Illustrated Textbook of Clinical Diagnosis in Farm Animals

Illustrated Textbook of Clinical Diagnosis in Farm Animals

Philip R. Scott

CRC Press is an imprint of the
Taylor & Francis Group, an **informa** business

First edition published 2022
by CRC Press
6000 Broken Sound Parkway NW, Suite 300, Boca Raton, FL 33487–2742

and by CRC Press
2 Park Square, Milton Park, Abingdon, Oxon, OX14 4RN

CRC Press is an imprint of Taylor & Francis Group, LLC

ISBN: 978-1-032-19750-0 (hbk)
ISBN: 978-0-367-61270-2 (pbk)
ISBN: 978-1-003-10645-6 (ebk)

DOI: 10.1201/9781003106456

Typeset in Times
by Apex CoVantage, LLC

Printed in the UK by Severn, Gloucester on responsibly sourced paper

Access the Companion Website: www.routledge.com/cw/scott

Contents

Author

Phil Scott has 43 years' experience of farm animal medicine and surgery in both first opinion practice and in a veterinary school teaching hospital. He has master's and doctoral degrees, is a Fellow of the Royal College of Veterinary Surgeons as well as a Diplomat of the European College of Small Ruminant Health Management and the European College of Bovine Health Management. He is the author of three other textbooks, many book chapters and over 150 refereed scientific publications.

Introduction

This book, and accompanying video recordings, result from a request by undergraduate students for a "5-minute" reference text that ranks common diseases showing particular clinical signs or syndromes such as weakness, weight loss, and abdominal distension. This book is not intended to replace reference veterinary textbooks but seeks to illustrate how common diseases and disorders typically present on commercial UK farms, what farmers expect, and what can be realistically achieved within a strict economic framework. Some UK veterinary schools do not operate a farm animal hospital, and those that have such facilities often do not have a realistic caseload for ever-expanding student numbers. As a result, extramural studies form the major component of an undergraduate's clinical experience, but finding an experienced mentor in farm animal practice has become increasingly difficult with the dramatic rise in demand. Developing clinical examination skills is not easy especially when much of farm animal veterinary medicine is strictly determined by agricultural economics and experienced clinicians adopt a pattern recognition approach with the diagnosis often based upon the telephone conversion with the client.

The author has 37 years' experience of undergraduate teaching and clinical instruction in first opinion practice and a university teaching hospital and has published textbooks in the traditional style on *Sheep Medicine* (2007; Second Edition 2014) and *Cattle Medicine* (2011). Over the past six years he has developed a veterinary consultancy and, as a consequence, is well aware of the challenges and demands faced by recent veterinary graduates arriving on a farm for the first time. Surveys reveal that some beef and sheep farms have fewer than one veterinary visit per six months and that engagement is often restricted to emergency events such as dystocia, uterine prolapse, hypomagnesaemic tetany etc or sudden high rates of mortality when circumstances are not always conducive to an easy working relationship.

In many farm animal practices, veterinary professional time on farm is billed at £120–180 per hour although some veterinary practices charge a standard professional fee for certain surgical procedures, such as caesarean operation and LDA, which often exceeds the *pro rata* hourly rate. Dairy farmers tend to have a more formal arrangement of routine veterinary visits, often the same practice partner every one–two weeks largely to maintain/improve the herd's fertility parameters, where fees may be charged per litre of milk produced. Dairy farmers are also more likely to employ nutritionists and other non-veterinary specialists as well as para-professional foot-trimmers although these people may be part of the vet-led team.

Farmers expect a benefit:cost in most situations which largely excludes veterinary care for many conditions affecting individual sheep. Septic pedal arthritis in a commercial value ewe is a good example of agricultural economics versus individual animal welfare whereby veterinary costs for digit amputation (undertaken while attending another problem therefore no visit charge, 15 minutes work plus drugs/materials) may exceed the sheep's market value (£50). This is difficult to accept for animal welfare reasons because amputation results in a sound sheep in 7–10 days whereas the consequences are more than three months' pain and weight loss until ankylosis eventually occurs (or not). Extradural injection to replace ovine vaginal prolapse is another good example of obvious animal welfare benefits restricted by perceived costs. The author has deliberately included a large number of such video recordings in this book. Unless veterinary undergraduates observe/participate in such procedures they will be less likely to propose/undertake such action to the farmer when the opportunity arises.

This book adopts a problem-based approach (e.g. chronic weight loss, poor production, frequent coughing/increased respiratory rate, weakness/recumbency) in response to requests from undergraduate students. The diseases that present with that characteristic are ranked in order with key clinical signs highlighted using a 7-point "traffic light" scale. No abnormality detected (NAD) is labelled in green, while mild changes are identified as amber with moderate and severe changes highlighted in red. This rather arbitrary scale is based upon the author's experience and interpretation and is offered as a guide only. These key clinical signs are highlighted in the images and in greater detail in the video recordings which emphasise lameness, respiratory signs, mental state etc.

+++ Rectal temperature ca. 41°C, heart rate >100–120 beats per minute, greatly distended/increased abdomen/viscus, tachypnoea with open-mouth breathing, seizure activity, opisthotonos.

++ Rectal temperature ca. 40°C, moderately distended/increased, markedly increased spinal reflexes, hyperaesthesia.

+ Rectal temperature ca. 39°C, slightly distended/increased, slightly increased spinal reflexes.

No abnormality detected (NAD) Within normal range. Animal sound.

– Reduced appetite, slightly decreased rumen fill, slightly increased spinal reflexes, unsteady on legs. Slightly lame.

–– Poor appetite, moderately decreased abdomen/viscus, moderately decreased spinal reflexes, weak—unable to stand. Moderately lame.

––– Not eating, greatly reduced/decreased abdomen/viscus, coma, weak—unable to maintain sternal recumbency. Severely lame, non-weight-bearing.

Locomotion scoring commonly adopts a 5-point system based on both gait and posture. This scale has been modified to fit the 7-point scale adopted earlier.

1) *Normal*: The cow is not lame; the back is flat. NAD within normal range. (NAD)
2) *Lame*: The back is slightly arched when walking. NAD within normal range. (+)
3) *Mildly lame*: The back is arched when both standing and walking. The cow walks with short strides in one or more legs. (+)
4) *Moderately lame*: The lame cow can still bear some weight on the affected foot. (++)
5) *Severely lame*: The back is arched; the cow refuses to bear weight on the affected foot and remains recumbent. Severely lame. (+++)

The important diagnostic tests are listed in an accompanying table. Images for each disease, diagnostic tests such as sonograms and radiographs, and pathognomonic findings are included. Key observations and clinical examination findings are highlighted in the image legends alongside likely key history events/predisposing factors. Video recordings are numbered in red and identified with the symbol 🎥. The video recordings are a critical component to each disease description expanding upon the images and clinical details. The video recordings feature extensively throughout the book with short captions to read before and after viewing. There is no narration to accompany the video recordings because the key points have already been summarised in the text and more information is retained after reading than listening. While the "disease characteristic" table is deliberately brief, the video recordings show sequentially: Disease presentation, diagnostic tests such as ultrasonography and radiography, then the animal after successful veterinary treatment or necropsy findings. The major value of a detailed necropsy is that the findings may be directly applicable to disease control for the whole flock such as fasciolosis.

One major criticism of a problem-based approach is that it simply matches the major presenting clinical sign(s) with a potential list of causes—veterinary medicine for the under-fives or pattern recognition at its most basic. To further develop clinical skills, and avoid undue criticism, the author has produced a short guide detailing his examination of each body system which emphasises key diagnostic tests in a cost-effective manner such as ultrasonography, lumbar cerebrospinal fluid collection and digital radiography. For example, published research clearly demonstrates that a sheep's chest can be ultrasound scanned within 20 seconds and provides much more accurate information than 5–10 minutes' auscultation and percussion. The author has used these techniques on farm for the past 30 years yielding many refereed scientific publications, and the important findings are reproduced here in the large number of diagnostic video recordings.

How to Use This Book

Undergraduate Years 1 and 2

There are two chapters specifically written on how to recognise sick cattle and sheep highlighting those clinical signs farmers identify before requesting veterinary attendance. The chapters on perinatal mortality and dystocia in sheep are especially relevant to the undergraduate animal husbandry courses and pre-clinical extramural studies/lambing placements.

Undergraduate Years 3 to 5

University lecture courses and reference textbooks traditionally adopt a systems-based approach detailing diseases in a standard format of aetiology, clinical signs, differential diagnoses, diagnosis, treatment, pathology, prevention and control. However, the presenting clinical signs are rarely pathognomonic for a particular disease and a problem-based approach is more relevant to general practice where time is money although it is much more difficult to teach. This book is intended to complement, not replace, systems-based lectures.

Examination of each body system focuses on the important features using images and video recordings with emphasis placed upon ultrasonography and to a lesser extent radiography. Correlating video recordings of sonographic changes with necropsy findings where available is the best way to learn how to apply this diagnostic facility to farm animal practice. Students are encouraged to challenge the findings reported here and compare ultrasound findings of the chest with auscultated sounds when they next have the opportunity. It is simply not possible to accurately assess the lungs and pleurae in cattle and sheep without ultrasonography (see published work).

The tables ranking the common complaints presented to veterinary practitioners are a guide with the relative importance/prevalence and will vary considerably between regions and seasons. These tables list the common diseases/disorders encountered in general practice, but they are not exhaustive. However, by following the recommended clinical examination approach for each organ/body system, and the ancillary tests, the reader should be able to establish a list of common differential diagnoses and be better placed to consult reference text books and senior colleagues in the practice.

Part 1

Veterinary Involvement on Farms

1.1

Flock Inspection

- One shepherd is usually responsible for 800–1,200 ewes.
- Be aware that annual ewe mortality is ca. 8–10 per cent in many flocks. The barren rate in hill flocks is typically 8–10 per cent. Veterinary help is usually requested only after several unexplained deaths. Farmers will make a financial judgement whether veterinary help is needed: Is there a likely benefit:cost?
- Veterinary time on farm is charged at £120–180 per hour plus visit charge. Ten minutes spent on the clinical examination of one sheep costs the farmer £20–30.
- Farmers/shepherds have ready access to a wide range of antibiotics, non-steroidal anti-inflammatory drugs, multivitamins, anthelmintics and flukicides.

1.1 and **1.2:** Veterinary help is usually requested only after numerous unexplained deaths. **1.3:** On-farm necropsy after five deaths in a group of 120 hoggs reveals acute respiratory disease; these sheep had not been vaccinated.

It is a legal requirement in the UK that intensively managed sheep and cattle are inspected daily and appropriate action taken to treat sick and/or diseased animals as directed in the veterinary health plans. However, individual sheep are rarely presented for veterinary examination unless they are financially valuable such as a pedigree ram or considered representative of a significant flock problem.

1.4, 1.5 and **1.6:** This sheep farmer paid £2,000 for this shearling ram (left), this ewe cost £120 as a gimmer (centre) and sold this store lamb for £50 (in most years).

DOI: 10.1201/9781003106456-2

How Does a Farmer Detect a Sick Sheep? Individual/Flock Behaviour

Sheep, even rams, are flock animals and graze and rest together. Isolation from the group is highly unusual and necessitates immediate investigation by the shepherd.

1.7 and **1.8:** Normal sheep, even rams, rest and graze together as a group. **1.9:** Isolation of this ewe from the group necessitates immediate investigation by the shepherd. 1.1

On most farms, sheep are inspected daily from a quad bike, but large groups of sheep are difficult to inspect once they have gathered into a tight group. 1.2, 1.3 Once sheep have stopped grazing, the appetite of an individual sheep is not known and abdominal fill is used as a proxy guide. Using "snackers" allows large numbers of sheep to be fed quickly and easily, but it can prove difficult for the shepherd to identify sick/lame sheep. 1.4

Similarly, it can prove difficult to identify sick sheep in cell/mob grazing systems with very high stocking density although walking sheep past you to the next grazing area, which occurs every 1–2 days, provides good opportunity (provided there are facilities to catch the sheep).

1.10: It can prove difficult to identify and catch sick sheep in very high stocking density situations where there are no portable handling facilities. **1.11:** Group of several hundred recently weaned ewes. Note that the sheep in the foreground have already stopped grazing as the shepherd approaches. 1.5

Typically, individual sheep are caught in the corner of the field using dogs but this is not always an easy process. Portable handling systems are used for group treatments in fields where sheep cannot be driven to central facilities. 1.6

1.12: Identifying sick sheep once gathered in large groups can prove difficult but many experienced shepherds can identify individual ewes. **1.13** and **1.14:** Portable sheep handling systems for remote fields.

Flock Inspection

Isolation from other sheep in the group, and lagging behind the group when gathered are common signs of severe lameness and/or illness. Normal sheep will generally rise and walk away when the quad bike is about 50 metres away although this distance is much reduced in sheep handled frequently. 🎥 1.7, 1.8, 1.9, 1.10

The ewe pictured next is easily identified because she is isolated from other sheep in the group and appears to be emaciated with a poor quality fleece. There appears to be little abdominal fill consistent with a poor appetite of several days' duration. Normal sheep are featured in the video recording for comparison. The affected sheep does not run off to join the other sheep in the group rather she assumes sternal recumbency.

Separation (**1.15** and **1.16**; same sheep) is a common sign of lameness and/or illness. This sheep appears dull, does not watch the person approaching nor attempt to join the other sheep. She has a much reduced abdominal fill and poorer fleece quality than other sheep in the group (**1.17**). 🎥 1.11

Separation from other sheep in the group (**1.18** and **1.19**; same sheep). **1.20:** Note that isolation is a normal behaviour during first stage labour.

1.21 and **1.22:** These two Greyface ewes have stopped grazing and are separated from other sheep in the group. **1.23:** Close-up image of ewe (**1.22**) reveals she is very dull, her ears are held back and the upper eyelids appear drooped.

1.24: Sick sheep are usually readily identified. **1.25:** Compare the attitude of these two weaned lambs to the person approaching. The sheep on the right has her head down, ears back and drooped upper eyelids. **1.26:** The sick ewe (left) is readily identified by her very dull demeanour, head down and drooped ears as well as her very low body condition score. 1.12a, 1.12b, 1.13, 1.14, 1.15, 1.16, 1.17

1.27: This ram is very dull with poor abdominal fill and requires immediate attention. **1.28:** This ewe appears dull with ears back and poor abdominal fill (1.18). **1.29:** This ewe appears reasonably bright but has poor abdominal fill and is emaciated with a poor quality fleece suggestive of chronic disease.

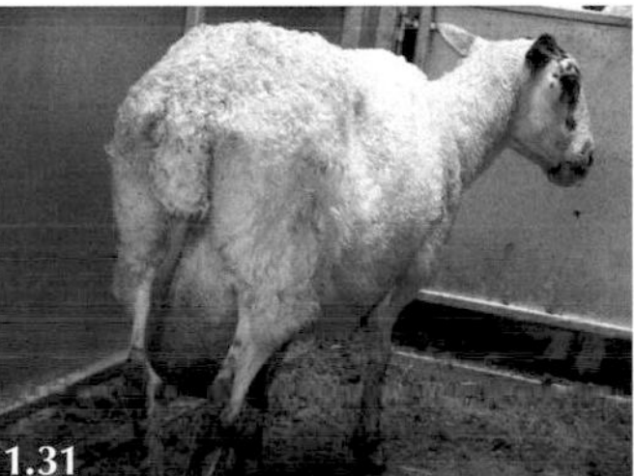

1.30: This nursing ewe has a low body condition score with obvious bony prominences, poor coat/wool quality and a sunken left flank. The left mammary gland and teat are enlarged (chronic illness). **1.31:** This Greyface appears very dull with its ears down. There is reduced abdominal fill indicating poor appetite over the past few days (acute illness).

1.32: The gimmer in the foreground is separated from others in the group and is not grazing. She appears dull with her ears back; there is evidence of chronic diarrhoea. **1.33:** View of same sheep from the side showing reduced abdominal fill and a more open, poorer quality, fleece. There is evidence of dermatophilosis on her muzzle. This sheep requires immediate attention. 1.19

Parasitic diseases generally affect a large proportion of sheep in the group although there can be a wide range of presenting signs. Occasionally, management errors such as failure to drench/inject individual sheep may result in diseases considered "group/flock problems" affecting only individual sheep e.g. failure to drench one/several sheep with a flukicide may allow chronic fasciolosis to cause disease in these sheep where normally it would be considered a group/flock problem. Production demands from a multiple pregnancy or rearing twin lambs with limited nutrition also affect large numbers within the group/flock and not individual sheep.

Poor body condition score can be identified by the prominent dorsal spines of the lumbar vertebrae giving a triangular outline to the sheep's dorsum when viewed from behind. Such sheep have a much reduced abdominal fill giving a flat-sided appearance rather than the full rounded nature of healthy sheep. 1.20, 1.21, 1.22, 1.23

Poor fleece quality is manifest as a dull open nature which readily breaks to the skin. Wool readily pulls out when sheep with chronic disease are incorrectly caught by grabbing the fleece. Shearing is undertaken after a "rise" in the wool allowing the shears to easily pass between the skin and old fleece. Sheep without this rise do not shear well ("sticky") and the remaining wool is striped by the shears.

1.34–1.36: Affected sheep on left hand side of all images show poorer body condition, poorer quality fleece and "slab-sided" appearance of abdomen when viewed from behind.

1.37 and **1.38:** Comparison with other sheep in the group identifies sheep with poorer body condition scores. **1.39**: Note the poor shearing job (right hand sheep) caused not by the shearer but lack of "rise" in the wool of a sheep in poor body condition. 1.24, 1.25

1.40: In this group of nursing ewes the ewe on the right is in very poor condition. It would be informative to compare her lamb(s) to other lambs in the group (only compare ewes with similar litter sizes; twins with twins etc).
1.41: On closer inspection this ewe has no udder (compare to ewe in background) and is not lactating; also note the poor fleece quality. This sheep has a significant chronic disease problem.

Veterinary Flock Inspection/Assessment

- Veterinary involvement with sheep can usually be divided into two broad categories either acute disease when a financially valuable sheep is found "sick" or chronic disease where the major manifestation is weight loss in several/numerous adult sheep over weeks/months or poor growth/weight loss in groups of lambs.

- Annual flock mortality rates are quoted as 5–10 per cent but there have been few long-term commercial farm studies. Carcase disposal costs are £18–20 per adult sheep. Cull ewes fetch £40–70 per head (Spring 2021 prices). Veterinary time on farm is charged at £120–180 per hour plus visit charge.
- Many farmers spend around £6 per ewe to protect against EAE and toxoplasmosis. Other good diseases candidates for veterinary control regimens include Johne's disease, fasciolosis and ovine pulmonary adenocarcinoma.

Assessing Flock Management/Disease Status

Farmers will happily report ewe pregnancy scanning percentages (when favourable and omitting barren ewes) but rarely flock profitability. Determining production-limiting parameters such as high involuntary culling rates proves difficult on many farms but this is essential information to assess potential problems. Necropsy examination of all deaths on farm is not realistic because of cost and lack of basic facilities. However, it should be possible to identify many of the affected sheep and the causes of illness before death/necropsy and implement effective preventive measures. Carcase disposal costs are £18–20 per adult sheep, and even poorly conditioned sheep have a disproportionately high economic value when culled; these savings could be used for structured veterinary investigations. Paratuberculosis, fasciolosis and ovine pulmonary adenocarcinoma are common diseases that can be controlled, but not necessarily eliminated, in a cost-effective manner.

1.42 and **1.43:** High rates of involuntary culling indicate a significant disease problem(s) on the farm and a good opportunity for veterinary involvement in flock health. **1.44:** Normal sheep on left compared to a sheep with chronic weight loss requiring veterinary investigation. 1.26, 1.27

The professional farm animal post-mortem and reporting service operated by Dr Ben Strugnell, in association with a knackery collection service, which investigates all deaths on farm on a contract basis is proving very successful in identifying and quantifying the major production losses on sheep farms. Veterinary practitioners can use this information to develop control programmes. Such ongoing independent monitoring allows the introduced control measures to be accurately evaluated.

Other information, such as comparing recently purchased sheep (rams are a good example) with older cohorts on the farm, gives an indication of the flock's management.

1.45: Suffolk shearling ram at purchase. **1.46** and **1.47:** Two Suffolk rams one year after purchase on the same farm suggest management deficiencies.

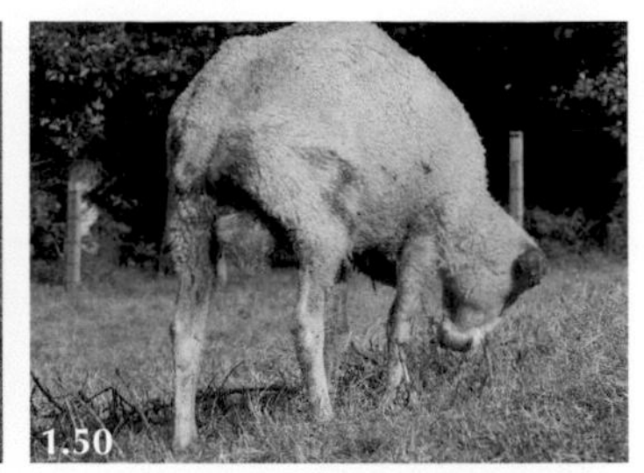

1.48: Blueface Leicester shearling ram at purchase. **1.49** and **1.50:** Two Blueface Leicester rams one year after purchase on the same farm. Appropriate management after the breeding season is a common problem. 1.28

The effects of improved nutrition are most often seen when sheep with molar dentition problems (**1.51**) have their ration supplemented with concentrates (**1.52**). 1.29

Clinical Examination of Sick Sheep

Direct visual comparison between the sheep causing concern and normal sheep in the group is not always possible because the shepherd will gather/catch the affected sheep and take it/them where it is more convenient for veterinary examination and subsequent treatment(s). Such gathering and isolation in a strange environment may temporarily mask some of the presenting signs and cause abnormalities such as hyperpnoea and tachypnoea. Sheep do not tolerate isolation well and it is preferable to confine it with another sheep from the same group but this may not be good biocontainment practice if the sheep is suspected of having a contagious disease. Examples of sick sheep are shown in the two images that follow and the video recordings; the benefits of normal sheep for comparative purposes are obvious.

1.53: The affected sheep (on right) shows no interest in the approaching person, has a much poorer body condition, markedly reduced abdominal fill, ears back, head down and lowered upper eyelids. **1.54:** Ewe on the right is very dull, has her ears down and there is a bilateral mucoid nasal discharge. 1.30, 1.31

The veterinarian will have to ask the shepherd what the presenting signs were and when the sheep was last inspected although such reports may be inaccurate particularly with respect to duration of signs because of irregular/cursory inspection.

1.2

General Examination: Toxaemia

Clinical examination of the "organ systems" is detailed in the respective chapters, but toxaemia is described here.

Toxaemia

Toxaemia is defined as "a condition in which the blood contains bacterial toxins disseminated from a local source of infection or metabolic toxins resulting from organ failure or other disease". Gram-negative bacteria, especially *Pasteurellae* and *Escherichia coli*, play a major role in the development of endotoxaemia in ruminants. The term "endotoxin" is mostly used synonymously with lipopolysaccharide (LPS) which is the major component of the outer membrane of gram-negative bacteria and released only after destruction of the bacterial cell wall. When endotoxin is liberated, it enters the general circulation which can lead to cardiovascular, respiratory and digestive system dysfunction and may result in death. The changes are manifested by an increased heart rate (often greater than 120 beats/minute), increased respiratory rate and effort, and rumen atony.

Examining Conjunctival Mucous Membranes

The eyelids are parted to check whether the eye is sunken (dehydration; right image that follows). The upper eyelid can be pinched to form a skin tent and the time for this fold to return to normal appearance is determined (not shown) but this is of little clinical use. The eye is very gently pushed back into its socket prolapsing the lower conjunctiva permitting examination (note the position of the right [top] thumb in the left and centre images).

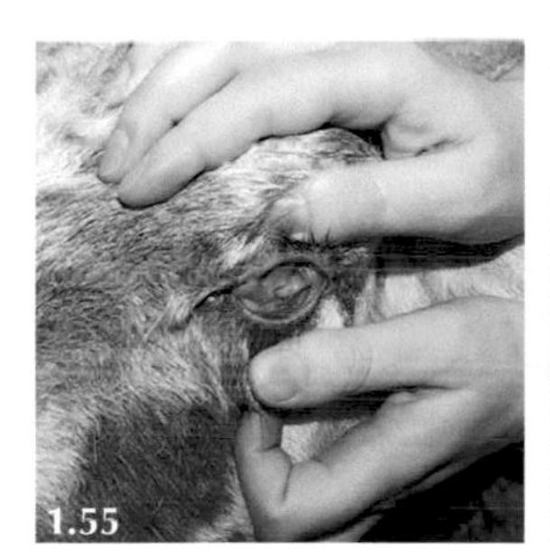

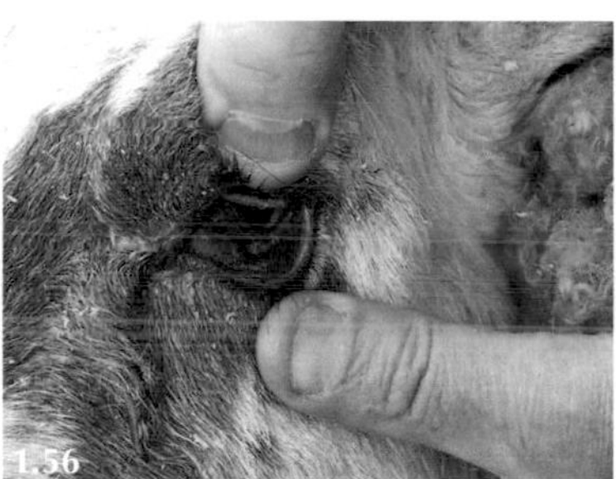

1.55: Healthy pink mucous membranes, toxic (purple) mucous membranes (**1.56**), toxic mucous membranes with sunken eye indicating dehydration (**1.57**).

DOI: 10.1201/9781003106456-3

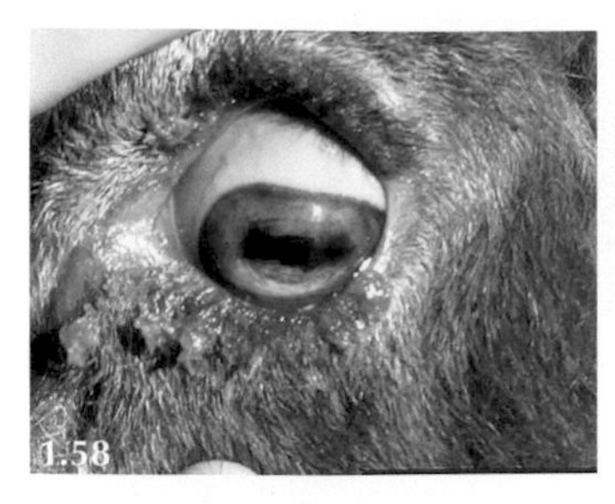
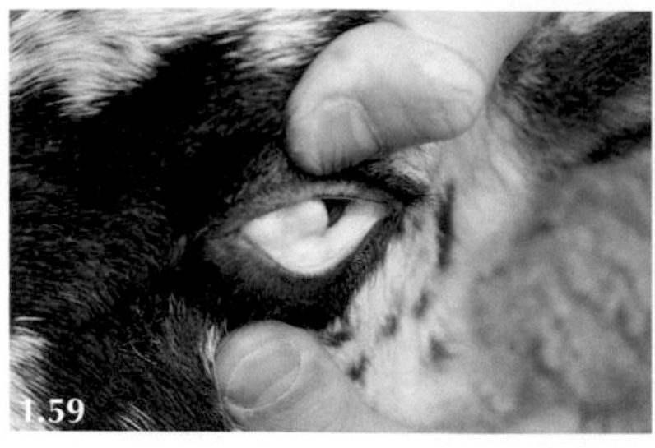
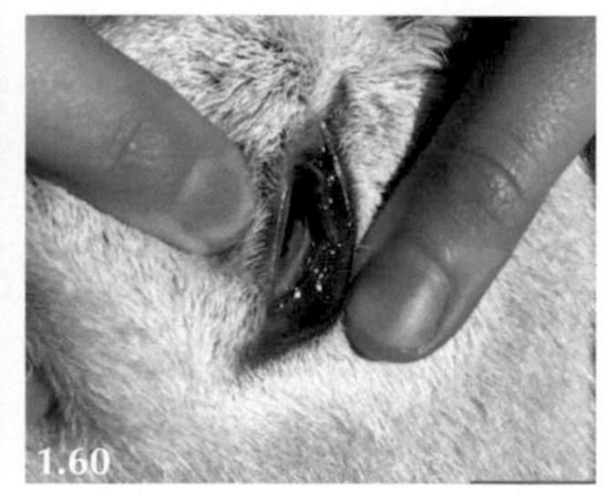

1.58: Jaundice (yellow appearance of sclera). **1.59**: Anaemia (very pale pink/white mucous membranes). **1.60:** Toxaemia (red/purple).

Identifying Toxaemia

Identifying a sheep suffering from toxaemia is much easier than determining the organ system involved because several organ systems are often affected.

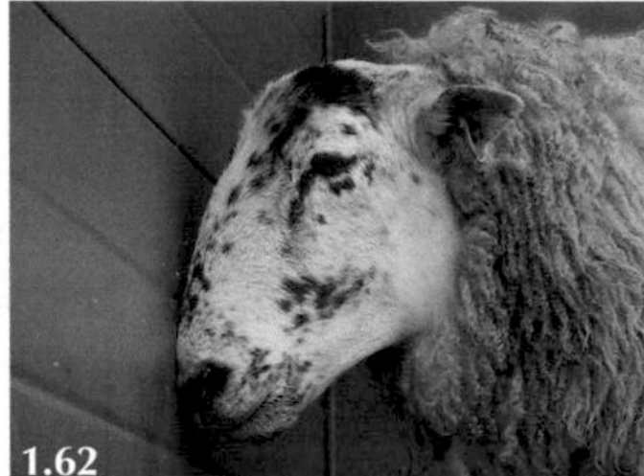
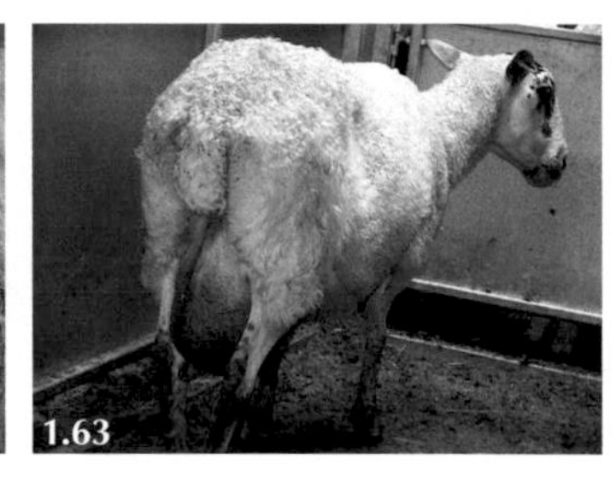

1.61: Toxaemia causing stupor, weakness leading to recumbency and profuse salivation associated with ruminal atony caused by acute mastitis 1.32. **1.62:** Stupor, drooped upper eyelids and ears caused by toxaemia secondary to OPA and per-acute *Pasteurella* pneumonia 1.33. **1.63:** Lowered head, ears back, reduced appetite/poor rumen fill caused by gangrenous mastitis. 1.34

Sheep can mask advanced pathology until the agonal stages; autolytic fetuses/lambs *in utero* is a good example of the apparent disparity of clinical signs and necropsy findings. The two ewes featured in the images that follow were euthanised immediately after veterinary examination. The necropsies revealed the necrotic purple/black uterus and advanced toxaemia yet these ewes were still ambulatory when presented for veterinary examination.

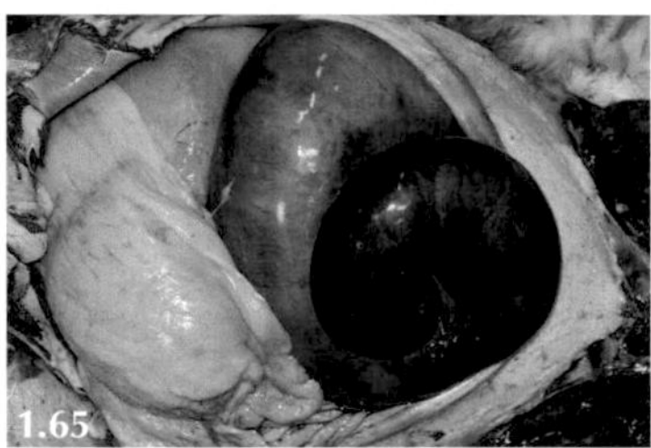

1.64: This ewe is dull, dehydrated, with a distended abdomen which masks her inappetence. The heart rate is >120 bpm with toxic mucous membranes. **1.65:** Immediate euthanasia, then necropsy, reveals necrosis of the uterine wall. See also the following. 1.35

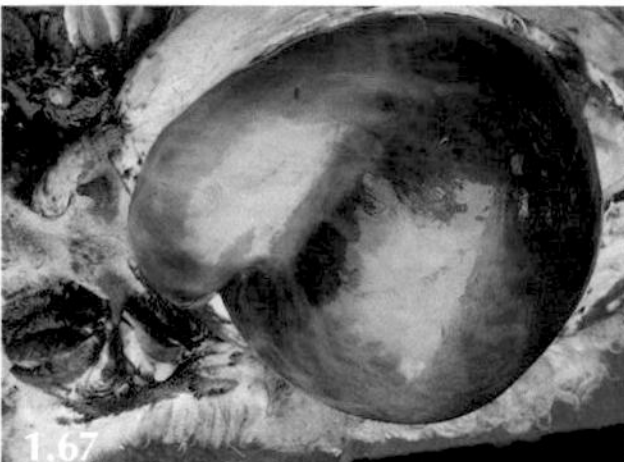

1.66: This ewe walked off the farmer's trailer but is weak, very dull and dehydrated, and has a distended abdomen. The heart rate is >120 bpm with toxic mucous membranes. **1.67:** Immediate euthanasia, then necropsy, reveals necrosis of the uterine wall.

1.3

Pain and Fear

Pain (FAWC report on farm animal welfare: health and disease 2012) is typically assessed by changes in behaviour, such as, reduced use of the affected part, and behavioural signs which include:

- Inappetence, decreased rumination rate.
- Dullness, depression, lethargy. Ears down, dropped upper eyelids.
- Greater time spent in sternal recumbency, head on ground.
- Teeth grinding.
- Increased respiratory rate.
- Increased sensitivity (hyperalgesia).
- Attention to site of pain (e.g. rubbing).
- Severe lameness when lesion affects foot/joint.

1.68 and **1.69:** These sheep suffering from endocarditis show many of the signs of pain listed earlier. 🎥 1.36

DOI: 10.1201/9781003106456-4

This six-month-old lamb (**1.70**) with polyarthritis is suffering from pain manifest by lateral recumbency when normal sheep rest in sternal recumbency. It is unable to flex its joints to maintain sternal recumbency (**1.71**), and has an unusual gait/stance indicating severe bilateral foreleg lameness (**1.72**). 1.37

1.73–1.75: Defining a "painful" expression proves very difficult but would be expected in these three weaned lambs suffering from polyarthritis where synovial membrane proliferation and articular cartilage erosion were identified at necropsy. 1.38

1.76 and **1.77:** These sheep are suffering from a septic joint. While there is obvious non-weight-bearing lameness, there is no convincing evidence of pain by facial expression. It is possible that pain has been over-ridden by fear during isolation for these photographs.

Clinical response to analgesia can be a useful way to assess pain if the animal's behaviour changes after an analgesic is given. Relief of pain is better appreciated after either high/low extradural injection of lignocaine in pelvic limb injuries and obstetrical procedures or IVRA for foot lesions because a total block of nerve transmission is achieved.

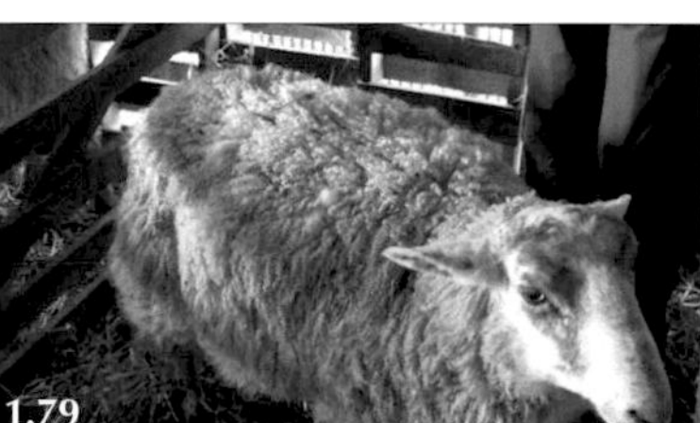

Relief of pain associated with dystocia (**1.78**) and immediately after low extradural injection of lignocaine and xylazine (**1.79**). 1.39

Sheep Pain Facial Expression Scale

A sheep pain facial expression scale (SPFES) has attempted to identify sheep suffering pain. Five traits are scored as 0 (not present), 1 (partially present) or 2 (present).

- *Orbital tightening*—there is a closing of the palpebral fissure by the eyelids and a narrowing of the eye aperture.
- *Cheek tightening*—there is a more convex shaping to the cheek in the area of the masseter muscle.
- *Abnormal ear posture*—ears become fully rotated ventrally and caudally.
- *Abnormal lip and jaw profile*—the jaw profile appears straight to concave.
- *Abnormal nostril and philtrum shape*—a "V" shape between nostril apertures is present. 🎥 1.40, 1.41

Fear

Fear is most commonly observed when animals are handled or isolated and can be a common side effect of treating a diseased animal. Fear is expressed by excitement, snorting, foot-stamping and frequent attempts to jump gates/fences to escape confinement. This abnormal behaviour often stops immediately when a companion is introduced into the pen. Fear may also result when the sheep is unable to avoid/flee an approaching person or sheep dog, but this is very much an anthropomorphic interpretation (see the following).

1.80: This sheep is severely lame and very reluctant to rise/walk. Her ears are directed forward and the head turned to the approaching person with "fear in her eyes".

1.4

Promoting Veterinary Services by Demonstrating a Benefit:Cost

Many diseases affecting flock profitability have an insidious onset with poor production and weight loss over three to six months resulting in involuntary culling or death. The farmer's vet is able to identify these sheep and investigate further to prevent such losses, but communicating such ideas to farmers is not always easy.

DOI: 10.1201/9781003106456-5

1.81: Scottish Blackface ewe with a low condition score and poor fleece consistent with chronic disease. No action was taken by the farmer (except for a "worm drench"). **1.82:** Same ewe eight weeks later—dead and awaiting knackery collection. The cause of death was likely Johne's disease (but not proven). Loss of cull ewe = £40 plus £18–20 knackery disposal fee; overall loss approximately £60 but, more importantly, serious animal welfare concerns. 1.42, 1.43, 1.44

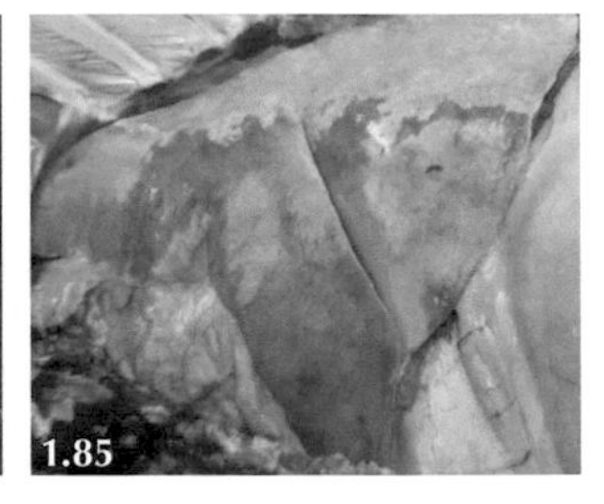

1.83: Four ewes from a 360-ewe flock identified with early stages of OPA—sold for slaughter at £80/head in September 2020. **1.84:** Advanced OPA on same farm euthanased for welfare reasons; carcase collection cost £18. **1.85:** Gross appearance of OPA at farm necropsy (sheep centre image). 1.45

A benefit:cost can be readily demonstrated in many commercial flocks for Johne's disease (**1.86**), fasciolosis (**1.87**) and ovine pulmonary adenocarcinoma (**1.88**).

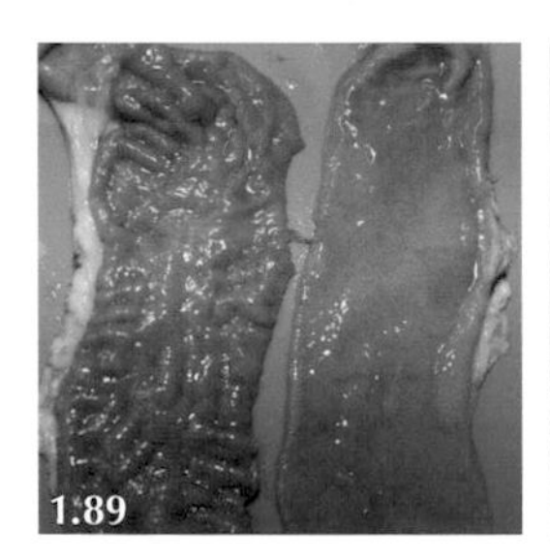
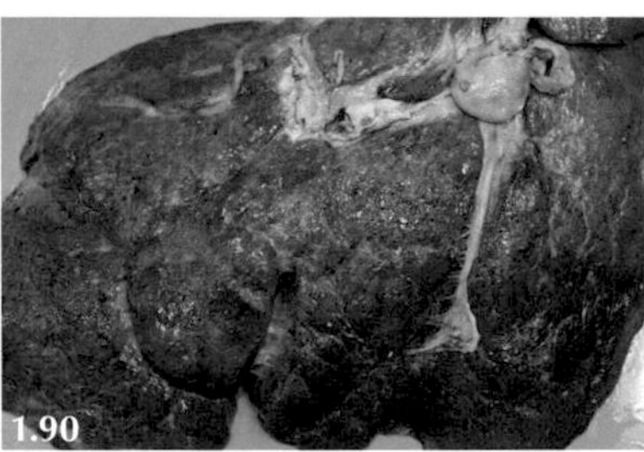
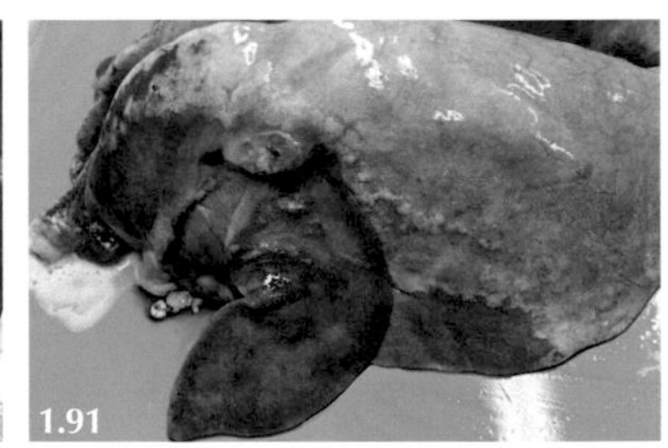

Characteristic gross necropsy findings.
1.89: Necropsy findings for Johne's disease (affected intestine on left side). **1.90:** Fasciolosis. **1.91:** Ovine pulmonary adenocarcinoma.

Promoting Biosecurity and Biocontainment

There is a very large range of infectious diseases and infestations that can be introduced with purchased sheep, and prevention is a major part of the veterinary flock health plan.

TABLE 1.1
Diseases That Can Be Introduced through Purchased Flock Replacements

Maedi visna virus
Ovine pulmonary adenocarcinoma (Jaagsiekte)
Orf (contagious pustular dermatitis)
Border disease
Scrapie (homozygous genotype for shortened incubation period)
Virulent footrot
Caseous lymphadenitis
Paratuberculosis (Johne's disease)
Chlamydia psittaci (enzootic abortion of ewes)
Campylobacter fetus fetus
Psoroptes ovis (sheep scab, including pyrethroid resistant strains)
Pediculosis (lice)
Melophagus ovinus (Keds)
Multiple anthelmintic resistant nematode strains (*Haemonchus contortus* and *Teladosargia* spp. in particular)
Fasciola hepatica (liver fluke)

1.5

Identification of Acute Disease in Cattle

Dairy Cattle

- Many dairy herds are housed all year with cows inspected multiple times per day and observed/handled/milked two to three times per day by the same team of skilled farm personnel who quickly identify behavioural changes in a cow. A dramatic reduction in milk yield accompanies acute disease in lactating cattle.
- Electronic transponders also record concentrate intake from out of parlour feeders and distance walked and time spent lying down. Significant deviations from normal trigger alerts.
- Most dairy farms have routine veterinary visits every one to two weeks largely to undertake (in)fertility work but also all other problems.
- Common disease problems such as mastitis are treated by the farmer using SOPs detailed in the herd health plan.
- Lameness is a significant problem on some dairy farms with many farmers using paraprofessional foot trimming services.

DOI: 10.1201/9781003106456-6

1.92 and **1.93:** Few lactating dairy cows graze during the summer months but are housed year-round. **1.94:** Grazing is restricted to youngstock, and occasionally dry cows to satisfy milk company marketing claims.

Beef Cattle

- Isolation from the group, lying down while all other cattle are grazing, must be investigated immediately however not all causes are urgent such as some causes of lameness. Lack of appropriate handling facilities at pasture frequently delays presentation/veterinary examination.
- The severity of acute diseases is often aggravated by poor detection of early clinical signs.
- Respiratory disease is the most common acute disease affecting significant numbers of beef calves.

While beef cattle must be examined daily as a legal requirement, lack of appropriate handling facilities for animals at pasture may delay veterinary examination.

1.95 and **1.96:** Cattle are gregarious animals and seek the company of others even in very extensive hill grazing (**1.97**).

1.98: This beef heifer is lying down and not grazing. Her ears are held down and the eyelids are drooped. **1.99:** This seven-month-old spring-born beef calf is very dull and disinterested in its surroundings. There is excess salivation and drooping of the ears and upper eyelids. There are marked mucopurulent nasal discharges with crusting around the nares. 🎥 1.46, 1.47

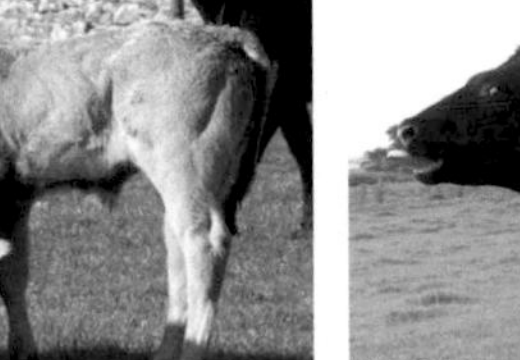

1.100: This six-week-old beef calf is dull with excess salivation and drooping of the ears. **1.101:** This beef cow at pasture is not grazing and presents with tachypneoa and frequent coughing. Both animals stand with their necks extended and heads lowered. 🎥 1.48, 1.49

1.102: This beef cow appears startled and is unsteady on her legs with a staggering gait and wide-based stance. **1.103:** There is an excitable and startled expression. **1.104:** Note the normal relaxed expression now that the cow has been sedated as part of the veterinary treatment. 1.50

1.6

Identification of Chronic Illness in Cattle

- Chronic illness results in poor milk production over several days/weeks and weight/body condition loss. In beef cows, there may be poor calf growth.
- Lameness is the major cause of body condition loss in dairy cows with national surveys reporting prevalence rates around 20 per cent although there are huge variations between herds.
- Recrudescence of chronic respiratory disease (bronchiectasis) is common in recently calved heifers but is rarely diagnosed and treated correctly.
- Paratuberculosis is a common cause of chronic weight loss in beef and dairy cows.
- Chronic illness in youngstock, often respiratory disease, results in poor growth, failure to gain body condition, and longer and poorly pigmented hair.

1.105: Dairy cow in late lactation with a target body condition score 3–3.5. **1.106:** Poor body condition caused by chronic lameness. **1.107:** Such poor condition in a beef cow would attract comment but is almost considered "normal" in lactating Holstein cows.

DOI: 10.1201/9781003106456-7

1.108 and **1.109:** Compare the animal under investigation (**1.109**) with other cattle in the group (**1.108**) to determine what is normal body condition for that production system.

1.110: During late summer this dry cow appears to be in reasonable body condition although the ribs are visible. **1.111:** Compared to other cows in this group (left) she is in much poorer condition.

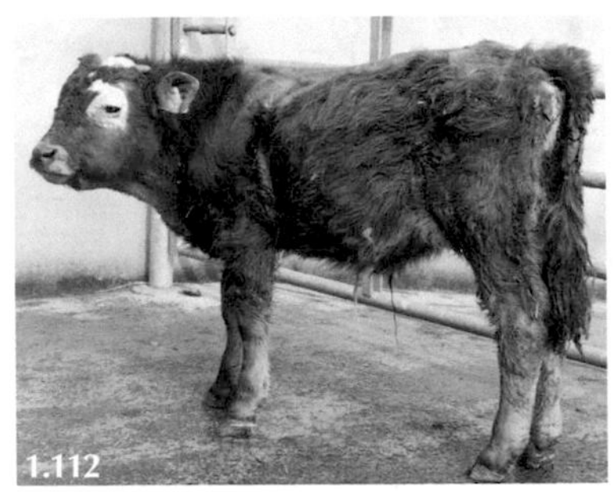

1.112 and **1.113:** Fecal staining of the perineum, long dull coat and poor condition score all indicate poor calf growth. **1.114:** The dull brown coat should be black.

Appetite/Feed Intake

Observation of feeding behaviour is more difficult in dairy cows where cattle are fed a total mixed ration (TMR) *ad libitum*. When a dairy cow is sick she often delays entry into the milking parlour until the end of the milking. Identification of chronic illness in lactating dairy cattle is most commonly identified, and quantified, by reduction in milk yield over several days. Many automated individual cow monitoring systems issue alerts when milk production drops significantly. Many dairy cows wear electronic transponders which record concentrate intake from out of parlour feeders and quantify distance walked and time spent lying down with changes in these behaviours helping to identify problems especially lameness. These parameters can also be used for oestrus detection.

During winter, housed beef cows and growing cattle are usually fed concentrates once or twice daily allowing appetite to be determined. Reduction/loss of appetite is also reflected in reduced abdominal fill although not all cows exhibiting chronic weight loss necessarily have a drawn-up abdomen.

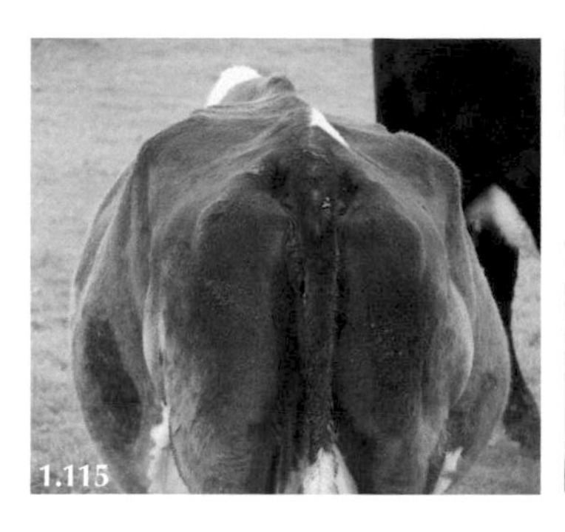

1.115 and **1.116:** Abdominal fill is greatly influenced by advanced pregnancy appearing much more pendulous in older beef cattle.

Loss of body condition score is a crude method of defining health because changes, such as more prominent ribs and lateral spines of the lumbar vertebrae, can only be identified after several weeks. Comparison with most other cattle in the management group serves to highlight these changes.

Several of the videos included next show normal healthy cattle for comparative purposes. Study the differences between these cattle and those presented to you for clinical examination.

Heifer (**1.117**) and cow (**1.118**) with chronic illness appear dull and disinterested in their surroundings. Note the stance with the neck often extended and head lowered. The left sublumbar fossa is sunken and the abdomen drawn-up consistent with a poor appetite. There are obvious bony prominences (ribs, wing of the ilium). The udder is flaccid for a lactating animal. The coat is longer than usual, dull and dirty.

These images feature the same heifer before treatment (**1.119**) and 10 days later (**1.120**). Note the much improved demeanour and abdominal fill (right). The coat is much cleaner (**1.120**) from self-grooming. 🎥 1.51, 1.52

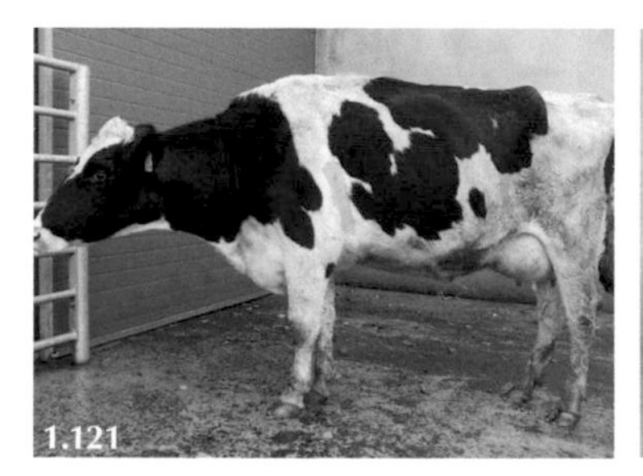

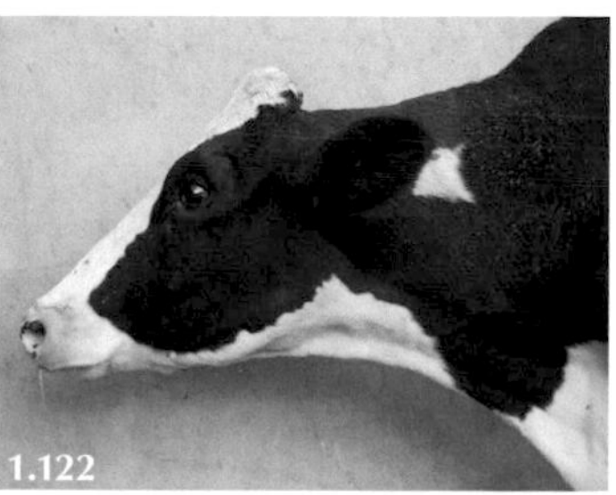

1.121: Dull and disinterested demeanour with poor body condition suggestive of chronic disease. **1.122:** Painful expression with ears down and nasal discharge.

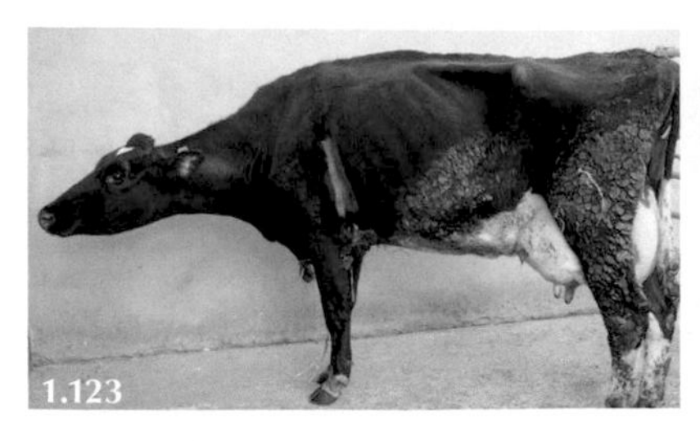

1.123 and **1.124:** Both Holstein cows are suffering from an acute flare-up of chronic suppurative pneumonia following the stress of calving. They are very dull and disinterested in their surroundings. The neck is outstretched with the head held lowered and the ears directed caudally. They have a painful expression. The cows are emaciated with prominent ribs and pelvis. The abdomen is very drawn-up with an obvious sunken triangle behind the rib cage caused by poor rumen fill. The coat lacks its normal sheen and is covered with dried feces (1.123). 1.53, 1.54, 1.55

1.125 and **1.126:** Chronic weight loss/emaciation. **1.127:** Lean cow (left) compared to normal cow on right—compare BCS and coat. 1.56, 1.57

1.128 and **1.129:** Extreme example of chronic disease caused by paratuberculosis (**1.129**). It may have taken six to nine months for this cow to become so emaciated. 1.58

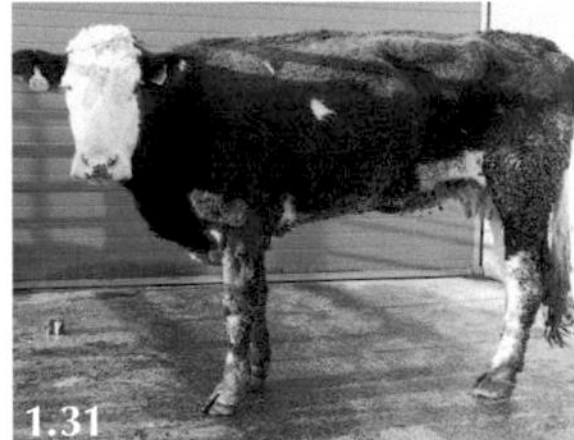

1.130: This beef cow with pyelonephritis is dull and disinterested as you approach. The cow is in very poor body condition with prominent ribs and pelvis. The coat is long and dull brown compared to the normal short sleek red coat of healthy cows. The abdomen is very drawn-up with an obvious sunken triangle behind the rib cage caused by poor rumen fill. **1.131:** Similar findings. It has likely taken three months for these cows to lose so much body condition. 1.59

Poor Growth in Calves

Dairy heifers are regularly weighed to check growth against production targets for mating and various stages of pregnancy, however this is uncommon in beef cattle where poorer growth and lower body condition score relative to their peer group is employed.

1.132: These beef steers are 18 months old and weigh ca. 550 kg except for the steer on the far left with horns which is considerably smaller. **1.133:** Poorly grown beef calf with poor body condition and long coat. **1.134:** Stirk with chronic respiratory disease with ribs cage showing and poor body condition. 🎥 1.60, 1.61, 1.62

Part 2

Examination of Organ Systems

2.1

Examination of the Reproductive System

Examination of the Female Reproductive Tract in Cattle and Sheep

Examination of the female reproductive tract can be largely divided into three main areas: Abortion, vaginal and uterine prolapses, and dystocia.

- Abortion is considered a major risk in intensively managed sheep where up to 90 per cent of sheep lamb within a two-week period. Beef cattle typically calve over three to four months, and many dairy herds calve all year round, resulting in a smaller percentage of cattle at a susceptible stage of pregnancy if challenged.

The minimum requirements for laboratory submissions for abortion investigation in cattle and sheep include:

- Fetus(es) or fetal stomach content
- Placenta
- Maternal serum sample

Instructions regarding safe packaging of pathological material (UN3373 Biological substances) can be obtained from the receiving veterinary laboratory. While the first submission may identify a recognised abortifacient agent, it is important to continue submitting aborted material during the outbreak as there may be more than one agent present within the flock/herd and such knowledge is essential when formulating treatment, control and prevention strategies.

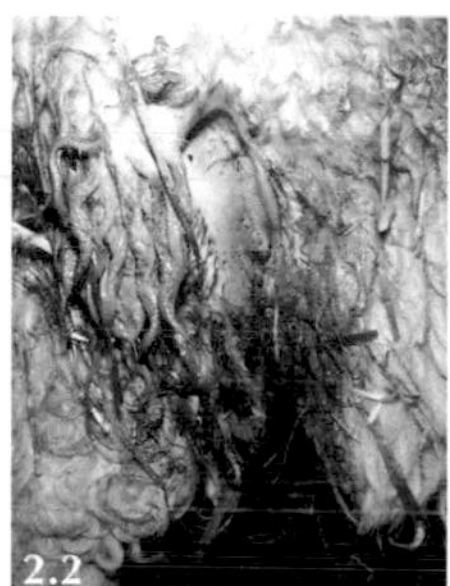

2.1: Retained placenta usually results after abortion in cattle. **2.2:** A purulent vaginal discharge may be the only sign of abortion in sheep. **2.3:** Various Salmonella serotypes can cause severe illness/death of the ewe due to fetal death and toxaemia before expulsion of the fetuses.

Abortion in Sheep

Infectious causes of abortion are most common after day 100 of pregnancy. While sporadic losses are variably attributed to handling procedures, overcrowding during housing/feeding, or movement, an abortion rate in excess of 2 per cent is suggestive of an infectious aetiology and laboratory investigation is strongly recommended. The farmer must isolate all suspect aborted sheep and dispose of all aborted material. The zoonotic potential of many abortifacient agents must be stressed to those attending sheep.

DOI: 10.1201/9781003106456-9

Appropriate hygiene precautions must also extend to the household where infection could arise from the farmer's contaminated clothing and footwear.

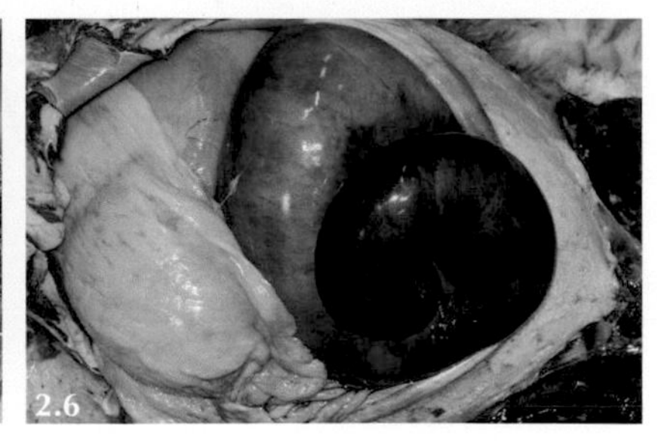

2.4: A vaginal discharge and much reduced abdominal fill suggest abortion in this ewe. **2.5:** Near term abortion. **2.6:** Death of the ewe can result before abortion.

Vaginal and Uterine Prolapses

- Vaginal prolapses are a common problem in sheep often occurring in low-ground sheep around 2 per cent, but this rate is highly variable between farms, breeds and years. Vaginal prolapse is often complicated by duration/state of the prolapsed tissues, haemorrhage from uterine vessels and prior unsuccessful attempts at replacement/retention by the farmer, and pending abortion/lambing.
- Vaginal prolapse is very occasionally seen in older beef cows but rarely in dairy cows. While often grossly contaminated with feces, such prolapses are easily replaced under low extradural block and present no further problem.
- Uterine prolapse typically occurs in sheep and beef cattle following prolonged second stage labour and delivery of an oversized lamb/calf in anterior presentation. Uterine prolapse is very occasionally encountered in dairy cows associated with hypocalcaemia. Positioning the cow in sternal recumbency with the hindlegs extended caudally greatly facilitates uterine replacement in cattle. Fatal haemorrhage from damaged caruncles and ruptured uterine vessels occurs not infrequently in cattle.

Vaginal prolapse in sheep, clinical examination should focus on:

- Potential abortion/lambing
- Clinical condition of the ewe
- Viability of the prolapsed tissues
- Content(s) of the prolapsed tissues which may aid replacement

Potential Abortion/Lambing

2.7: Healthy allantois during normal first stage labour. **2.8** and **2.9:** Abortion may cause vaginal prolapse; conversely, vaginal prolapse may cause abortion. In either case the appearance of the necrotic placenta and discoloured fetal fluids in both sheep indicate fetal death.

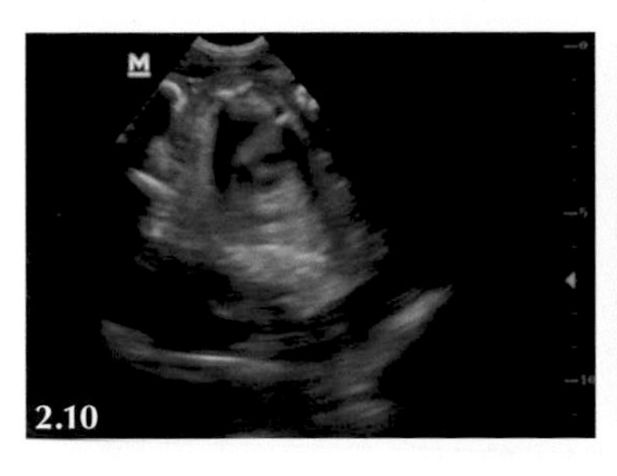
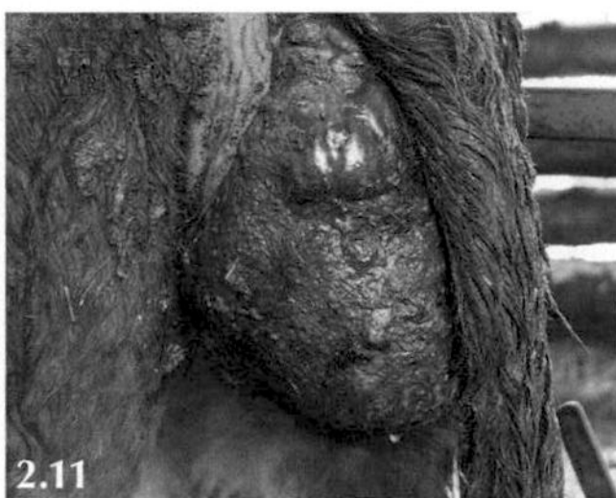

2.10: Where doubts arise about fetal viability in sheep, ultrasound examination using a microarray probe can be used to check for the beating heart and/or limb movement.
2.11: Vaginal prolapse in beef cows is easily replaced and rarely presents future problems. 2.1, 2.2, 2.3, 2.4, 2.5, 2.6

Clinical Condition of the Ewe with Vaginal Prolapse

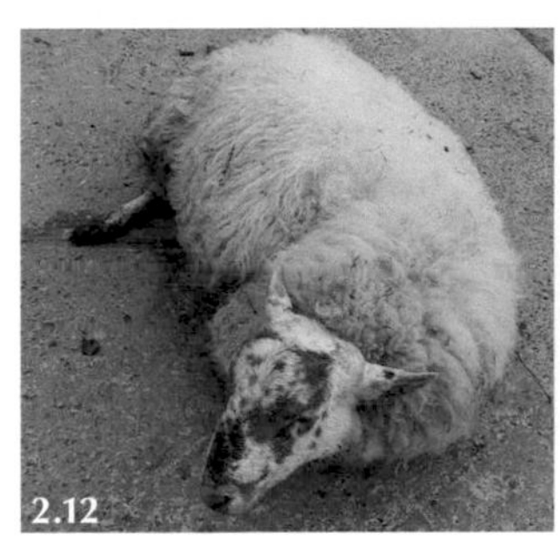

2.12: Profound toxaemia is common in ewes with vaginal prolapse and dead/autolytic lambs *in utero*. **2.13** and **2.14:** Very pale mucous membranes as a consequence of internal haemorrhage from middle uterine artery in a ewe with vaginal prolapse. 2.7, 2.8

Viability of the Prolapsed Tissues

Oedema of the vaginal wall and vulva is common when the tissues have been prolapsed more than 6 hours. Tenesmus may cause rupture of uterine vessels with the prolapsed tissue (and ewe mucous membranes) appearing very pale. Compromised blood supply and increased pressure from the prolapse's contents over several days result in a devitalised vaginal wall often with adherent dried feces.

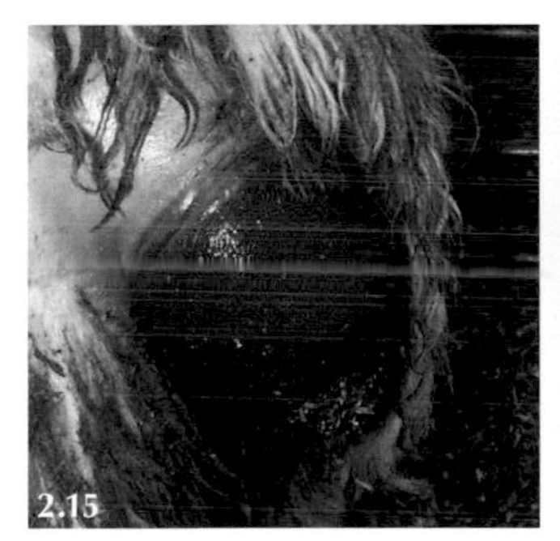
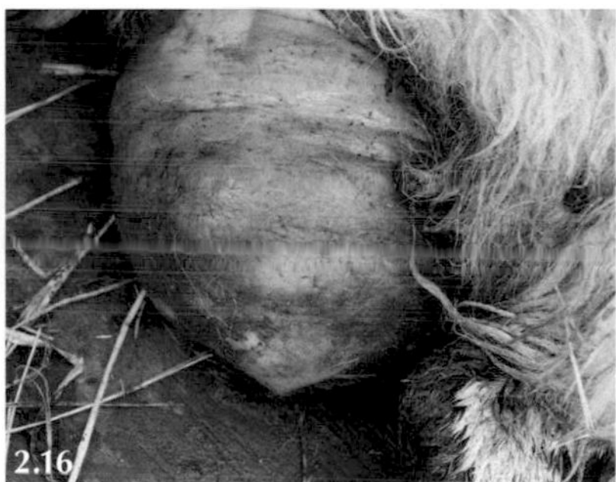
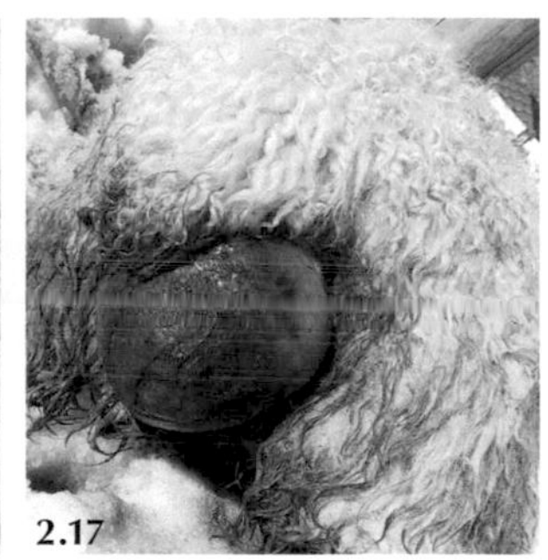

2.15: Oedematous wall of the prolapsed vagina and vulva. **2.16:** Pale oedematous vaginal wall. **2.17:** Devitalised, almost translucent (oedematous) vaginal wall covered by adherent dried feces.

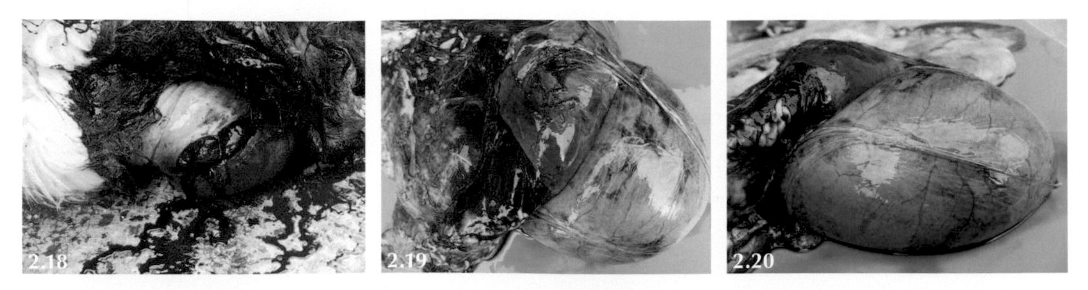

2.18: Rupture of the vaginal mucosa with arterial bleeding. **2.19** and **2.20:** Necropsy examination of the ewe (left) reveals internal haemorrhage and the urinary bladder within the prolapse.

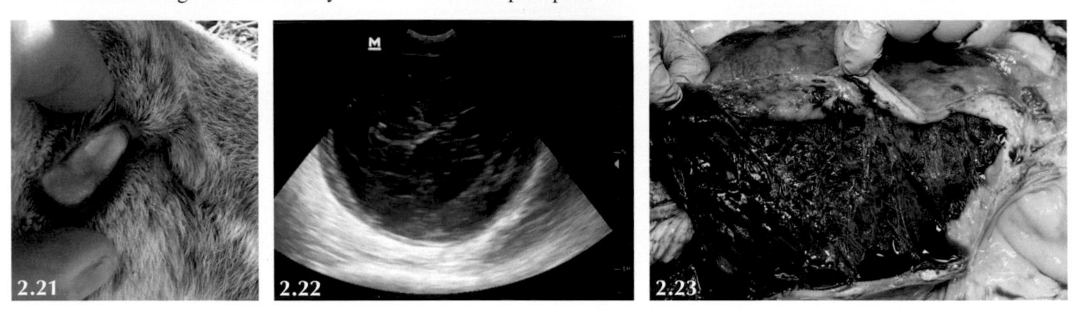

2.21: Marked anaemia as a consequence of haemorrhage associated with vaginal prolapse. **2.22:** Ultrasound examination of 7 cm diameter blood clot cranial to the pelvic inlet. **2.23:** Confirmation of haemorrhage at necropsy.

Content(s) of the Prolapsed Tissues (Sheep)

The contents of a vaginal prolapse are either the urinary bladder (much more common) or uterine horn. Raising the prolapse after low extradural block corrects the kink in the urethra and urine drains freely from the bladder (see next). Needle decompression of the contents of the vaginal prolapse is not needed nor advised.

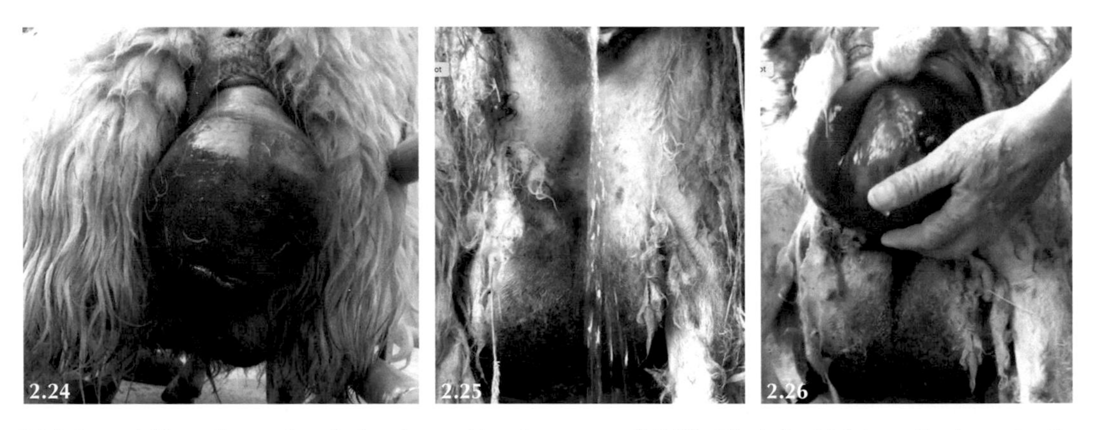

2.24: Large (>25 cm diameter) vaginal prolapse with oedematous wall. **2.25:** Alleviating kink in urethra by raising the bladder/prolapse after low extradural block releases a steady flow of urine. **2.26:** Same case as left after draining the urinary bladder. Note the swelling now largely comprises very oedematous vaginal wall and vulva. 2.9, 2.10

Dystocia

Absolute fetal oversize is common in beef cattle where many bulls are still purchased on phenotypic appearance. Caesarean operations are common in certain purebred beef breeds. Mal-posture is the most common cause of dystocia in dairy herds where much greater use is made of estimated breeding values for dystocia risk and gestation length. Veterinary involvement with ovine dystocia is largely determined by farm economics and often restricted to valuable pedigree ewes with a single lamb in posterior presentation. Paravertebral anaesthesia takes 5–10 minutes to be effective; surgery time is around 20 minutes for sheep and 40 minutes for cattle although more experienced surgeons may be quicker. Total costs are around £90–180 for sheep and £240–400 for cattle.

Examination of the Female Reproductive Tract—Dystocia in Sheep

- Always administer a combined low extradural block of lignocaine and xylazine before commencing your obstetrical examination.
- Administer a NSAID intravenously before investigating the problem rather than afterwards.
- Make the decision to undertake a caesarean operation early (within 5 minutes).
- Vaginal delivery of a large lamb in posterior presentation risks serious damage to the lamb's rib cage and/or liver.

First stage labour is represented by cervical dilation which takes 3–6 hours but is more rapid in multiparous ewes. Second stage labour is represented by expulsion of fetus(es), and typically takes about 1 hour. Third stage labour is completed by expulsion of fetal membranes which usually occurs within 2–3 hours of the end of second stage labour. Retained fetal membranes and subsequent metritis are rare in sheep.

2.27: Early first stage labour—the ewe tries to find solitude (not easy in a crowded lambing shed). **2.28:** Ewe during first stage labour with appearance of the allantois. **2.29:** Frequent forceful abdominal contractions during early second stage labour.

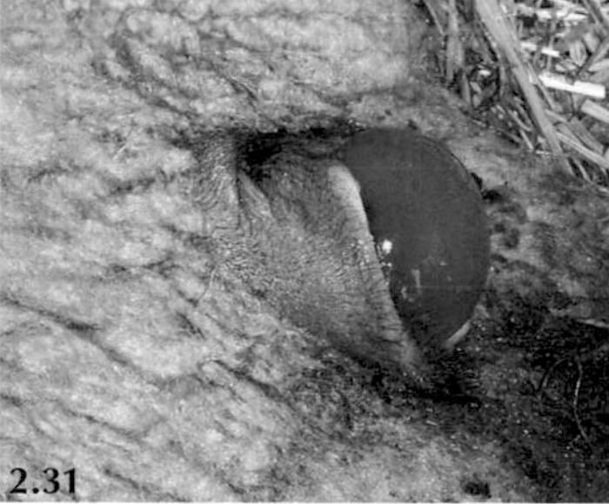

Sequential images of normal lambing. **2.30:** Frequent forceful abdominal contractions during early second stage labour. **2.31:** Appearance of the amnion. **2.32:** Head and forefeet (anterior presentation) at the vulva.

It is essential to assess the patient in detail before attempting to correct any problem. Lateral recumbency with frequent abdominal straining and vocalisation may result from engagement of the lamb within the pelvis or excessive manual interference leading to trauma of the posterior reproductive tract. Bruxism (teeth grinding) and an elevated respiratory rate with an abdominal component (panting) may indicate more serious concerns such as uterine rupture.

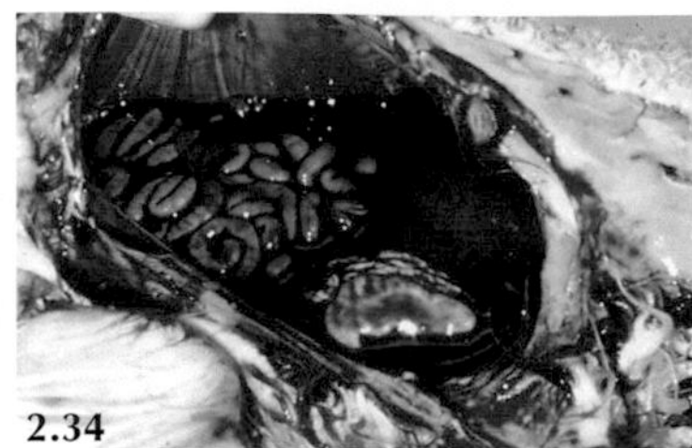

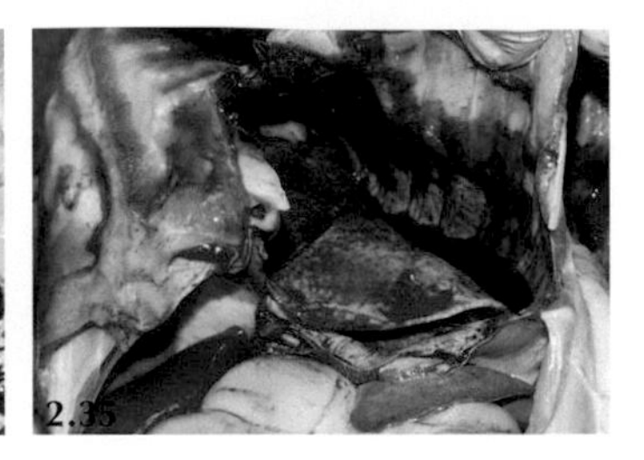

2.33: Haemorrhage with blood contamination of the fleece arising from unskilled attempt to correct dystocia. **2.34:** Liver rupture as a consequence of forced delivery of posterior presentation. **2.35:** Rib fractures and associated sub-pleural haemorrhage caused by forced delivery of posterior presentation.

Attempted delivery by an unskilled shepherd frequently results in oedema, reddening and bruising of vulval labiae. The ewe's mucous membranes must be checked for evidence of pallor. Blockage of the ewe's reflex abdominal contractions by low extradural block greatly assists corrections/manipulations, especially bilateral shoulder flexion, with obvious animal welfare benefits.

Make the decision to undertake a caesarean operation as soon as possible to achieve the best outcome. Discuss with the farmer beforehand if you think that there has been trauma to the posterior reproductive tract that could compromise the surgical outcome. 2.11, 2.12

Parturition in Cattle

Gestation length is around 285 days but is highly variable both within and between breeds. Imminent parturition can be detected by udder development, accumulation of colostrum and slackening of the sacro-iliac ligaments. Where possible, cows will isolate themselves from the group. The birth process is divided into three stages.

First Stage Labour

First stage labour is represented by cervical dilation which takes 3–6 hours but is more rapid in multiparous cows. There are various behavioural changes including separation into a corner of the field or barn, and alternatively lying/standing. A thick string of mucus is often observed hanging from the vulva. The bouts of abdominal straining occur more frequently, usually every 2–3 minutes. At the end of first stage labour the cervix is fully dilated. 2.13

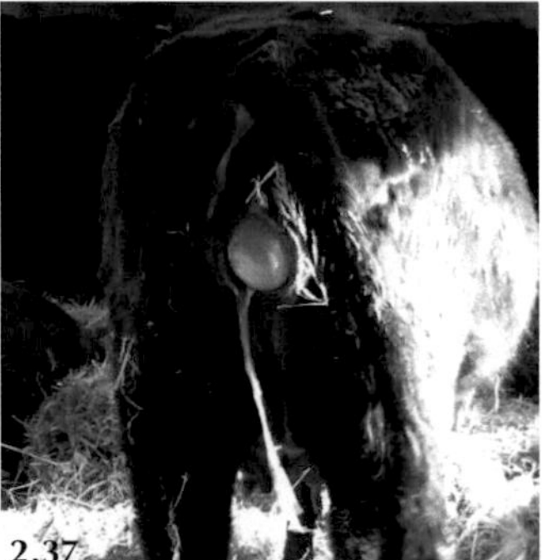

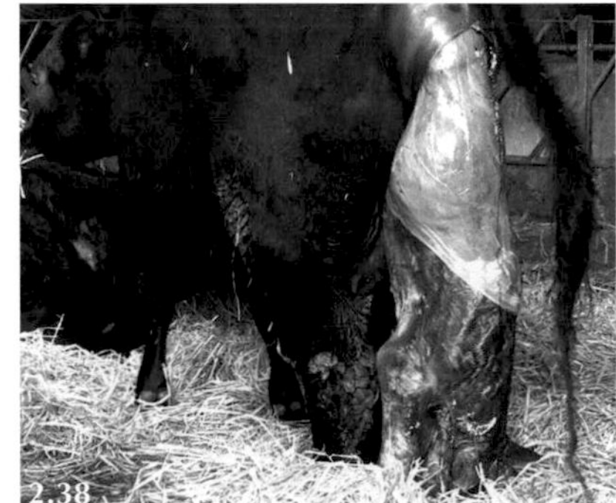

2.36: Slackening of the sacro-iliac ligaments with a thick string of mucus observed hanging from the vulva. **2.37:** Appearance of the allanto-chorion. **2.38:** End of second stage labour. 2.14

Second Stage Labour

Second stage labour is represented by expulsion of fetus(es), and takes from 5 minutes to several hours. There is rupture of allanto-chorion with a sudden rush of fluid. The amnion and fetal parts are then engaged in pelvic inlet. The amniotic sac appears at the vulva and ruptures at this stage. Powerful reflex and voluntary contractions of abdominal muscle and diaphragm ("straining") serve to expel the fetus. Intravenous NSAIDs are indicated before a veterinary-assisted calving. 🎥 2.15

Third Stage Labour

Third stage labour is completed by expulsion of fetal membranes which usually occurs within 2–3 hours of the end of second stage labour.

Facilities

Many farmers routinely calve cows in cattle stocks; every farmer has a calving jack with calving ropes stored hanging on a nearby gate. However, wherever safe to do so, the cow should be haltered then released into a clean calving pen with the shank end tied low down to a substantial post allowing approximately 1.5 m of lead rope because delivery of the calf is greatly facilitated when the cow is able to lie in lateral recumbency and strain using her abdominal muscles. Be aware that some beef cows can become very aggressive after parturition; make sure you can safely exit the pen.

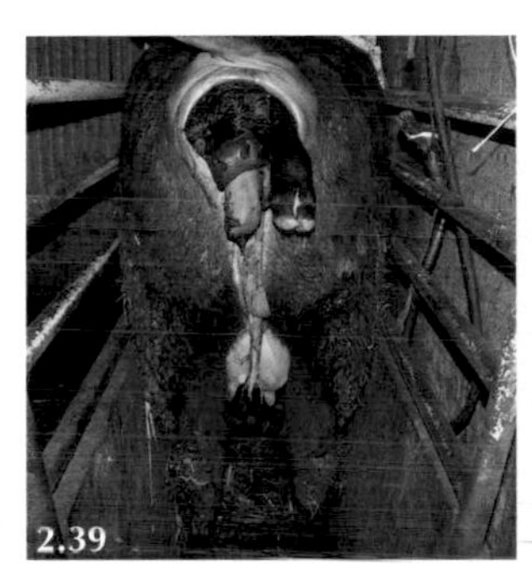

2.39: If safe to do so, release the cow from the stocks once the malposture has been corrected.
2.40: Cows naturally lie down to calve.

General Guidelines

The general guidelines that follow can be applied when presented with suspected fetal oversize whether absolute or relative.

Anterior Presentation

Calf's front fetlock joints must protrude more than one hand's breadth beyond the cow's vulva after a maximum period of 10 minutes' traction (two people; yourself and farmer) to safely proceed (calf's shoulders fully engaged in the pelvis); reconsider option if greater traction necessary. A calving rope placed around the calf's poll helps guide the head into the pelvic canal and is especially useful in delivering large calves; it should be applied before the leg ropes.

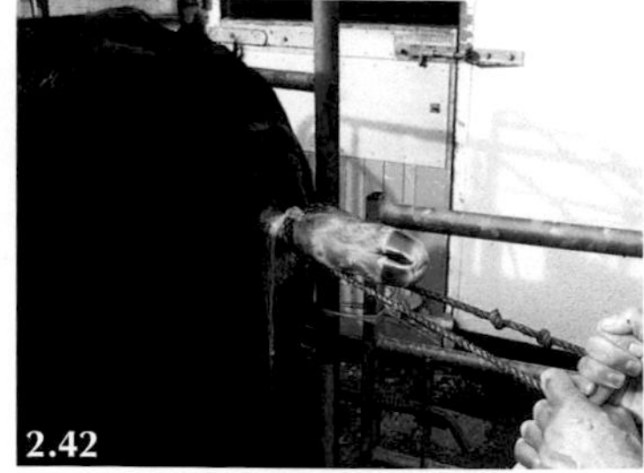

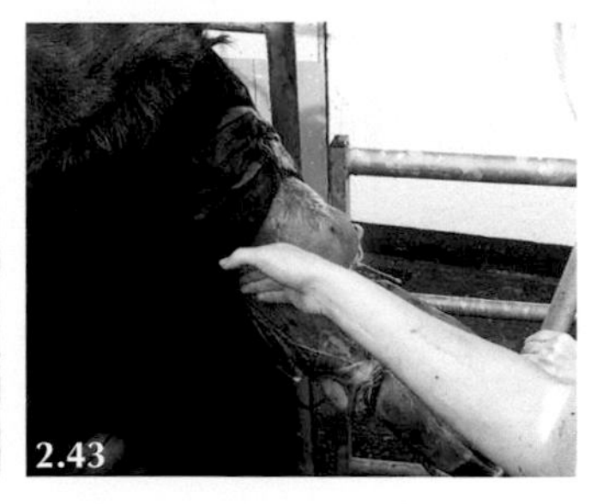

2.41: Calf in anterior presentation. **2.42:** Calf's front fetlock joints now protrude more than one hand's breadth beyond the cow's vulva with two people pulling. **2.43:** Calf's head through vulva and delivered safely.

Posterior Presentation

Two strong people pulling on calving ropes should be able to extend the hocks more than one hand's breadth beyond the cow's vulva within 10 minutes (calf's hindquarters now fully within the pelvic inlet). With experience it is possible to apply greater traction than the forces described here and still achieve a successful resolution, but there are occasional doubts when the calf becomes lodged. Long bone fractures can occur when calves in posterior presentation are delivered in cattle stocks if the cow lies down and the calving jack cannot be immediately released.

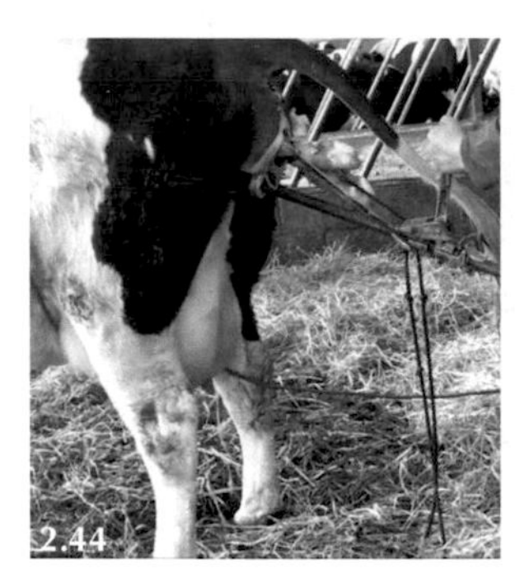

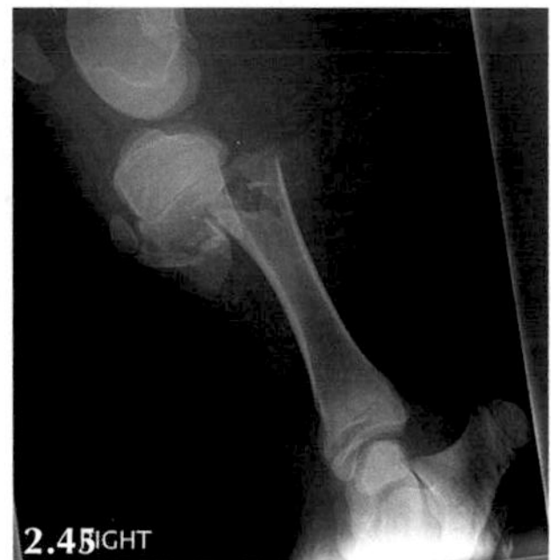

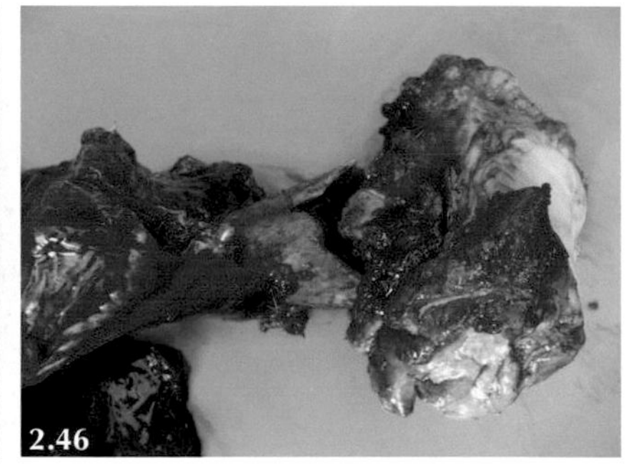

2.44: Note that the calf's hocks are not beyond the cow's vulva and considerable traction is being applied. **2.45:** Lateral radiograph revealing fracture through the proximal tibial growth plate caused by cow attempting to lie down when delivering calf in posterior presentation in cattle stocks. **2.46:** Necropsy findings of fracture (**2.45**).

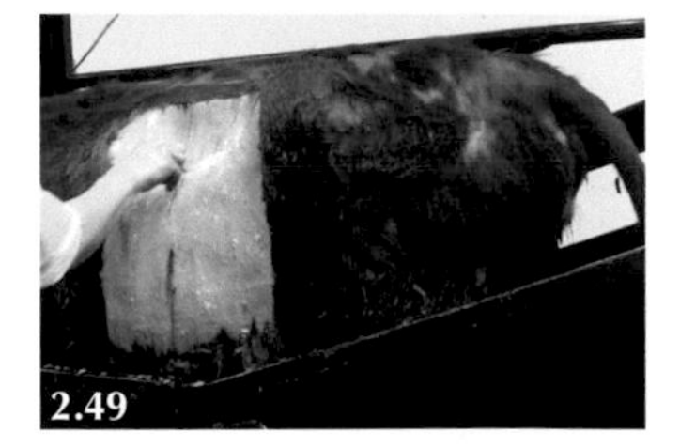

2.47: Many cattle stocks do not allow access to the left flank for a caesarean operation. **2.48:** Close-up of cattle stocks featured (left). **2.49:** A different crate design allows excellent safe access to the left flank.

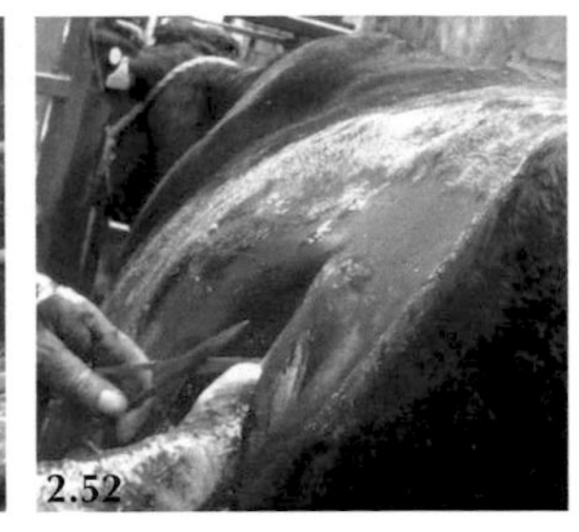

2.50 and **2.51:** Before and after caesarean operation where the cow was restrained only with a rope halter. **2.52:** Intra-op image showing cow restrained only with a rope halter. Operator safety must be carefully considered in these situations however good your paravertebral anaesthetic technique. 2.16, 2.17

Uterine Prolapse

Uterine prolapse is readily identified by the presence of caruncles and often adherent placenta. Uterine prolapse most commonly occurs immediately after assisted delivery of an over-sized lamb/calf in anterior presentation. Low extradural block is essential for replacement and preventing tenesmus and re-prolapse.

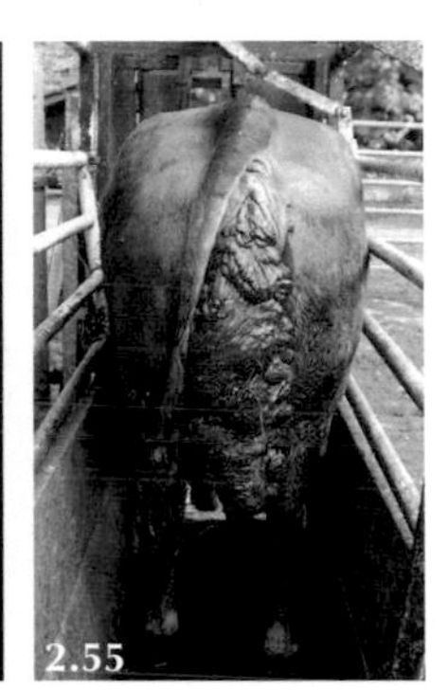

2.53: Uterine prolapse occurred immediately after delivery of an over-sized lamb in anterior presentation—note the caruncles and attached placenta. **2.54:** Incomplete return of uterine prolapse by farmer has resulted in re-prolapse 24 hours later after prolonged tenesmus—note the mattress suture in the vulva in an attempt to hold the uterus in place. The uterus is oedematous with superficial necrosis of the caruncles. **2.55:** Replacement of uterine prolapse in cattle is most easily achieved with the cow resting on its brisket and the hindlegs extended caudally although this is not always possible in fractious animals.

Examination of the Male Reproductive System

- Note ram/bull body condition, conformation and any lameness.
- Measure scrotal circumference.
- Palpate scrotum/testicles for consistency, symmetry, heat, swelling and free movement within scrotum. Ultrasound scan all palpable scrotal abnormalities.
- Collect semen sample using either natural service and an artificial vagina or electro-ejaculation. Determine progressive motility and sperm morphology.
- Be aware of the limitations of a single semen sample collected by electroejaculation in rams especially if the sample is abnormal when there are no palpable abnormalities—seek advice and consider referral to a facility that uses an artificial vagina for semen collection.
- Examine the prepuce/penis.
- Observe natural service (bulls).

Physical Examination

Rams should have a minimum scrotal circumference >36 cm for shearlings and >32 cm for ram lambs; animals with measurements below these threshold values should be rejected (>34 cm for bulls more than two years old, and >32 for eighteen-month-old bulls). Determine symmetry of the testicles and whether there is free movement within the scrotal sac. The semen sample should have >60 per cent progressive motility and >70 per cent normal sperm morphology. 🎥 2.18, 2.19

Bull Breeding Soundness Examination

The reader is directed to the article appearing in The *Veterinary Record* in 2010 by Dr Colin Penny which details the bull breeding soundness examination. The BCVA's bull pre-breeding examination certificate. *Veterinary Record* (2010) 167: 551–554 doi: 10.1136/vr.c5216

Ultrasound Examination

Ultrasound examination allows differentiation between a transudate within the scrotum (inguinal hernia) and an inflammatory exudate separating the vaginal tunics (orchitis). Ultrasound examination differentiates abscesses from testicles when surrounded by thick fibrous tissue as occurs in epididymitis and sperm granuloma cases in rams.

Ultrasonographic examination of the scrotum is undertaken in the standing animal using a 5 MHz linear scanner connected to a real-time, B-mode ultrasound machine (bulls are restrained in cattle stocks). Sequential examination of the pampiniform plexus, head of the epididymis, testicle, then body and tail of the epididymis is undertaken as the transducer head is moved distally over the lateral aspect of the scrotum. The pampiniform plexus reveals a matrix of hyperechoic (bright white) lines throughout the conical anechoic area. The normal testicle appears as a uniform hypoechoic area with a hyperechoic mediastinum clearly visible. The sonograms shown next feature rams but the pathology in bulls would show similar ultrasonographic changes.

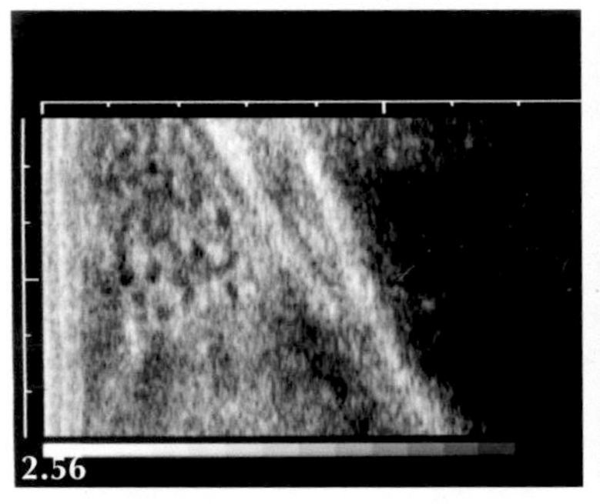

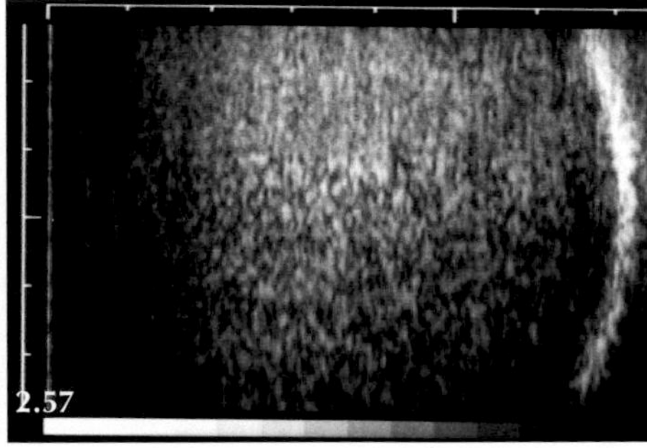

2.56: Junction between pampiniform plexus dorsally (top) and proximal pole of the testicle in a ram. **2.57:** Normal uniform echotexture of 7 cm diameter testicle in a ram (cm scale top of image).

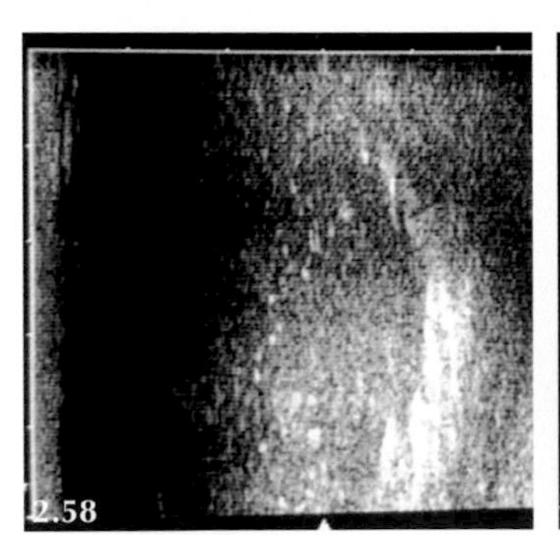

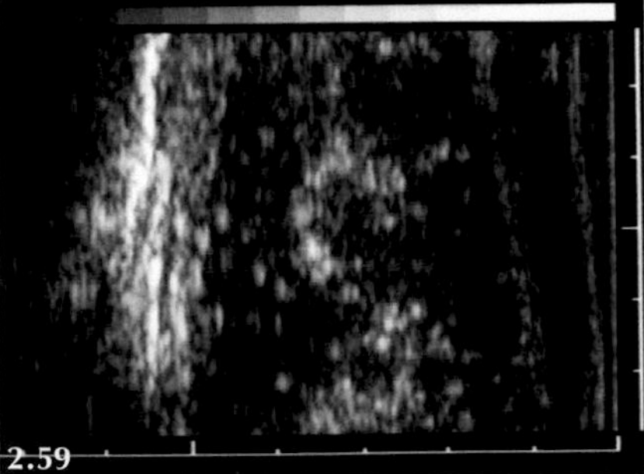

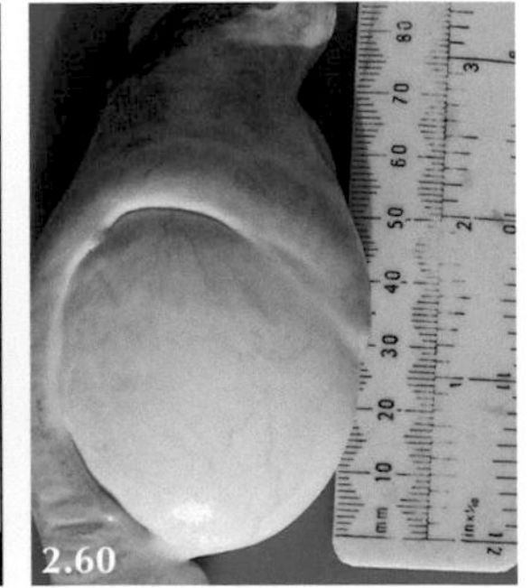

2.58 and **2.59:** Testicular atrophy/hypoplasia in rams—diameter is less than 5 cm diameter and more hypoechoic containing multiple hyperechoic dots. **2.60:** Testicle diameter less than 5 cm.

Testicular atrophy/hypoplasia is diagnosed when the diameter is less than 5 cm and the testicles appear more hypoechoic and contain multiple hyperechoic spots. These hyperechoic dots are thought to represent the fibrous supporting architecture now more obvious due to atrophy of the seminiferous tubules. In some cases, the hyperechoic dots with shadowing may represent calcification. The tail of the epididymis is distinct from the testicle and considerably smaller in diameter (2 to 3 cm compared to 6 to 7 cm) with a distinct capsule. It may prove difficult to obtain good contact between the linear probe head and the smaller diameter tail of the epididymis.

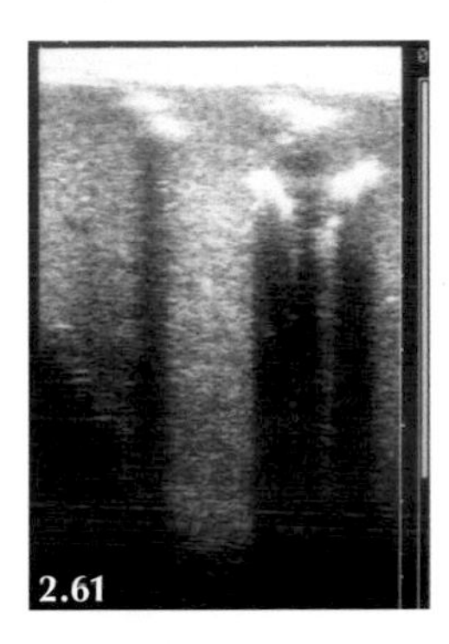

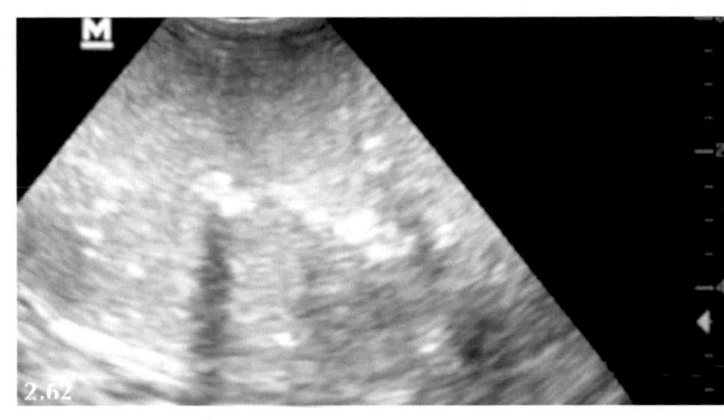

Testicular atrophy/hypoplasia with the testicle appearing more hypoechoic and containing multiple hyperechoic spots with shadowing in a bull (**2.61**) and ram (**2.62**).

Orchitis

There are variable amounts of fibrinous exudate between the vaginal tunics and changes in echotexture of the testicles with hyperechoic areas throughout.

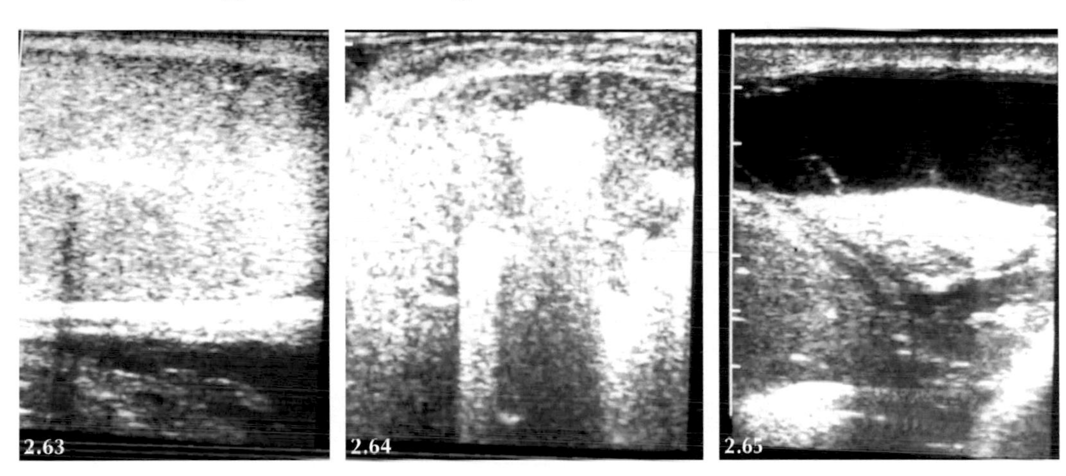

Unilateral orchitis and fibrinous exudate in a ram. The image quality is poor as they are reproduced from screen grabs of a video recording but are nonetheless useful for comparative purposes. **2.63:** Testicular atrophy contralateral to orchitis. **2.64:** Numerous large hyperechoic areas within testicle in case of orchitis. **2.65:** Distal pole of testicle showing 2–3 cm of inflammatory exudate (anechoic area containing numerous fine fibrin strands). 2.20

Epididymitis

Acute epididymitis results in a swollen oedematous scrotum with exudate within the epididymes (see next). Ultrasonographic examination of rams with chronic epididymitis (three to six months' duration) reveals several 2 to 8 cm diameter well-encapsulated abscesses (anechoic areas containing many bright spots, "snowstorm" appearance). It may prove difficult to differentiate abscesses from atrophied testicles during ultrasound examination because the large amount of fibrous tissue in the abscess capsule walls reduces image quality.

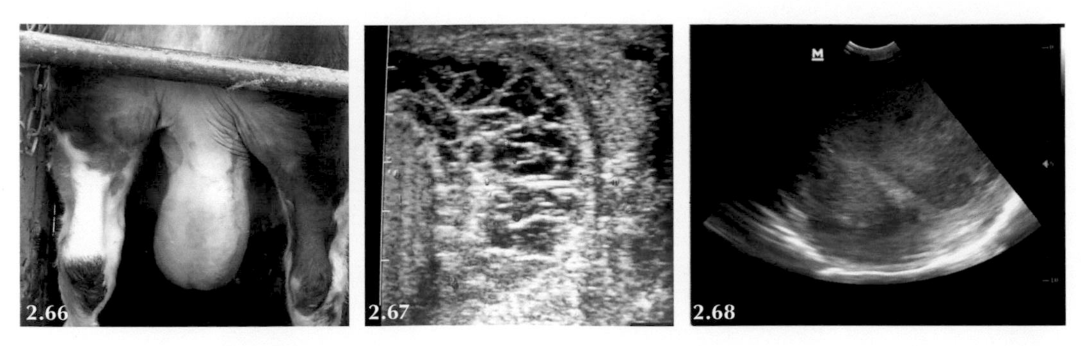

2.66: Enlarged oedematous scrotum in a young bull. **2.67:** Ultrasound examination of bull (left) reveals inflamed tail of the epididymis. **2.68:** Chronic epididymitis (four to six months) in a ram forming 8 cm diameter abscesses within the tail of the epididymis with 1 cm fibrous capsule. 2.21, 2.22, 2.23

Sperm Granuloma

With sperm granuloma the scrotal contents have a bi-lobed appearance due to testicular atrophy and significant abscessation of the tail of the epididymis.

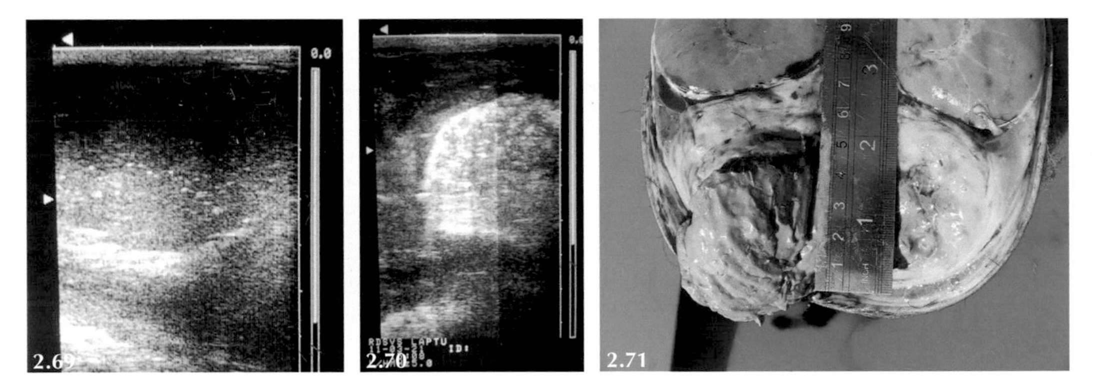

2.69: Testicular atrophy/hypoplasia—diameter is less than 5 cm diameter and more hypoechoic containing multiple hyper-echoic dots. **2.70:** 4 cm diameter abscess with 1 cm fibrous capsule. **2.71:** Findings confirmed at necropsy. 2.24

Inguinal Hernia

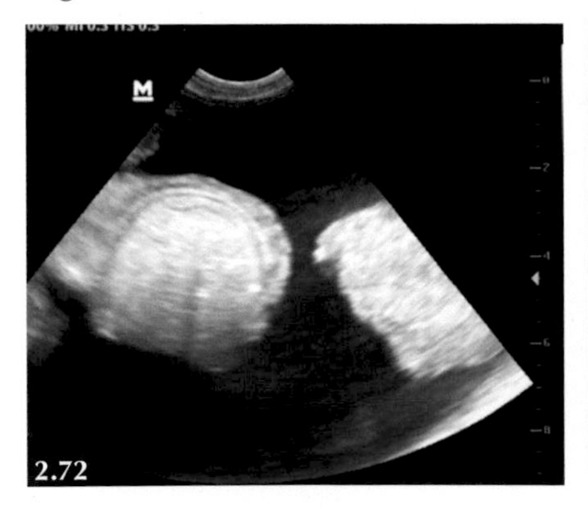

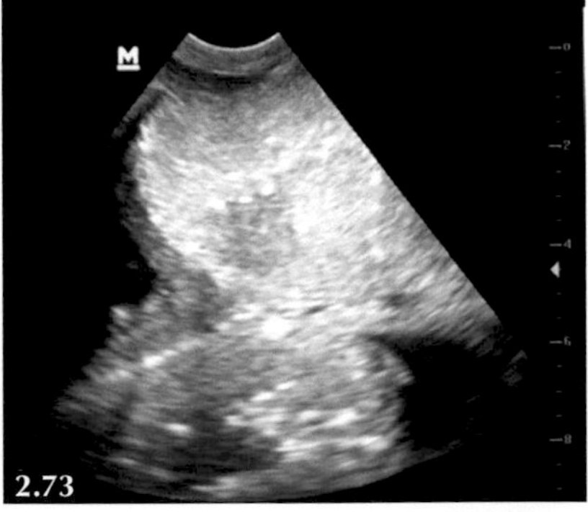

2.72: Ultrasound examination of an inguinal hernia reveals a large volume of peritoneal fluid and cross section of small intestine and omentum within scrotum. **2.73:** Testicular atrophy with heterogenous appearance of the testicle. 2.25

Semen Evaluation

Despite disadvantages with regard to assessment of rams' libido, occasional collection of an unrepresentative poor sample and animal welfare concerns, electroejaculation remains the most common method for checking individual suspect rams for breeding soundness on commercial farms. However, semen sample(s) collected from an artificial vagina are increasingly required for many infertility claims. Fresh semen is much preferred for artificial insemination programmes.

The ram is positioned in lateral recumbency and the penis extruded by extending the sigmoid flexure. A gauze swab is wrapped around the penis proximal to the glans to prevent retraction into the prepuce. A warmed 7 ml plastic collection tube is held over the glans penis. A Ruakura type electro-ejaculator (possibly modified to a greater diameter probe) is introduced into the rectum to a pre-determined length just caudal to the pubic symphysis. The handle of the probe is gently raised positioning the stimulatory electrodes on the probe tip adjacent to the accessory sex glands. The ram is stimulated for 4 seconds, the probe is then switched off for 4 seconds. Stimulation often causes vocalisation and sudden muscular spasm with resultant arching of the back and extension of the hindlegs. Rigid extension of one hindleg during stimulation indicates that the tip of the probe is not positioned in the midline but has moved toward the side of the extended hindleg. While the electrical current is switched off the probe is slowly moved to massage the accessory sex glands.

Note that the ram will ejaculate during massage of the accessory sex glands while the probe is switched *off*. A colourless/pale yellow watery sample of 0.5 ml probably represents pre-ejaculatory fluid and a semen sample, thick creamy white 0.7 to 2.0 ml sample, will be collected after the next cycle of electrical stimulation/accessory sex gland massage. This sequence can be repeated for a maximum of three stimulations. If no sample is collected the ram should be released and a further attempt made later.

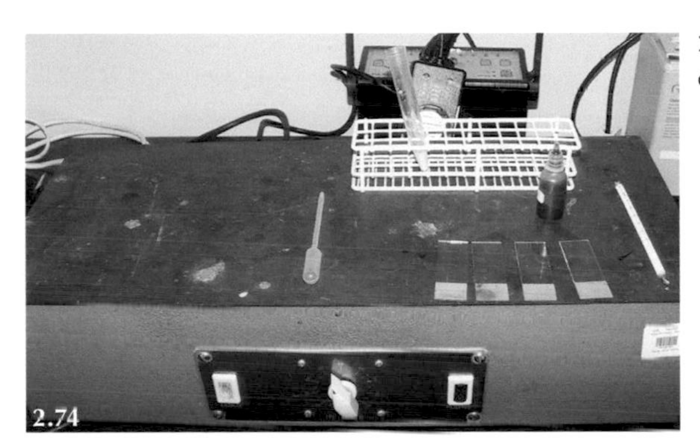

2.74: A heated plate and microscope stage are essential for semen examination.

Progressive Motility and White Blood Cells

One drop of semen is transferred to the centre of a warm microscope slide. A small amount of semen is then collected on the corner of a cover slide and transferred to 5 drops of warm phosphate buffered saline (PBS). After mixing, the cover slip is placed on top of the now-diluted semen sample and examined ×100 under dark ground microscopy. Vigorous forward motility of individual sperm can be readily identified in normal semen samples. White cells, whose presence is grossly abnormal, appear as round transparent cells somewhat smaller than the sperm heads.

Sperm Morphology

Five drops of nigrosin/eosin stain are now added to the drop of semen in the centre of the slide and thoroughly mixed. A thin smear is then made by picking up a small amount of semen sample/stain on the corner of a microscope slide and pushing this slide at a shallow angle along the original slide. When dry the sample is examined under oil immersion ×400. The presence of white cells is a significant finding and indicates inflammation of the urino-genital system. Primary spermatozoan abnormalities involve the head and acrosome and are associated with serious testicular conditions. Tail abnormalities are often associated with less severe problems and disease of the epididymis. Sperm abnormalities should not total more than 30 per cent of the 100 spermatozoa examined.

2.2

Examination of the Digestive System

Whilst both species are ruminants, the dietary management of sheep and cattle varies from sheep grazing fibrous hill pasture all year to dairy cows permanently housed and fed very high energy rations with limited roughage content *ad libitum*. As a consequence, dairy cows present with a wide range of digestive tract disorders rarely seen in sheep.

Adult Cattle

- Lactating dairy cattle are fed high quality, high energy rations *ad libitum* to maximise dry matter intake whereas spring-calving beef cows are often fed restricted high fibre rations necessitating mobilisation of body reserves.
- Abdominal shape and rumen motility are influenced by diet; both are increased with a fibre-based ration. Advanced pregnancy can cause considerable abdominal distension especially in old beef cows fed fibre-based diets.
- Normal lactating dairy cow feces contacting the ground has been described as "a slow handclap" whereas beef cow feces on a straw-based ration can be used to build eco-friendly huts. Melaena is common in dairy cows following abomasal ulceration. The passage of only mucus suggests intestinal obstruction.

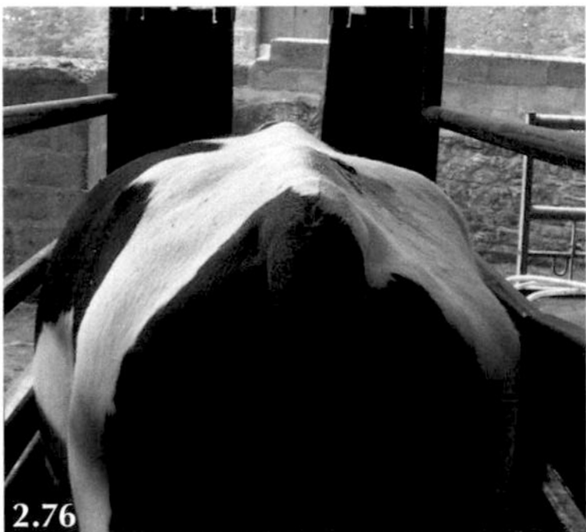

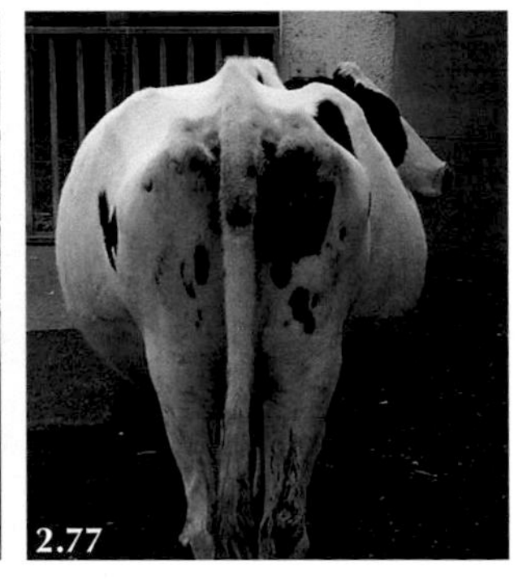

Abdominal contours **2.75:** advanced pregnancy in an old beef cow. **2.76:** Choke causing visible distension of the left sublumbar fossa in a dairy cow. **2.77:** Vagus indigestion affecting a non-pregnant dairy cow.

DOI: 10.1201/9781003106456-10

2.78 and **2.79:** Older, heavily pregnant beef cows produce a range of interesting abdominal contours. The passage of only mucus (**2.80**) suggests intestinal obstruction.

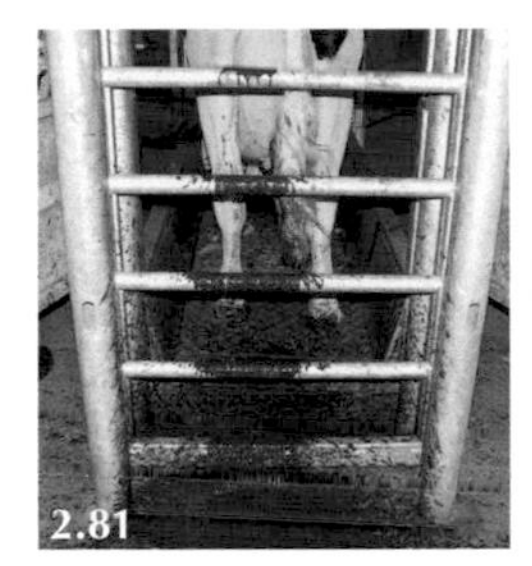

2.81: Melaena in a Holstein cow. **2.82:** Normal feces from another cow on same ration. **2.83:** Close-up of melaena, slightly "tarry" feces (no filters used on either fecal sample allowing direct comparison). It is important to recognise this change—see 2.26.

Adult Sheep

- Profuse diarrhoea is an abnormal finding under all sheep feeding/management systems.
- With the exception of parasitic gastroenteritis, digestive tract problems are much less common in sheep compared to cattle.
- Molar dental problems cause significant mastication problems (quidding) and chronic weight loss in older sheep.

Sheep pass pelleted feces except when grazing very lush pastures when soft stools are passed; diarrhoea is an abnormal finding under all feeding systems. The presence of dried fecal staining the tail and perineum reveals a prior episode of diarrhoea and is abnormal.

Diarrhoea is an abnormal finding under all sheep management systems. Episodes involving a large percentage of the group suggests endoparasites (**2.84** and **2.85**) while diarrhoea in individual sheep suggests an underlying health problem (**2.86**). 2.27, 2.28

Dentition

Dental problems are very uncommon in cattle whereas correct dentition is of critical importance to the maintenance of body condition/weight gain in sheep. The necessity to fully examine the ovine mouth, with particular reference to the molar teeth, cannot be over-emphasised. Overgrown, worn and absent molar teeth cause serious problems with mastication of fibrous feeds. Sheep have 32 permanent teeth with a dental formula of 2 (incisors 0/4, premolars 3/3, and molars 3/3). The temporary incisor teeth erupt sequentially at approximately weekly intervals from birth. The three temporary premolars erupt within two to six weeks. The first permanent molar erupts at three and five months in the lower and upper jaws, respectively. The second permanent molar erupts at 9 to 12 months, and the third permanent molar and permanent premolars erupt between 18 and 24 months.

Incisor teeth alignment is examined by running an index finger along the dental pad and incisor teeth with the sheep's mouth closed. This examination will reveal any teeth projecting forward of the normal contact area on the dental pad (overshot jaw or prognathia) or behind (undershot jaw or brachygnathia). This examination must be undertaken with the mouth closed and the head held in the normal resting position.

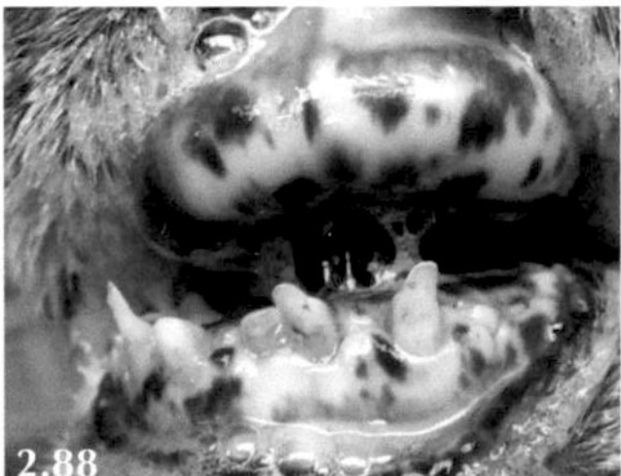

2.87: Lateral view of sheep with undershot jaw or brachygnathia. 2.29. **2.88:** Missing incisor teeth often referred to as "broken mouth". **2.89:** Elongated and mal-aligned central incisor which does not contact the dental pad; two incisor teeth are missing.

Molar tooth problems can best be identified by impaction of food in the cheeks and short jerky jaw movements with the mouth held slightly open resulting from excessive tooth growth or cheek lesions causing pain during mastication. The sheep will often raise its head to assist movement of the food bolus over the dorsum of the tongue and swallowing. Affected sheep often have pieces of fibrous feed protruding from the commissures of the mouth with frequent quidding. Careful palpation of the dental arcade through the cheek reveals the sharp irregular ridges of the labial aspect of the upper cheek teeth, and any lost upper cheek teeth. There is no enlargement of submandibular lymph nodes associated with molar tooth loss. Examination of the molar teeth with a gag and torch is essential but lengthy examination is resisted by the sheep. Radiographs can provide useful information of the jaw and cheek teeth, but it proves difficult not to superimpose the cheek teeth of the contralateral jaw even using oblique views (see section on radiography).

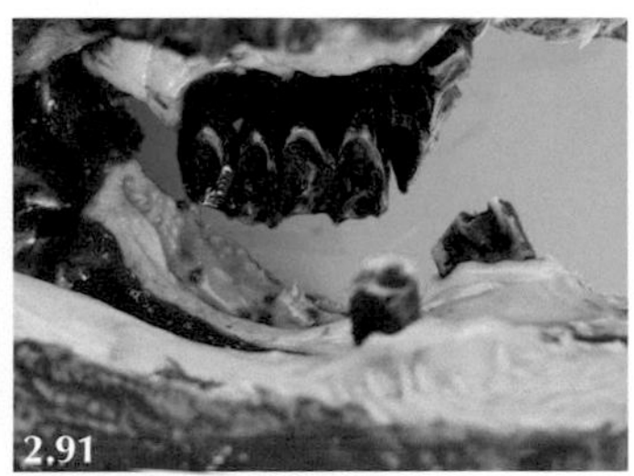

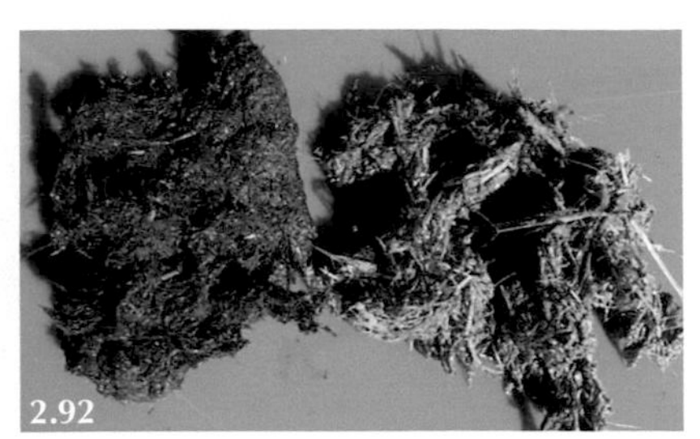

2.90: Molar tooth problems can best be identified by impaction of food in the cheeks and excessive salivation. Note also the reduced abdominal fill and poor body condition of this ewe. **2.91:** Missing cheek teeth with overgrown opposite teeth. **2.92:** Normal rumen contents (left in image) compared to more fibrous contents (less masticated) in sheep with molar dentition problems. 🎥 2.30, 2.31

Examination of the Forestomachs

When viewed from behind the rumen pushes the lower left flank beyond the outline of the costal arch. Bloat causes visible distension of the left sublumbar fossa. Choke caused by potatoes/turnips is common in cattle but rarely seen in sheep. A probang must be used carefully to move an obstruction because oesophageal perforation can result from excessive force.

Auscultation of the rumen is performed in the upper left sublumbar fossa with pressure applied to the stethoscope head to achieve good contact between the stethoscope/flank/rumen wall. Note that a gap between the rumen wall and abdominal wall will reduce transmitted sound while percussion will elicit a high-pitched "pinging sound" referred to as "rumen void". There are two independent reticulo-ruminal contraction sequences, The primary biphasic contraction cycle of the reticulum is followed by rumen contractions, occurring approximately once a minute, mixing ingesta and forcing small particles into the omasum. The secondary contraction does not involve the reticulum, but rumen activity pushes the gas cap into the cardia region with resultant eructation. Typically, one secondary cycle follows two primary cycles such that three cycles occur every 2 minutes. Cattle fed fibrous diets have a more distended rumen than those fed solely concentrates with more frequent contractions.

Rumen Fluid Analysis

Percutaneous rumen fluid collection and pH determination can be used to investigate the possibility of subacute ruminal acidosis (SARA) in cattle fed high energy dense rations, but recent work using continuous rumen fluid pH monitors has questioned the value of such investigations.

Abdominocentesis

Abdominocentesis is most frequently attempted from the ventral midline immediately caudal to the xiphisternum in cattle with suspected reticulo-peritonitis however such results rarely add any significant additional information to that already gathered by ultrasonography.

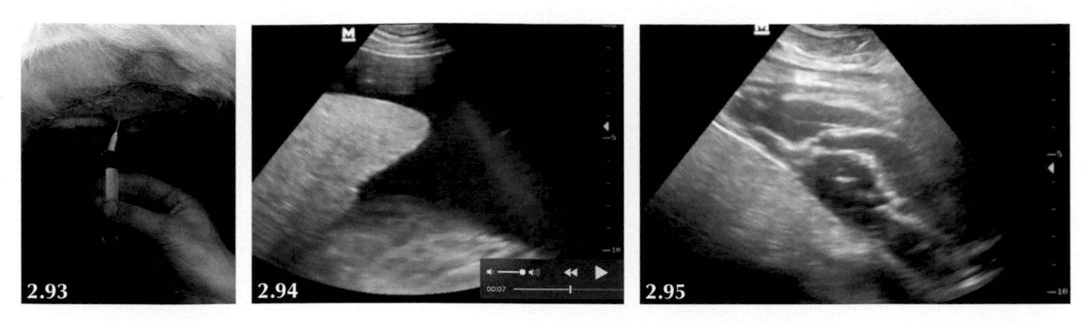

2.93: Abdominocentesis is attempted from the ventral midline immediately caudal to the xiphisternum. **2.94:** Ultrasound examination reveals >10 cm transudate within the abdomen with dorsal displacement of viscera (liver and greater omentum). **2.95:** Trans-abdominal ultrasonography from the same site in another cow reveals 6–7 cm organised fibrinous exudate—note the hyperechoic fibrin tags within the anechoic fluid (scale right hand margin). 2.32

Radiography

Radiography of the bovine digestive system can be undertaken to investigate potential mandibular fractures although these are usually identified by a step in the dental arcade when palpated with a mouth gag placed in the contralateral side of the mouth. With the exception of tooth root abscesses in growing cattle, molar teeth problems are very uncommon in cattle which is very different to sheep.

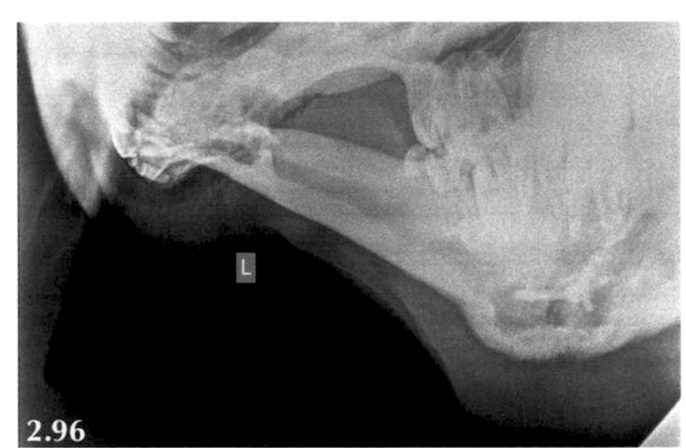

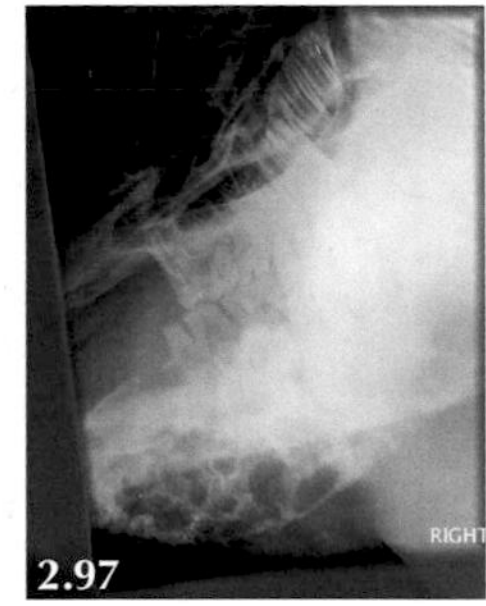

2.96: Radiography is helpful in the investigation of potential molar teeth lesions; there is chronic tooth root infection extending into the horizontal ramus in this bull. **2.97:** The extent of new bone proliferation in this case of actinomycosis of the right mandible is revealed by radiography.

Radiography can be usefully employed to investigate potential mandibular fractures and molar dentition problems although the latter can generally be quantified using a mouth gag and torch.

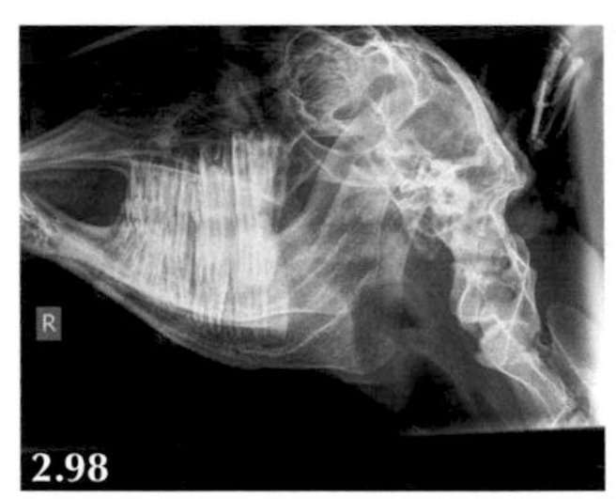

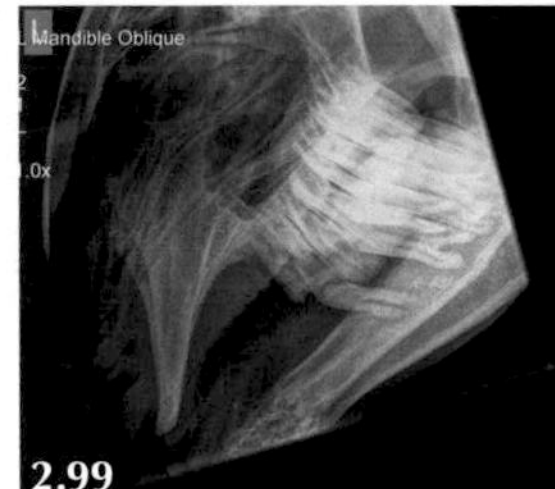

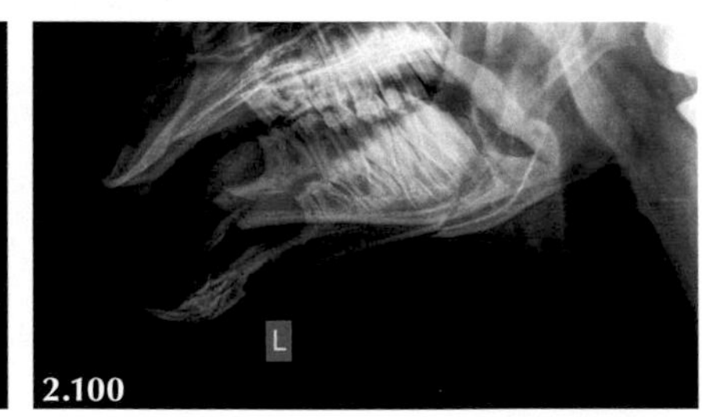

2.98: Normal dentition. **2.99:** Missing premolar tooth, abnormal wear and angulation of cheek teeth. **2.100:** Mandibular fracture in valuable stud ram.

Historically, radiography has been used to identify metallic objects penetrating the reticulum in cattle (traumatic reticulo-peritonitis) but this pathology is much more accurately quantified using trans-abdominal ultrasonography including 5 MHz linear scanners.

Ultrasonographic Equipment

A 5.0 MHz linear transducer connected to a real-time, B-mode ultrasound machine can be used for abdominal ultrasonographic examination except examination of the right kidney where a 5–6.5 MHz microarray transducer ensures better contact with the concave flank of the right sublumbar fossa. The field setting of

10 cm on the linear scanner is appropriate for most common abdominal disorders but will not allow the normal contraction of the reticulum to be fully assessed. Occasionally, the 30 cm field depth afforded by many ultrasound machines/microarray probes more accurately determines the size of the liver and the extent of fluid accumulation, but these data do not significantly alter the clinical diagnosis and prognosis.

The scanning site can be either shaved or clipped and ultrasound gel applied. The abdominal wall is 2 to 3 cm thick in cattle depending upon the site and body condition score (subcutaneous fat). There is scant peritoneal fluid in normal cattle and sheep which is not visualized during ultrasonographic examination.

Cranial Ventral Abdomen

With cattle adequately restrained in stocks, the reticulum is imaged from the ventral midline immediately caudal to the xiphisternum.

Right Side—Right Kidney, Liver and Small Intestine/Caecum

Examination of the right kidney necessitates preparing an area over the last intercostal space and right sublumbar fossa immediately caudal to the last rib (see photos that follow). The microarray transducer head is firmly held against the skin to ensure good visualisation of the right kidney juxtaposed the caudal lobe of the liver. The bovine liver is readily identified in the 9th to 11th intercostal spaces on the right side at the level of a line joining the right wing of the ilium with the right shoulder. In cases of hepatomegaly, the liver can be imaged immediately caudal to the costal arch at this level with the 5–6.5 MHz microarray probe head pointed toward the opposite elbow.

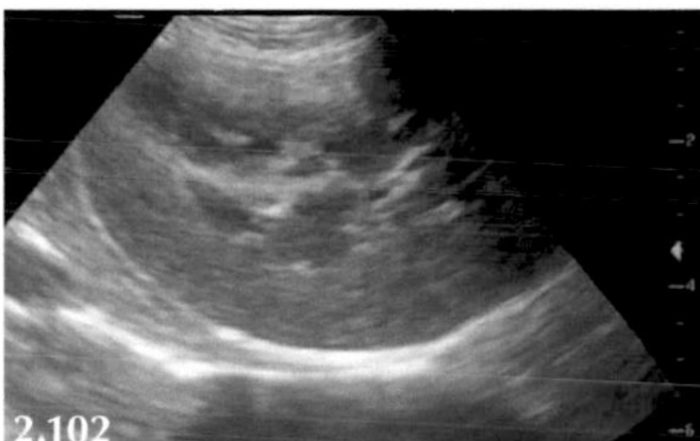

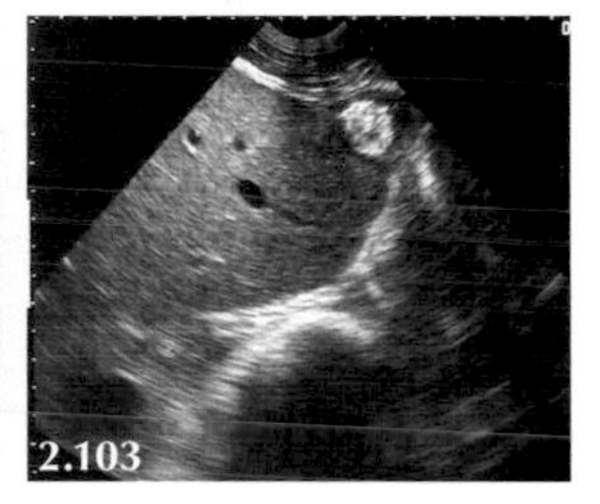

2.101: Caudal shaved area over the right kidney; the larger more cranial shaved area is over the liver. **2.102:** Ultrasonographic appearance of normal bovine kidney. **2.103:** Ultrasonographic appearance of normal bovine liver (note use of different field depth; kidney 6 cm versus liver 16 cm).

The liver in sheep can readily be visualised from half way down the right chest wall at the 8th to 10th intercostal spaces with the microarray probe head angled towards the left shoulder.

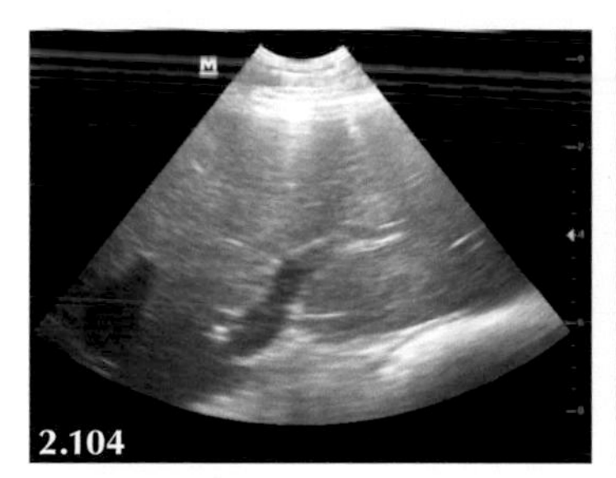

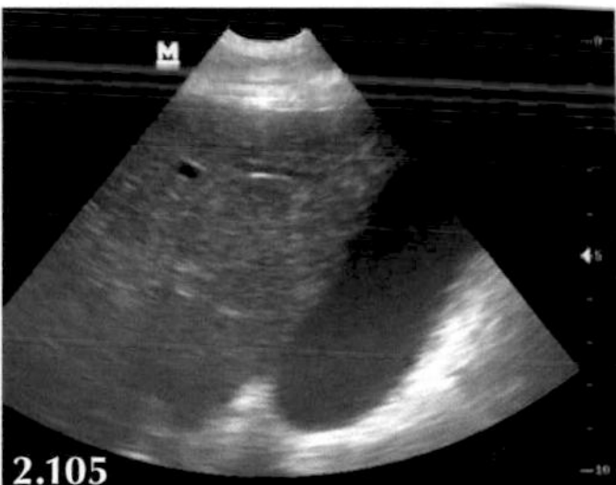

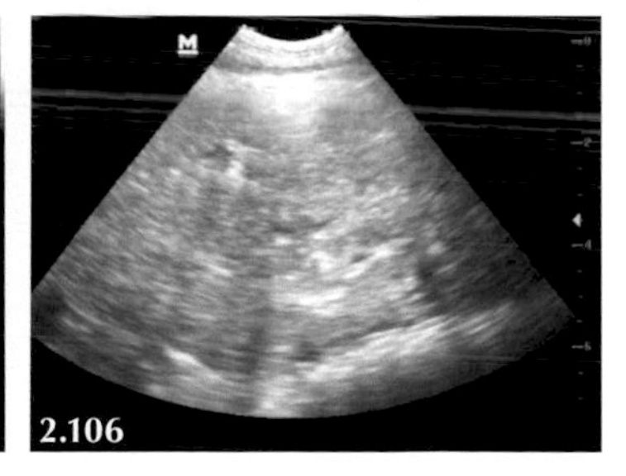

2.104: Ultrasonographic appearance of normal ovine liver. 6.5 MHz microarray probe. **2.105:** Chronic fasciolosis in a sheep. There is a loss of normal liver architecture and massively distended (10 cm) gall bladder ventrally (to the right). **2.106:** Chronic ovine fasciolosis. There is loss of normal liver architecture and thickened bile ducts with distal shadowing. 🎥 2.33, 2.34, 2.35

Small intestine and caecum are examined from the lower right sublumbar fossa. The pylorus is found in the cranioventral area of the right paralumbar fossa. The intestines are clearly outlined as broad hyperechoic (white) lines/circles (1–2 cm diameter in sheep, 2 cm diameter in cattle) containing material of varying echogenicity. By maintaining the probe head in the same position for 10 to 20 seconds, digesta can be visualised as multiple small dots of varying echogenicity forcibly propelled within the intestines. Such movement of intestinal contents prevents confusion with other structures such as an abscess.

Left Side—Left Displaced Abomasum (LDA)

Examination for a LDA necessitates shaving an area over the 9th–12th intercostal spaces two-thirds the way up the left body wall. The distended abomasum occupies the craniodorsal area of the left side of the abdominal cavity (under the rib cage) but differentiation from the rumen is not simple; auscultation and succussion revealing high-pitched metallic "pinging" sounds is the usual diagnostic approach. Normal rumen movements can be heard caudally in the sublumbar fossa.

Transrectal Examination

Transrectal examination of the bladder in adult cattle uses a 5 MHz linear probe.

Ultrasound Findings—Ascites

Ascites is occasionally observed in cattle and sheep with similar ultrasonographic findings although the causes differ; right-sided heart failure is the common reason in cattle whereas the major cause in sheep is severe hypoalbuminaemia. In standing cattle examined from the ventral midline, ascitic fluid appears as an anechoic area with abdominal viscera displaced dorsally. The intestines are clearly outlined as hyperechoic (bright white) tubes/circles containing material of varying echogenicity. By maintaining the probe head in the same position digesta can be visualised as multiple small dots of varying echogenicity forcibly propelled within the intestines. Hepatomegaly with rounded liver edges is consistent with chronic venous congestion.

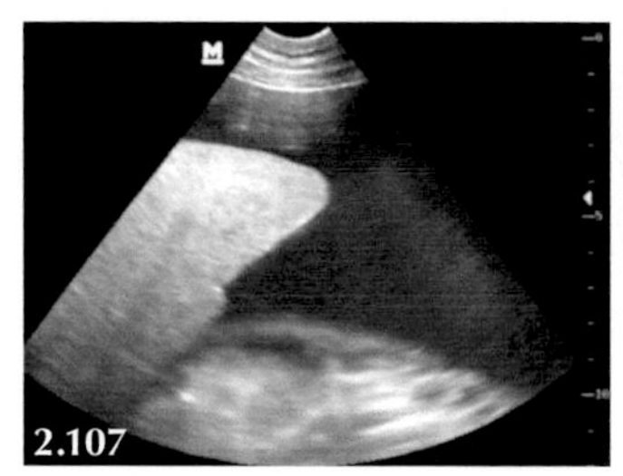

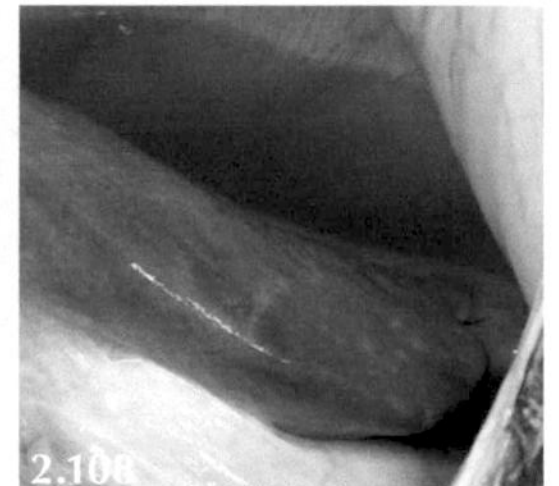

2.107: Ascitic fluid in this cow with right-sided heart failure appears as an anechoic area with the liver and small intestines/omentum displaced dorsally. The edges of the liver are enlarged and rounded. **2.108:** Necropsy reveals a large accumulation of transudate in the abdominal cavity with the rounded edges of the enlarged liver clearly shown. 🎥 2.36

In sheep, large accumulations of transudate can be observed in some cases of paratuberculosis and bacterial endocarditis. Trans-coelomic spread of intestinal adenocarcinoma causes impaired lymphatic drainage and accumulation of transudate.

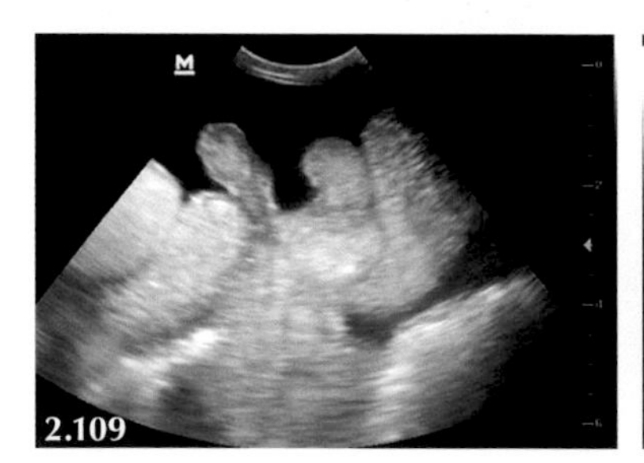

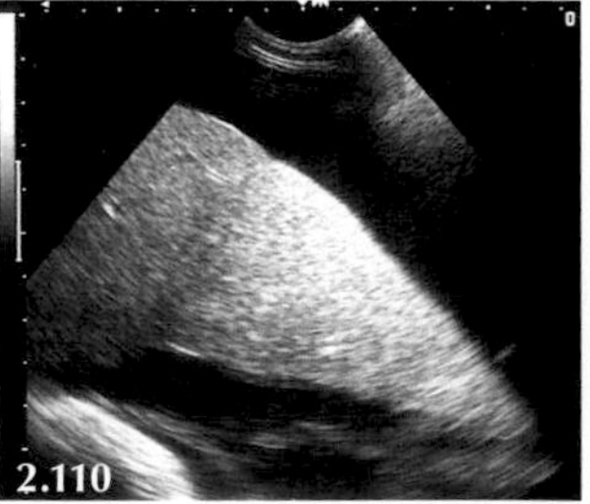

2.109: Ascitic fluid (transudate) appears as an anechoic area with abdominal viscera displaced dorsally in the standing animal. The intestines are clearly outlined as hyperechoic (bright white) lines/circles containing material of varying echogenicity. **2.110:** Large accumulation of transudate with the liver clearly outlined in this ram with endocarditis. **2.111:** Necropsy reveals ascites associated with an intestinal adenocarcinoma. 🎥 2.37

Traumatic Reticulitis and Associated Localised Peritonitis

The normal reticulum is in contact with the diaphragm and ventral body wall and is examined from the ventral midline site immediately caudal to the xiphisternum. Two forceful biphasic reticular contractions lasting 3–5 seconds should be observed over a 2–3-minute period in normal cattle; absence represents reticular impairment.

Traumatic reticulitis is common in cattle where ultrasonography identifies, in chronological order of the disease process, reduced/absent reticular motility, increasing volumes of peritoneal fluid in the anterior abdomen, the development of fibrinous adhesions between the reticulum and abdominal wall, and abscess formation and enlargement. 🎥 2.38

Significant quantities of peritoneal exudate, often up to 8–10 cm between peritoneum and reticular wall and thick fibrin deposits on the reticular wall, are commonly observed in traumatic reticulitis when (beef) cows have been sick more than one week. The hyperechoic latticework appearance of the fibrinous reaction in the anterior abdomen in these cattle contrasts with the anechoic peritoneal fluid. Extension of the fibrinous peritonitis to involve the spleen is commonly seen in advanced cases.

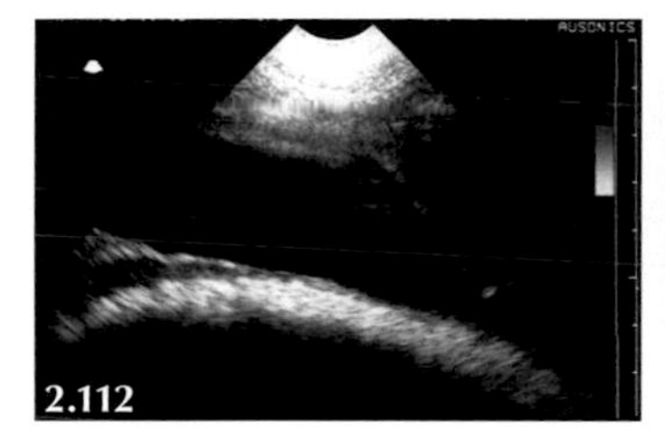

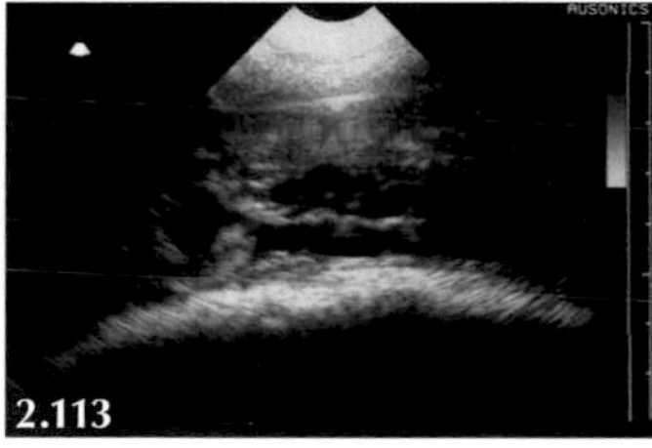

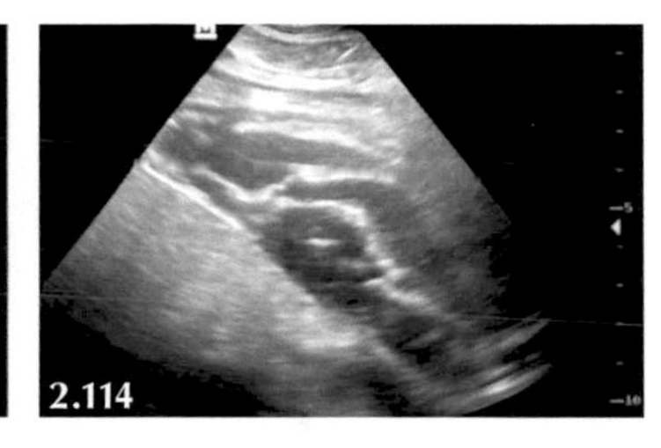

2.112–2.114: Trans-abdominal ultrasound scans of the cranial abdomen (6.5 MHz microarray probe) showing displacement of the reticulum (broad hyperechoic semicircle near the bottom of **2.112** and **2.113**) from the abdominal wall by increasing amounts of inflammatory exudate containing large fibrin tags.

Ultrasonographic examination is strongly recommended before considering surgical retrieval of the penetrating foreign body because the normal vigorous reticular contractions are greatly impaired by large fibrinous adhesions. While the wire may be successfully removed in cattle with fibrinous peritonitis, prognosis for a return to full production is guarded. Indeed, return to normal milk production may take up to four weeks even in those dairy cattle which have been ill for only one day. 🎥 2.39, 2.40, 2.41 recorded using microarray probe. Same case: 🎥 2.42 recorded using a linear probe.

Localised Peritonitis

Occasionally, the peritoneal reaction is limited to a few fibrinous adhesions causing constriction of intestinal lumen which cannot be imaged. In this situation, the intestines proximal to the lesion are grossly distended with fluid (anechoic appearance) rather than containing normal digesta (anechoic appearance containing multiple bright dots) and there are no propulsive contractions. Omental bursitis, viewed at necropsy in images that follow, may not be imaged because it is separated from the probe head by abdominal wall and several centimetres of greater omentum.

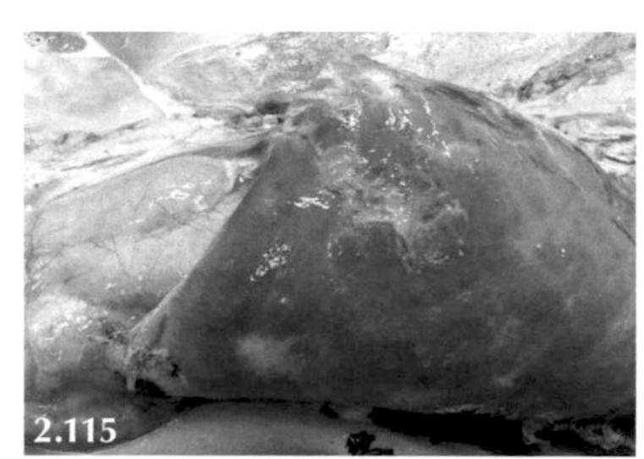

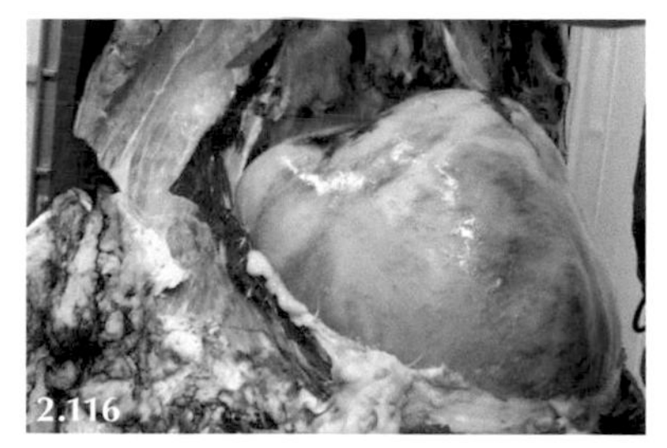

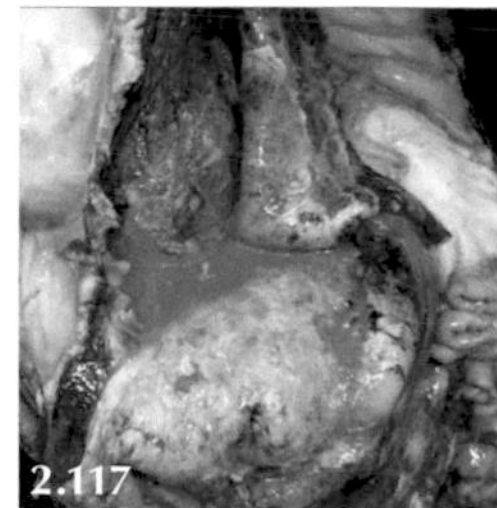

2.115: Necropsy specimen shows fibrinous exudate on the liver capsule with adherent small intestine. These adhesions may not be imaged, but impaired gut motility as a consequence would cause distended loops of intestines proximally. **2.116** and **2.117:** Ruptured viscus where digesta has become enveloped by the greater omentum (omental bursitis). Ultrasound examination failed to identify any pathology because it was separated from the abdominal wall by several centimetres of greater omentum which greatly reduced sound wave transmission.

Generalised Septic Peritonitis

Cattle with acute septic peritonitis are dull, depressed, with an increased pulse often >100 beats per minute due to circulatory compromise and anorexic with a painful expression. They are slow to rise and reluctant to move. Peritonitis involving small intestine results in abdominal distension due to fluid sequestration within the intestines. Excessive accumulation of inflammatory exudate and fibrin deposition over several days causes abdominal distension despite a poor appetite.

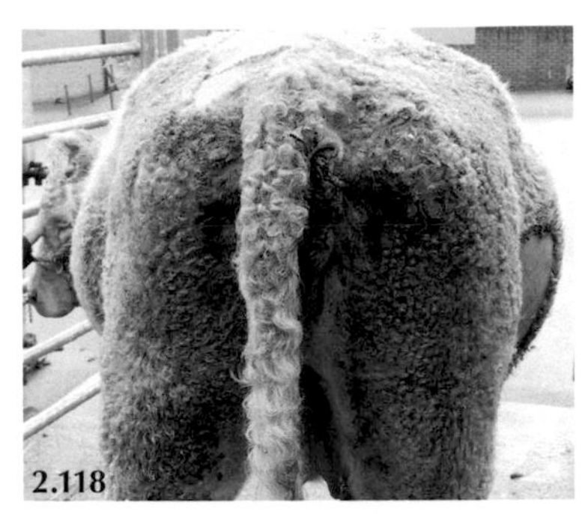

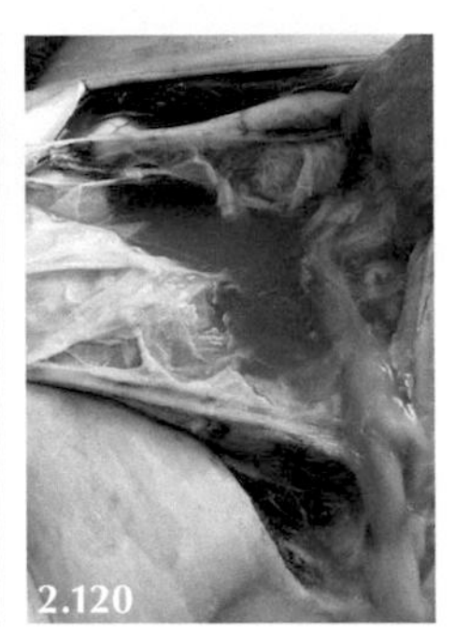

2.118: Simmental cow with abdominal distension due to large volume of exudate and ileus. **2.119:** Painful expression with sunken eyes and ears back. **2.120:** Extensive accumulation of inflammatory exudate/pus within the ventral abdominal cavity revealed at necropsy. 2.43

Peritonitis in Sheep

Unlike cattle, where ingestion of sharp metallic objects causes traumatic reticulo-peritonitis, peritonitis is uncommon in adult sheep and more often associated with sub-acute liver fluke, and very occasionally septicaemia. The ultrasonographic appearance of acute fibrinous peritonitis shows very distended loops of intestine with poor propulsion of digesta and increasing accumulations of inflammatory exudate appearing as anechoic pockets with bridging hyperechoic fibrin tags. With advancement of the condition, fibrin is apparent on serosal surfaces and the tags become thicker and more organised forming a latticework.

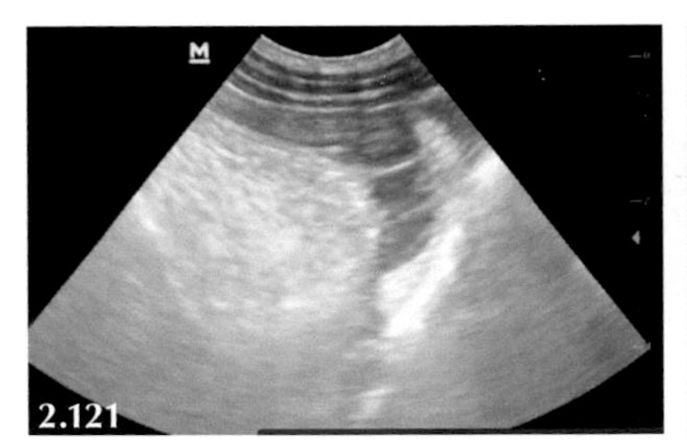
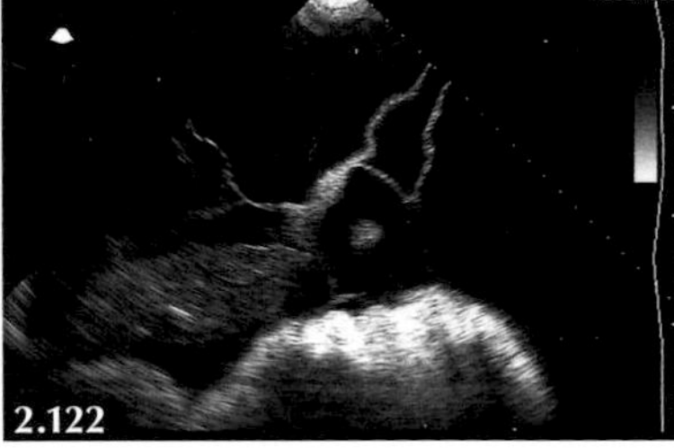

2.121: Distended loops of small intestine bounded by hyperechoic (white) circles representing the intestinal wall. There is an accumulation of fluid (exudate) to the right with hyperechoic bands (fibrin strands) connecting loops of intestines. **2.122:** Extensive peritoneal exudate extending to 16 cm from the abdominal wall with hyperechoic fibrin strands bridging the liver and peritoneum. Such extensive changes are uncommon and caused by sub-acute fasciolosis in this sheep. 2.44

Intra-Abdominal Abscess(es)

Infection within the peritoneal cavity often becomes enveloped and localised by the greater omentum. The peritoneal abscess shown in the images that follow appears as a circular anechoic area containing multiple hyperechoic dots surrounded by an anechoic (fibrous) capsule; the contents are static (view video recording that follows) which differentiates this circular structure from a cross section of intestine containing digesta.

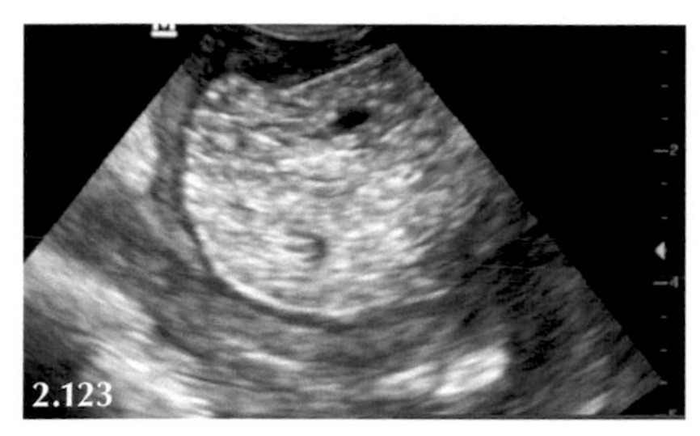

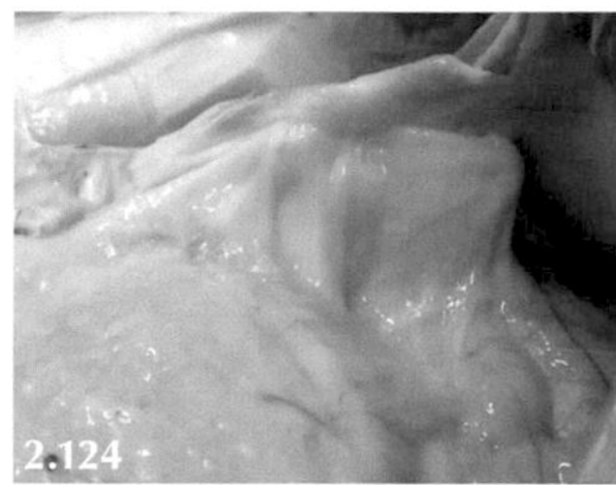

This 4 cm circular intra-abdominal abscess (**2.123**) in this ewe has been encapsulated by the omentum and is shown at necropsy adherent to the abdominal wall (**2.124**). **2.125:** Incision into the inspissated abscess. 🎥 2.45

Small Intestine

The intestines are clearly outlined as broad 1–2 cm diameter (sheep) and 2–4 cm diameter (cattle) hyperechoic (white) lines/circles containing material of varying echogenicity. By maintaining the probe head in the same position for 10 to 20 seconds, digesta can be visualised as multiple small hyperechoic dots within anechoic fluid forcibly propelled within the intestines. Such movement of intestinal contents prevents confusion with other structures such as an abscess or fluid accumulation within the uterine horns (metritis/pyometra) which may present with a similar heterogeneous appearance but the contents remain static. 🎥 2.46

Ileus

Ileus most commonly results from peritonitis with resultant irregular fluid distension of some sections of the intestine up to 6–10 cm diameter containing far fewer hyperechoic dots and no peristalsis. In the case of an obstruction, such as intussusception, intestine proximal to the lesion is dilated while loops distal to the lesion are empty.

Intestinal Torsion

Ultrasound examination of the caudal abdomen is again approached from the ventral midline. In the case of intestinal torsion there are multiple fluid-distended loops of small intestine with little/no peristaltic movement.

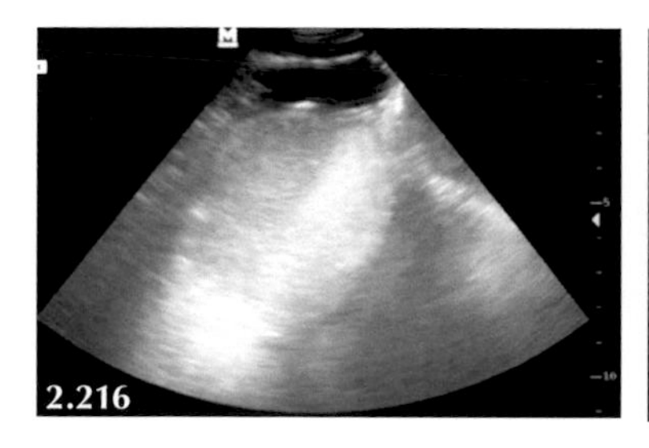

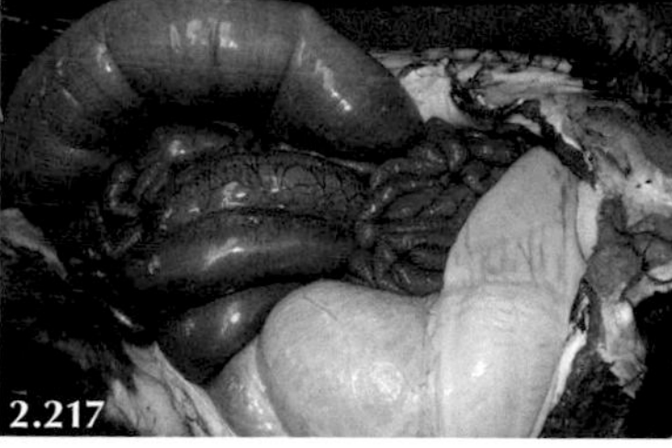

2.126: Ultrasound examination of the abdominal cavity of this 150 kg calf reveals 8 cm diameter loops of distended small intestine with uniform echogenicity and little/no peristaltic movement. **2.127:** Necropsy confirms a torsion around the root of the mesentery. 🎥 2.47

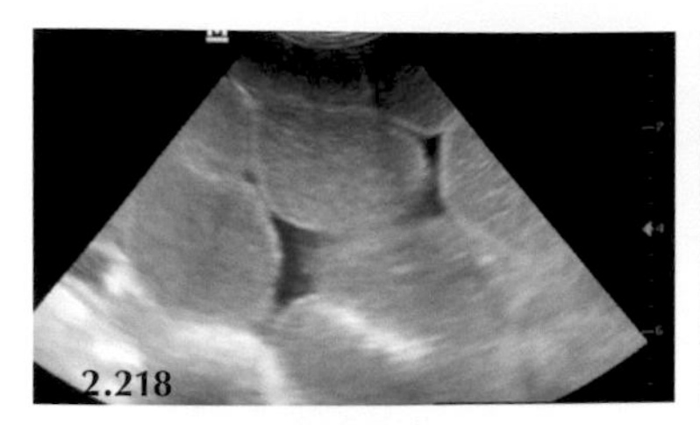
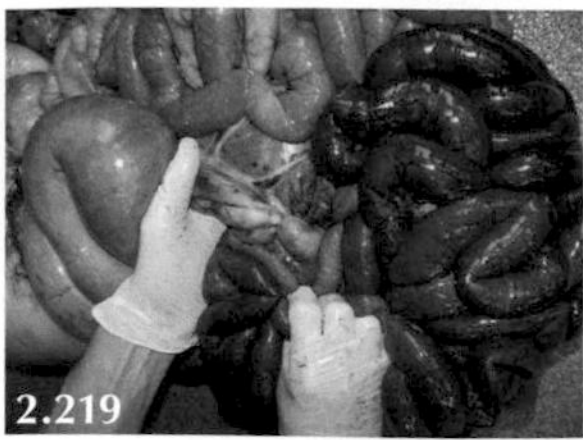

2.128: Ultrasound examination of the abdominal cavity of this 65 kg Blueface Leicester ewe lamb reveals 3–4 cm fluid-distended loops of small intestine with little/no peristaltic movement (see video recording next). A small amount of fluid is present within the abdominal cavity which is abnormal although no fibrin is yet visible. **2.129:** Necropsy reveals very congested fluid-distended loops of intestine caused by a torsion around the root of the mesentery. 🎥 2.48

Caecum

The distended caecum in cattle can readily be palpated per rectum as a blind-ended sac except for the rare occasion when it becomes retroflexed dorsally. Ultrasound examination reveals only the wall of the gas-filled caecum adjacent to the right flank as far forward as the 10th intercostal space. The caecum may also be fluid-filled in the ventral aspect of the abdomen when ventrally retroflexed.

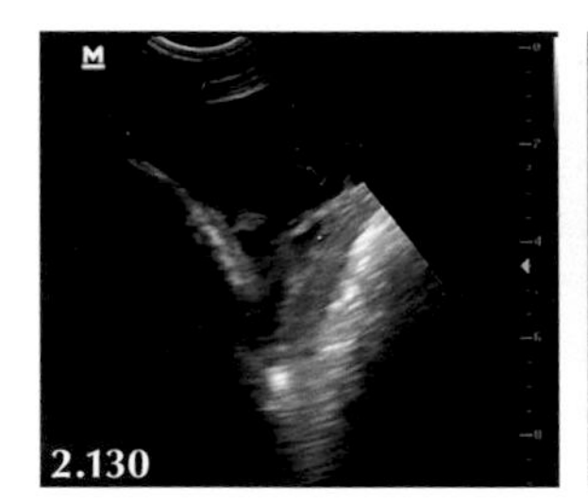
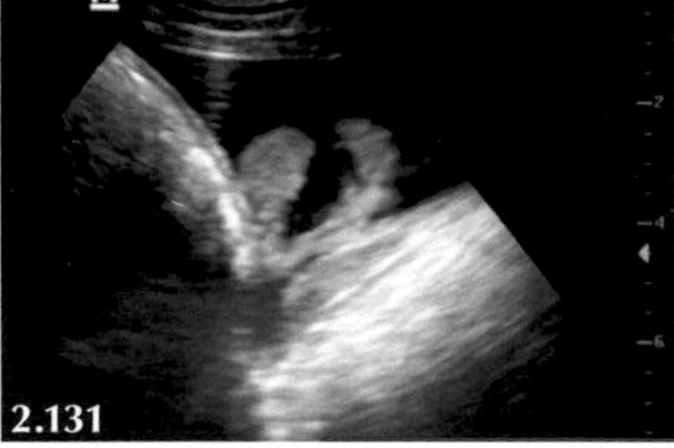
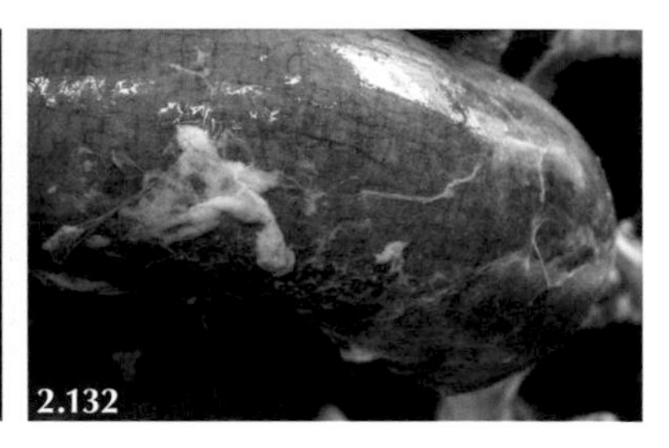

2.130 and **2.131:** Trans-abdominal ultrasound examination of the right flank reveals inflammatory exudate within the peritoneal cavity extending to 4 cm with large fibrin deposits on the serosal surface of the distended caecum. These sonographic findings were confirmed at necropsy **(2.132).** 🎥 2.49

Fecal Worm Egg Count Method

A fecal worm egg count can be undertaken within a couple of minutes requiring only a microscope and McMaster slide; interpretation of the result can be altogether more difficult. Parasite control should be part of a structured veterinary flock health plan. 🎥 2.50

TABLE 2.1

Fecal Worm Egg Count Method
Weigh 2 g of feces and place into tea strainer in a plastic bowl.
Add 28 ml of saturated salt (sodium chloride) solution.
Mix (stir) the contents thoroughly with a teaspoon.
While stirring the filtrate take a sub-sample with a Pasteur pipette.
Fill both sides of the McMaster counting chamber with the sub-sample.
Allow the counting chamber to stand for 5 minutes (this is important).
Examine the sub-sample of the filtrate under a microscope at 10 x 10 magnification.
Count all eggs and coccidia oocytes within the engraved area of both chambers.
The number of eggs per gram of feces can be calculated as follows: Add the egg counts of the two chambers together. Multiply the total by 50.

2.3

Examination of the Respiratory System

The occurrence and prevalence rates of respiratory diseases, as well as their lung pathology and consequent clinical signs, differ markedly between cattle and sheep. However, the principles of ultrasound examination remain the same for both species although the clinical significance of various pathologies are very different.

- Outbreaks of respiratory disease are very common in growing cattle but with the exception of systemic pasteurellosis are uncommon in sheep.
- During respiratory disease outbreaks in growing cattle, rectal temperature >39.6°C is frequently used as the sole indicator of infection necessitating antibiotic treatment. More than 100 calves can be examined per hour. A fall in rectal temperature <39.2°C after 48 hours is considered a positive treatment response.
- 5 MHz linear ultrasound probes, used in bovine fertility work, can be successfully employed for all bovine respiratory tract ultrasound examinations.
- Ultrasound examination of the chest takes less time than auscultation and is much more accurate in assessing lung and pleural pathologies. However, it is important to appreciate that there may be no sonographic changes in septicaemia cases.
- Lung consolidation with abscessation and bronchiectasis are common in chronic respiratory disease in growing and adult cattle but rare in sheep.
- Pleural exudate with the development of large pleural abscesses is common in sheep but rare in cattle. Pleural effusion is seen occasionally in cattle with right-sided heart failure but rarely in sheep.

Acknowledging the Limitations of Chest Auscultation in Cattle and Sheep

In veterinary textbooks wheezes are described as prolonged musical sounds that usually occur during inspiration, and occasionally throughout the breath cycle, resulting from vibration of airway walls caused by air turbulence in narrowed airways. Crackles are loud, explosive, short duration (typically 10–30 ms), non-musical and "rattling or bubbling" sounds. Crackles are thought to be caused by air bubbling through, and causing vibrations of, respiratory secretions within the larger intrathoracic airways, including those that are pooling within the dependent part of the rostral thoracic trachea.

It is important to recognise that attenuation of heart and lung sounds results from lung consolidation and significant pleural effusion in cattle, and large (>4 cm diameter) pleural abscesses and >1–2 cm fibrinous pleurisy which are common pathologies in sheep. In many cases of respiratory disease/pathology, reduced/absent lung sounds are clinically more significant than increased adventitious lung sounds. Auscultation can lead to an overestimate of lung pathology when there is muco-pus within the larger dorsal airways which generates loud crackles audible over much of the lung field. Simultaneous ultrasonography and auscultation highlights the limitations of the latter in identifying and defining lung pathology. Ultrasound examination of the chest takes less time than auscultation and is much more accurate in assessing lung and pleural pathologies.

DOI: 10.1201/9781003106456-11

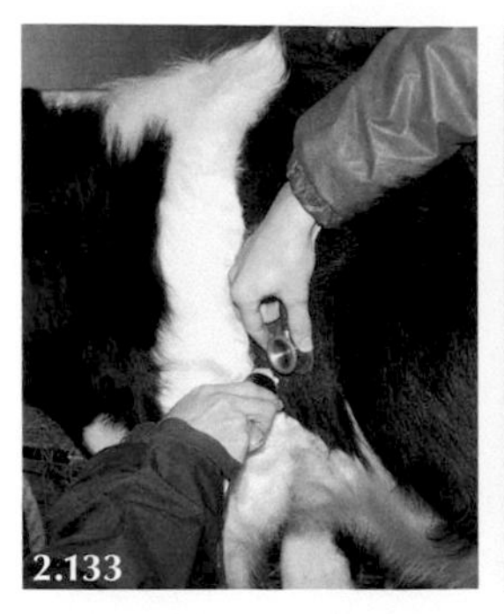
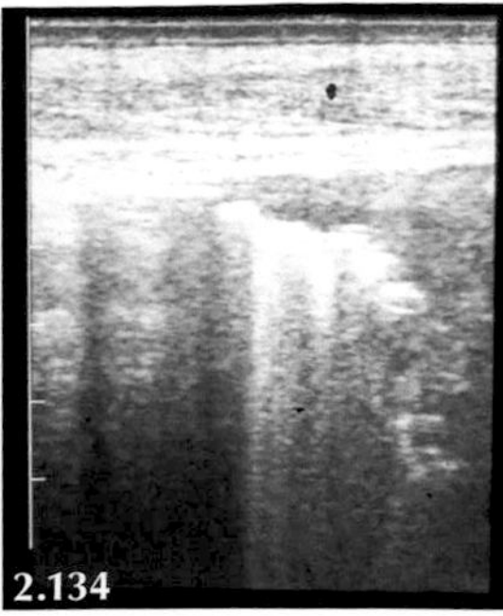
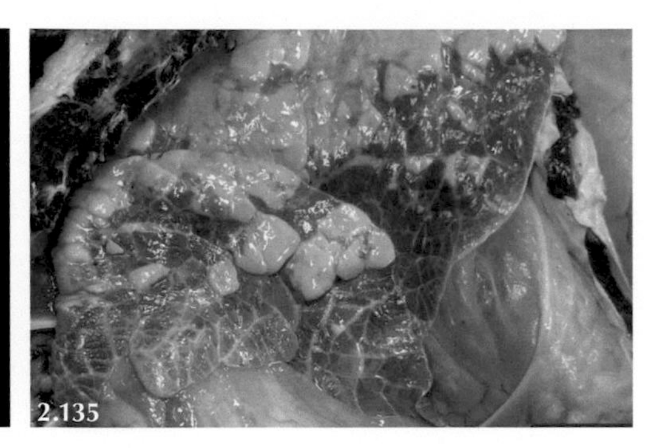

2.133: Simultaneous ultrasonography and auscultation highlights the limitations of the latter in identifying and defining lung pathology. **2.134:** A 5 MHz linear probe demonstrating the dorsal extent of lung pathology (bronchiectasis) in a cow. **2.135:** Sharply demarcated ventral lung consolidation typical of severe bronchiectasis in adult cattle.

Respiratory Disease Outbreaks in Calves

Respiratory disease in calves and growing cattle usually affects a large proportion of the group often after a combination of stressful events such as weaning, sale, transport, co-mingling and housing at a new farm (beef cattle), and weaning and formation of larger management groups (dairy cattle). Respiratory viral infections are commonly involved.

2.136: Low stocking density and excellent ventilation. **2.137:** Ventilation of this building housing beef cattle only occurs through open doors. **2.138:** Purchased beef cattle turned out to pasture for four to six weeks after purchase and before housing separating stressful events.

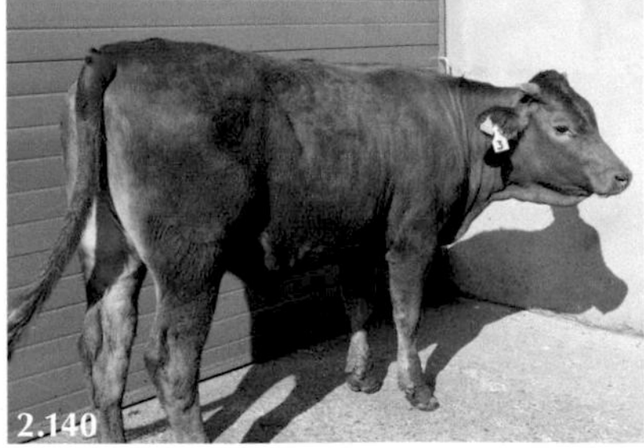

2.139–2.141: Identification of sick cattle suffering from acute respiratory disease is simple; dealing with the remainder of the group in a logical manner using antibiotics responsibly is altogether more challenging.

Treatment Strategies for Groups of Growing Cattle

The approach to an outbreak of respiratory disease in a large group of growing cattle is very different to individual adults. With a large number of calves detailed veterinary examination of each calf is not possible for time/cost reasons and the very limited application of auscultation. Studies which selected calves for antibiotic treatment based solely upon rectal temperature greater than 39.6°C ("temp and treat") were as effective as mass treatment (whole group metaphylaxis). In most cattle handling systems two stockmen can check the rectal temperature and treat more than 100 cattle per hour. Reasons against regular rectal temperature monitoring of calves during disease outbreaks have included fears regarding further stress-induced disease or exacerbation of existing problems, availability and cost of farm labour, and perceived additional veterinary costs but are unfounded. 🎥 2.51, 2.52

Antibiotic Re-Treatment

Antibiotic re-treatment becomes necessary when bacterial infection recurs during the period of viral-induced compromise of host defence mechanisms in the upper respiratory tract. By definition, the interval between antibiotic treatments has to be greater than the period of maintenance of effective minimum inhibitory concentrations of antibiotic in lung and bronchial secretions; otherwise the antibiotic has not been fully effective and/or the bacterial pathogens were not susceptible to the antibiotic selected by the veterinary surgeon. Recurrence of respiratory disease with pyrexia greater than 39.6°C after four or more days' interval has been treated with the same antibiotic in many field studies with very good results. This is a very important principle; the antibiotic was effective but because of the impaired physical defences caused by viruses, bacterial infection has recurred and there is every likelihood that the same antibiotic would again be effective. Such recurrence of pyrexia is rarely the result of the rapid development of resistance and the term relapse should be used with caution because it may prove misleading especially to farmers. Whole group metaphylactic antibiotic injection is contrary to responsible antibiotic use.

Clinical Examination of the Respiratory System

- Measure rectal temperature and observe respiratory rate and effort. Determine case history. Determine whether other sick animals/deaths occurred in recent days/weeks.
- Carry out clinical examination noting limitations of chest auscultation.
- For individuals/small groups scan chest and record any significant pathology to laptop using Elgato software (1–2 minutes per animal). Be aware that septicaemic cases may show no sonographic changes.
- Post-mortem any recent deaths (sheep) and always submit histology samples. Growing and adult cattle would likely be examined at regional veterinary laboratory allowing detailed investigations.

Most textbooks report a rectal temperature ca. 41.0°C in cases of per-acute respiratory disease. In sheep, the respiratory rate may be increased by gathering, handling stresses and body condition score. The presence of a full fleece during hot weather can rapidly induce panting which may further complicate interpretation of auscultation findings. Painful conditions, especially lameness, can also have a considerable influence on observed respiratory parameters. Toxaemia causing an abnormal respiratory rate and effort may originate from a body system other than the lungs. Confining a sheep from the group causes distress often masking abnormal clinical signs. It would generally be advised that a sheep has a companion for welfare reasons but biocontainment would be a concern for contagious diseases such as OPA.

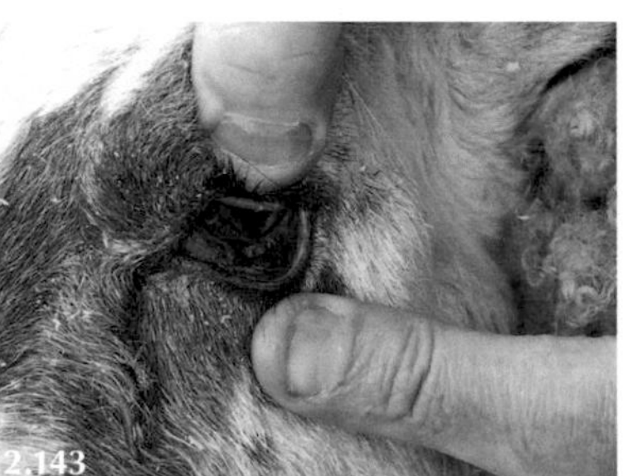

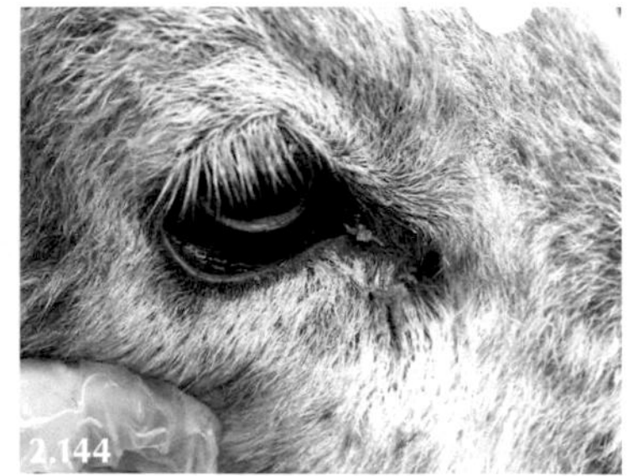

2.142: Panting can be caused by gathering and handling stresses during hot weather. **2.143** and **2.144:** Toxaemia can result from diseases other than respiratory infections. 🎥 2.53, 2.54, 2.55, 2.56, 2.57

Case History

It is important to recognise that the frequency and thoroughness of flock supervision will influence the accuracy, and thereby usefulness, of the sheep's given history. Respiratory disease generally develops 10–20 days after housing growing cattle where reduced appetite is often the first sign of illness.

Body condition score relative to other animals in the group may provide some indication of likely duration although two or more conditions may co-exist. One disease may precede and/or predispose to another; for example, ovine pulmonary adenocarcinoma frequently predisposes to a per-acute episode of illness caused by *Mannheimia haemolytica*. Note that the extent of lung pathology predisposing to pasteurellosis (following, right) is much less that the OPA alone (following, left).

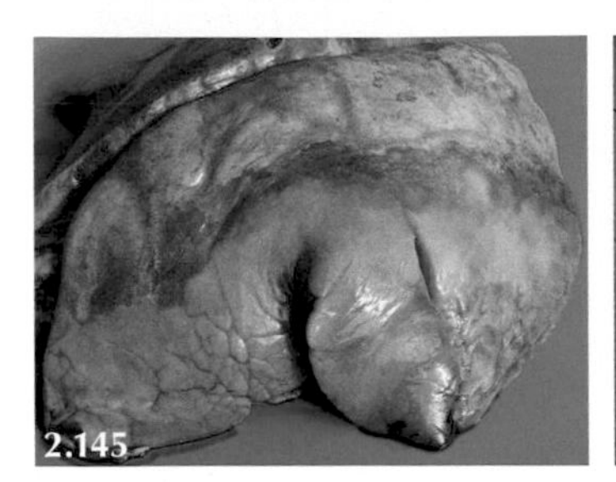

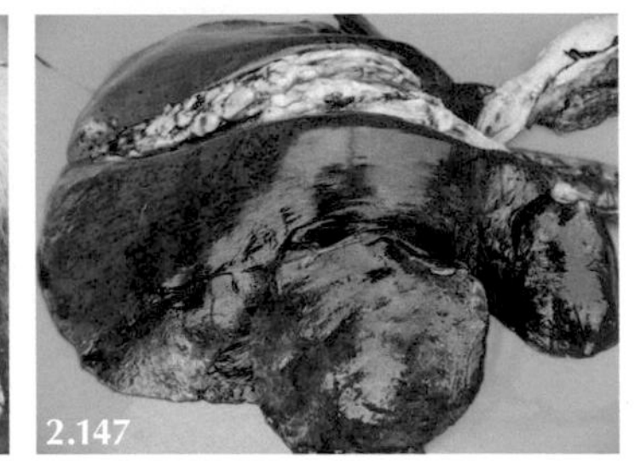

2.145: Extensive OPA affecting 75% of the left lung although the ewe showed no respiratory signs. **2.146:** Moribund ewe with septicaemic pasteurellosis secondary to OPA. **2.147:** Necropsy of ewe shown in centre—OPA affects only the ventral half of the cardiac lobe, but there are widespread septicaemic changes. 2.58

Respiratory Disease Prevalence

It is important to appreciate that the true prevalence of respiratory diseases in sheep is not known because few sheep are examined by veterinary practitioners and such clinical examination does not accurately identify lung and pleural pathologies. Respiratory diseases are widely regarded by farmers and veterinary practitioners as the most common causes of sudden death in growing lambs and adult sheep, but few animals are subjected to expert post-mortem examination supported by histopathology.

Routine whole flock trans-thoracic ultrasonography of more than 0.3 million adult sheep as part of an OPA elimination programme identified lung pathology in 0.5 per cent of sheep which had survived significant acute respiratory disease as evidenced by pathology comprising >2 cm thick exudate extending over the majority of the ventral lung field on one or both sides of the chest (0.15 per cent), >4 cm diameter well-encapsulated abscess(es) within the pleural space (0.14 per cent), and irregular >3 cm deep consolidation often containing abscesses and with >2 mm overlying pleurisy classified as bacterial pneumonia (0.11 per cent). It is clear from these field data that sheep can survive severe respiratory disease without antibiotic treatment.

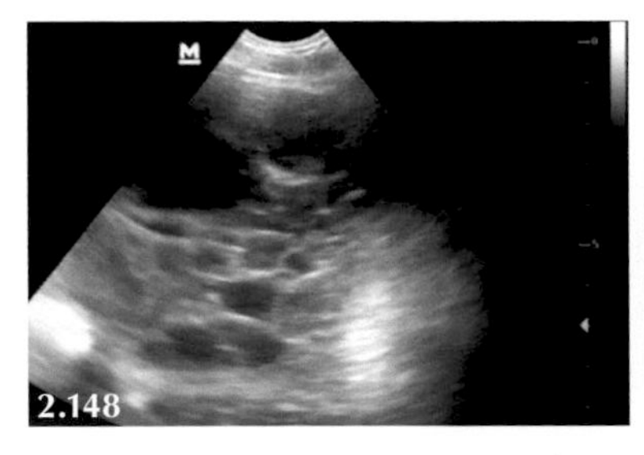

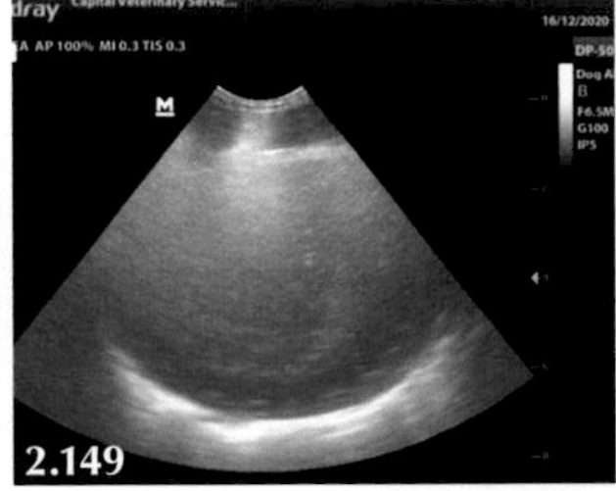

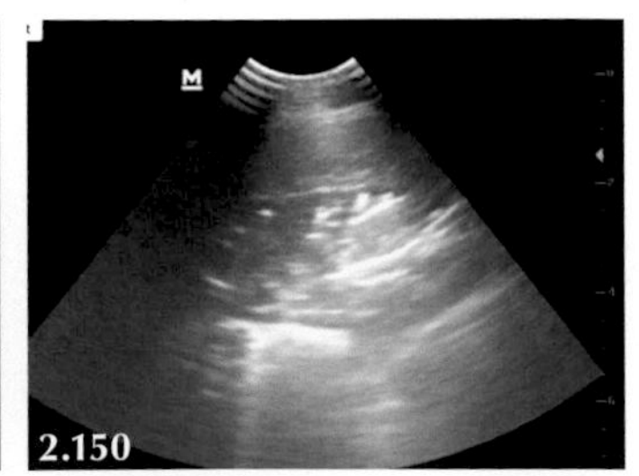

2.148: 8 cm thick inflammatory exudate in the chest cavity. **2.149:** 6 cm diameter pleural abscess. **2.150:** 3 cm deep consolidation showing distinct bronchial pattern with 2 cm overlying pleural exudate (organised fibrin). None of these sheep was classified as "sick" by the farmer when identified during whole flock screening for OPA.

Ultrasonographic Examination

Ultrasound machines are readily portable and have batteries allowing remote use. Always record all abnormal findings using Elgato software or similar to a tablet or laptop for subsequent reference.

A 6.5 MHz microarray (sector) transducer connected to a real-time, B-mode ultrasound machine is recommended with an initial depth setting of 4–6 cm for sheep and 6–8 cm for adult cattle which is increased as necessary to image the full extent of pathology. Do not use the full field depth (25–30 cm) for routine examinations because this greatly reduces image definition of lesions which are often 2–8 cm deep. The focus is set at the depth of the visceral pleura usually 1–2 cm in sheep and 2–3 cm in cattle depending upon body size and condition score. It is important to appreciate that lesions within the lung parenchyma, and not in contact with the visceral pleura, will not be imaged however such distribution is uncommon.

A 5 MHz linear transducer, commonly used trans-rectally for early bovine pregnancy diagnosis, works perfectly well to examine the pleurae and lungs in both sheep and cattle. While field depth is restricted to 10 cm with a linear probe, and lesions such as pleural abscesses may extend beyond 16–20 cm from the chest wall, this limitation does not adversely affect decisions relating to either diagnosis or prognosis.

When interpreting published ultrasound images always refer to the scale on the image margin otherwise the size of the lesion(s) cannot be appreciated; scans without a clear scale attached are worthless. Note that field depth has frequently been increased in the video recordings in this chapter, often during the examination, to fully define lesions. It is much easier to appreciate ultrasound changes in the video recordings than images.

Sheep

The sheep is restrained in the standing position against a solid partition by an assistant. The operator flexes the ipsilateral foreleg at the elbow and draws it forward to gain access to the cranial chest to image the ventral lung field at the level of the base of the heart. The ultrasound machine is sited behind the sheep and viewed from over its back (see images earlier). The transducer head is firmly held at a right angle against the skin and systematically moved over the 4th to 7th intercostal spaces. The starting point is approximately 5 cm above the point of the sheep's elbow. This area of skin has no fleece and only sparse hairs such that shaving is not necessary; ultrasound gel is applied only to the probe head. Applying surgical spirit to the skin is irritant and unnecessary.

The ultrasound examination first recognises the junction between the lung and base of the heart; the beating heart and ventricle(s) are quickly identified. The probe head is then moved cranially and caudally over adjacent intercostal spaces at that level. The probe head is then lowered by 5 cm and moved into adjacent intercostal spaces to further examine the ventral lung margins especially caudally over the diaphragmatic lung lobe (see photos that follow).

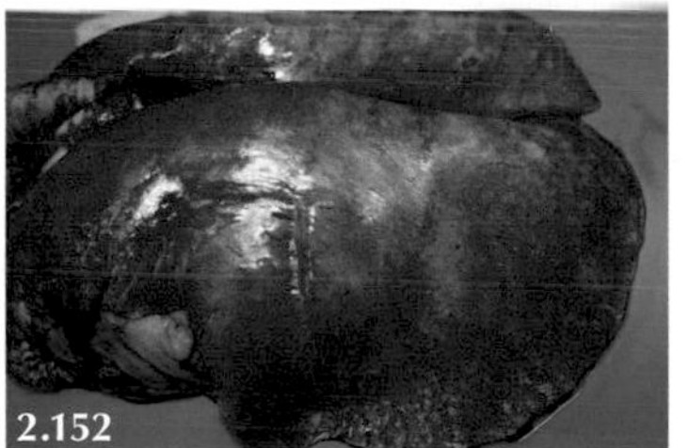

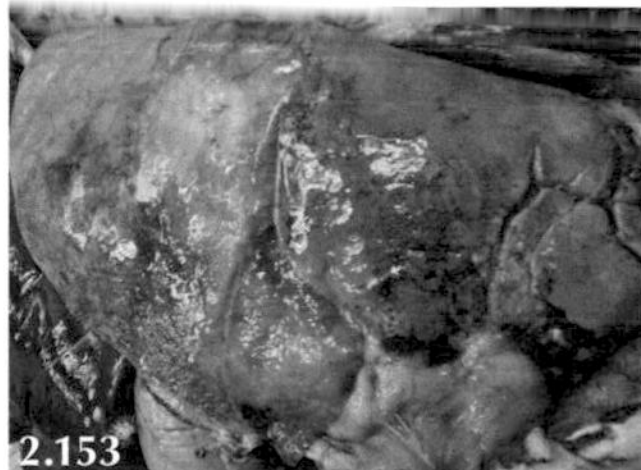

2.151: The sheep is held against the solid partition and the ultrasound machine viewed from over the sheep's back. **2.152:** Necropsy specimen of the left lung showing the ventral margins of the apical and cardiac lung lobes at the level of the heart base. By lowering the probe 5 cm the ventral portion of the diaphragmatic lobe is fully examined. **2.153:** Necropsy specimen reveals the right lung "*in situ*" showing the anatomical relationship between diaphragmatic lung lobe, diaphragm and liver.

It is essential to examine both sides of the chest because lesions are often unilateral and/or there can be significant differences in size/severity. The full extent of all pathology can be determined by increasing field depth. Images of the diaphragm and liver are produced when the probe head is positioned too far caudally over the right hand side of the chest but this ensures that the whole lung field has been checked. Do not mistake liver for hepatoid change in the diaphragmatic lobe (see photos that follow). Similarly, images of the diaphragm and rumen/reticulum, and occasionally spleen, are produced when the probe head is positioned too far caudally over the left hand side of the chest.

Video recordings should be captured as mp4 files using Elgato Video Capture software (www.elgato.com) which permits monitoring of lesions and their response to treatment, and reflective learning should necropsy results become available later. Where necessary, such recorded files can readily be e-mailed for a second opinion with some farm animal telemedicine services provided *gratis*. All ovine video images and recordings in this book have been taken on farm with no skin preparation.

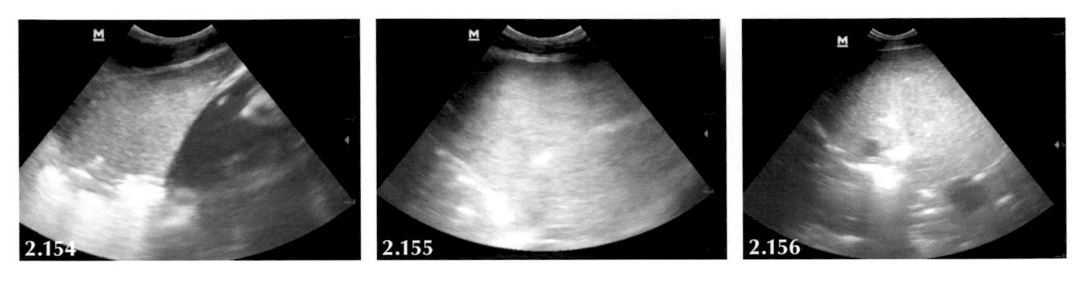

2.154: Image quality is better in low BCS sheep because sound wave penetration is not reduced by a thick body wall. **2.155:** OPA tumour fills the screen with depth setting at 4 cm. **2.156:** The full extent of the tumour can be determined by increasing the field depth to 10 cm. This recording took less than 1 minute on farm and shows all the important sonographic changes reported in the literature for OPA. 🎥 2.59

Ultrasound examination allows the pleurae/lungs to be examined logically. The ultrasound image next shows consolidation of the ventral lung with a bronchial pattern. There is 2–3 cm fluid in the chest cavity which looks more like pus than fibrin. Also, pleurisy tends to extend dorsally and not stop abruptly. The lesion was unilateral which would likely rule out an effusion. It is easier to appreciate these changes in the video recording. 🎥 2.60

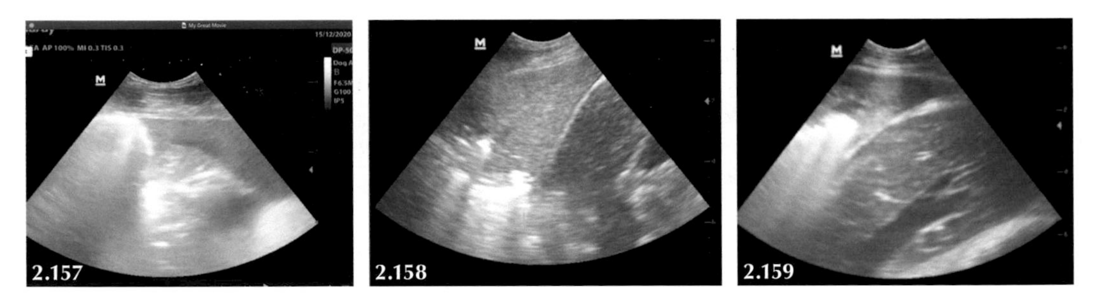

2.157: Sharply demarcated consolidation of the ventral lung with a bronchial pattern. There is 2–3 cm fluid in the chest cavity which looks more like pus than fibrin. **2.158:** OPA tumour dorsally (to the left), the semi-circular hyperechoic line represents the diaphragm, with the liver ventrally. **2.159:** 1–2 cm exudate within the pleural space with an irregular hyperechoic lung surface. The liver parenchyma is more readily identifiable in this image because of the prominent vessels.

Cattle

A 5–7 cm wide strip of hair is shaved from both sides of the thorax extending in a vertical plane from the caudal edge of the scapula to below the point of the elbow. Shaving the skin over the whole lung field is not necessary. The skin is soaked with warm tap water and ultrasound gel liberally applied to the wet skin to ensure good contact; alcohol is not needed and can be irritant. The transducer head is firmly held against the skin overlying the 6th or 7th intercostal spaces. The probe head is moved from the dorsal (normal) lung ventrally to below the point of the elbow. This examination routine shows normal lung then pathology which has a cranio-ventral distribution in almost all cases. In bronchiectasis, normal lung dorsally is followed by lobular consolidation then widespread consolidation of the ventral cranial and cardiac lung lobes. This scanning approach also readily identifies pleural exudate (fibrinous pleurisy) and pleural effusion.

Normal Lung—Sheep

The chest wall is approximately 1 cm thick in 20 to 40 kg lambs extending up to 2 cm in adult sheep in good body condition (>90 kg; body condition score 3 or greater, scale 1 to 5). The surface of a normal lung (visceral or pulmonary pleura) is characterised by a bright white linear echo. Image quality is adversely affected in fat/very fat sheep. Equally spaced reverberation artefacts appear beneath the visceral pleural line when sheep are in very poor body condition. The lung surface moves several millimetres in the vertical plane during respiration. There is no detectable pleural fluid in normal sheep.

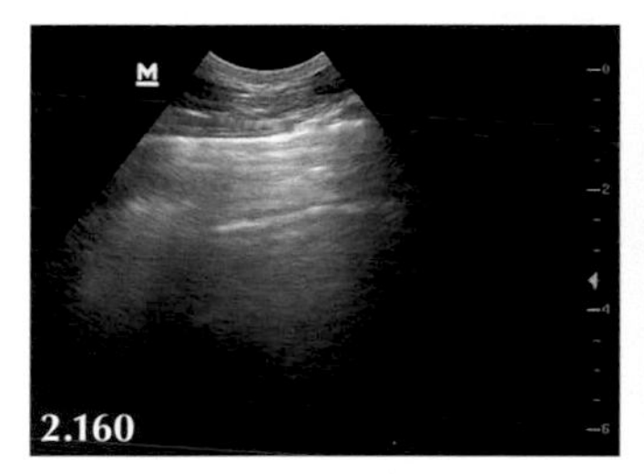

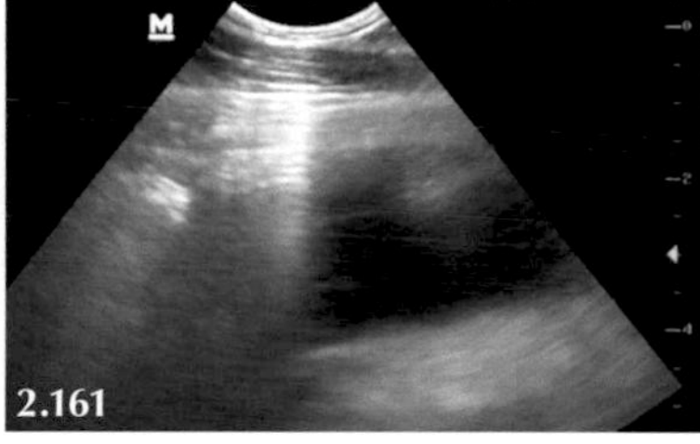

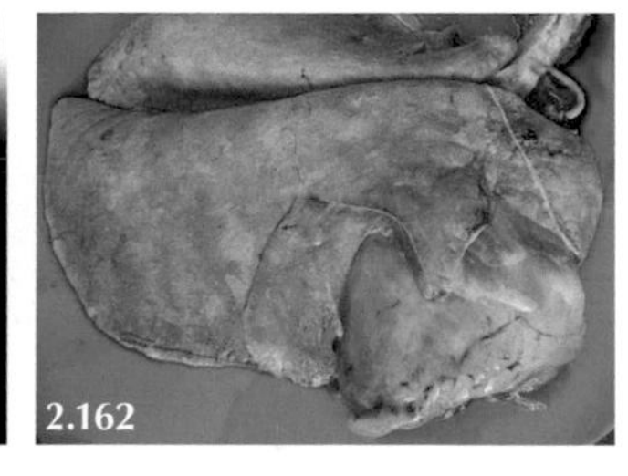

2.160: The surface of normal aerated lung (visceral or pulmonary pleura) is characterised by the uppermost white linear echo; 6.5 MHz (micro-convex) sector scanner. The probe head is at the top of the image, dorsal is to the left, centimetre markers in the right hand margin. The field depth is set at 4–6 cm including chest wall. **2.161:** The ventral margins of the apical and cardiac lobes extend to the level of the base of the heart which is in contact with the chest wall. Blood in the ventricle appears anechoic; the ventricular wall is approximately 8 mm thick in this adult sheep. **2.162:** Normal left lung showing position of the heart base in relation to the ventral lung margins. 🎥 2.61, 2.62, 2.63, 2.64

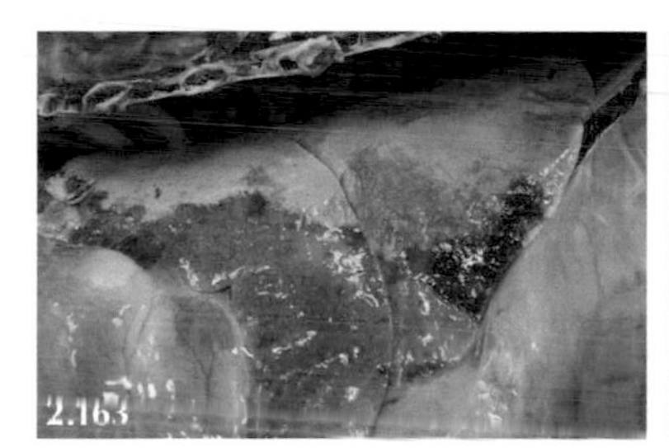

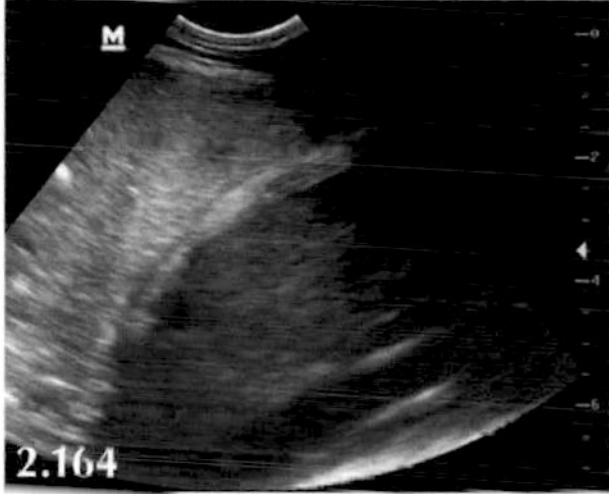

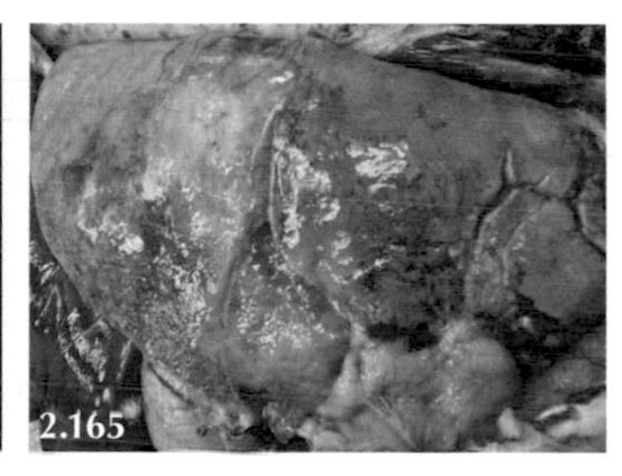

2.163: The base of the heart (note that this refers to the top of the heart as imaged here "*in situ*") is shown in relation to the left lung lobes. **2.164:** The diaphragm, imaged here as a slightly hyperechoic 5 mm band, separates OPA lung dorsally (to the left) from the liver ventrally (to the right). **2.165:** This necropsy image of the right chest cavity and abdomen (diaphragm reflected) shows the relative positions of the right diaphragmatic lobe of the lung and liver. 🎥 2.65, 2.66, 2.67, 2.68

Normal Cattle

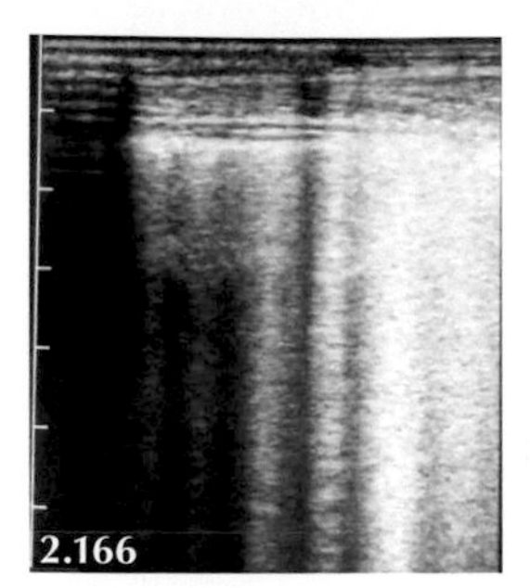
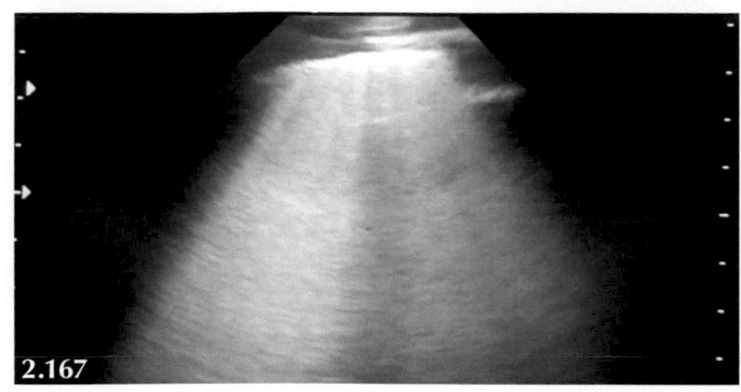

2.166: 5 MHz linear scanner. The surface of normal aerated lung (visceral or pulmonary pleura) is characterized by the continuous hyperechoic (white) linear echo 1.5 cm (chest wall) from the top of the image. **2.167:** Unlike in sheep, "comet tails" are commonly observed in cattle arising from the lung surface and are of no clinical significance.

Pleural Effusion

Pleural effusion is not uncommon in cattle with right-sided heart failure. An increasing depth of fluid (anechoic area) separates the pleurae as the ultrasound probe head is moved down the chest wall. At the ventral margin of the lung field the effusion may extend up to 15–20 cm. There is dorsal displacement of the lung lobes and collapse of the ventral lung margins.

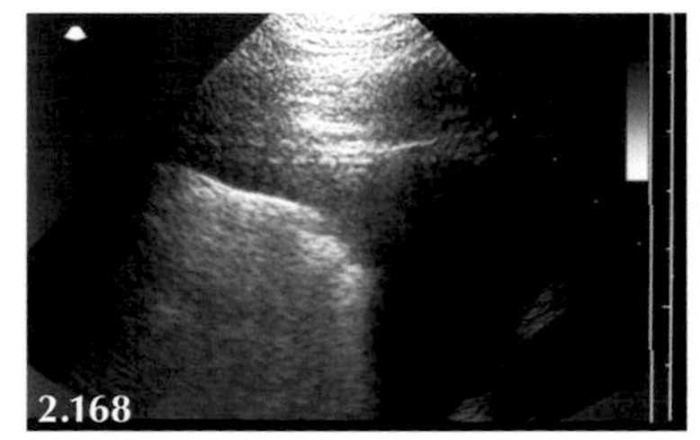
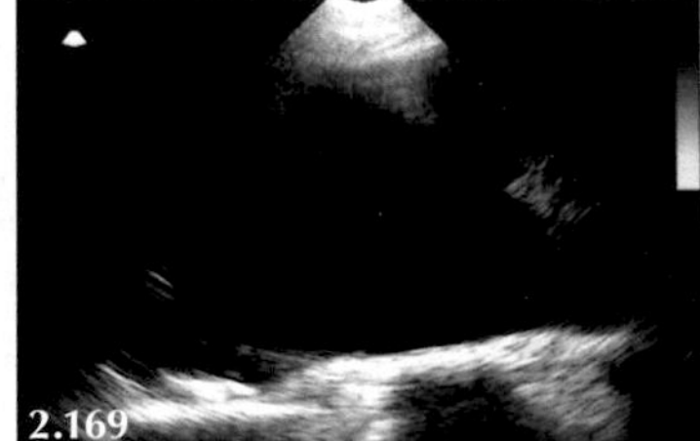

Dorsal margin of pleural effusion (**2.168**) which extends to 15 cm (**2.169**) as the ultrasound probe moves ventrally. The lung surface appears more hyperechoic than normal due to acoustic enhancement (sound waves pick up energy travelling through the pleural fluid before striking aerated lung). **2.170:** Extent of pleural effusion revealed at necropsy.

Fibrinous Pleurisy

With significant accumulations of inflammatory exudate, pleurisy causes attenuation of lung and heart sounds; there are no audible "pleural friction rubs". With unilateral lesions, an increased volume of breath sounds can often be heard on the normal side of the chest caused by the increased respiratory rate due to reduced functional lung capacity.

Fibrinous pleurisy is common in sheep but rare in cattle. The nature of fibrinous pleurisy in sheep varies from a 2–5 mm layer covering much of the visceral pleura to 60–80 mm exudate occupying much of the pleural space. Well-encapsulated lesions can measure up to 20 cm from the chest wall with little or no functional lung on that side of the chest. Sheep with such large pleural lesions show tachypnoea because of reduced functional lung volume but may not be as sick as the clinician would have expected. Contrary to the situation with sheep, large volumes of pleural exudate/fibrinous pleurisy in growing cattle are associated with a very poor prognosis.

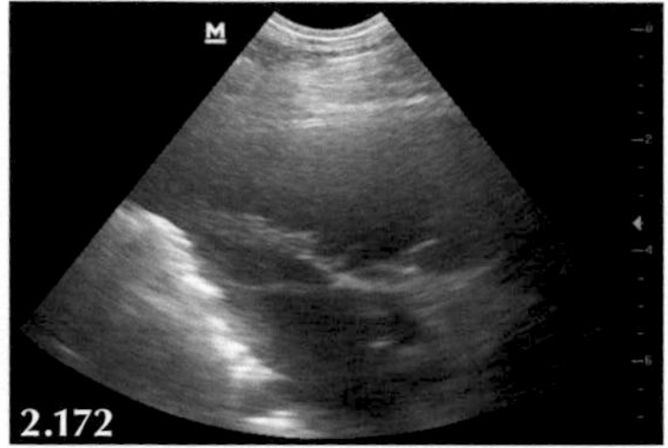

2.171: This apparently healthy ram had 8 cm fibrinous exudate in the left chest (see centre). **2.172:** Extensive fibrinous exudate in the pleural space extending to 8 cm. **2.173:** This heifer was severely ill with extensive fibrinous pneumonia.

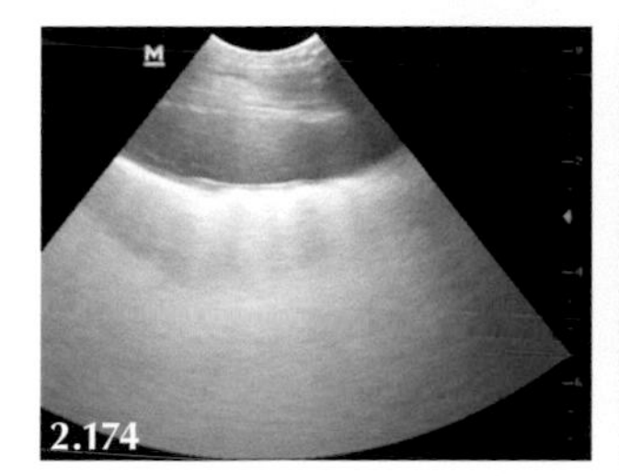

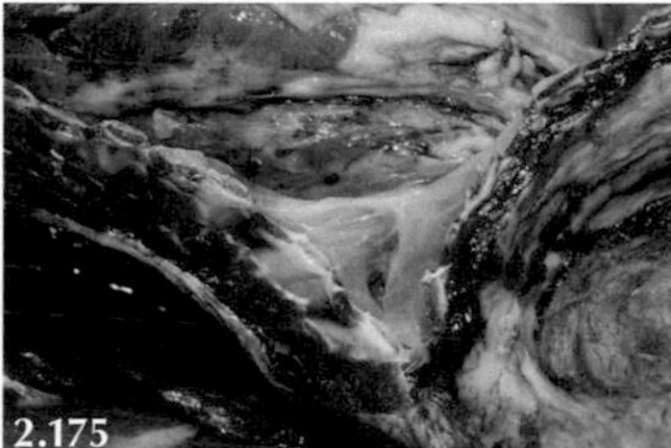

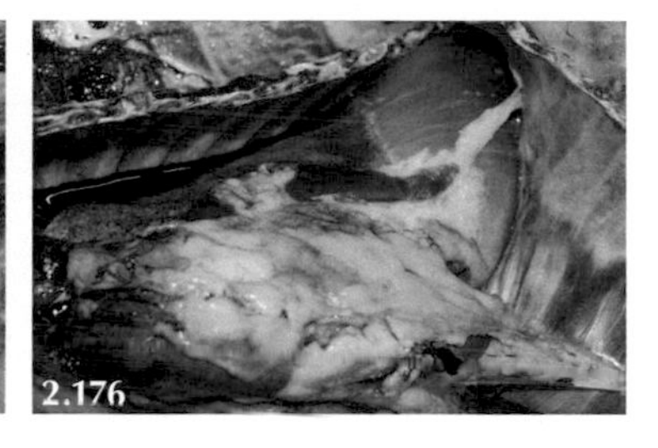

2.174: 10–15 mm thick hypoechoic layer, containing a fine hyperechoic matrix representing organised fibrinous pleurisy, separates the lung surface from the chest wall in this sheep. The hyperechoic line, representing the lung surface, is thicker and more hyperechoic than normal (acoustic enhancement). **2.175:** Necropsy reveals organised fibrin within the pleural space covering much of the lung surface. **2.176:** Left lung of sheep *in situ* with chest wall reflected, exudate has gravitated to the most ventral margin which is adjacent to the diaphragm.

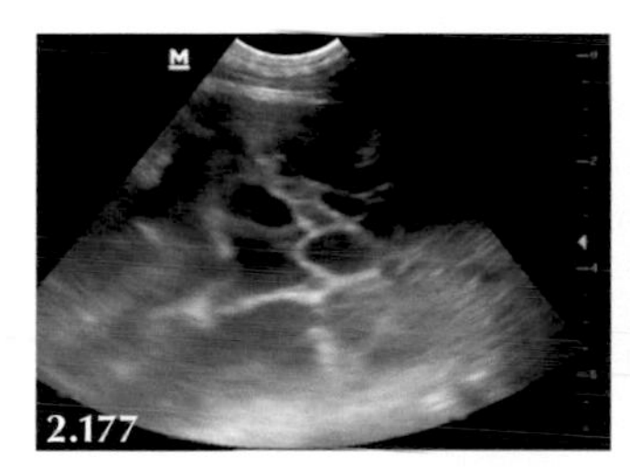

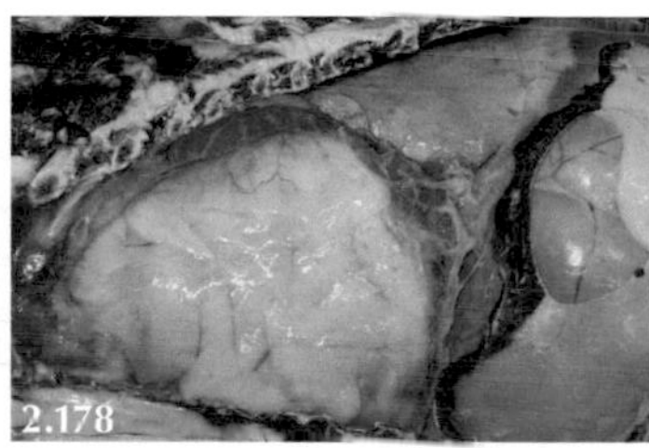

2.177: Organising fibrinous exudate in this ram extends to 6 cm separating the pleurae. **2.178:** Necropsy confirms accumulation of exudate in the ventral pleural space adjacent to the diaphragm (right).

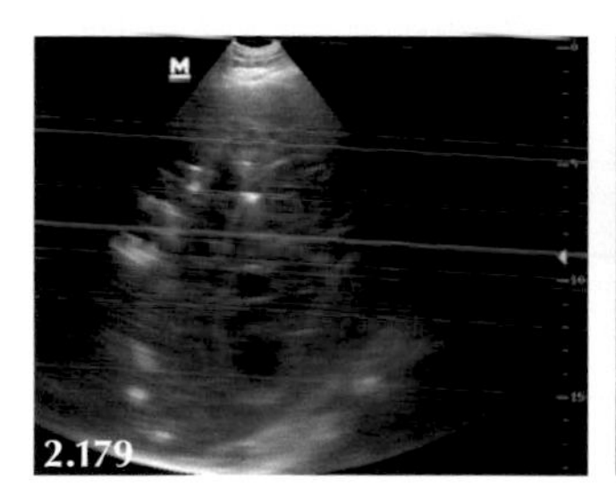

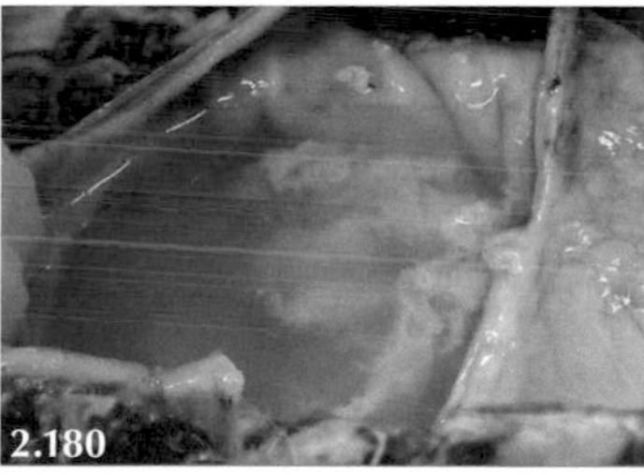

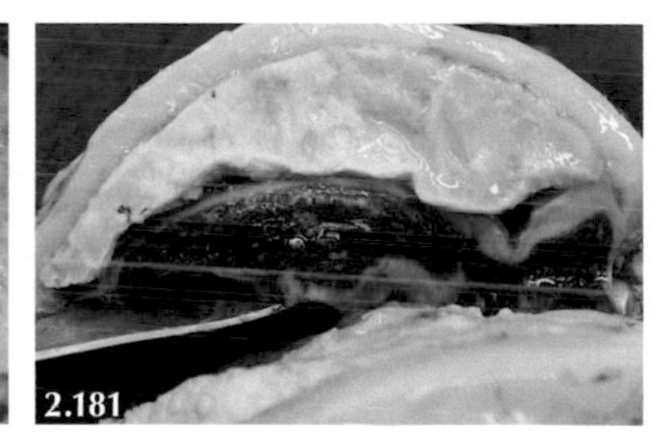

2.179: Sonogram showing unilateral exudate extending to 16 cm from the chest wall in this ewe (note the scale on the right hand margin of ultrasound image). **2.180:** Necropsy of case (left) revealing the fibrinous exudate is well-encapsulated and therefore of no clinical significance apart from as a space-occupying lesion reducing lung capacity. **2.181:** Extensive chronic fibrinous pleurisy which has caused collapse of the left lung of this ewe (supported by knife and looks like sectioned liver). 2.69, 2.70, 2.71, 2.72, 2.73, 2.74, 2.75, 2.76, 2.77

Acute fibrinous pleurisy is uncommon in cattle but is generally associated with severe illness. With large accumulations of inflammatory exudate, pleurisy causes attenuation of lung and heart sounds but there are no audible friction rubs. With unilateral lesions, an increased volume of breath sounds can often be heard on the normal side of the chest caused by increased air flow due to the overall reduced functional lung capacity. Fibrous adhesions are occasionally observed at necropsy and are of little significance.

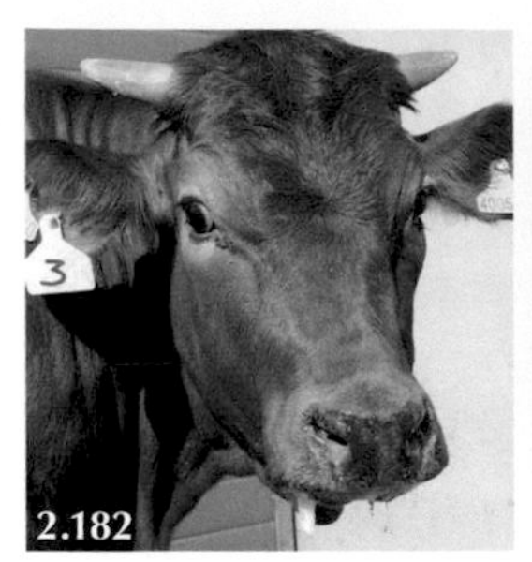

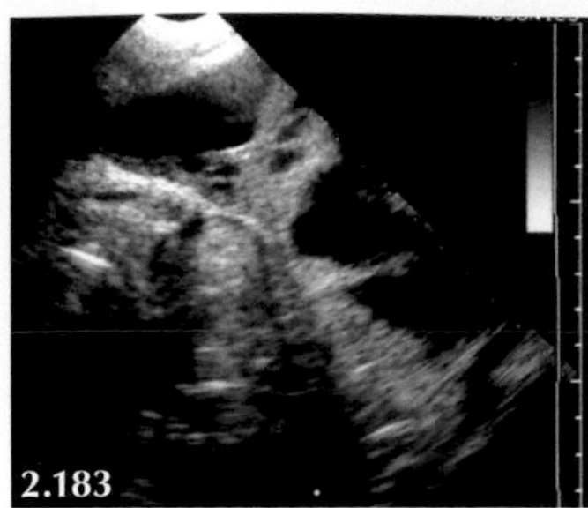

2.182: Unlike in sheep, fibrinous pleurisy is associated with serious illness in growing cattle. **2.183:** The image quality initially looks poor, but this is caused by the large amounts of fibrin deposited on both the parietal and visceral pleurae. 2.78 Fibrin tags can be seen bridging these two surfaces which are 7–8 cm apart. **2.184:** Necropsy findings reveal extensive fibrinous pleurisy. 2.79

Resolution of Fibrinous Exudate—Sheep

Many apparently healthy sheep have been detected with considerable fibrinous exudate within the pleural space during routine whole flock scans for OPA control. These sheep have been treated for 10 consecutive days with procaine penicillin or a single injection of long-acting oxytetracycline as directed by the farm's veterinary health plan. In some cases when re-scanned 4–6 weeks later there was only approximately 3 mm pleural exudate/pleurisy. In other cases, resolution was much slower with little difference when the sheep was re-scanned four weeks later but after several months there was only a 2–3 mm band of pleurisy. Such lesion resolution is shown in videos: 2.80, 2.81, 2.82.

Cattle

Repeated ultrasound examinations showed resolution of fibrinous pleurisy in this heifer where the diaphragm can be used as a landmarks (see the following).

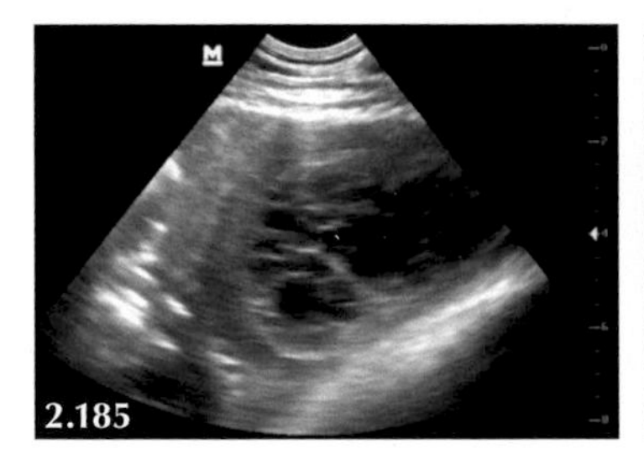

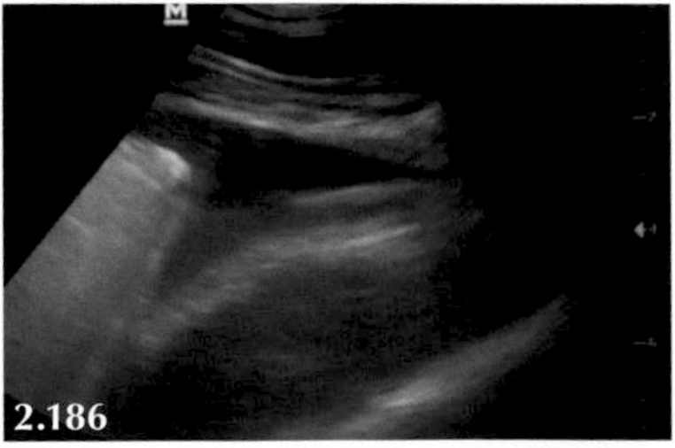

2.185: There is 5–6 cm organised fibrinous exudate in the right caudo-ventral pleural space. The diaphragm appears as a broad hyperechoic band in the lower right hand side of the image with the liver beneath (bottom right). There is consolidated lung dorsally showing "hepatoid" change. **2.186:** Same animal taken two weeks later and after antibiotic therapy. The pleural exudate appears only 5–10 mm thick; the diaphragm and liver are clearly visible showing that the same lung area has been imaged. 2.83

Development of Pleural Abscesses in Sheep

Pleural abscesses are common in sheep but cause no adverse clinical signs. They likely develop following pyogenic infection of organising pleural exudate as illustrated in the sequential figures and video recordings shown next. Pleural abscesses are generally single and large to very large (4–20 cm diameter), occupy the ventral pleural space and are well-encapsulated. Pyaemic spread to the lungs results in multiple small (1–3 mm diameter) abscesses in the dorsal lung field and other parenchymatous organs but such lesions were not observed in sheep with large pleural abscesses.

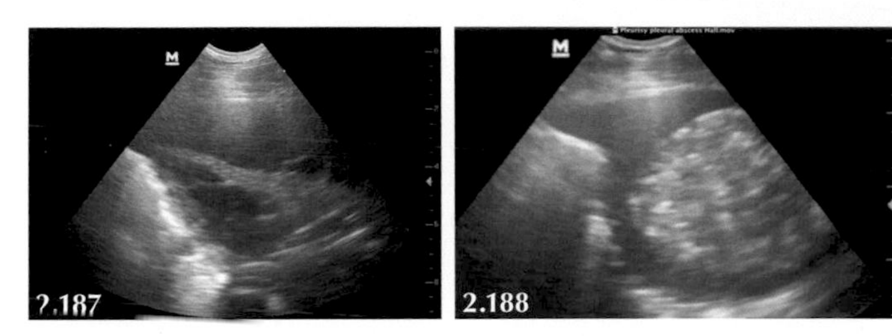

2.187: Ultrasound examination reveals >8 cm fibrinous exudate filling one side of the chest. Note acoustic enhancement of the lung surface with adherent fibrin (left) and the hypoechoic lung lobe consistent with consolidation. **2.188:** Four weeks later the pleural exudate has become organised. There is a 5 cm diameter circular hypoechoic lesion containing multiple hyperechoic dots surrounded by an anechoic periphery approximately 1 cm thick which extends dorsally separating the pleurae. It is not possible at this stage to distinguish whether this circular lesion is an organised fibrin clot or an abscess however necropsy of similar cases (see next) has revealed these lesions are abscesses.

With large pleural abscesses auscultation of the lung field on the affected side fails to detect lung sounds and only transmitted gut sounds, particularly rumen contraction sounds over the left thorax. Auscultation of the contralateral (normal) chest reveals increased volume of normal lung sounds. Heart sounds are also increased over the contralateral chest due to displacement of the heart towards the unaffected side.

Pleural abscesses appear as anechoic circles containing multiple hyperechoic dots representing minute gas bubbles giving a "snowstorm" appearance. The surrounding capsule appears as a hyperechoic circle varying in thickness up to 10 mm. Typically, there is adjacent pleurisy and acoustic enhancement of the surface of normal lung tissue dorsal to the lesion. 2.84, 2.85, 2.86. 2.87

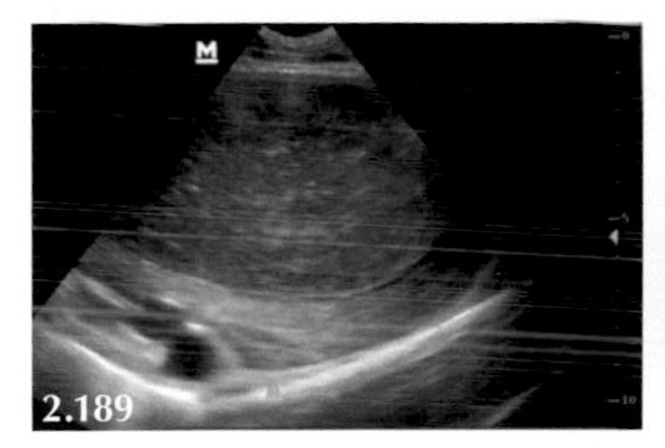

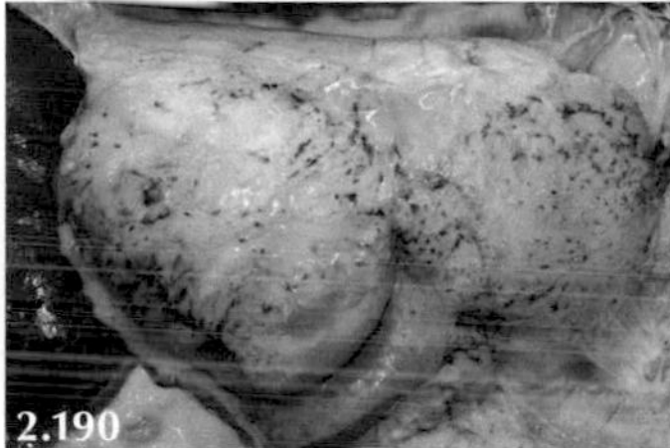

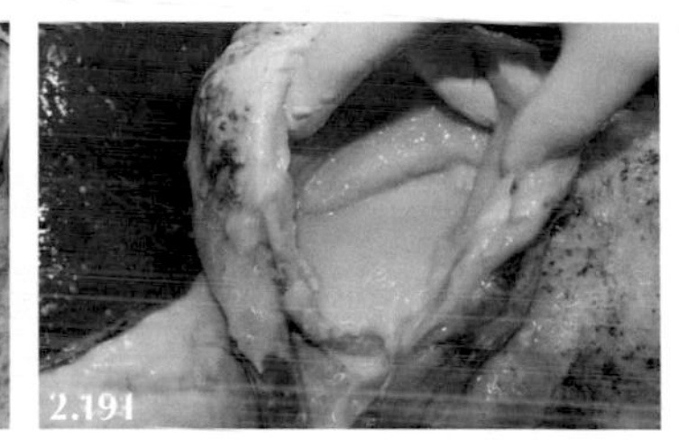

2.189: Sonogram of a 6 cm diameter pleural abscess with a 1 cm thick hyperechoic capsule in the ventral pleural space. **2.190** and **2.191:** Lesion confirmed at necropsy.

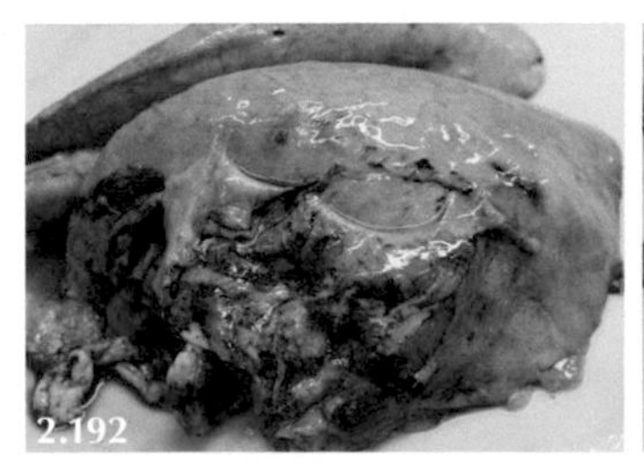

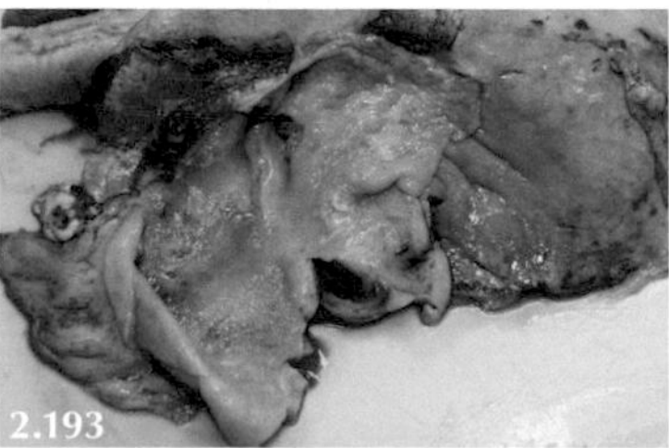

2.192: Extensive fibrinous pleurisy and pleural abscess at necropsy. **2.193:** The pleurisy has been peeled off the lung surface to reveal the abscess capsule (the contents have been drained). Note that there are no abscesses in the lung parenchyma.

Sequential ultrasound recordings over several months (shown earlier) demonstrate that abscesses likely develop from infected pleural exudate and that this exudate becomes encapsulated over several months.

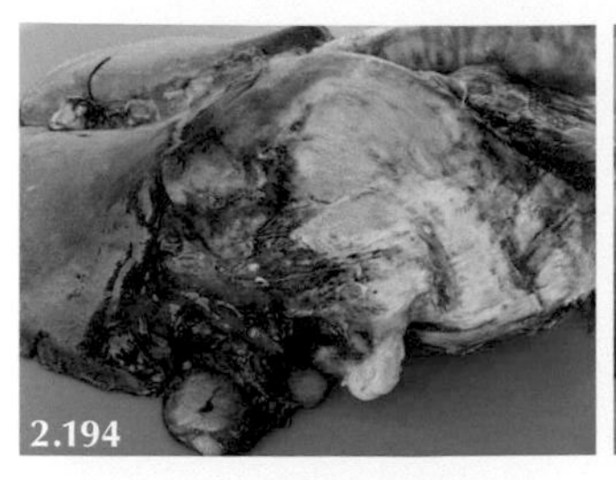

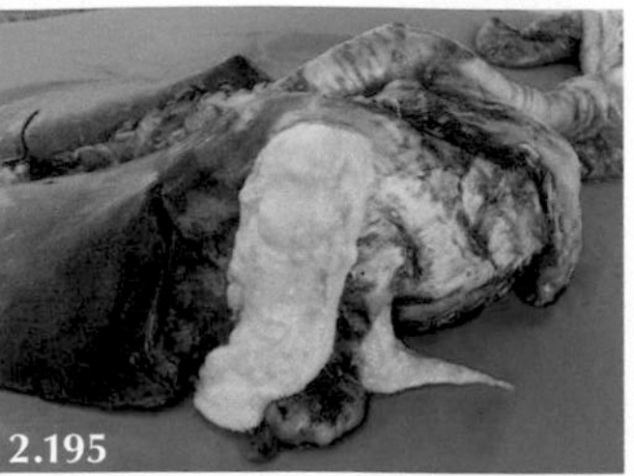

2.194 and **2.195:** Careful removal of the lungs at necropsy shows no evidence of pyaemic spread to the caudo-dorsal lung only encapsulation of infection by fibrous tissue within the pleural space.

Lung abscesses appear as anechoic circles containing multiple hyperechoic dots.

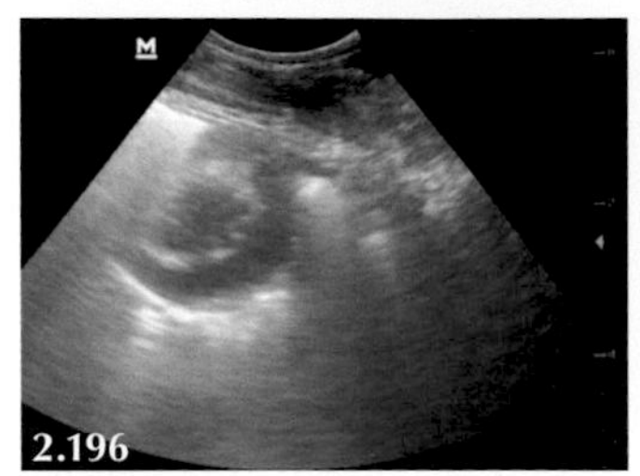

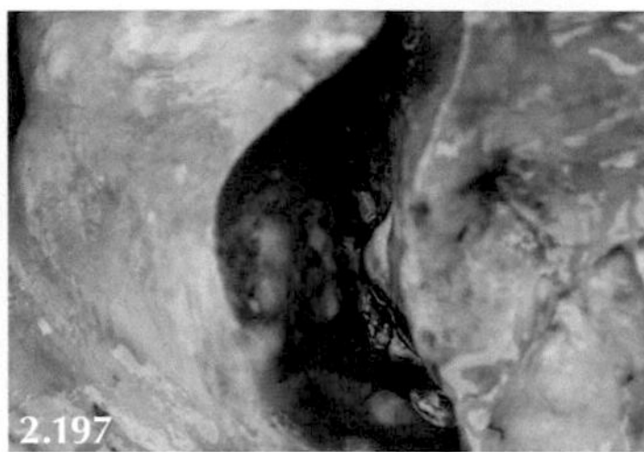

2.196: Ultrasound image of 3 cm diameter lung abscess revealed at necropsy (**2.197**). 2.88, 2.89, 2.90, 2.91

Lung Consolidation

Lung consolidation is most commonly observed in housed sheep during their first winter often caused by *Mycoplasma* infection (cuffing pneumonia). Pleurisy can often be observed in these sheep with the sharply demarcated consolidated lung, often showing a marked bronchial pattern, displaced off the chest wall by organising exudate. 2.92, 2.93, 2.94, 2.95

Secondary abscessation, consolidation, fibrosis and pleurisy occur occasionally in adult sheep. Such extensive lung pathology means the affected sheep will be unproductive and should be culled for welfare reasons. These lung lesions cannot be differentiated on gross examination from secondary bacterial infection/necrosis of OPA tumour masses and it is still essential to undertake histopathology to ascertain the role of JSRV should control measures for this disease prove necessary.

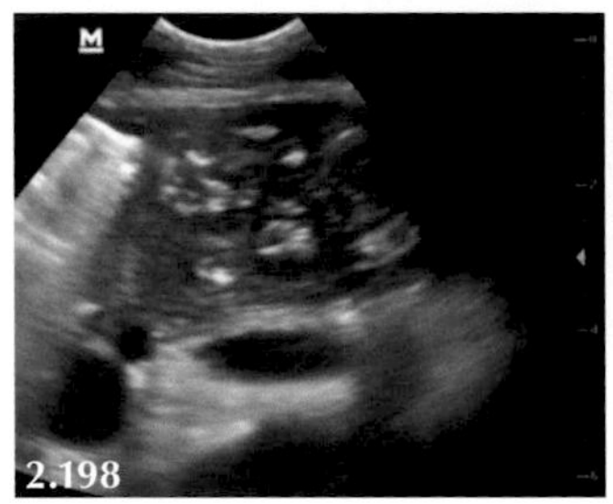

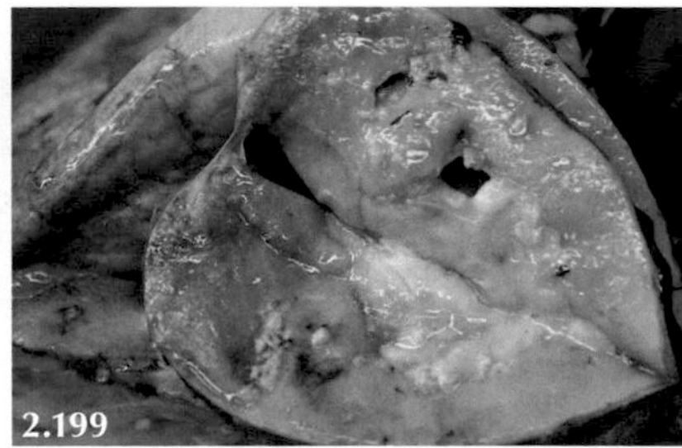

2.198: Multiple well-encapsulated abscesses within consolidated (fibrotic) lung. There is 3–5 mm pleurisy more noticeable dorsally displacing the hyperechoic visceral pleura. These changes would suggest a bacterial aetiology (**2.199**) but histopathology is necessary to rule out the possible role of JSRV. 2.96, 2.97, 2.98

Chronic Suppurative Pneumonia/Bronchiectasis in Cattle

As the probe head is advanced ventrally from normal lung tissue present in the dorsal lung field, the first ultrasonographic changes attributed to chronic pneumonia in cattle are hypoechoic "columns" extending 2 to 6 cm from the visceral pleura bordered distally by bright hyperechoic lines representing lobular consolidation.

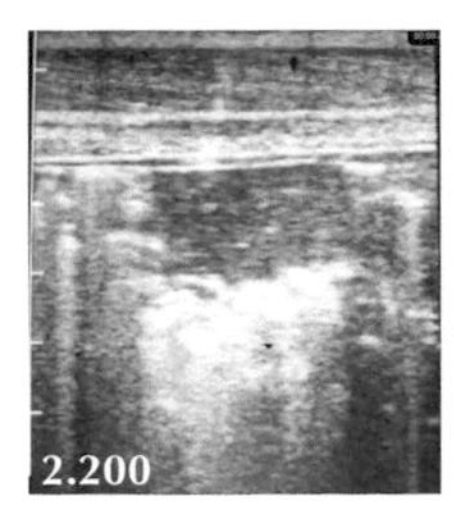
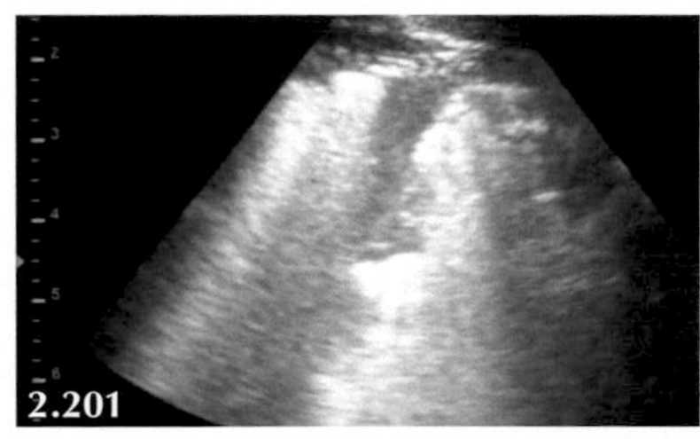
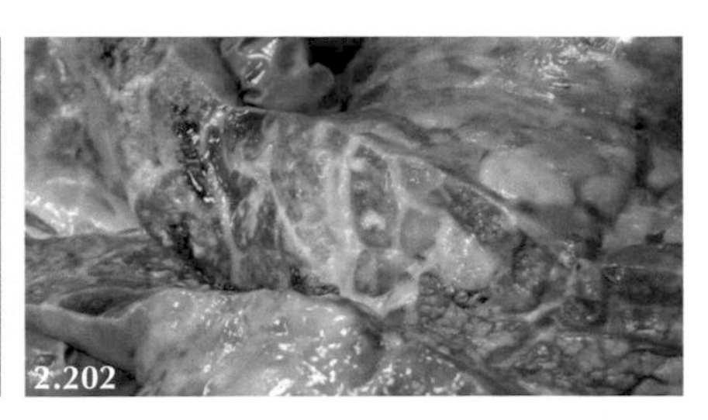

2.200: A hypoechoic "column" extends 2 cm from the visceral pleura bordered distally by bright hyperechoic lines (5 MHz linear scanner). This area represents lobular consolidation. **2.201:** A hypoechoic "column" extends 2 cm from the visceral pleura bordered distally by bright hyperechoic lines (6.5 MHz microarray scanner). There is normal lung either side of this affected lobule. **2.202:** Necropsy reveals a lobular pattern of lung consolidation (collapsed dark purple areas contrasts with normal inflated pink lobules). 🎥 2.99, 2.100

Bronchiectasis is the term given to permanently dilated small bronchi and bronchioles located in the ventral parts of the lungs which contain a wide range of microorganisms, predominantly *Trueperella pyogenes*, causing chronic disease. In bronchiectasis cases, the columnar appearance of lung pathology in the dorsal lung field changes abruptly to large hypoechoic areas of lung consolidation extending to the ventral lung margins. This consolidated lung extends for 6–10 cm (deep) from the chest wall.

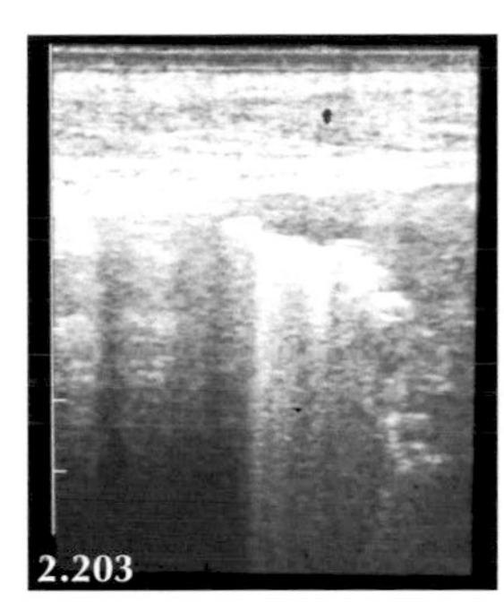
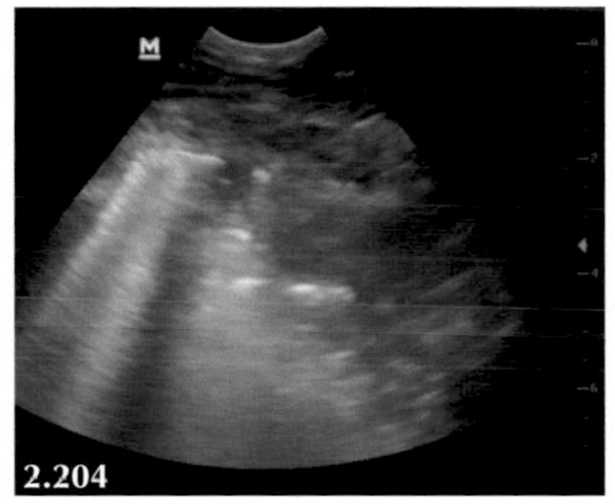
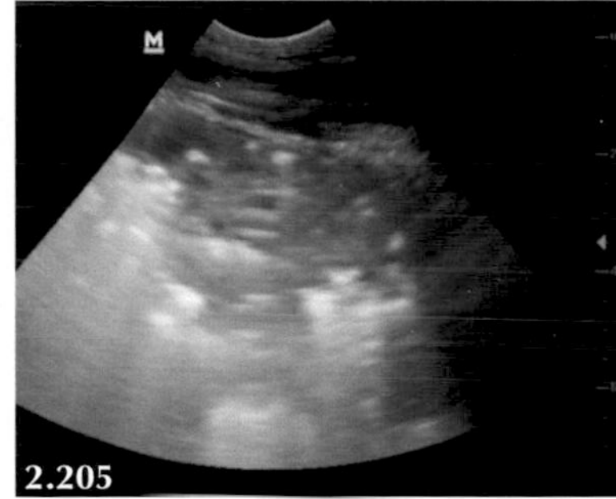

2.203: There is sharp demarcation in this image between the normal hyperechoic (bright white) visceral pleura dorsally (to the left) and lung consolidation ventrally (to the right) represented by the hypoechoic area bordered distally by a broad hyperechoic line running diagonally left to right. 5 MHz linear scanner. **2.204:** There is sharp demarcation between normal visceral pleura dorsally and 5–6 cm lung consolidation ventrally ("hepatoid appearance"). 6.5 MHz microarray scanner. **2.205:** There is lung consolidation appearing as a large (>7 cm deep) hypoechoic area containing numerous hyperechoic dots that probably represent small airways. 6.5 MHz microarray scanner.

The sonographic change to a large hypoechoic area affecting the ventral lung lobes representing lung consolidation is clearly illustrated at necropsy (see photos that follow). The dorsal margin of this sudden change can be measured from the point of the olecrannon to quantify lung pathology.

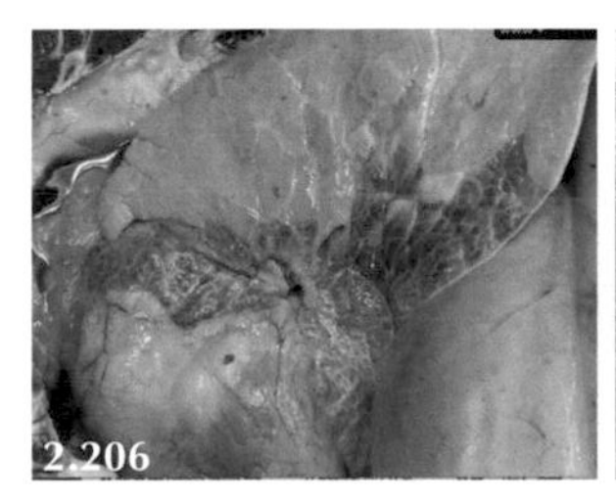
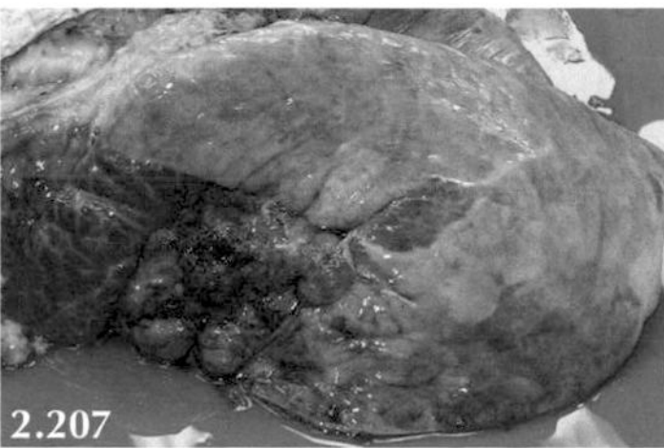

2.206: Lungs *in situ* at necropsy. Sharp horizontal demarcation between normal lung dorsally and ventral lung consolidation. **2.207:** Lungs removed from chest cavity. While the sharp demarcation between diseased lung ventrally is obvious, the horizontal demarcation cannot be appreciated.

It is often possible to image the right lung and liver in the same image with diseased lung having the sonographic appearance of liver hence the term "hepatoid change".

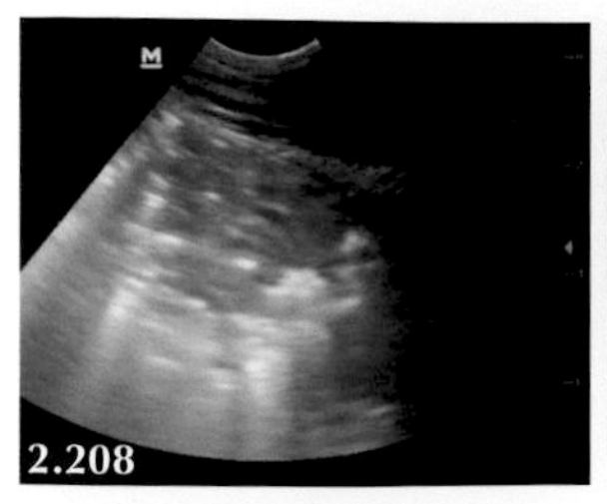

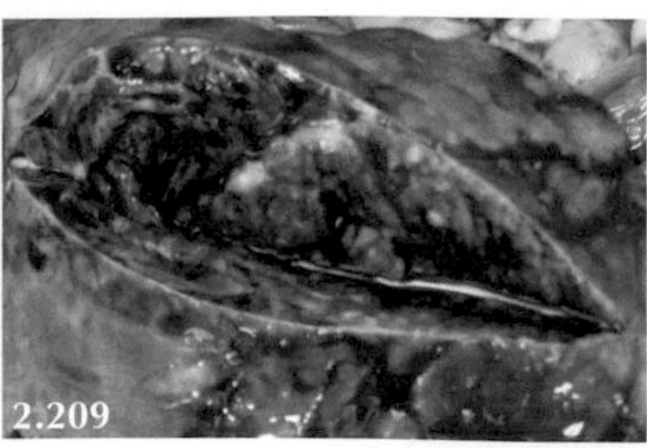

2.208: The lung is hypoechoic (consolidated) with the sonographic appearance of liver (hepatoid change). 6.5 MHz microarray scanner. **2.209:** Necropsy findings reveal extensive lung consolidation on cut surface with the gross appearance of liver. 2.101, 2.102

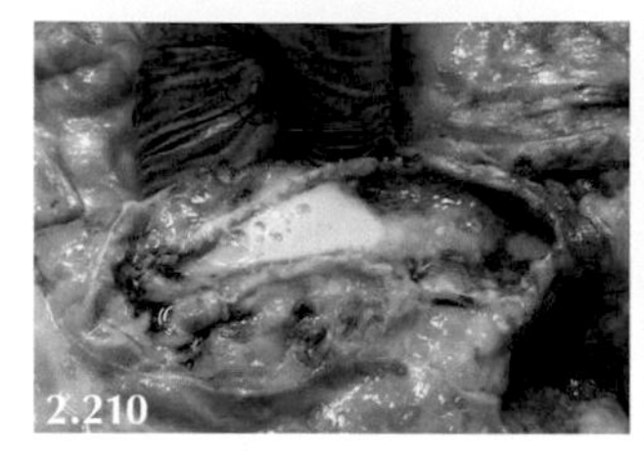

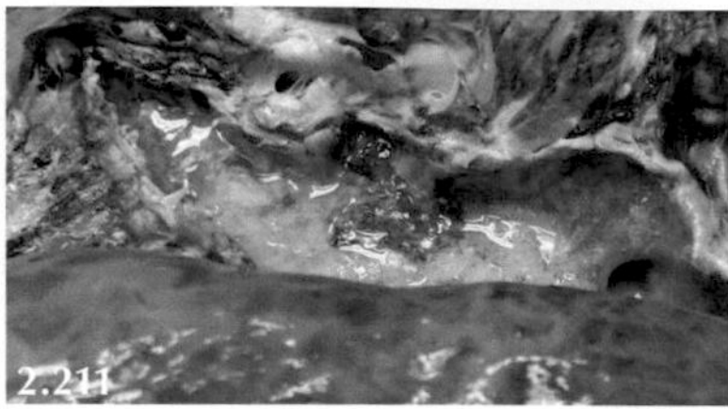

2.210 and **2.211:** Bronchiectasis demonstrated at necropsy—consolidated lung with pus expressed from the airways in cut sections of affected ventral lung lobe.

The distance from the point of the elbow joint (olecrannon process of the ulna) to the most dorsal area of lung consolidation (hypoechoic/heterogenous area) is measured using a flexible steel tape to quantify the extent of lung pathology. As a general guide, the prognosis is considered poor when the sonographic changes representing chronic pneumonia extend more than 40 cm above the level of the olecranon of the elbow when values for both sides of the chest are added together. Be aware that lung pathology defined ultrasonographically may change little during antibiotic treatment despite marked clinical improvement; antibiotic treatment in bronchiectasis cases is treating disease within the airways and resolution of associated lung consolidation may not occur.

Using ultrasonography, experienced veterinarians could screen groups of cattle (e.g. heifers being selected for breeding programme) for chronic respiratory disease at a rate >30 per hour. Those heifers with significant lung pathology should be fattened and not be retained in the breeding herd. 2.103

Lung Abscessation

Multiple 5–10 mm abscesses are readily imaged within consolidated lung (see the following) and usually accompany bronchiectasis in cattle.

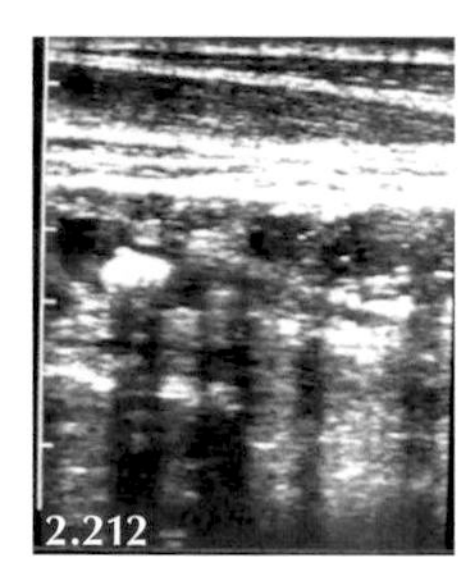

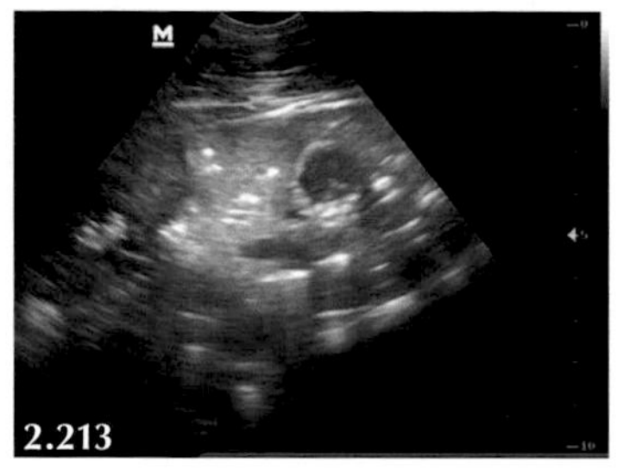

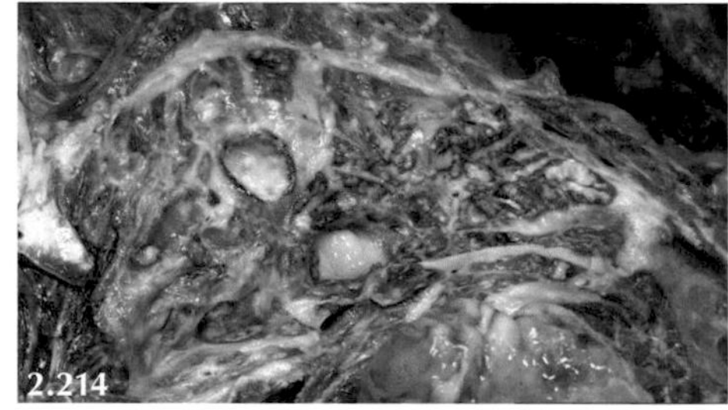

2.212: 5 MHz linear scanner. Chronic suppurative pneumonia/bronchiectasis with extensive consolidation containing numerous 5–15 mm diameter abscesses. **2.213:** 6.5 MHz microarray scanner. Consolidated lung containing multiple 10–20 mm diameter abscess. **2.214:** Chronic suppurative pneumonia with extensive lung consolidation/bronchiectasis and numerous 10–20 mm diameter abscesses.

OPA

Many farmers with an OPA problem in their flocks have seen a benefit:cost after whole flock scanning with important animal welfare benefits achieved by selling sheep before disease becomes apparent. Biannual ultrasound examination of lungs has been adopted by progressive pedigree sheep breeders to reduce the likelihood of selling rams with OPA lesions however ultrasonographic examination of the lungs cannot guarantee freedom from this disease.

2.215 and **2.216:** Author's set up on a commercial farm to allow rapid ultrasound scanning for OPA with 150 sheep examined per hour (over 1,200 sheep per day is possible). Note the laptop computer connected to the ultrasound machine to record all pathologies for future reference.

It is essential to examine both sides of the chest because OPA lesions commonly involve only one lung. Widespread B lines and thickening of the hyperechoic line representing the visceral pleura alert the clinician to the possibility of very early OPA with re-scanning often necessary four to eight weeks later. However, these sonographic changes can also be observed in sheep with mild pleurisy. Several 2 mm to 5 mm hypoechoic/anechoic circles within the visceral pleural line can be identified during ultrasonographic examination of the lungs alongside B lines, but it may still not be possible to differentiate whether these hypoechoic areas are early OPA lesions, abscesses, consolidation or other pathologies such as aberrant lungworm lesions, and these sheep must be re-assessed one to two months later when OPA would have progressed whereas other pathologies will likely have remained unchanged (abscesses) or reduced/resolved (consolidation/lungworm lesions).

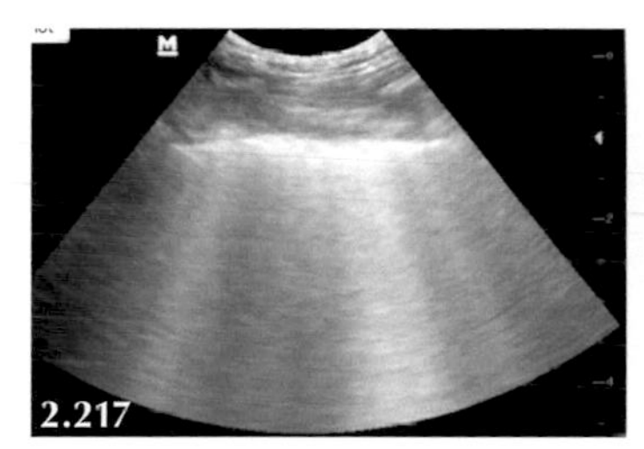

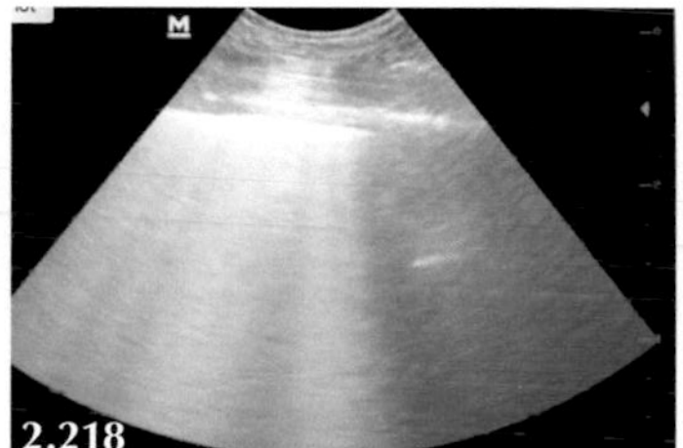

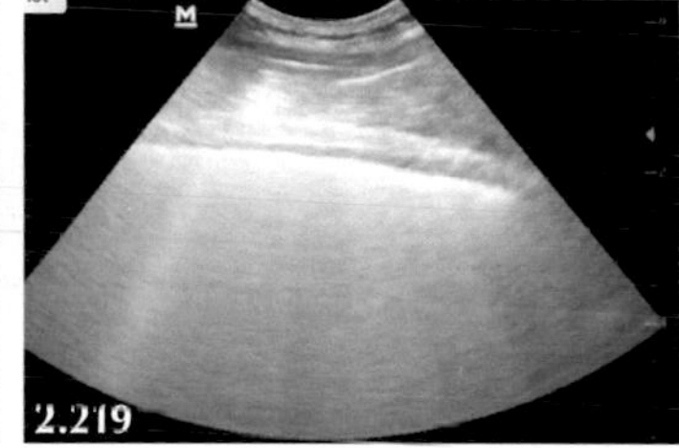

2.217–2.219: B lines arising from mild pleurisy with 1–2 mm separation of the pleurae. There are no anechoic visceral pleural lesions. 2.104, 2.105, 2.106

Widespread B lines and 2–8 mm diameter hypoechoic dots at the visceral pleura can also be imaged in aberrant cases of verminous pneumonia and re-scanning in six to eight weeks is necessary to avoid errors. 2.107

A broader and more hyperechogenic white line representing the visceral pleura and containing multiple 2–8 mm diameter hypoechoic dots is typically seen in sheep with early signs of OPA. These sonographic changes probably represent very small tumours with a lobular distribution. Broad hyperechoic lines (B lines) originate from the visceral pleural likely caused by inter-lobular oedema.

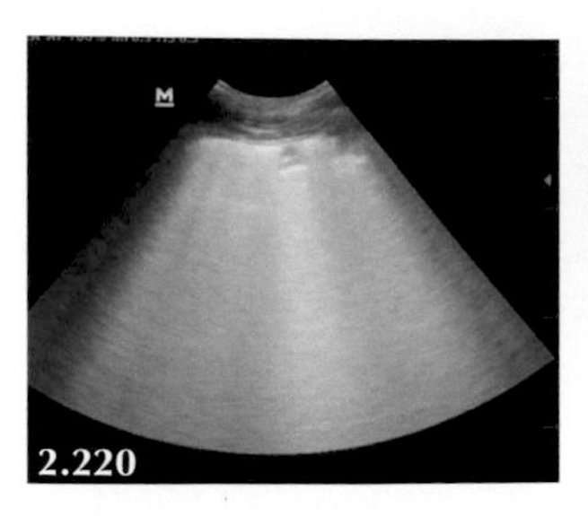

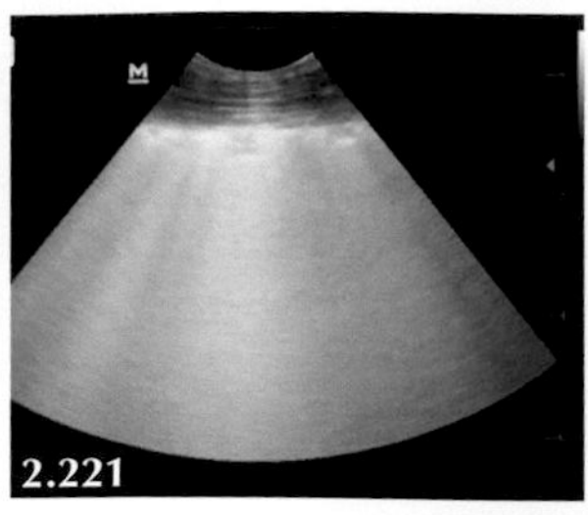

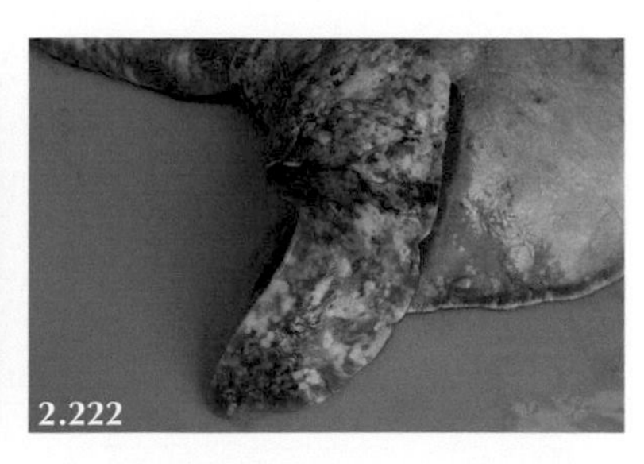

2.220 and **2.221:** Very early sonographic changes in OPA. The hyperechoic line representing the lung surface is thicker and more hyperechoic than normal and punctuated with several 2–3 mm anechoic dots. **2.222:** Necropsy finding revealing many small OPA tumours arising from multiple sites of JRSV virus infection; note that this is not metastatic spread from a single tumour source.

B lines can be caused by a variety of pathologies, most often interlobular oedema, but in combination with multiple 2–8 mm anechoic lesions at the visceral pleura should alert the clinician to the likelihood of very early OPA lesions. In the video recordings featured next the clinician is suspicious of the B lines and pleural lesions on the first side of the chest (and starts the recording) with confirmation of OPA on the other side of the chest by identifying more typical advanced lesions. 🎥 2.108, 2.109, 2.110

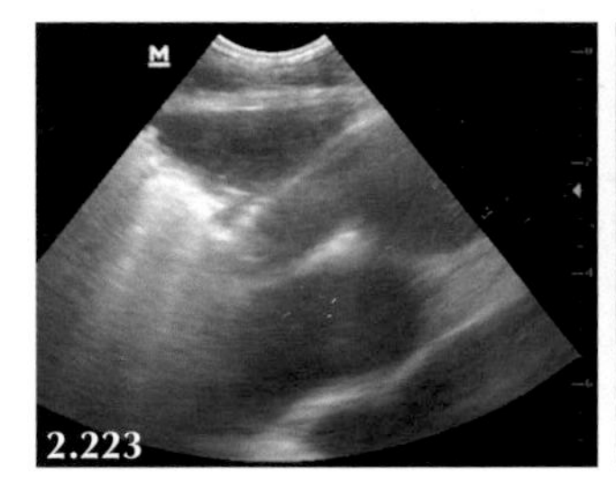

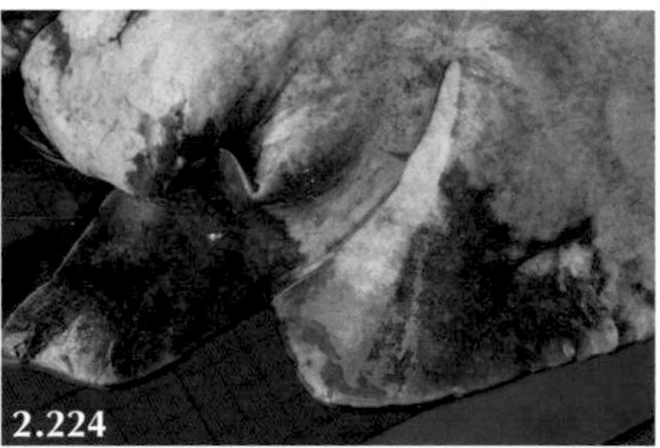

2.223: Sonogram showing displacement of the heart from the chest wall by a hypoechoic triangular-shaped area of consolidation (up to 20 mm wide; to the left) at the ventral lung margin is one of the earliest reliable indications of OPA in adult sheep. More dorsal examination would likely reveal widespread B lines and several 2–5 mm hypoechoic pleural lesions (shown in video recordings next). Blood in the heart chambers appears anechoic; the chambers are displaced off the chest wall by consolidated lung. **2.224:** Necropsy of case (left) confirms the presence of sharply demarcated lung consolidation revealed as OPA on histopathological examination. 🎥 2.111

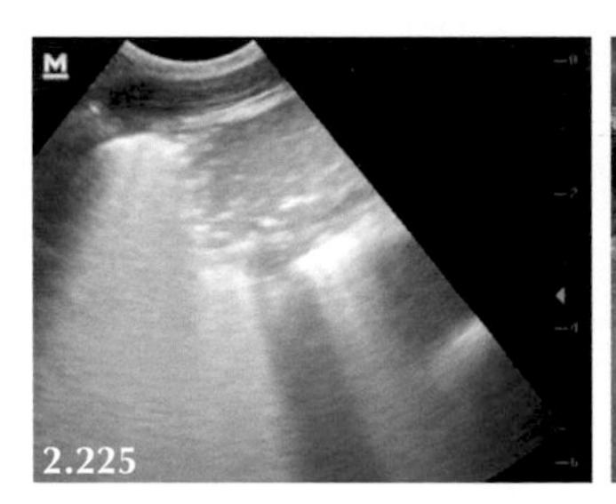

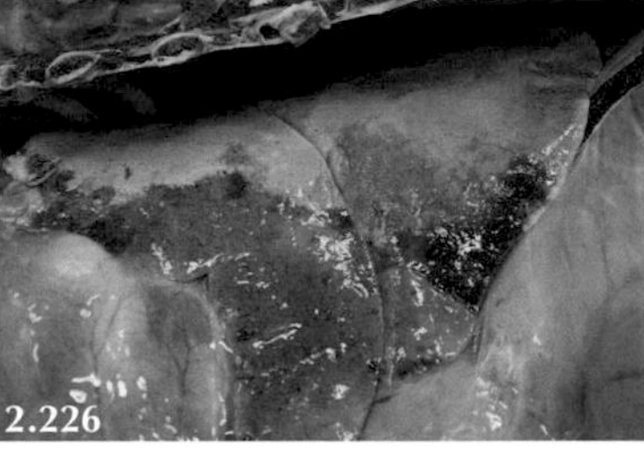

2.225: Sonogram showing displacement of the heart from the chest wall by a 20 mm deep, sharply demarcated hypoechoic area at the ventral lung margin. **2.226:** Sharply delineated lung pathology ventrally involving the cardiac and diaphragmatic lung lobes confirmed as OPA on histological examination. 🎥 2.112

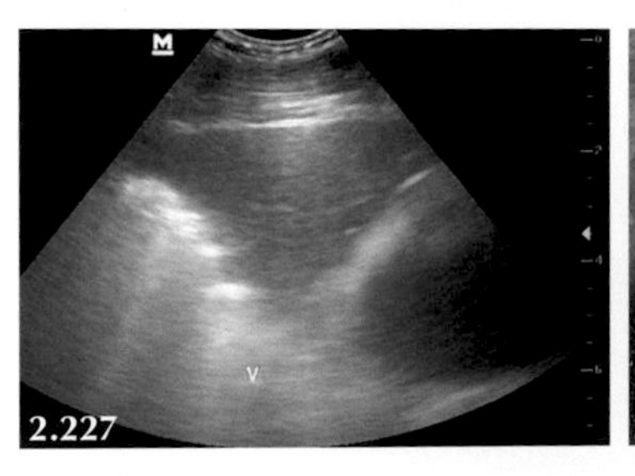

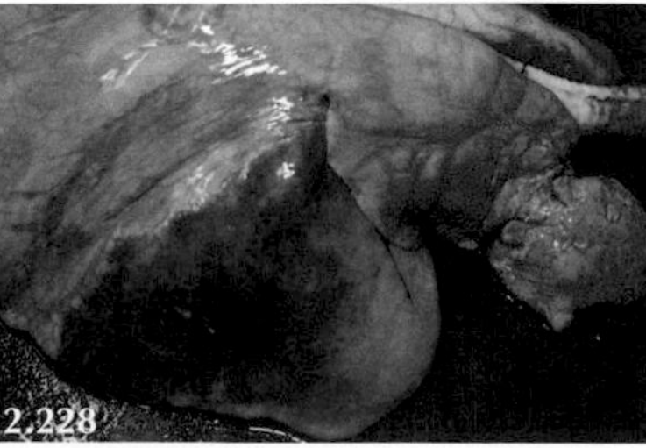

2.227: Sonogram showing displacement of the heart from the chest wall by a 4–5 cm deep, hypoechoic, triangular-shaped area of consolidation of the ventral lung margin. **2.228:** Necropsy, then histology, confirm OPA in this adult sheep.

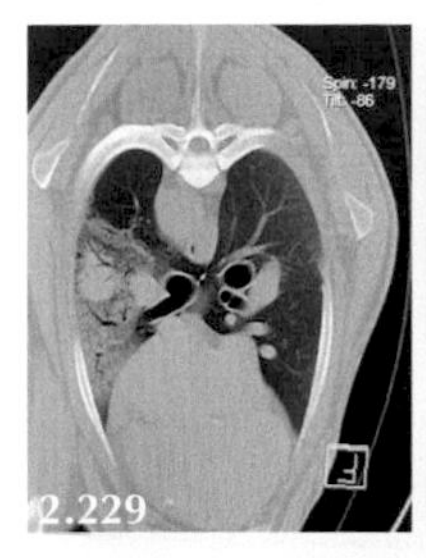
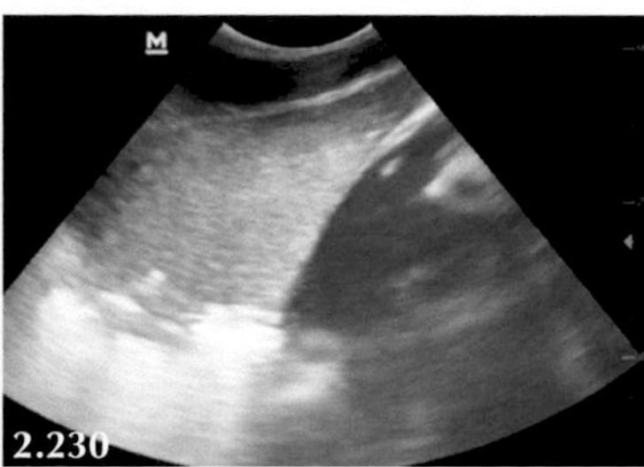

2.229: CT showing the presence of a large OPA tumour in the ventral left lung—note how the medial aspect of the tumour mass is shaped by the heart. **2.230:** Sonogram showing the characteristic semi-circular shape of the medial aspect of the OPA mass which is formed by the heart.

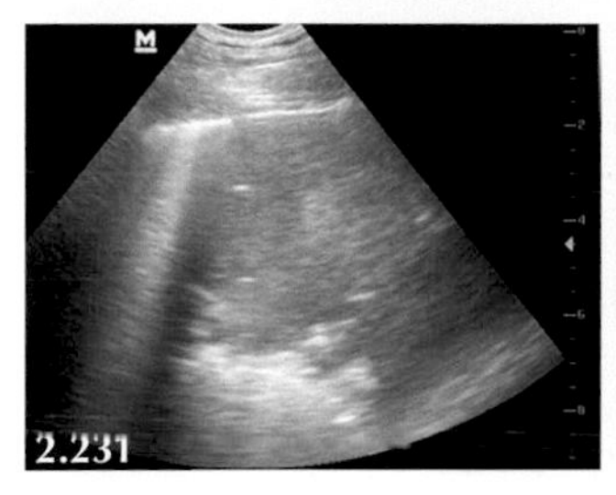
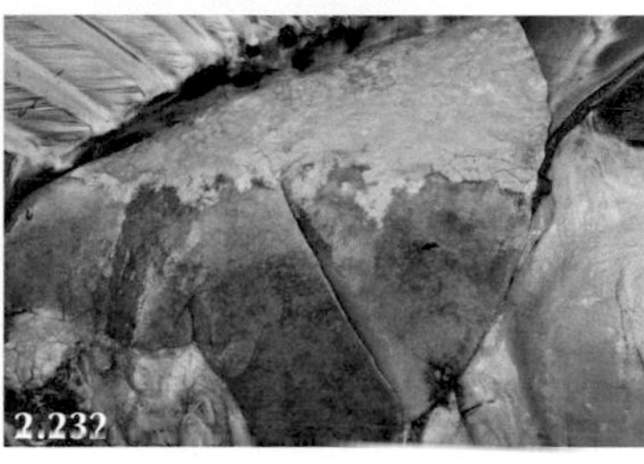

2.231: Extensive hypoechoic area of consolidation (up to 7 cm wide) with a semi-circular distal (medial) surface moulded by the heart. The heart is visible in the lower right of the sonogram. **2.232:** Necropsy then histology confirm OPA in this adult sheep. 2.113, 2.114, 2.115, 2.116, 2.117

The liver extends as far forward as the 7th–8th intercostal spaces on the right hand side of the chest and presents just above the level of the elbow at its most cranial position. The marked convex nature of the diaphragm means that the chest (dorsally) and abdomen (ventrally) are present underneath the same intercostal spaces. As the probe head moves ventrally over the same intercostal space it moves from lung to liver. Care is needed not to mistake the liver for consolidated ventral lung. There is an abrupt loss of the hyperechoic line which represents the lung surface as the probe head moves ventrally from the chest over the abdominal cavity with blood vessels and bile ducts clearly visible in the liver.

The "hepatoid" echogenic appearance of OPA tumours is highlighted where both organs are imaged together with OPA pathology dorsally and liver ventrally separated by the diaphragm which appears as a broad convex hyperechoic line in the centre of the image.

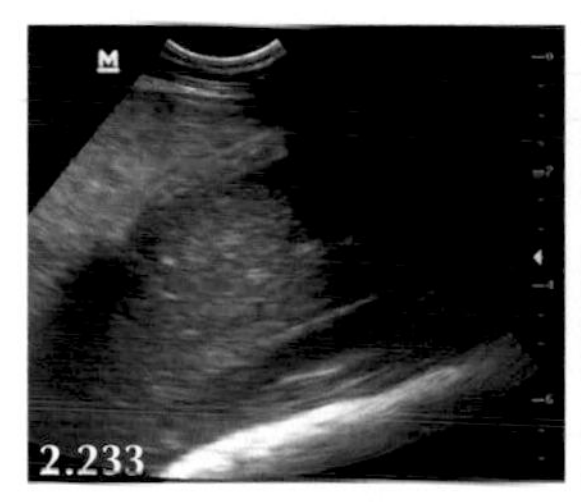
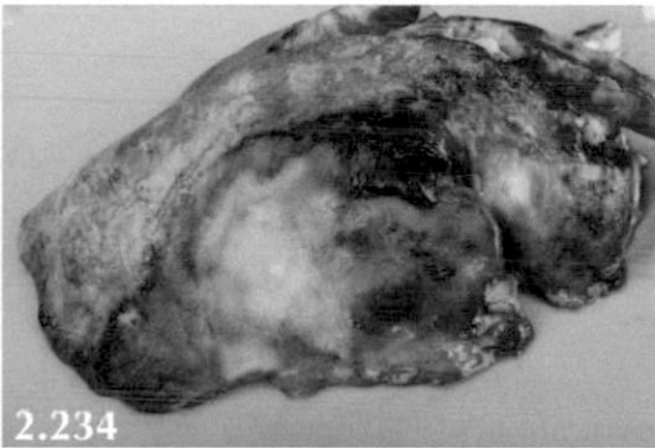

2.233: The "hepatoid" appearance of this OPA tumour is highlighted in this sonogram when both organs are imaged together with OPA pathology dorsally (to the left) and liver ventrally (to the right) separated by the diaphragm which appears as a broad convex hyperechoic line in the centre of the image. **2.234:** Necropsy reveals involvement of the right diaphragmatic lung lobe.

Necropsy

Necropsy studies of knackery submissions list ovine pulmonary adenocarcinoma, chronic bronchopneumonia and pasteurellosis as three of the six most common causes of adult sheep deaths in the UK. However, gross inspection of lungs during a quick on-farm post-mortem examination often leads to erroneous conclusions especially where there is some degree of autolysis; histology is essential especially if significant decisions are to be made on these findings.

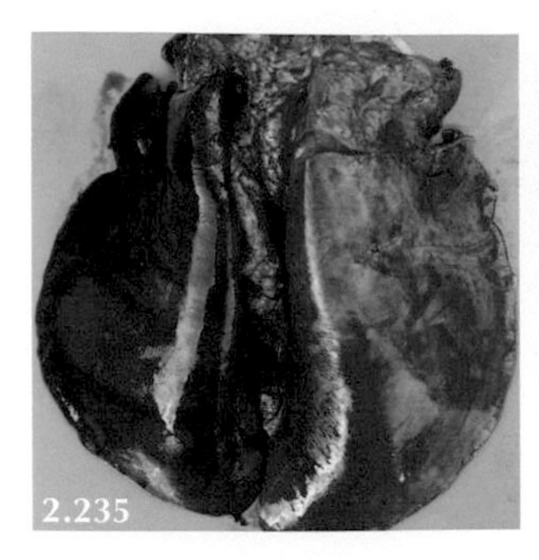

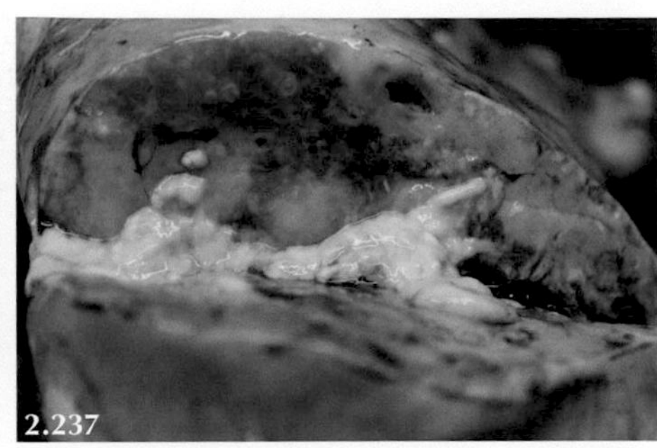

2.235: Hypostatic congestion of the left lung. **2.236:** Normal lung floats, lung pathology sinks; a simple but very useful field test. **2.237:** Gross examination cannot differentiate lung fibrosis/abscessation from an OPA lesion with a necrotic centre.

Examination of the lungs is most easily achieved by placing the dead sheep on its back and incising down through the skin in the midline onto the sternum then using a hand saw to cut through the bony sternum. Using a 10–15 cm block of wood or similar, the split sternum is forced apart revealing the pericardium and ventral margins of the lungs. The lungs can then easily be removed and photographed with representative samples taken for histopathology. Interpreting changes in lung appearance following death is difficult but immersion in water where diseased lung sinks and normal, aerated lung floats is a reliable guide. Lung samples for histopathology should include the margin between normal and diseased lung and represent no more than 10 per cent of the formalin volume.

Necropsy examination of growing and adult cattle is generally undertaken at regional veterinary laboratories (noting anthrax regulations where applicable).

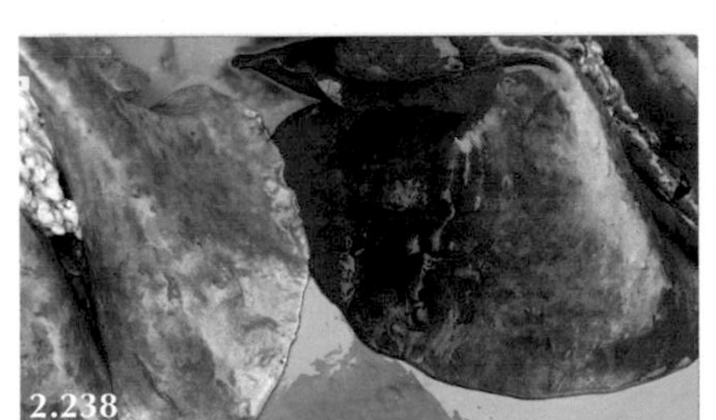

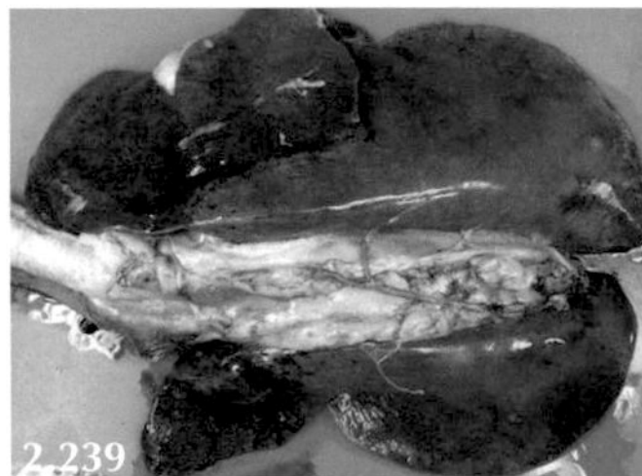

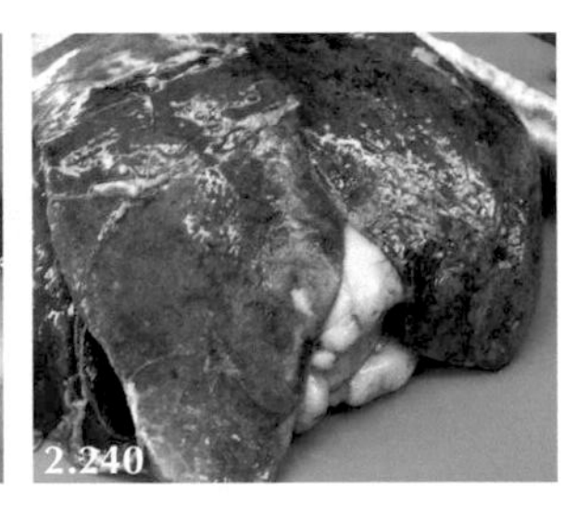

2.238: Comparison of normal lungs (left) and lung pathology (right)—note the purple discolouration and swollen diseased lungs. **2.239:** Swollen "wet" lungs (especially right lung) with widespread petechiae. **2.240:** Widespread petechiae with fibrin tags adherent to the lung surface.

Thoracocentesis

Pleural exudate, effusion and pericarditis in cattle are readily differentiated during trans-thoracic ultrasound examination. Ultrasound-guided thoracocentesis allows analysis of the pleural fluid, and would provide bacteriological culture results but rarely yields critically important information. Examination of the images that follow from various ovine pathologies illustrates why blockage of the sample needle with fibrin is very common.

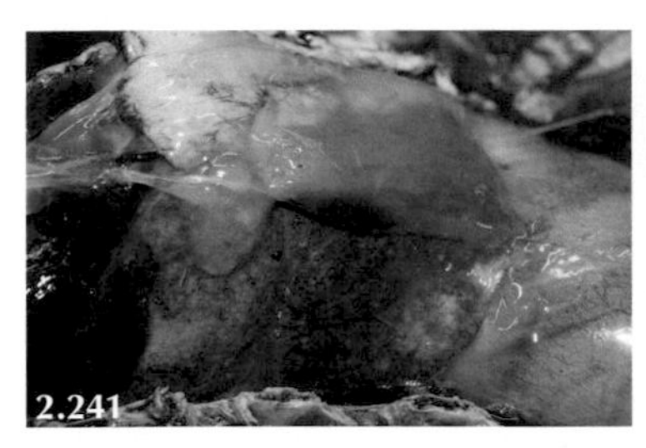

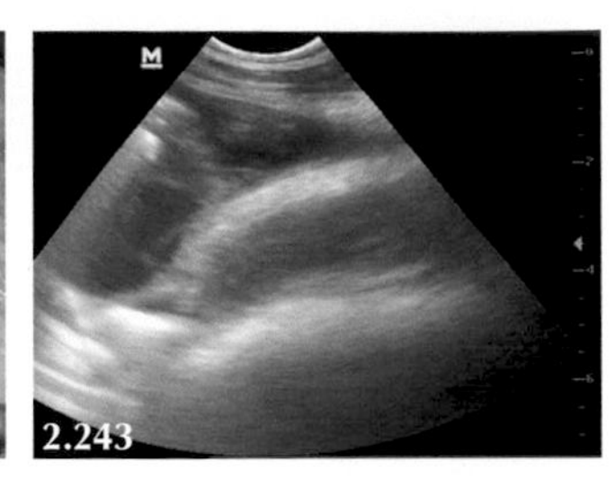

2.241: Fine needle aspiration would not collect any material from this fibrin exudate in this sheep with septicaemia. **2.242:** Culture of aspirate from this abscess in the pleural space of a sheep is of academic interest only because the infection is so well-encapsulated. **2.243:** Most exudate gravitates to the lowest point of the pleural space; on the right side even ultrasound-guided needle aspiration risks puncturing the adjacent liver in this sheep.

Bacteriology

Daily veterinary monitoring of respiratory disease outbreaks in calves largely overcomes the necessity for bacteriology and antibiotic sensitivity testing. There is little convincing evidence of widespread antibiotic resistance in sheep to oxytetracycline which is the antibiotic most commonly used by veterinary surgeons for suspected respiratory disease.

Broncho-Alveolar Lavage

Broncho-alveolar lavage (BAL) is not routinely undertaken in respiratory diseases because bacterial isolates are generally similar to those collected from deep naso-pharyngeal swabs. However, respiratory pathogens can frequently be isolated from the major airways of healthy cattle and sheep with no respiratory disease. Delays of several days before bacteriology results become available prevent immediate guidance on antibiotic selection. BAL samples are much more useful than nasal swabs for isolating BRSV during the early stages of disease outbreaks in young cattle.

Identification of Jaasiekte sheep retrovirus (JSRV) in BAL fluid by PCR has been used for OPA screening, but not all JSRV-infected sheep develop clinical OPA.

Radiography

Problems with access to the cranial chest, presence of a thick fleece, potential for aggravating dyspnoea while restraining the sheep in lateral recumbency, cost/time, and health and safety concerns mean that radiography is rarely used to investigate respiratory disease in sheep. While radiography could be used in growing cattle <200 kg, it is never employed for many of the reasons listed for sheep.

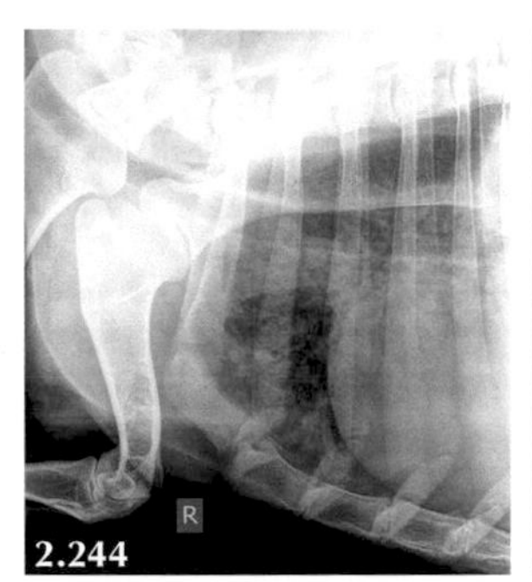

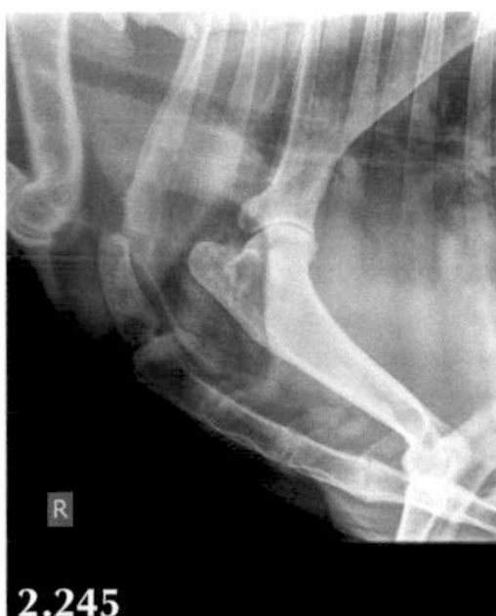

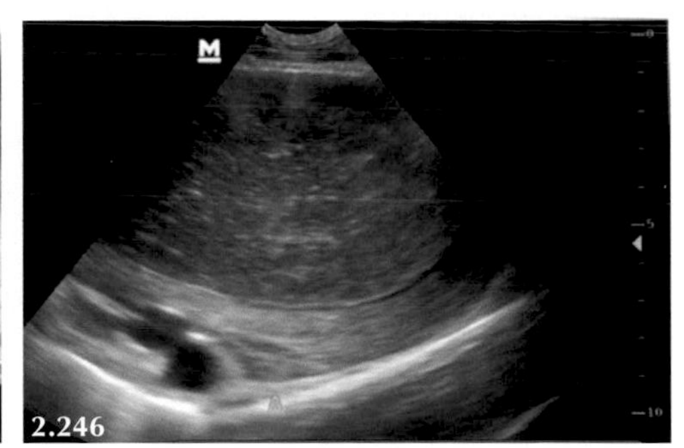

2.244 and **2.245:** Lateral radiographs showing a single large pleural abscess in the anterior chest. Without a DV view it is not possible to determine whether the abscess is within the pleural space or mediastinum. **2.246:** Trans-thoracic ultrasonography shows a well-encapsulated 6 cm diameter pleural abscess in the right chest (same sheep as left radiograph).

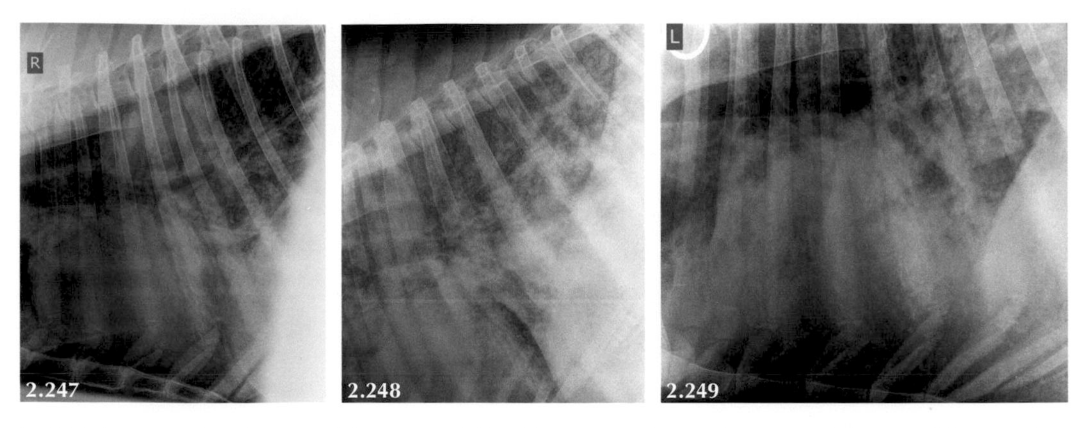

2.247–2.249: Lateral radiographs of three cases of OPA. The heart shadow has been lost and there is a diffuse mottling. The detail obtained using radiography is much less precise than ultrasonography and would not permit a diagnosis of OPA.

2.4

Examination of the Cardiovascular System

- While splashing and tinkling sounds are typically auscultated in septic pericarditis cases in cattle, large vegetative heart valve lesions (endocarditis) do not usually cause an audible murmur in sheep and cattle.
- While field depth of a 5 MHz linear scanner is limited to 10 cm, determining dorsal displacement of the base of the heart in cattle is a proxy measure of pericardial distension in cattle.
- Ascites secondary to right-sided heart failure in adult cattle can be readily demonstrated by trans-abdominal ultrasonography.
- Calves with small ventricular septal defects, yet intense murmurs, may remain asymptomatic throughout life.

Coccygeal, femoral and brachial pulses are easily found in cattle. The rate, rhythm and intensity of heart sounds are determined by auscultation over the chest in the region immediately beneath, and cranial to, the elbow joints. It is essential to listen to both sides of the chest because unilateral effusion/exudate and space-occupying lesions in the cranial thorax frequently displace the heart leading to marked disparity in intensity and origin of heart sounds. The heart rate of neonates may approach 120–160 beats per minute; adult cattle and sheep have a heart rate around 60 beats per minute. Handling and other stresses may temporarily increase the heart rate by more than 50 per cent, but it returns to normal within 5 to 10 minutes.

Cattle and sheep are sedentary animals and exercise intolerance is not assessed; lethargy is more easily appreciated. Heart failure is commonly observed in cattle but rarely in sheep. Clinical signs include brisket, submandibular and ventral abdominal oedema, and jugular distension or pulse. Heart rate is generally increased above 100 beats per minute and the mucous membranes are either normal or pale.

DOI: 10.1201/9781003106456-12

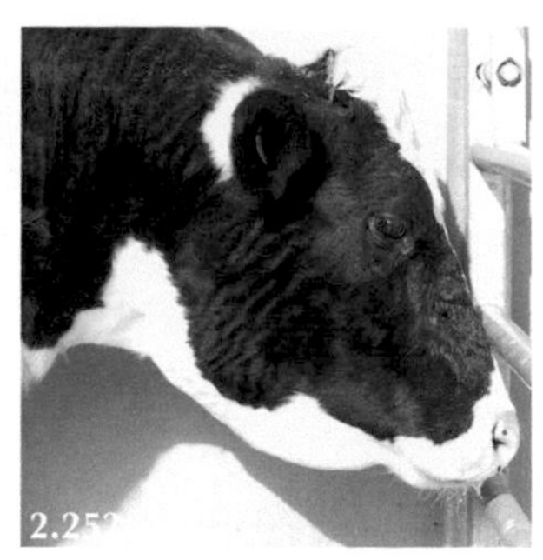

2.250–2.252: Signs of right-sided heart failure in a lethargic cow with marked brisket and submandibular oedema. 🎥 2.118, 2.119, 2.120

Right-sided heart failure can lead to ascites in severe cases. Chronic venous congestion results in hepatomegaly with rounded edges, and a "nutmeg" appearance at necropsy in the cow featured next.

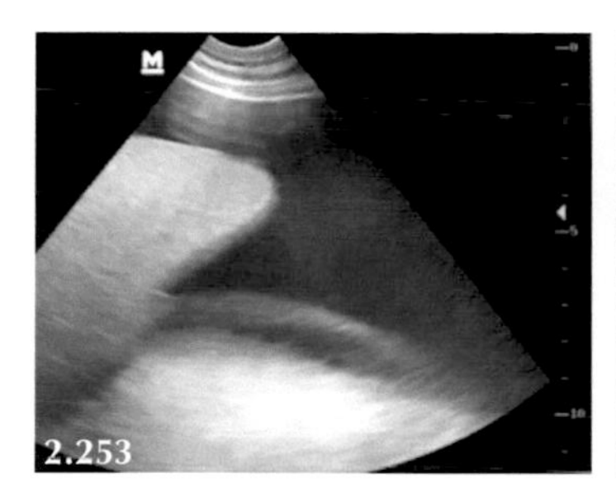
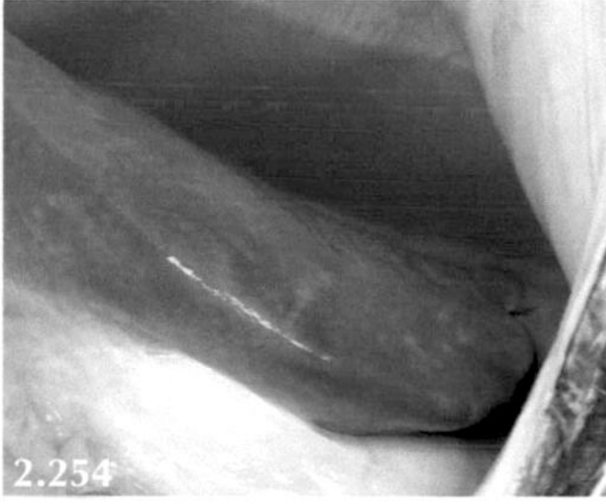
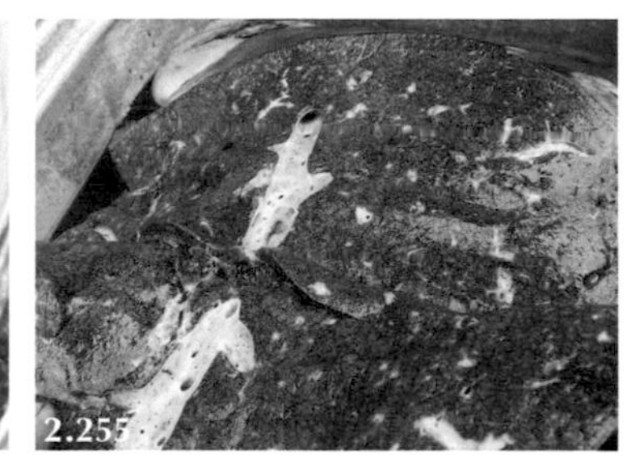

2.253: Ultrasound scan (6.5 MHz microarray probe) reveals a transudate (ascites) and hepatomegaly with rounded liver edges. **2.254:** Necropsy confirms rounded liver edges/hepatomegaly and ascites. **2.255:** Nutmeg liver indicative of chronic venous congestion. 🎥 2.121, 2.122

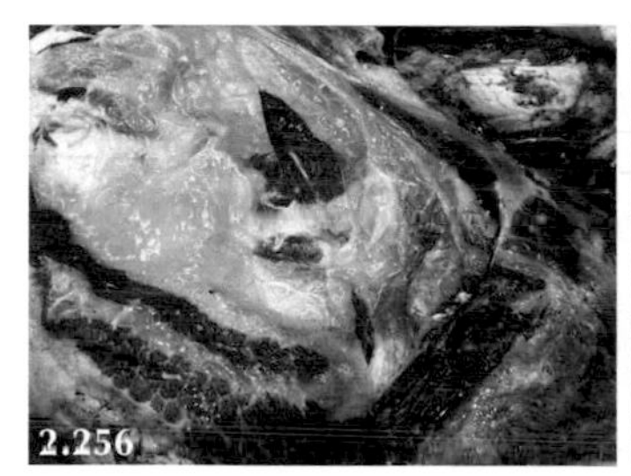
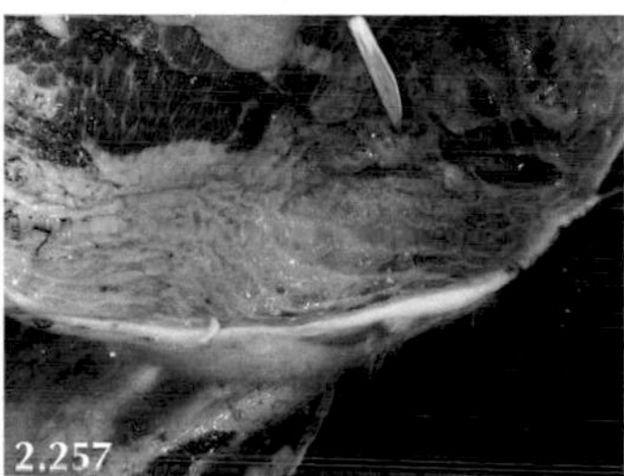

2.256 and **2.257:** Necropsy of this cow with right-sided heart failure reveals the extent of subcutaneous oedema in the brisket region extending to 15 cm in the right hand image.

Ultrasonography is particularly useful in cattle to differentiate pleural/pericardial effusion from exudate/pus within the pericardial sac. Unlike in sheep, extensive pleural exudate/abscesses are very uncommon in cattle unless secondary to traumatic reticulitis when the left side of the chest is affected. Ultrasound images/recordings of septic pericarditis often appear to be poor quality, and this can be attributed to the large amount of fibrin present on the epicardium and lining the pericardial sac. Furthermore, purulent material does not transmit sound waves well, appearing as a "snowstorm". Finding vegetative endocarditis lesions is time-consuming in adult cattle and a microarray probe is necessary to gain ready access between the rib spaces and achieve sufficient field depth. A right-sided approach is preferred with the foreleg drawn forward and placed on a 30 cm high block to facilitate access in adult cattle.

Ultrasound examination of the chest (pleurae, pleural space, lungs, pericardium and heart) is particularly informative in adult cattle with suspected right-sided heart failure and takes only 5 minutes. The ultrasound findings in a cow shown in the video recording that follows, acquired using a 5 MHz linear probe (rectal scanner), demonstrate the pathology well and allow an informed diagnosis. This investigative approach is much more rewarding than trying to fit a label to some otherwise poorly defined clinical signs. 🎥 2.123

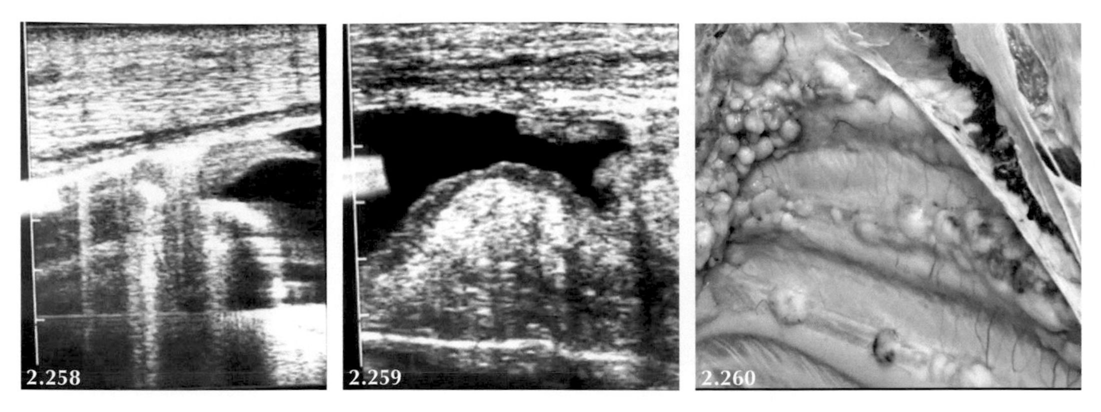

2.258: Examination of ventral chest of a cow reveals the dorsal margin of pleural effusion with consolidation of the ventral lung margin. **2.259:** 4–5 cm multi-lobulated mass on the surface of the pericardial sac. There are several 1–2 cm diameter and 5 mm high masses attached to the parietal pleural. 5 MHz linear (rectal) scanner. **2.260:** Necropsy reveals these lesions attached to the parietal pleura.

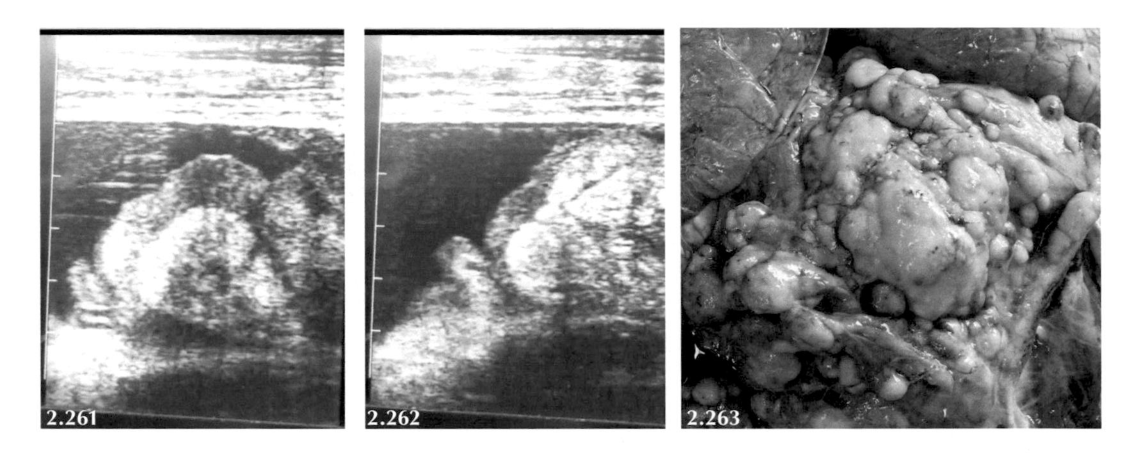

2.261 and **2.262:** Examination of the ventral chest reveals 4–5 cm deep multi-lobulated mass on the surface of the pericardial sac. There are several 2 cm diameter, 5 mm high masses attached to the parietal pleura. A heart chamber is represented by the semi-circular anechoic area at the bottom of these two images. 5 MHz linear (rectal) scanner. **2.263:** Necropsy reveals the multi-lobulated mass attached to the pericardial sac.

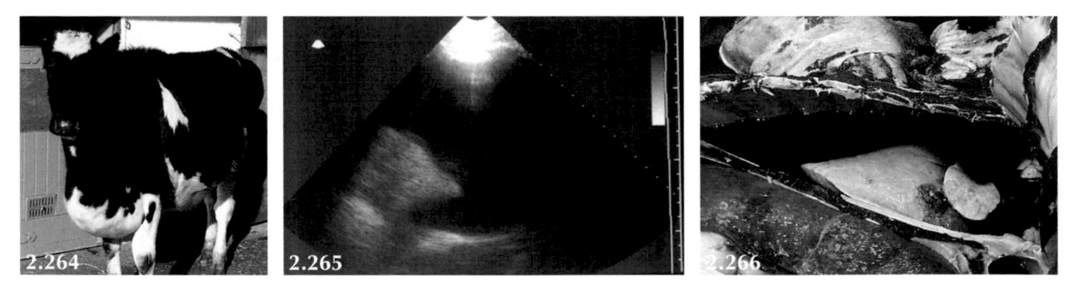

2.264: Extensive brisket oedema in a Holstein heifer secondary to dilated cardiomyopathy. **2.265:** Extensive pleural effusion extending to more than 20 cm with consolidation of the ventral lung lobe. **2.266:** Necropsy shows the amount of pleural effusion, note also consolidation of the ventral margin of the diaphragmatic lung lobe (purple area at ventral margin of lung).

Pleural and pericardial effusions can be readily demonstrated in cattle using a 6.5 MHz microarray probe and differentiated on appearance from septic pericarditis (effusion versus pus). 🎥 2.124

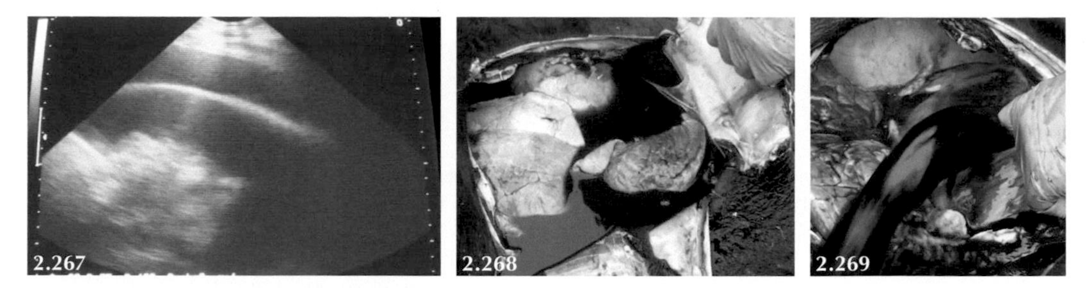

2.267: Ultrasound examination taken halfway up the chest wall using a 6.5 MHz microarray probe identifies 5–6 cm fluid within the pleural space and 6–8 cm fluid in the pericardial sac. There is also a large >10 cm diameter multi-lobulated mass attached to the heart base. The amount of fluid in both compartments is likely to be much greater ventrally. The lungs have been displaced dorsally. **2.268:** Necropsy reveals a very large volume of blood-stained fluid within the pleural space. **2.269:** Heavily blood-contaminated fluid released from the pericardial sac at necropsy caused by bleeding from the heart base tumour. 2.125

The extent of pericardial effusion/exudate can be readily identified using a 5 MHz linear probe by locating the junction between the displaced-dorsally lung and the pericardial sac ventrally (see the following). The video recordings that follow compare linear and microarray probes; pericardial sac distension can readily be appreciated by noting the dorsal displacement of the lungs.

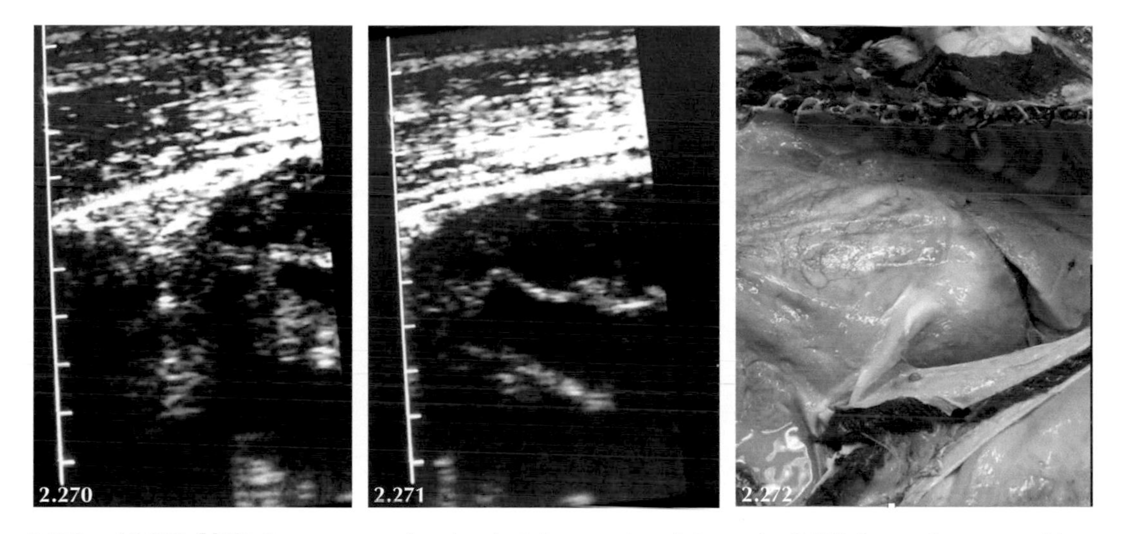

2.270 and **2.271:** 5 MHz linear scanner, dorsal to the left, cm scale on left margin. **2.270:** Image shows normal lung dorsally (hyperechoic line to the left extending halfway across the screen) with the distended pericardial sac ventrally containing several broad hyperechoic fibrin strands. **2.271:** Probe moved more ventrally reveals only distended pericardial sac with wavy hyperechoic fibrin strands. **2.272:** Dorsal displacement of the lung indicates the size of the pericardial sac shown here at post-mortem examination (carcase in right lateral recumbency, cranial to the left). 2.126, 2.127, 2.128

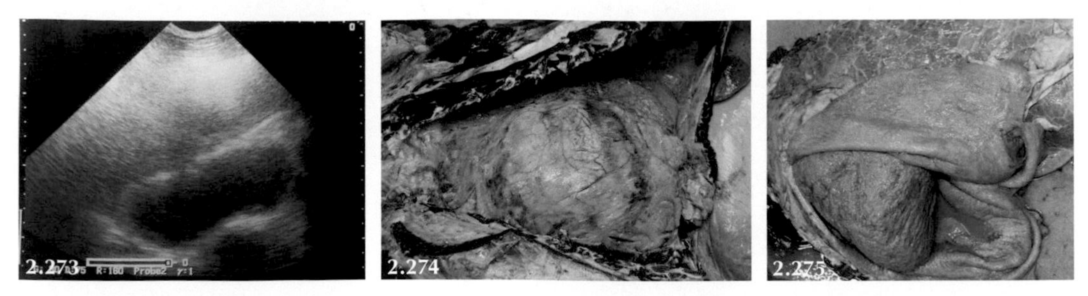

2.273: Ultrasound examination of the pericardial sac with a 6.5 MHz microarray probe reveals an anechoic area containing multiple hyperechoic dots ("snowstorm") consistent with pus extending up to 10 cm from the probe head. **2.274:** Necropsy with the carcase in right lateral recumbency with the head to the left shows the extent to which the distended pericardium fills the chest. **2.275:** Pus has been drained from the pericardial sac revealing fibrin deposition on the epicardium and lining the pericardial sac. 2.129, 2.130

2.276: Ultrasound examination of the pericardial sac with a 6.5 MHz microarray probe reveals fibrin tags lining the pericardial sac but especially on the epicardium. **2.277:** Release of pus from the pericardial sac at necropsy. **2.278:** Fibrin on the epicardium. 2.131, 2.132

In some cases where the wire remains within the heart chamber wall, fibrin deposition on the epicardium can be identified as a broad irregular hyperechoic band with oedema of the myocardium appearing as a narrow anechoic band underlying the fibrin deposit.

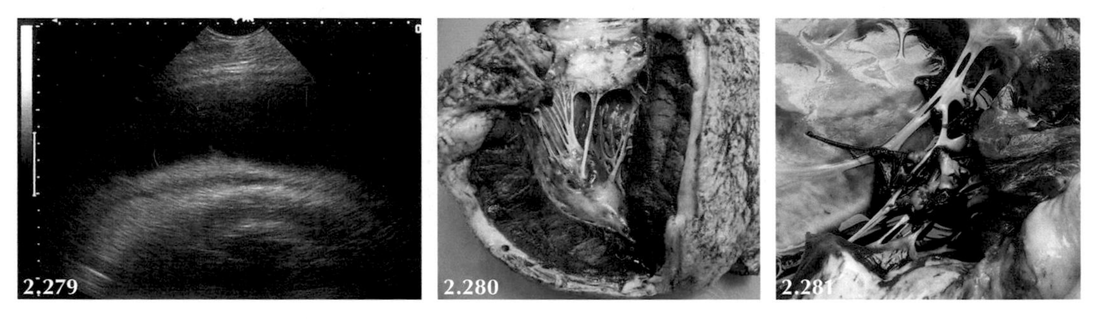

2.279: Fibrin deposited on the epicardium imaged as a broad irregular hyperechoic band with oedema of the myocardium appearing as a narrow anechoic band beneath. There is 6–8 cm pus within the pericardial sac. **2.280:** 1–2 cm organised fibrin on the epicardium. **2.281:** Close-up view of the ventricular wall with an 8 cm length of thin wire embedded in the myocardium (centre of image).

Large volume of pericardial fluid is a very uncommon finding in sheep and was observed in only two animals scanned as part of whole flock screens for OPA; neither sheep showed any signs of illness.

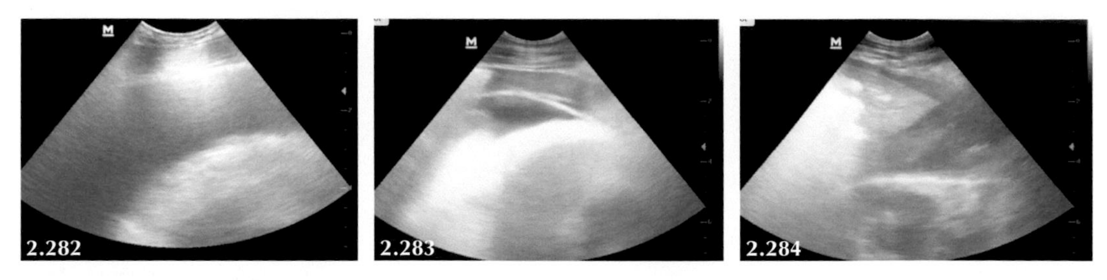

2.282: 2–4 cm fluid within the pericardial sac displacing the ventricle from the chest wall in a sheep. There are several fibrin tags on the epicardium (video recording 2.133). **2.283:** Small volume of fluid present within the pericardial sac (video recording 2.134 shows this to be an exudate). **2.284:** Same case as centre—3–4 cm pleural exudate with consolidated ventral lung lobe. The pericardial sac is represented by the hyperechoic line in the centre of the sonogram.

Vegetative Endocarditis

The major clinical signs of endocarditis include reduced appetite, chronic weight loss, lethargy and jugular distension. Pain is often, but not always, related to secondary joint infection causing a painful expression and an arched back with the head lowered; polyarthritis is more common in sheep than cattle. Subcutaneous oedema is not a common finding in endocarditis. 2.135, 2.136, 2.137, 2.138

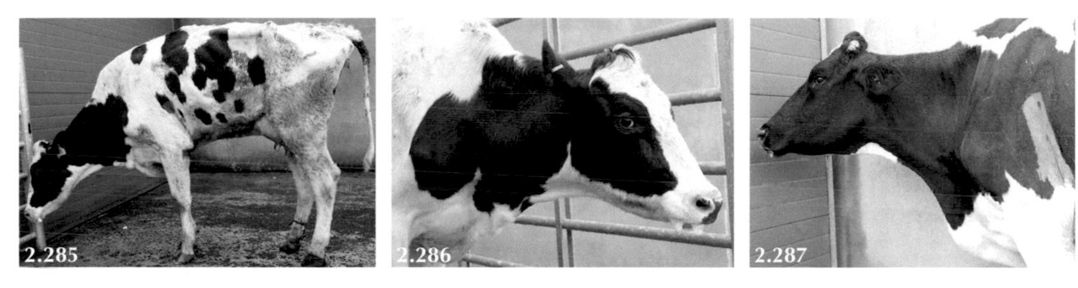

2.285: Cow with endocarditis, head lowered, arched back and painful expression. **2.286:** Ears directed caudally, painful expression. **2.287:** Ears down, painful expression and (left) jugular distension.

Ultrasonographic confirmation of vegetative endocarditis in adult cattle can be time-consuming and necessitates use of a microarray probe. Necropsy reveals cardiac enlargement with ventricular hypertrophy/dilation.

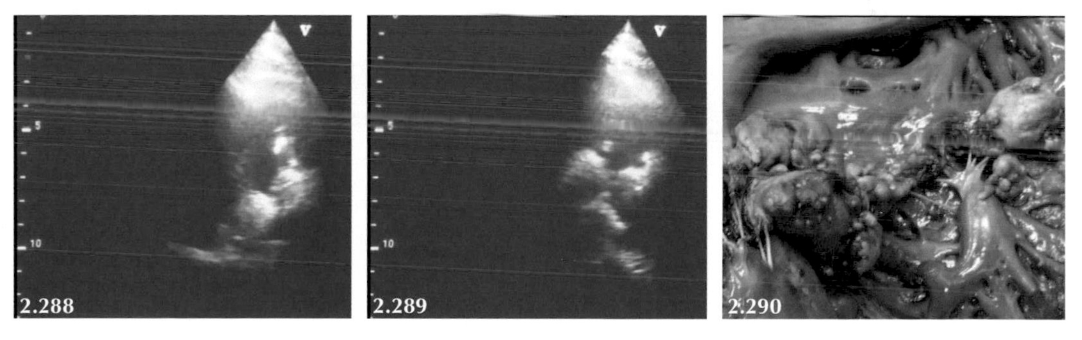

2.288 and **2.289:** 6.5 MHz microarray probe reveals vegetative lesions on the tricuspid valve of a 300 kg steer (images have not been cropped in order to show scale on left margin). **2.290**: Necropsy confirms lesions on the tricuspid valve.

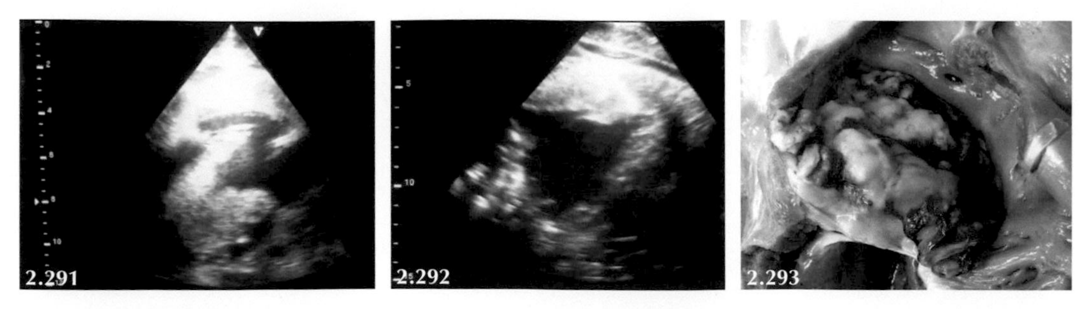

2.291: 6.5 MHz microarray probe reveals vegetative lesions on the mitral valve (images have not been cropped in order to show scale on left margin). **2.292:** Lesions on the tricuspid valve (to the left of image). **2.293:** Necropsy shows lesions on the tricuspid valve. 2.139, 2.140, 2.141

Necropsy findings also include bacteraemic spread to the joints and kidneys.

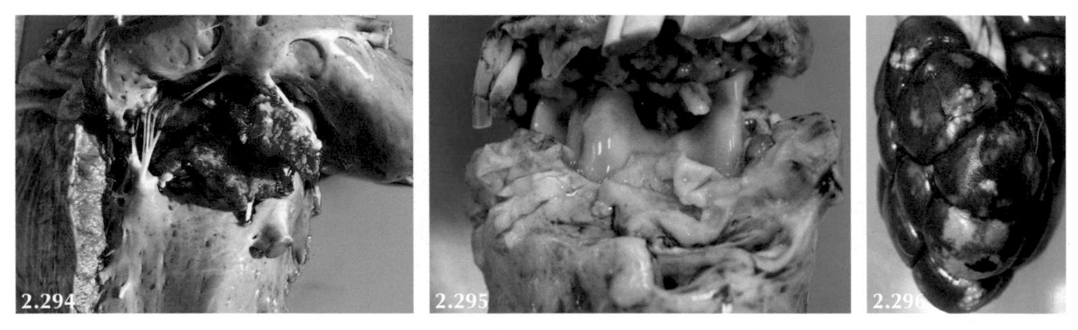

2.294: Vegetative endocarditis lesion at necropsy. **2.295:** Pannus in the hock joint causing marked lameness (8/10) in a cow with endocarditis. **2.296:** Pyogenic foci in the kidney.

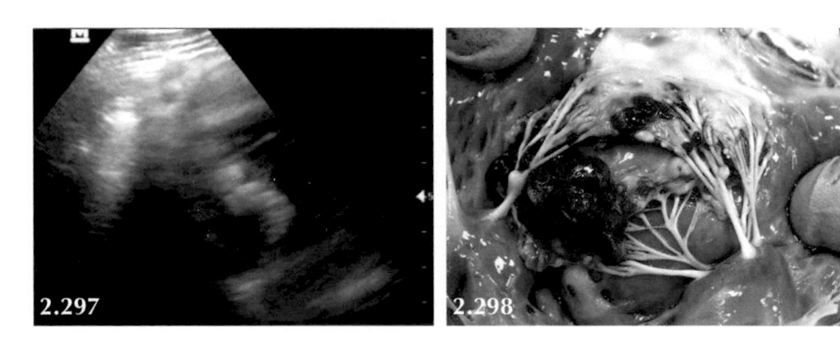

2.297: Ultrasound demonstration of vegetative lesion in a ram. **2.298:** Demonstration of the heart valve lesion at necropsy. 2.142, 2.143

Congenital Cardiac Defects

Of the congenital cardiac defects, only ventricular septal defect is common in calves. Affected calves with large defects present with poor appetite and reduced growth since birth. The rectal temperature is normal. The calf is typically dull and lethargic. Auscultation may reveal a normal heart rate but a harsh pansystolic murmur in the tricuspid valve area louder on the right than the left side. A palpable cardiac thrill is present. The respiratory rate may be elevated with a slight abdominal component. Calves with small defects, yet intense murmurs, may remain asymptomatic throughout life.

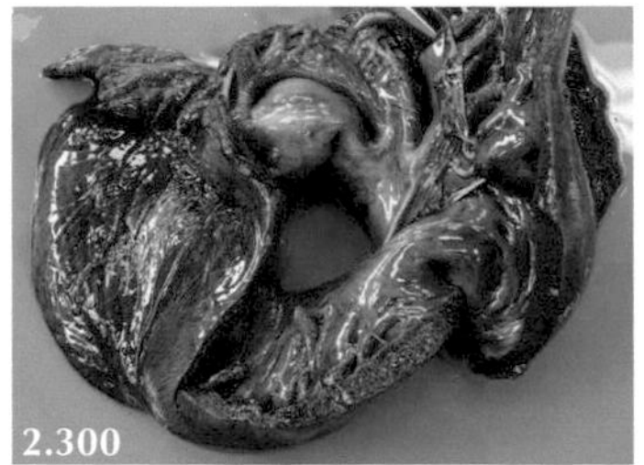

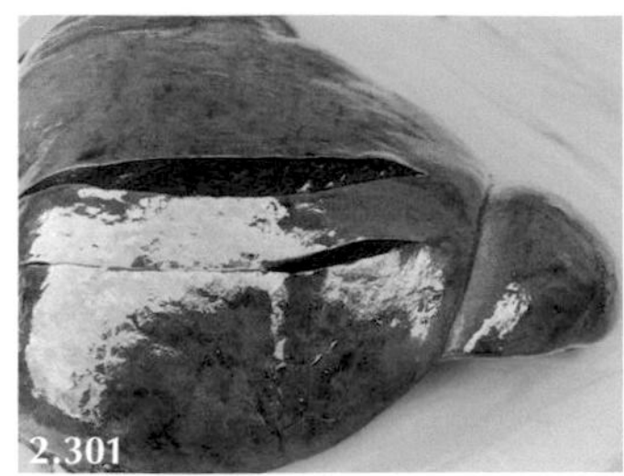

2.299: Lethargic calf with an elevated respiratory rate. **2.300:** Very large ventricular septal defect. **2.301:** Chronic venous congestion of the liver with enlargement and rounded edges.

Hepatocaval Thrombosis

Several case series have reported ultrasonographic findings of hepatocaval thrombosis in cattle, but this is not a simple diagnosis and there are no specific premonitory signs before the fatal haemorrhage to suggest detailed sonographic examination.

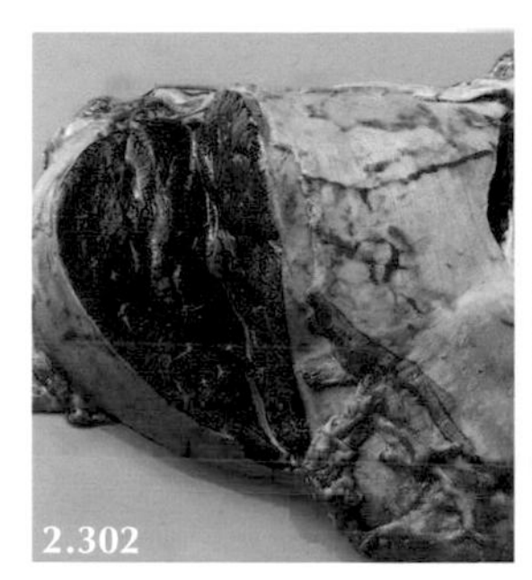

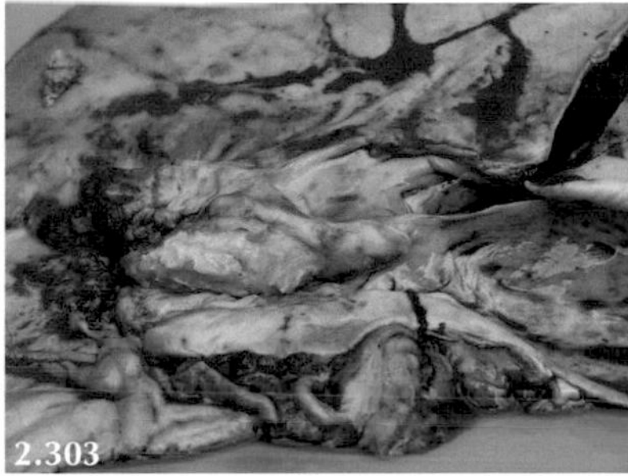

2.302: Chronic venous congestion secondary to a thrombus/abscess adjacent to the caudal vena cava (see centre image). **2.303:** Liver abscess (incised) adjacent to the caudal vena cava in a cow. **2.304:** Fatal haemorrhage following rupture of major pulmonary vessel (a lot of blood is swallowed).

2.5

Examination of the Nervous System

- Attempt to assign the major presenting neurological signs to one area of the brain (neurological syndrome); the neurological signs are the same for cattle or sheep.
- Carefully check for other neurological signs caused by diseases of this area of the brain.
- Clinical signs of many neurological diseases may change over time e.g. cerebral diseases typically progress from depression and coma to hyperaesthesia, seizure activity and opisthotonos.
- Collect lumbar CSF under local anaesthesia in unusual/problem cases and submit for laboratory examination (total protein, white cell count and differential).

DOI: 10.1201/9781003106456-13

The brain is conveniently divided into six "areas", each with a recognised neurological "syndrome" although some overlap in clinical signs may result from the complex pathways within the brain. Five neurological syndromes; cerebral, cerebellar, pontomedullary (brainstem), vestibular and hypothalamic syndromes are encountered in farm animal practice; the midbrain syndrome is very uncommon in ruminants.

	Cerebral	Pontomedullary	Vestibular	Cerebellar	Hypothalamic
Mentation	---/+++	---/+	NAD	NAD	NAD/--
Vision	---	NAD	NAD	NAD	NAD/---
Responses	---/+++	NAD	NAD	NAD	NAD/--
Cranial nerve deficits	NAD	V and VII	Peripheral VII	NAD	Variable II–VII
Balance	NAD	NAD	-/---	NAD	NAD
Head tilt	NAD	NAD/+	+++	NAD	NAD
Ataxia	NAD	NAD	+/+++	+/+++	NAD
Dysmetria	NAD	NAD	NAD	-/+	NAD

Cerebral Syndrome

Cerebral dysfunction is the most common neurological syndrome encountered in ruminants. The cerebrum is concerned with mental state, behaviour and, in conjunction with the eye and optic nerve (II), vision. Clinical signs that suggest cerebral dysfunction include:

- Blindness, but with normal pupillary light reflex.
- Circling, constant walking forward, teeth grinding.
- Initial depression and dementia typically progress within hours/days to hyperaesthesia, seizure activity and opisthotonos.

Note that the clinical signs typically progress when treatment is delayed from aimless wandering, walking forward, dullness and reduced responses to hyperaesthesia, seizure activity and opisthotonos, often over 24 hours. The clinical signs therefore depend on when in the disease process the animal is presented for veterinary examination.

Approximately 90% of efferent nerve fibres cross at the optic chiasma, therefore animals with a left-sided space-occupying lesion would be blind in the right eye. The pupillary light reflex would be normal. The menace response is not always reliable in cases of unilateral space-occupying lesions and this test can be supplemented with unilateral blindfolding when safe to do so.

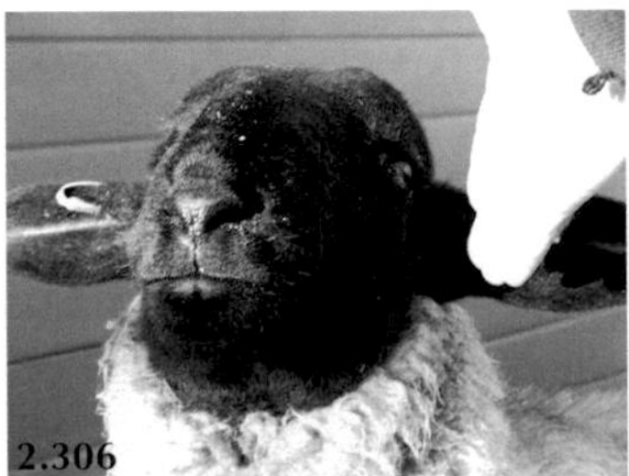

2.305: Early signs of PEM (a diffuse cerebral lesion): Isolation, aimless wandering and blindness. **2.306:** Dorsiflexion of neck, lack of menace response, and dorso-medial strabismus. **2.307:** Untreated sheep with PEM quickly progress to seizure activity and opisthotonos. Note that these clinical signs present in the same disease process but depend upon duration (see also Meningitis next).

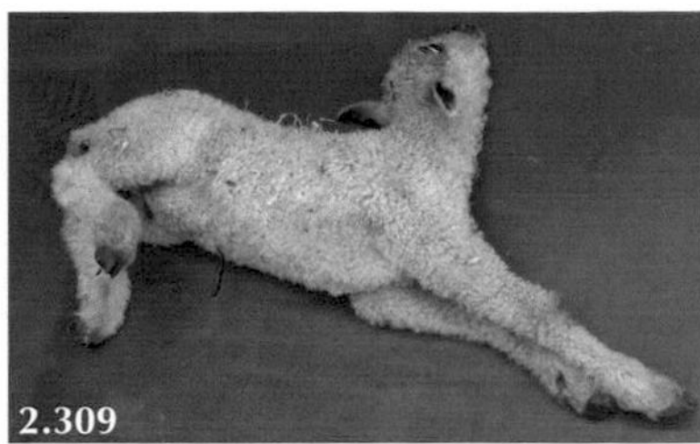
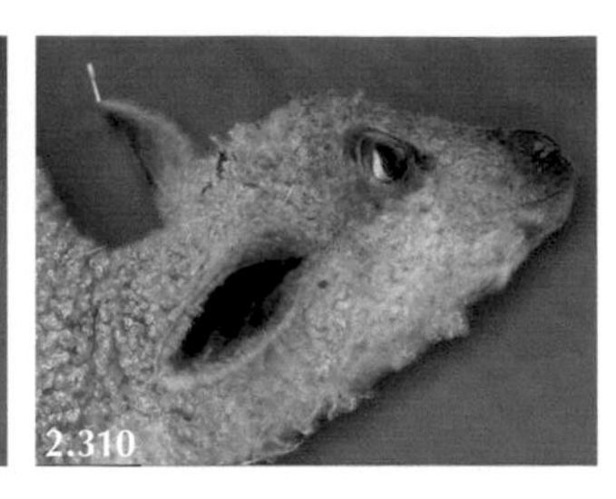

2.308: Obtunded lamb with early meningitis. **2.309** Advanced meningitis with lamb showing seizure activity and opisthotonos. **2.310:** Dorsiflexion and dorso-medial strabismus. Same disease process but different duration of infection. 🎥 2.144, 2.145, 2.146, 2.147, 2.148

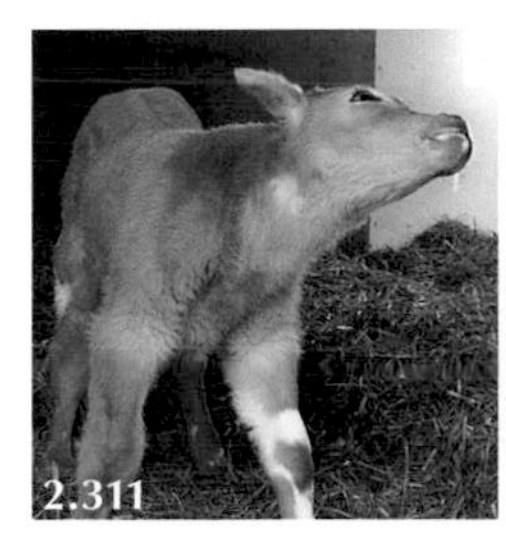
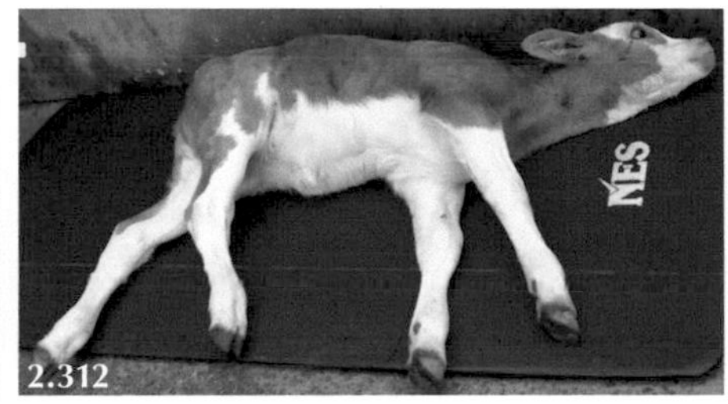
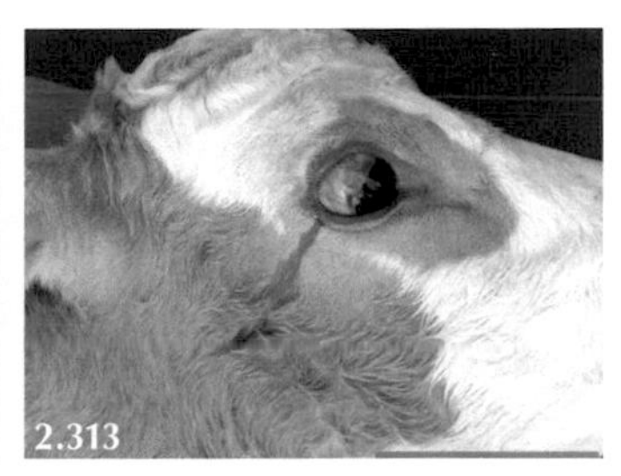

2.311: Obtunded calf with early meningitis. **2.312** Advanced meningitis with calf showing seizure activity and opisthotonos. **2.313:** Dorsiflexion and dorso-medial strabismus. 🎥 2.149

Cerebellar Syndrome

- The cerebellum is primarily concerned with fine coordination of voluntary movement (gait).
- In cerebellar disease all limb movements are spastic (rigid), clumsy and jerky.
- Initiation of movement is delayed and may be accompanied by tremors.

Cerebellar disease is uncommon in sheep but is characterised by a wide-based stance and ataxia (incoordination), particularly of the pelvic limbs with preservation of normal muscle strength. In addition to ataxia, dysmetria (problems associated with stride) may be observed. Hypermetria, or overstepping, is seen as a Hackney-type gait. With hypometria, the animal will frequently drag the dorsal aspect of the hoof along the ground. Hypermetria is the more common form of dysmetria observed in cerebellar disease. Hypermetria may also be observed in spinal cord disease or affecting the contralateral limb in vestibular disease.

Cerebellar disease may result in jerky movements of the head, especially when the animal is aroused or at feeding times when affected animals will often overshoot the feed bowl. This condition is often referred to as "intention tremors". This clinical feature is most commonly seen in cases of congenital Border disease virus infection affecting neonatal lambs.

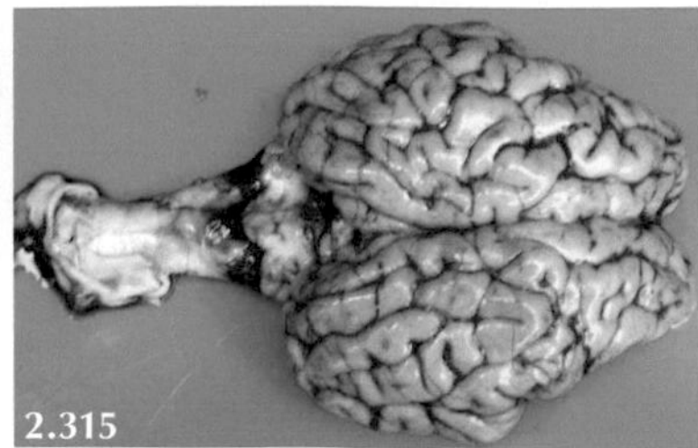

2.314: Intention tremors in lamb with Border disease. **2.315:** Cerebellar hypoplasia confirmed at necropsy. **2.316:** Wide-based stance in this ram with cerebellar abscess. 2.150

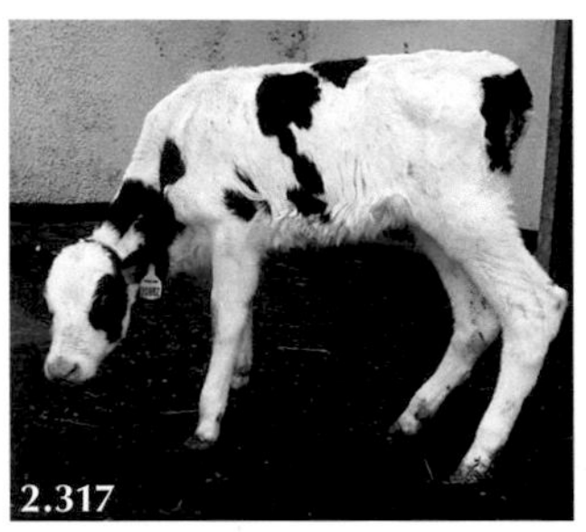

2.317: Wide-based stance in this calf with cerebellar hypoplasia. **2.318:** Wide-based stance in this calf with a cerebellar abscess. **2.319:** Necropsy reveals the extent of the cerebellar abscess (centre). 2.151

Vestibular Syndrome

- The major clinical sign associated with a vestibular lesion is a head tilt.
- Ipsilateral facial nerve paralysis is common in peripheral vestibular disease.
- Circling may also be observed in vestibular disease.

Rapid vertical and horizontal movements of the head in normal animals can induce positional nystagmus with the fast phase in the direction of head movement. Positional nystagmus may be depressed or absent in animals with a vestibular lesion when the head is moved towards the side of the lesion (head tilt). Resting nystagmus is present and permits differentiation between the two forms of vestibular disease: Peripheral vestibular disease—fast phase away from side of lesion; central vestibular disease—fast phase in any direction including dorsal or ventral. In addition, Horner's syndrome and facial nerve paralysis is common in peripheral vestibular disease (see section on cranial nerves for further information on facial nerve paralysis) since both facial and sympathetic nerves fibres pass close to the middle ear.

Determination of a head tilt is best made when the animal is viewed from a distance of 5 to 10 metres. The head is normally held in the vertical plane but in vestibular disease a 5° to 10° tilt is present. The poll is tilted down to the affected side and may be exaggerated by blindfolding the animal. A head tilt must be differentiated from an altered head position or aversion.

2.320–2.322: Loss of balance and head tilt caused by otitis media (vestibular syndrome). 2.152

Pontomedullary Syndrome

- The pontomedullary syndrome is characterised by cranial nerve V and VII deficits. Depression is attributed to a specific lesion in the ascending reticular activating system. Ipsilateral hemiparesis and proprioceptive defects are common.
- Propulsive circling may result from involvement of the vestibulocochlear nucleus. Involvement of the facial nucleus results in ipsilateral facial nerve paralysis evident as drooped ear, drooped upper eyelid (ptosis) and flaccid lip.
- Involvement of trigeminal nerve or the trigeminal motor nucleus results in paralysis of the cheek muscles and decreased facial sensation.
- Abnormal respiratory patterns may result from damage to the respiratory centre in the medulla.

2.323: Walk forward behaviour in this ewe (she is not caught in the fence but will not reverse out). **2.324:** Close-up of ewe (left) showing obtunded behaviour and right-sided drooped ear, ptosis and flaccid lip. **2.325:** Dullness affecting this heifer with left sided facial and trigeminal nerve paralysis. 2.153, 2.154

Hypothalamic Syndrome

There are multiple cranial nerve deficits (II to VII) depending upon which nerve roots are affected by the lesion which is typically an abscess arising from the circle of Willis.

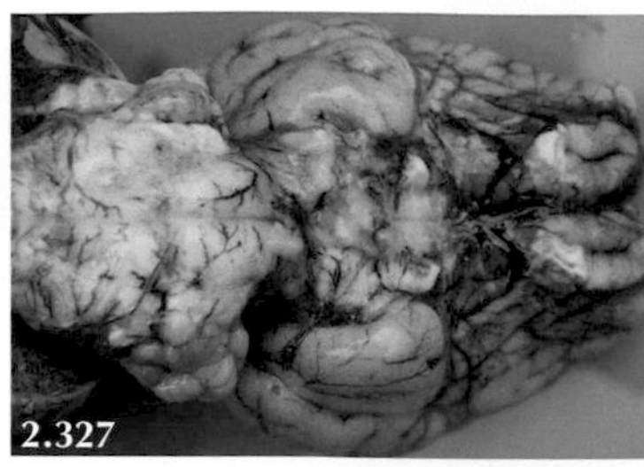

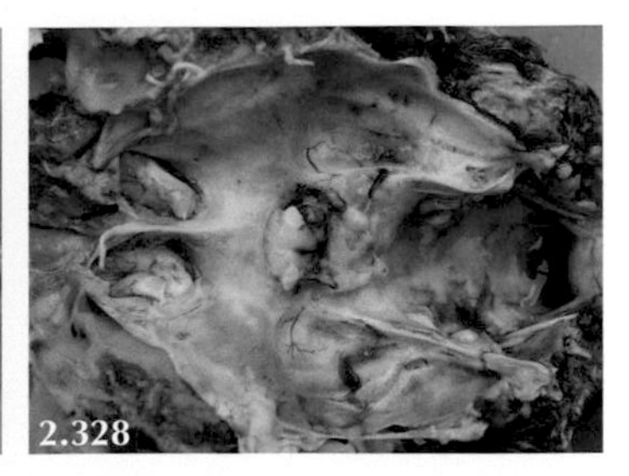

2.326: Multiple cranial nerve II–VII deficits. **2.327:** Brain reflected to show ventral aspect with pus surrounding the roots of cranial nerves II–VII. **2.328:** Floor of the cranial vault showing infection spreading caudally from the pituitary foramen and Circle of Willis (basilar empyema). 🎥 2.155

Spinal Lesions

	Muscle tone forelegs	Reflex arc forelegs	Muscle tone hindlegs	Reflex arc hindlegs
C1-C6	–/––	+/++	–/––-	+/++
C6–T2	––/–––	––/–––	–/––	+/++
T2-L3	NAD	NAD	––/–––	+/+++
L4-S2	NAD	NAD	–/–––	–/–––

Cervical Spinal Lesions (C1–C6)

In mild cervical lesions (C1 to C6) there is ataxia and weakness involving all four limbs, but the pelvic limbs are more severely affected. There are hopping, placing and conscious proprioceptive deficits. The thoracic and pelvic limb reflexes are increased (upper motor neuron signs). Severely affected sheep may not be able to maintain sternal recumbency and must be supported.

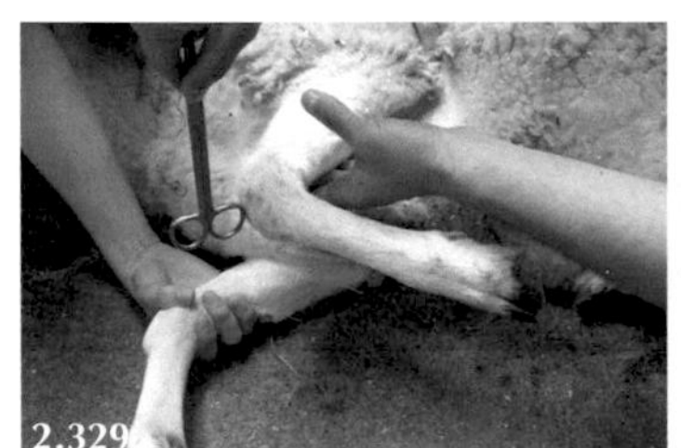

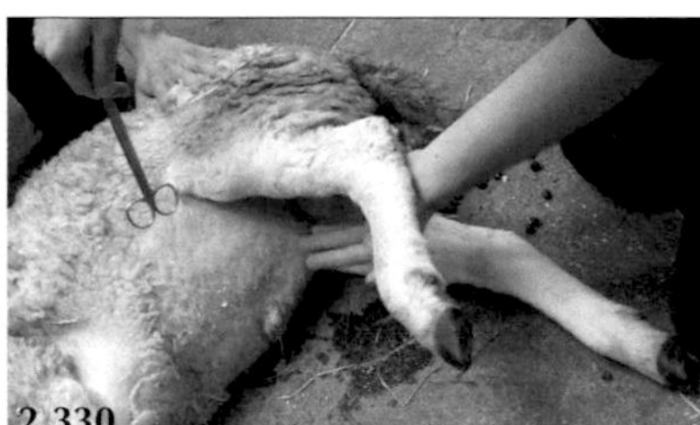

Testing the reflex arcs of the forelegs (**2.329**) and hindlegs (**2.330**).

2.331: Weakness of all four legs resulting from *S. dysgalactiae* infection of the atlanto-occipital joint in a five-day-old lamb. **2.332:** Cervical vertebral body fracture caused by fighting injury causing weakness of all four legs. **2.333:** Weakness caused by cervical vertebral malformation in a yearling ram. 🎥 2.156, 2.157, 2.158, 2.159

Brachial Intumescence (C6–T2)

The term intumescence refers to the concentration of neurons in the grey matter of the spinal cord at the level of the thoracic and pelvic limbs. Lesions affecting this length of spinal cord cause reduced thoracic limb reflexes (lower motor neuron signs) and increased pelvic limb reflexes (upper motor neuron signs). Thoracic limb weakness is judged by the resistance to lateral movement of the animal by pushing the animal's shoulder away.

2.334 and **2.335:** Weakness of the forelegs caused by a lesion in the region C6 to T2. The hindlegs have increased reflex arcs. 🎥 2.160

T2–L3

The presence of a spinal cord lesion in the thoraco-lumbar region T2 to L3 results in upper motor neuron signs of increased tendon jerk and withdrawal reflexes in the pelvic limbs. There are conscious proprioceptive deficits and weakness of the pelvic limbs.

Animals with a spinal lesion caudal to T2 have normal thoracic limb function. These animals frequently adopt a dog-sitting posture with the hips flexed and the pelvic limbs extended alongside the abdomen rather than the normal flexed position underneath the body. The clinician's attention is immediately drawn to this abnormal posture because, unlike horses, ruminants always raise themselves by the pelvic limbs first.

2.336: Weakness of the hindlegs with knuckling over at the fetlock joints. **2.337:** Empyema with destruction of T5 and infection (pus) tracking into the vertebral canal causing spinal cord compression. **2.338:** Characteristic "dog sitting" posture of cow with a T/L lesion. 🎥 2.161, 2.162

L4–S2

Lesions in the region L4 to S2 result in flaccid paralysis of the pelvic limbs with reduced or absent reflexes.

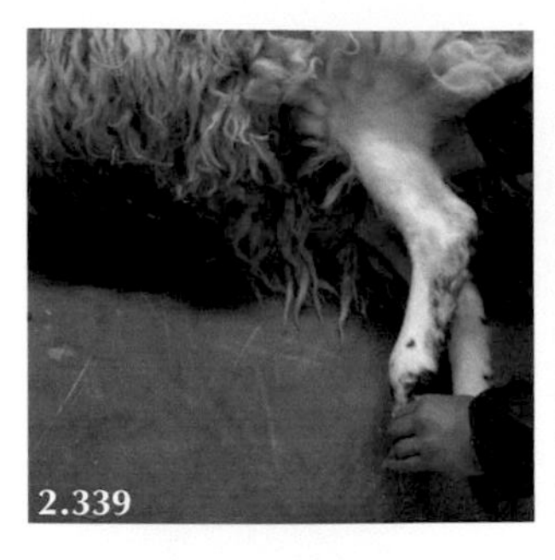

2.339: Pinch test of coronary band to test pelvic limb reflex arc. **2.340:** Hindleg weakness with the sheep using the wall for support. **2.341:** After assuming sternal recumbency (centre) the hindlegs are not drawn under the sheep as normal. 2.163, 2.164

S1–S3 (Cauda Equina)

Lesions affecting S1 to S3 cause hypotonia of the bladder and rectum resulting in distension with urine and feces, respectively.

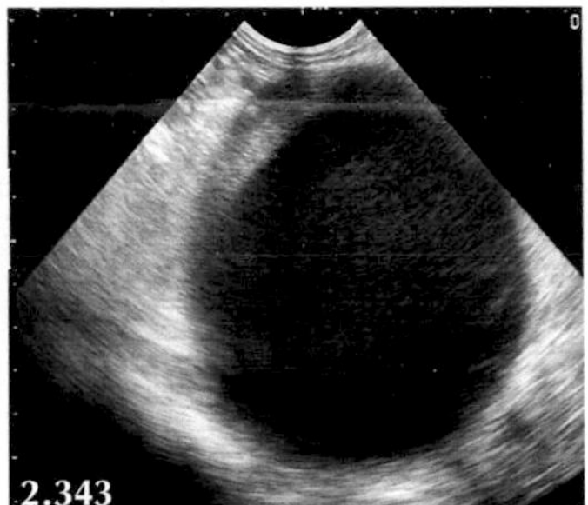

2.342: Checking tail tone. **2.343:** Urinary bladder distension can be easily and quickly checked using trans-abdominal ultrasonography in sheep. **2.344:** Sacral fracture causing paralysis of the tail, bladder and rectal atony/distension detected on rectal examination.

Lumbar Cerebrospinal Fluid (CSF) Collection

When correctly performed under local anaesthesia, lumbar CSF collection in cattle and sheep is a safe procedure. Lumbar cerebrospinal fluid (CSF) collection and subsequent analysis is particularly useful when investigating new/unusual diseases and more commonly, potential compressive lesions of the spinal cord. Familiarity with the technique aids high extradural injection of lidocaine and/or xylazine which has many applications in farm animal clinical practice. There are few, if any, indications for cisternal CSF collection in farm animal practice.

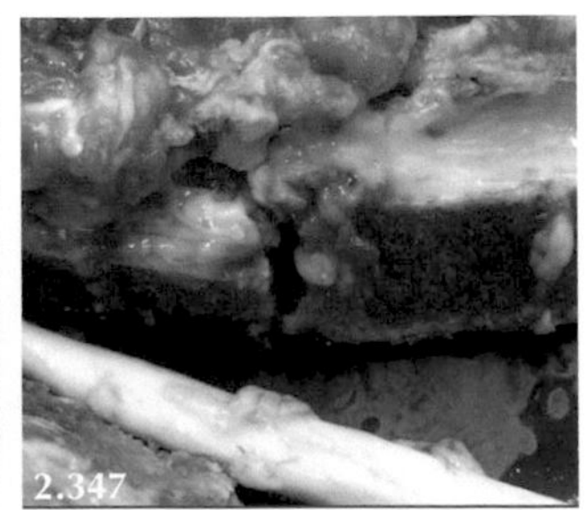

2.345 and **2.346:** Pelvic limb weakness caused by vertebral empyema and delayed swayback, respectively. **2.347:** Vertebral empyema causing cord compression can be identified by an elevated lumbar CSF protein concentration.

When possible, the animal is positioned in sternal recumbency with the hips flexed and the pelvic limbs extended alongside the abdomen. The animal's head is averted against the animal's flank. Sedation of the animal is not usually necessary but intravenous injection of 10 μg/kg detomidine can be used in sheep and intravenous xylazine (0.04 to 0.07 mg/kg bodyweight) in cattle.

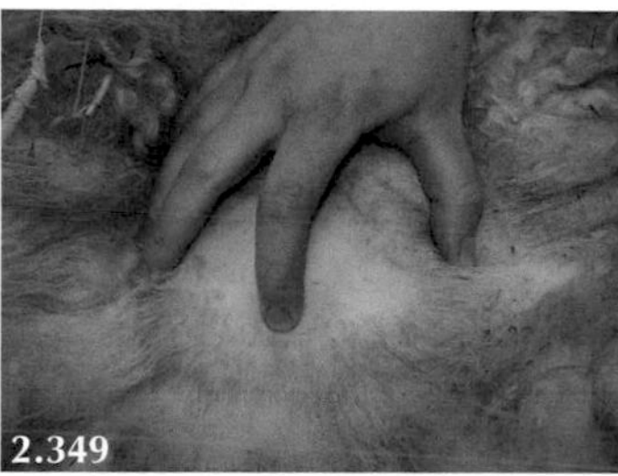

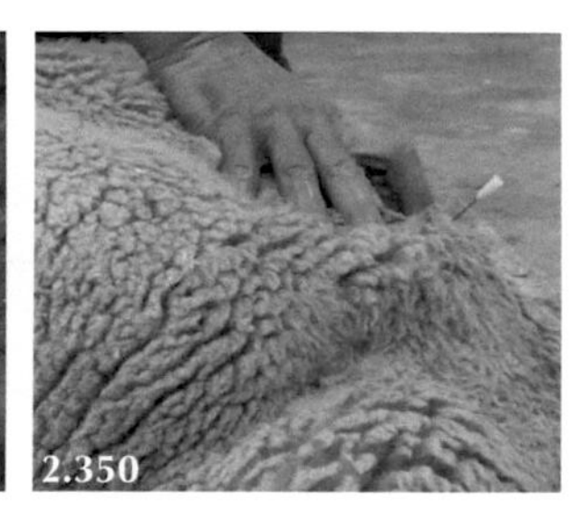

2.348: This lamb has been positioned in sternal recumbency with the hips flexed and the pelvic limbs extended alongside the abdomen. **2.349:** The site for lumbar CSF collection is the midline depression between the last palpable dorsal lumbar spine (L6) and the first palpable sacral dorsal spine (S2) (sheep's head top of image). **2.350:** The needle is positioned at a right angle (90°) to the vertebral column (sheep's head left of image).

The site for lumbar CSF collection is the midpoint of the lumbosacral space identified as the midline depression between the last palpable dorsal lumbar spine (L6) and the first palpable sacral dorsal spine (S2). This site is approximately 2 to 3 cm caudal to an imaginary line joining the wings of the ilium in adult sheep. The thumb and middle finger of one hand are placed over the wings of the ilium and the index finger used to palpate the midline depression between L6 and S2. The site must be clipped, surgically prepared and between 1–2 ml of local anaesthetic injected subcutaneously. Hypodermic needles are preferred for sheep and cattle <200 kg because, unlike spinal needles, they are very sharp and CSF wells up as soon as the needle point enters the dorsal subarachnoid space without the need to remove a stylet. Spinal needles with internal stylets are necessary for cattle >200 kg.

The hypodermic needle (Table 2.2) is slowly advanced over 10 seconds at a right angle to the plane of the vertebral column or with the hub directed 5° to 10° caudally. It is essential to appreciate the changes in tissue resistance, as the needle point passes sequentially through the subcutaneous tissue, interarcuate ligament then the sudden "pop" due to the loss of resistance as the needle point exits the ligamentum flavum into the extradural space. Once the needle has been advanced fractionally farther and the point has penetrated the dorsal subarachnoid space CSF will well up in the needle hub within 2 to 3 seconds. Care must be taken not to advance the needle point from the dorsal subarachnoid space into the conus medullaris when the syringe is attached to the needle hub. The needle hub must be firmly stabilised between the thumb and index finger with the wrist resting on the dorsal spines of the vertebral column. The seal on the syringe should be broken before it is connected to the needle hub. Selection of the correct needle length ensures that the needle hub is close to the skin thereby assisting stabilisation. Between 1 and 2 ml of CSF is sufficient for laboratory analysis and while the sample can be collected by free flow over 1 to 2 minutes, it is more convenient to employ very gentle syringe aspiration over 20 seconds. There is no justification to collect from the ventral subarachnoid space.

TABLE 2.2

Guide to Needle Length and Gauge for Lumbar CSF Sampling

Neonatal lambs	1 cm	21–23 gauge
Lambs <30 kg	2.5 cm	21 gauge
Ewes 40–80 kg	4 cm	19 gauge
Rams >80 kg	5 cm	19 gauge
Calves <40 kg	2.5 cm	19 gauge
Calves <100 kg	4 cm	19 gauge
Calves 100–200 kg	5 cm	19 gauge
Cattle >200 kg	10 cm	18 gauge + internal stylet

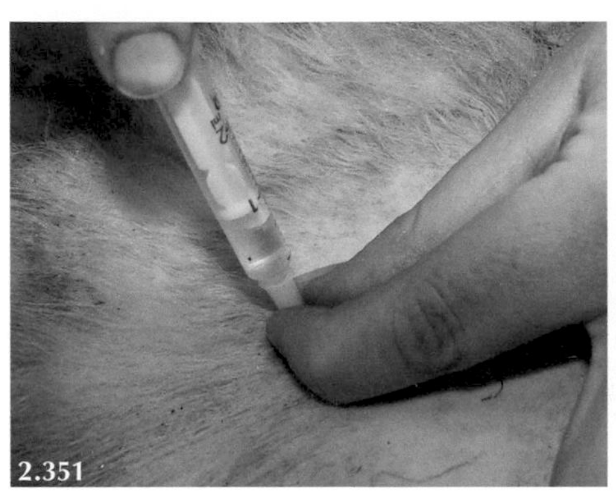

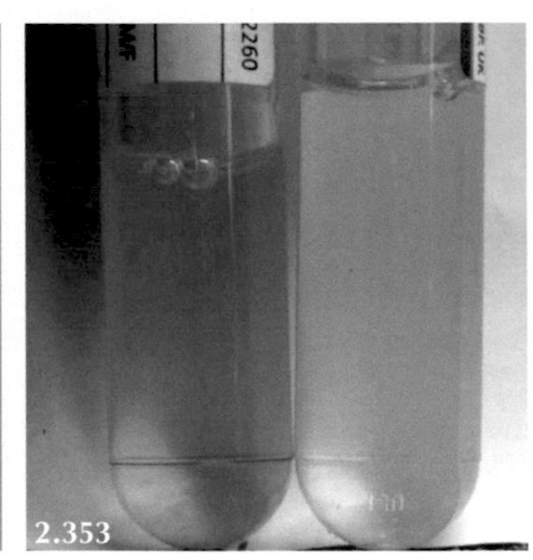

2.351: The needle hub is firmly stabilised between the thumb and index finger with the wrist resting on the dorsal spines of the vertebral column. **2.352:** Iatrogenic haemorrhage with red blood cells having gravitated to the bottom of the tube after 1–2 hours. **2.353:** Xanthochromia sample (left as viewed). 2.165, 2.166, 2.167, 2.168

Cerebrospinal fluid is clear and colourless. A stable foam is indicative of an increased protein concentration but agitation of the sample will cause damage to cell morphology. Cloudy or turbid CSF is indicative of a markedly elevated (>100-fold) white cell concentration. Red blood cells may be present in the CSF following haemorrhage into the subarachnoid space. Much more commonly, the presence of red blood cells in the CSF sample results from needle puncture of blood vessels on the dura mater, particularly the leptomeninges, during the sampling procedure. Accidental trauma resulting in intrathecal haemorrhage is more common following repeated attempts to obtain CSF. Haemorrhage caused by the sampling technique appears as streaking of blood in clear fluid which may gradually disappear as more CSF is allowed to flow freely from the needle hub. Turbidity caused by recent haemorrhage into the CSF will clear after centrifugation to leave a transparent supernatant. Alternatively, if the sample is left to stand for approximately 2 hours, the red blood cells gravitate and form a small plug at the bottom of the collection tube. A yellow discoloration of the CSF, referred to as xanthochromia, appears within a few hours after subarachnoid haemorrhage and may persist for two to four weeks. Xanthochromia is caused by release of pigment, following lysis of red blood cells, which results from red blood cell membrane fragility as a consequence of the low CSF protein concentration.

Radiography

Radiography of the vertebral column is not commonly undertaken in farm animal practice because early bony changes are difficult to detect and more information can be gathered from lumbar CSF collection.

Myelography

Myelography has be superseded by CT scan in the investigation of spinal/vertebral lesion in valuable sheep.

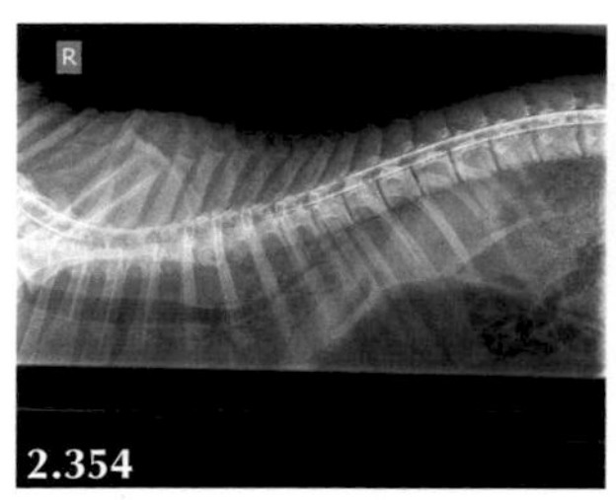

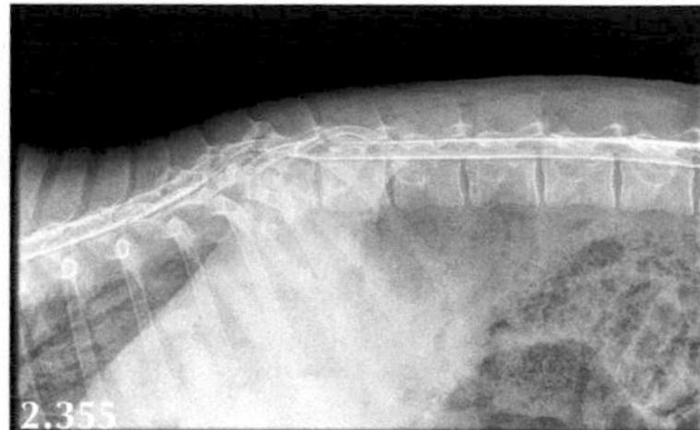

2.354 and **2.355:** Myelography has been used to confirm the presence of early (left) and advanced (right) cases of vertebral empyema in valuable two–three-month-old ram lambs.

Cranial Nerves

Cranial nerves leave the forebrain and brainstem and have a variety of specialised functions.

TABLE 2.3

Assessment of Cranial Nerve Function

Assessment of normal function	Tests the undernoted cranial nerves and associated centres
• Vision	Eye, II, cerebrum (contralateral)
• Pupillary light response (pen torch)	II, III
• Pupil size and symmetry (pen torch)	II, III, brainstem, sympathetic nervous system
• Menace response (rapidly approaching object)	II, VII, cerebrum, brainstem, cerebellum
• Eyeball position	III lateral strabismus IV dorsal and medial VI medial strabismus
• Normal head/cheek muscle tone	V
• Touch cornea, eyeball retracts	V, VI
• Touch medial canthus eye closes	V, VII
• Ears held in normal position	V, VII
• Nostrils—normal sensation	V
• Eyelids in normal position	III, VII, sympathetic nervous system
• Hearing	VIII
• Normal head position	VIII, cerebrum
• Deglutition, tongue movement	IX, X, XII

Olfactory Nerve (I)

Assessment of the olfactory nerve has little clinical application in ruminant species.

Olfactory Nerve (II)

The visual pathway can be tested by observing the sheep encountering obstacles such as gateways, and by the menace response when the eyelids close quickly in response to a rapidly approaching object. The menace response can be difficult to evaluate in depressed animals and should be interpreted with caution.

Up to 90% of optic nerve fibres decussate at the optic chiasma therefore vision in the left eye is perceived in the contralateral (right) cerebral hemisphere. The signal generated by the menace response in left eye travels along the left optic nerve to the optic chiasma then crosses to the right optic tract and right occipital cortex. The motor (efferent) pathway is from the right visual cortex to the left facial nucleus resulting in closure of the left eye.

Lesions of the eye and optic nerve result in ipsilateral blindness. Lesions of the optic tract or nucleus cause contralateral blindness. Therefore, an abscess in the right cerebral hemisphere affecting the nucleus will result in blindness of the left eye.

Oculomotor Nerve (III)

Pupillary diameter is controlled by constrictor muscles innervated by the parasympathetic fibres in the oculomotor nerve and dilator muscles innervated by the sympathetic fibres from the cranial cervical ganglion.

The normal response to light directed in one eye is constriction of both pupillary apertures with a direct response in stimulated eye and a consensual response in contralateral eye.

A dilated pupil in an eye with normal vision (menace response) would suggest a lesion in the oculomotor nerve (III). The contralateral eye with normal oculomotor nerve (III) function will respond to both direct and consensual stimulation. If a lesion involves primarily one cerebral hemisphere increased pressure to one oculomotor nerve presents as different pupillary aperture diameters (anisocoria) with the affected side displaying pupillary dilation.

Horner's Syndrome

Horner's syndrome refers to the clinical appearance of damage to the sympathetic nerve supply to the eyeball causing slight ptosis (drooping of upper eyelid), miosis (constriction of the pupil) and slight protrusion of the nictitating membrane. The menace response (vision) and pupillary light response are normal.

Oculomotor (III), Trochlear (IV) and Abducens (VI) Nerves

These cranial nerves are responsible for normal position and movement of the eyeball within the bony socket. An abnormal eyeball position is referred to as strabismus. Abnormal position of the eyeball is rarely seen as an acquired syndrome in sheep.

- Paralysis of the oculomotor nerve—lateral strabismus
- Paralysis of the trochlear nerve—dorsomedial strabismus
- Paralysis of the abducens nerve—medial strabismus

Many cerebral lesions can result in strabismus. If there is a unilateral cerebral lesion the strabismus is directed to the ipsilateral side. Dorsomedial strabismus is classically seen in PEM and acute bacterial meningitis of neonates as a reflection of cerebral oedema involving upper motor neuron pathways.

Trigeminal (V) Nerve

As the name implies, there are three branches: Mandibular, maxillary and ophthalmic which supply the motor fibres to the muscles of mastication and sensory fibres to the face.

Loss of motor function of the mandibular branch of the trigeminal nerve results in rapid atrophy of the temporal and masseter muscles which are responsible for mastication. Unilateral lesions result in deviation of the lower jaw and muzzle away from the affected side. Responses to stimulation of the skin around the face are mediated through sensory fibres in the trigeminal nerve and motor fibres in the facial nerve. These reflexes require intact nerves V and VII, trigeminal and facial nuclei and brainstem.

Abducens (VI) Nerve

Lesions of the abducens nerve result in constant medial strabismus and loss of the ability to retract the eyeball into the bony socket (corneal reflex).

Facial (VII) Nerve

The facial nerve is concerned primarily with motor supply to the facial muscles. The facial nerve contains the lower motor neurons for movement of the ears, eyelids, nares and muzzle and the motor pathways of the menace and palpebral reflexes. When the periocular skin is touched the normal reflex is that the animal will close the palpebral fissure.

The lack of the palpebral reflex may indicate a lesion in:

- The trigeminal nerve or nucleus (sensory pathway)
- The facial nerve or facial nucleus (motor pathway)
- Both nerves or nuclei involved

Facial nerve paralysis is characteristically seen as drooping of the upper eyelid and ear, and with a unilateral lesion deviation of the muzzle towards the unaffected side due to loss of facial muscle tone in the affected side.

Differentiation between central or peripheral facial nerve involvement can be attempted by identifying involvement of other central structures such as the trigeminal and vestibulocochlear nuclei.

Vestibulocochlear (VIII) Nerve

If the facial (VII) nerve is involved, deafness in sheep may be difficult to determine. The vestibular system controls orientation of the head, body and eyes. Nystagmus refers to movement of the eyeball within the bony socket. Normal vestibular nystagmus refers to horizontal movement of the eyeball as the head is turned laterally with the fast movement phase toward the side to which the head is turned. Pathological change causing nystagmus originates in the vestibular system.

Spontaneous nystagmus refers to nystagmus when the head is held in the normal position. Positional nystagmus results when the head is held in various abnormal positions.

Glossopharyngeal (IX) and Vagal (X) Nerves

Damage to these nerve nuclei results in dysphagia and associated salivation. Affected animals cannot swallow or drink.

Accessory (XI) Nerve

In ruminants, the accessory nerve appears to have little specific function.

Hypoglossal (XII) Nerve

The hypoglossal nerve provides motor supply to the muscles of the tongue. With a unilateral lesion there is atrophy of musculature but the sheep is still able to retract the tongue within buccal cavity. In the case of a bilateral lesion the sheep is unable to prehend and masticate food and the tongue remains protruded.

2.6

Examination of the Musculoskeletal System

The musculoskeletal system comprises the skeleton, joints, ligaments, tendons and muscles. Together with the nervous system, it is responsible for the animal's stance and gait. Clinical involvement of the musculoskeletal system is manifest as lameness and much less commonly as weakness.

Lameness is probably the major welfare concern in sheep and dairy cows. Advances in knowledge have not resulted in significant reductions in lameness prevalence with figures similar to those reported 20–30 years ago. The very large variations in lameness suggest significant management effects. In both species, immediate intervention and treatment of the cause(s) greatly reduce duration and flock/herd lameness prevalence.

Clinical Examination of the Musculoskeletal System

- Lameness is considered one of the most serious cattle and sheep welfare concerns.
- Severity of lameness on a farm = intensity × duration × number of animals affected.
- Government advisory bodies give lameness prevalence targets of 2 per cent for sheep but no annual incidence figures. Foot lesions are especially common in housed dairy cattle with prevalence values for lameness around 20 per cent.
- Control measures for dairy cow lameness include routine foot trimming undertaken every four to six months by para-professionals. Routine foot trimming is actively discouraged in sheep.
- Polyarthritis occurs occasionally in calves and lambs secondary to a heavily contaminated environment and failure of passive transfer.

Veterinary examination of a lame sheep or cow:

- Where possible, walk the animal along a flat non-slip surface for 10–20 yards and identify leg(s) affected scoring them 1–10 (1 being barely lame and 10 being severely lame/not bearing any weight).
- Determine/compare extent of muscle wastage of affected leg(s) (indicates likely duration).
- Restrain animal (in cattle stocks).
- Palpate drainage lymph node(s).
- Clean foot thoroughly.
- Examine the interdigital space (for foreign body/material/infection) and coronary band.
- Pare overgrown and under-run horn in cattle. Only pare sheep's foot/feet if suspect a white line abscess.
- Re-shape bovine foot (five step Dutch method) as necessary.

DOI: 10.1201/9781003106456-14

- There are several specific conditions where radiographic examination is needed to confirm the diagnosis.
- Joint ultrasonography is of very limited use. Arthrocentesis rarely yields a diagnostically useful sample of synovial fluid.

Lameness is best defined when the animal is made to walk with another animal from the same group on a non-slip level surface in a confined area. Isolation stresses many sheep and beef animals and this may over-ride mild/moderate lameness.

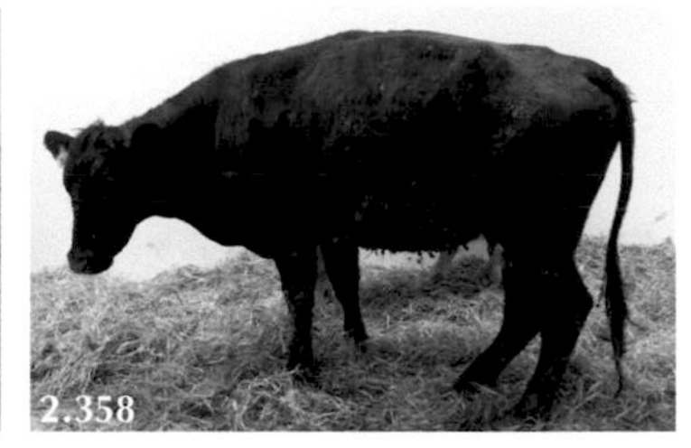

2.356–2.358: Painful expression, arched back and lowered head carriage in these three lame cows.

The extent of the lameness is subjectively scored on a 10-point scale; 1 being very slight lameness to 10 which is non-weight-bearing even at rest with the animal unwilling to take even one or two steps forward (a five-point scale 1–5 is used by some authors with 0.5 points added by other observers). Typically, long bone fractures and septic joints result in severe (10/10) lameness but so too can white line abscesses, especially those infections which track to the coronary band. Therefore, the degree of lameness does not necessarily determine prognosis and a detailed examination is essential. Painful foreleg lesions result in the hind feet being drawn forward under the body to bear more weight on the hind feet. 🎥 2.169, 2.170, 2.171, 2.172, 2.173, 2.174, 2.175, 2.176, 2.177

2.359: This ram's lameness, caused by interdigital dermatitis, with normal horn wear indicates acute severe pain but does not determine outcome. **2.360:** Grossly overgrown forefeet hooves indicate one to two months' lameness. **2.361:** Severe pain and overgrown hoof horn of left hind foot; lameness reduces horn wear, overgrown feet do not cause lameness *per se*. 🎥 2.178, 2.179, 2.180, 2.181, 2.182, 2.183, 2.184

Painful lesions affecting the foot and distal joints, but also elbow arthritis, cause sheep to graze on their knees, leading to abrasions, thickening of the skin overlying the knees and carpal hygromas.

2.362: Carpal hygromas with white hair caused by skin trauma. **2.363** and **2.364:** Large carpal hygromas with thickened skin as a consequence of chronic foreleg lameness. The swelling does not involve the carpal joints. 2.185

Prolonged periods in sternal recumbency on unyielding surfaces may result in large swellings (bursitis) over the lateral aspect of the hocks and dorsal aspect of the carpi in cattle. Such lesions are common in dairy cows and result from poor cubicle design and maintenance.

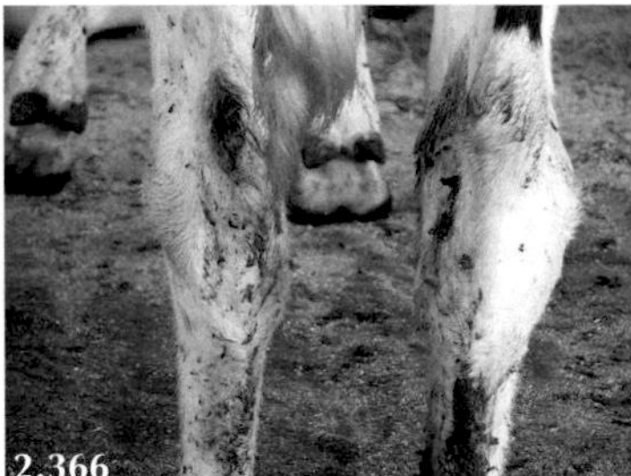

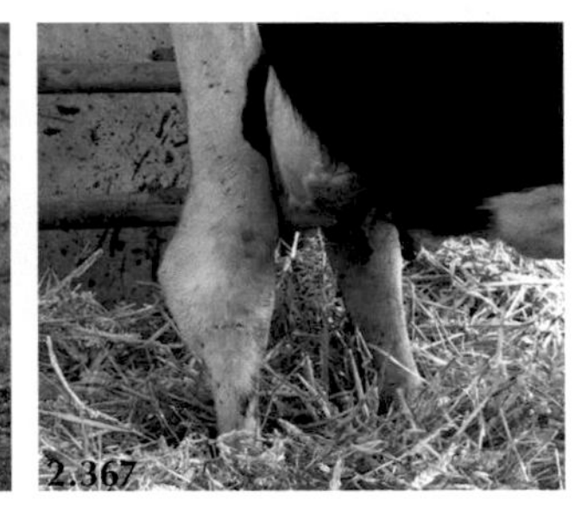

2.365: Marked carpal bursitis of right foreleg. **2.366:** Moderate right tarsal bursitis. **2.367:** Extensive swelling of capped right hock. The deep straw bedding was not the usual housing situation.

Muscle Wastage

The severity of muscle wastage depends upon both the intensity and duration of lameness. Muscle wastage can be reliably detected after five to seven days' moderate to severe lameness by careful palpation over bony prominences such as the spine of the scapula and head of the femur for forelegs and hindlegs, respectively. Comparison of changes with the contralateral limb, if sound, is recommended.

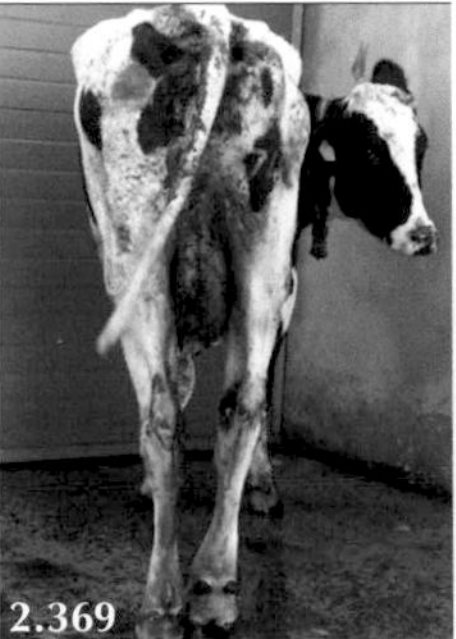

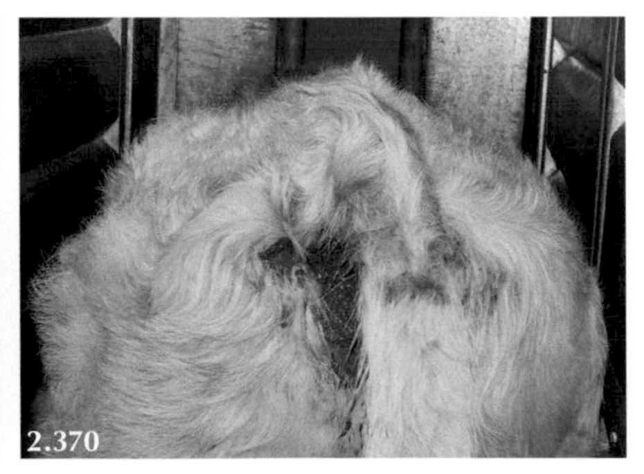

2.368: Muscle wastage over right hip area **2.369:** Muscle wastage over left hip area. Also note how dirty these cattle are which reflects their housing conditions. **2.370:** Muscle wastage over left hip area. 2.186

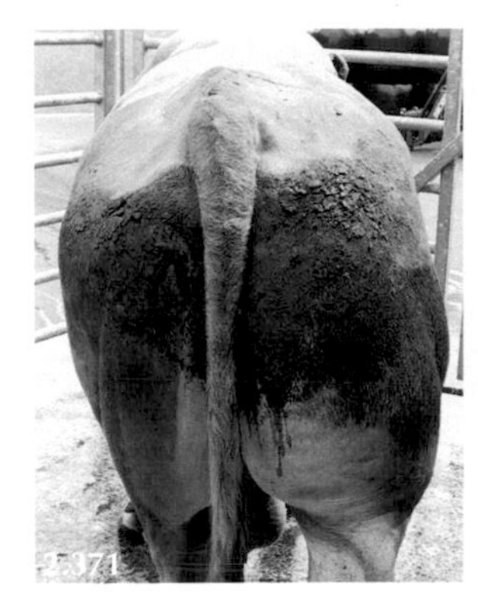

2.371 and **2.372:** Considerable muscle wastage over left hip area in these lame bulls. **2.373:** The physical size of many bulls can present problems with clinical examination and may render on-farm radiography impossible.

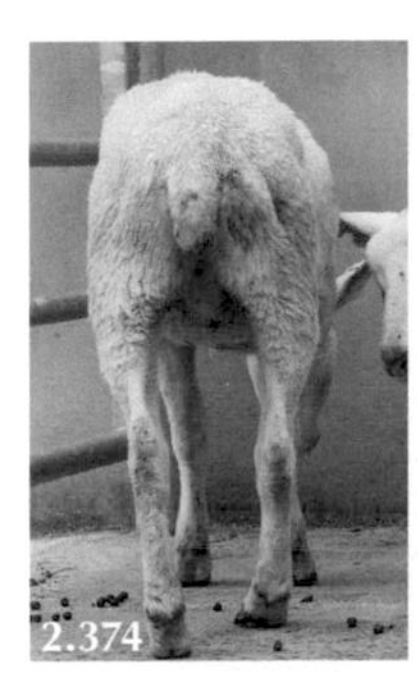
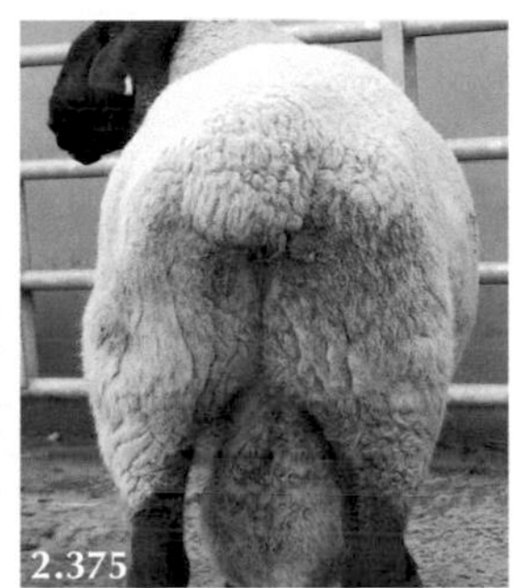

2.374 and **2.375:** Marked muscle of the left hindleg due to chronic lameness in these sheep.

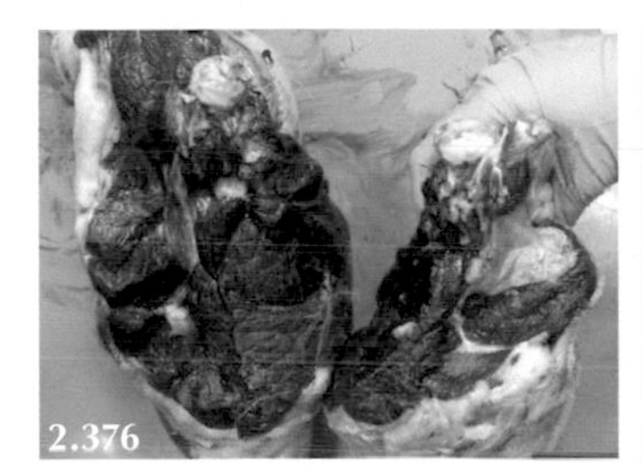
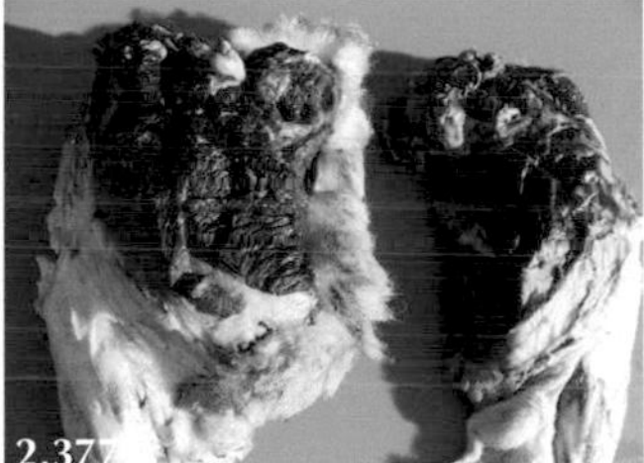
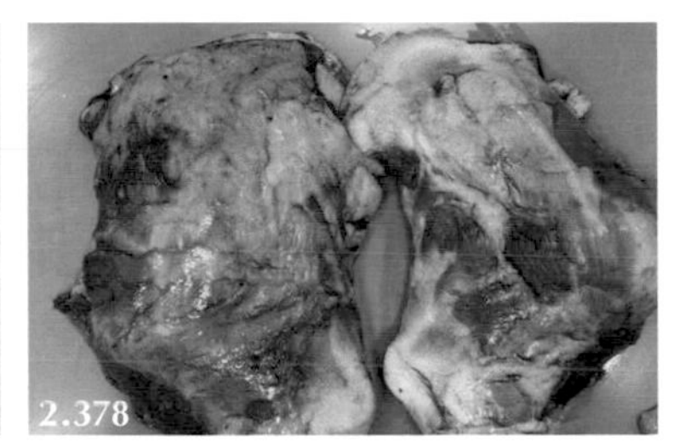

2.376 and **2.377:** Muscle wastage of (right) affected hindleg compared to normal (left) leg revealed at necropsy. **2.378:** Muscle wastage over the right (affected) shoulder showing reduced musculature with prominent spine of the scapula. 🎥 2.187

Lymph Nodes

Enlargement of the prescapular lymph node (two to 10 times normal size) can be readily appreciated within three to seven days of bacterial infection of the forelimb joints and cellulitis lesions. White line and sole abscesses, and digital dermatitis lesions do not usually cause such obvious drainage lymph node enlargement. Infected lesions distal to the stifle joint cause enlargement of the popliteal lymph node but this node is not readily palpable unless there is considerable muscle atrophy because it is deep to the gastrocnemius muscle. Infection proximal to the stifle joint results in enlargement of the deep inguinal lymph nodes within the pelvic canal.

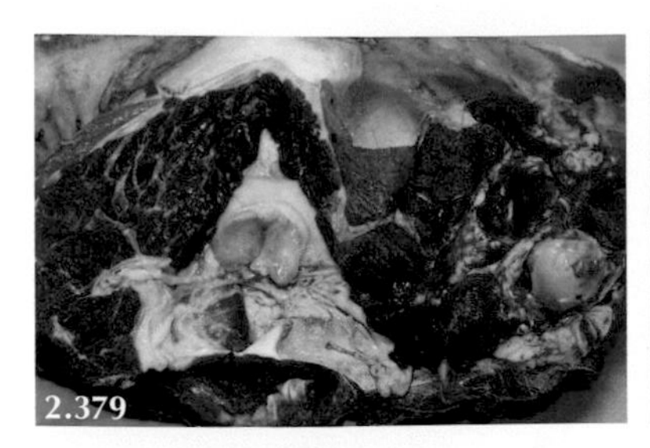

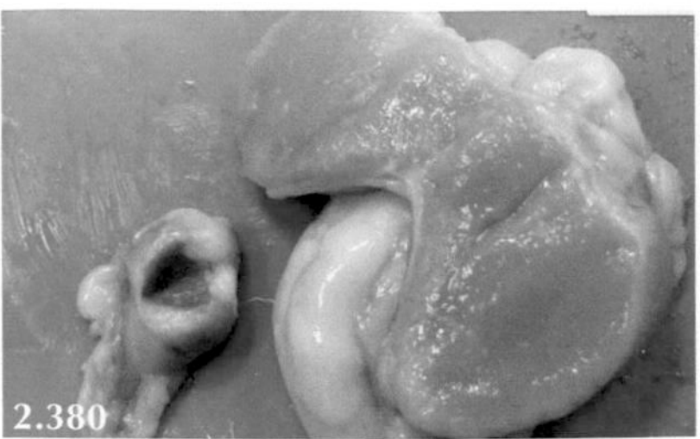

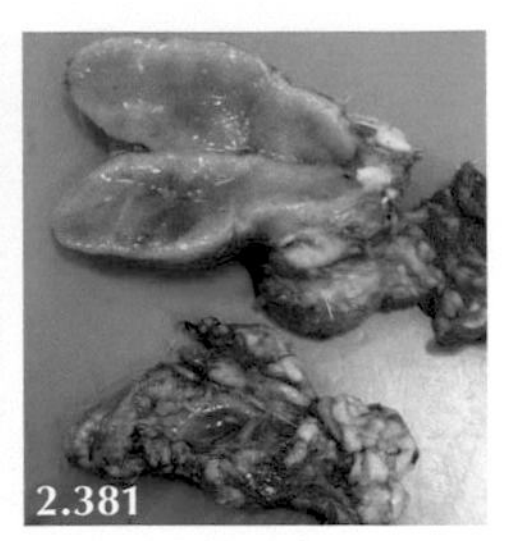

2.379: Position of popliteal lymph node (enlarged) deep to the gastrocnemius muscle renders palpation difficult. **2.380:** Dissection reveals marked enlargement of (right) affected popliteal lymph node compared to normal (left) node revealed at necropsy. **2.381:** Enlargement and cortical reaction of the left prescapular lymph node (top) compared to normal size (below).

Be aware that swellings caused by abscesses/haematomas, cellulitis lesions and fractures can cause lameness such that there appears to be "more muscle" over the affected leg.

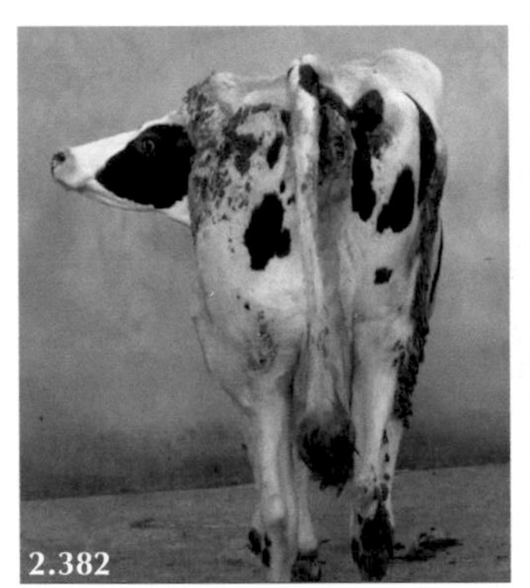

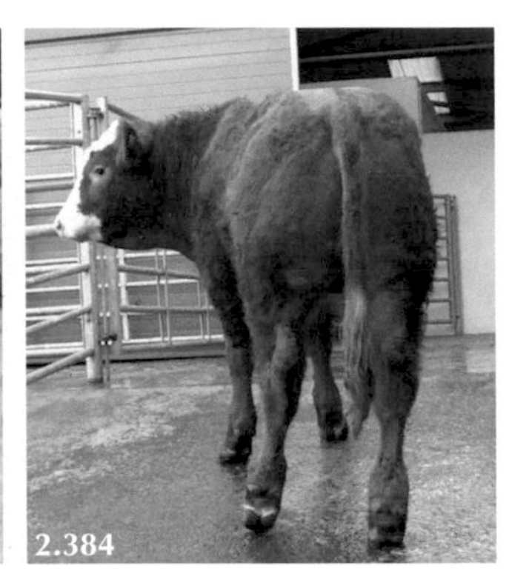

Soft tissue swelling associated with lameness (left side in all three animals). **2.382:** Cellulitis. **2.383:** Oedema associated with blackleg. **2.384:** Haemorrhage from mid-shaft femoral fracture. 2.188, 2.189, 2.190

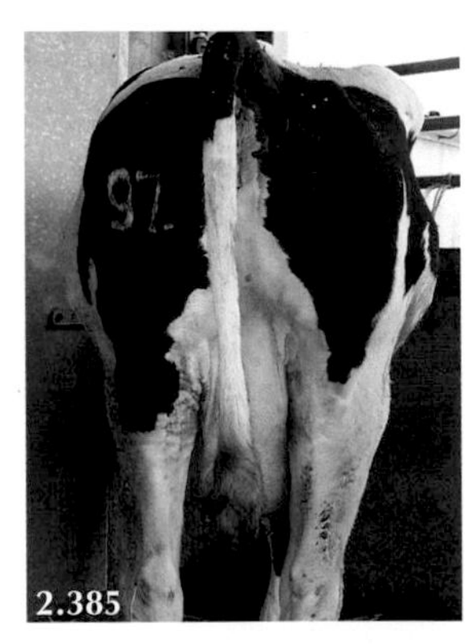

2.385: Deep cellulitis of the left hindleg causing swelling of the leg compared to the normal right hindleg.

The clinician must always remember that lameness originates from a painful lesion and that manipulations should be kept to a minimum and undertaken with care and empathy. In particular, joint lesions are especially painful; manipulations to elicit crepitus are not usually necessary. The clinical examination must not exacerbate the degree of lameness. 2.191

Restraint

There are various means of raising feet to allow detailed examination and most farmers have specialised "cattle foot crates". Raising the foot is undertaken as the last component of the examination investigation but should not be undertaken if there is a painful joint lesion or suspected fracture. A broad webbing strap is much more preferable to rope when raising the leg. Do not over-flex the hindleg joints, especially the stifle joint, when examining the hind feet.

2.386 and **2.387:** Examination of the feet is often difficult in conventional cattle stocks especially with large bulls. Note there is no access to the left forefoot with the cattle stocks sited next to a wall. **2.388:** Deep sedation may be necessary in some situations.

2.389: Wopa box with broad support band around chest wall to support animal when a foot is raised. **2.390:** Damage to the brachial plexus from the narrow (3–4 cm wide) support band can happen when a cow accidentally sits down (**2.391**).

Foot Examination

The interdigital space is a common site for impacted stones and must be carefully examined using a torch where necessary. In cattle, any overgrown horn from the abaxial walls and toes is removed with a sharp hoof knife noting the quality of the horn removed and presence of any sole haemorrhages or black marks (in unpigmented horn). When foot paring, continually check for thickness of the sole to avoid making the sole too thin with consequent pressure necrosis of the corium.

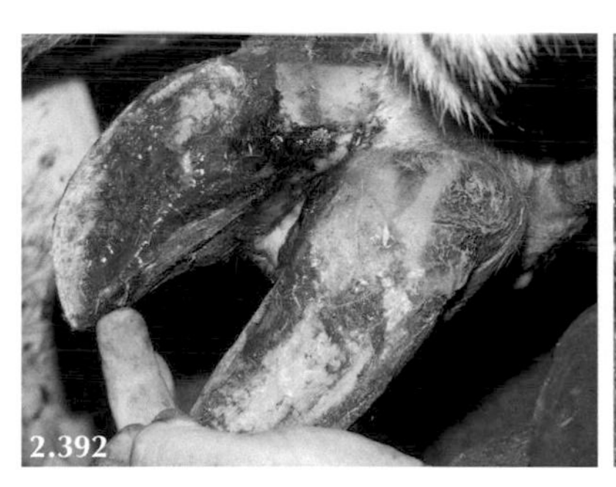

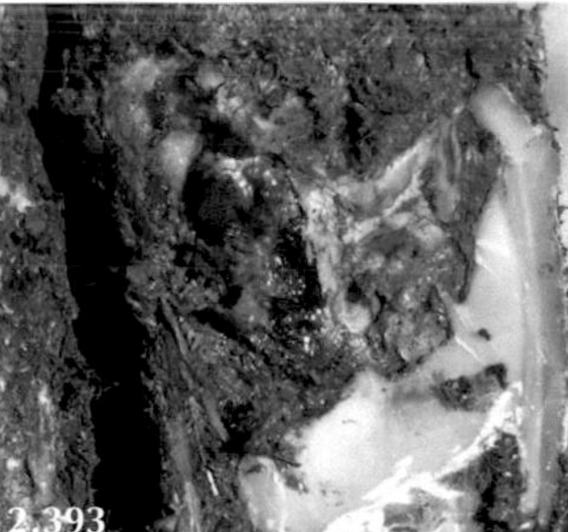

2.392: The interdigital area is carefully examined for foreign material. **2.393:** Careful foot paring to remove sole horn underrun with foreign material/pus. Sole ulcer has resulted from pressure to the corium.

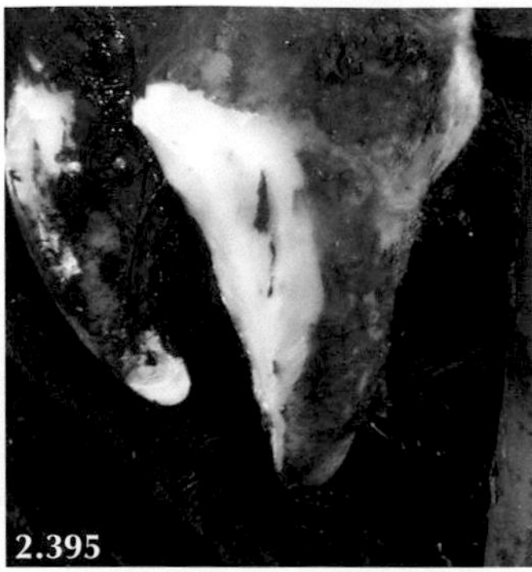
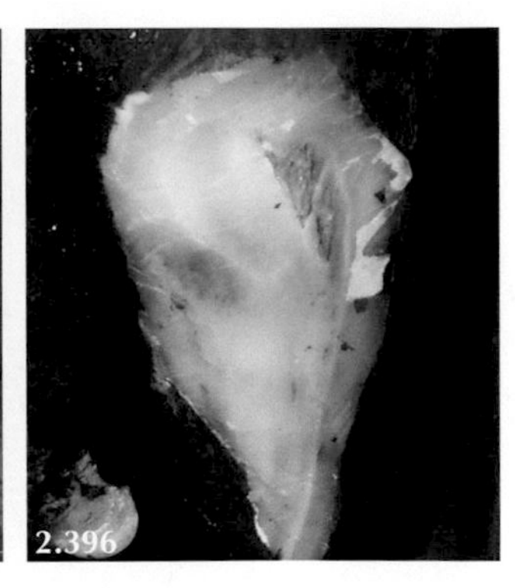

2.394: Limousin bull with overgrown lateral claw of right hind foot at presentation. **2.395:** Black mark (impacted dirt) detected in the abaxial white line. **2.396:** Careful foot paring releases pus from a white line abscess. Note paring has caused no damage to the corium and no bleeding. There is considerable bruising of sole horn caused by overgrown horn with abnormal weight distribution putting pressure on the corium.

Any under-run horn commencing at the axial margin of the sole should be removed. It is important not to damage the sensitive corium as this will lead to delayed regeneration of epithelium and extended healing time. Exposure of the sensitive corium to irritant chemicals, such as formalin, may result in exuberant granulation tissue. 🎥 2.192, 2.193

2.397: Use of tipping crate and angle grinder to remove hoof horn by para-professional hoof trimmer. **2.398:** Over-paring causing bleeding from the corium; pressure to the thinned corium is now likely causing bruising.

Foot Paring in Sheep

Grossly overgrown horn can be carefully removed from the abaxial wall and toe of the foot of the lame leg with shears or a sharp hoof knife to inspect the foot and check for a white line abscess. Paring all but excessive hoof horn of the wall must be avoided because it simply transfers weight to the sole which is abnormal. Foot paring must never be undertaken where there is separation of the horn from the corium. Under-running of the horn and exposure of the corium caused by footrot or CODD is very painful and is best treated with topical and parenteral antibiotics. 🎥 2.194

2.399: Foot paring undertaken to investigate under-run wall and possible abscess. The sole should not have been pared nor the wall of the other hoof. **2.400:** This case of footrot should have been treated with antibiotics and not hoof paring. **2.401:** Treat this case of footrot with antibiotics and re-examine in seven to 10 days when the lesion will be much less painful and the grossly overgrown hoof horn could be cut away with shears.

Joint Effusion/Thickening

Gentle digital palpation will reveal joint effusion. Distension of the tarso-metatarsal joint is common in bulls often with no signs of lameness and is a useful learning exercise. Distension of the stifle joint is more difficult to appreciate. With the exception of neonatal calves, significant accumulation of joint effusion is uncommon even in acute infections.

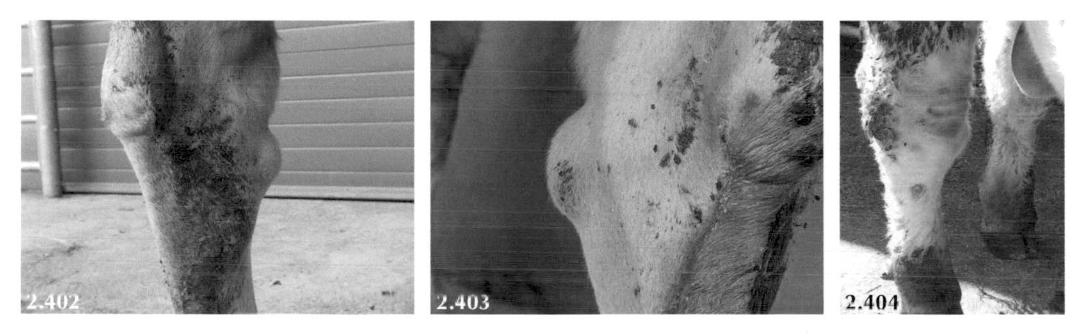

2.402–2.404: Distension of the tibio-tarsal joint is common in bulls often with no signs of lameness.

Thickening of the fibrous joint capsule reflects chronic joint trauma/infection. Comparing changes in the joint capsule (thickness) with the contralateral limb, where sound, gives a useful guide to likely duration of the pathology. Joint manipulation will reveal reduced joint excursion but attempting to elicit crepitus rarely gives significant prognostic information, causes unnecessary and avoidable pain, and should not be routinely performed.

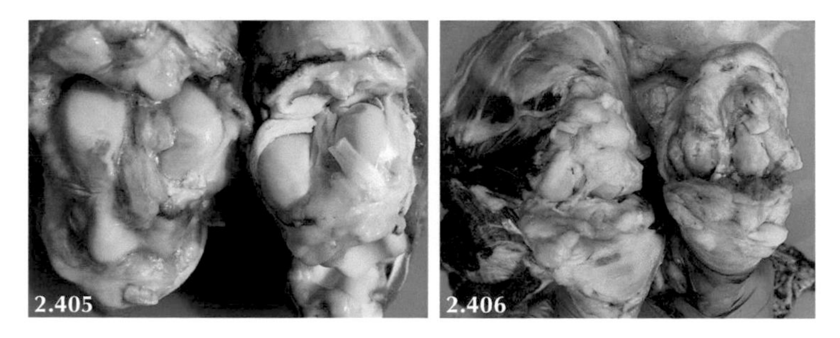

2.405: Synovial proliferation and thickening of the fibrous joint capsule in the chronically infected left stifle joint of a sheep noting the overall pink colour caused by synovial membrane hyperaemia. **2.406:** Thickening of the fibrous joint capsule and pronounced synovial proliferation in a chronically infected ovine stifle joint (right) with the articular surface barely visible as a consequence. Note also severe muscle atrophy in this leg.

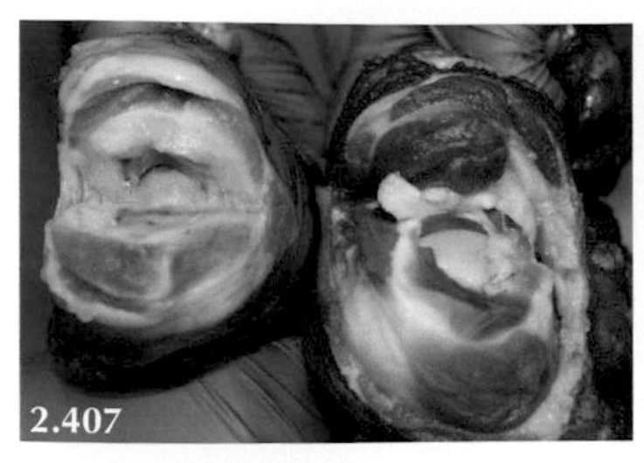
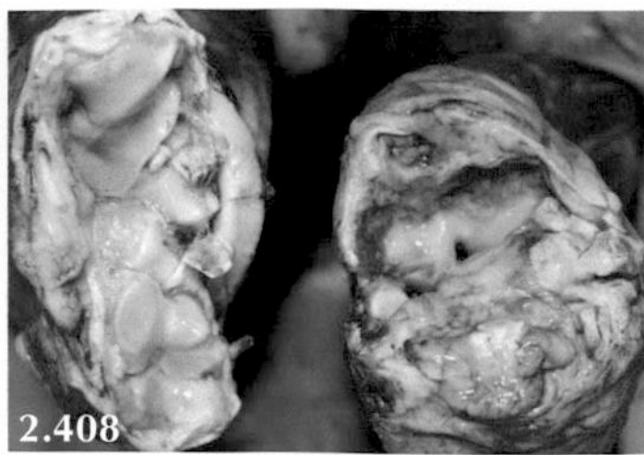

2.407: Fibrous tissue proliferation surrounding the infected shoulder joint (left) as viewed. Note also the resultant muscle atrophy. **2.408:** Synovial and fibrous tissue proliferation and pannus involving the right carpal joint (right as viewed). 2.195, 2.196, 2.197, 2.198, 2.199, 2.200

Arthrocentesis

Arthrocentesis is not commonly undertaken because joint infections with the common bacterial pathogens such as *E. coli* and *Salmonella* spp. quickly cause pannus rather than marked joint effusion. Pannus is defined as an abnormal layer of fibrovascular tissue or granulation tissue over a joint surface.

Attempts can be made to collect synovial fluid from distended joints under local anaesthesia, but the anaesthetic solution can only be given subcutaneously; the infected joint capsule/synovial membrane cannot be readily de-sensitised. Intravenous regional anaesthesia can be used to anaesthetise the fetlock and phalangeal joints. A high extradural block can be used when sampling/flushing the stifle and more distal joints.

The arthrocentesis site is shaved and aseptically prepared. The approach depends upon the particular joint, avoiding tendons and ligamentous structures. In general, the joint capsule is punctured where it is most distended as this "pouching" occurs away from joint structures such as ligaments and tendons. The calf's age and marked joint distension (next) are not typical of a septic joint but demonstrate the procedure as well as case selection for joint lavage.

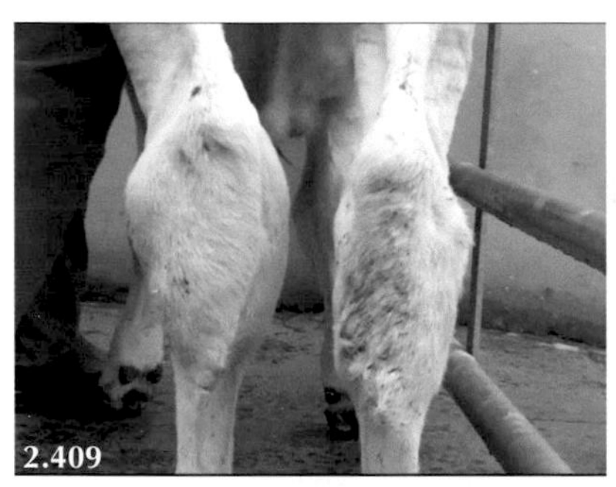
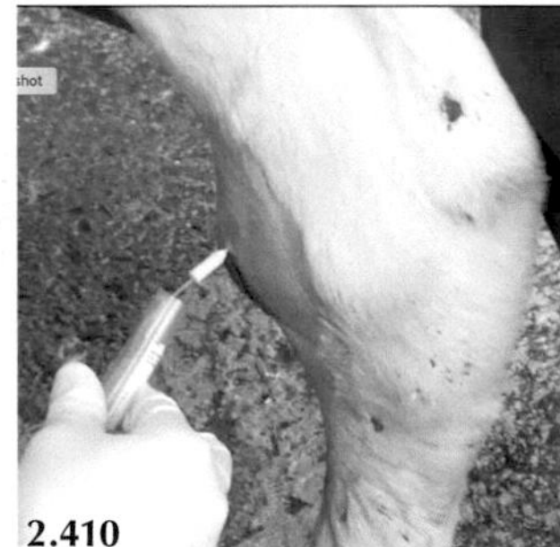
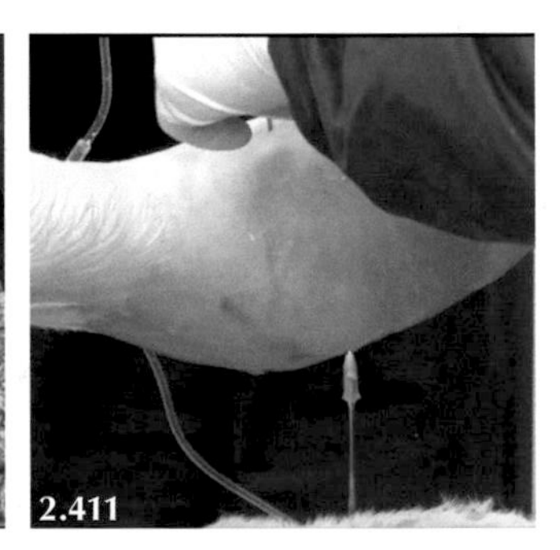

2.409: Obvious distension of the left tibio-tarsal joint in a two-month-old calf. **2.410:** Arthrocentesis site is shaved and aseptically prepared. **2.411:** Joint lavage under high extradural block. 2.201

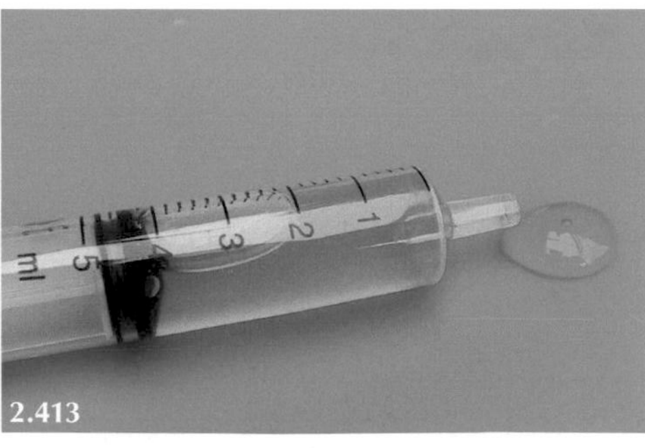

2.412 and **2.413:** Excessive quantities of slightly turbid but less viscous joint fluid suggests infection—laboratory examination of the arthrocentesis sample is necessary. The sample shown (right) was collected at necropsy.

Arthrocentesis is not commonly undertaken in sheep because joint infection with the common bacterial pathogens *Streptococcus dysgalactiae* and *Erysipelothrix rhusiopathiae* rarely causes marked joint effusion.

Normal synovial fluid is pale yellow, viscous, clear and does not clot. The protein concentration is less than 18 g/l with a low white cell concentration comprised mainly of lymphocytes. Septic arthritis is characterised by a turbid sample caused by increased white cell concentration which is comprised almost exclusively of neutrophils. The protein concentration is increased above 40 g/l. Samples are especially difficult to collect from chronically infected joints and often fail to grow bacteria. Direct smears of the aspirate can be made onto a glass slide and stained with Gram's stain to gain some information of the potential pathogen(s) involved. Synovial membrane collected at necropsy from an untreated animal provides the best chance of successful bacterial growth.

Radiography

Radiography is most useful in the investigation of long bone fractures where doubts exist over the diagnosis. This is especially useful in potentially highly valuable neonatal calves where fractures result from dystocia. Deep sedation, or preferably general anaesthesia, may be required to allow correct positioning for radiography of the humerus. A high extradural block can be used when investigating suspected femoral fractures. Radiography is very useful in less common clinical situations such as growth plate fractures.

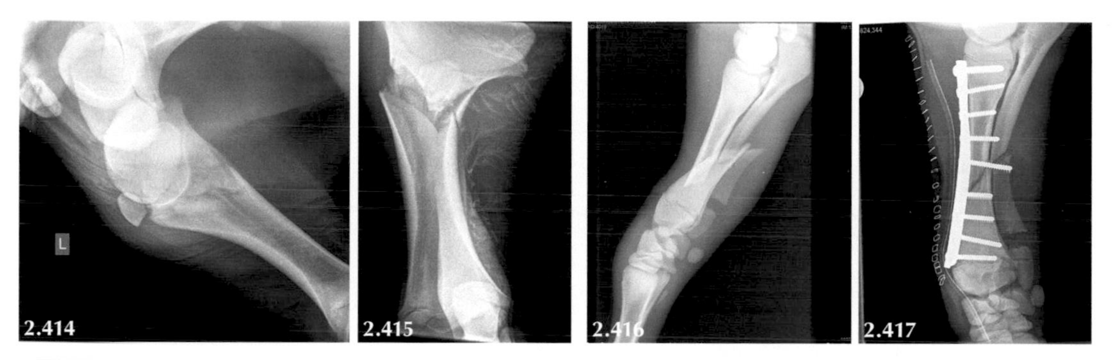

2.414 and **2.415:** Tibial fracture in a pedigree bull calf following assisted delivery. **2.416** and **2.417:** Repair of mid-shaft radial fracture in a neonatal calf. 🎥 2.202

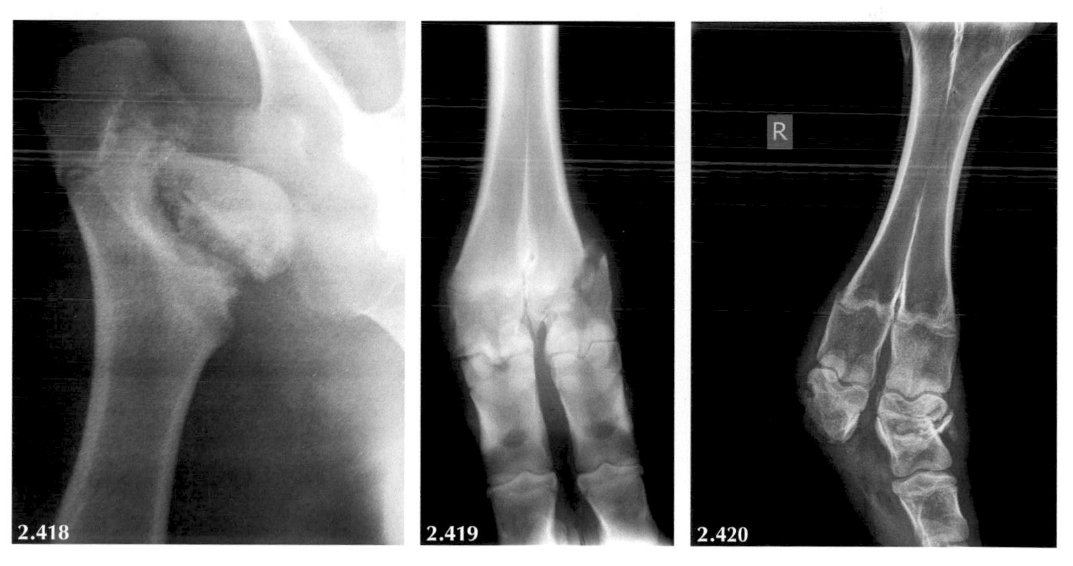

2.418: Radiographic confirmation of pathological fracture of the proximal femoral growth plate. **2.419:** Salter-Harris type III fracture of the distal third metatarsal bone. **2.420:** Amputation of the medial digit for septic pedal arthritis was followed several weeks later by fracture through the proximal growth plate of P1 of the lateral digit (which healed well after casting).

Radiography is useful in the identification of greenstick fractures in young lambs, and the investigation of suspected long bone fractures, in particular, the humerus and femur. Enthesophyte formation is common in the elbow joint of adult sheep with best radiographic results obtained from an oblique view.

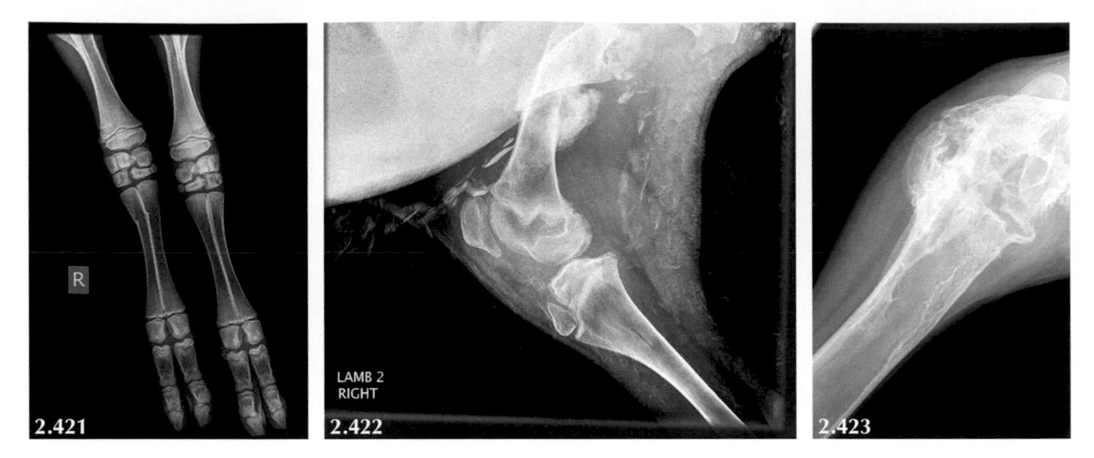

2.421: Radiography is helpful in the identification of greenstick fractures in young lambs (left as viewed). **2.422:** Investigation of suspected femoral fracture in six-month-old ram. **2.423:** Oblique view of the elbow joint of a ram showing extensive enthesophyte formation involving the lateral co-lateral ligament.

Radiography may prove useful in the investigation of chronic foot infections in cattle, where infection may involve either the proximal or distal interphalangeal joints, and possibly both distal interphalangeal joints.

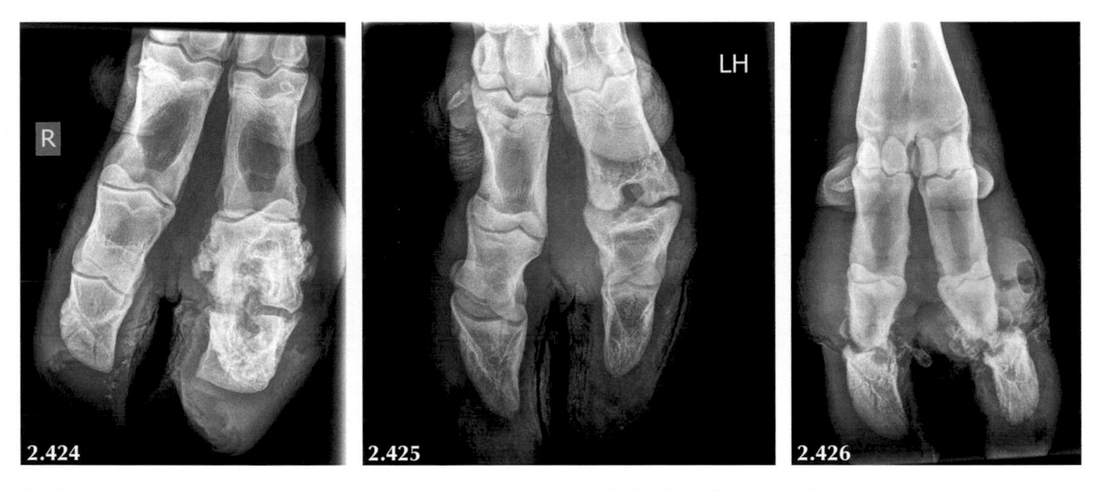

Radiographs of the bovine foot used to determine prognosis. **2.424:** Extensive osteophyte formation but amputation remains the better option. **2.425:** Infection of the proximal inter-phalangeal joint. **2.426:** Infection/destruction of both distal interphalangeal joints.

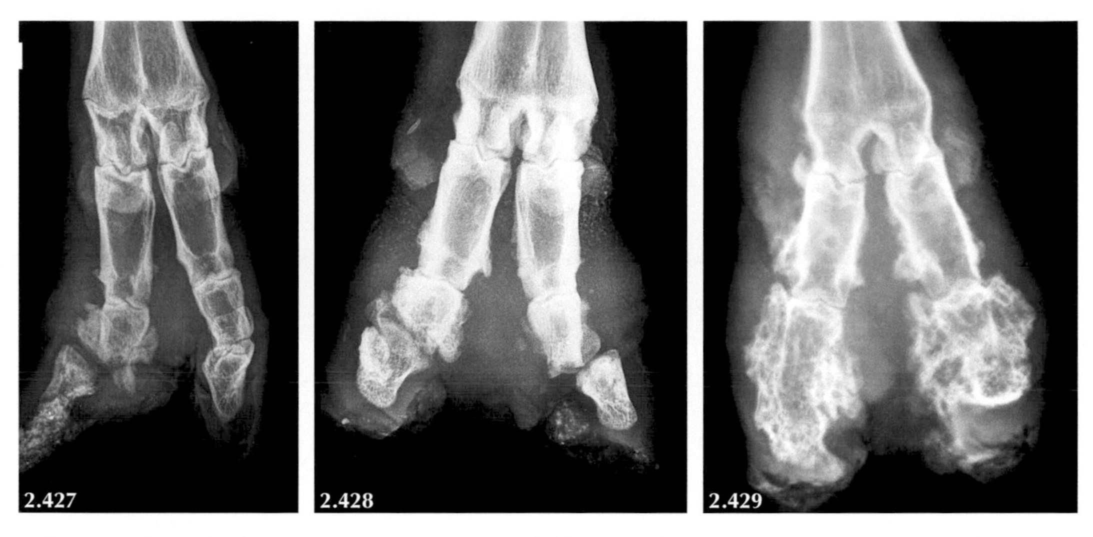

Radiographs of the ovine foot used to determine prognosis. **2.427:** DP radiograph shows soft tissue swelling of interdigital area and dislocation of P2/P3 joint (left as viewed). **2.428:** DP radiograph shows infection of both distal interphalangeal joints with considerable new osteophyte formation of the lateral joint (left as viewed) and dislocation of the medial joint. **2.429:** DP radiograph of the right hind foot reveals destruction of both distal interphalangeal joints with ankylosis and extensive osteophyte formation involving the proximal interphalangeal joints and extending to involve the fetlock joint.

Radiography adds little critical information to the investigation of most cases of septic arthritis other than to reveal widening of the joint space and osteophyte formation in neglected cases. Radiography of a septic joint during the early stages of infection in a heifer (see the following) reveals only widening of the joint space which may be mistakenly interpreted that there is little pathology present.

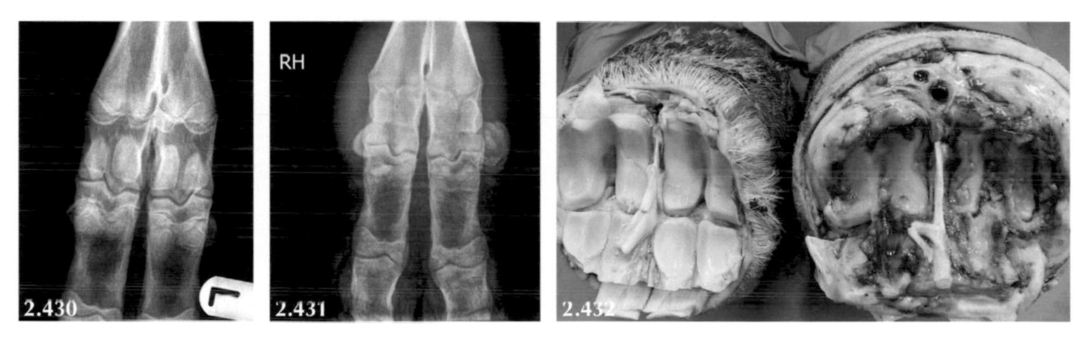

2.430 and **2.431:** Yearling Simmental heifer. DP radiographs of normal left fetlock joint and septic right fetlock joint; the only difference is the large amount of soft tissue swelling of the right hind fetlock. **2.432:** Normal (left) and septic fetlock joint revealing pannus (right).

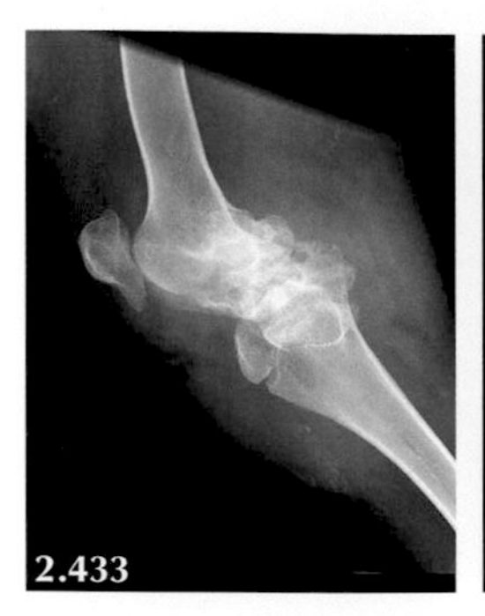

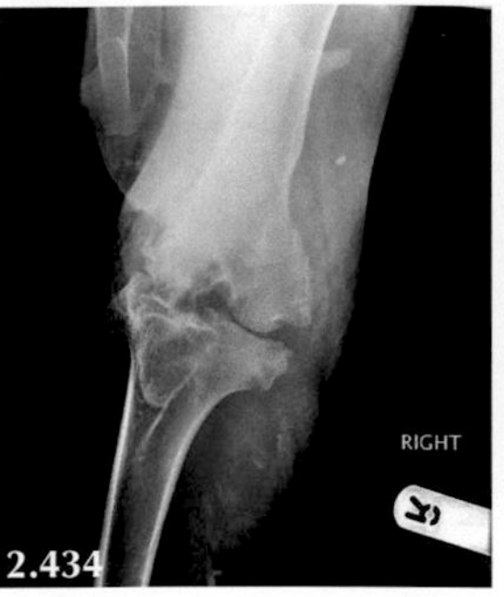

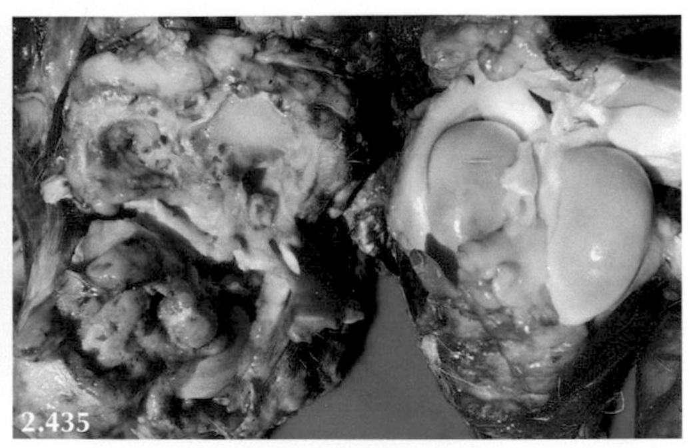

Lateral view (**2.433**) and DP view (**2.434**) of an infected stifle joint of a ewe showing erosion of articular surfaces, widening of joint space (by pannus/synovial membrane hypertrophy), and extensive osteophyte formation. **2.435:** Necropsy examination reveals extensive erosion of articular surfaces of the stifle joint extending into subchondral bone with pronounced synovial membrane hypertrophy.

Ultrasonography

Ultrasonography using a 5 MHz linear array scanner (preferably higher frequency) can provide details of the thickness of the joint capsule and extent of joint effusion/exudate but this information is usually determined by palpation. The skin overlying the joint is shaved to ensure good contact; a stand-off may be required for examination of smaller joints, and in sheep. Affected joint(s) can be compared to the contralateral joint where normal. Interpreting the nature of joint effusion is difficult because chronic joint infection comprises pannus and not large volumes of pus. Fibrin tags must not be mistaken for synovial membrane hypertrophy.

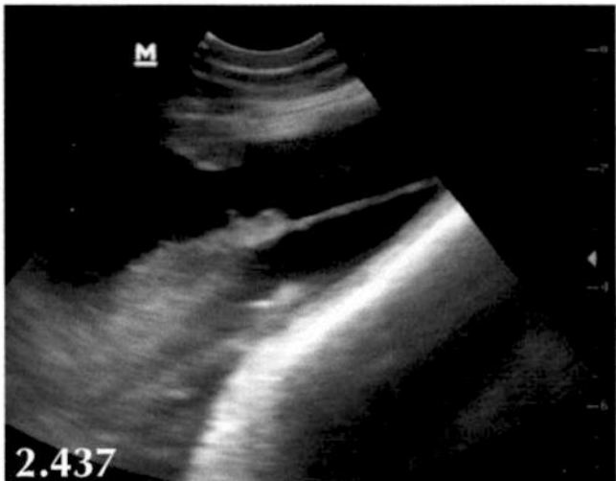

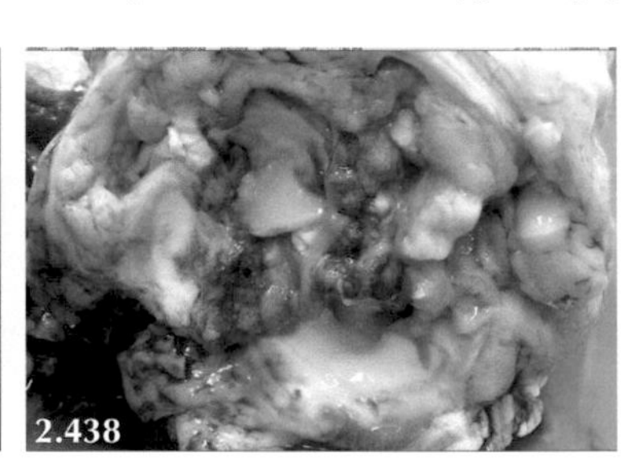

2.436: Severe muscle wastage over the left gluteal region of this Holstein heifer. The left stifle joint capsule was markedly thickened. **2.437:** Ultrasound examination reveals increased fluid within the left stifle joint containing fibrin. **2.438:** Inflamed and proliferative synovial membrane and intra-articular fibrin clot (fluid was released on incision into joint). 🎥 2.203

Joint Lavage

Joint lavage, undertaken under high extradural block for hindleg joints or general anaesthesia for forelegs joints, is occasionally successful in neonatal calves where only one joint is affected. It is rarely, if ever, successful in sheep because of the chronic nature of lesion(s) at presentation with extensive synovial membrane hypertrophy and inability to flush out pannus. 🎥 2.204, 2.205, 2.206

Boiling Out

The extent (severity and chronicity) of bony pathology in joint disease is most graphically demonstrated in "boiled out" preparations (next).

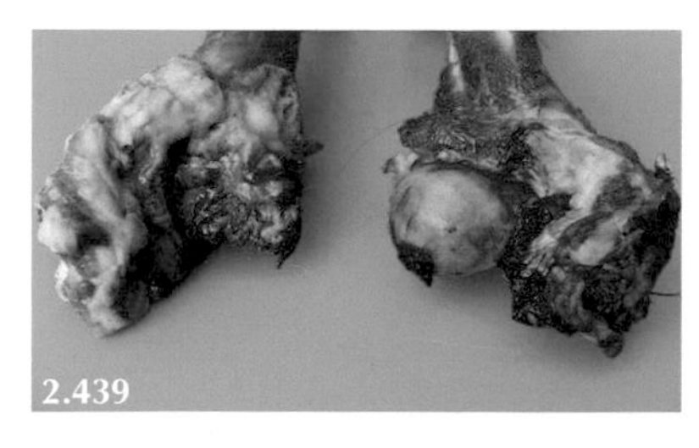

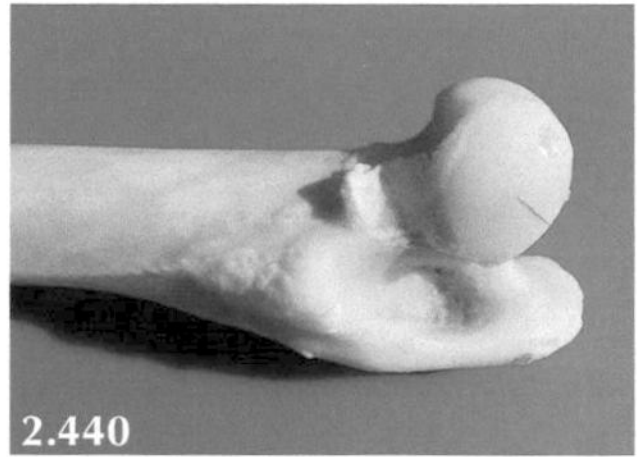

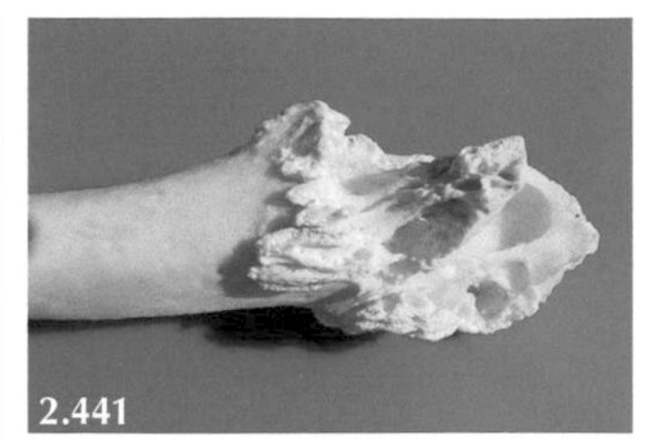

2.439: The effects of chronic hip joint infection on bone are clearly shown in the femurs from the same sheep after "boiling out"; normal femoral head (**2.440**), chronic hip infection (**2.441**).

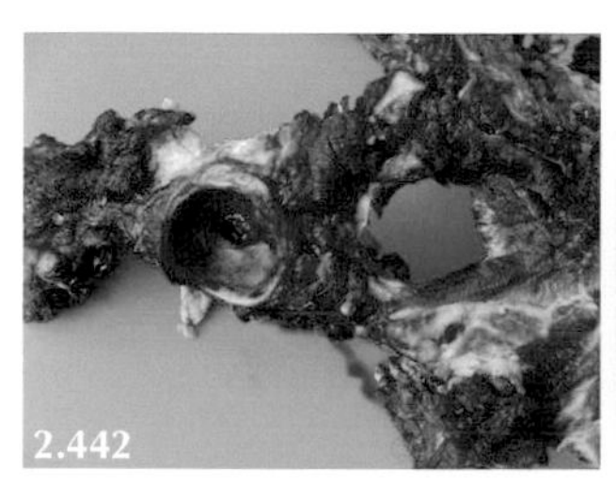

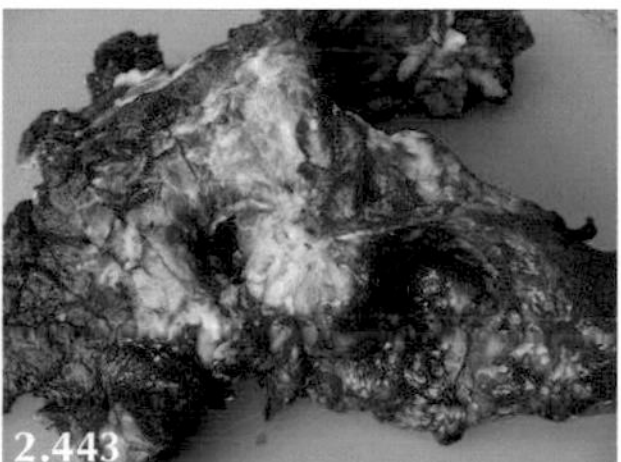

2.442 and **2.443:** Necropsy reveals the effects of chronic hip joint infection causing extensive bone re-modelling (right) of the acetabulum.

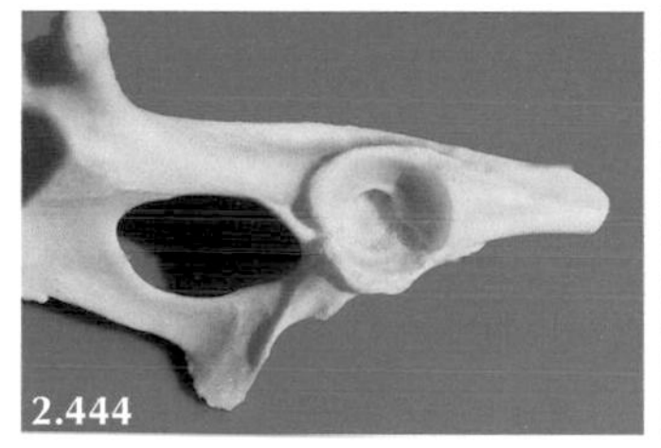

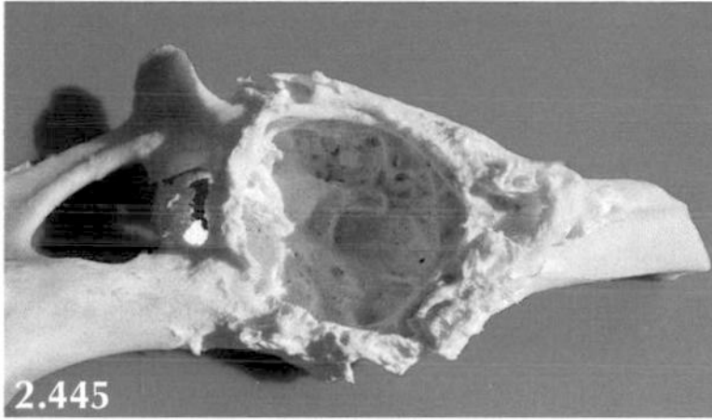

2.444 and **2.445:** The effects of chronic hip joint infection causing extensive bone re-modelling (right) of the acetabulum are highlighted after "boiling out". 🎥 2.207

Nerve Blocks

Unlike in horses, nerve blocks and intra-articular anaesthesia are seldom used in cattle and sheep.

Intravenous Regional Anaesthesia—Method

In most situations, the animal is restrained in cattle stocks/Wopa box and the affected hindleg raised. In an adult dairy cow weighing 600–700 kg, 25 to 30 ml of 2 per cent lignocaine solution (or equivalent) is injected into a superficial vein after application of a tourniquet either above or below the hock. A soft calving rope can be used for this purpose. The recurrent metatarsal vein runs on the cranio-lateral aspect of the mid region of the third metatarsal bone and is readily palpable unless the leg is oedematous. Insertion of the 18 gauge 25 mm needle into the distended superficial vein releases a small quantity of blood under pressure, and blood flow then quickly reduces to the occasional drop if the tourniquet is tight enough. The anaesthetic solution is injected over 10–15 seconds, but it is not unusual for the cow to kick when only two-thirds of the solution has been injected. Analgesia is effective within 2 minutes and is tested by needle pricking the coronary band. After amputation and application of the bandage the tourniquet is removed around 10–20 minutes after it is first applied. A similar approach is employed for sheep using 5 to 7 ml of 2 per cent lignocaine solution (or equivalent) and a 20 gauge 25 mm needle. 🎥 2.208, 2.209

2.7

Examination of the Urinary System

- Obstructive urolithiasis is common in rams and castrates, much less so in bulls.
- It can be difficult to induce urination and assess whether a ram or bull can urinate normally.
- Trans-abdominal ultrasound examination readily identifies bladder distension in rams while rectal palpation is undertaken in bulls.
- Rectal examination and transrectal ultrasonography identify bladder distension in bulls, and thickening of the ureters in pyelonephritis cases.
- It is important to examine the (right) kidney in rams for hydronephrosis before tube cystotomy surgery.
- Pyelonephritis is not uncommon in beef cows but rarely seen in dairy cows and never in sheep (except after sub-ischial urethrostomy surgery).

Observation

Normal urination in male sheep produces a steady flow lasting at least 20 seconds. Tenesmus producing only drops of blood-tinged urine is highly abnormal and must be investigated immediately. Compare the two video recordings: 🎥 2.210, 2.211. Chronic weight loss and frequent urination containing flecks of fresh blood and pus, often found sticking to the tail, are common in cows with pyelonephritis.

Clinical Examination

Blockage to the free flow of urine leads to bladder distension and abdominal pain expressed by frequent tail swishing and long periods in sternal recumbency. Frequent vocalisation is common in lambs suffering from urolithiasis, while bruxism is more common in adult rams and bulls. Rupture of the urethra causes swelling of the prepuce and subcutaneous skin extending caudally to the scrotum.

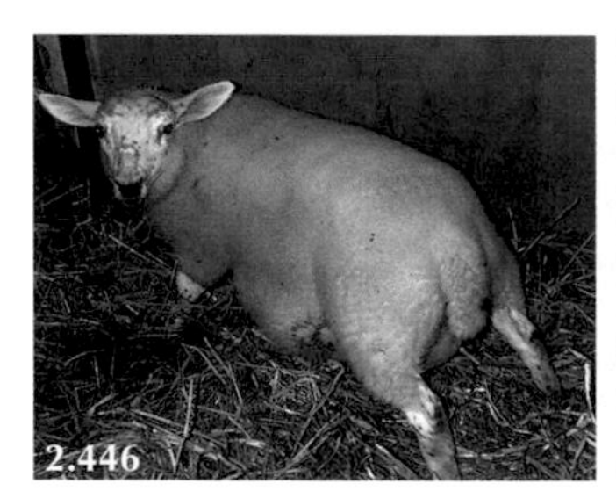
2.446

2.447

2.448

2.446: Wide-based stance in ram lamb with urolithiasis. **2.447:** Isolated castrated male lamb with colic signs arising from bladder distension. **2.448:** Scottish Blackface ram not grazing with other sheep in the group and curled upper lip/expression suggest pain. 🎥 2.212, 2.213

DOI: 10.1201/9781003106456-15

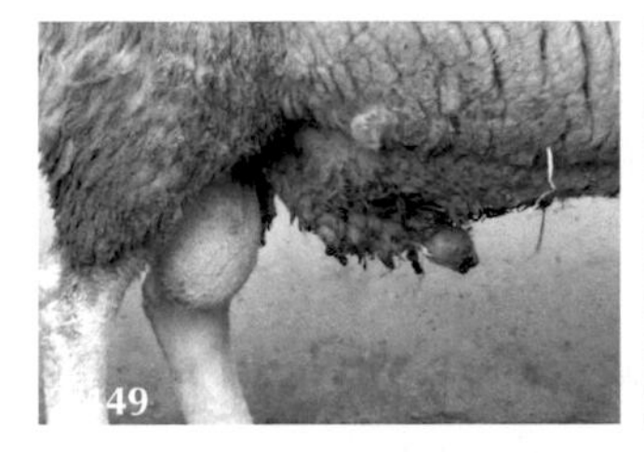
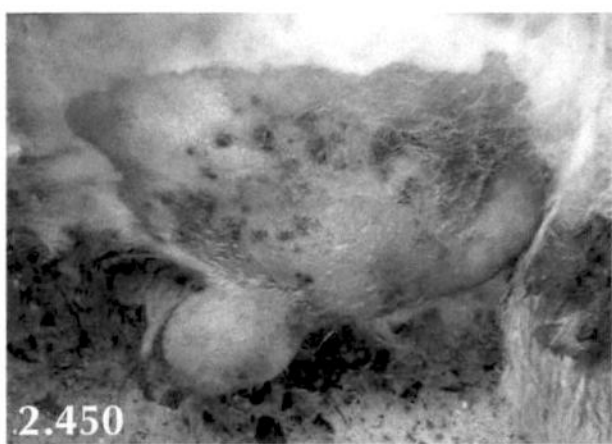
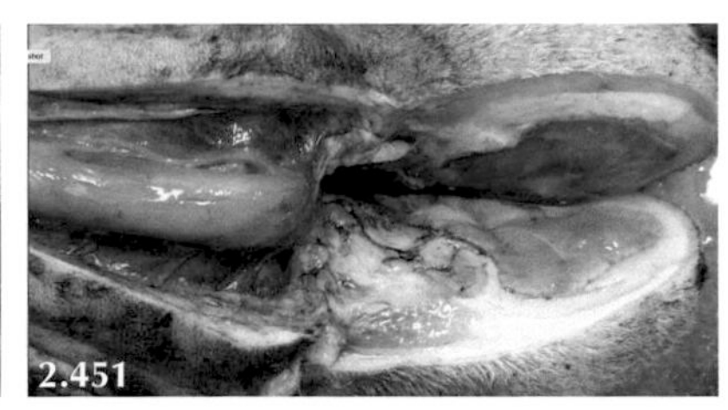

2.449: Urethral rupture causing swollen prepuce extending caudally to involve the scrotum. **2.450:** Subcutaneous urine causing swollen prepuce with surrounding sharply demarcated purple/black skin necrosis. **2.451:** Subcutaneous urine accumulation extending into the scrotum following ruptured urethra revealed at necropsy. 2.214

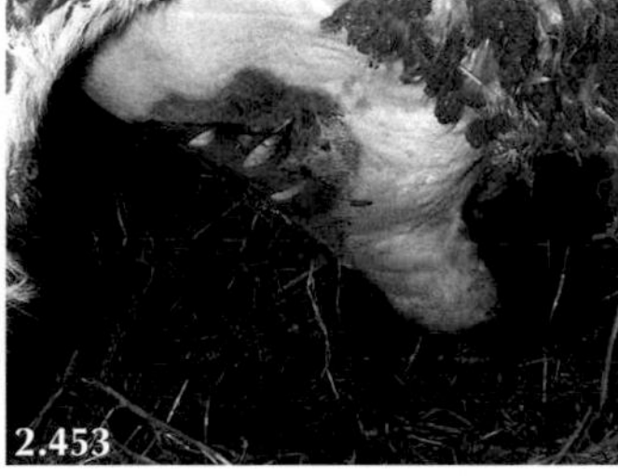

2.452: Extensive ventral subcutaneous urine accumulation following urethral rupture. **2.453:** Extensive subcutaneous urine accumulation surrounding the prepuce with overlying skin necrosis. Stab incisions in the necrotic skin are an attempt to drain urine (not undertaken by this author).

Crystals can be felt when urethral obstruction occurs within the vermiform appendage in rams; the portion distal to the obstruction often appears black and necrotic. The penis can be exteriorised in sexually mature male sheep to allow examination of the glans and vermiform appendage (see the following photographs). The ram is sat on its rump and the penis is held through the sheath and the sigmoid flexure extended at the same time as retracting the prepuce. A gauze swab is then wrapped around the penis proximal to the glans. This procedure is demonstrated in a cadaver specimen in the video recording that follows. In bulls, sedation usually results in protrusion of the penis allowing examination; a pudendal nerve block is rarely attempted.

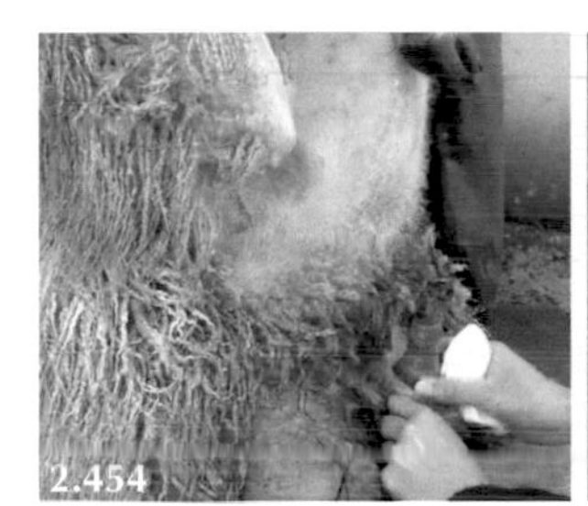

2.454: The ram is sat on its hindquarters and the penis extruded. **2.455:** The penis is grasped with a gauze swab allowing examination of the glans and vermiform appendage. 2.215, 2.216

Urinalysis

Pyelonephritis in cows in caused by *Corynebacterium renale*; sheep rarely develop ascending urinary tract infections. Urine specific gravity ranges from 1.015 to 1.045 in normal sheep with pH values from 7.4 to 8.0. Urine can be checked for the presence of sediment, in particular struvite crystals. Glucose, ketones, protein, blood and bilirubin should be absent from urine. Serum creatinine and urea nitrogen concentrations can be used to determine renal function, but such laboratory examination results in delays compared to the immediate diagnostic information from rectal/ultrasonographic examination of the bladder/kidneys.

Ultrasonography

Ultrasonography has great practical application in the field investigation of suspected urolithiasis in male sheep. Bladder distension, uroperitoneum and subcutaneous urine accumulation post-urethral rupture can readily be determined using both linear array and microarray probes. Examination of the right kidney can be undertaken with a 6.5 MHz microarray probe. There is no requirement to scan the left kidney because hydronephrosis secondary to urethral obstruction affects both kidneys. Hydronephrosis rarely develops in cattle, and ultrasonography of the right kidney is usually undertaken in suspected cases of pyelonephritis in (beef) cows.

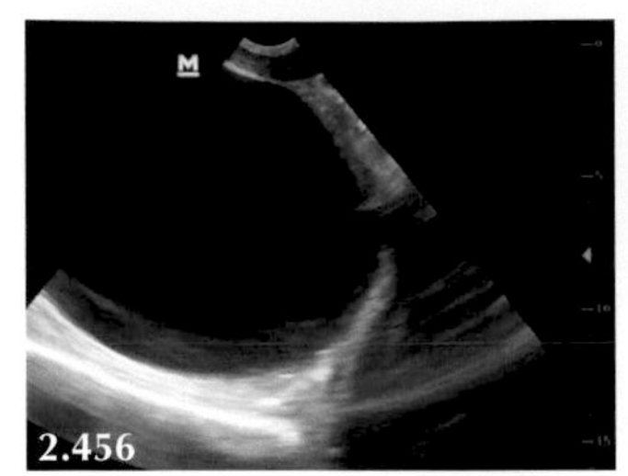

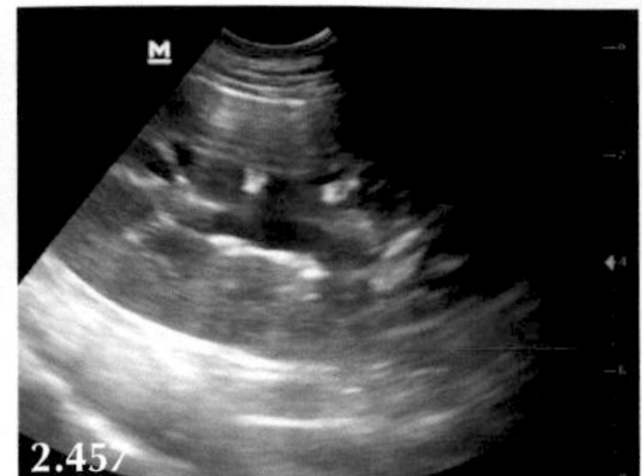

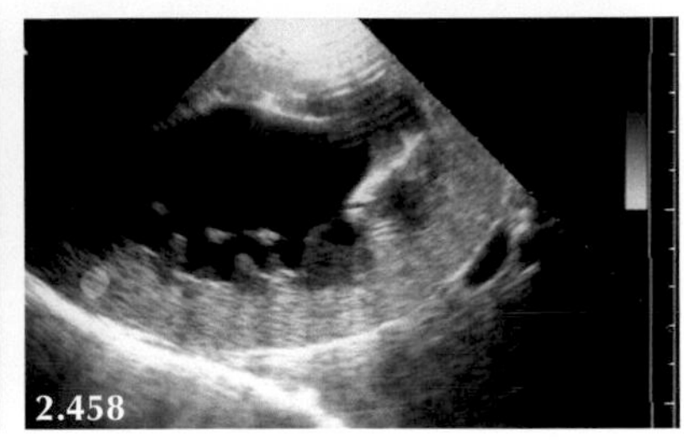

2.456: Urinary bladder is >14 cm diameter in this 40 kg ram lamb. **2.457:** Ultrasound examination shows slight enlargement of the kidney due to increased distension of the renal pelvis, but these changes are not clinically significant. **2.458:** Enlarged kidney with dilated renal pelvis consistent with hydronephrosis. 2.217, 2.218, 2.219, 2.220

Necropsy

Necropsy reveals that bladder rupture is very uncommon in sheep with urolithiasis except for neglected cases; instead, back pressure causes hydroureters and hydronephrosis.

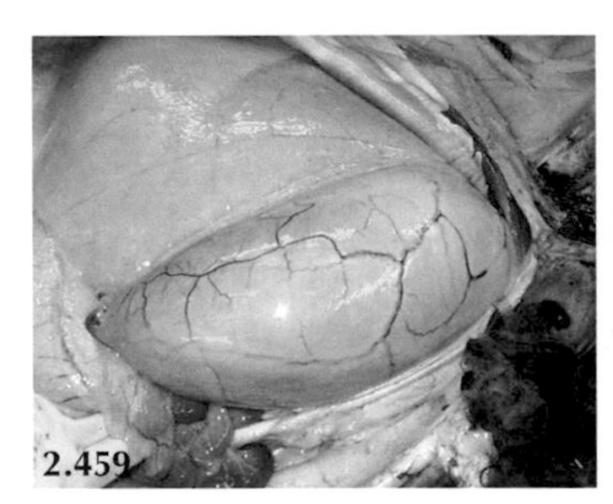

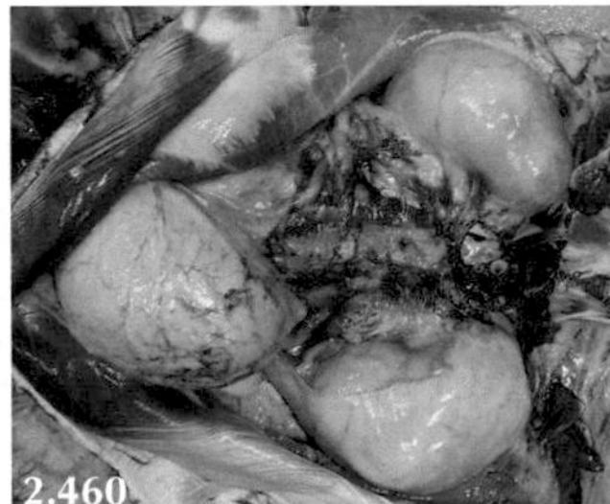

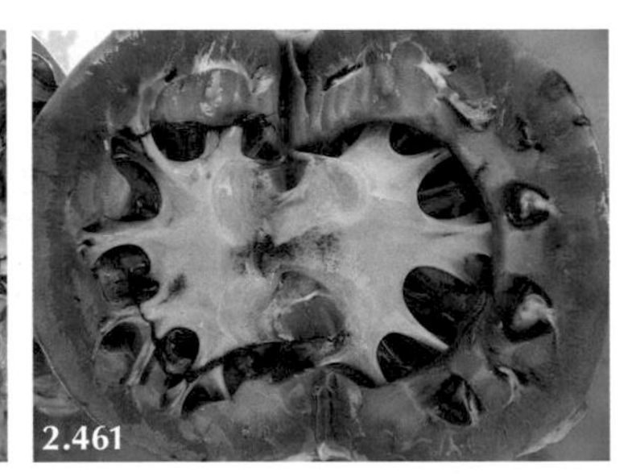

At necropsy. **2.459:** Ram positioned on its back reveals a massively distended but intact urinary bladder. **2.460:** Distended urinary bladder, bilateral hydroureters and hydronephrosis. **2.461:** Longitudinally sectioned kidney reveals marked hydronephrosis (distension reduced following release of urine on sectioning). 2.221

Examination of the Penis

- Subcutaneous urine accumulation extending caudally to the scrotum is observed following rupture of the penis in cattle and sheep.
- Balanoposthitis is very common in rams on lush pasture causing painful lesions.
- Paraphimosis occurs during the breeding period in rams due to wrapped wool.

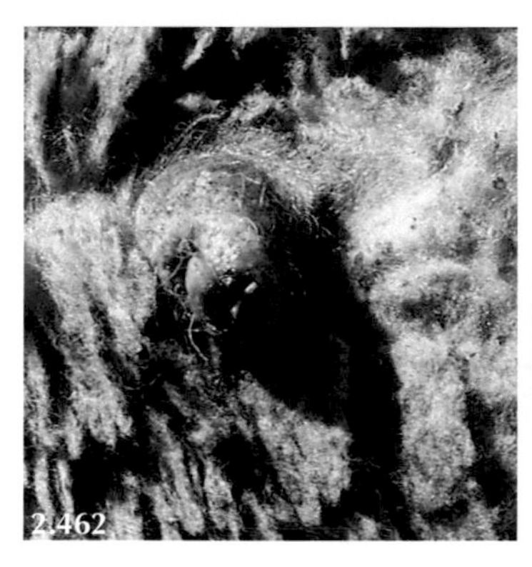

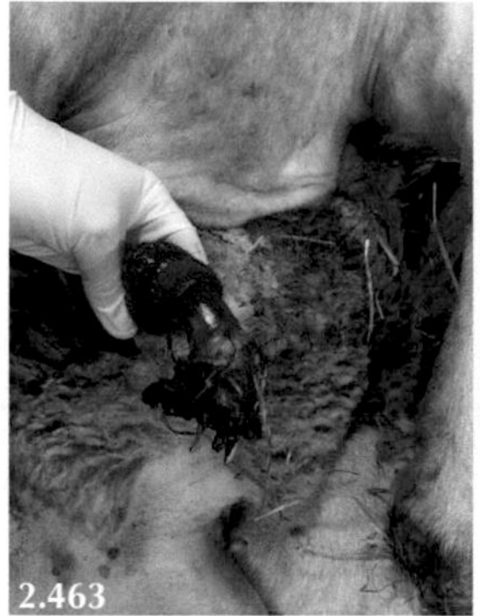

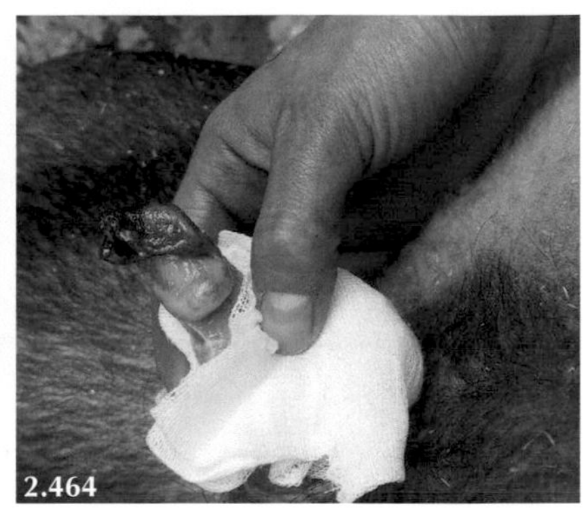

2.462: Ulcerated and oedematous preputial skin in a ram. **2.463:** The ram is positioned on its rump and the sigmoid flexure extended to examine the penis. **2.464:** The sigmoid flexure is extended and the penis held proximal to the glans using a gauze swab. 2.222, 2.223

2.8

Examination of the Skin/Fleece

Ectoparasite infestations are very common in sheep with sheep scab and pediculosis causing serious welfare issues despite simple, cheap and effective control measures. The common observation of herd-wide severe louse infestation in beef cattle suggests a similar attitude problem by some cattle farmers.

- Inspect all groups on the farm; hair/wool loss and pruritus are the key signs of ectoparasite infestations.
- Lice on cattle and sheep skin are visible to the naked eye.
- Skin scraping for sheep scab should be collected from the periphery of lesions; take plenty of samples.
- Be aware of the behavioural signs suggestive of cutaneous myiasis and headfly in sheep.
- During group inspection, fleece and hair quality ("coat") can be a good indicator of overall body condition especially in sheep.

Diseases of the skin are important in sheep for numerous reasons; wool yield and quality are adversely affected which is the major product of the sheep industry in some countries, and generalised infections and infestations such as sheep scab and cutaneous myiasis can lead to debility and possibly death. Despite highly effective control programmes, ectoparasite infestations are common in the UK where they cause economic loss and serious sheep welfare concerns. Ticks are also vectors for bacterial, rickettsial and viral infections in sheep which may cause serious disease outbreaks. Superficial bacterial infection of the skin (dermatophilosis) is common in debilitated sheep.

DOI: 10.1201/9781003106456-16

2.465: Sheep scab. **2.466:** Cutaneous myiasis. **2.467:** Pediculosis are common causes of serious animal welfare concerns in sheep.

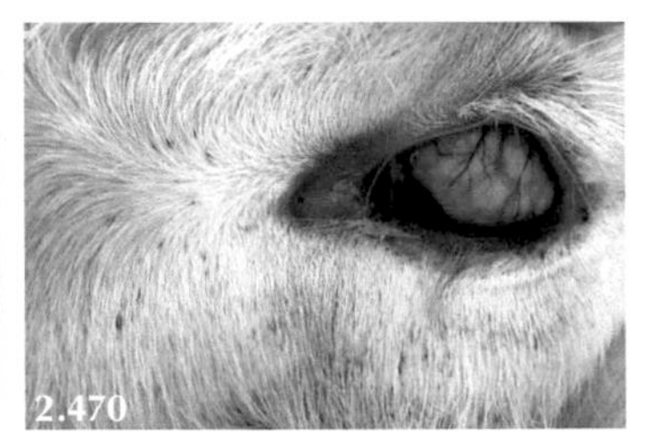

2.468 and **2.469:** Hair loss caused by rubbing in cattle heavily infested with lice. **2.470:** The short grey objects in the hair at the medial canthus of this calf are chewing lice.

Observation of Group

Grazing behaviour is disrupted by ectoparasite infestations. Affected sheep often isolate themselves and remain in shade where available. Sheep affected by headfly stand with the head held lowered with frequent head shaking and ear movements. Kicking at the head often greatly exacerbates damage caused by headflies. Alternatively, sheep adopt a submissive posture in sternal recumbency with the neck extended and the head held on the ground.

Flock inspection will reveal behavioural signs characteristic of ectoparasite infestations. **2.471:** Advanced sheep scab. **2.472:** Lice. **2.473:** Headfly. 2.224

Sheep with blowfly strike appear agitated and are frequently observed away from the group. They only graze for 5 to 10 seconds then suddenly trot away for 10–20 metres with frequent tail swishing before re-commencing grazing. They frequently turn their head and nibble at the wool over the tailhead. 2.225

Infestation with *Bovicola ovis* may cause disrupted feeding patterns, with chewing at the fleece over the flanks causing "wool plucks" and rubbing against fences causing fleece damage/loss. 2.226

During the early stages of sheep scab infestation sheep have disturbed grazing patterns and are observed kicking at their flanks with their hind feet and/or rubbing themselves against fence posts which leads to a dirty ragged appearance to the fleece and loss of wool. There is serum exudation which gives the fleece overlying the skin lesions a moist yellow appearance. The skin lesions are most commonly observed on the flanks but may extend to involve the whole body in neglected cases. Typically, after eight weeks' infestation or so the hair loss on the flanks may extend to 20–30 cm diameter surrounded by an area of hyperaemia and serum exudation at the periphery. The skin is thrown into thickened corrugations in advanced cases. 🎥 2.227

Skin Inspection

Blowfly eggs are deposited on soiled areas of the fleece or adjacent to traumatised skin. Lesions may range from centimetre diameter areas of skin hyperaemia with a small number of maggots when detected during the first days of infestation to extensive areas of traumatised skin surrounded by moist, foul-smelling wool.

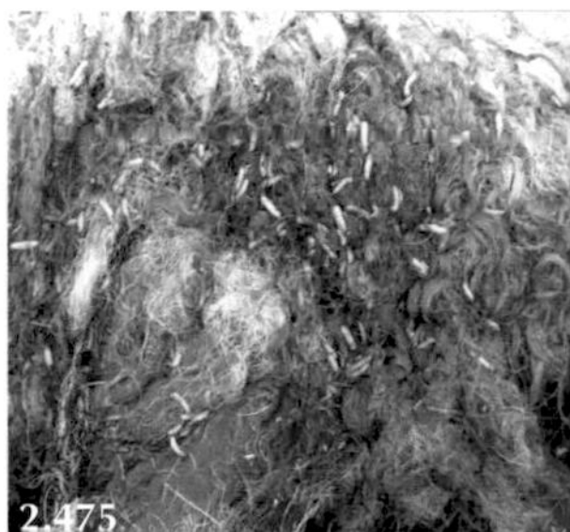

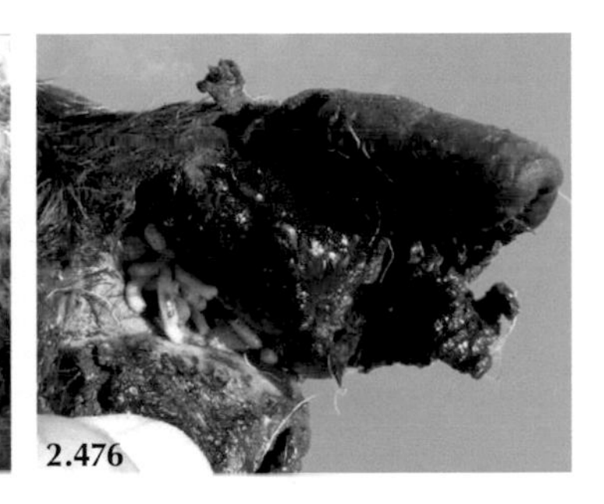

2.474: Presence of bluebottles and large numbers of eggs on the skin. **2.475:** Cutaneous myiasis with blackened skin surrounded by moist, foul-smelling wool. **2.476:** Myiasis affecting the interdigital space of ewe with footrot.

Careful inspection of the fleece over the tailhead and along the dorsum using a magnifying glass will identify significant *B. ovis* populations. Lice in cattle are most commonly observed on the face, neck and scrotum.

Skin Scrapings

Cutaneous stimulation of sheep with scab during handling procedures may precipitate seizure activity which lasts for 2 to 5 minutes before animals recover fully. Skin scrapings taken using a scalpel drawn held at a right angle to the skin surface at the periphery of active lesions demonstrate large numbers of mites under ×100 magnification. Liquid paraffin can first be applied to the skin to aid collection of the scrapings. A similar technique is used for chorioptic mange affecting the tailhead of cattle. Prior digestion of the sample in potassium hydroxide is not usually necessary.

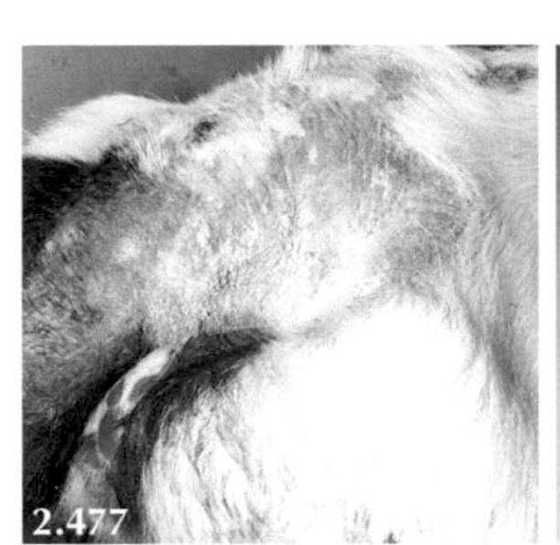

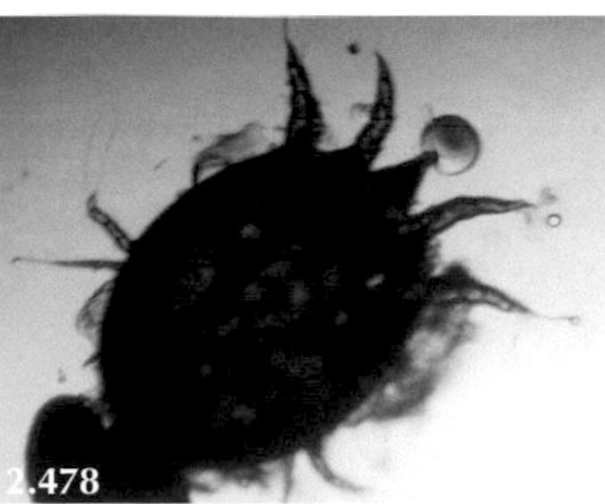

2.477: Chorioptic mange affecting the skin at the tailhead of this dairy cow.
2.478: *Psoroptes ovis* mite from a case of sheep scab.

Palpation

Examination/palpation of the skin pays particular attention to the superficial lymph nodes including the parotid, submandibular, prescapular, popliteal and precrural lymph nodes. Caseous lymphadenitis, caused by *Corynebacterium pseudotuberculosis*, is occasionally seen affecting the parotid lymph node especially in rams following fighting injury to the poll.

Punch Biopsy and Histopathology

Skin and tissue samples can be readily collected under local anaesthesia although skin tumours are uncommon in sheep and cattle.

Season

The life cycle of lice is the egg, three nymph stages and the adults spend their entire time on the sheep. The slow reproductive capacity of *Bovicola ovis* during the colder months results in a gradual build-up of louse numbers during late autumn/winter. This seasonal peak during late winter also applies to cattle. Sheep scab, caused by *Psoroptes ovis*, is typically encountered during the autumn/winter months from September to April. In the UK, headflies and blowflies are generally active from May to September depending upon daytime temperature.

2.479: Lice in winter months. **2.480:** Sheep scab in winter on lambs grazing stubble turnips. **2.481:** Headfly in mid-summer.

Poor Wool Quality in Individual Ewes

Wool quality is directly affected by nutrition, and poor wool quality is commonly encountered in sheep with low condition scores.

2.482: Ewe with low body condition and poor quality "carpet-like" fleece.
2.483: Dermatophilosis skin lesions on face and ears are common in emaciated sheep.
🎥 2.228, 2.229, 2.230, 2.231

2.9

Examination of the Udder (Sheep)

- The udder is routinely checked for gross lesions (abscesses) before the start of the breeding season.
- Milk production is only determined indirectly by observing lamb growth rate during the first 4–6 weeks but is affected by many factors not least ewe nutrition.
- Gangrenous mastitis is a major cause of ewe mortality and premature culling.
- Somatic cell counts, California mastitis test and bacteriology are rarely undertaken.
- The role of MVV in the aetiology of indurative mastitis is considered by some to be a major problem in the UK, but there are insufficient clinical data at present to be certain that a seropositive test represents significant pathology and production loss.

Examination of the udder is rarely undertaken except to confirm clinical mastitis which is indicated by gaunt hungry lambs making frequent attempts to suck and, in severe cases, ipsilateral lameness with the ewe dragging the leg.

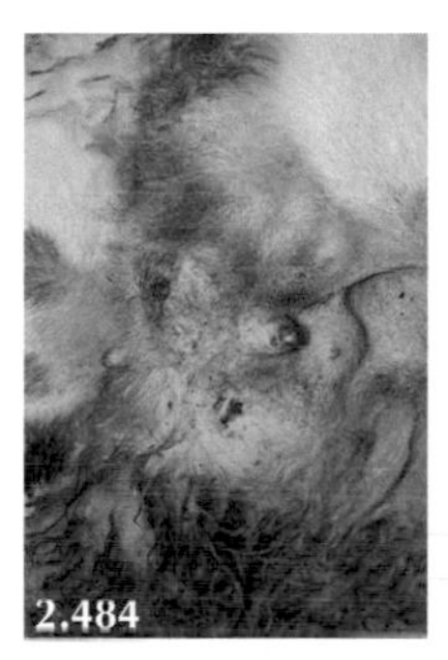

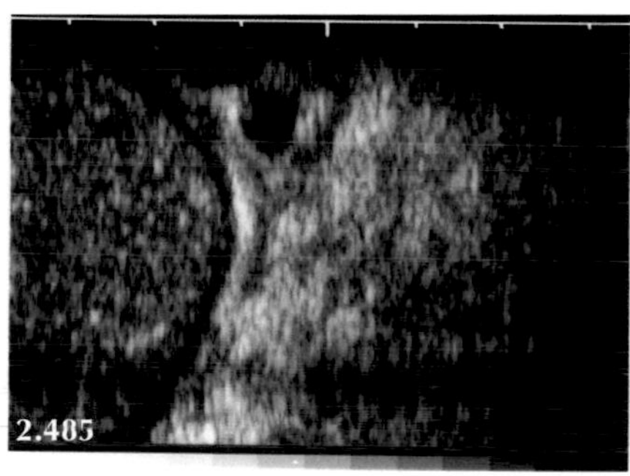

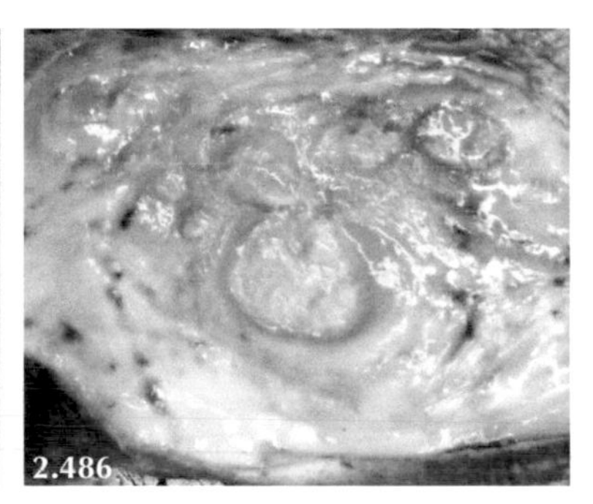

2.484: Several 2 cm diameter firm swelling within the udder palpated before the breeding season. **2.485:** 3 cm diameter well-encapsulated udder abscess demonstrated ultrasonographically (probe head at right hand margin of image). **2.486:** Confirmation of the sonographic findings at necropsy.

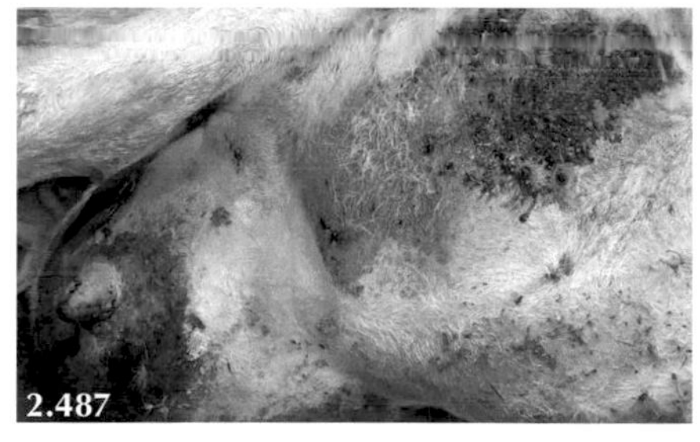

2.487: Gangrenous mastitis with thrombophlebitis of the drainage mammary vein. **2.488:** Fibrous pegs left after the gangrenous gland has been sloughed. 2.232, 2.233, 2.234

DOI: 10.1201/9781003106456-17

Mastitis Detection in Dairy Cows

Stripping foremilk to detect mastitis is an integral part of the milking routine.

The generally accepted somatic cell count threshold (mostly leucocytes) for a healthy cow is <100,000; >200,000 cells/ml of milk indicates mastitis, with higher cell counts used as an indicator of the severity of infection.

Individual cow somatic cell counts are usually performed monthly.

There are a number of computer programmes designed to interpret somatic cell count and mastitis data.

Automated mastitis detection (AMD) systems use in-line sensors to monitor milk from individual cows.

Part 3

Clinical Problems

3.1

Blindness

Common Causes of Blindness in Adult Sheep

- Pregnancy toxaemia is common in underfed multiparous, twin- and triplet-bearing ewes.
- Polioencephalomalacia is common in weaned lambs and young adult sheep.
- Bilateral ocular lesions are common arising from storm weather conditions (IKC) and silage feeding (iritis).

TABLE 3.1

Common Causes of Blindness in Adult Sheep

	Mentation	Pupillary light reflex
Pregnancy toxaemia	–/––	NAD
Polioencephalomalacia	–/+++	NAD
Infectious kerato-conjunctivitis (severe)	NAD/+	
Iritis	NAD/+	
Closantel toxicity	+/++	–––
Sulphur toxicity	–/––	NAD
Pituitary tumour	–/––	–––
Localised infection at optic chiasma	+/++	–––

TABLE 3.2

Ancillary Tests for Common Causes of Blindness in Adult Sheep

Disease	Ancillary tests
Pregnancy toxaemia	Serum 3-OH butyrate >4.0 mmol/l. Rothera's test positive
Polioencephalomalacia	Response to vitamin B1. Necropsy/histopathology
Infectious kerato-conjunctivitis	Response to topical/parenteral antibiotics
Iritis	Response to topical/parenteral antibiotics
Closantel toxicity	Histopathology of optic nerve
Sulphur toxicity	Necropsy/histopathology. Feed/water analysis
Tumour	Necropsy/histopathology
Localised infection at optic chiasma	Necropsy

DOI: 10.1201/9781003106456-19

Pregnancy Toxaemia

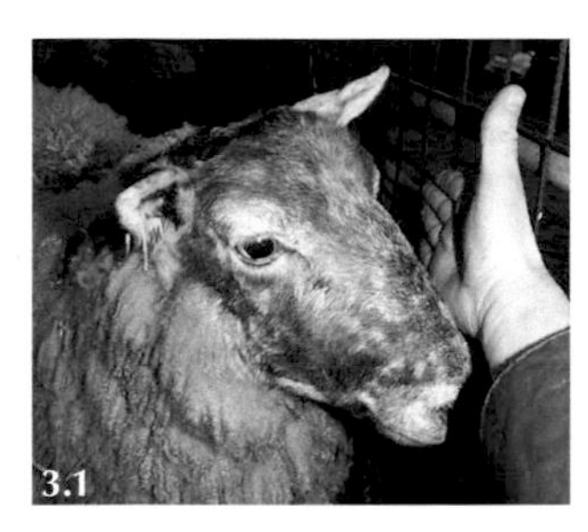

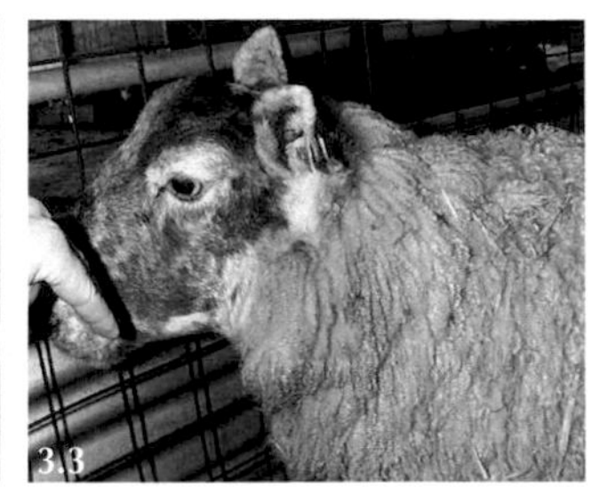

3.1–3.3: All sheep show reduced mentation and absence of menace response (approaching hand).

- **Key observations**: Multiparous, >2 lambs, last 2–3 weeks of pregnancy.
- **Clinical examination**: Often poor BCS caused by chronic energy deficiency, distended abdomen due to large fetal load (often triplets).
- **Key history events**: Several ewes affected—flock problem of inadequate energy supply, sudden adverse weather, recent housing. Individual sheep—problem with access to feed such as severe lameness, poor dentition etc. 🎥 3.1

Polioencephalomalacia

3.4–3.6: All sheep show lack of menace response but normal pupillary light reflexes (not shown). 🎥 3.2

- **Key observations**: Dorsiflexion of neck, hyperaesthetic to touch/sound, rapid progression over 12–24 hours to seizure activity and opisthotonos. 🎥 3.3
- **Clinical examination**: Dorsomedial strabismus. Normal pupillary light reflexes.
- **Key history events**: Pasture/feeding change 2–3 weeks previously.

Infectious Kerato-Conjunctivitis

3.7: Early case of IKC showing blepharospasm, photophobia and a serous ocular discharge. Vision would be reduced, but this ewe would not be blind. **3.8:** IKC is more common after winter storms ("snow blindness"). **3.9:** Marked scleral vessels; developing keratitis/opacity impairs vision with blindness in severe cases.

- **Key observations**: Blindness in (bilateral) severe cases, blepharospasm, photophobia, serous ocular discharge.
- **Clinical examination**: Marked conjunctivitis, keratitis.
- **Key history events**: Adverse weather during winter. Close contact at feeding troughs. 🎥 3.4, 3.5, 3.6

Iritis

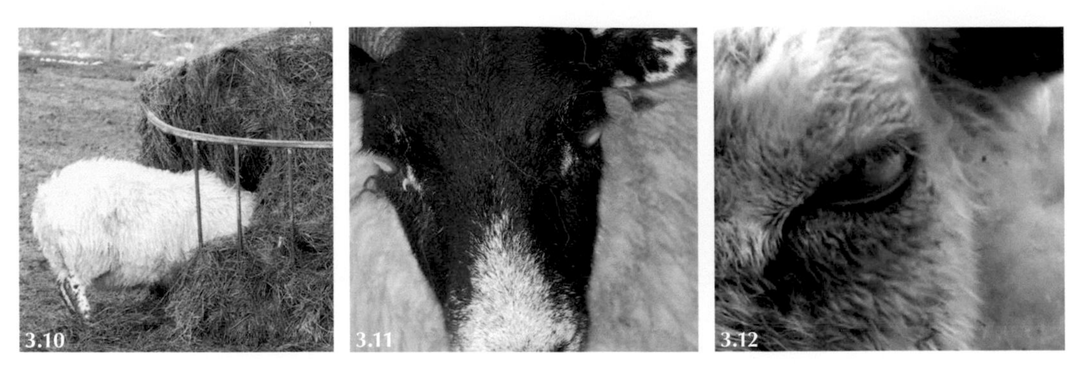

3.10: Iritis is associated with (big bale) silage feeding. **3.11** and **3.12:** Blepharospasm, photophobia, serous ocular discharge and iritis.

- **Clinical examination**: Anterior uveitis.
- **Key history events**: Often winter months associated with big bale silage feeding. 3.7, 3.8, 3.9

Closantel Toxicity

3.13 and **3.14:** Normal mentation, but sheep are blind and easily startled by sound and touch. **3.15:** Dilated pupil showing absence of pupillary light reflex.

- **Key observations**: Often several animals affected. Easily startled—normal mentation but blind. For animal welfare reasons—pen together with other sheep.
- **Clinical examination**: Normal appetite. No ocular lesions but pupillary light reflex absent.
- **Key history events**: Overdose with closantel 3–10 days previously (low safety index). 3.10

Sulphur Toxicity

3.16–3.18: Suffolk shearling ram being fed intensively for sale. The ram is blind and wanders aimlessly.

- **Key observations**: Dull, blind, wanders aimlessly. No response to vitamin B1 treatment but arrests progress of clinical signs.
- **Key history events**: Several sheep affected. High sulphur content in concentrate feed or ground water supply.

Pituitary Tumour

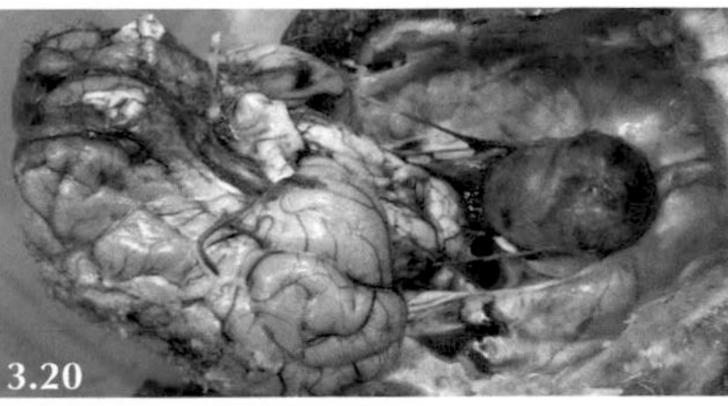

3.19: Aged Cheviot ewe with bilateral II and III cranial nerve deficits caused by a pituitary tumour. **3.20:** Necropsy reveals the pituitary tumour pressing on cranial nerves II and III.

- **Key observations**: Single blind older sheep, poor appetite and weight loss.
- **Clinical examination**: No pupillary light reflexes. Dullness may be caused by compression of the cerebrum from the tumour mass.
- **Key history events**: Slow progression over several/many weeks. Chronic weight loss. 🎥 3.11

Abscess at the Optic Chiasma

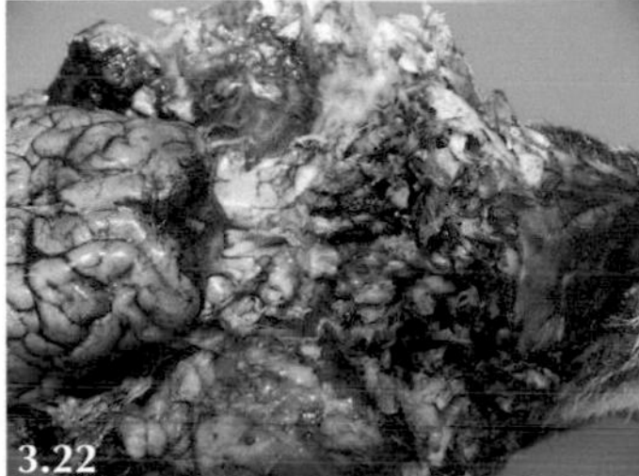

3.21: Affected ewe in foreground—note startled expression compared to normal sheep. **3.22:** Abscess causing bone lysis (black area in centre of image) in region of the optic chiasma.

- **Key observations**: Blind, easily startled.
- **Clinical examination**: No pupillary light reflexes.
- **Key history events**: Likely infection tracking from frontal sinuses. 🎥 3.12

Common Causes of Blindness in Lambs and Growing Sheep

- Polioencephalomalacia is common in weaned lambs.
- Entropion is common in some breeds; lack of attention leads to physical trauma to the eye causing severe conjunctivitis and possible rupture of the eyeball.
- Ocular lesions are common arising from storm weather conditions (IKC) and silage feeding (iritis).

TABLE 3.3

Common Causes of Blindness in Lambs and Growing Sheep

	Age	Mentation	Pupillary light reflex
Polioencephalomalacia	>4 months	+/+++	NAD
Entropion (severe, untreated)	At birth	NAD	
Closantel toxicity	After drenching	+/++	−−−
Meningitis	2–3 weeks	−−/+++	NAD
Microphthalmia	At birth	NAD	
Sulphur toxicity	From ration	−/−−	NAD

TABLE 3.4

Ancillary Tests for Common Causes of Blindness in Lambs and Growing Sheep

Disease	Ancillary tests
Polioencephalomalacia	Response to vitamin B1, necropsy/histopathology
Entropion (severe, untreated)	Clinical examination
Closantel toxicity	Histopathology of optic nerve
Meningitis	Increased lumbar CSF protein, neutrophilic pleocytosis
Microphthalmia	Progeny of one ram

Polioencephalomalacia

3.23–3.25: Dorsiflexion of neck, lack of menace response but normal PLRs.

- **Key observations**: Dorsiflexion of neck, hyperaesthetic to touch/sound, progression to seizure activity after 1–2 days.
- **Clinical examination**: Dorsomedial strabismus. Normal PLRs.
- **Key history events**: Pasture/feeding change 2–3 weeks previously. 🎥 3.13, 3.14, 3.15

Entropion (Severe, Untreated)

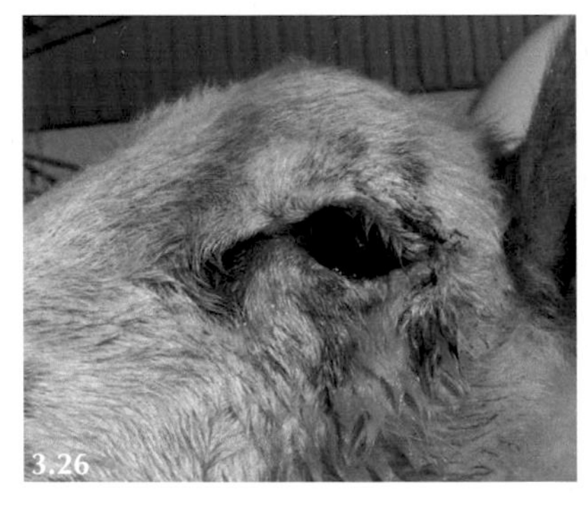

3.26–3.28: Unilateral blindness in these three lambs with increasing severity/duration of pathology; corneal ulceration, hypopyon, then rupture of the anterior chamber.

- **Key observations**: Usually present at birth or within 24 hours. Ocular discharge, tear-staining of face.
- **Clinical examination**: Exposure keratitis. Injected scleral vessels.
- **Key history events**: Progeny of particular ram. Common in certain breeds. 🎥 3.16, 3.17, 3.18, 3.19

Closantel Toxicity

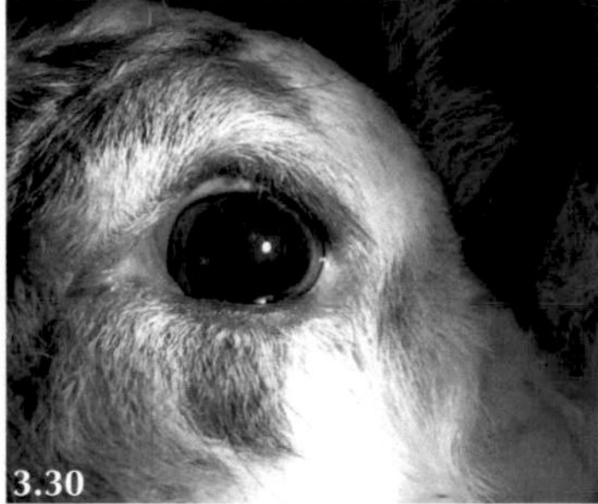

3.29: Normal mentation, but sheep are blind and easily startled by sound and touch.
3.30: Absence of pupillary light reflex.

- **Key observations**: Often several animals affected. Easily startled—normal mentation but blind (pen together as a group).
- **Clinical examination**: Normal appetite. No ocular lesions, pupillary light reflexes absent.
- **Key history events**: Overdose with closantel 3–10 days previously (low safety index).

Meningitis

3.31–3.33: Three-week-old lambs with altered mentation and blindness before rapid progression to opisthotonos and seizures.

- **Key observations**: Often unusual gait, observed walking backwards. Initially dull, not sucking, separated from dam.
- **Clinical examination**: Hyperaesthesia, normal PLRs, rapid deterioration in clinical signs.
- **Key history events**: Around 2–3 weeks old.

Microphthalmia

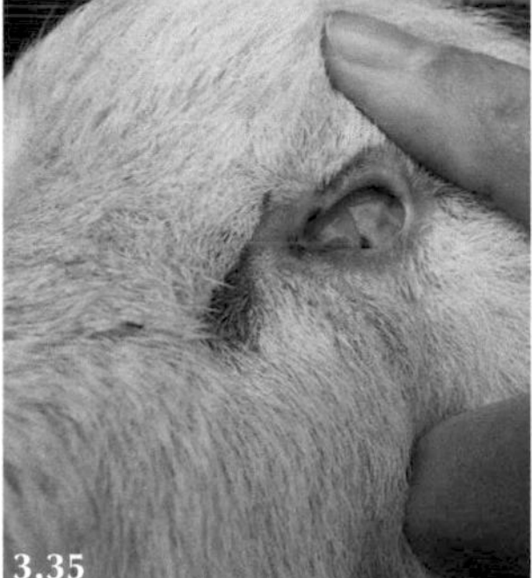

3.34–3.36: Microphthalmia, domed skulls in three neonatal Texel lambs.

- **Key observations**: Blind, microphthalmia, slightly domed skulls.
- **Key history events**: Autosomal recessive condition of Texel breed, progeny of one sire. 🎥 3.20

Common Causes of Blindness in Growing and Adult Cattle

- Housed cattle with vitamin A deficiency have often been blind for several months and it is only noted when the cattle leave their pen and encounter a new environment.

TABLE 3.5

Common Causes of Blindness in Growing and Adult Cattle

	Mentation	Pupillary light reflex
Lead poisoning	−−/+++	NAD
Vitamin A deficiency	NAD	−−−
Polioencephalomalacia	+/++	NAD
Iritis	NAD/+	

TABLE 3.6

Ancillary Tests for Common Causes of Blindness in Growing and Adult Cattle

Disease	Ancillary tests
Lead poisoning	Blood lead >0.35 ppm, >10 ppm in kidney cortex or liver
Vitamin A deficiency	Plasma concentration < 5µg/dl
Polioencephalomalacia	Response to vitamin B1. Necropsy/histopathology
Iritis	Response to topical/parenteral antibiotics

Lead Poisoning

3.37 and **3.38:** Yearling steer dull with vacant expression and head pressing behaviour. **3.39:** Discarded car batteries found in field grazed by this steer.

- **Key observations**: May also show of salivation, colic, constipation/tenesmus.
- **Clinical examination**: Blindness, with disease progressing to bellowing, convulsions and seizures in untreated cases.
- **Key history events**: Grazing youngstock with access to lead source such as discarded car batteries. 🎥 3.21

Optic Nerve (Vitamin A Deficiency)

3.40: Intensively reared bull beef animals; the bull on the left is blind and has walked into the gate.

- **Key observations**: Blindness often not noted when in pen. Walk into gates/fences when moved into strange environment.
- **Clinical examination**: Blind, absent PLRs, papilloedema.
- **Key history events**: Intensive cereal ration, insufficient vitamin A supplementation. 3.22

PEM

3.41: This intensively reared young beef bull is blind and presents in sternal recumbency. **3.42:** After treatment the animal was assisted to its feet but wandered aimlessly in the pen.

- **Key observations**: Aimless wandering, walk into corners, incoordination.
- **Clinical examination**: Cortical blindness. Normal PLRs.
- **Key history events**: Often cattle fed intensive rations. 3.23

Iritis (Silage Eye)

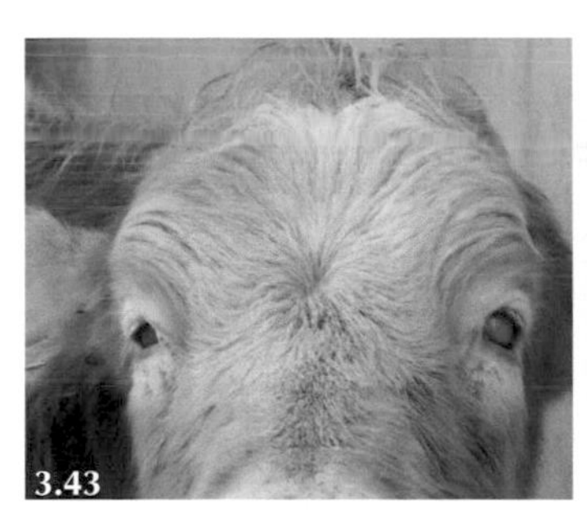

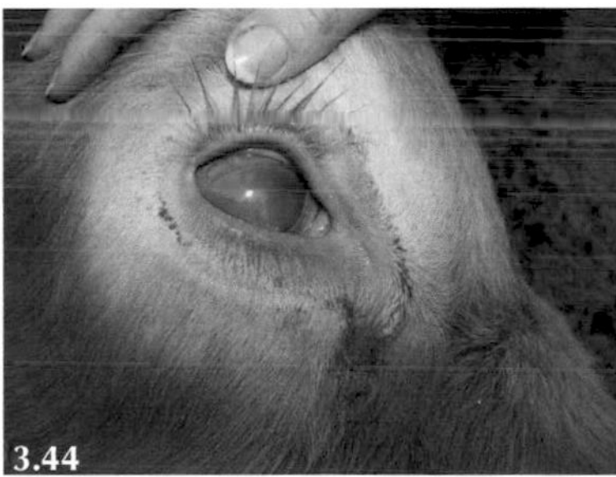

3.43 and **3.44:** Marked anterior uveitis.

- **Key observations**: "Blue eye".
- **Clinical examination**: Anterior uveitis with fibrinous material in anterior chamber.
- **Key history events**: Usually associated with big bale silage feeding during winter months.

3.2

Behavioural Changes, Gait Abnormalities, Cranial Nerve Deficits, Head Tilt

Cerebral Syndrome in Sheep

- The common clinical signs of diffuse cerebral dysfunction include behavioural changes which range from coma to seizure activity and opisthotonos, and often blindness. The likely duration of clinical signs is indicated in the table that follows to illustrate how the clinical presentation changes over time especially for PEM and bacterial meningitis.
- Clinical causes of signs attributable to a cerebral lesion localised to one cerebral hemisphere, including circling and contralateral blindness, are commonly caused by an abscess or coenurosis.

TABLE 3.7

Common Differential Diagnoses of Behavioural Changes in Sheep

	Likely duration of illness	Mentation	Responses	Seizures
PEM (early)	Hours	–/––	+	NAD
PEM (late)	Days	++/+++	++/+++	++/+++
Pregnancy toxaemia	Days	–/––	–/––	NAD
Hypocalcaemia	Hours	–/––	–/––	NAD
Meningitis (early)	Hours	–/–––	NAD	NAD
Meningitis (late)	Days	++/+++	++/+++	++/+++
Sulphur toxicity	Days	–/––	–/––	NAD
Sarcocystosis	Hours/days	+/++	+/+++	+/+++

TABLE 3.8

Laboratory Tests for Behavioural Changes in Sheep

Disease	Laboratory tests
PEM (early)	Response to vitamin B1
PEM (late)	Response to vitamin B1, UV autofluorescence at necropsy. Histology
Pregnancy toxaemia	Serum 3-OH butyrate concentration >3–4.0 mmol/l. Rothera's test
Meningitis	Increased lumbar CSF protein concentration, neutrophilic pleocytosis
Sulphur toxicity	Brain histology
Sarcocystis	Increased CSF eosinophil number and percentage, histology. Immunohistochemistry

DOI: 10.1201/9781003106456-20

PEM (Early Stages)

3.45 and **3.46:** Suffolk gimmer isolated from group and shows no menace response. **3.47:** Six-month-old Scottish Blackface lamb with dorsi-flexion of the neck "star gazing" and lack of menace response.

- **Key observations**: Isolated, aimless wandering, blind.
- **Clinical examination**: Bilateral lack of menace response, normal pupillary light reflexes, dorsomedial strabismus, dorsi-flexion of neck ("star gazing").
- **Key history events**: Often follows weaning, change in pasture/other management two to three weeks previously. 🎥 3.24, 3.25, 3.26, 3.27, 3.28

PEM (Late Stages)

3.48: Blindness, seizure activity, opisthotonos affecting gimmer two weeks after pasture change. **3.49** and **3.50:** Scottish Blackface ram lamb before and after treatment for PEM.

- **Key observations**: Lateral recumbency, opisthotonus, seizure activity.
- **Clinical examination**: Hyperaesthetic to touch/sound precipitating seizures.
- **Key history events**: Often follows weaning, change in pasture/other management. 🎥 3.29

Pregnancy Toxaemia

3.51: Walk forward behaviour. **3.52:** Lack of menace response. **3.53:** Adverse weather conditions preventing feeding may precipitate pregnancy toxaemia in underfed ewes.

- **Key observations**: Isolated, dull, unresponsive.
- **Clinical examination**: Blind, normal PLR.
- **Key history events**: Multiple pregnancy, recent housing, extreme weather interruption to feeding. Individual sheep—lameness, vaginal prolapse. 3.30

Hypocalcaemia

3.54–3.56: 4-crop ewes due to lamb within two weeks. The ewes are dull, unresponsive, weak and unable to stand; bloat has developed (centre).

- **Key observations**: Dull and unresponsive, sternal recumbency, weak/unable to stand.
- **Clinical examination**: Afebrile, normal menace response and PLRs. No ruminal activity.
- **Key history events**: Late gestation, handled/housed (stressed) one to two days previously. 3.31, 3.32.

Meningitis (Lambs)

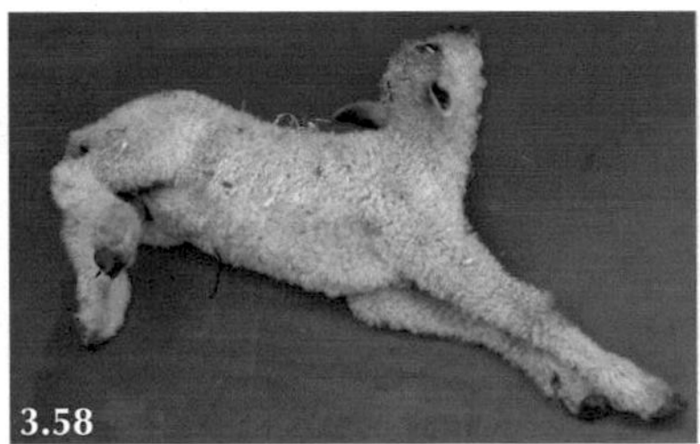

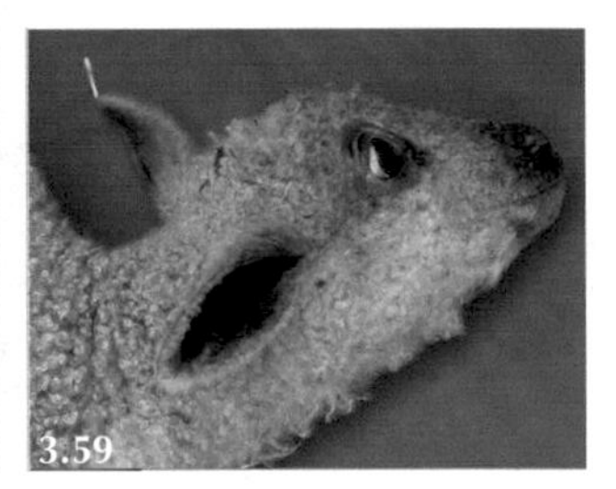

3.57: Obtunded three-week-old lamb. **3.58:** Seizure activity/opisthotonos. **3.59:** Dorsi-flexion of the neck and dorso-medial strabismus.

- **Key observations**: Short period of obtunded behaviour quickly progresses to seizures/opisthotonos.
- **Clinical examination**: Blind, normal PLR, hyperaesthetic to touch/sound precipitating seizures.
- **Key history events**: None. 3.33, 3.34

Sulphur Toxicity

3.60: Blind Suffolk shearling ram fed a diet high in sulphur.

- **Key observations**: Dull, unresponsive, blind, normal PLR. Signs do not progress unlike PEM.
- **Clinical examination**: Signs consistent with diffuse cerebral lesion.
- **Key history events**: Home-mixed rations with feed high in sulphur. Bore hole water high in sulphur.

Sarcocytosis

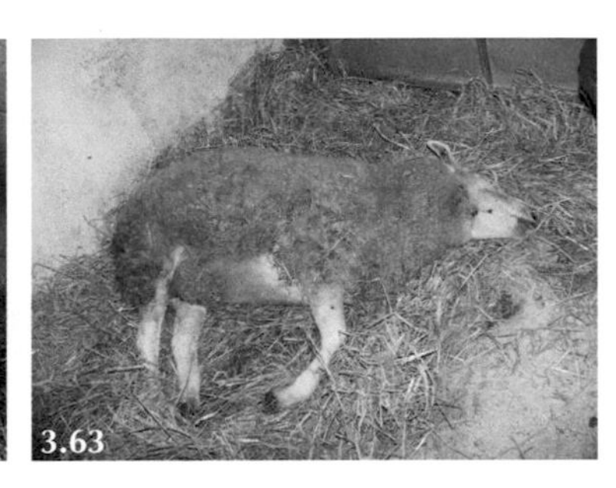

3.61 and **3.62:** Pelvic limb weakness. **3.63:** Seizure activity in a ewe.

- **Key observations**: Very variable ranging from pelvic limb weakness to seizures activity.
- **Clinical examination**: Signs consistent with spinal cord/diffuse cerebral lesions.
- **Key history events**: Recent access to pasture/roughage heavily contaminated by dog feces. 3.35

Scrapie

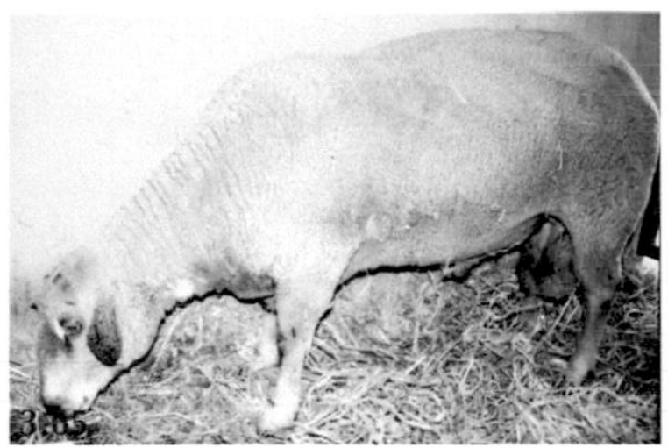

3.64–3.66: Head down, dull, detached behaviour in sheep with scrapie.

- **Key observations**: Dull, isolated from group, will drop into sternal recumbency when stressed returning to normal several minutes later. Chronic weight loss, rubbing leads to fleece loss, positive nibble response.
- **Clinical examination**: Behavioural signs consistent with diffuse cerebral lesion but normal menace and PLR.
- **Key history events**: Susceptible genotype.

Unilateral Cerebral Lesions

TABLE 3.9

Common Differential Diagnoses of Unilateral Cerebral Lesions in Sheep

	Duration	Mentation	Responses	Circling
Cerebral abscess	Weeks	-	NAD	+/++
Coenurosis	Weeks	-	NAD	+/+++

TABLE 3.10

Ancillary Tests for Unilateral Cerebral Lesions in Sheep

	Ancillary tests
Cerebral abscess	Increased CSF protein and white cell concentrations
Coenurosis	Softening of overlying skull, surgery

Cerebral Abscess (Affecting One Cerebral Hemisphere)

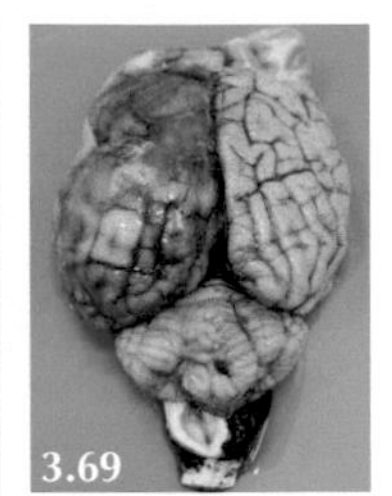

3.67: Ram does not see feed bucket from right eye. **3.68:** Normal vision in left eye and approaches feed bucket. **3.69:** Necropsy reveals abscess in left cerebral hemisphere.

- **Key observations**: Circling to one side.
- **Clinical examination**: Contralateral blindness and proprioceptive deficits but normal pupillary light reflexes.
- **Key history events**: None. 3.36

Coenurosis

- **Key observations**: Compulsive circling behaviour. Dullness, head-pressing behaviour.
- **Clinical examination**: Contralateral blindness and proprioceptive deficits but normal pupillary light reflexes if affects one cerebral hemisphere.
- **Key history events**: Contamination of pastures grazed by sheep with dogs infested with the tapeworm *Taenia multiceps*.

Cerebral Syndrome in Cattle

Cattle with a specific disease caused by cerebral dysfunction may show the full range of mentation changes from coma to seizure activity depending upon duration. Cattle may wander aimlessly and are often blind. See video recording which shows the typical clinical presentation of a cerebral lesion. 3.37

- The common conditions that present with diffuse cerebral dysfunction include metabolic diseases such as hypocalcaemia, nervous ketosis and hypomagnesaemic tetany. Other causes include PEM, lead poisoning and bacterial meningitis of neonates.
- Clinical signs attributable to a cerebral lesion localised to one cerebral hemisphere include abscessation.

TABLE 3.11

Common Differential Diagnoses of Behavioural Changes in Cattle

	Likely duration at presentation	Mentation	Responses	Seizures
Hypocalcaemia	Hours	–/––	–––	NAD
Hypomagnesaemia	Hours	+/+++	+/+++	+/+++
Nervous ketosis	Days	–/––	NAD	NAD
Lead poisoning	Hours	––/+++	+/++	+/+++
Meningitis	Hours	–/+++	++/+++	+/+++
PEM	Days	––/+	+/++	+/++
BSE	Week/months	–/+++	+/+++	NAD

TABLE 3.12

Laboratory Tests for Behavioural Changes in Cattle

Hypocalcaemia	Serum calcium concentration <1.2 mmol/l
Hypomagnesaemia	Serum magnesium concentration <0.4 mmol/l
Nervous ketosis	Serum 3-OH butyrate concentration >4.0 mmol/l
Lead poisoning	Blood lead >0.35 ppm, >10 ppm in kidney cortex or liver
Meningitis	Lumbar CSF increased protein concentration, massive neutrophilic pleocytosis
PEM	Response to vitamin B1, UV autofluorescence at necropsy
BSE	Brain histology—vacuolar change

Hypocalcaemia

3.70: Dairy cow with advanced milk fever, dull, weak/unable to stand and unresponsive. **3.71:** Beef cow with early sign of weakness and reduced awareness—compare with her calf's alert attitude. **3.72:** Twenty minutes after treatment of beef cow (centre).

- **Key observations**: Brief initial period of mild ataxia, hypersensitivity and excitability progressing to dullness and eventually coma. Head averted against flank.
- **Clinical examination**: Afebrile, generalized paresis, and circulatory collapse.
- **Key history events**: Immediately prior to, and within 24 hours, of parturition. 🎥 3.38, 3.39, 3.40

Hypomagnesaemic Tetany—Early Clinical Signs

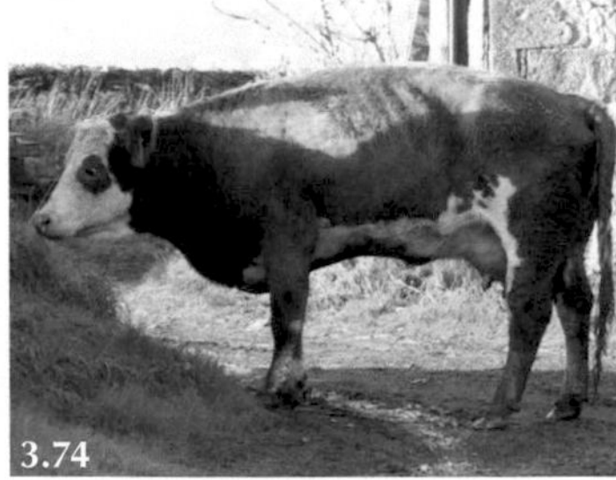

3.73: During the early stages of hypomagnesaemia cows can be aggressive—this cow charged the farmer and vet but was ataxic and fell over (shown after treatment). The frothy salivation was caused by frequent teeth grinding. **3.74** and **3.75:** Behavioural changes; this cow has a startled expression (best appreciated in Video 🎥 3.41).

Hypomagnesaemic Tetany—Advanced Clinical Signs

3.76–3.78: Serial images of same cow. **3.76:** Beef cow with hypomagnesaemic tetany; opisthotonus with seizure activity. **3.77:** Seizures controlled 10 minutes after veterinary treatment. **3.78:** Following day cow eating mineralised barley from a bucket.

- **Key observations**: Lateral recumbency. Often present with seizures.
- **Clinical examination**: Opisthotonus. Hyperaesthetic—approach/restraint for intravenous injection may precipitate seizure.
- **Key history events**: Beef cattle, lush pasture, stormy weather, no supplementation. Often four to eight weeks after calving. 🎥 3.42

Nervous Acetonaemia

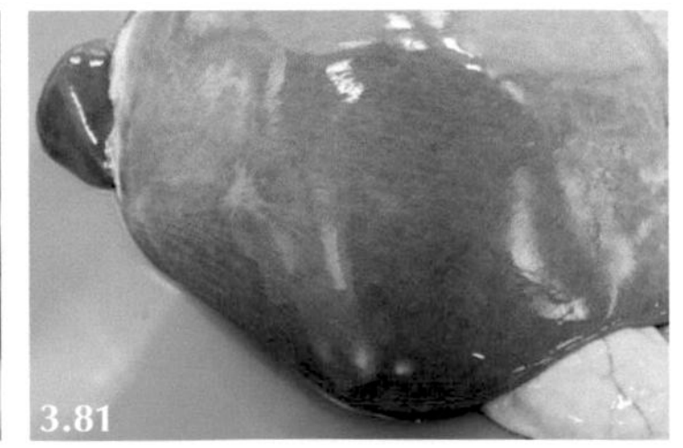

3.79: Aimless wandering/constant licking at flank. **3.80:** Severe fatty liver. **3.81:** Normal healthy liver contrasts with fatty liver (3.80).

- **Key observations**: Reduced appetite/milk yield, dull, unresponsive. Pica, manic licking/chewing of inanimate objects.
- **Clinical examination**: Ketotic smell on breath.
- **Key history events**: Two to four weeks post-calving. 🎥 3.43

Pregnancy Toxaemia in Beef Cattle

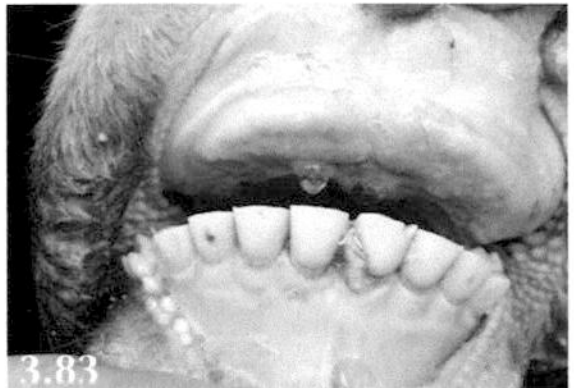
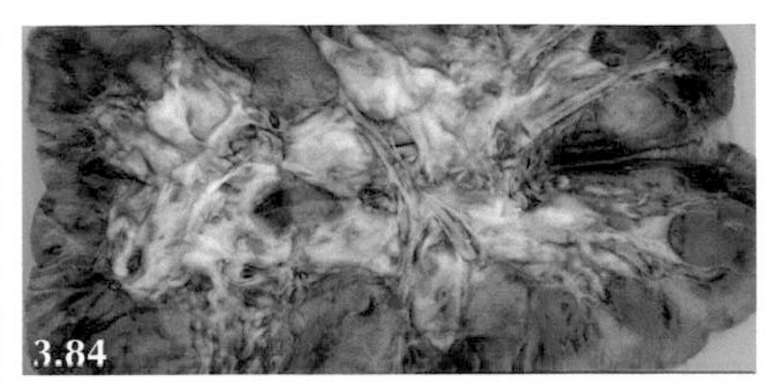

3.82: Weakness/recumbency in beef cow. **3.83:** Jaundice of the oral mucous membranes. **3.84:** Fatty change of the parenchymatous organs including the kidneys.

- **Key observations**: No appetite, dull, unresponsive, weak.
- **Clinical examination**: Jaundiced mucous membranes, developing peripheral oedema.
- **Key history events**: Sudden dietary energy restriction in twin-bearing beef cows. 3.44

Meningitis—Early Signs

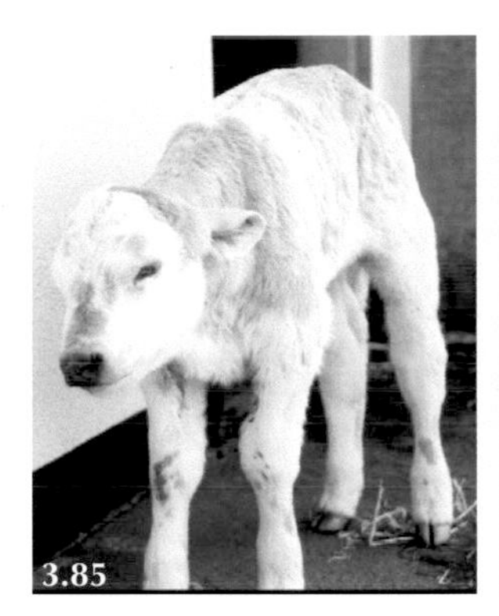
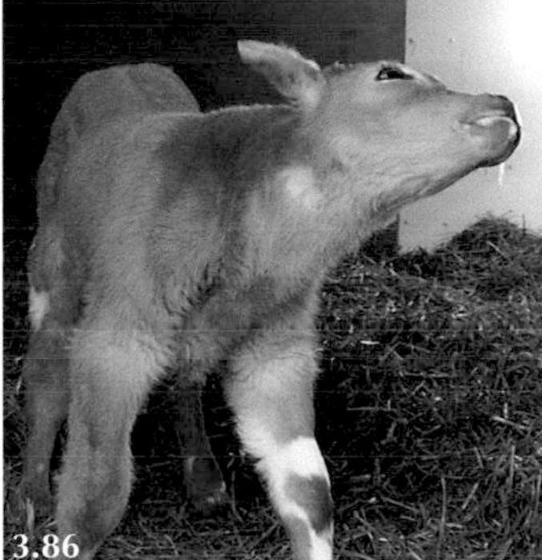
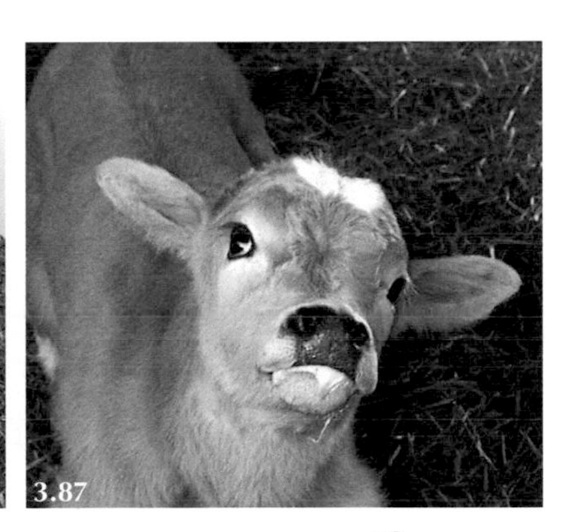

3.85–3.87: Stupor, blindness and dorsi-flexion of the neck during the early stages of bacterial meningitis. 3.45

Meningitis—Late Stage Signs

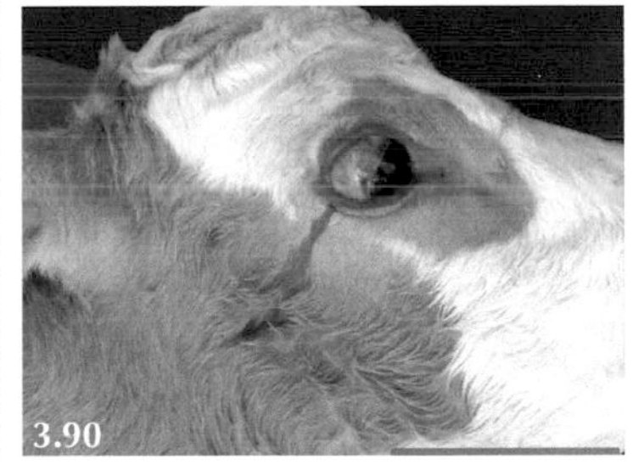

3.88 and **3.89:** Seizure activity during the latter stages. **3.90:** Dorso-medial strabismus. Injected scleral vessels are suggestive of septicaemia.

- **Key observations**: Short period of obtunded behaviour quickly progressing to seizures/opisthotonos.
- **Clinical examination**: Blind, normal PLR, hyperaesthetic to touch/sound precipitating seizures.
- **Key history events**: Failure of passive antibody transfer, dirty calving environment. 3.46, 3.47

Lead Poisoning

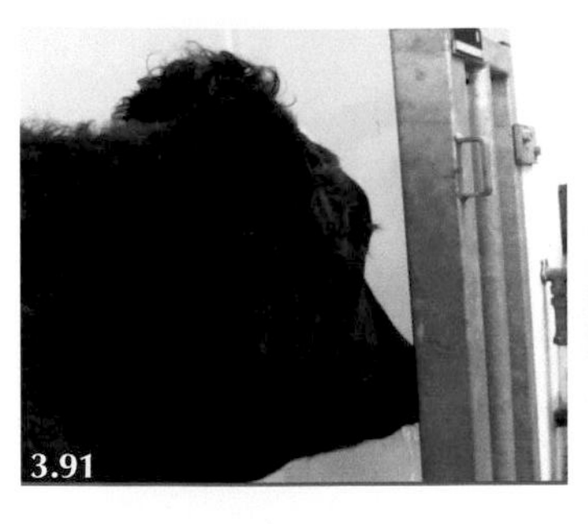
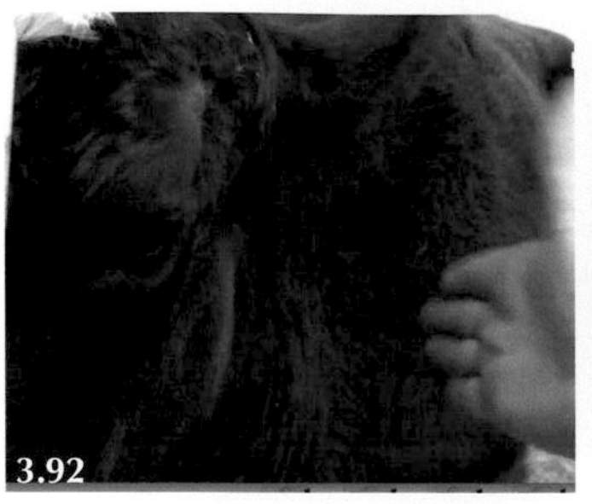

3.91 and **3.92:** Dull yearling steer with lack of menace response. **3.93:** Discarded car batteries in field grazed by this steer.

- **Key observations**: May also show GI tracts signs of salivation, colic, constipation/tenesmus.
- **Clinical examination**: Blindness, with disease progressing to bellowing, convulsions and seizures in untreated cases.
- **Key history events**: Grazing youngstock with access to lead source such as discarded car batteries. 3.48

PEM

3.94: This intensively reared young beef bull is blind and presents in sternal recumbency. **3.95:** After treatment the animal was assisted to its feet but wandered aimlessly in the pen.

- **Key observations**: Aimless wandering, walk into corners, incoordination.
- **Clinical examination**: Cortical blindness, normal PLRs.
- **Key history events**: Often cattle fed intensive rations. 3.49

Bovine Spongiform Encephalopathy

3.96: Cow in centre, head down, dull, detached behaviour not reacting to observer like other cows. **3.97:** Cow with arched back and not interested in observer unlike her calf.

- **Key observations**: Dull, isolated from group, altered behaviour, kicking at head, can become aggressive to other cows and when handled (syn. “mad cow disease”). Chronic weight loss, arched back.
- **Clinical examination**: Ataxia. Aggressive behaviour when handled can limit examination.
- **Key history events**: Access to prion-contaminated feed as youngstock. 3.50

Unilateral Cerebral Lesion

TABLE 3.13

Common Differential Diagnoses of Unilateral Cerebral Lesions in Cattle

Disease	Duration	Mentation	Responses	Circling
Cerebral abscess	Weeks/months	–/–––	NAD	+/++

TABLE 3.14

Ancillary Tests for Unilateral Cerebral Lesions in Cattle

Disease	Ancillary tests
Cerebral abscess	Increased CSF protein and white cell concentrations. Necropsy.

- **Key observations**: Circling to one side.
- **Clinical examination**: Contralateral blindness and proprioceptive deficits but normal pupillary light reflexes.
- **Key history events**: None.

Ataxia, Dysmetria and Wide-Based Stance—Cerebellar Lesions in Sheep

- Cerebellar disease is characterised by a wide-based stance and ataxia (incoordination), particularly of the pelvic limbs, but with preservation of normal muscle strength.
- Cerebellar lesions are uncommon in sheep except for congenital lesions caused by Border disease and other viruses (Schmallenburg and Bluetongue).
- Dysmetria (problems associated with stride) may be observed. Hypermetria, or overstepping, is seen as a Hackney-type gait. With hypometria, the animal will frequently drag the dorsal aspect of the hoof along the ground.

TABLE 3.15

Common Differential Diagnoses for Gait Abnormalities in Lambs and Growing Sheep

	Duration	Intention tremors
Border disease (lambs)	Present at birth	+/+++
Cerebellar abscess	Weeks	NAD
Cerebellar abiotrophy	Develops slowly from 2–3 months old	NAD

TABLE 3.16

Ancillary Tests for Cerebellar Lesions in Lambs and Growing Sheep

Border disease (lambs)	Virus isolation, necropsy
Cerebellar abscess	Increased CSF protein and white cell concentrations
Cerebellar abiotrophy	Necropsy/histology

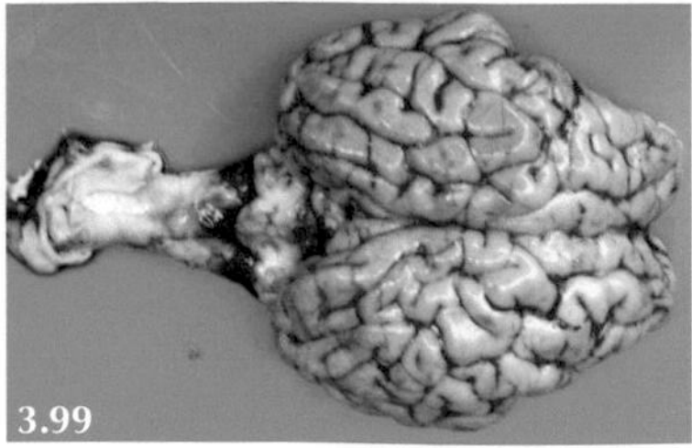

3.98: Neonatal lamb showing wide-based stance. **3.99:** Cerebellar hypoplasia revealed at necropsy. **3.100:** Poorly grown lamb persistently infected with Border disease virus.

- **Key observations**: Intention tremors precipitated on arousal.
- **Clinical examination**: Wide-based stance.
- **Key history events**: Several new-born lambs affected. Poor flock biosecurity. 3.51, 3.52

Cerebellar Abiotrophy

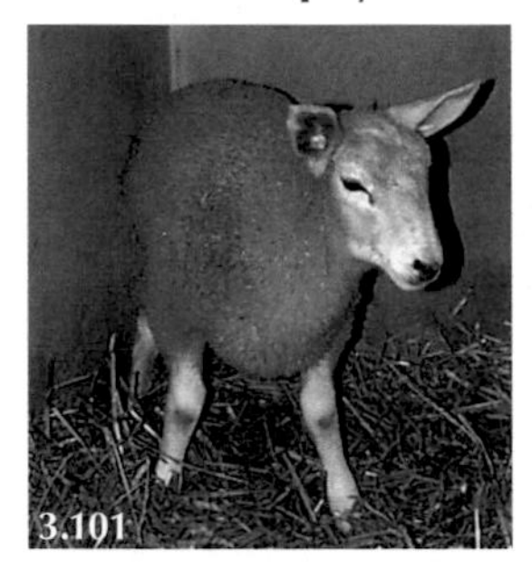

3.101: Wide-based stance in this growing lamb.

- **Key observations**: Insidious onset gait abnormalities.
- **Clinical examination**: Wide-based stance and ataxia.
- **Key history events**: Several lambs affected related to certain sire.

Cerebellar Abscess

3.102: Wide-based stance in this adult ram. Lowered head carriage cannot be appreciated in this image.

- **Key observations**: Insidious onset gait abnormalities.
- **Clinical examination**: Wide-based stance and ataxia.
- **Key history events**: None. 3.53, 3.54

Ataxia, Dysmetria and Wide-Based Stance—Cerebellar Lesions in Cattle

- Cerebellar disease is characterised by a wide-based stance and ataxia (incoordination), particularly of the pelvic limbs, but with preservation of normal muscle strength.

- Cerebellar lesions are common in calves where congenital lesions of cerebellar hypoplasia and hydranencephaly are caused by BVD and other viruses worldwide (Akabane).
- Occasional cases of cerebellar abscess are encountered.
- Dysmetria (problems associated with stride) may be observed. Hypermetria, or overstepping, is seen as a Hackney-type gait. With hypometria, the animal will frequently drag the dorsal aspect of the hoof along the ground.

Common Causes of Cerebellar Dysfunction

TABLE 3.17

Common Differential Diagnoses for Gait Abnormalities in Calves and Growing Cattle

	Onset	Wide-based stance	Ataxia
Cerebellar hypoplasia	Present at birth	++/+++	+/++
Cerebellar abscess	Weeks/months old	++/+++	+/++
Cerebellar abiotrophy	2–3 months old	+/+++	+/++

TABLE 3.18

Ancillary Tests for Cerebellar Lesions in Calves and Growing Cattle

Disease/disorder	Ancillary tests
Cerebellar hypoplasia (BVD)	Pre-colostrum antibody, virus isolation. Necropsy. Histology
Cerebellar abscess	CSF increased protein and white cell concentrations Necropsy
Cerebellar abiotrophy	Histology

Cerebellar Hypoplasia

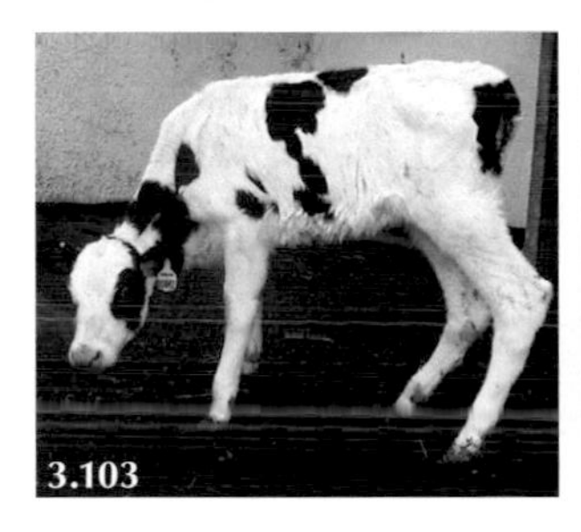

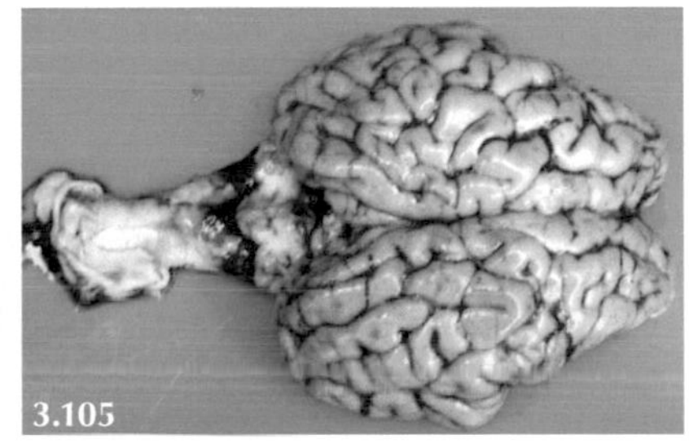

3.103 and **3.104:** Wide-based stance and lowered head carriage in two neonatal calves with cerebellar hypoplasia. **3.105:** Cerebellar hypoplasia revealed at necropsy.

- **Key observations**: Signs present since birth.
- **Clinical examination**: Low head carriage, wide-based stance and ataxia.
- **Key history events**: Congenital (possibly vector-borne) viral infection. Often several/many new-born calves affected over several weeks. Failed herd biosecurity/vaccination. 🎥 3.55

Cerebellar Abscess

3.106 and **3.107:** Six-month-old cattle showing wide-based stance caused by a cerebellar abscess. **3.108:** Cerebellar abscess revealed at necropsy.

- **Key observations**: Insidious onset gait abnormalities, falls over often.
- **Clinical examination**: Wide-based stance and ataxia. Normal strength. Slowly progressive over several months.
- **Key history events**: None. 🎥 3.56

Cerebellar Abiotrophy

3.109: Slow progression of wide-based stance, ataxia and low head carriage from 2–3 months old.

- **Key observations**: Normal for first two to three months then progression over weeks/months.
- **Clinical examination**: Ataxia, low head carriage, wide-based stance.
- **Key history events**: Hereditary, all affected calves in herd progeny of one bull.

Hydranencephaly

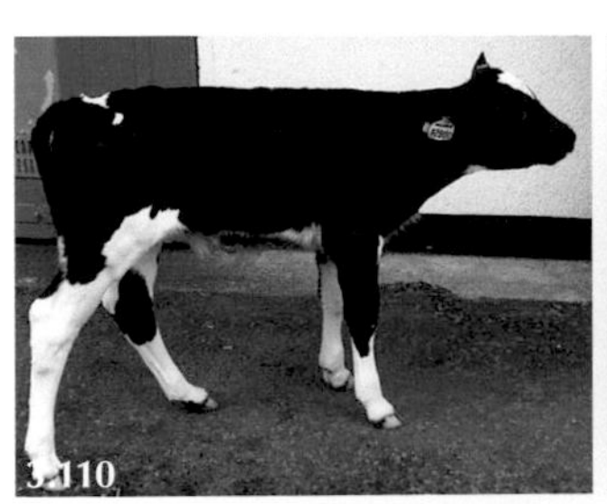
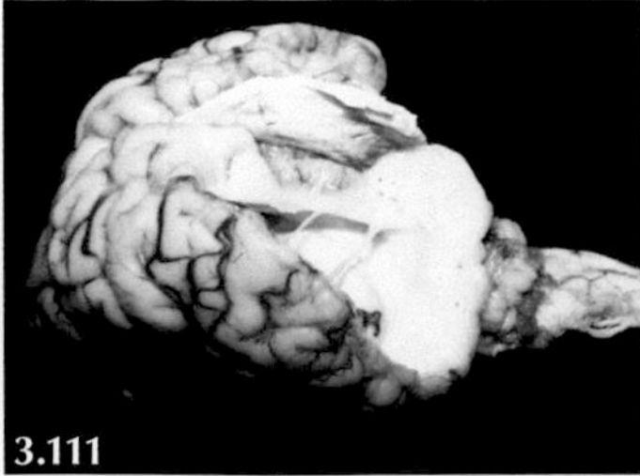

3.110: Two-week-old calf blind since birth with low head carriage and a wide-based stance. **3.111:** Necropsy confirms hydranencephaly.

- **Key observations**: Signs present since birth.
- **Clinical examination**: Blind, low head carriage, wide-based stance and ataxia.
- **Key history events**: Congenital lesion caused by a range of viruses including BVDv, Akabane and Bluetongue.

Multiple Cranial Nerve Deficits—Hypothalamic Syndrome in Sheep

- This is a relatively uncommon syndrome in sheep. There are multiple cranial nerve deficits (II to VII) depending upon which nerve roots are affected by the lesion which is typically an abscess (basilar empyema) or tumour of the pituitary gland.

TABLE 3.19

Hypothalamic Syndrome in Sheep

		Menace response	PLR	Cranial nerves IV–VII
Basilar empyema	Days	–/–––	–/–––	––/–––
Pituitary tumour	Days/weeks	–––	–––	NAD

TABLE 3.20

Ancillary Tests for Multiple Cranial Nerve Deficits in Sheep

Basilar empyema	Increased CSF protein and white cell concentrations
Pituitary tumour	Necropsy, histology

Basilar Empyema

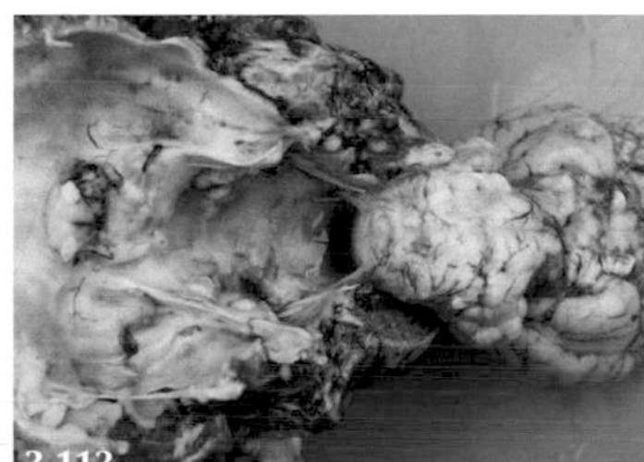

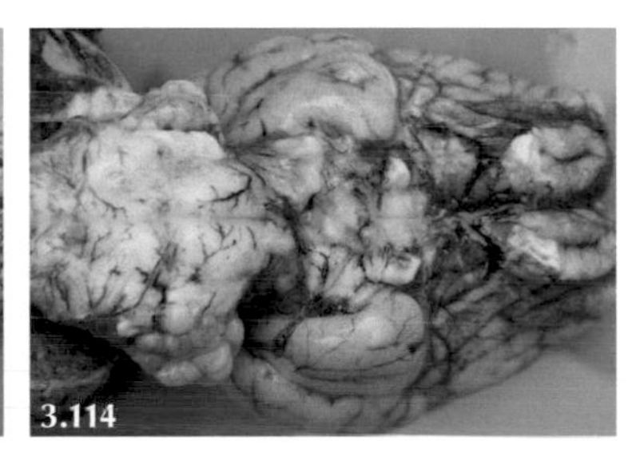

3.112: Multiple cranial nerve deficits II–VII. **3.113:** Brain reflected to show deposition of yellowish purulent material extending from the pituitary gland over the basisphenoid bone. **3.114:** Ventral view of brain showing purulent material surrounding nerve roots II–VII.

- **Key observations**: Very variable clinical presentation depending upon extent of lesion.
- **Clinical examination**: Multiple cranial nerve deficits II–VII.
- **Key history events**: Often related to purulent infection of frontal sinuses. 🎥 3.57

Pituitary Tumour

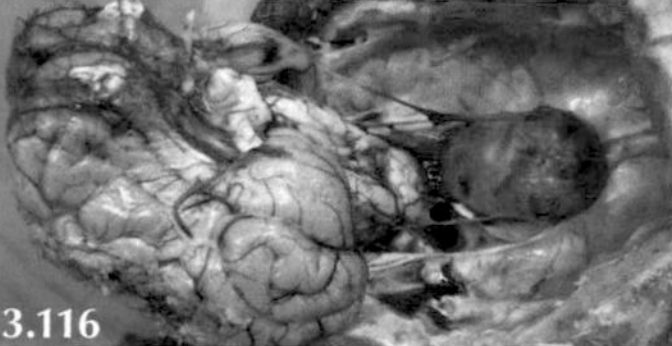

3.115: Aged ewe with bilateral loss of menace response. **3.116:** Necropsy reveals pituitary tumour compressing nerve roots II, III and the cerebral cortex.

- **Key observations**: Insidious onset, often treated for PEM by farmer.
- **Clinical examination**: Bilateral loss of menace and PLR responses.
- **Key history events**: None. 🎥 3.58

Multiple Cranial Nerve Deficits—Hypothalamic Syndrome in Cattle

- Basilar empyema is not uncommon in cattle secondary to sinus infections after dehorning and insertion of nose rings.
- There are multiple cranial nerve deficits (II to VII) depending upon which nerve roots are involved.
- Often good response to early antibiotic treatment.

TABLE 3.21

Hypothalamic Syndrome in Cattle

		Menace response	PLR	Cranial nerves IV–VII
Basilar empyema	Days	–/–––	–/–––	––/–––

TABLE 3.22

Ancillary Tests for Multiple Cranial Nerve Deficits in Cattle

Disease	Ancillary tests
Basilar empyema	Increased CSF protein and white cell concentrations. Necropsy

Basilar Empyema

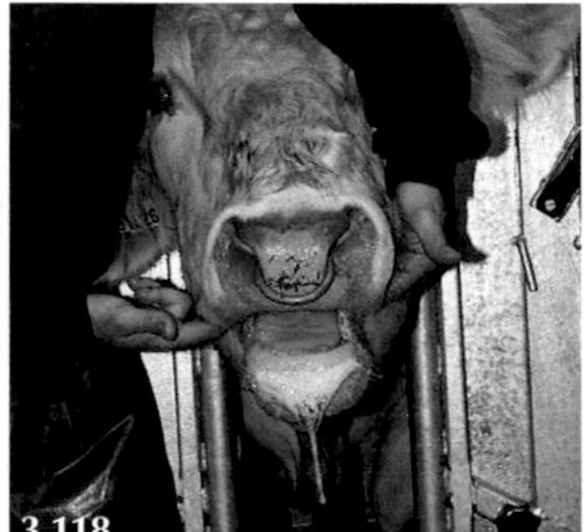

3.117–3.119: Animal appears dull with multiple cranial nerve deficits including bilateral V causing passive protrusion of the tongue.

- **Key observations**: Highly variable clinical presentation depending upon extent of lesion within cranial vault.
- **Clinical examination**: Multiple cranial nerve deficits II–VII.
- **Key history events**: Often related to purulent infection of frontal sinuses following dehorning. Can occur two to three weeks after inserting nose ring in bulls.

Trigeminal and Facial Nerve Deficits—Pontomedullary Syndrome—Sheep

- Dysfunction referred to as the pontomedullary syndrome is characterised by multiple cranial nerve deficits.
- Involvement of the facial nucleus results in ipsilateral facial nerve paralysis evident as drooped ear, drooped upper eyelid (ptosis) and flaccid lip.
- Involvement of the trigeminal motor nucleus results in paralysis of the cheek muscles and decreased facial sensation.

- Depression is attributed to a specific lesion in the ascending reticular activating system of the brainstem.
- Ipsilateral hemiparesis and proprioceptive defects are common.
- Propulsive circling may result from involvement of the vestibulocochlear nucleus.

TABLE 3.23

Common Differential Diagnoses for Trigeminal and Facial Nerve Deficits in Sheep

	Duration	Mentation	Cranial nerves	Walking forward
Listeriosis	Hours/days	--/+	V and VII (unilateral)	NAD/+++

TABLE 3.24

Ancillary Tests for Trigeminal and Facial Nerve Deficits in Sheep

	Ancillary tests
Listeriosis	Increased CSF protein and mononuclear cell concentrations. Brain culture. Histology

Listeriosis

3.120: This twenty-two-month-old sheep has walked forward into the corner of a fence and will not reverse out. **3.121:** Close-up view reveals right-sided drooped ear, lowered eyelid and flaccid lip. **3.122:** This lamb with listeriosis has walked forward into a barrier and stopped; this is not forceful head-pressing.

3.123: Left-sided hemi-paresis, V and VII cranial nerve deficits. **3.124:** Obtunded ewe caused by ARAS involvement. **3.125:** Left-sided V and VII cranial nerve deficits.

- **Key observations**: Walking forward into fences/gates etc and not reversing, salivation, drooped upper eyelid and ear on one side, dull.
- **Clinical examination**: Unilateral V and VII cranial nerve deficits, obtunded caused by ARAS involvement, hemiparesis.
- **Key history events**: Usually access to poor quality silage about two weeks previously. Most common in eighteen–twenty-four-month-old sheep. 🎥 3.59, 3.60, 3.61, 3.62, 3.63, 3.64, 3.65, 3.66, 3.67, 3.68

Trigeminal and Facial Nerve Deficits—Pontomedullary Syndrome—Cattle

- In listeriosis, unilateral cranial nerve V and VII deficits in cattle are much less pronounced than in sheep and are easily overlooked.
- Listeriosis in cattle progresses much more slowly than the disease in sheep.
- In listeriosis, walking forward into objects is common, often resulting in animals becoming caught under gates and fences, and even pushing into the pit of the milking parlour.

	Duration	Mentation	Cranial nerves	Walking forward
Listeriosis	Often several days	-/--	Unilateral V and VII	NAD/++

Disease	Ancillary tests
Listeriosis	Increased CSF protein and mononuclear cell concentrations. Histology. Brain culture

Listeriosis

3.126: Obtunded cow caused by ARAS involvement. **3.127** and **3.128:** Left-sided V and VII cranial nerve deficits. Note deficits in centre image appear mild.

- **Key observations**: (Very) dull, walking forward against objects, can become lodged across cubicles. Ocular discharge.
- **Clinical examination**: Mild unilateral V and VII cranial nerve deficits. Exposure keratitis.
- **Key history events**: Usually access to poor quality silage about two weeks previously. 🎥 3.69, 3.70

Head Tilt—Vestibular Syndrome in Sheep

- The major clinical signs associated with a vestibular lesion are loss of balance and ipsilateral head tilt.
- Ipsilateral facial paralysis is often observed.
- Circling may also be observed in vestibular disease.
- Ascending infection of the eustachian tube is a common cause in growing lambs.

TABLE 3.25

Common Differential Diagnoses for Head Tilt and Facial Nerve Deficits in Sheep

	Duration	Head tilt	VII
Otitis media	Days/weeks	++/+++	-/---

TABLE 3.26

Ancillary Tests for Head Tilt and Facial Nerve Deficits in Sheep

	Ancillary tests
Otitis media	Good response to treatment, necropsy

Otitis Media

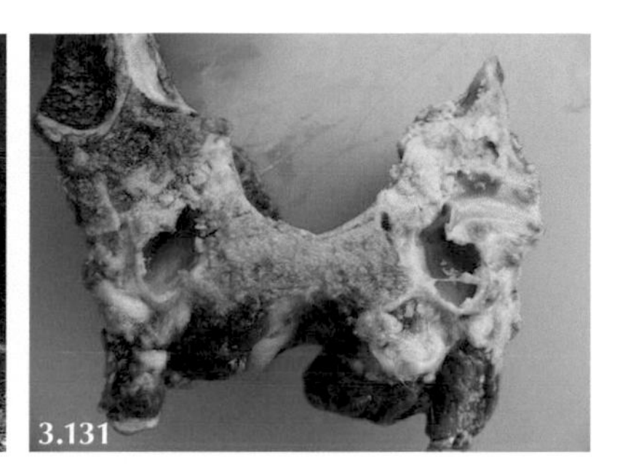

3.129: Loss of balance/sternal recumbency, head tilt to right, right-sided facial paralysis. **3.130:** Head tilt to left, left-sided facial paralysis. **3.131:** Necropsy reveals pathology of the left middle ear chamber (left as viewed).

- **Key observations**: Unsteady/loss of balance.
- **Clinical examination**: Head tilt and ipsilateral facial paralysis.
- **Key history events**: None. 🎥 3.71, 3.72

Head Tilt—Vestibular Syndrome in Cattle

- The major clinical signs associated with a vestibular lesion are loss of balance and ipsilateral head tilt.
- Ipsilateral facial nerve paralysis is often observed.
- Circling may also be observed in vestibular disease.
- Ascending infection of the eustachian tube is commonly caused by *Mycoplasma bovis*.

TABLE 3.27

Common Differential Diagnoses for Head Tilt and Facial Nerve Deficits in Cattle

	Likely duration at presentation	Head tilt	VII
Otitis media	Days/weeks	++/+++	–/––

Otitis Media

3.132: Head tilt to right, right-sided facial paralysis. **3.133:** Head tilt to left, left-sided facial paralysis. **3.134:** Steer featured in centre image shows loss of balance to the left (viewed from behind).

- **Key observations**: Unsteady/loss of balance.
- **Clinical examination**: Head tilt and ipsilateral facial paralysis.
- **Key history events**: Several calves in group affected and history of respiratory disease when caused by *Mycoplasma bovis*. 🎥 3.73

3.3

Weakness—Spinal Lesions

Spinal Lesions in Sheep

- Vertebral empyema is common in two–four-month-old lambs particularly affecting the region T2-L3.
- Neonatal infection of the atlanto-occipital joint with *Streptococcus dysgalactiae* is common and frequently misdiagnosed.
- Lumbar CSF protein concentration is a very useful indicator whether an abscess (vertebra empyema) is causing spinal cord compression.

The categorisation of the common spinal cord lesions is detailed next based upon the presence or absence of spinal reflex arcs. Lesions are listed in anatomical location starting at the head and progressing caudally to aid understanding, and in order of prevalence for that anatomical location. When considering all spinal lesions, A/O joint infection and thoracolumbar empyema are common; the other conditions listed much less so.

DOI: 10.1201/9781003106456-21

TABLE 3.28

Common Differential Diagnoses for Spinal Lesions in Sheep

	Likely duration at presentation	Muscle tone forelegs	Reflex arc forelegs	Muscle tone hindlegs	Reflex arc hindlegs
A/O infection	Hours/days	–/––	+/++	–/––	+/++
Empyema C1-C6	Days	–/––	+/++	–/––	+/++
Infection tracking to spinal canal C1-C6	Days	–/––	+/++	–/––	+/++
Cervical vertebral malformation	Weeks/months	–/––	NAD/+	–	NAD/+
Empyema C6-T2	Days	––/–––	––/–––	NAD	+/++
Empyema T2-L3	Days	NAD	NAD	––/–––	+/+++
Delated swayback	Weeks	NAD	NAD	–/––	NAD
L4-S2 (sarcocystosis)	Days	NAD	NAD	–/–––	–/–––
Cauda equina	Days	NAD	NAD	NAD	NAD

TABLE 3.29

Ancillary Tests for Common Spinal Lesions in Sheep

	Ancillary tests
A/O infection	Mild increase in lumbar CSF protein concentration and white cell concentration. Rapid response to penicillin and dexamethasone
Empyema C1–C6	Mild increase in lumbar CSF protein concentration and white cell concentration
Infection tracking to spinal canal C1–C6	Localised painful swelling in the neck musculature from bite wound or contaminated injection site abscess
Empyema C6–T2	Mild increase in lumbar CSF protein concentration and white cell concentration
Empyema T2–L3	Moderate to large increase in lumbar CSF protein concentration and white cell concentration. Radiography/myelography has been superseded by CT for valuable individual sheep
Delayed swayback	Lumbar CSF normal
L3-S2 (sarcocystosis)	Definitive diagnosis not possible in live sheep. Specialised immuno-staining necessary but rarely undertaken. Absolute increase in lumbar CSF eosinophil count if lesion involves T/L cord
Cauda equina	Radiography if caused by sacral fracture

A/O Infection

3.135: The lamb (right hand side) is weak on all four legs with an arched back and lowered head carriage compared to twin. **3.136:** This five-day-old lamb is weak on all four legs. **3.137:** Same lamb as centre image 6 hours after dexamethasone and penicillin injections.

- **Key observations**: Bright and alert but weak on all four legs with lowered head carriage.
- **Clinical examination**: Increased reflex arcs in all four legs.
- **Key history events**: High prevalence of *Streptococcus dysgalactiae* polyarthritis in flock. 3.74, 3.75, 3.76, 3.77, 3.78, 3.79

Empyema C1-C6

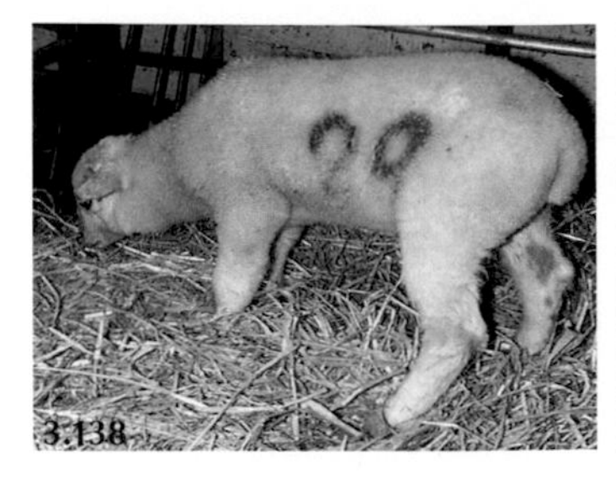

3.138: Painful expression, lowered head, weak on all four legs. **3.139:** Painful expression, neck held in rigid extension. **3.140:** Weakness causing lateral recumbency.

- **Key observations**: Weakness, pain in neck region, neck held rigidly.
- **Clinical examination**: Increased reflex arcs in all four legs.
- **Key history events**: Weakness progressing to recumbency over four to seven days.

Localised Soft Tissue Infection Tracking to Spinal Canal C1–C6

3.141: Lateral recumbency in a ten-day-old Scottish Blackface lamb where a fox bite has tracked to the cervical vertebral canal.

- **Key observations**: Lamb weak on all four legs/lateral recumbency.
- **Clinical examination**: Painful soft tissue swelling/abscess tracking to vertebral canal.
- **Key history events**: Fox/dog bite several days earlier. Contaminated injection site. 🎥 3.80

Cervical Vertebral Malformation (Wobbler)

3.142: Weakness and stumbling on the forelegs in a Texel ram with CVM. **3.143:** Wide-based stance and proprioceptive deficit of the left hind foot.

- **Key observations**: Insidious onset stumbling and weakness.
- **Clinical examination**: CVM is a stenotic myelopathy of the vertebral canal causing variable compression of the cervical spinal cord and weakness of all four legs.
- **Key history events**: Prevalent in certain bloodlines of Texel sheep and other meat breeds. 🎥 3.81, 3.82

Empyema C6–T2

3.144 and **3.145:** Lambs show weakness of both forelegs with normal hindleg function.

- **Key observations**: Weakness of both forelegs. UMN to pelvic limbs.
- **Clinical examination**: Much reduced/absent reflex arcs involving the forelegs.
- **Key history events**: Usually vertebral empyema but origin of bacteraemia rarely confirmed. 🎥 3.83

Empyema T2–L3

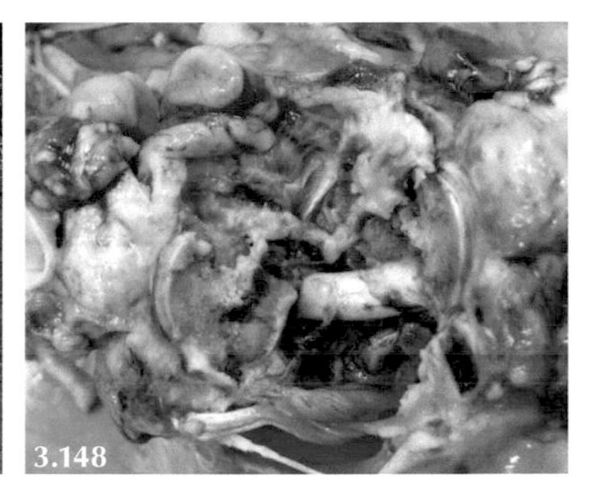

3.146: Weakness of the hindlegs in a ten-week-old lamb with a T/L spinal cord lesion. **3.147:** Characteristic "dog sitting" posture of animals with a T/L lesion. **3.148:** Empyema with bone destruction and infection (pus) tracking into the vertebral canal causing spinal cord compression.

- **Key observations**: Weak on hindlegs, frequently adopt "dog sitting" posture.
- **Clinical examination**: Normal foreleg reflexes, upper motor neuron signs in hindlegs.
- **Key history events**: Extensive bone destruction before weakness observed; consequently, sheep never improve with treatment. 🎥 3.84, 3.85, 3.86, 3.87

Delayed Swayback

3.149: Hindleg weakness caused by delayed swayback.

- **Key observations**: Observed weak on hindlegs by shepherd for several weeks.
- **Clinical examination**: Often difficult to appreciate any possible upper motor neuron signs in hindlegs.
- **Key history events**: Often only one or two lambs affected during one summer in flock with no history of swayback. 3.88, 3.89

L4–S2

3.150: Hindleg weakness with this six-month-old Scottish Blackface ram using the wall for support. **3.151:** Failure to draw hindlegs under body when lying down. **3.152:** Despite these Pyrenean Mountain guard dog puppies being reared in intimate contact with pedigree Suffolk lambs there was no history of spinal cord disease.

- **Key observations**: Insidious onset hindleg weakness.
- **Clinical examination**: Lesions in the region L4 to S2 result in flaccid paralysis of the pelvic limbs with reduced or absent reflexes.
- **Key history events**: Exposure of housed young lambs to feces from litter of puppies. Lambs grazing fields frequently used by dog walkers. 3.90, 3.91

Cauda Equina

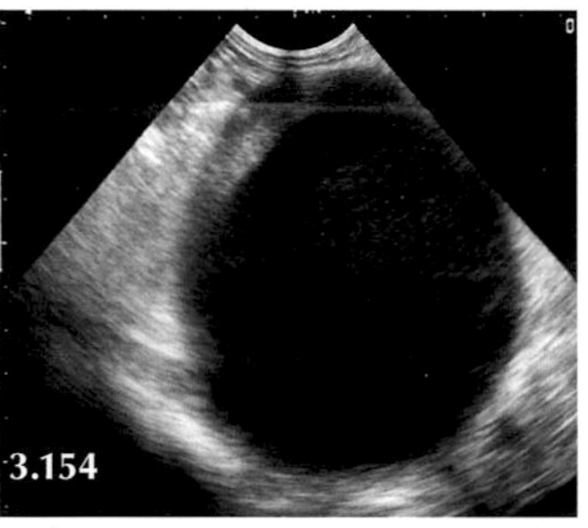

3.153: Lack of tail tone. **3.154:** Urinary bladder distension can be readily identified using trans-abdominal ultrasound (6.5 MHz microarray probe shown but 5 MHz linear probes work equally well).

- **Key observations**: Loss of tail tone.
- **Clinical examination**: Distended bladder, flaccid impacted rectum.
- **Key history events**: Sacral fractures following fall in other species but rare in sheep.

Spinal Lesions in Cattle

- Vertebral empyema is uncommon in cattle but occurs most often in two–six-month-old calves, particularly *Salmonella Dublin* lesions, affecting the lower cervical region.
- Lumbar CSF protein concentration is useful in determining whether an abscess or haemorrhage is causing spinal cord compression.
- Assessing limb muscle tone and reflex arcs is difficult in adult cattle.

TABLE 3.30

Common Differential Diagnoses for Spinal Deficits in Cattle

	Likely duration at presentation	Muscle tone forelegs	Reflex arc forelegs	Muscle tone hindlegs	Reflex arc hindlegs
C1–C6	Days	−−/−−−	+/++	−−/−−−	+/++
C6–T2	Days	−−−	−−−	+/++	+/++
T2–L3	Days	NAD	NAD	−−/−−−	+/++
L3–S2	Days	NAD	NAD	−−−	−−−
Cauda equina	Hours/Days	NAD	NAD	NAD	NAD

TABLE 3.31

Ancillary Tests for Common Spinal Lesions in Cattle

Disease	Ancillary tests
Empyema C1–C6	Radiography (calves). Lumbar CSF protein concentration
Empyema C6–T2	Radiography (calves). Lumbar CSF protein concentration
Empyema T2–L3	Radiography (calves). Lumbar CSF protein concentration
L3–S2	Lesion may be caudal to lumbar CSF sampling site
Cauda equina	Atonic bladder and rectum. Palpate vertebral fracture site per rectum

C1–C6

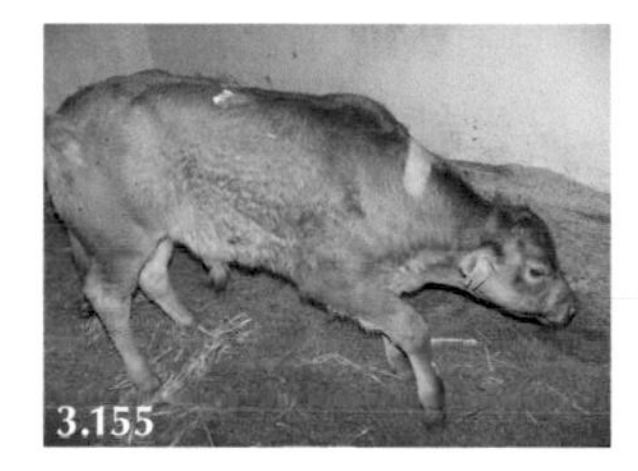

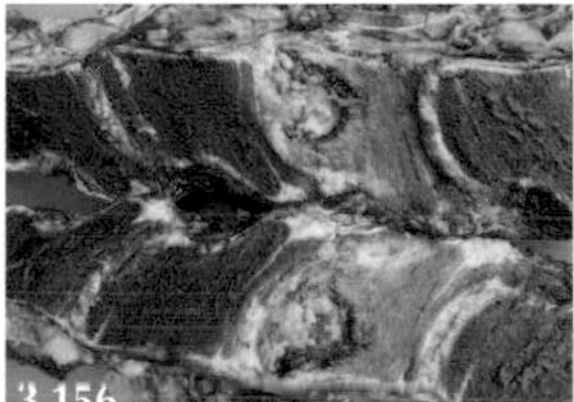

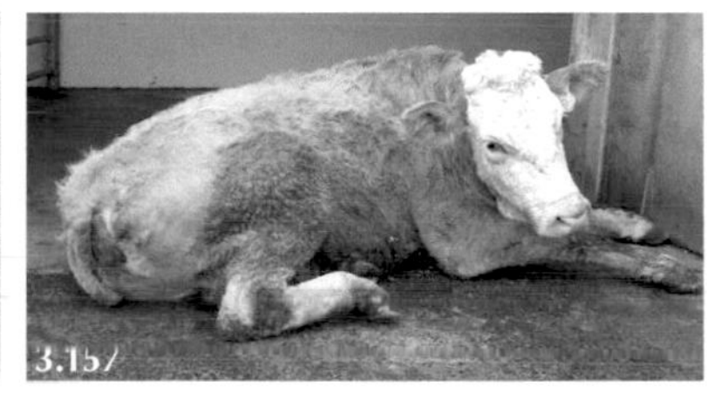

3.155: Calf weak on all four legs, low head carriage. **3.156:** Sagittal section of C5 reveals empyema caused by *Salmonella Dublin*. **3.157:** Abnormal foreleg position caused by cow falling; cervical cord compression has caused severe weakness affecting all four legs.

- **Key observations**: Weak on all four legs. Lowered head carriage.
- **Clinical examination**: Cervical pain. Increased spinal reflexes all four legs.
- **Key history events**: Insidious onset unless caused by trauma/vertebral fracture. 🎥 3.92, 3.93, 3.94

T2–L3

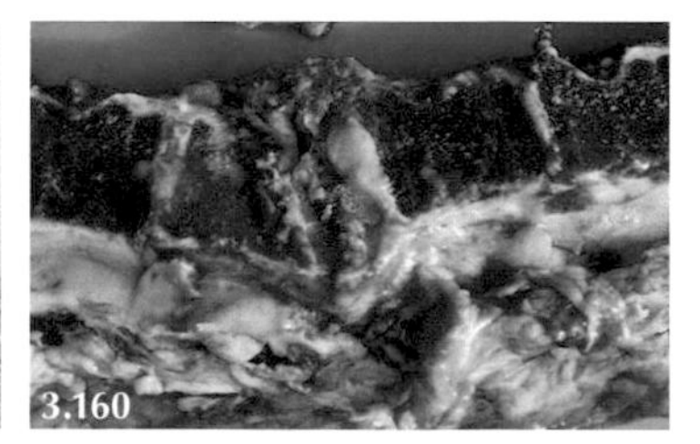

3.158 and **3.159:** Hindleg weakness. Cattle usually raise themselves hindquarters first, so this "dog-sitting" posture indicates a serious problem. **3.160:** Sagittal section of thoraco-lumbar vertebral column/spinal canal showing abnormal angulation caused by vertebral body infection (empyema) and subsequent collapse compressing the spinal cord (removed).

- **Key observations**: Weak hindlegs. Dog-sitting posture.
- **Clinical examination**: Upper motor neurons signs in hindlegs can be difficult to assess.
- **Key history events**: Often insidious onset unless caused by trauma (e.g. served by bull). 3.95

L3–S2

3.161 and **3.162:** Damage to the sciatic (L4–S2 spinal segments) and obturator nerves (L2-L4) after intrapelvic parturient trauma may cause hindleg weakness and recumbency.

- **Key observations**: Weak hindlegs.
- **Clinical examination**: LMN of hindlegs.
- **Key history events**: Damage to the sciatic (L4–S2 spinal segments) and obturator nerves (L2–L4) after intrapelvic parturient trauma may cause recumbency after calving.

Cauda Equina

3.163 and **3.164:** Abnormal shape of vertebral column following sacral fracture. **3.165:** Sacral fracture with haemorrhage into the spinal canal causing paralysis of the rectum, bladder and tail (see centre image).

- **Key observations**: Painful expression, abnormal vertebral column in pelvic area, flaccid tail.
- **Clinical examination**: Atonic tail, bladder and rectum. Fracture may be palpable per rectum.
- **Key history events**: Bulling activity, slippery flooring. 3.96

CSF Collection

The site for lumbar CSF collection is the midpoint of the lumbosacral space which can be identified as the midline depression between the last palpable dorsal lumbar spine (L6) and the first palpable sacral dorsal spine (S2). Collection of lumbar CSF is facilitated if the animal can be positioned in sternal recumbency with the hips flexed and the pelvic limbs extended alongside the abdomen (typically calves and recumbent adults). The site has been shaved, surgically prepared and between 1–2 ml of local anaesthetic injected subcutaneously. Sterile surgical gloves should be worn for the collection procedure (but not in 3.97).

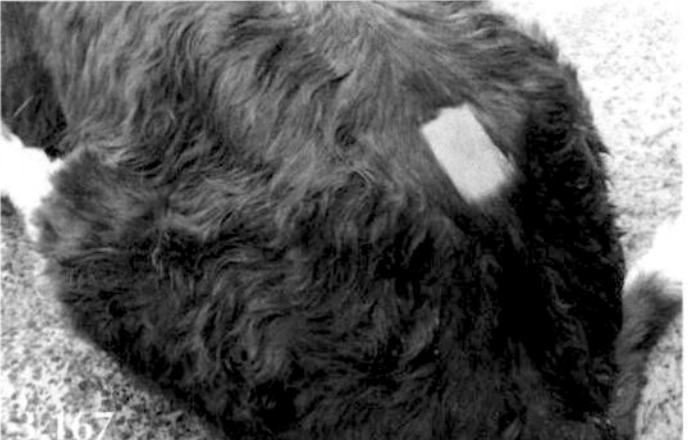
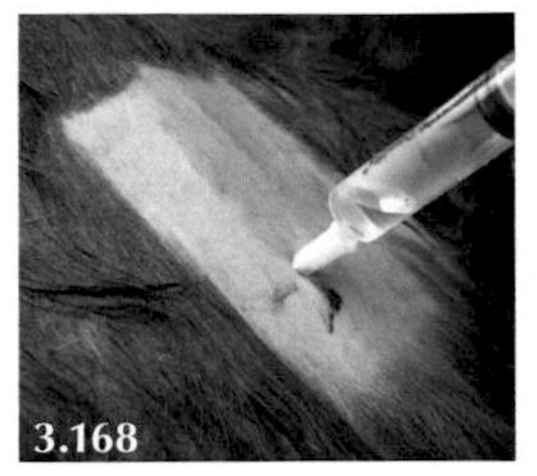

3.166: The animal is positioned in sternal recumbency with the hips flexed and the pelvic limbs extended alongside the abdomen. **3.167:** The site has been shaved, surgically prepared and 2 ml of local anaesthetic injected subcutaneously. **3.168:** Normal CSF is clear and colourless.

The 50 mm 19 gauge needle (see Table 1 below) is slowly advanced (over 10 seconds) at a right angle to the plane of the vertebral column. It is essential to appreciate the changes in tissue resistance as the needle point passes sequentially through the subcutaneous tissue, interarcuate ligament and then the sudden "pop" due to the loss of resistance as the needle point exits the ligamentum flavum into the extradural space. Use only your index finger pressure to advance the needle. Once the needle point has penetrated the dorsal subarachnoid space CSF will well up in the needle hub within 2–3 seconds. Failure to appreciate the change in resistance to needle travel may result in needle puncture of the conus medullaris. This may elicit an immediate pain response and cause unnecessary discomfort to the animal which must be avoided at all times. Between 1 and 2 ml of CSF is sufficient for laboratory analysis and while the sample can be collected by free flow over 1 to 2 minutes, it is more convenient to employ very gentle syringe aspiration over 10 to 20 seconds. Note that the needle hub is securely anchored when attaching the syringe so as not to advance the needle point into the conus medullaris (this grip was then released to show CSF in the syringe for demonstration purposes only).

Guide to Needle Length and Gauge for Lumbar CSF Sampling

Calves <100 kg	2.5 cm 19 gauge
Calves 100–250 kg	5 cm 19 gauge
Cattle >250 kg	10 cm 18 gauge + internal stylet

The normal range for CSF protein concentration quoted for cattle is <0.3 g/l. Normal CSF contains less than 10 cells/ml which are predominantly lymphocytes with an occasional neutrophil. As a general rule, a predominantly polymorphonuclear intrathecal inflammatory response is found in acute CNS bacterial infections whereas a mononuclear response is seen in viral CNS infections. 🎥 3.97

Leg Weakness—Peripheral Nerve Injuries in Sheep

- Damage to a peripheral nerve typically causes weakness affecting one leg.
- Except for sciatic and obturator nerve paralysis associated with dystocia, both forelegs or both hindlegs affected more likely indicates a spinal cord lesion.

An attempt has been made to rank the prevalence of peripheral nerve lesions observed by the author; this ranking may differ from other veterinary practices.

TABLE 3.32

Common Clinical Presentation of Peripheral Nerve Lesions in Sheep

	Likely duration at presentation	Lameness	Strength	Muscle wastage
Peroneal nerve	Weeks/months	+/++	–/––	NAD
Radial nerve	Weeks/months	+/++	–/–––	–/––
Brachial plexus	Days/weeks	++/+++	––/–––	––/–––
Obturator nerve and sciatic nerve	Days	++/+++	–/–––	NAD
Visna	Months	+	–/––	––/–––

TABLE 3.33

Ancillary Tests for Common Peripheral Nerve Deficits in Sheep

Disease	Ancillary tests
Visna	Serology, histology

Peroneal Nerve

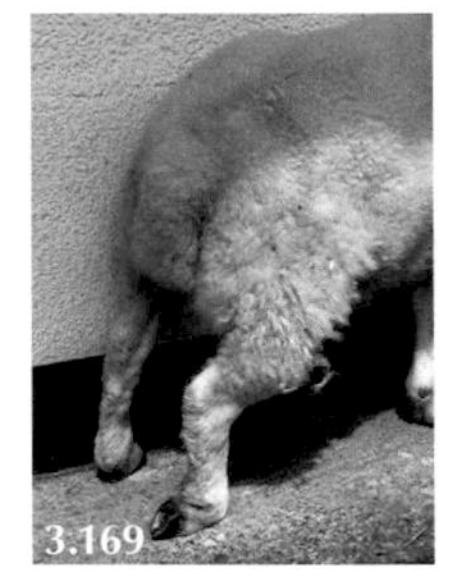

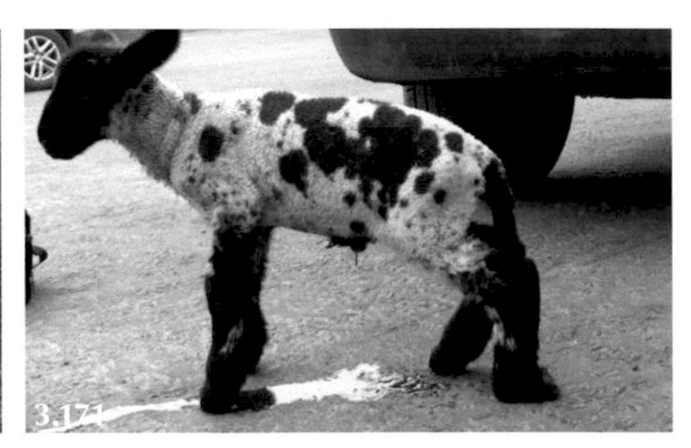

3.169 and **3.170:** Soft tissue swelling in the stifle area can lead to signs of peroneal nerve paralysis. **3.171:** Peri-neural injection is a common temporary cause of sciatic/peroneal nerve paralysis in neonatal lambs.

- **Key observations**: The dorsal surface of the hoof contacts the ground. With chronic lesions there is a plantigrade stance of contralateral foot.
- **Clinical examination**: Often related to stifle joint swelling in older lambs.
- **Key history events**: Often caused by perineural injection in neonatal lambs. 🎥 3.98, 3.99

Brachial Plexus/Radial Nerve Injury

It can prove difficult to differentiate radial nerve paralysis from damage to the brachial plexus.

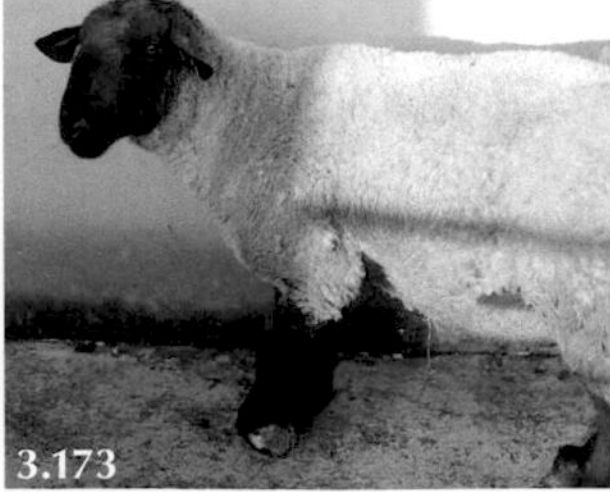

3.172–3.174: Weak left foreleg with a dropped elbow and the dorsum of foot resting on the ground.

3.175 and **3.176:** Ewes are unable to bear weight on the left leg. **3.177:** This ewe shows only mild weakness of the left foreleg with knuckling of the fetlock joint.

- **Key observations**: Unable to fully extend the elbow, carpus and fetlock, and, in severe cases, bear weight on the affected limb.
- **Clinical examination**: Appreciable muscle wastage after more than five days.
- **Key history events**: In neonates occurs following excessive traction of posterior presentation delivery. Trauma to the mid/distal humeral region can result in radial nerve paralysis. 3.100, 3.101, 3.102, 3.103

Obturator Nerve and Sciatic Nerve Injuries

3.178: Hindlegs not positioned correctly when ewe in sternal recumbency. **3.179:** Hindlegs weakness with ewe unable to lift her hindquarters. **3.180:** Peroneal nerve damage (knuckling of right fetlock joint) still evident two days later.

- **Key observations**: Hindleg weakness but makes frequent attempts to stand.
- **Clinical examination**: May show signs of extended parturition—vulval swelling.
- **Key history events**: Obturator nerve and sciatic nerve injury is rare in sheep but can result from dystocia where there is protracted second stage labour. Usually recover within days. 3.104

Kangaroo Gait

3.181: Sudden onset foreleg weakness in young nursing ewe at pasture. **3.182:** Young ewe with kangaroo gait now housed. **3.183:** Much improved 10 days later after supplementary feeding.

- **Key observations**: Sudden onset foreleg weakness. Sheep bright and alert.
- **Clinical examination**: No obvious cause of weakness.
- **Key history events**: Typically affects young ewes nursing twin lambs at pasture without supplementary feeding. Resolution after housing and improved energy supply. 3.105, 3.106

Visna

3.184: Emaciated ewe with lameness and muscle wastage of the left hindleg.

- **Key observations**: Chronic weight loss over months, weakness affecting one hindleg.
- **Clinical examination**: No obvious cause of weakness, specifically no chronic stifle joint pathology.
- **Key history events**: Other sheep with signs of MVV infection in flock—indurative mastitis, chronic pneumonia.

Leg Weakness—Peripheral Nerve Injuries in Cattle

Damage to a peripheral nerve typically causes weakness affecting one leg and must be differentiated from other causes of gait abnormality/lameness. An attempt has been made next to rank the prevalence of peripheral nerve lesions observed by the author in predominantly beef cattle practice. This ranking may differ in other veterinary practices.

TABLE 3.34

Common Clinical Presentation of Peripheral Nerve Lesions in Cattle

	Likely duration at presentation	Lameness	Strength	Muscle wastage
Femoral nerve	Weeks/months	++/+++	––/–––	–/–––
Obturator nerve	Days/weeks	++/+++	––/–––	–/––
Tibial nerve	Weeks/months	+	NAD/–	NAD
Radial nerve	Weeks/months	+	–	–/––
Brachial plexus	Days/weeks	++/+++	––/–––	––/–––
Peroneal nerve	Weeks/months	+	–	NAD

Femoral Nerve

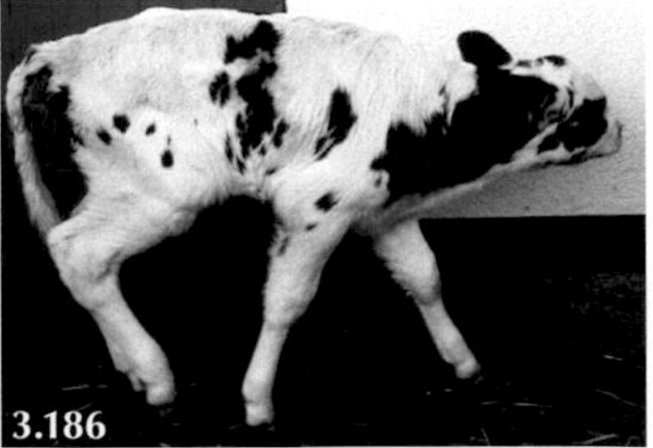

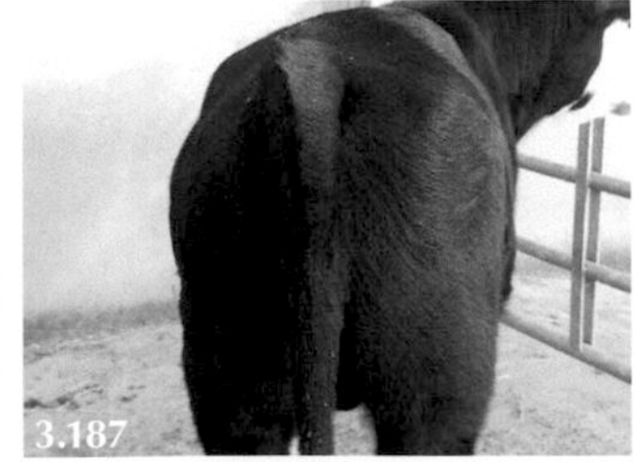

3.185: Bilateral, but predominantly left femoral nerve paralysis causing recumbency following assisted delivery. **3.186:** Right hindleg extensor weakness caused by femoral nerve paralysis. **3.187:** Recovery from femoral nerve damage (left image). The animal shows considerable muscle atrophy of the left hindleg even after six months.

- **Key observations**: Unable to extend the stifle joint and bear weight. Calves with bilateral femoral paralysis are unable to stand and adopt a dog-sitting posture.
- **Clinical examination**: Atrophy of the *Quadriceps femoris* muscle group.
- **Key history events**: Common when assisted delivery of calf in anterior presentation becomes hip-locked. 🎥 3.107, 3.108, 3.109

Obturator Nerve

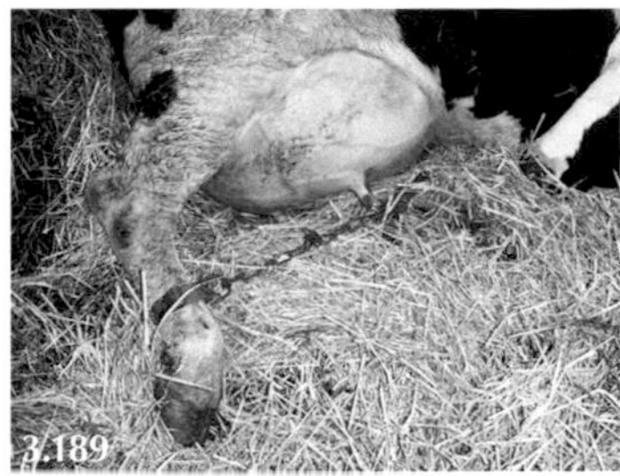

3.188: Excessive abduction of the left hindleg from adductor paresis. **3.189** and **3.190:** Use of hobbles to limit further muscle damage (should be placed above fetlock joints).

- **Key observations**: Adductor paresis in newly calved cows.
- **Clinical examination**: Difficult to assess and offer accurate prognosis.
- **Key history events**: Obturator nerve injuries following "hip lock" calves result in adductor paresis.

Sciatic and Obturator Nerves Injury

3.191: Cow standing on its forelegs because she cannot force herself up using her hindlegs. **3.192:** Cow is able to stand for a few minutes. **3.193:** Muscle weakness from sciatic nerve damage results in loss of stifle and hock extensor muscle tone.

- **Key observations**: Cow unable to rise, attempts to stand with forelegs first.
- **Clinical examination**: Difficult to afford accurate prognosis.
- **Key history events**: Obturator nerve and sciatic nerve injuries following "hip lock" calves. Cow loses her footing on wet, slippery surfaces. 🎥 3.110, 3.111, 3.112

Tibial Nerve

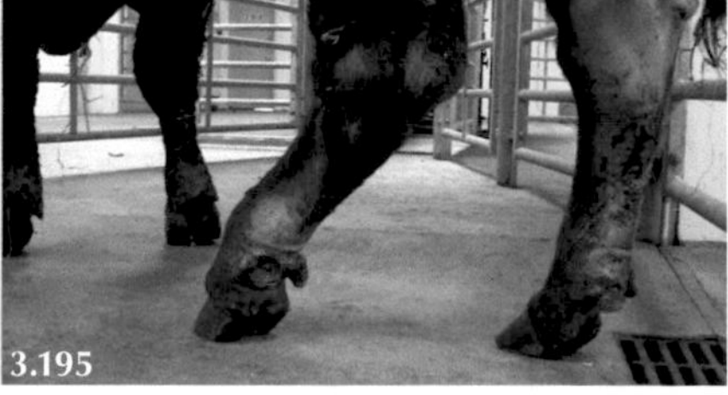

3.194: Slight knuckling over of both hind fetlock joints. **3.195:** Over-flexion of the left hock and knuckling over on the left fetlock joint.

- **Key observations**: Injury results in dropped hock and knuckling of the fetlock joint.
- **Clinical examination**: Findings as earlier.
- **Key history events**: May be caused by dystocia. 🎥 3.113

Peroneal Nerve

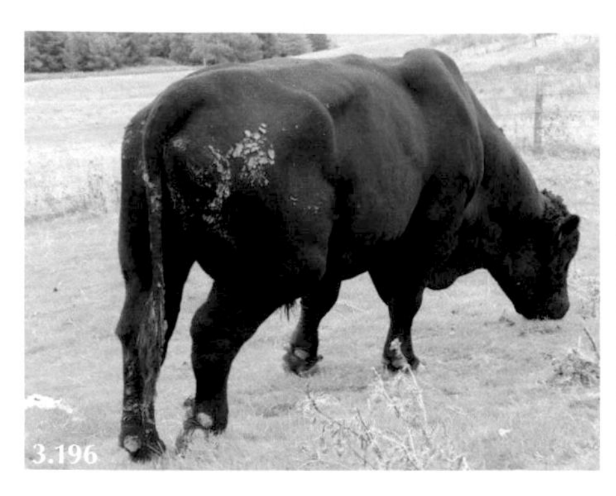

3.196 and **3.197:** Peroneal nerve damage with the dorsal surface of the right hind foot touching the ground. **3.198:** Affecting the left hindleg in this bull.

- **Key history events**: Trauma/swelling over the lateral aspect of the stifle region, often following prolonged recumbency on an unyielding surface.

Radial Nerve Injury

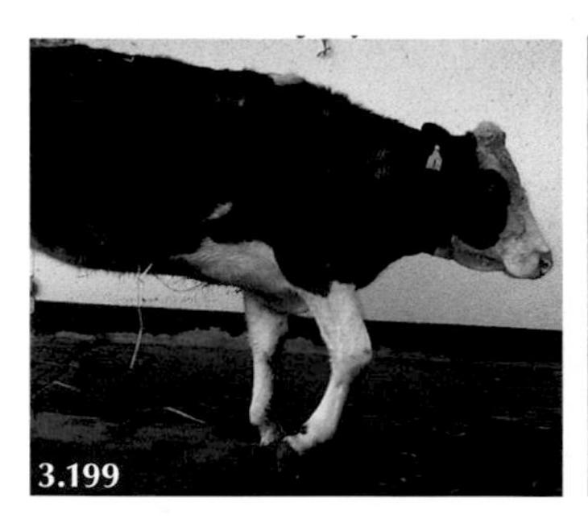

3.199–3.201: These cows are unable to fully protract the right foreleg and bear weight.

- **Key observations**: Unable to extend the elbow, carpus and fetlock, and drags the foot.
- **Clinical examination**: Appreciable muscle wastage after more than five days.
- **Key history events**: Blunt trauma to the mid/distal humeral region causing soft tissue swelling. 3.114

Brachial Plexus Avulsion/Injury

3.202: This calf is weak on the right foreleg with the leg resting on dorsum of the fetlock joint. The calf has been like this since assisted delivery by the farmer.

- **Key observations**: Unable to extend the elbow, carpus and fetlock and, in severe cases, bear weight on the affected limb.
- **Clinical examination**: Appreciable muscle wastage over the foreleg after five days.
- **Key history events**: Occurs following excessive traction to correct an anterior presentation dystocia.

3.4

Common Causes of Diarrhoea

Adult Sheep

- Diarrhoea with dried feces on the perineal wool and tail affects many adult sheep.
- The refugia concept in endoparasite control is not understood by most farmers whose standard response to diarrhoea in their sheep is more anthelmintic treatments.
- The repeated use of long-acting moxidectin injections every spring to control PGE is a serious concern.

TABLE 3.35

Common Clinical Presentation of Adult Sheep with Diarrhoea

	Shape	Appetite	Demeanour	Feces
Parasitic gastroenteritis	NAD	NAD	NAD	Diarrhoea
Trichostrongylosis	NAD	NAD	NAD	Profuse diarrhoea
Grain overload	+/++	–/–––	–/–––	Whole grains in foul-smelling feces

TABLE 3.36

Ancillary Tests to Determine Cause of Diarrhoea in Sheep

	Ancillary tests
Parasitic gastroenteritis	Fecal worm egg count. Response to effective anthelmintic
Trichostrongylosis	Usually very high FWEC; (thousands of epg)
Grain overload	Rumen fluid pH<4.5

Parasitic Gastro-Enteritis

3.203–3.205: Diarrhoea with fecal staining on the perineal wool and tail caused by endoparasite infestation.

DOI: 10.1201/9781003106456-22

- **Key observations**: Fecal staining of the tail/perineum. Mid-summer onwards.
- **Clinical examination**: NAD, check for cutaneous myiasis.
- **Key history events**: No effective parasite control programme including quarantine treatments (especially rams) in flock health plan. Contaminated grazing. Be aware that sheep with paratuberculosis can have very high FWEC indeed (>30,000 epg) including *N. battus* and lungworm larvae. 3.115

Trichostrongylosis

3.206: Profuse diarrhoea affecting eighteen-month-old sheep during late autumn/early winter. **3.207:** Large percentage of group of eighteen-month-old sheep present with profuse diarrhoea during late autumn/early winter affected. **3.208:** Marked condition score loss caused by Trichostrongylosis during late autumn.

- **Key observations**: Most often eighteen-month-old sheep during late autumn/early winter. Large percentage of group affected.
- **Clinical examination**: NAD.
- **Key history events**: No effective parasite control programme. Contaminated grazing. 3.116

Grain Overload

3.209: Dull and inappetant sheep with foetid diarrhoea. **3.210:** Evidence of foetid diarrhoea on the wool of the perineum and tail.

- **Key observations**: Weak animal with fluid-distended rumen, foetid diarrhoea may contain whole grains.
- **Clinical examination**: Rumen stasis, diarrhoea, dehydration, toxic mucous membranes.
- **Key history events**: Sudden unaccustomed access to excessive amount of grain. 3.117, 3.118

Lambs

- Farmers are often unaware of reliable models to predict Nematodiriasis risk.
- The use of clean grazing (silage ground) post-weaning is well-established farming practice, but the refugia concept is astrophysics to most farmers.
- Parasite control is the most important component of the veterinary flock health plan.

TABLE 3.37

Common Causes of Diarrhoea in Lambs

	Age	Shape	Appetite	Demeanour	Feces
Parasitic gastroenteritis	>2 months	NAD	NAD	NAD	Profuse diarrhoea
Nematodiriasis	1–3 months and 7–9 months	NAD/–	–	–	Profuse diarrhoea
Trichostrongylosis	7–9 months	NAD	NAD	NAD	Profuse diarrhoea
Coccidiosis	2–8 weeks	–/––	–	–	Mucoid diarrhoea, occasional fresh blood
Watery mouth disease	12–36 hours	+/++	––/–––	––/–––	Agonal diarrhoea
Cryptosporidiosis	7–21 days	–	–	–	Mucoid diarrhoea
Grain overload		+/++	–––	––/–––	Foul–smelling containing grain
Nephrosis	2–4 months	––/–––	–/––	–/––	Chronic diarrhoea

TABLE 3.38

Ancillary Tests for Common Causes of Diarrhoea in Lambs

	Ancillary tests
Parasitic gastroenteritis	Fecal worm egg count >400–1,000 epg. Response to effective anthelmintic. Check for concurrent trace element deficiencies
Nematodiriasis	Pre-patent infestations—negative FWEC. Anthelmintic treatment, move pasture. Necropsy
Trichostrongylosis	Usually very high FWEC; (thousands epg)
Coccidiosis	Fecal smear; oocyst count and speciation
Watery mouth disease	High lactate and BUN, low plasma glucose concentrations, leucopaenia. Bacteriology. Necropsy
Cryptosporidiosis	Fecal smear, histology of gut sections at necropsy
Grain overload	Rumen fluid pH<4.5. Necropsy
Nephrosis	Increased BUN, hypoalbuminaemia, necropsy

Parasitic Gastro-Enteritis

3.211–3.213: Profuse diarrhoea staining the tails and perineum of growing lambs.

- **Key observations**: Large percentage of group affected. Fecal staining of the tail/perineum. Midsummer onwards.
- **Clinical examination**: Check for cutaneous myiasis.
- **Key history events**: No effective parasite control programme. Contaminated grazing. 3.119, 3.120

Nematodiriasis

3.214 and **3.215:** Profuse diarrhoea affecting six-week-old lambs. **3.216:** Profuse diarrhoea affecting six-week-old twin lambs; note the dam is unaffected.

- **Key observations**: Rapid weight/condition losses. Fecal staining of the tail/perineum. Possible "sudden" deaths.
- **Clinical examination**: Empty abdomen. Variable dehydration. Ewes not affected.
- **Key history events**: Four- to ten-week-old lambs grazing contaminated grazing (grazed by lambs previous year). 3.121, 3.122, 3.123, 3.124

Trichostrongylosis

3.217–3.219: Profuse diarrhoea staining the tails and perineum of nine-month-old lambs.

- **Key observations**: Large percentage of group affected. Fecal staining of the tail/perineum. Late autumn/early winter.
- **Clinical examination**: NAD.
- **Key history events**: No effective parasite control programme. Contaminated grazing. 3.125

Coccidiosis

3.220: Five-month-old lamb with tenesmus and chronic diarrhoea. **3.221** and **3.222:** Rectal prolapse may result from coccidiosis.

- **Key observations**: Two- to eight-week-old lambs most commonly affected, tenesmus, mucoid diarrhoea and weight loss.
- **Clinical examination**: Tenesmus, mucoid diarrhoea, possible rectal prolapse in some cases.
- **Key history events**: Possible infection build-up in creep areas. Following removal of coccidiostat medication from ration. 3.126

Watery Mouth Disease

3.223: Excess salivation in one-day-old lamb. **3.224:** Abdominal distension in one-day-old lamb. **3.225:** Endotoxaemia causing recumbency in one-day-old lamb with watery mouth disease.

- **Key observations**: Salivation, increasing abdominal distension, end stage diarrhoea.
- **Clinical examination**: Weak, dehydration, sequestration of fluid in abomasum/intestines.
- **Key history events**: Large numbers of neonatal lambs affected towards end of lambing period.

Grain Overload

3.226: Weak, dehydrated, Scottish Blackface lamb. **3.227:** Change to *ad libitum* feeding using hoppers may cause grain overload.

- **Key observations**: Weak/sternal recumbency, fluid-distended rumen. May die before diarrhoea signs appear.
- **Clinical examination**: Dehydration, toxic mucous membranes.
- **Key history events**: Sudden unaccustomed access to grain.

Nephrosis

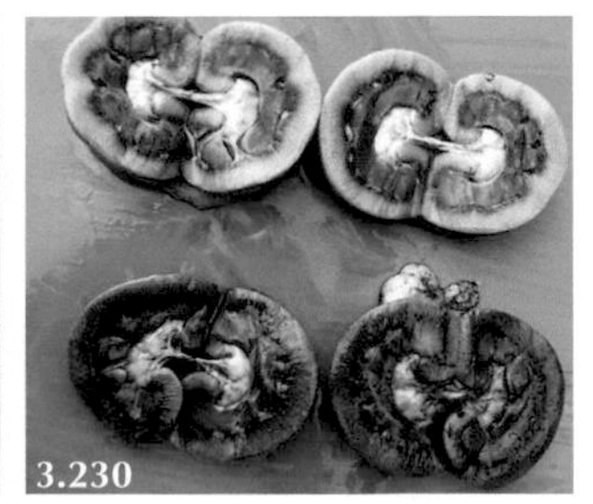

3.228 and **3.229:** Chronic weight loss and diarrhoea following nematodiriasis. **3.230:** Pale nephrotic kidneys (top), normal kidneys (bottom).

- **Key observations**: Individual lamb, chronic diarrhoea, severe weight loss.
- **Clinical examination**: Often dull, emaciated, variable dehydration.
- **Key history events**: Previous episode of coccidiosis/nematodiriasis. 🎥 3.127, 3.128

Adult Cattle

- Paratuberculosis is a major disease problem in many beef and dairy herds.
- Flukicide treatment at dry-off in dairy herds is a useful strategy to reduce levels of exposure and increase milk production in the subsequent lactation. Control measures are simpler in beef herds where milk withhold does not apply.

TABLE 3.39

Common Clinical Presentation of Adult Cattle with Diarrhoea

	Likely duration at presentation	Shape	Appetite	Demeanour	Feces
Paratuberculosis	Weeks/months	––/–––	–/––	–	Profuse diarrhoea
Salmonellosis	Hours	–	––/–––	––/–––	Dysentery
Fasciolosis	Weeks/months	–	NAD	NAD	Diarrhoea
Abomasal ulcers	Days	–/––	–/–––	–/––	Melaena
Grain overload	Hours	+/++	–/––	–/––	Whole grains in foul–smelling feces
Acorn poisoning	Days	NAD	––/–––	–/––	Foetid diarrhoea

TABLE 3.40

Ancillary Tests to Determine Cause of Diarrhoea in Adult Cattle

Disease	Ancillary tests
Paratuberculosis	ELISA, fecal PCR, culture, histology
Salmonellosis	Fecal culture
Fasciolosis	Fecal egg count, coproantigen ELISA
Abomasal ulcers	Melaena, necropsy
Grain overload	Rumenocentesis with fluid pH< 4.5
Acorn poisoning	Circumstantial evidence. Necropsy with acorns in rumen

Paratuberculosis

3.231: Profuse "hose pipe" diarrhoea in beef cow during agonal stage of paratuberculosis. **3.232:** "Hose pipe" diarrhoea. **3.233:** Normal jejunum (bottom) with ridged paratuberculosis gut (top).

- **Key observations**: Chronic weight loss over three to six months. Profuse "pea soup" diarrhoea latterly.
- **Clinical examination**: NAD during early stages; afebrile, normal appetite.
- **Key history events**: Purchased animal although generally herd problem. 3.129, 3.130, 3.131, 3.132, 3.133, 3.134, 3.135

Salmonellosis

- **Key observations**: Foetid dysentery containing mucosal casts.
- **Clinical examination**: Fever, toxaemia, dehydration, weakness progressing to sternal recumbency, death.
- **Key history events**: Often outbreak of disease from external source e.g. watercourse. Clinical signs often appear following parturition.

Fasciolosis

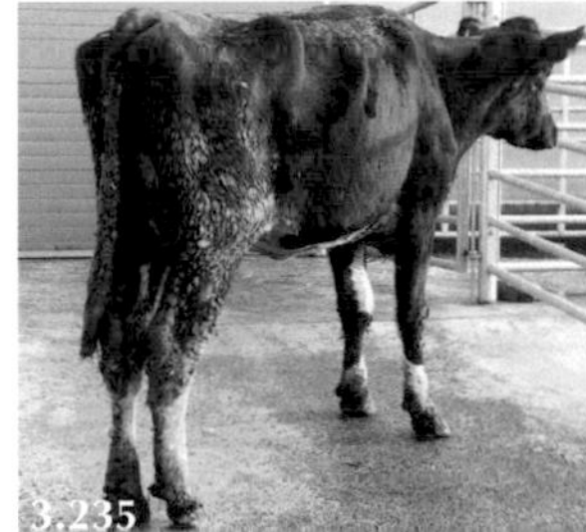
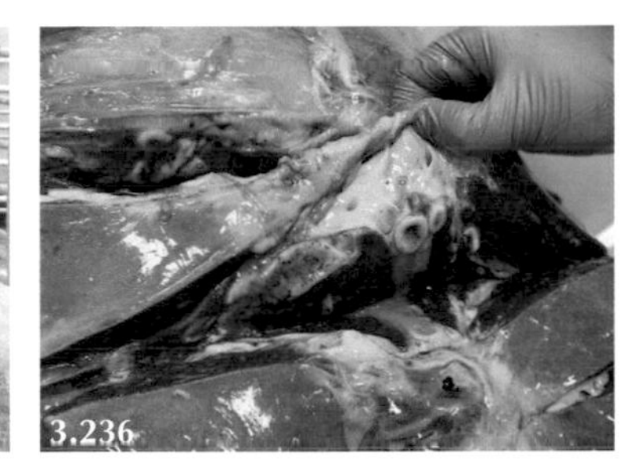

3.234: Chronic weight loss, diarrhoea and submandibular oedema in a beef cow in winter. **3.235:** Chronic weight loss and diarrhoea in a dairy heifer. **3.236:** Considerable biliary fibrosis caused by chronic fasciolosis.

- **Key observations**: Chronic weight loss and diarrhoea.
- **Clinical examination**: No specific clinical signs.
- **Key history events**: Several/many cattle in group affected. Difficult to implement control in dairy herds due to drug withdrawal times for milk. 3.136, 3.137

Abomasal Ulceration

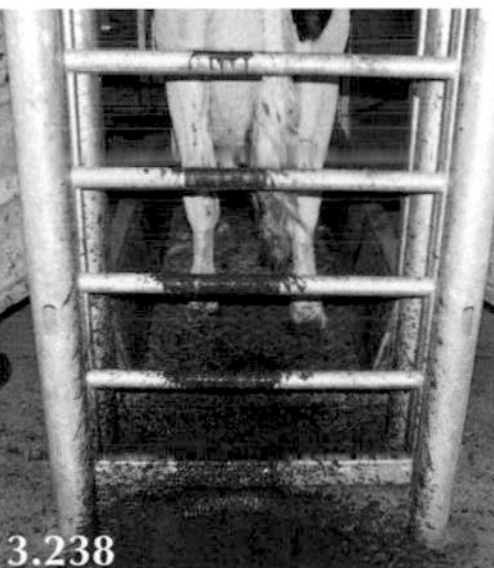

3.237: Painful expression, ears back. **3.238:** Profuse, often foetid, diarrhoea/melaena.

- **Key observations**: Chronic weight loss, poor milk production and melaena (dark sticky feces containing partly digested blood, as a result of internal bleeding).
- **Clinical examination**: Variable anaemia, abomasal distension/right dorsal displacement.
- **Key history events**: Dairy cows in first two months after calving. 3.138

Abomasal Perforation/Acute Fibrinous Peritonitis

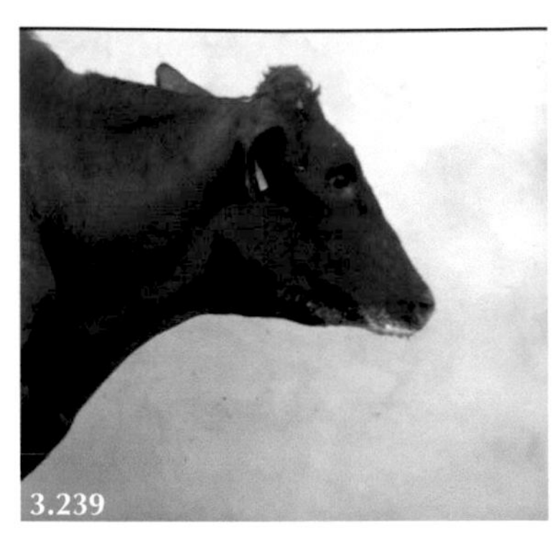
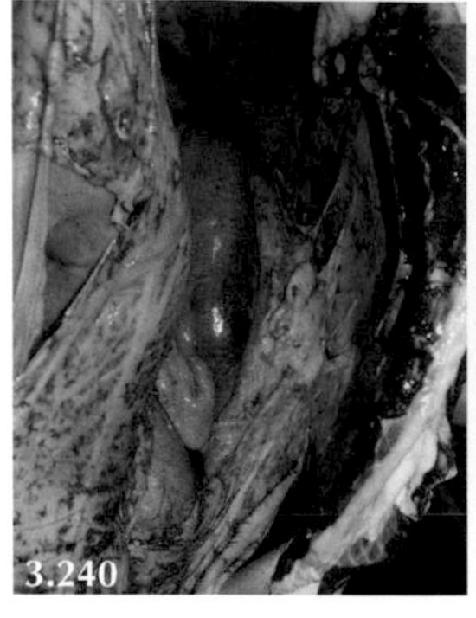
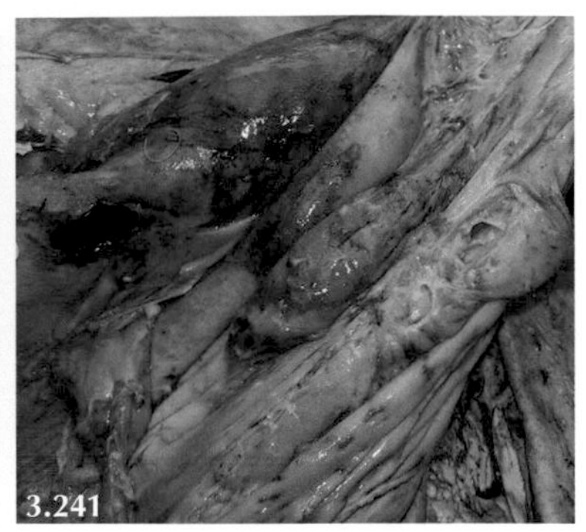

3.239: Sick cow with sunken eyes showing severe pain caused by abomasal perforation/acute fibrinous peritonitis. **3.240:** Necropsy reveals fibrinous exudate especially between abomasum and right flank. **3.241:** Abomasal perforation (upper left), adjacent inflamed abomasal wall with fibrin on serosal surface.

- **Key observations**: Sick cow, slow, dehydrated, very little milk yield, no appetite, painful expression/ears back.
- **Clinical examination**: Signs of peritonitis—arched back, painful abdomen, no ruminal activity, rapid pulse >100 beats per minute.
- **Key history events**: Rapid deterioration after several days' poor appetite and melaena. 🎥 3.139

Grain Overload

3.242: Recumbent heifer with ruminal atony, bloat and foetid diarrhoea caused by grain overload. **3.243:** Cattle breakouts with access to stored feed are the common cause of grain overload.

- **Key observations**: Sick cattle, foetid diarrhoea often containing whole grain.
- **Clinical examination**: Distended atonic rumen, dehydrated.
- **Key history events**: Often accidental access to large quantities of grain 12–24 hours earlier. 🎥 3.140, 3.141

Jejunal Haemorrhage

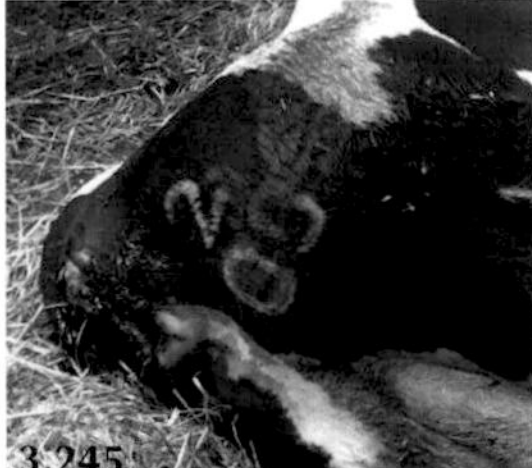

3.244: Weak, recumbent cow with painful expression. **3.245** and **3.246:** Passage of digested blood/melaena.

- **Key observations**: Sudden onset weakness, colic, partially digested blood/melaena.
- **Clinical examination**: Pain, cold extremities, anaemia, rapid pulse, severe dehydration, subnormal rectal temperature.
- **Key history events**: None. 🎥 3.142

Acorn Poisoning

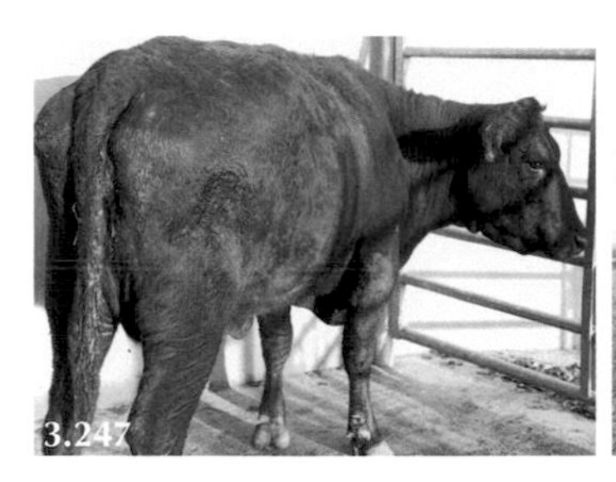

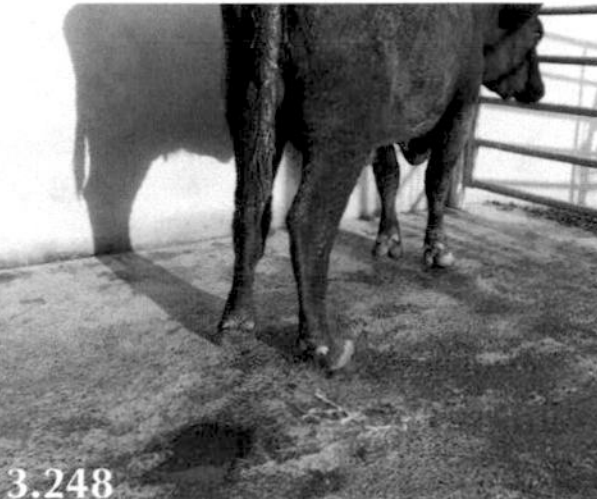

3.247: Rumen atony causing initial bloat, colic. **3.248:** Foetid mucoid diarrhoea.

- **Key observations**: Sudden onset, initial bloat, sick cattle, colic, foetid mucoid diarrhoea.
- **Clinical examination**: Atonic rumen, rapid dehydration, afebrile.
- **Key history events**: (Beef) cattle at pasture in late autumn; access to acorns follows storms.

Young Calves

- ETEC can cause sudden death without premonitory signs in susceptible one–three-day-old calves (non-vaccinated dams).
- Correction of severe metabolic acidosis in calves with viral-induced diarrhoea is readily achievable on farm.
- Salmonellosis is often quoted as a common cause of dysentery, but chronic pneumonia and septic joints may be more common.
- Improved hygiene, reduced stocking density and clean dry bedding will reduce most perinatal infections. In spring, cows and calves are turned out to pasture as soon as possible after calving.

TABLE 3.41

Common Causes of Diarrhoea in Calves

	Shape	Appetite	Demeanour	Pain	Feces
Milk scour	NAD	NAD/–	NAD/–	NAD	Diarrhoea
Viral entero–pathogens	+/++	–/––––	–/–––	+	Diarrhoea
Cryptosporidiosis	–/––	–	–	NAD	Pasty diarrhoea
Coccidiosis	–/––	–	–	NAD	Mucoid diarrhoea
ETEC	NAD	NAD	–/–––	NAD	Agonal diarrhoea
Salmonellosis	–/––	–/–––	–/–––	+/++	Dysentery
Intussusception	–	––/–––	–/–––	+/++	–––

TABLE 3.42

Ancillary Tests to Determine Cause of Diarrhoea in Calves

Disease/disorder	Ancillary tests Viral Entero-Pathogens
Milk scour	Circumstances—beef cow dam too much milk
Viral entero-pathogens	Calf-side tests
Cryptosporidiosis	Microscopic examination of gut specimens
Coccidiosis	Fecal smear, oocyst count and speciation. Response to treatment
ETEC	Calf-side tests
Salmonellosis	Bacteriology
Intussusception	Ultrasound confirmation is not easy. Surgery

Milk Scour

3.249: Diarrhoea but beef calf bright and alert.

- **Key observations**: Often beef calves within first few weeks of life with mild diarrhoea.
- **Clinical examination**: Calves bright and alert and not easily caught.
- **Key history events**: Beef cows have too much milk. Incorrect mixing/feeding milk substitutes in dairy calves. 3.143

Viral Entero-Pathogens

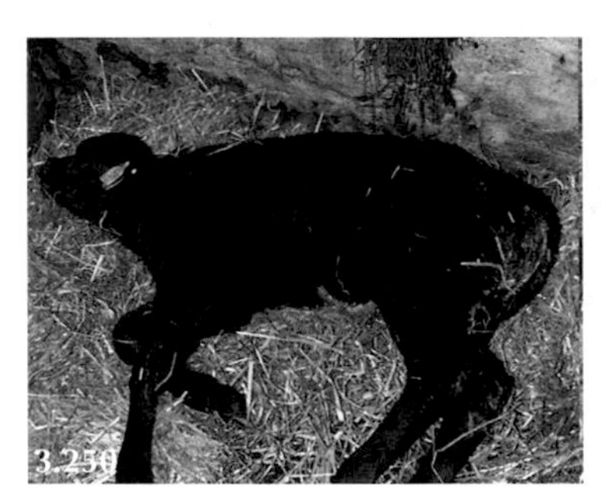

3.250 and **3.251:** Comatose 8–12-day-old beef calves with 2- to 3-day history of history of profuse diarrhoea. Often normal hydration status—weak due to acidosis. **3.252:** Successful treatment of metabolic acidosis within 6 hours (calf centre).

- **Key observations**: 8–12 days old, persistent diarrhoea, reluctance to suck cow, fluid-distended atonic abomasum/intestines (ileus).
- **Clinical examination**: Variable dehydration (administration of oral electrolyte solutions by farmer). Progress to recumbency and coma (acidosis).
- **Key history events**: Build-up of infection indoors towards end of calving period. Unvaccinated dams and/or failure of passive antibody transfer. 3.144

Cryptosporidiosis

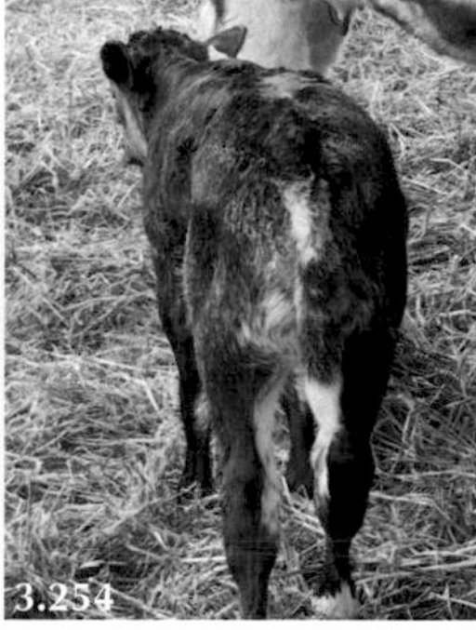

3.253 and **3.254:** 14-day-old calves showing persistent diarrhoea and condition loss. Affected calves have a poor appetite and may fail to suck their dam.

- **Key observations**: Seven to twenty-one days old, persistent diarrhoea, reluctance to suck cow, condition loss.
- **Clinical examination**: Variable dehydration (administration of oral electrolyte solutions by farmer).
- **Key history events**: Build-up of infection indoors towards end of calving period.

Coccidiosis

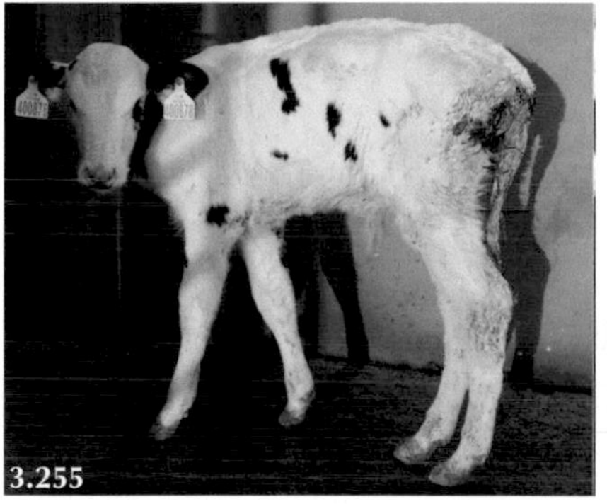
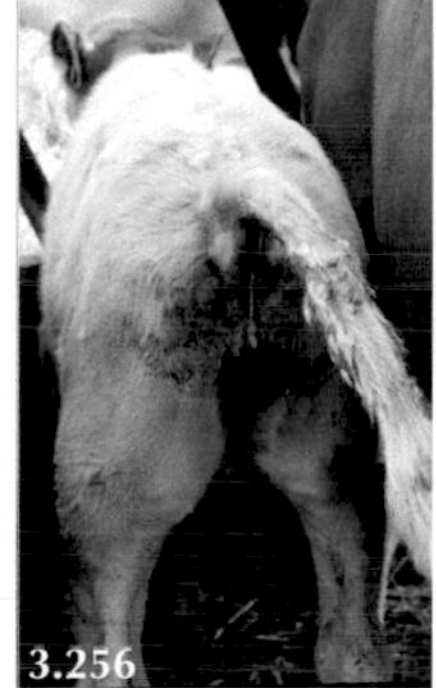

3.255: Poorly grown six-week-old dairy calf with chronic mucoid diarrhoea caused by coccidiosis. **3.256:** Tenesmus with mucoid diarrhoea. **3.257:** Two-month-old beef calves at pasture (housed for treatment) with tenesmus and fecal staining of the perineum caused by *Eimeria alabamensis*.

- **Key observations**: Rapid condition loss. Tenesmus, mucoid diarrhoea with flecks of blood.
- **Clinical examination**: Afebrile, gaunt appearance, tenesmus
- **Key history events**: Contaminated surface water in beef calves at pasture. 3.145

ETEC

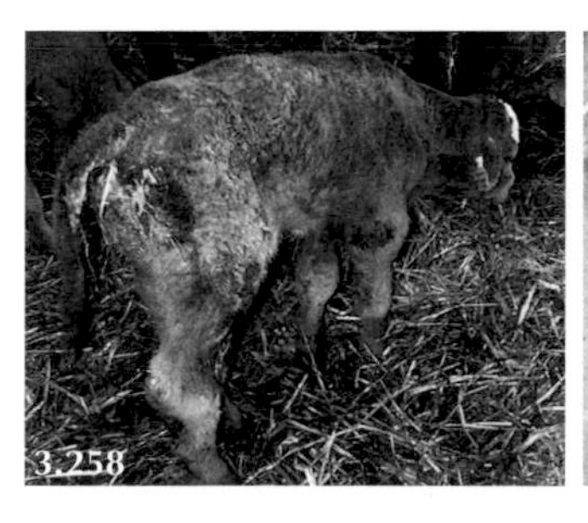
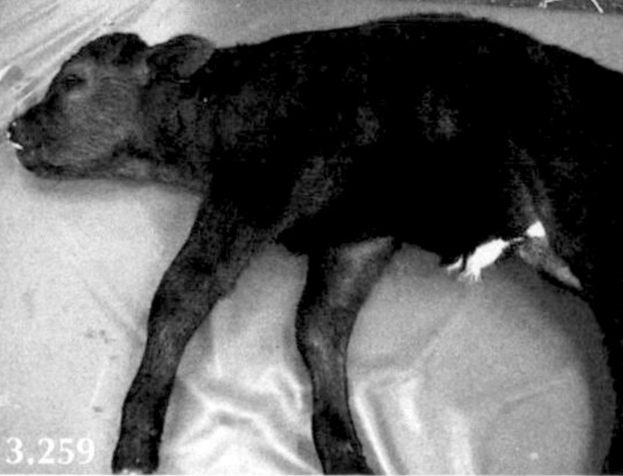

3.258: Weakness affecting a 24-hour-old calf. **3.259:** Collapse in a 36-hour-old beef calf.

- **Key observations**: Apparent sudden death in one–three-day-old calves before diarrhoea appears.
- **Clinical examination**: Very rapid dehydration, ileus.
- **Key history events**: Unvaccinated dam.

3.5

Common Causes of Poor Appetite and Low Body Condition

Adult Sheep

The common causes of poor appetite, reduced abdominal shape/content and poor body condition in adult sheep are ranked in order of annual incidence in the tables that follow to provide a guide. Severe lameness was the most common cause of poor body condition on the majority of 90 farms, totalling over 40,000 adult sheep, where all adults were handled as part of an OPA control programme. However, it is clearly acknowledged that disease incidence varies a great deal between farms; some farms monitored by the author had an annual mortality rate of 3–5 per cent caused by OPA before veterinary intervention whilst others never had a single case. Similarly, paratuberculosis caused 4–5 per cent premature culls/death per annum in some flocks and was rarely diagnosed in others.

- Severe lameness is the most common cause of reduced grazing, poor abdominal fill and condition loss in adult sheep regardless of management.
- Fasciolosis may cause poor pregnancy rate due to fetal resorption as well as severe weight/condition loss in high-risk years. Farmers rarely use proven fluke risk forecasts based on summer rainfall.
- Serology tests for paratuberculosis are not sensitive.
- Assessment of suspected respiratory diseases necessitates trans-thoracic ultrasound examination.
- Without laboratory confirmation, many "while you are here" farm necropsies are a waste of time and could be counter-productive.

TABLE 3.43

Common Differential Diagnoses of Poor Body Condition in Sheep

	Abdomen	Appetite	Demeanour	Toxaemia	Pain	Feces	Rectal temperature
Severe lameness	––/–––	–	NAD	NAD	++/+++	NAD	NAD/+
Parasitic gastro-enteritis	–/––	NAD	NAD	+	NAD	Diarrhoea/or anaemia	NAD
Fasciolosis	–/–––	–/–––	–/–––	NAD	NAD/++	NAD	NAD
Poor dentition	–/––	NAD	NAD	NAD	NAD	Fibrous	NAD
Paratuberculosis	––/–––	–	NAD/–	NAD	NAD	NAD/ Agonal diarrhoea	NAD
Abortion	–/––	NAD/––	NAD/––	NAD/++	NAD	NAD/–	NAD/+
Gangrenous mastitis	–/––	––/–––	––/–––	+++	++	––	++/+++
Chronic mastitis	–	NAD/–	NAD	NAD/+	NAD/+	NAD	NAD/+
OPA	–/––	NAD/–	NAD/–	NAD	NAD	NAD	NAD
Chronic suppurative pneumonia	–	–	NAD/–	NAD	NAD	NAD	NAD/+
Intestinal tumours	–/––	–	NAD	NAD	NAD/+	–/–––	NAD

DOI: 10.1201/9781003106456-23

Table 3.44

Diagnostic Tests for Common Causes of Poor Body Condition in Sheep

Condition	Diagnostic tests
Severe lameness	Examination foot/joint. Radiography
Parasitic gastro-enteritis	Fecal worm egg count (may not be proportional to parasite burden)
Fasciolosis	Liver enzymes; GLDH. Coproantigen ELISA. Hypoalbuminaemia, hyperglobulinaemia. Necropsy. Fecal egg count if patent
Poor dentition	Palpation, inspect teeth using gag/torch. Radiography—oblique angle
Paratuberculosis	Hypoalbuminaemia and normal/low globulin concentration. Serum ELISA. Fecal examination/PCR. Necropsy with histology
Abortion	Vaginal discharge. Abdominal palpation. Ultrasonography to confirm abortion
Gangrenous mastitis	Clinical examination
Chronic mastitis	Palpation. Ultrasonography
OPA	Ultrasonography. Necropsy. Histology
Chronic suppurative pneumonia	Ultrasonography. Necropsy
Intestinal tumours	Necropsy. Histology

Severe Lameness

Depending upon farm management, severe lameness/pain can be the most important reason for reduced grazing causing emaciation.

3.260–3.262: Severe, non-weight bearing lameness leading to reduced abdominal fill and marked loss of condition score. The overgrown forefeet (**3.260**) indicates much reduced hoof horn wear (lameness) over four to six weeks (minimum).

- **Key observations**: Severe lameness, prolonged periods lying down, grazing on knees/hygromas. Overgrown hoof horn of affected foot/plantigrade stance of contralateral foot.
- **Clinical examination**: Foot lesions are readily identifiable but often difficult to resolve.
- **Key history events**: Sudden lameness problems may result from purchased sheep not quarantined (e.g. introduction of CODD from dairy farm) but most often inadequate and delayed treatment. 🎥 3.146, 3.147, 3.148

Parasitic Gastro-Enteritis

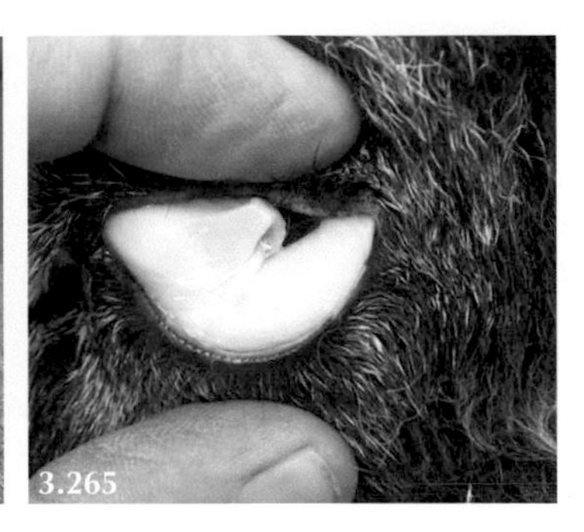

3.263 and **3.264:** Diarrhoea, poor abdominal fill, poor coat and low condition score. **3.265:** Emaciated ewe with anaemia caused by haemonchosis.

- **Key observations**: Diarrhoea (not haemonchosis), low condition score, several/many sheep affected. May have concurrent fasciolosis.
- **Clinical examination**: Usually NAD except diarrhoea or anaemia in case of haemonchosis.
- **Key history events**: No farm parasite control plan. Purchased sheep not quarantine drenched. Anthelmintic resistance. 🎥 3.149, 3.150

Fasciolosis

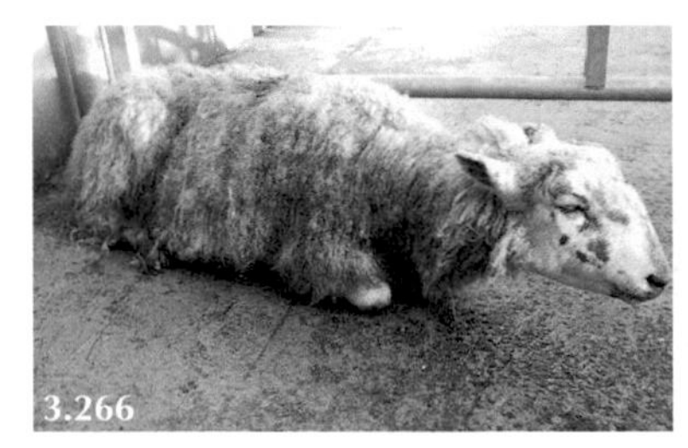
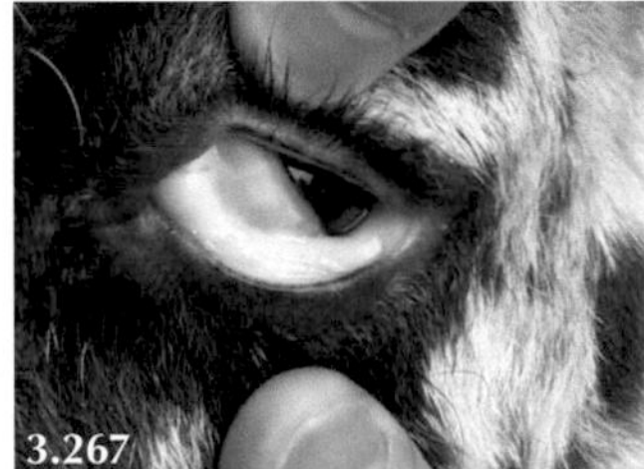
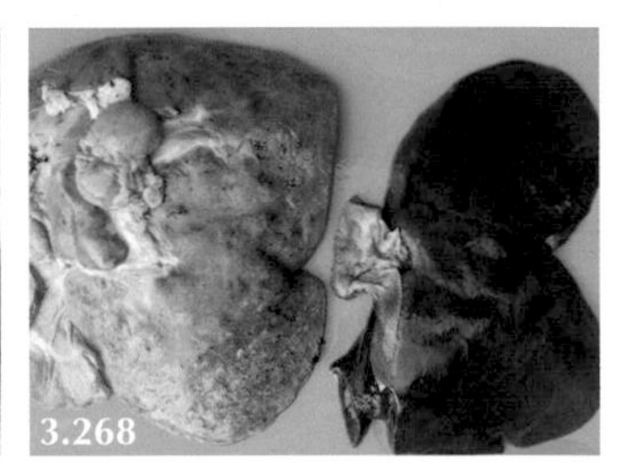

3.266: Ewe with subacute fasciolosis. **3.267:** Pale ocular mucous membranes representing anaemia. **3.268:** Necropsy findings showing subacute fluke-affected liver on left compared to normal liver on right.

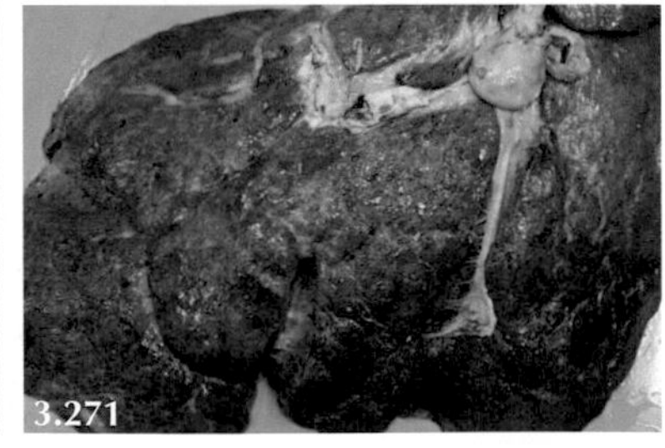

3.269: Ewe with subacute fasciolosis. **3.270:** Submandibular oedema in chronic fasciolosis. **3.271:** Necropsy findings of chronic fasciolosis.

- **Key observations**: Weak, poor body condition/emaciated. Several/many sheep affected. May have concurrent PGE.
- **Clinical examination**: Anaemia. Bottle jaw in some cases of chronic fasciolosis.
- **Key history events**: No parasite control plan. No quarantine drench. Inappropriate flukicide for developing fluke stages. 🎥 3.151, 3.152, 3.153, 3.154, 3.155

Poor Dentition

3.272 and **3.273:** Ewes with poor abdominal fill caused by molar dentition problems. **3.274:** Difficulty masticating roughages with food pouching in cheeks.

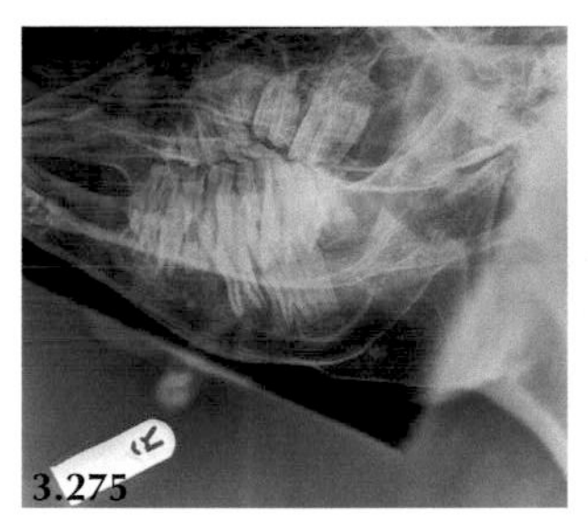
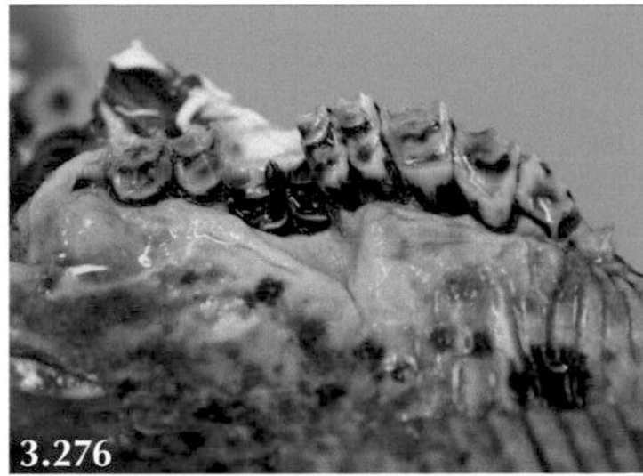

3.275: Oblique radiographic view of head revealing molar dentition problems. **3.276:** Necropsy reveals a very irregular maxillary dental arcade plus excessive wear of second right molar tooth. **3.277:** Lower jaw of another sheep showing missing lower premolars and grossly overgrown molars.

- **Key observations**: Quidding, food pouching in cheeks.
- **Clinical examination**: Palpate dental arcade through cheek. Inspect teeth with mouth gag in place—often resented by sheep. Radiography employing oblique view.
- **Key history events**: Chronic weight loss but normal appetite. 3.156, 3.157

Paratuberculosis

3.278–3.280: Emaciated ewes with paratuberculosis. Poor fleeces. Diarrhoea is often an agonal finding. Secondary dermatophilosis is commonly seen on the ears and face caused by debility.

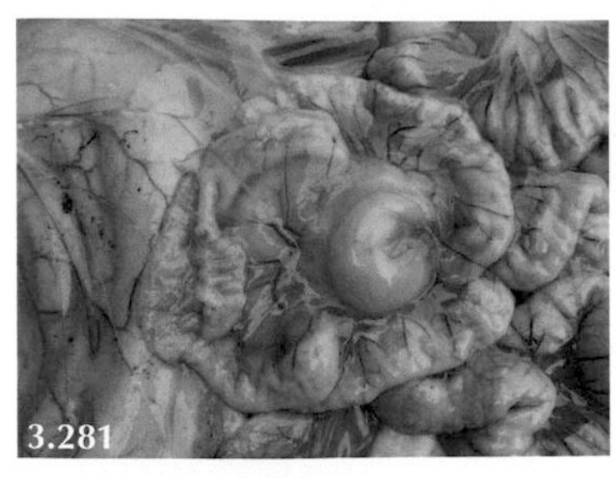
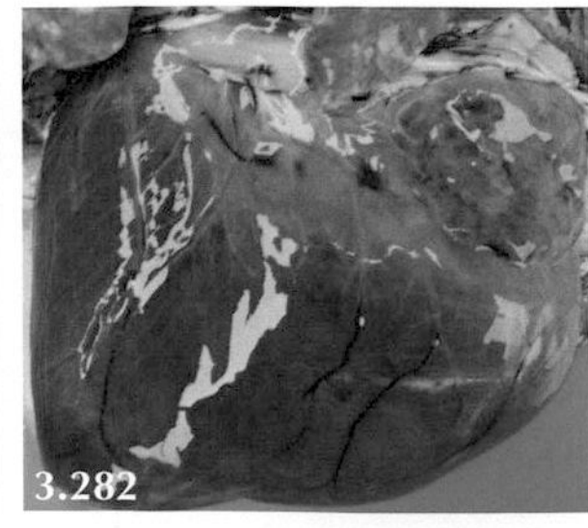
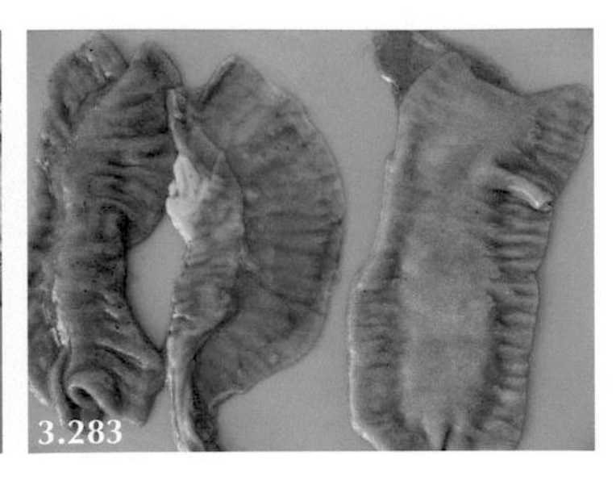

3.281: Serous atrophy of omental fat with prominent lacteals and grossly enlarged mesenteric lymph node (centre of image). **3.282:** Serous atrophy of fat in epicardial groove. **3.283:** Middle sample is a section of incised gut from normal ewe flanked by grossly thickened small intestine from ewe with paratuberculosis.

- **Key observations**: Chronic loss of body condition over four to six months.
- **Clinical examination**: NAD apart from emaciation. May show diarrhoea. McMaster slide often shows very high number (many thousands) of strongyle eggs including *N. battus* and lungworm larvae. Serology tests have poor sensitivity.
- **Key history events**: Previous cases in group/farm. No paratuberculosis vaccination programme. 3.158, 3.159, 3.160, 3.161, 3.162, 3.163

Abortion

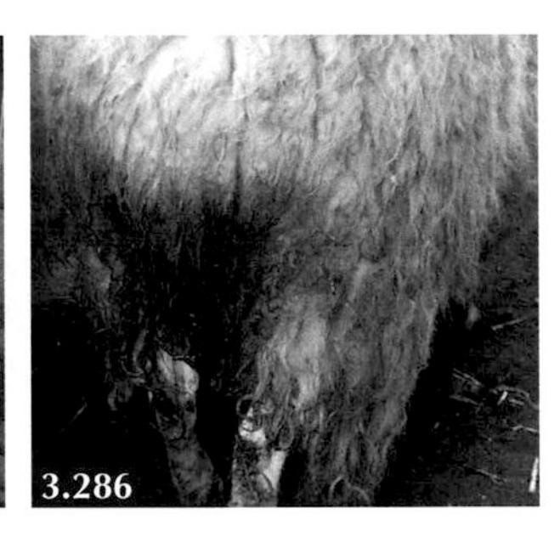

3.284 and **3.285:** Sudden reduction of abdominal size after abortion. **3.286:** Foetid vaginal discharge visible on wool of perineum.

- **Key observations**: Sudden loss of abdominal contents. Vaginal discharge. Several abortions/storm over past week.
- **Clinical examination**: No palpable fetus. Ultrasound fails to find significant uterine fluid and fetus.
- **Key history events**: Final trimester, often last three weeks of pregnancy.

Gangrenous Mastitis

3.287: Acute toxaemia. **3.288:** Swollen mammary gland with sharply defined gangrene which often causes ipsilateral hindleg lameness.

- **Key observations**: Acute toxaemia, swollen mammary gland, sharply defined gangrene, often shows ipsilateral hindleg lameness. Hungry lambs.
- **Clinical examination**: Toxic mucous membranes, dehydration. Swollen, oedematous then gangrenous, mammary gland(s) with serum-like secretion.
- **Key history events**: Inadequate nutrition while nursing four–six-week-old twin lambs. 3.164, 3.165

Chronic Mastitis

3.289 and **3.290:** Reduced abdominal fill but very swollen left mammary gland and associated wool slip. **3.291:** Necropsy reveals extensive abscessation and surrounding fibrosis.

- **Key observations**: Often show ipsilateral hindleg lameness. Poor lamb(s).
- **Clinical examination**: Swollen, oedematous mammary gland(s).
- **Key history events**: Acute episode of mastitis several months previously. 3.166

OPA

3.292 and **3.293:** Emaciated ewes with poor abdominal fill; the ewe in the centre shows woolslip. **3.294:** Advanced OPA encountered in emaciated ewes.

- **Key observations**: Poor abdominal fill and emaciation during agonal phase.
- **Clinical examination**: Increased respiratory and effort after gathering.
- **Key history events**: History of condition on farm. Often first encountered in purchased ram. 3.167, 3.168

Chronic Suppurative Pneumonia

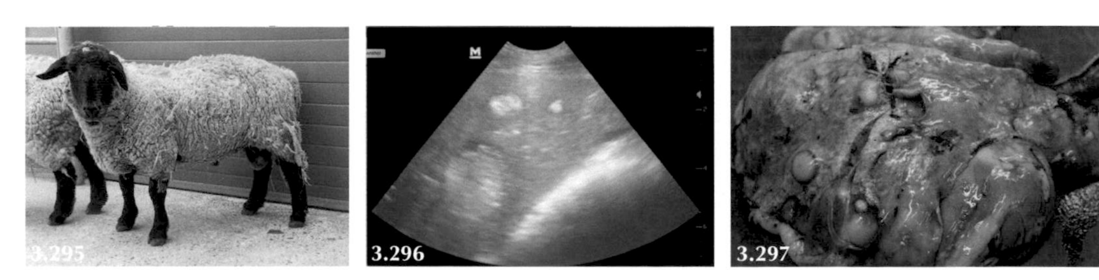

3.295: Suffolk ram in poor body condition. **3.296:** Sonogram showing lung fibrosis with widespread abscessation. **3.297:** Lung abscessation and ventral consolidation.

- **Key observations**: Weight loss over several months.
- **Clinical examination**: No adventitious sounds on auscultation.
- **Key history events**: Often rams—uncertain whether management issue or financial value. 3.169

Cud-Spilling

Cud-spilling is an uncommon cause of weight loss more often seen in pedigree rams and is clearly illustrated in videos 3.170, 3.171.

Intestinal Tumours

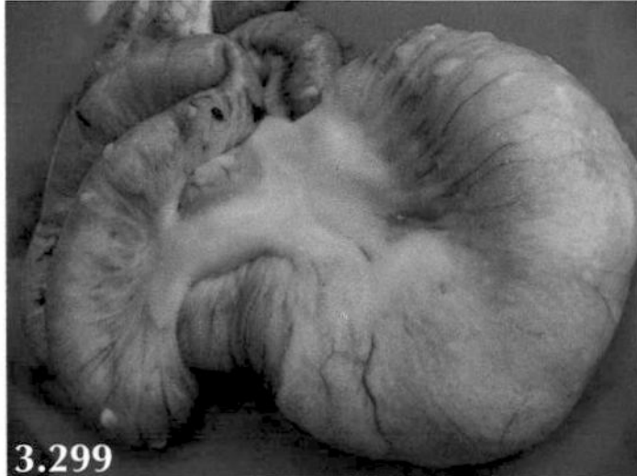
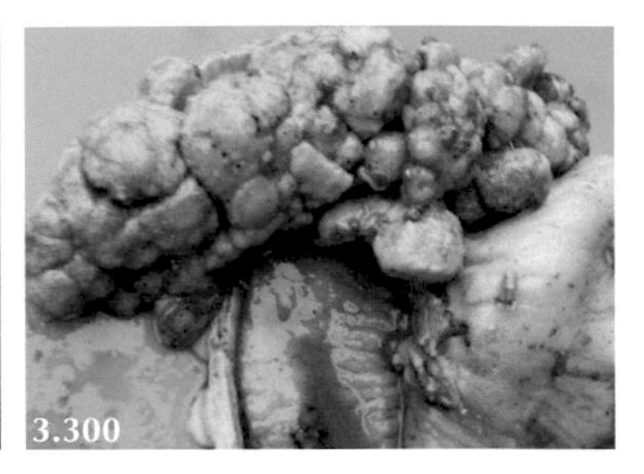

3.298: Reduced contents of abdominal viscera. **3.299** and **3.300:** Tumour arising from the intestinal mucosa occluding the lumen greatly restricting passage of digesta.

- **Key observations**: Poor abdominal fill and emaciation during agonal phase.
- **Clinical examination**: Reduced fecal output. Ascites may not be present if no trans-coelomic spread.
- **Key history events**: None. 3.172

Lambs

TABLE 3.45

Common Causes of Poor Body Condition in Young Lambs

	Shape	Appetite	Demeanour	Pain	Feces
Starvation	–/–––	NAD	NAD/–	NAD	NAD
Coccidiosis	–/––	–	–	Tenesmus	Mucoid diarrhoea
Peritonitis (chronic)	–/––	––/–––	––/–––	+/++	–/––
Hepatic necrobacillosis	–/––	–/––	–/––	+/++	–

TABLE 3.46

Ancillary Tests for Common Causes of Poor Body Condition in Young Lambs

	Ancillary tests
Starvation/rejection	Palpate lamb's abdomen. Ultrasound scan abomasum. Inspect dam's udder for mastitis. Observe dam's behaviour/attitude to lamb
Coccidiosis	Fecal smear (speciation). Response to treatment
Peritonitis (chronic)	Ultrasonography; abdominocentesis of any exudate but often only small quantities. Necropsy. Response to antibiotic therapy (penicillin) is poor.
Hepatic necrobacillosis	Ultrasonography of liver. Response to antibiotic therapy (penicillin) is generally poor

Starvation

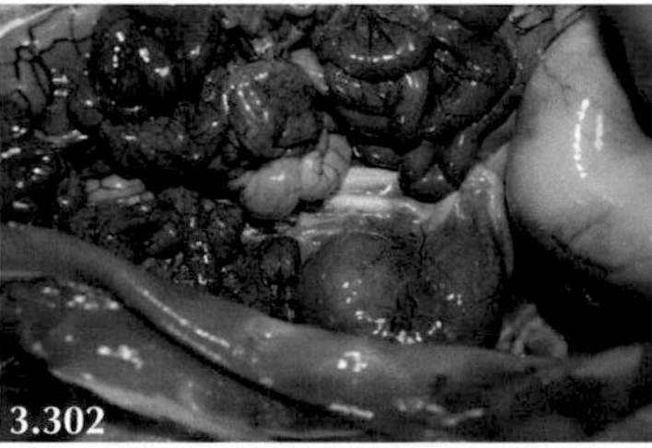

3.301: Dull, arched back, skin folds visible. Age consistent with rejection by dam. **3.302:** Mobilisation of fat reserves leaving no omental or peri-renal fat.

- **Key observations**: Dull, skin folds visible, arched back. Frequent attempts to suck but dam walks away.
- **Clinical examination**: Dehydrated, empty abdomen/abomasum. Check ewe's udder.
- **Key history events**: Rejection by dam (often fostered lamb), ewe has mastitis or other reason for poor milk supply. 🎥 3.173, 3.174, 3.175

Coccidiosis

3.303: Mucoid diarrhoea, poor abdominal fill. **3.304:** Chronic weight loss, poor abdominal fill, mucoid diarrhoea.

- **Key observations**: Several/many lambs affected, dull, occasional tenesmus, mucoid diarrhoea with flecks of fresh blood.
- **Clinical examination**: Empty abdomen.
- **Key history events**: Contaminated feed area. No coccidiosis medication. 🎥 3.176, 3.177, 3.178

Chronic Peritonitis

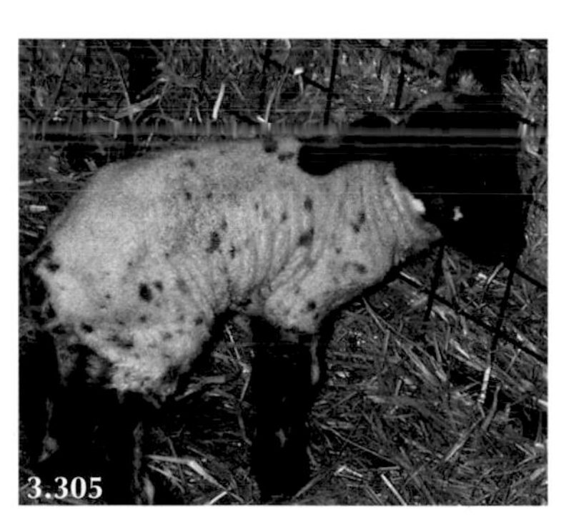

3.305: Dull, head down with an arched back.

- **Key observations**: Dull, arched back, painful expression (adhesions), does not suck.
- **Clinical examination**: Empty abdomen/abomasum unless significant accumulations of peritoneal exudate/adhesions blocking digesta.
- **Key history events**: Ineffective navel dressing, dirty lambing environment.

Hepatic Necrobacillosis

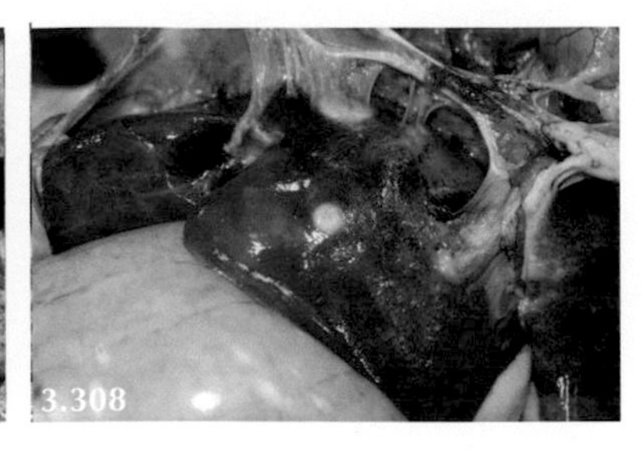

3.306: Left hand lamb affected; normal twin. **3.307:** Dull, head down with an arched back. **3.308:** Necropsy reveals liver abscesses and widespread adhesions to the diaphragm.

- **Key observations**: Dull, arched back, painful expression (adhesions), does not suck.
- **Clinical examination**: Empty abdomen/abomasum unless adhesions block digesta.
- **Key history events**: Ineffective navel dressing, dirty lambing environment. 🎥 3.179

Intussusception

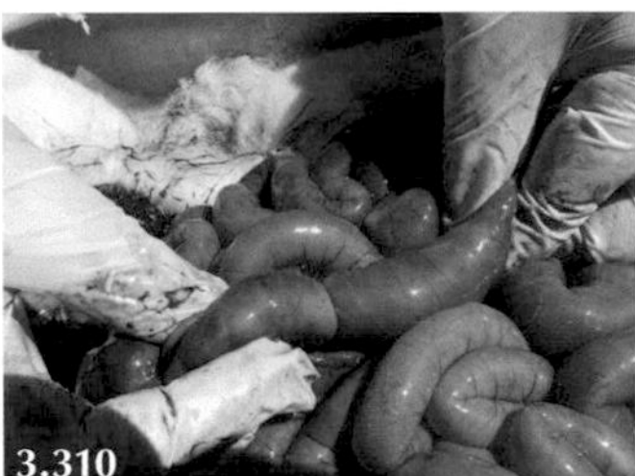

3.309: Lamb showing an arched back suggestive of abdominal pain. **3.310:** Necropsy confirms the intussusception.

- **Key observations**: Dull, arched back, painful expression, does not suck.
- **Clinical examination**: Initially, there is reduced abdominal fill due to the lamb not sucking but this changes with advancing disease and fluid sequestration proximal to lesion.
- **Key history events**: Often none identified. 🎥 3.180

Adult Cattle

This section includes conditions and diseases that have resulted in poor appetite and reduced abdominal fill. While adult cattle with lameness of various causes, paratuberculosis, endocarditis, liver abscessation and pyelonephritis present with much reduced abdomen fill and poor appetite, chronic weight/body condition score loss has occurred over the previous weeks or months therefore the main presenting sign is chronic weight loss. There will be considerable overlap between these two common presenting complaints of poor appetite and chronic weight loss; the main difference is duration which may not be known by the clinician. 🎥 3.181

- Severe lameness is the most common cause of reduced abdominal fill and rapid condition loss in dairy cows.
- Chronic bacterial pneumonias are commonly under-diagnosed because there are no adventitious sounds—accurate diagnosis necessitates ultrasound examination.
- Fasciolosis may be a problem in dairy herds because of extended milk withhold times after treatment.
- Herd control of paratuberculosis is difficult; serological tests are not sensitive.
- Ultrasound examination is essential to diagnose localised peritonitis associated with traumatic reticulitis, liver abscesses, endocarditis and pyelonephritis; routine haematology results, acute phase proteins and various enzymes will not differentiate between them.

TABLE 3.47

Common Causes of Poor Appetite and Low Body Condition Score in Adult Cattle

	Abdomen Volume	Appetite	Demeanour	Toxaemia	Pain	Feces	Rectal temperature
Severe lameness	–/––	–/––	–/––	NAD	++/+++	NAD	NAD
Coliform mastitis	–/––	–/––	–/–––	+/+++	NAD	Diarrhoea	NAD/–
Chronic pneumonia	–	–	–	NAD/+	NAD/+	NAD	NAD/+
Metritis	–/––	–/––	–/––	+/++	NAD/+	Diarrhoea	+/++
Ketosis	–/––	–/––	Nervous ketosis	NAD	NAD	Constipated	NAD
Paratuberculosis	––/–––	NAD/–	NAD	NAD	NAD	Profuse diarrhoea	NAD
LDA	–/––	––	–	NAD	NAD	–/––	NAD
Peritonitis	–/–––	–––	–/–––	++/+++	+/+++	–/–––	NAD/+
Summer mastitis	–/––	–/––	–/––	+/++	+/++ Joints	NAD	++
Endocarditis	–/––	/––	–/––	NAD/+	+/++ Joints	–	NAD/+
Liver abscesses	–/––	–	NAD/–	NAD	NAD/+	NAD	NAD/+
Traumatic reticulitis	+/––	–––	––	+	+++	Nil	+
Pyelonephritis	–/––	––	–/––	+	NAD	–	NAD/+

TABLE 3.48

Ancillary Tests for Common Causes of Poor Appetite and Low Body Condition Score in Adult Cattle

Severe lameness	Clinical examination. Radiography
Mastitis	California mastitis test. Bacteriology
Pneumonia	Ultrasonography
Metritis	Retained placenta? Vaginal examination
Ketosis	Serum 3-OH butyrate >4.0 mmol/l. Rothera's reagent positive
Paratuberculosis	Serology. Fecal PCR. Culture
Peritonitis	Ultrasonography. Abdominocentesis
LDA	Surgical correction
Endocarditis	Ultrasonography
Liver abscesses	Ultrasonography. Liver enzymes. Serum proteins
Pyelonephritis	Rectal palpation of ureters and left kidney, ultrasonography

Acute Severe Lameness

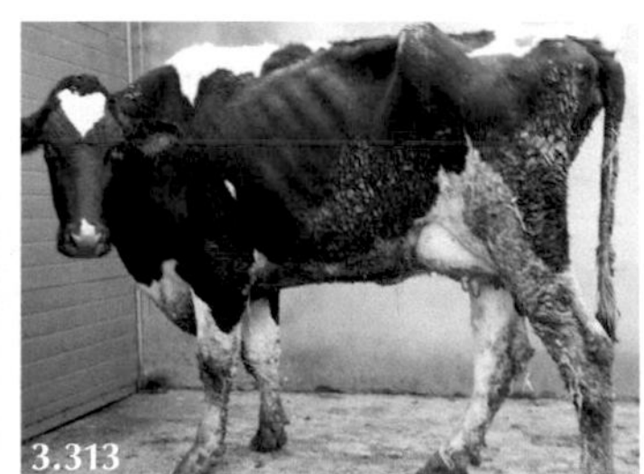

3.311: Simmental bull with poor appetite/reduced rumen fill as a consequence of severe left hindleg lameness. **3.312** and **3.313:** Dairy cows in early lactation with sudden onset lameness and poor appetite.

- **Key observations**: Severe sudden onset lameness. Long periods in sternal recumbency.
- **Clinical examination**: White line abscesses and interdigital necrobacillosis commonly cause acute lameness.
- **Key history events**: New roadways/walking surfaces often lead to lameness problems. 🎥 3.182, 3.183, 3.184

Mastitis

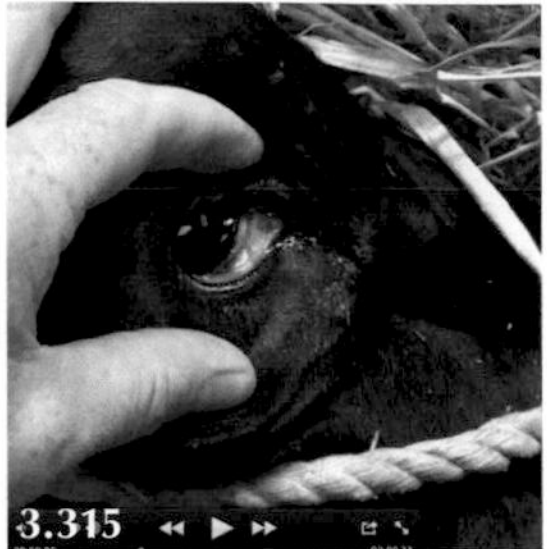

3.314: Inappetance, sternal recumbency in dairy cow with coliform mastitis.
3.315: Marked dehydration.

- **Key observations**: Off food, weakness, recumbency.
- **Clinical examination**: Dehydrated, possible diarrhoea. Serum-like mammary secretion.
- **Key history events**: Sudden onset, often within several days of calving. 🎥 3.185

Chronic Pneumonia

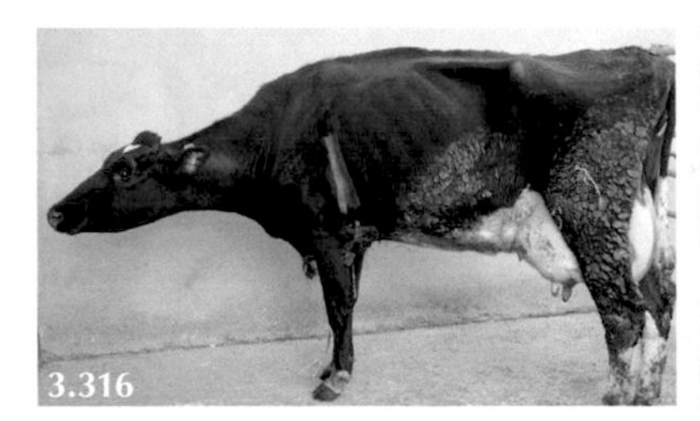

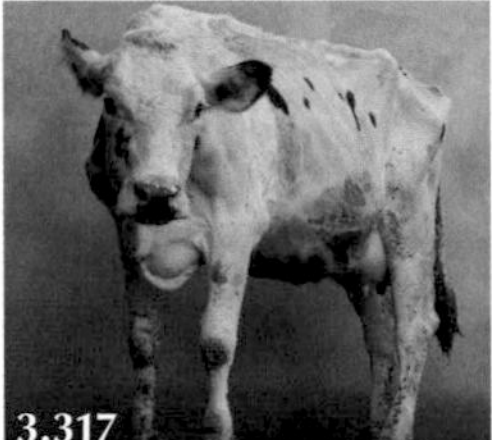

3.316–3.318: Images show much reduced abdominal fill in dairy cows with bronchiectasis.

- **Key observations**: Reduced appetite and poor abdominal fill.
- **Clinical examination**: Often no discernible adventitious sounds. Neck extended, head lowered. Occasional nasal discharge, coughing, halitosis.
- **Key history events**: Often flare-up of chronic bronchiectasis following calving. 🎥 3.186, 3.187

Metritis

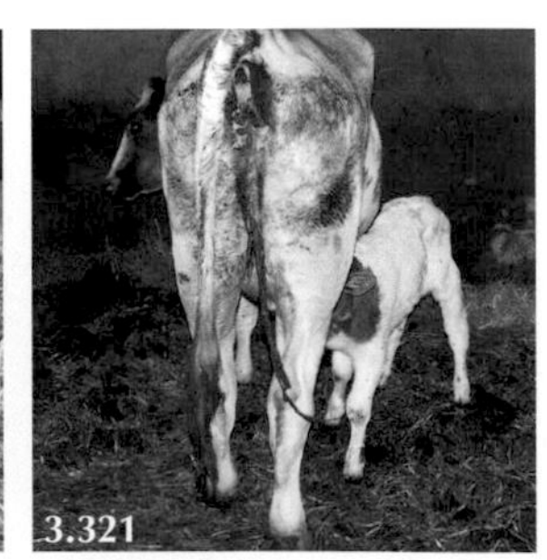

3.319: Much reduced appetite abdominal fill in a dairy cow with retained placenta. **3.320** and **3.321:** Poor appetite in a beef cow with retained placenta and metritis.

- **Key observations**: Retained placenta for three to five days, foetid vaginal discharge.
- **Clinical examination**: Foetid "oxtail soup" uterine contents.
- **Key history events**: Twins, hypocalcaemia, dystocia all predispose to metritis.

Ketosis

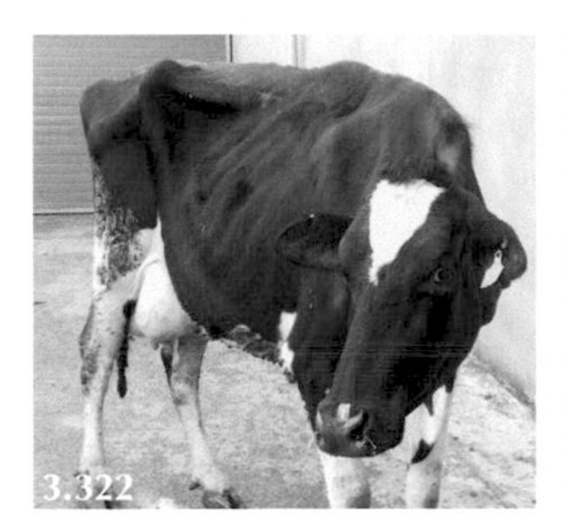
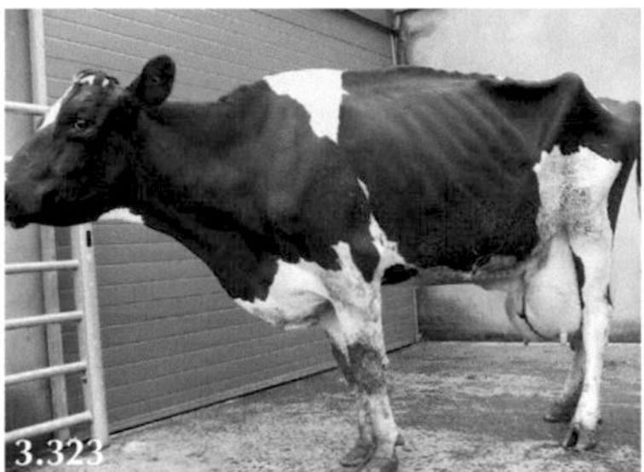

3.322 and **3.323:** Inappetance with very poor abdominal fill in Holstein cow. **3.324:** Cow showing behavioural signs of nervous ketosis by continually licking her flank.

- **Key observations**: Stop eating concentrates, rapid fall in milk yield.
- **Clinical examination**: Ketotic smell on breath. Varying behavioural signs including licking inanimate objects.
- **Key history events**: Two to four weeks post-calving. Check for LDA. 3.188

Paratuberculosis

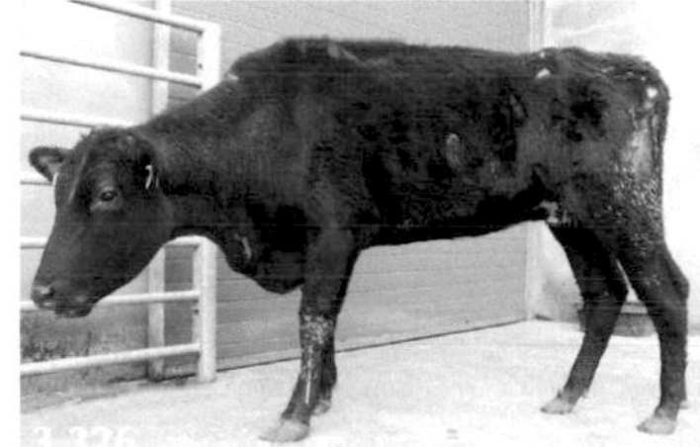

3.325–3.327: All three beef cows show reduced abdominal fill but also early weight loss.

- **Key observations**: Reduced abdominal fill. Diarrhoea.
- **Clinical examination**: NAD.
- **Key history events**: Reduced abdominal fill before rapid loss of body condition. 3.189, 3.190, 3.191

Peritonitis

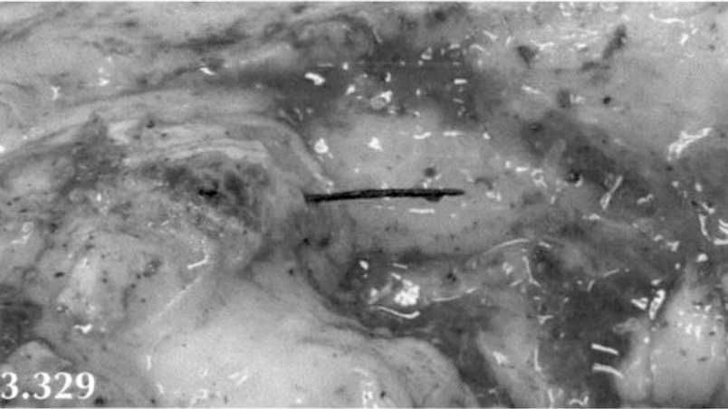

3.328 and **3.329:** Cow presents with poor appetite, painful expression and arched back. Foreign bodies occasionally become walled off causing chronic signs of poor production and weight loss.

- **Key observations**: Arched back, painful expression, reduced appetite.
- **Clinical examination**: Reduced rumen motility. Positive wither's pinch test but ultrasonography much more useful.
- **Key history events**: Sudden fall in milk yield to 2–5 litres. 3.192, 3.193, 3.194

LDA

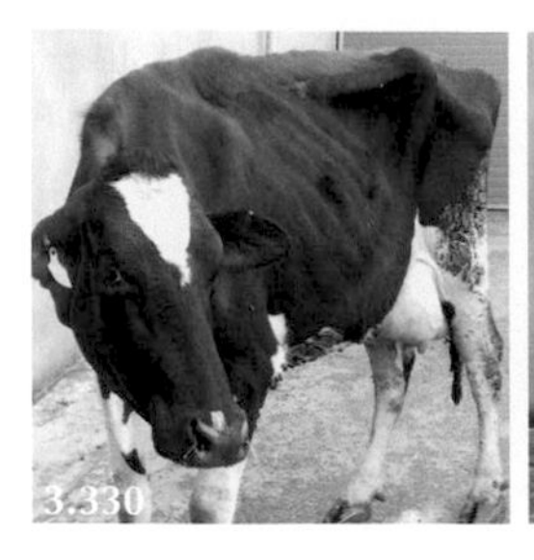

3.330 and **3.331:** Both dairy cows show poor appetite and much reduced abdominal fill.

- **Key observations**: Poor appetite although eat roughage.
- **Clinical examination**: High-pitched resonant "pings" audible in left sublumbar fossa.
- **Key history events**: Two to four weeks post-calving. 3.195

Summer Mastitis

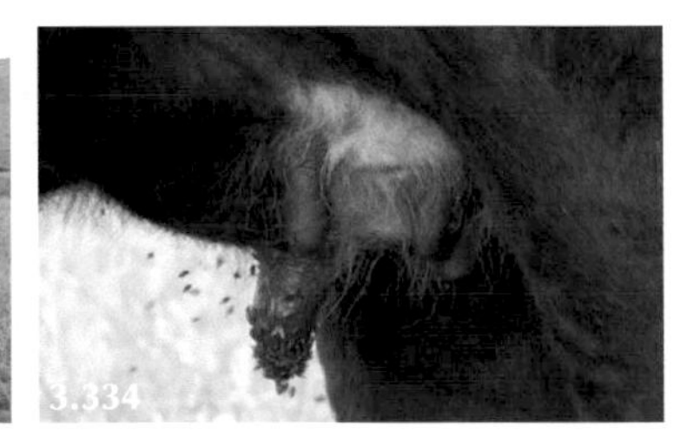

3.332: Sick beef cow isolated in field, swollen right fore mammary gland and teat. **3.333:** Lameness (arched back) caused by distended painful hock joints. Note swollen teat. **3.334:** Close-up of udder reveals a swollen right fore teat with many flies around the teat orifice.

- **Key observations**: Isolated in field. Swollen teat/quarter. Large number of flies around teat orifice.
- **Clinical examination**: Sick cow, fever (ca. 39.7°C). Oedematous, painful teat/quarter, secretion foul-smelling "grains of rice" in yellow watery liquid. Often have multiple joint effusions, especially hocks, causing pain and marked lameness.
- **Key history events**: Summer months, dry period. 3.196

Endocarditis

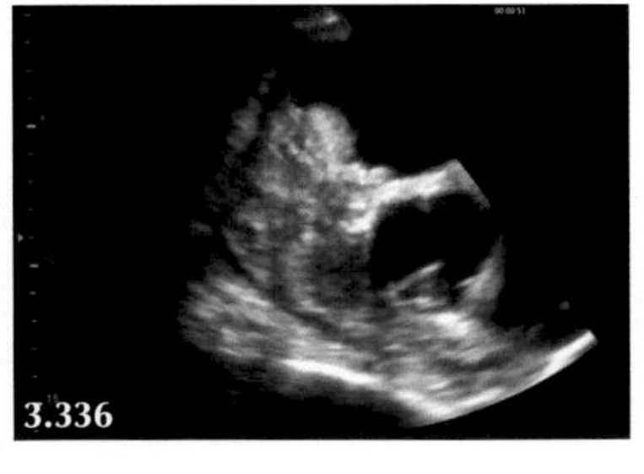

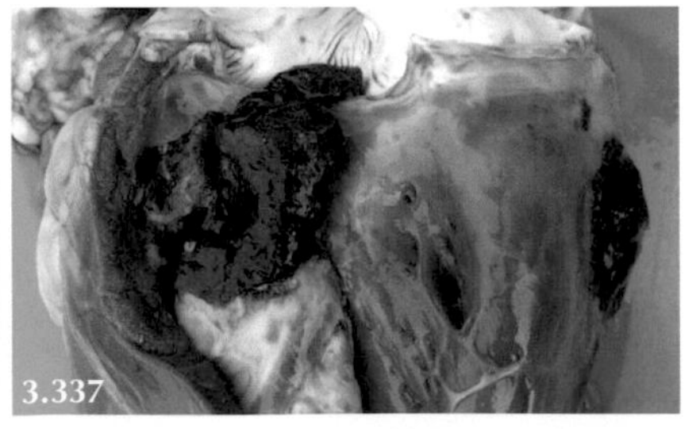

3.335: Cattle with endocarditis often present with lameness and joint effusions. **3.336:** Ultrasound examination can identify vegetative lesion on tricuspid heart valve. **3.337:** Necropsy finding of vegetative lesion.

- **Key observations**: Poor appetite, often moderate lameness arising from several joints.
- **Clinical examination**: Heart murmur not often audible. Joint effusions.
- **Key history events**: Chronic vague lameness. 3.197, 3.198

Liver Abscesses

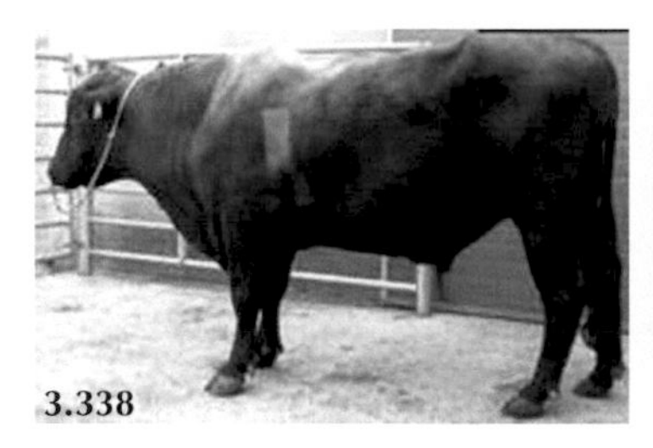
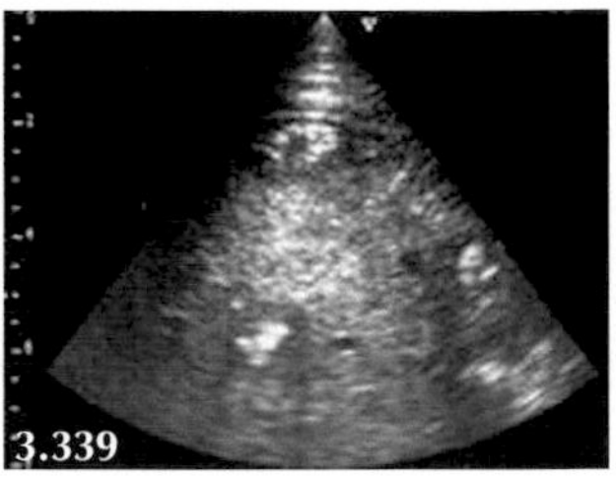
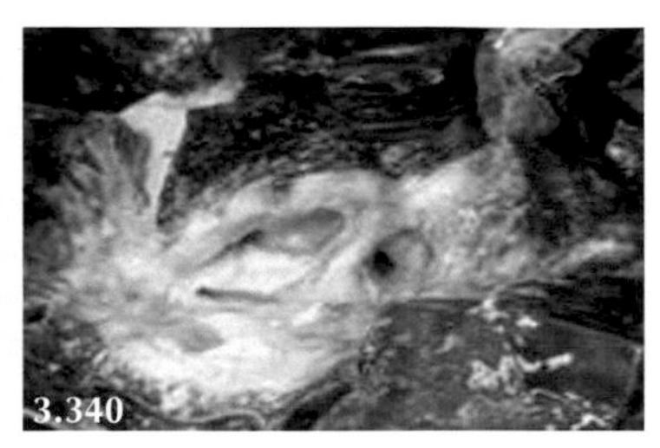

3.338: Beef bull with poor appetite and reduced abdominal fill. **3.339:** Ultrasound examination identifies several 2–3 cm diameter abscesses throughout the liver parenchyma. **3.340:** Necropsy confirms the presence of liver abscesses.

- **Key observations**: History of poor appetite leading to weight loss.
- **Clinical examination**: No specific clinical signs.
- **Key history events**: Associated with high concentrate feeding—dairy cows and beef bulls. 🎥 3.199

Pyelonephritis

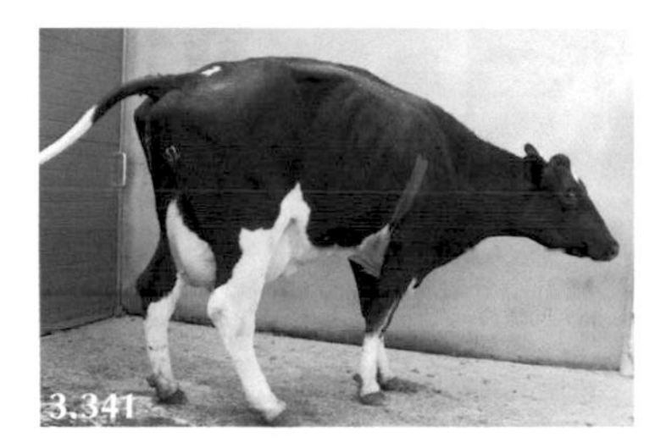
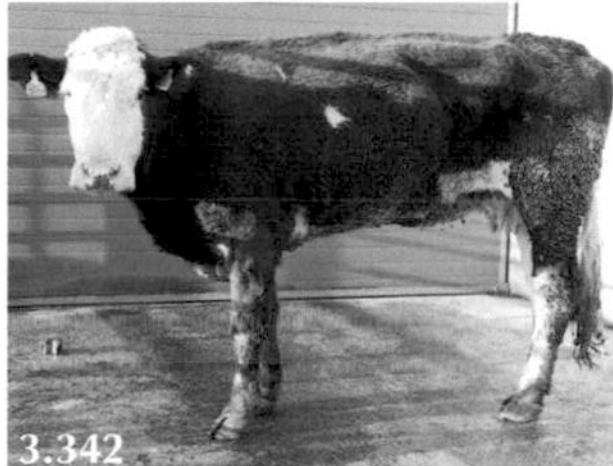
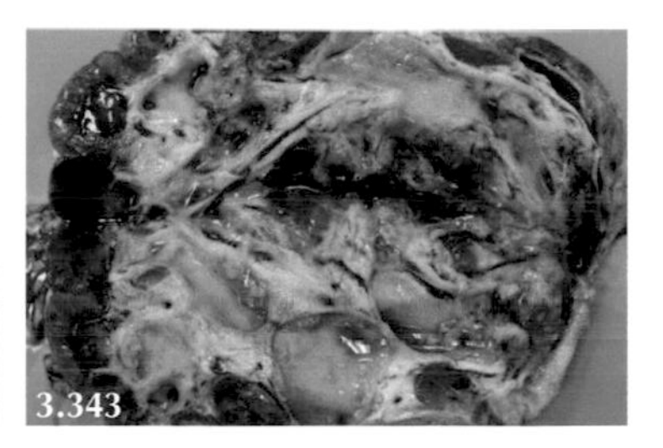

3.341: Frequent urination. **3.342:** Beef cow with poor appetite and reduced abdominal fill. **3.343:** Necropsy confirms diagnosis of pyelonephritis (Centre).

- **Key observations**: Frequent urination, small flecks of blood and pus on tail/perineum.
- **Clinical examination**: Rectal examination reveals thickened bladder wall and thickened ureters.
- **Key history events**: Often appear as small clusters in beef cows several months after calving. 🎥 3.200

Growing Cattle

- Severe lameness is the most common cause of reduced abdominal fill and rapid condition loss in growing cattle.
- The clinical significance of lung pathology caused by bacterial pneumonia (bronchiectasis) can only be determined by ultrasonography.
- Grazing management errors occasionally result in lungworm and type I ostertagiosis outbreaks but should be covered by the veterinary herd health plan.
- Paratuberculosis can cause clinical disease as early as 15 months old.

TABLE 3.49

Common Causes of Poor Appetite and Low Body Condition Score in Growing Cattle

	Volume	Appetite	Demeanour	Toxaemia	Pain	Feces	Rectal temperature
Acute lameness	–/––	–/––	NAD	NAD	++/+++	NAD	NAD/+
Type I Ostertagiasis	–	NAD	NAD	NAD	NAD	Diarrhoea	
Chronic pneumonia	–	–	–	NAD/+	NAD	NAD	NAD/+
Liver abscesses	–/––	–	NAD/–	+/++	NAD/+	NAD	NAD/+
Necrotic enteritis	–/––	–/––	–/––	+/++	+/++	Diarrhoea	++
Paratuberculosis	–/–––	NAD/–	NAD	NAD	NAD	Diarrhoea	NAD
Endocarditis	–/––	–/––	––/–––	+	++/+++	–	NAD/+

TABLE 3.50

Ancillary Tests for Common Causes of Poor Appetite and Low Body Condition Score in Growing Cattle

Severe lameness	Clinical examination. Radiography in chronic cases
Type I Ostertagiosis	Fecal worm egg count
Pneumonia	Ultrasonography
Liver abscesses	Ultrasonography
Necrotic enteritis	Necropsy
Paratuberculosis	Serology. Fecal PCR. Culture. Necropsy
Endocarditis	Ultrasonography

Acute Lameness

3.344

3.345

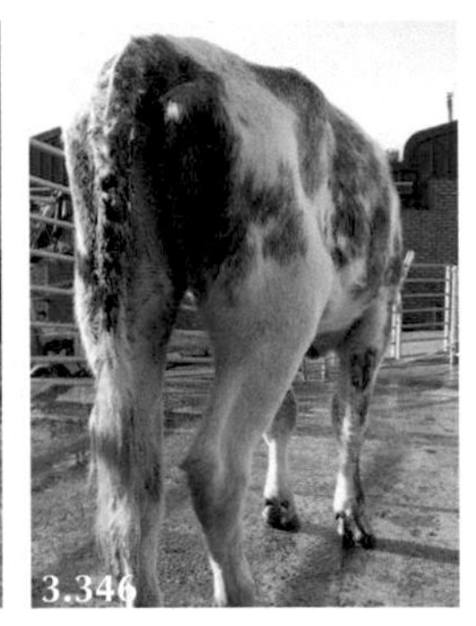
3.346

3.344–3.346: All three calves show severe lameness with poor appetite and much reduced abdominal fill.

- **Key observations**: Sudden severe lameness persisting over several days.
- **Clinical examination**: Often interdigital lesions such as necrobacillosis.
- **Key history events**: Can appear as a cluster in housed cattle. 🎥 3.201, 3.202

Type I Ostertagiosis

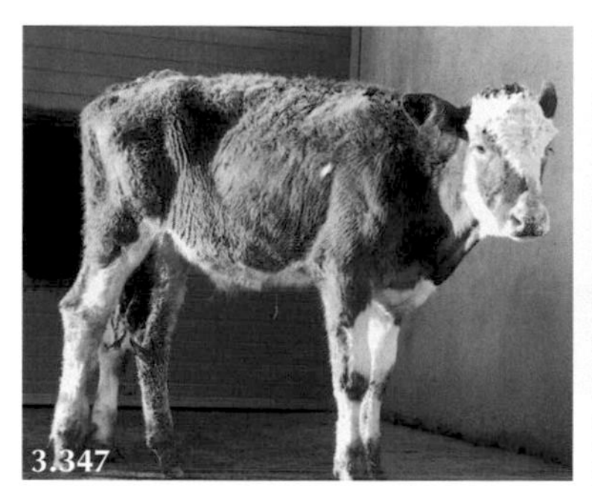
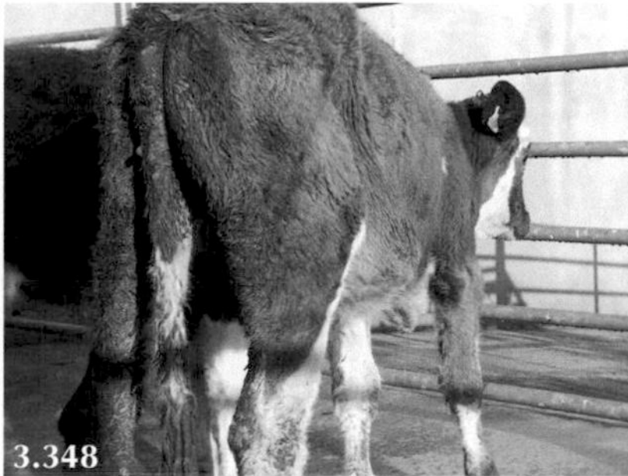
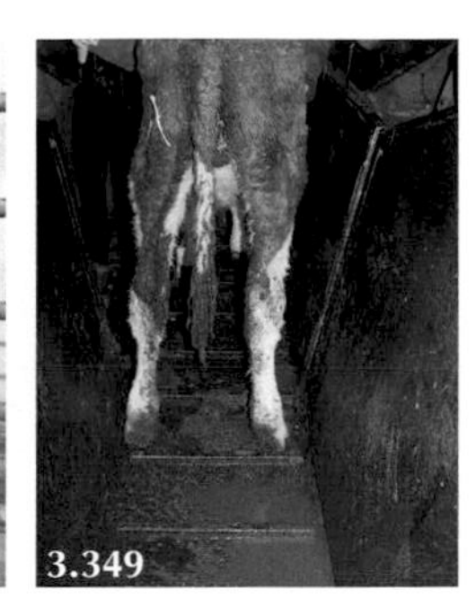

3.347–3.349: Profuse diarrhoea and rapid weight loss in grazing yearling.

- **Key observations**: Profuse diarrhoea, rapid weight loss.
- **Clinical examination**: Afebrile.
- **Key history events**: Several/many animals in group affected during late summer grazing contaminated pasture.

Chronic Pneumonia

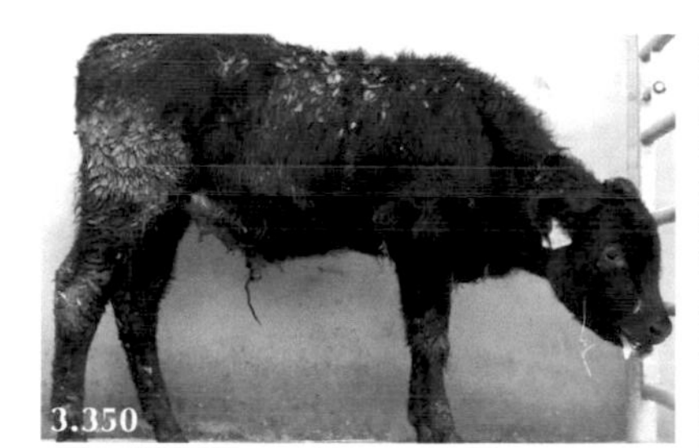
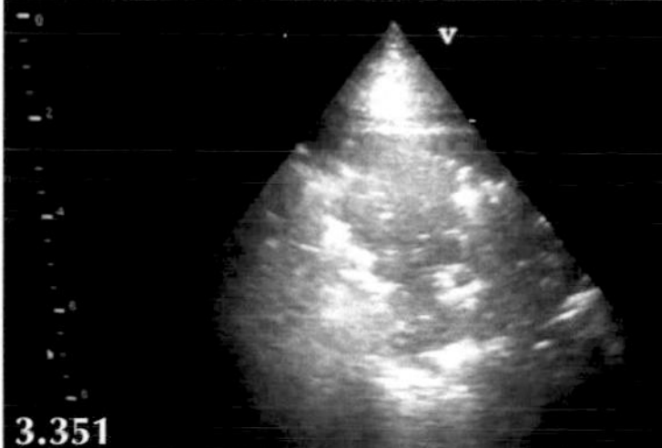
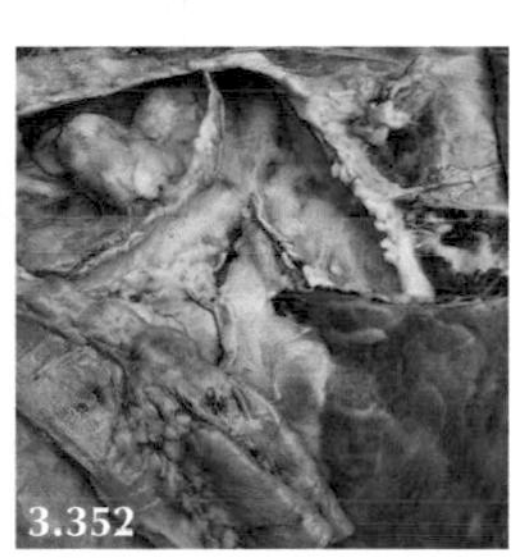

3.350: Beef calf with poor appetite and occasional coughing. **3.351:** Thoracic auscultation fails to determine nature and extent of lung pathology compared to ultrasonography. **3.352:** Pus in the larger airways—often clinical signs caused by recrudescence of chronic infection (bronchiectasis).

- **Key observations**: Reduced appetite over several days.
- **Clinical examination**: Usually no significant adventitious sounds.
- **Key history events**: Often recrudescence of chronic infection. 3.203

Liver Abscesses

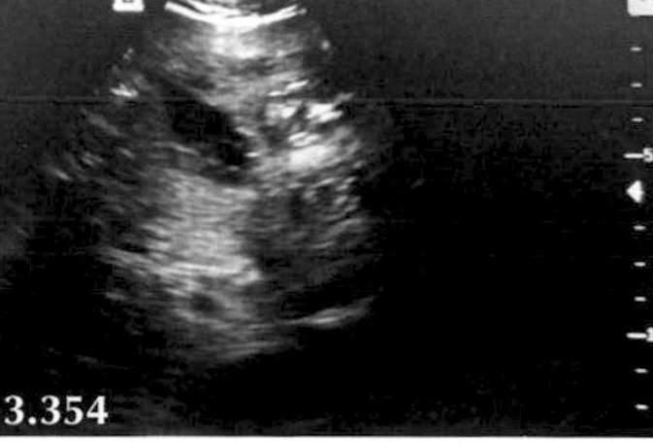

3.353: Reduced appetite and gaunt appearance. **3.354:** 2–3 cm diameter anechoic areas containing hyperechoic dots consistent with abscesses present throughout the liver parenchyma.

- **Key observations**: Reduced appetite and gaunt appearance have resulted over several days.
- **Clinical examination**: Febrile 39°C, no specific clinical findings.
- **Key history events**: Usually intensively reared "bull beef" animals. 3.204

Necrotic Enteritis

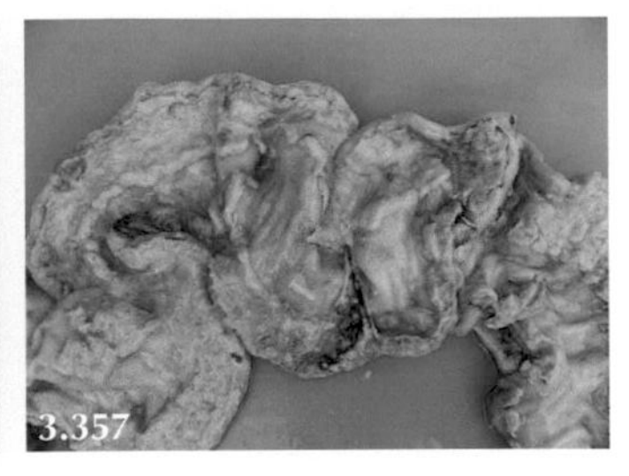

3.355: Gaunt appearance, tenesmus (raised tail). **3.356:** Purulent nasal discharges, sunken eyes. **3.357:** Necropsy reveals diphtheritic membrane overlying Peyer's patches in small intestine.

- **Key observations**: Foetid diarrhoea, rapid weight loss, tenesmus, gaunt appearance.
- **Clinical examination**: Lymphadenopathy, oral ulceration, dehydrated, febrile >40°C, occasional stertor.
- **Key history events**: Often single beef calf two–four months old. Guarded prognosis. 🎥 3.205

Paratuberculosis

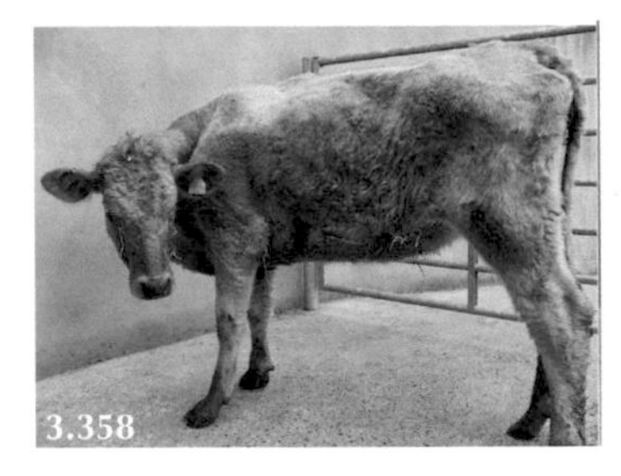

3.358 and **3.359:** Very gaunt appearance in this fifteen-month-old steer. **3.360:** Necropsy confirms paratuberculosis.

- **Key observations**: Chronic severe weight loss.
- **Clinical examination**: Dehydrated, afebrile. No specific clinical findings.
- **Key history events**: Unusual to have fifteen-month-old animal affected; possibly the progeny of clinical case with congenital/neonatal infection. 🎥 3.206

Endocarditis

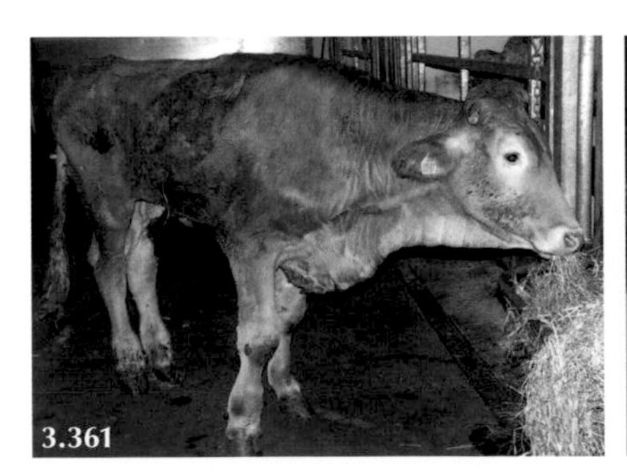
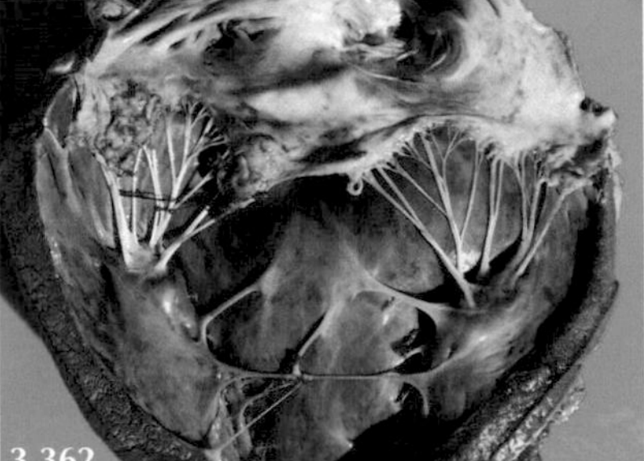
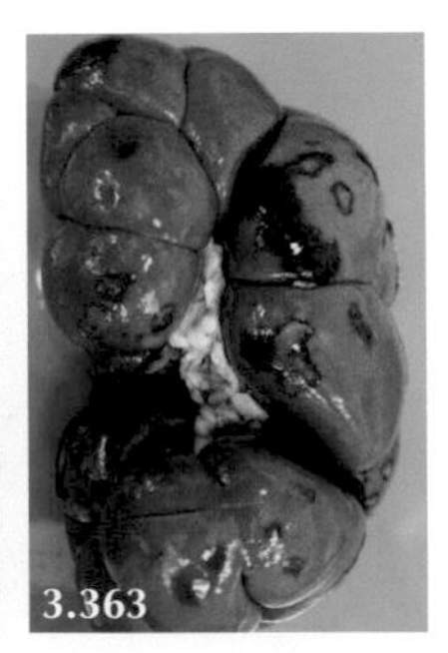

3.361: Chronic weight loss, gaunt appearance. **3.362:** Vegetative endocarditis involving the tricuspid valve. **3.363:** Renal infarcts develop from bacteraemia.

- **Key observations**: Chronic severe weight loss, arched back, variable moderate/severe lameness caused by joint effusion, pannus and/or erosion of articular cartilage.
- **Clinical examination**: Usually multiple joint effusions, often no discernible heart murmur.
- **Key history events**: Origin of bacteraemia rarely identified. 🎥 3.207

3.6

Common Causes of Increased Abdominal Size

Adult Sheep

- Increased abdominal size is an uncommon presentation in sheep with the most common conditions of intestinal torsion and dead lambs in utero necessitating immediate euthanasia.

TABLE 3.51

Common Differential Diagnoses for Increased Abdominal Size in Adult Sheep

	Likely duration at presentation	Abdominal size	Appetite	Demeanour	Toxaemia	Pain	Feces	Rectal temperature
Intestinal torsion	Hours	+++	−−−	−−−	+++	+++	−−−	NAD
Ruptured prepubic tendon	Weeks	+/++	NAD/−	NAD	NAD	NAD	NAD	NAD
Dead lambs in utero	Hours	++/+++		−/−−−	+++	+	−	+
Urolithiasis (intact bladder)	Days	+	−−−	−−−	+	+++	NAD/−	NAD
Peritonitis	Days	+/+++	−−−	−−/−−−	++/+++	+/+++	−−−	NAD/+
Bloat	Days	+/+++	−	−	NAD	NAD	NAD	NAD
Intestinal adenocarcinoma	Weeks	+/++	NAD/−	NAD	NAD	NAD	NAD/−	NAD
Abomasal emptying defect	Weeks/months	+++	−/−−	−	NAD	NAD	−	NAD
Urolithiasis (ruptured bladder)	Days	+/++	−−−	−−−	+/+++	+	NAD/−	NAD

TABLE 3.52

Ancillary Tests for Common Causes of Increased Abdominal Size in Adult Sheep

	Ancillary tests
Intestinal torsion	Ultrasonography. Grossly dilated and atonic intestinal loops. Necropsy
Ruptured prepubic tendon	Ultrasonography reveals thinned body wall, normal abdominal contents
Dead lambs in utero	Ultrasonography reveals poorly defined caruncles and heterogenous uterine content. Necropsy
Urolithiasis	Ultrasonography—grossly distended bladder. 10–20x increase in serum BUN and creatinine
Peritonitis	Ultrasonography. Necropsy. Abdominocentesis rarely successful
Bloat	Great care needed when passing orogastric tube
Intestinal adenocarcinoma	Ultrasonography—large volumes of abdominal transudate. Smear of centrifuged abdominocentesis sample. Necropsy
Abomasal emptying defect	Rumen fluid chloride concentration unreliable. Necropsy
Urolithiasis (ruptured bladder)	Ultrasonography—large volumes of urine in abdomen although bladder wall still appears intact. Abdominocentesis. Necropsy

DOI: 10.1201/9781003106456-24

Intestinal Torsion

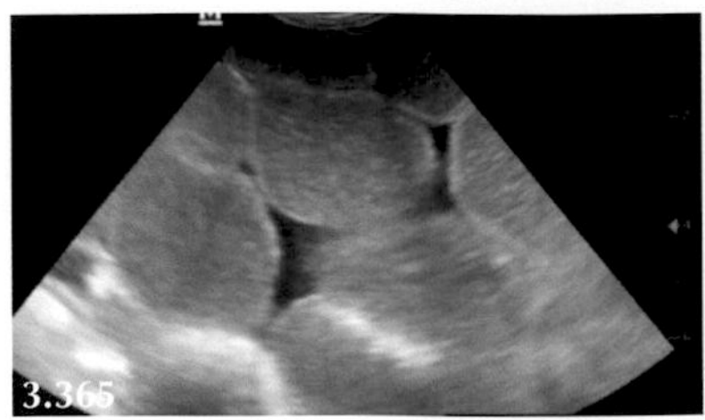
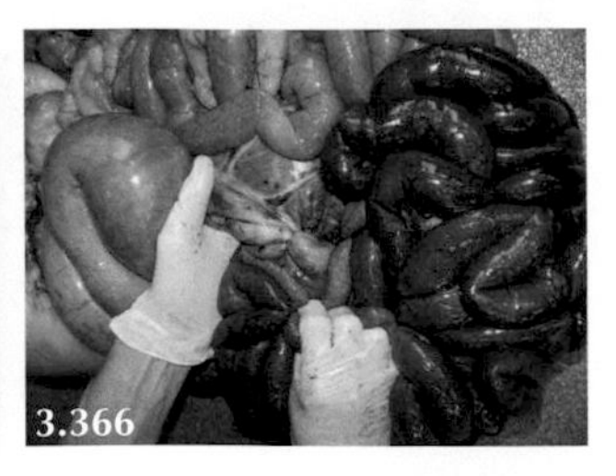

3.364: Ewe with arched back, distended abdomen, painful expression. **3.365:** Distended atonic loops of small intestine. **3.366:** Necropsy reveals torsion around root of mesentery.

- **Key observations**: Very painful (arched back), taut abdomen. No feces.
- **Clinical examination**: Toxic mucous membranes. Variable dehydration. No rumen activity. Heart rate >120 bpm.
- **Key history events**: Sudden onset (several hours). 3.208

Ruptured Prepubic Tendon

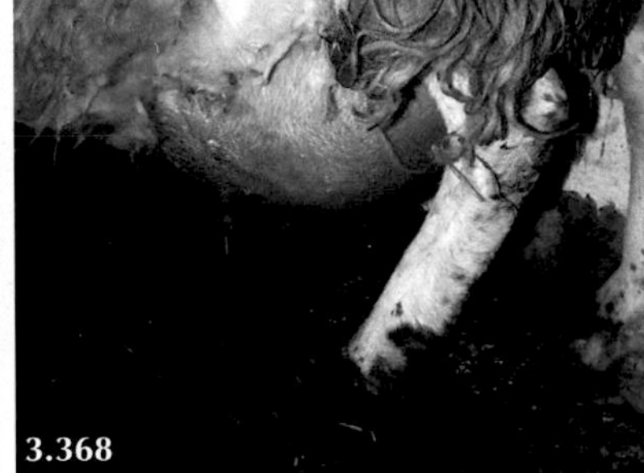

3.367: Pendulous abdomen in heavily pregnant multiparous ewe.
3.368: Large swelling on left side immediately cranial to the udder with pitting subcutaneous oedema.

- **Key observations**: Large swelling immediately cranial to the udder.
- **Clinical examination**: Thin, often oedematous, ventral abdominal wall.
- **Key history events**: Late gestation, multiple pregnancy, older ewe.

Dead Lambs *in Utero*

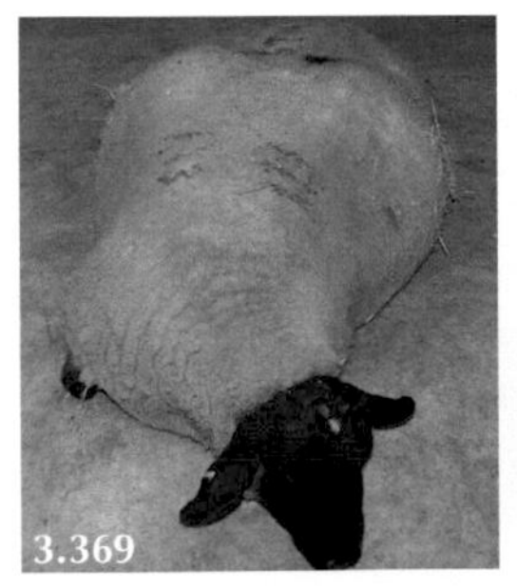
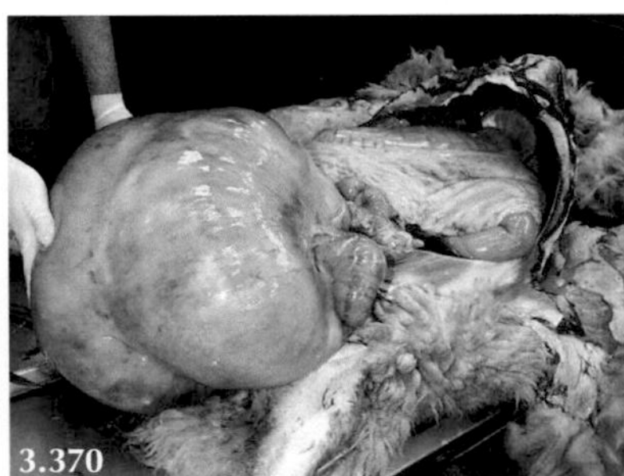

3.369: Sick Suffolk ewe with gross abdominal distension. **3.370** and **3.371:** Abdominal distension caused by dead (emphysematous) lambs *in utero*.

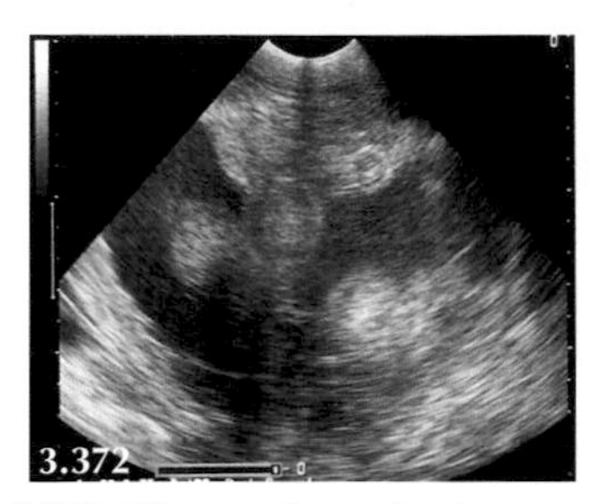
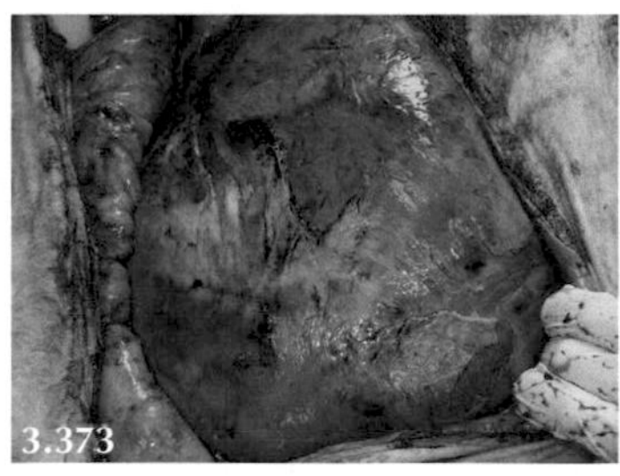
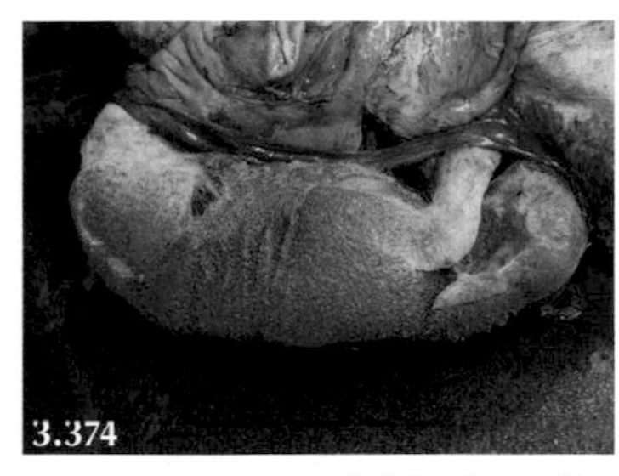

3.372: Ultrasound examination reveals poorly defined cotyledons and heterogenous uterine content. **3.373:** Devitalised uterine wall with fibrin on serosal surface. **3.374:** Dead lamb with meconium staining.

- **Key observations**: Grossly enlarged abdomen, sick/moribund ewe.
- **Clinical examination**: Toxic mucous membranes. Often no vaginal discharge.
- **Key history events**: Various abortifacient agents. Second stage labour/malpresentation not detected/ corrected. 3.209, 3.210

Urolithiasis

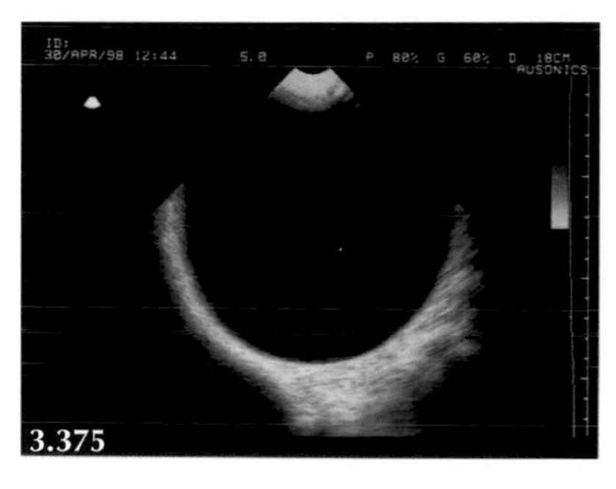
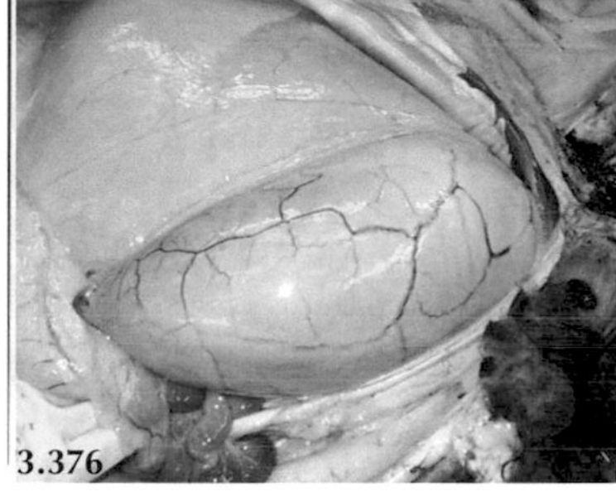

3.375: Bladder measures 16 cm diameter causing moderate abdominal distension. **3.376:** Necropsy confirms the extent of bladder distension.

- **Key observations**: Sternal recumbency, inappetant, intermittent tenesmus.
- **Clinical examination**: Often crystals palpable within vermiform appendage, ultrasonography.
- **Key history events**: Feeding ewe minerals to rams, lack of water supply over several days. 3.211

Peritonitis

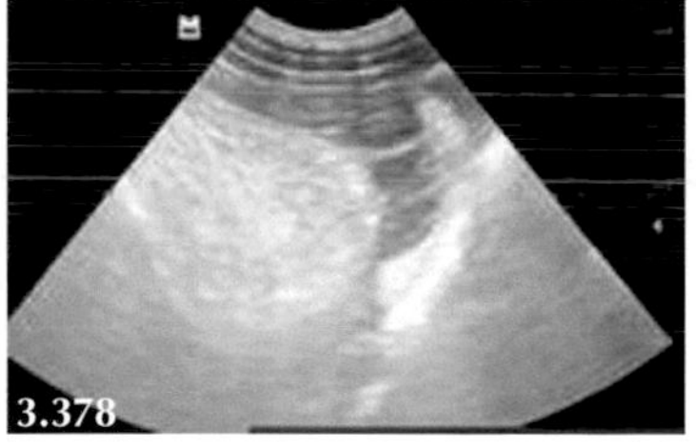
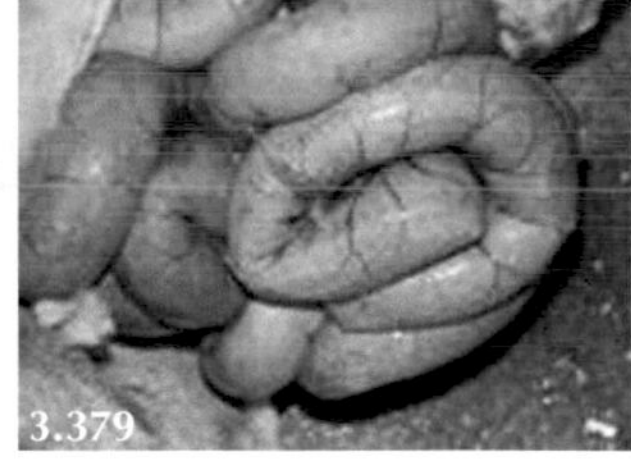

3.377: This Scottish Blackface ram is dull, bloated, with toxic mucous membranes and a heart rate >100 bpm. **3.378:** Ultrasonography reveals grossly dilated loops of small intestine with little propulsion of digesta and pockets of inflammatory exudate appearing as anechoic areas with bridging hyperechoic fibrin strands. **3.379:** Ultrasonographic findings of fibrin tags (inflammatory exudate) confirmed at necropsy.

- **Key observations**: Pain/colic, isolated from group.
- **Clinical examination**: Toxic mucous membranes, dehydrated, heart rate >100 bpm, no ruminal motility.
- **Key history events**: None. 3.212, 3.213

Bloat

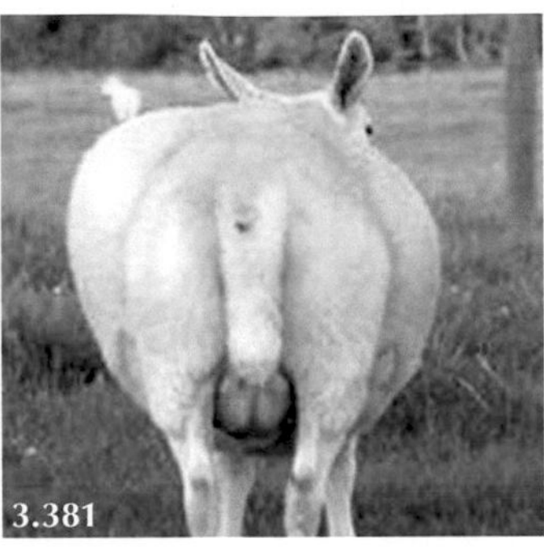

3.380 and **3.381:** High left-sided abdominal distension in this Cheviot ewe.

- **Key observations**: High left-sided abdominal distension. Salivation.
- **Clinical examination**: Take great care when attempting to pass oro-gastric tube.
- **Key history events**: Access to root crops, beet pulp etc causing oesophageal obstruction. 3.214

Intestinal Adenocarcinoma

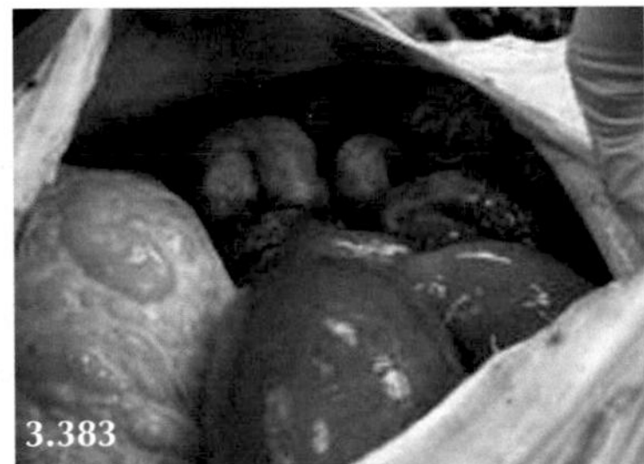
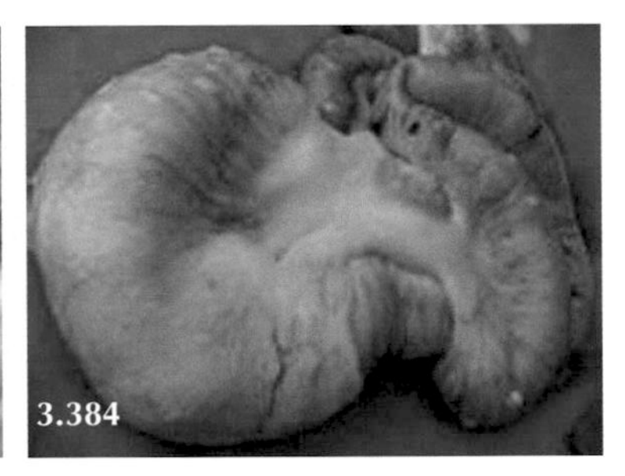

3.382: Abdominal distension caused by trans-coelomic tumour spread and impaired lymphatic drainage. **3.383:** Accumulated transudate within the abdominal cavity at necropsy. **3.384:** Intestinal adenocarcinoma with tumour nodules on serosal surface.

- **Key observations**: Poor BCS, fluid-distended abdomen (trans-coelomic tumour spread).
- **Clinical examination**: Fluid thrill across abdomen. Ultrasonography.
- **Key history events**: None.

Abomasal Emptying Defect

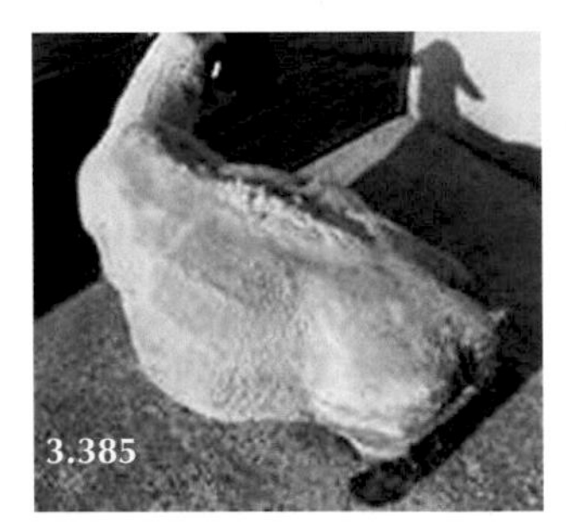
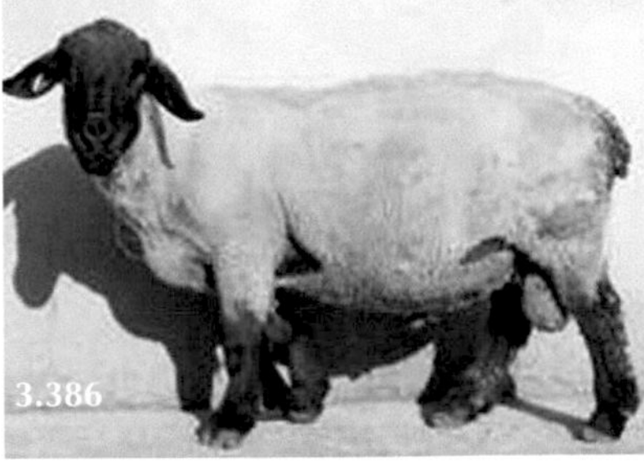
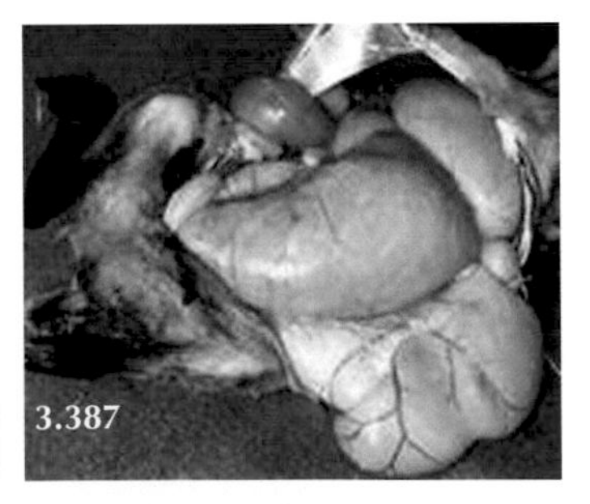

3.385 and **3.386:** Massively distended abdomen. **3.387:** Necropsy confirms massive abomasal distension.

- **Key observations**: Poor BCS, massively distended abdomen, soft feces.
- **Clinical examination**: Rumen motility usually increased 3–4 times normal.
- **Key history events**: None—genetic effect? 3.215

Lambs

- Quantifying disease prevalence in lambs is difficult because there have been few studies. Veterinary assessment could be biased by cases of atresia ani brought to the surgery.

TABLE 3.53

Common Causes of Increased Abdominal Size in Lambs

	Age at presentation	Size	Appetite	Demeanour	Pain	Feces
Watery mouth disease	12–36 hours	+/++	−−−	−−/−−−	+/++	Agonal diarrhoea
Atresia ani	24–48 hours	++/+++	−−−	−/−−	+/++	Nil
Intestinal torsion	4–8 months	++/+++	−−−	−−/−−−	++/+++	Nil
Peritonitis	2–5 days	+/++	−−/−−−	−−/−−−	+/++	−
Abomasal distension/torsion	3–14 days	+++	−−−	−−/−−−	+++	NAD
Atresia coli	24–72 hours	++/+++	−−−	−/−−	+/++	Nil
Intussusception	3–14 days	+/++	−−−	−/−−	+/++	−−−
Obstructed bladder		+/++	−−−	−−/−−−	+/++	NAD

TABLE 3.54

Ancillary Tests for Common Causes of Increased Abdominal Size in Lambs

Watery mouth disease	Increased BUN, creatinine. Leucopaenia, low plasma glucose. Culture of *E. coli* of limited diagnostic value
Atresia ani	Skin bulge under tail, high extradural injection, puncture, release mucus/meconium/gas.
Intestinal torsion	Ultrasonography—grossly distended and atonic gut loops. Necropsy
Peritonitis	Abdominocentesis more likely punctures distended gut. Ultrasonography; dilated intestines but only small amount of exudate.
Abomasal distension/ torsion	Symptomatic treatment for suspect abomasal distension—hyoscine, NSAID. Needle decompression rarely successful. Euthanase for welfare reasons
Atresia coli	No diagnostic test, confirm at necropsy
Intussusception	Palpation, ultrasonography shows fluid-filled intestines proximally, confirm at necropsy
Obstructed bladder	Ultrasonography. If male lamb check whether elastrator ring constricting urethra

Watery Mouth Disease

3.388: Salivation in early clinical stage. **3.389:** Recumbent with distended abdomen causing pain in latter stages.

- **Key observations**: 12–36-hour-old not sucking, profuse salivation, colic signs, rapid progression to distended abdomen.
- **Clinical examination**: Dehydration, sequestration of fluid in abomasum/intestines.
- **Key history events**: Multiple litter, poor hygiene, failure passive antibody transfer. 3.216

Atresia Ani

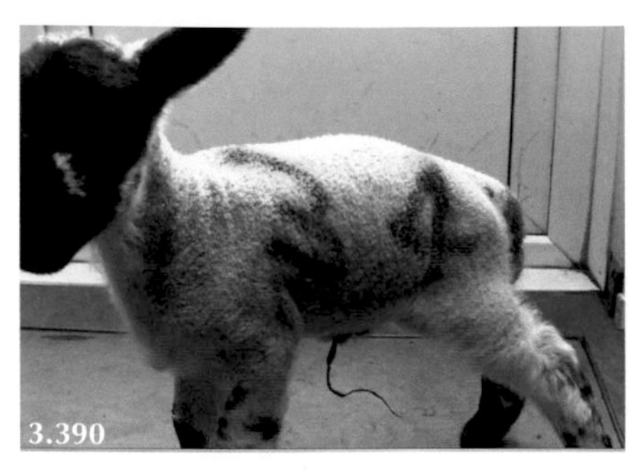

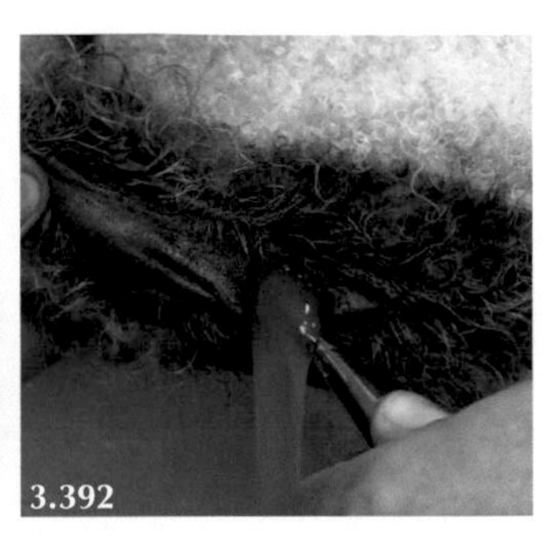

3.390: Abdominal distension despite not sucking. **3.391:** No anus. **3.392:** Correction under low extradural block.

- **Key observations**: Squatting/tenesmus, distended abdomen, not sucking, 24–48 hours old.
- **Clinical examination**: No anus. Correct surgically under low extradural block.
- **Key history events**: None. 🎥 3.217, 3.218

Intestinal Torsion

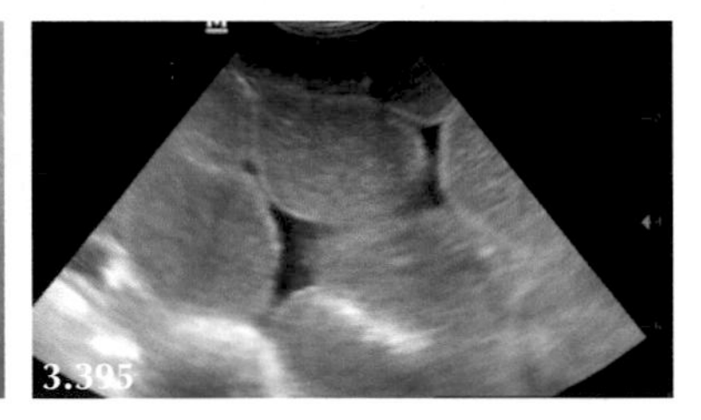

3.393: Distended abdomen. **3.394:** Depressed demeanour, ears held back, painful expression. **3.395:** Dilated atonic loops of intestine.

- **Key observations**: Colic—depressed demeanour, ears held back, painful expression.
- **Clinical examination**: Increased heart and respiratory rates. Ultrasonography reveals dilated atonic loops of intestine.
- **Key history events**: None. 🎥 3.219

Peritonitis

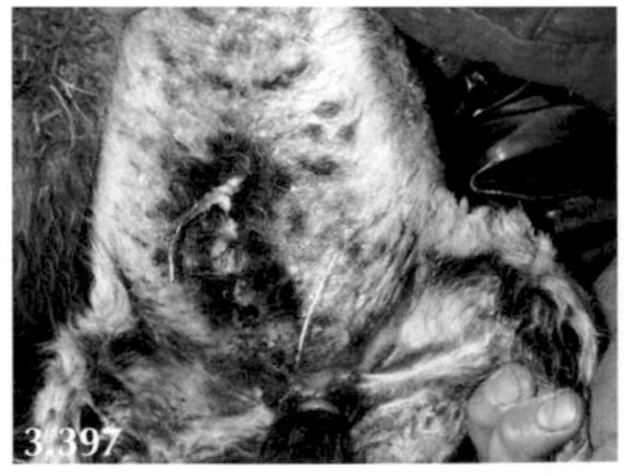

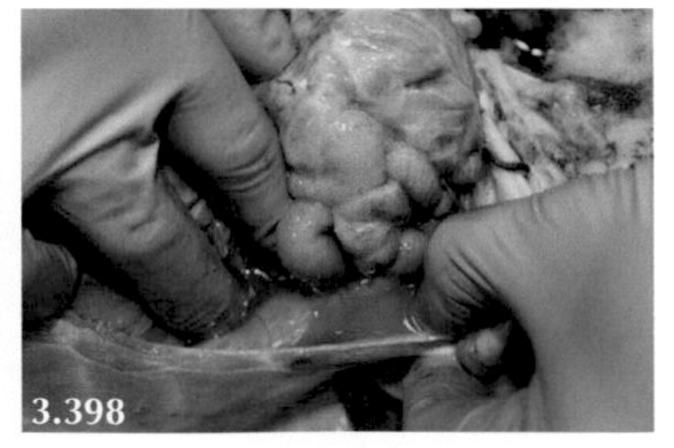

3.396: Five-day-old lamb with arched back and taut painful abdomen. **3.397:** Abdominal distension with umbilicus still "wet". **3.398:** Extensive fibrin covering dilated intestines with small quantity of pus.

- **Key observations**: Weak, distended abdomen contrary to not sucking, often moribund at presentation.
- **Clinical examination**: Dehydrated, toxic mucous membranes, thickened painful umbilicus.
- **Key history events**: Peritonitis secondary to umbilical infection.

Intussusception

3.399: Very dull lamb with arched back; demeanour suggests abdominal pain.
3.400: Necropsy findings of intussusception (centre of image).

- **Key observations**: Arched back. Distended intestines/abdomen, not sucking.
- **Clinical examination**: No feces, only mucus.
- **Key history events**: Normal until acute onset of illness. 🎥 3.220

Abomasal Bloat/Torsion

- **Key observations**: Sudden onset after milk feeding, very distended abdomen, colic signs.
- **Clinical examination**: Taut abdomen, often gas-filled abomasum. Needle decompression usually unsuccessful in long term—may indicate torsion rather than bloat.
- **Key history events**: Orphan lambs, irregular milk feeding.

Urethral Obstruction

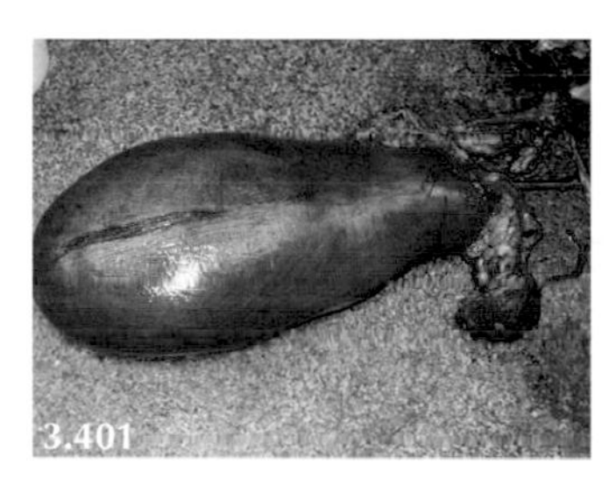

3.401: Necropsy finding of urinary bladder distension.

- **Key observations**: Colic signs, not sucking, increasing abdominal distension.
- **Clinical examination**: Distended viscus in caudal abdomen.
- **Key history events**: Congenital abnormalities are rare, but bladder distension can occur after incorrect rubber ring castration incorporates the penis.

Adult Cows

TABLE 3.55

Common Causes of Increased Abdominal Size in Adult Cows

	Volume	Appetite	Demeanour	Toxaemia	Pain	Feces	Rectal temperature
Grain overload	+/++	–/––	–/––	+/++	+	Foetid	NAD/+
Bloat (choke)	++/+++	––/–––	–	–	+	NAD	NAD
Frothy bloat	++/+++	–/––	–	–	+	NAD	NAD
Peritonitis	+/++	–––	–/–––	++/+++	+/+++	–/–––	NAD/+
Advanced/twin pregnancy	+/++	NAD	NAD	NAD	NAD	NAD	NAD
Ascites/right-sided heart failure	+/++	–	–	NAD	NAD	NAD	NAD
Vagus indigestion	+/++	––/–––	–	NAD	NAD	–	NAD
Hydrops allantois	++/+++	NAD	NAD	NAD	NAD	NAD	NAD
Obesity	+/++	NAD	NAD	NAD	NAD	NAD	NAD

TABLE 3.56

Ancillary Tests for Common Causes of Increased Abdominal Size in Adult Cows

Grain overload	Rumen pH<4.5
Bloat (choke)	Oro-gastric tube—obstruction. Probang/extractor to relieve
Frothy bloat	No oesophageal obstruction
Peritonitis	Ultrasonography (exudate). Abdominocentesis?
Advanced/twin pregnancy	NAD on clinical, rectal examination advanced pregnancy
Ascites	Ultrasonography (transudate). Abdominocentesis? Identify primary cause
Vagus indigestion	Variable rumen motility but often greatly increased
Hydrops allantois	Stage of gestation. Rectal examination
Obesity	BCS 5

Grain Overload

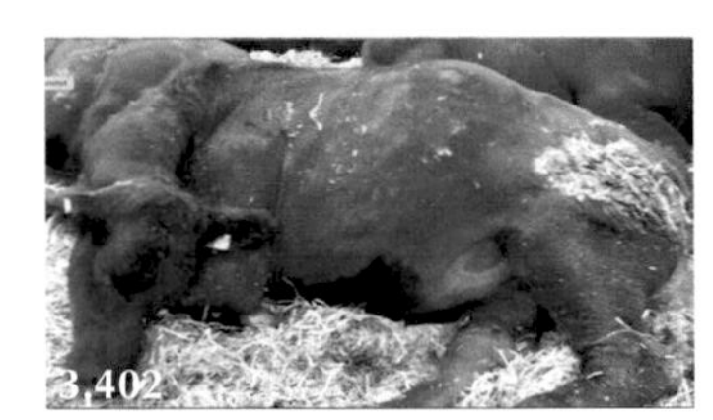
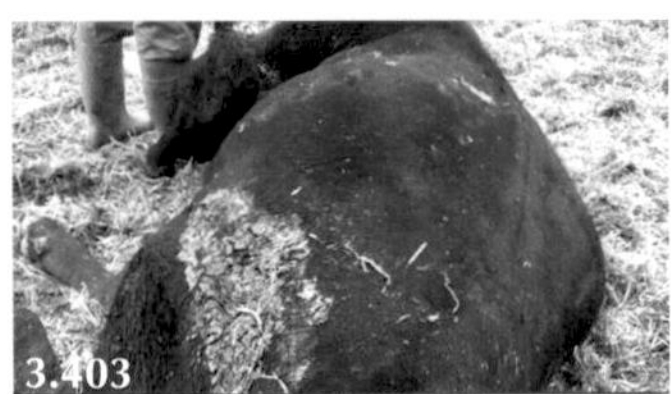

3.402–3.404: Beef heifers with barley poisoning/grain overload. Profound dullness/toxaemia with abdominal distension during early stages due to fluid sequestration.

- **Key observations**: Several/many in group of cattle. Rapid (within hours) abdominal distension. Dull, weak, toxaemic, "drunken" appearance in most severely affected.
- **Clinical examination findings**: Dehydration, no rumen activity, later profuse foetid diarrhoea.
- **Key history events**: Sudden unaccustomed access to large quantities of grain/concentrates. 🎥 3.221, 3.222

Bloat (Choke)

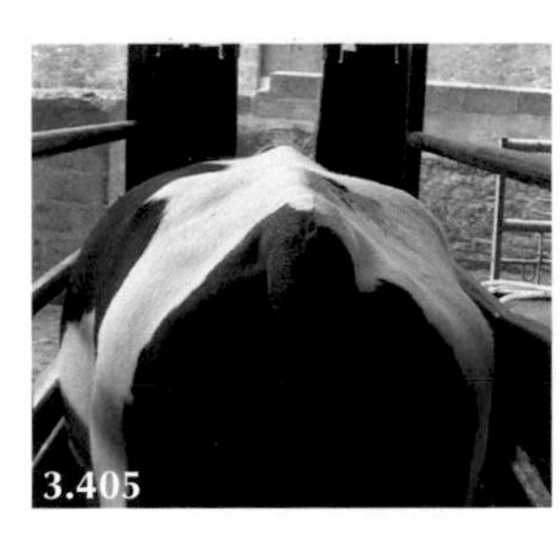

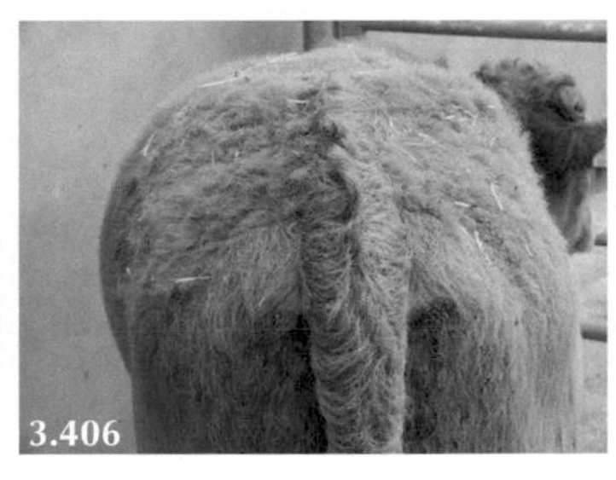

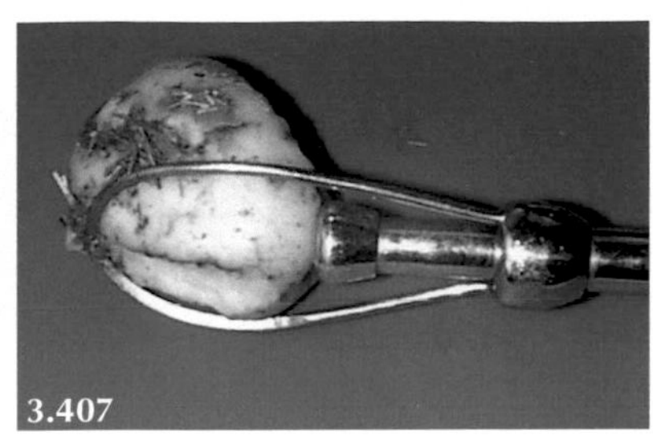

3.405 and **3.406:** High left-sided abdominal distension. **3.407:** Removal of potato from oesophagus at the thoracic inlet.

- **Key observations**: Individual animal. Rapid (within hours) high left-sided abdominal distension. Profuse salivation.
- **Clinical examination findings**: Unable to pass orogastric tube (typically at level of base of heart).
- **Key history events**: Access to potatoes, turnips, apples etc.

Frothy Bloat

- **Key observations**: Often several/many animals affected. Rapid (within hours) high left-sided abdominal distension.
- **Clinical examination findings**: No oesophageal obstruction. Unable to relieve rumen distension with oesophageal tube due to stable foam.
- **Key history events**: Recent (within hours) movement to lush pasture.

Peritonitis

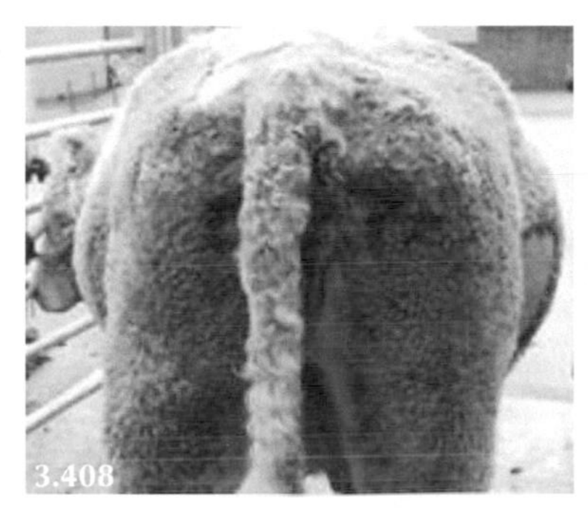

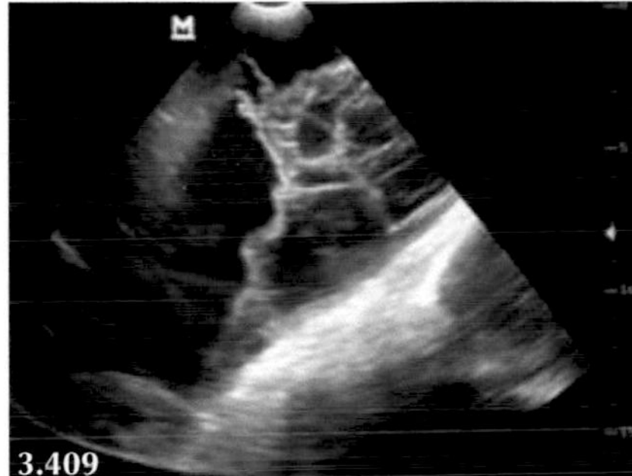

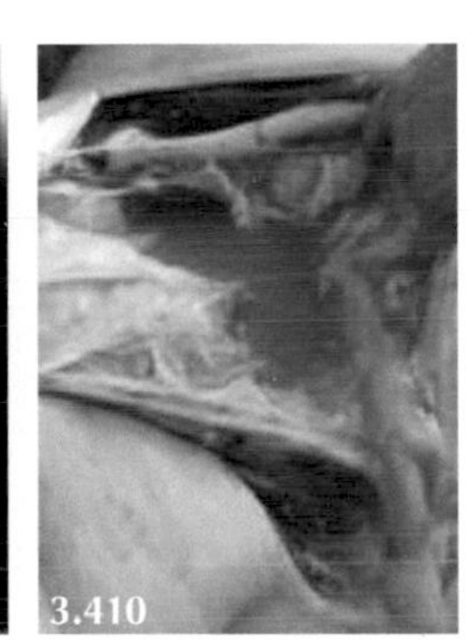

3.408: Abdominal distension caused by unusually extensive peritoneal exudate. **3.409:** Sonogram showing peritoneal exudate up to 15 cm (scale on right hand margin). **3.410:** Necropsy reveals organised inflammatory exudate with widespread fibrin deposition.

- **Key observations**: Painful expression. Distended abdomen despite very poor appetite. Dehydration—sunken eyes.
- **Clinical examination findings**: 7–10 per cent dehydration, sequestration of fluid within intestines, toxaemia, no ruminal sounds, taut, painful abdomen. Variable exudate (up to 20 cm) detected ultrasonographically.
- **Key history events**: Recent dystocia—iatrogenic uterine rupture. Ruptured viscus e.g. perforated abomasal ulcer.

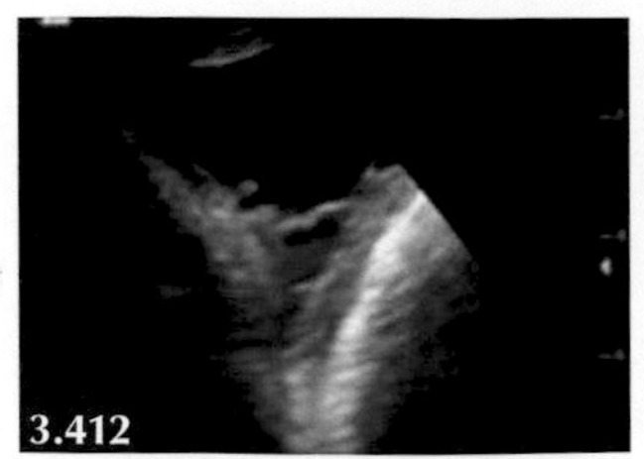
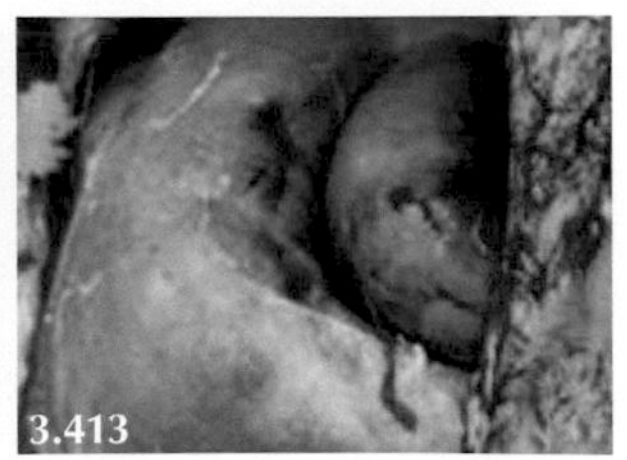

3.411: Beef cow with distended abdomen. **3.412:** Trans-abdominal ultrasonography in mid right flank region reveals inflammatory exudate within the abdomen; fibrin tags also present on the serosal surface of the distended caecum. **3.413:** Necropsy confirms distended caecum and widespread inflammatory exudate/peritonitis.

- **Key observations**: Distended abdomen despite inappetence.
- **Clinical examination findings**: Percussion of right flank yields high-pitched "pings" resulting from gas distension of caecum. 30 cm diameter blind-ended sac palpable per rectum. Ultrasound scan via the right flank reveals peritoneal exudate and distended caecum.
- **Key history events**: No recognised risk factors. 3.223, 3.224

Advanced (Twin) Pregnancy

3.414 and **3.415:** Distended abdomen especially lower right side in older beef cows.

- **Key observations**: Distended abdomen, often older beef cow. Fibrous diet.
- **Clinical examination**: Twin pregnancy identified during earlier ultrasound pregnancy diagnosis between nine and 16 weeks pregnant.
- **Key history events**: More common in spring-calving beef cows. 3.225

Ascites/Right-Sided Heart Failure

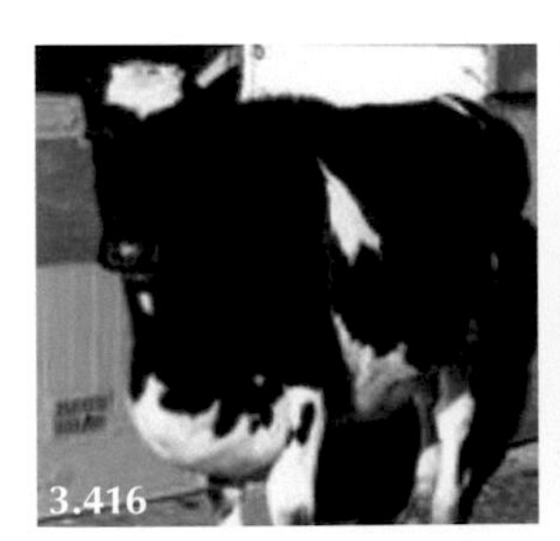
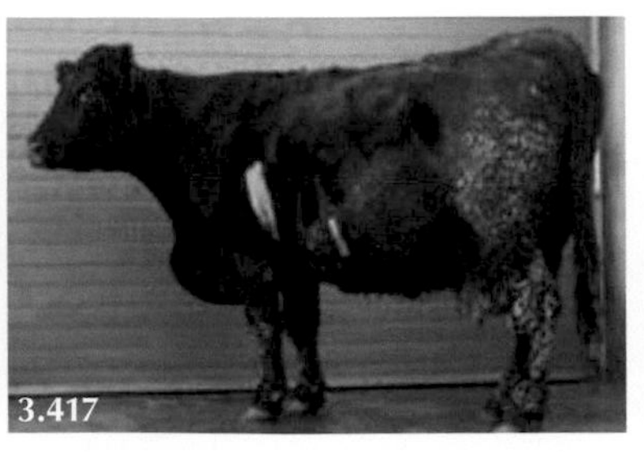
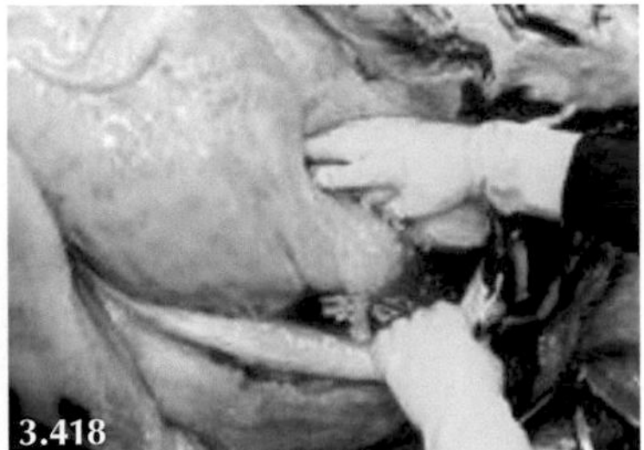

3.416 and **3.417:** Abdominal distension caused by fluid accumulation. **3.418:** Transudate in the abdominal cavity (ascites).

- **Key observations**: Distended abdomen, brisket oedema.
- **Clinical examination**: Primary disease may be identified e.g. septic pericarditis.
- **Key history events**: Insidious abdominal distension often combined with condition loss. 3.226

Vagus Indigestion

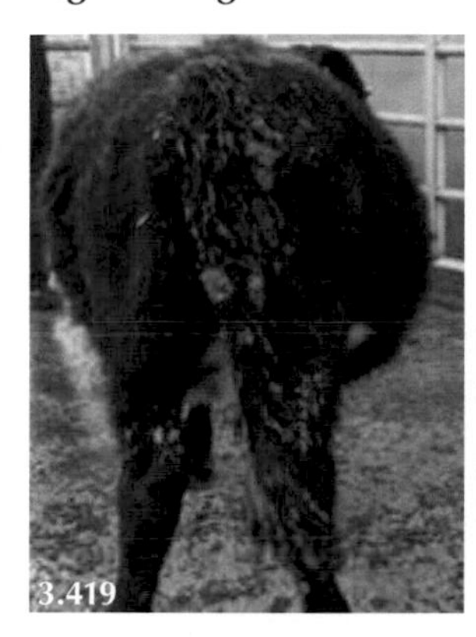

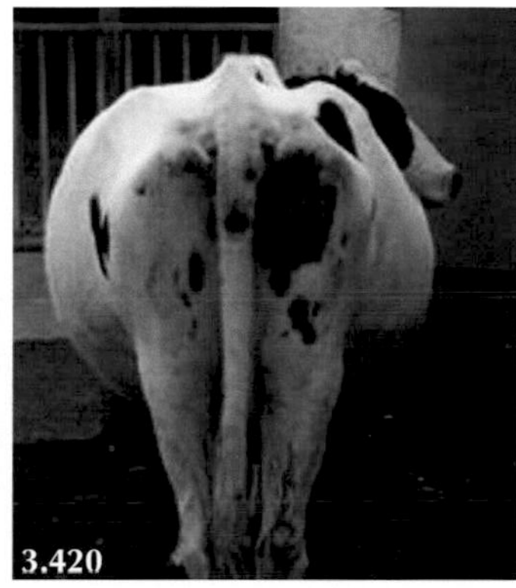

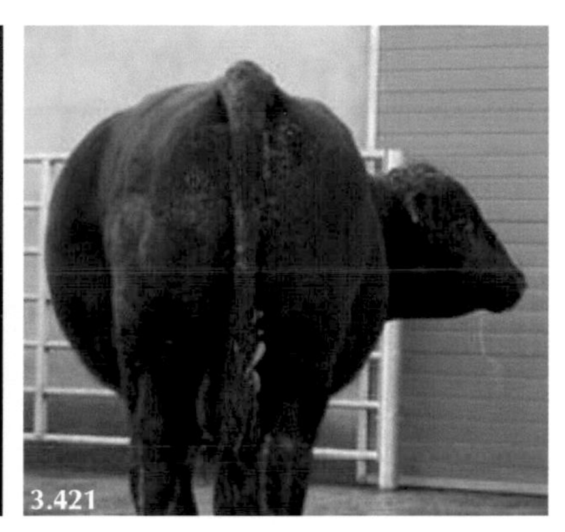

3.419–3.421: Markedly distended abdomen "10 to 4" abdominal silhouette ("papple") when viewed from behind.

- **Key observations**: Markedly distended abdomen "10 to 4" abdominal silhouette ("papple") when viewed from behind. Often chronic weight loss.
- **Clinical examination**: Often markedly increased frequency of ruminal contractions.
- **Kcy history events**: None. 3.227

Hydrops Allantois

3.422: Very large fluid-distended abdomen of pregnant cow.

- **Key observations**: Large fluid-distended abdomen.
- **Clinical examination**: Four to five months pregnant; scanned for single fetus.
- **Key history events**: Rapid increase in abdominal size over 10–14 days.

Obesity

3.423 and **3.424:** Beef cows with excessive body condition scores.

- **Key observations**: BCS 5.
- **Clinical examination**: NAD.
- **Key history events**: Overfed during prolonged dry period. Often pet cows which are not bred.

Calves

- Recurrent bloat occurs sporadically in growing cattle but generally resolves after placing a rumen trochar for four to eight weeks.
- Abomasal perforation typically affects the best four–eight-week-old beef calf in the herd.

TABLE 3.57

Common Causes of Increased Abdominal Size in Calves

	Volume	Appetite	Demeanour	Toxaemia	Pain	Feces	Rectal temperature
Bloat	+/+++	NAD/–	–	NAD	+	NAD	NAD
Torsion of small intestine	++/+++	–––	–––	+++	++/+++	Nil	NAD/–
Abomasal perforation	+/++	––/–––	–––	+++	++/+++	NAD	Often subnormal
Thymic lymphosarcoma	+/++	NAD/–	–	NAD	NAD	–	NAD

TABLE 3.58

Ancillary Tests for Common Causes of Increased Abdominal Size in Calves

	Ancillary tests
Bloat (choke)	Oro-gastric tube—effective rumen decompression. Place trochar in recurring cases
Torsion of small intestine	Ultrasonography reveals greatly distended SI loops. Severity of clinical signs precludes attempted GA and surgical correction. Necropsy
Abomasal perforation	Necropsy
Thymic lymphosarcoma	Necropsy. Pleural effusion often present due to right-sided heart failure identified ultrasonographically

Bloat

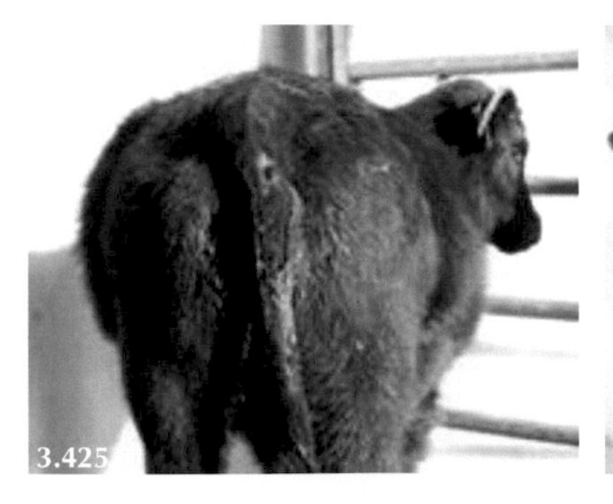

3.425 and **3.426:** Distended upper left quadrant.

- **Key observations**: Individual animal. High left-sided abdominal distension.
- **Clinical examination findings**: Able to decompress via an orogastric tube.
- **Key history events**: Often NAD. Chronic pneumonia with enlarged mediastinal lymph nodes.

Recurrent Bloat

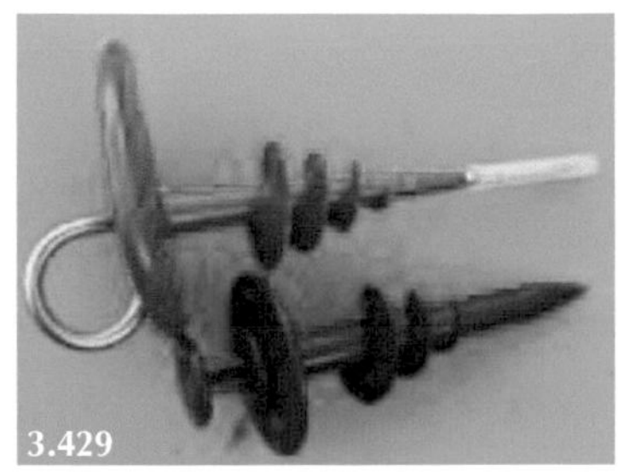

3.427: Distended upper left quadrant. **3.428:** Several days following rumen trochar. **3.429:** Two common types of trochar.

- **Key observations**: Individual animal. High left-sided abdominal distension.
- **Clinical examination findings**: Able to pass orogastric tube with release of gas.
- **Key history events**: NAD. 3.228, 3.229

Torsion of Small Intestine

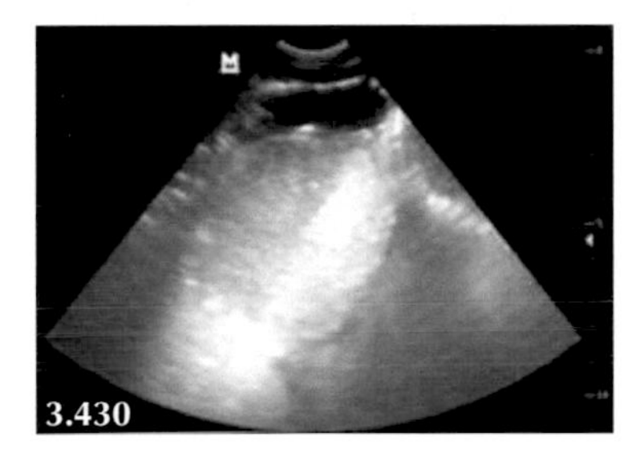
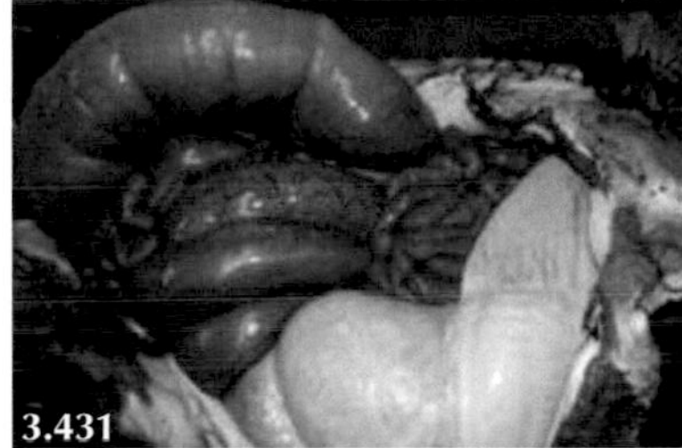

3.430: Ultrasound examination reveals greatly distended >8 cm diameter and static small intestine. **3.431:** Small intestinal torsion confirmed by on-farm necropsy.

- **Key observations**: Individual animal. Sudden onset colic/collapse.
- **Clinical examination findings**: Severely dehydrated, rapid heart rate, distended painful abdomen.
- **Key history events**: NAD. 3.230

Abomasal Perforation

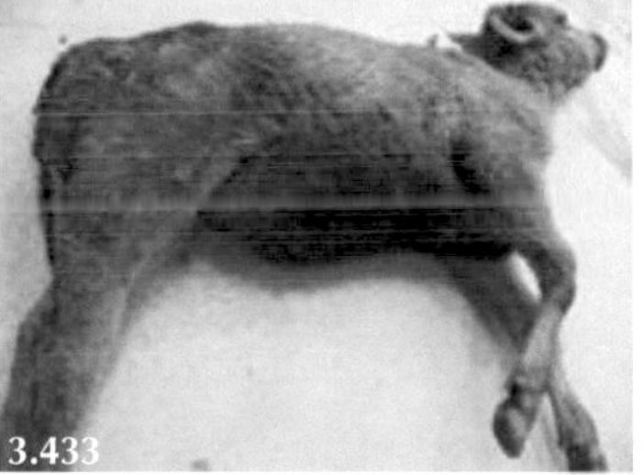
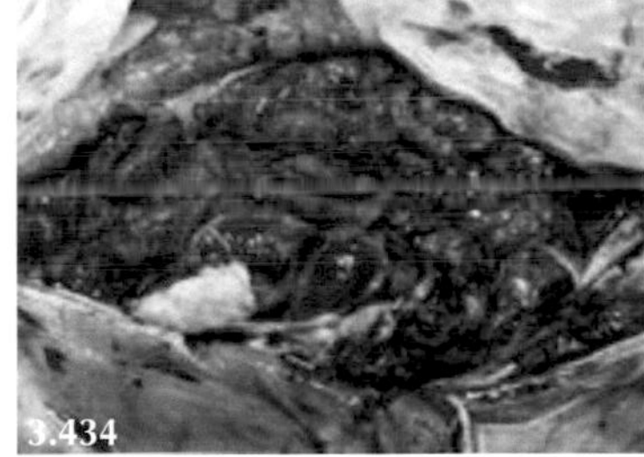

3.432: Collapsed six-week-old beef calf with distended abdomen. **3.433:** Calf with distended abdomen presented for necropsy. **3.434:** Digesta leaked into the peritoneal cavity from a large abomasal perforation.

- **Key observations**: Individual animal—often best calf in the beef herd. Sudden onset colic/collapse.
- **Clinical examination findings**: Severely dehydrated, rapid heart rate, distended painful abdomen.
- **Key history events**: Often six–eight-week-old beef calves. 3.231

Thymic Lymphosarcoma

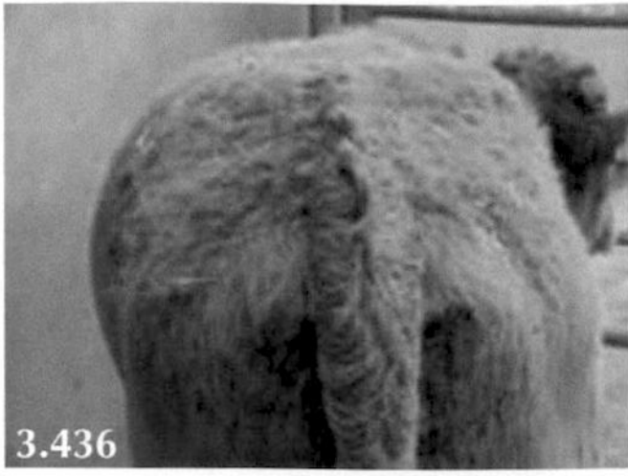
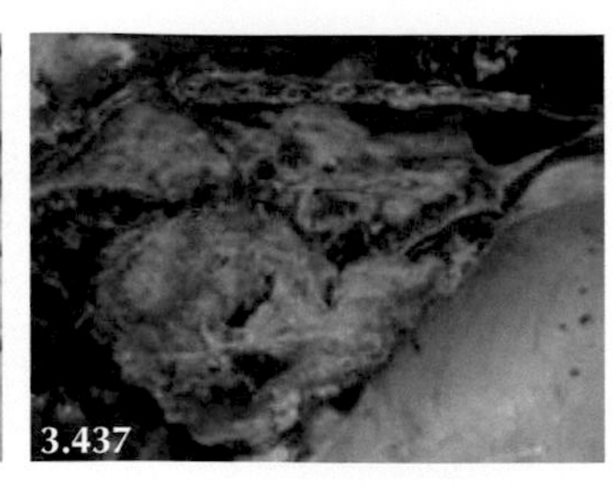

3.435 and **3.436:** Moderate bloat and brisket oedema in steer with thymic lymphosarcoma. **3.437:** The tumour occupies much of the chest cavity with extensive fibrin deposits.

- **Key observations**: Individual animal. Chronic left-sided abdominal distension—free gas bloat.
- **Clinical examination findings**: Able to pass orogastric tube with release of gas.
- **Key history events**: Insidious onset. Development of peripheral oedema due to right-sided heart failure.

Neonatal Calves

- Abomasal distension and ileus occur secondary to metabolic acidosis.
- Fibrin affecting gut motility causes greatly distended small intestinal loops.
- Atresia coli presents in three- to four-day-old calves with colic and abdominal distension.

TABLE 3.59

Common Causes of Increased Abdominal Shape/Content in Neonatal Calves

	Shape	Appetite	Demeanour	Pain	Toxaemia	Feces
Abomasal distension/acidosis	+/++	–––	–––	NAD	NAD	Diarrhoea
Peritonitis	+/++	––/–––	–––	++	+++	––
Atresia coli/ani	++/+++	–––	––/–––	+/++	NAD	Nil
Intussusception	+/++	––	–/––	+/++	+	Nil

TABLE 3.60

Ancillary Tests for Common Causes of Increased Abdominal Shape/Content in Neonatal Calves

Disease/disorder	Ancillary tests
Abomasal distension/acidosis	Succussion—fluid distended loops. Significant acidaemia; Base deficit 20 mEq/L
Peritonitis	Ultrasonography. Necropsy
Atresia coli	Necropsy
Atresia ani	Simple stab incision correction
Intussusception	Ultrasonography. Confirm during surgery

Abomasal Distension/Acidosis

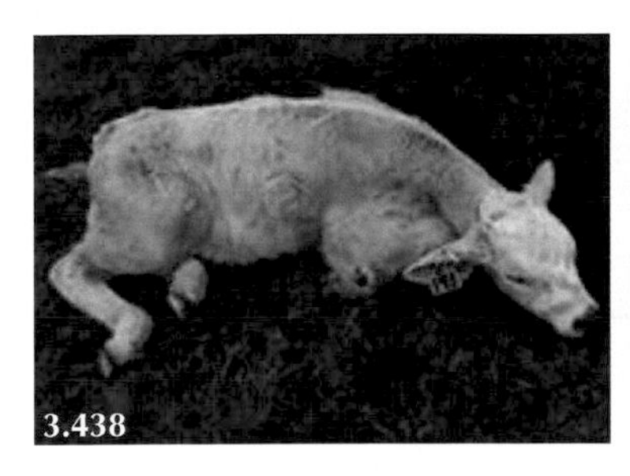

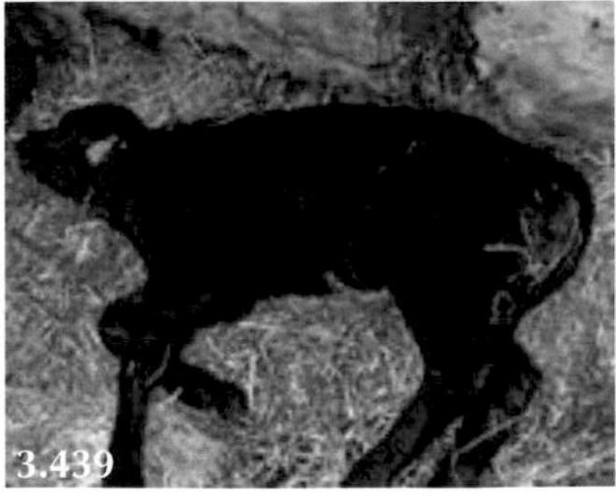

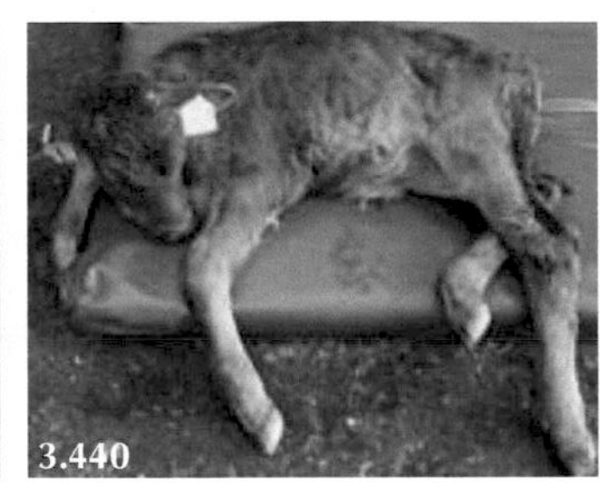

3.438–3.440: These three calves are recumbent, weak and have fluid-distended abomasum/intestines associated with acidosis.

- **Key observations**: Weakness/recumbency. Fluid-distended abomasum/intestines which "slosh" on gentle palpation.
- **Clinical examination**: Usually have normal hydration status; weakness results from acidaemia.
- **Key history events**: Diarrhoea for several days. No dam vaccination. 3.232.

Peritonitis

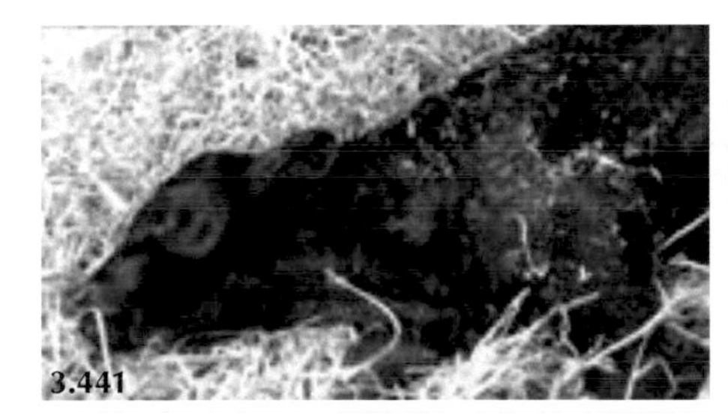

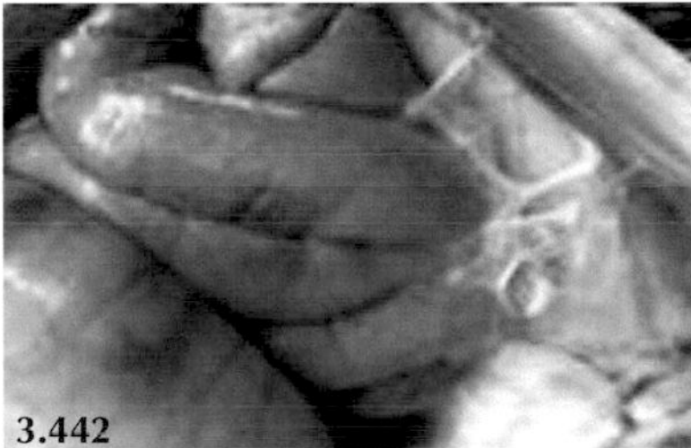

3.441: Recumbent calf with painful distended abdomen. Intravenous fluid therapy proved unsuccessful.
3.442: Passage of digesta prevented by fibrinous adhesions.

- **Key observations**: Painful swollen abdomen.
- **Clinical examination**: Swollen painful umbilicus with no evidence of preventive iodine dip.
- **Key history events**: Usually 3–5 days old. Normal for first two days then stop sucking. Born in contaminated environment, often no navel dressing. 3.233, 3.234

Atresia Ani/Coli

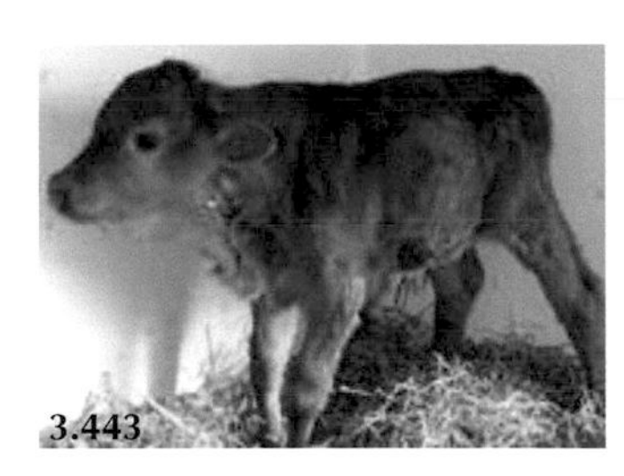

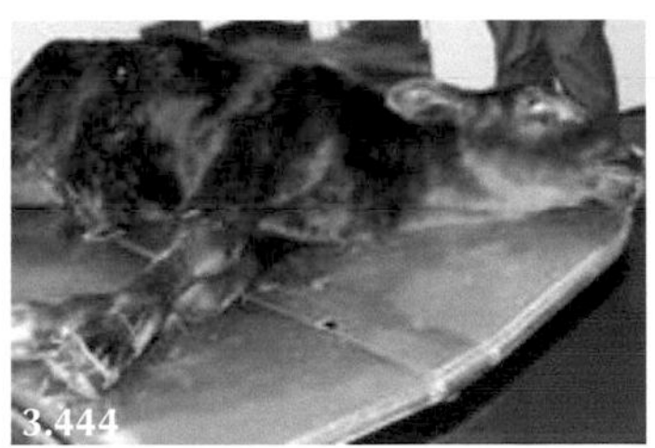

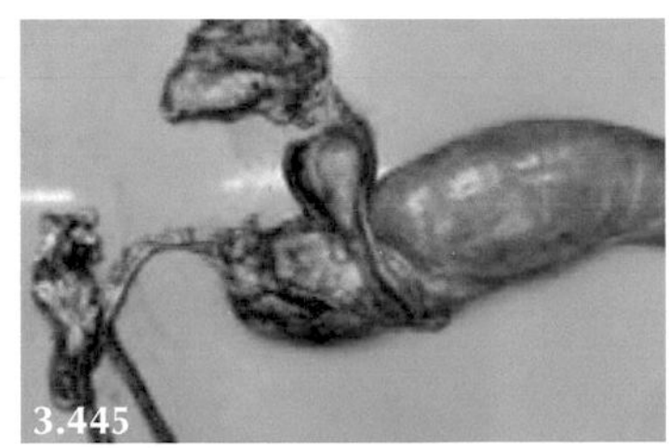

3.443: Two-day-old calf not sucking but with increasing abdominal distension. **3.444:** Recumbent calf with distended abdomen. **3.445:** Necropsy confirmation of atresia coli showing distended, blind-ended large intestine.

- **Key observations**: Increasing abdominal distension since birth. Colic.
- **Clinical examination**: Fluid-distended intestines, absence of feces.
- **Key history events**: Cases of atresia coli may occur as small clusters—genetic effect? 3.235, 3.236, 3.237

Intussusception

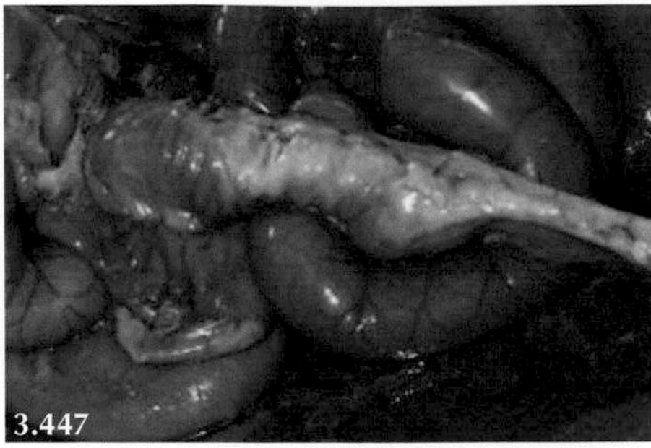

3.446: Two-week-old calf with distended abdomen; wide-based stance suggests abdominal pain.
3.447: Intussusception at necropsy.

- **Key observations**: Increasing abdominal distension over several days. Colic.
- **Clinical examination**: Fluid-distended intestines, absence of feces. 3.238

3.7

Common Causes of Abortion in Sheep

- Start investigations after several abortions.
- Always collect dam serum sample and vaginal swab; collect and bag all aborted material.
- Always detail biocontainment measures to farmer. Re-iterate possibility of zoonotic infection with appropriate precautions.

TABLE 3.61

Common Causes of Abortion in Sheep

	Ewe demeanour	Rectal temperature	Fetus/lambs
EAE	–	NAD/+	Fresh, may be born alive but weak
Toxoplasmosis	NAD	NAD	Mummified fetus and one normal lamb at full term
Campylobacter	–	NAD/+	NAD
Listeria	NAD	NAD	NAD
Salmonella	––/–––	++/+++	Varying stages of autolysis. Foetid uterine discharge. Metritis. High mortality
Q fever	NAD	NAD	Aerosol infection from goats

DOI: 10.1201/9781003106456-25

TABLE 3.62

Definitive Ancillary Tests for Abortion (Where Appropriate/Necessary)

EAE	Ewe often not sick, lambs may be born near-term. Elementary bodies in placental smears stained by the modified Ziehl-Nielsen method. Serology
Toxoplasmosis	Dam serological testing merely indicates past infection. Histopathological examination of placentae
Campylobacter	Culture of fetal stomach contents
Listeria	Culture of fetal stomach contents
Salmonella	Culture of fetal stomach contents/ewe vaginal swab
Q fever	Paired serology. Culture, immunohistochemical and PCR tests can be used to identify *C burnetii* in tissues

EAE

3.448: Two dead fetuses aborted 10 days before full term.

- **Key observations**: Fresh fetuses aborted during last three weeks.
- **Clinical examination**: NAD apart from vaginal discharge.
- **Key history events**: No vaccination policy, purchased unvaccinated sheep.

Toxoplasmosis

3.449: Delivery of mummified fetus at full term. **3.450:** Open feed stores are a risk factor for contamination by vermin.

- **Key observations**: Mummified fetus at full term.
- **Clinical examination**: Other lamb usually normal. Ewe normal.
- **Key history events**: Cat feces contaminating feed/grazing but rarely observed/detected.

Campylobacter

3.451 and **3.452:** Surface water, unhygienic and overcrowded grazing/feeding conditions are risk factors for Campylobacter abortion.

- **Key observations**: Fresh fetuses.
- **Clinical examination**: Ewe often shows diarrhoea.
- **Key history events**: Unhygienic and overcrowded grazing/feeding conditions. Contaminated water supply.

Listeria

3.453: Poor quality silage with wrap removed several days before feeding. **3.454** and **3.455:** Centre of bale may not be eaten for several weeks.

- **Key observations**: Poor quality silage. Poor storage, long exposure to air before eaten.
- **Clinical examination**: No evidence of encephalitis.
- **Key history events**: Soil contamination of conserved grass silage. Poor storage/perforated wrap.

Salmonella

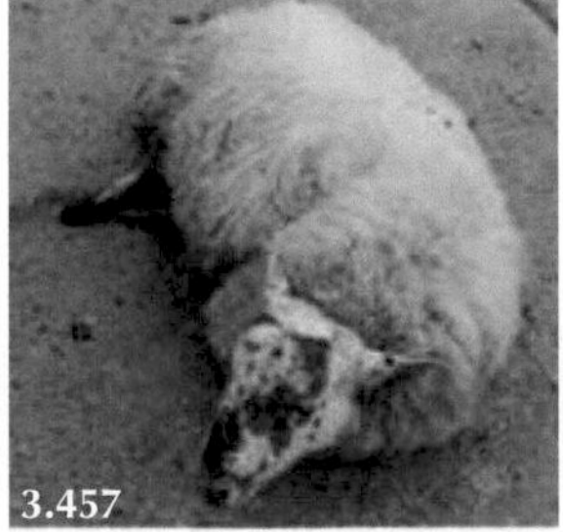

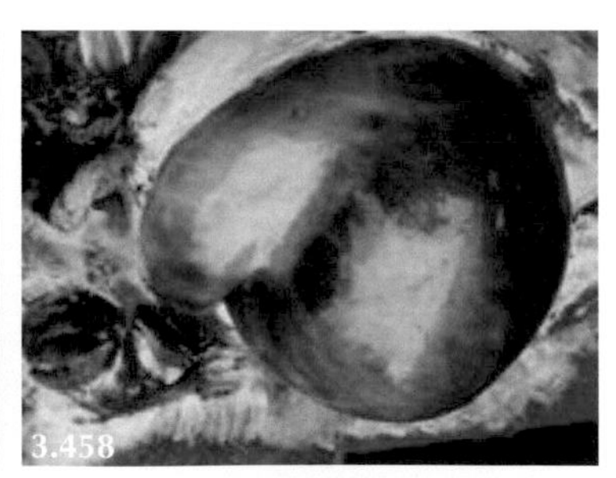

3.456: Seabirds are considered an important vector for *Salmonella montevideo*. **3.457:** Toxaemia and abdominal distension following fetal death. **3.458:** Necrotic uterus following fetal death caused by *Salmonella montevideo*.

- **Key observations**: Outbreak of abortions, may be confined to one group/field.
- **Clinical examination**: Ewe profound toxaemia if lambs not expelled, high mortality.
- **Key history events**: Outdoors, role of seagulls/carrion (*S. montevideo*).

3.8

Common Causes of Dystocia

Sheep

- Always administer a low extradural block of lignocaine before proceeding. Add xylazine for analgesic effect lasting 24–36 hours.
- Intravenous NSAID injection is also indicated.
- Disposable gloves should be worn.
- Take great care when applying traction to a pedigree single meat breed lamb in posterior presentation—consider performing a caesarean operation because of the risk of trauma.
- If events are not progressing after 5 minutes—perform a caesarean operation. 🎥 3.239, 3.240, 3.241, 3.242, 3.243

TABLE 3.63

Common Causes of Dystocia in Sheep

	Dam behaviour	Fetal extremities
Bilateral shoulder flexion	May not strain	Head only presented through vulva
Posterior presentation	May not strain	Hind feet, hooves pointed up
Breech (bilateral hip flexion)	May not strain	Tail only at vulva
Absolute fetal oversize	Straining	Fore feet, hooves pointed down. Muzzle at vulva
Simultaneous presentation	Straining	Various limbs presented from two lambs
Incomplete cervical dilation	First stage labour behaviours	Only fetal membranes at vulva
Uterine torsion	Some evidence of first stage labour behaviours	Unable to detect torsion by vaginal examination

TABLE 3.64

Action to Correct Dystocia

Bilateral shoulder flexion	Low extradural block. Repel fetus, extend both forelegs. Traction
Posterior presentation	Traction—caution if single meat breed lamb as risk of rib fractures and/or liver rupture
Breech (bilateral hip flexion)	Low extradural block. Repel fetus, extend both hindlegs to resemble a posterior presentation. Traction
Absolute fetal oversize	Traction, if this does not deliver lamb—caesarean operation
Simultaneous presentation	Low extradural block. Ensure both legs/head belong to same lamb
Incomplete cervical dilation	Attempt digital dilation of cervix for up to 5–10 minutes. If no progress consider caesarean operation
Uterine torsion	Caesarean operation

DOI: 10.1201/9781003106456-26

Bilateral Shoulder Flexion

3.459 and **3.460:** Lamb's swollen head presented through vulva.

- **Key observations**: Lamb's swollen head presented through vulva.
- **Clinical examination**: Bilateral shoulder flexion.
- **Key history events**: Normal first stage labour. 🎥 3.244, 3.245

Posterior Presentation

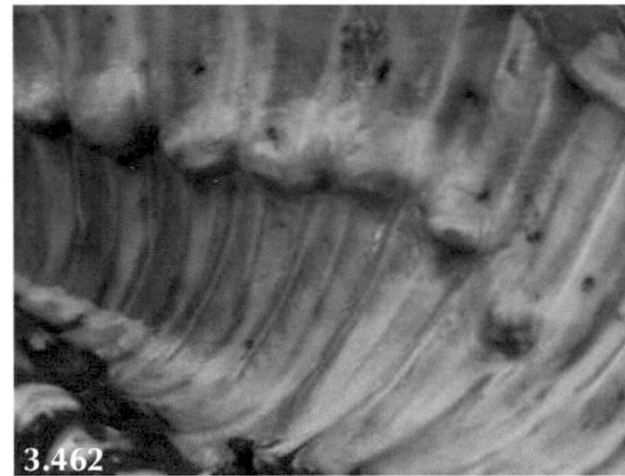

3.461: Hind feet at vulva, hooves pointed up. **3.462:** Healed rib fractures several weeks after assisted posterior presentation delivery of large single lamb.

- **Key observations**: Hind feet at vulva, hooves pointed up.
- **Clinical examination**: Possible absolute fetal oversize especially if singleton and meat breed such as Texel. A caesarean operation is likely to be necessary in these circumstances.
- **Key history events**: Normal first stage labour.

Breech (Bilateral Hip Flexion)

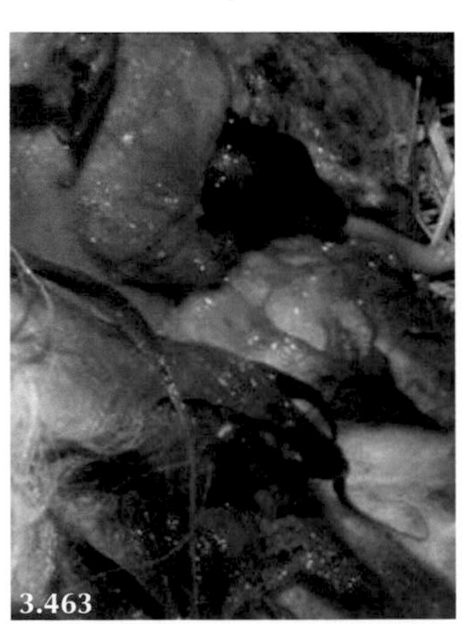

3.463: The lamb's tail is sometimes presented through vulva.

- **Key observations**: Lamb's tail sometimes presented through vulva.
- **Clinical examination**: Lamb's pelvis has entered maternal pelvis but both hips fully flexed.
- **Key history events**: Normal first stage labour.

Absolute Fetal Oversize

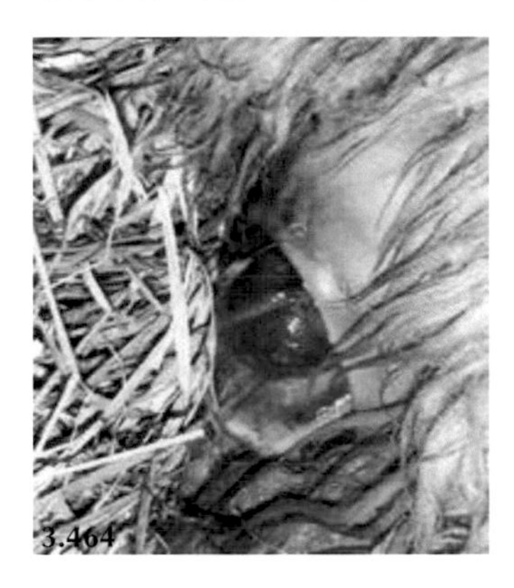

3.464: Lamb's muzzle and forefeet presented at vulva.
3.465: Elbows extended—delivery will continue normally.

- **Key observations**: Lamb's muzzle and forefeet presented at vulva.
- **Clinical examination**: Lamb firmly lodged in pelvis, elbows not yet extended.
- **Key history events**: Normal first stage labour. 🎥 3.246

Simultaneous Presentation

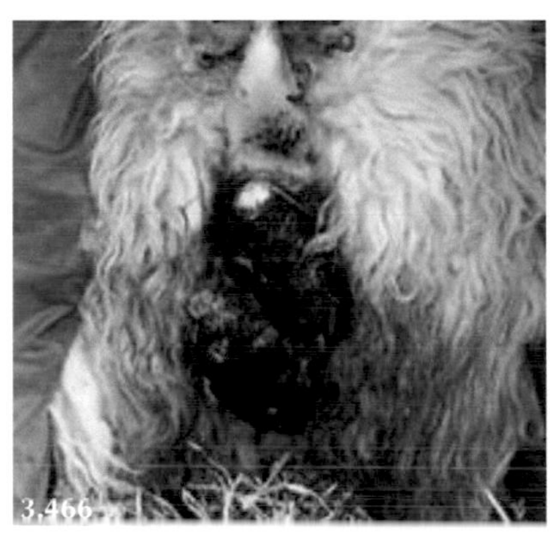

3.466: Two lambs presented simultaneously.

- **Key observations**: Head/limbs from two lambs presented simultaneously.
- **Clinical examination**: Trace one foreleg to chest, neck, head then reverse process to lamb's other foreleg. Apply traction—lamb should be delivered easily because multiple litter.
- **Key history events**: Normal first stage labour.

Dead Lambs *in Utero*

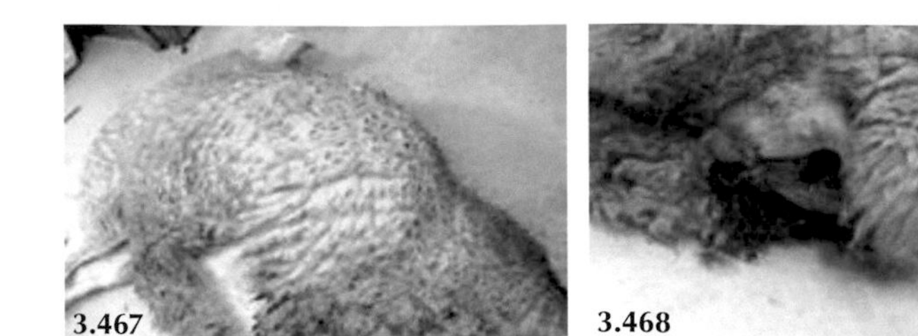
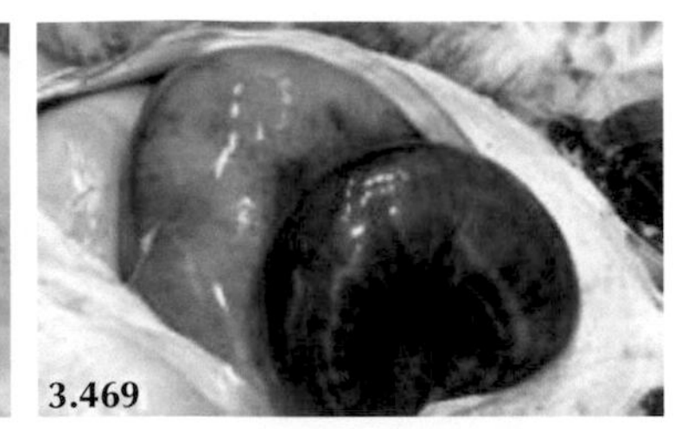

3.467: Markedly distended abdomen. **3.468:** Foetid vaginal discharge, cervix could not be palpated—note vulval oedema. **3.469:** Autolytic lambs *in utero*.

- **Key observations**: Markedly distended abdomen and foetid vaginal discharge.
- **Clinical examination**: Toxaemia, cervix can not be palpated because the vagina is not dilated.
- **Key history events**: Inadequate supervision of lambing ewes. Abortion near/at term. 3.247

Caesarean Operation on Farm

A caesarean operation is indicated in this Greyface ewe presenting in first stage labour with a large oedematous vaginal prolapse. The facilities on farm are a gate placed on top of a lamb heat box and an old oil drum for the tray of instruments.

3.470 and **3.471:** Greyface ewe presenting in first stage labour with a large oedematous vaginal prolapse. **3.472:** Facilities on farm comprise a gate placed on top of a heat box and an old oil drum for the tray of instruments.

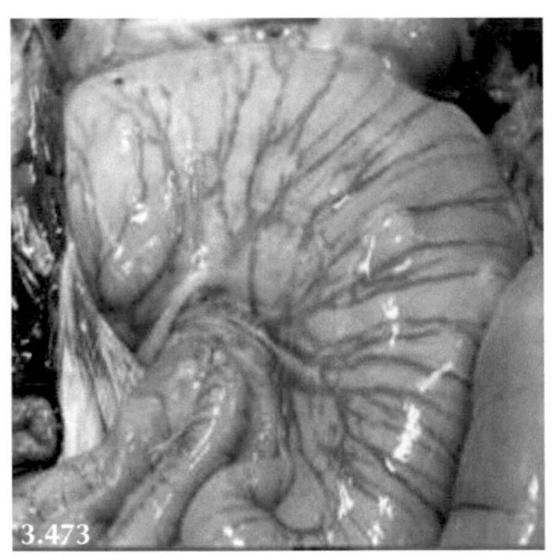
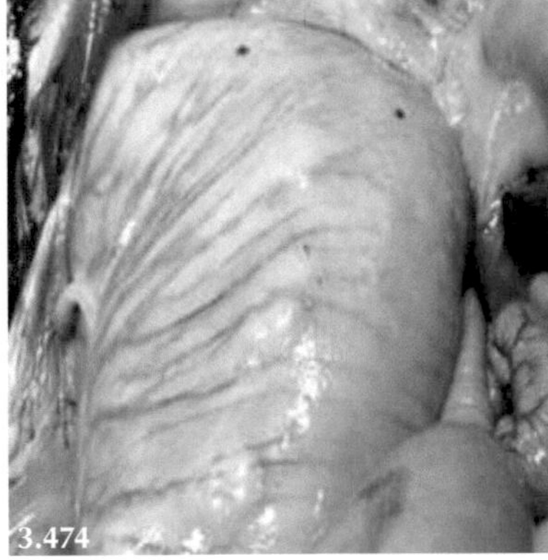

3.473 and **3.474:** Necropsy specimens of heavily pregnant ewes show blood supply to the uterine wall/caruncles and the very much narrower diameter of these vessels on the greater curvature (to the right hand side of images). 3.248, 3.249, 3.250, 3.251, 3.252

Cattle

- Wherever safe, calve cows in pen not cattle stocks. 🎥 3.253
- Ensure you can successfully block a cow's flank for surgery (paravertebral anaesthesia) before your first night on call for a farm animal practice.
- Administer a NSAID intravenously before surgery.
- Consider whether a low extradural block is necessary.
- Many veterinary practices have two vets attend each caesarean operation as a routine.
- There is no need to give antibiotics into the abdominal cavity and along the flank wound edges.

TABLE 3.65

Common Causes of Dystocia in Cattle

	Dam behaviour	Fetal extremities at vulva
Absolute fetal oversize	Straining	Fore feet—hooves pointed down. Forelegs crossed in vagina. Calf's muzzle at vulva, very swollen tongue
Uterine torsion	Prolonged first stage labour behaviour	Able to detect torsion by vaginal examination
Posterior presentation	May not strain	Hind feet—hooves pointed up
Breech (bilateral hip flexion)	Does not strain	Tail only at vulva
Unilateral shoulder flexion	Straining	Head and one leg presented. Swollen head and tongue
Simultaneous presentation	First stage labour behaviours	Various limbs/heads from two calves at pelvic inlet
Hypocalcaemia	Recumbent, prolonged first stage labour	Dilated cervix but no fetal extremities within the pelvic canal
Incomplete cervical dilation	First stance labour behaviour	Only fetal membranes protrude through cervix

TABLE 3.66

Action to Correct Dystocia

Condition	Action
Absolute fetal oversize	Calf's foreleg fetlock joints protruding one hand's breadth beyond the vulva—traction generated by two people delivers calf in 10 minutes—if not then consider caesarean operation
Uterine torsion	Often absolute fetal oversize. Attempt to correct torsion. May require caesarean operation
Posterior presentation	Calf's hock joints protruding one hand's breadth beyond the vulva—traction generated by two people delivers calf in 10 minutes—if not then consider caesarean operation
Breech (bilateral hip flexion)	Low extradural block, repel calf, extend both hip joints then follow preceding instructions for posterior presentation. Take care not to wrap umbilical cord around one hindleg
Unilateral shoulder flexion	Low extradural block. Repel fetus, extend other foreleg. Traction
Simultaneous presentation	Ensure both forefeet and head belong to same calf
Hypocalcaemia	400 ml 40% calcium borogluconate intravenously over 5–10 minutes
Incomplete cervical dilation	Attempt manual dilation. Caesarean operation if no progress within 15 minutes

Normal Parturition

3.475: Calf's muzzle and forefoot presented at vulva. **3.476:** Calf's fore fetlock joints protruding much more than one hand's breadth beyond the vulva—no intervention necessary. **3.477:** The cow has stood up for the final stage of delivery of the calf. 3.254

Absolute Fetal Oversize

Deciding how much traction to apply to a calf in anterior presentation is not easy.

3.478: Calf's forefeet presented at vulva, fetlock joints level with vulva. **3.479:** Two people pulling—calf's fore fetlock joints protruding more than one hand's breadth beyond the vulva. **3.480:** Traction applied when cow strains, head almost out, uneventful delivery. Note that the cow should have been released from the stocks but this was not possible in this situation.

3.481: Two people pulling—calf's fore fetlock joints protruding more than one hand's breadth beyond the vulva. **3.482:** Calf delivered. Note that the cow should be released from the stocks wherever possible (not considered safe in this situation). **3.483:** Considerable force can be applied using a calving jack (cow released from stocks but has not lain down to strain). 3.255

Uterine Torsion

3.484 and **3.485:** Extended first stage labour with occasional colic in cow with a uterine torsion. **3.486:** Cow cast using Rueff's method.

- **Key observations**: Cow does not progress beyond behavioural changes of early first stage labour. Possible mild colic—lying down, occasional kick at belly.
- **Clinical examination**: Vaginal examination reveals obvious corkscrewing of vaginal wall which is also palpable per rectum.
- **Key history events**: Extended first stage labour. 3.256

Absolute Fetal Oversize, Hip Lock, Corrected by Embryotomy

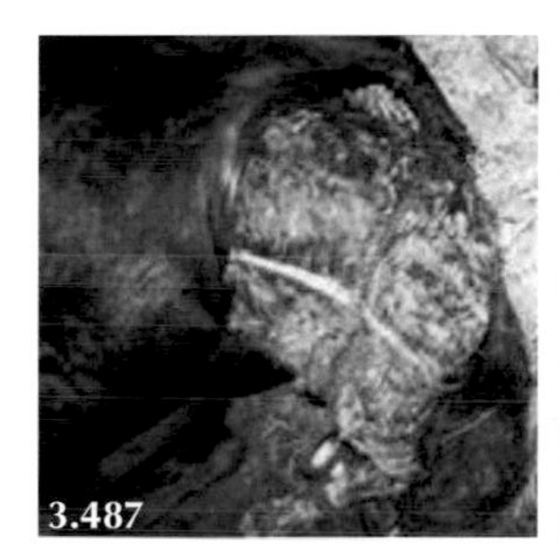

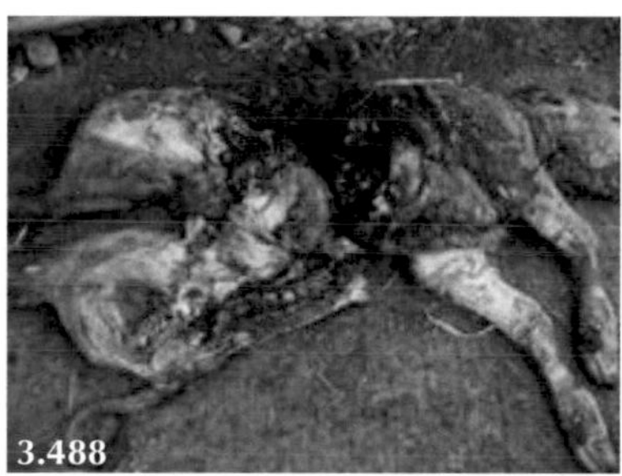

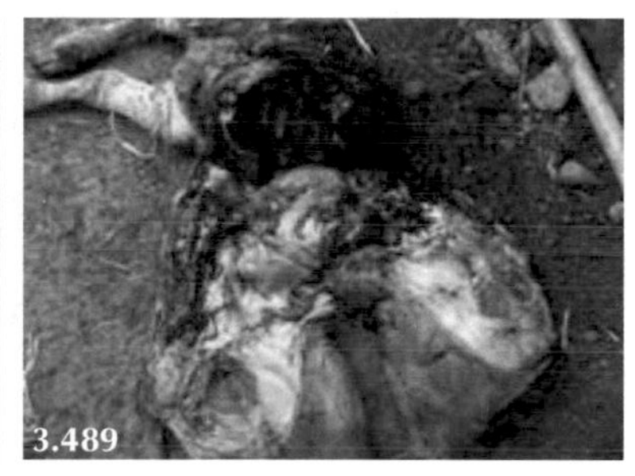

3.487: Calf is locked at the hips because excessive traction has been applied when a caesarean operation was necessary. **3.488** and **3.489:** The calf's forequarters are removed as close to the vulva as possible by cutting through the skin, caudal rib cage, then splitting the caudal vertebral column and pelvis. 3.257, 3.258

Posterior Presentation

Two strong people pulling on calving ropes should be able to extend the hocks more than one hand's breadth beyond the cow's vulva (calf's hindquarters now fully within the pelvic inlet) within 10 minutes. With experience it is possible to apply greater traction than the forces described here and still achieve a successful resolution, but there are occasional doubts when the calf becomes lodged. Is delivery of a live calf which dies within 24 hours due to trauma sustained during forced extraction a successful resolution to a "difficult calving"?

3.490: Hind feet at vulva, hooves pointed up. Calf's hock joints are not visible. **3.491:** Calf's hock joints are not protruding one hand's breadth beyond the vulva—continue traction but with considerable caution. Note that the cows have been released into a pen. **3.492:** Excessive traction by the farmer has ruptured the middle uterine artery and a large volume of arterial blood is visible.

- **Key observations**: Hind feet at vulva, hooves pointed up.
- **Clinical examination**: Possible absolute fetal oversize—assess risk of damage to calf.

Breech (Bilateral Hip Flexion)

3.493: The calf's tailhead is presented at the pelvic inlet with the tail protruding from the vulva. 3.259

Unilateral Shoulder Flexion

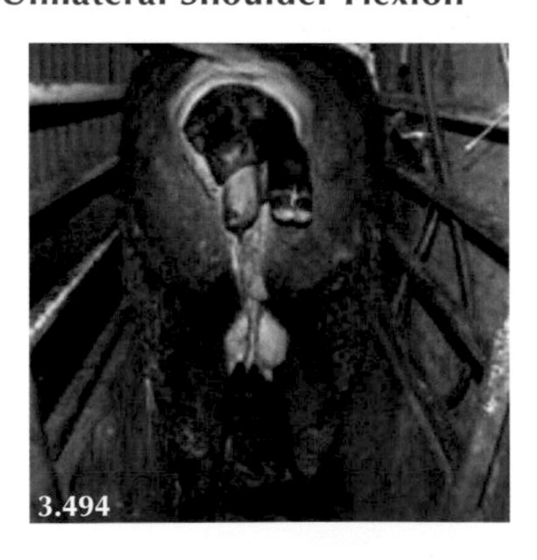

3.494: Unilateral shoulder flexion. **3.495:** A very swollen oedematous tongue is common when the calf's head is stuck in the cow's pelvic canal.

- **Key observations**: Calf's muzzle but only one forefoot presented at vulva.
- **Clinical examination**: Calf's foreleg alongside the chest wall. Repel head/foreleg under low extradural block and extend other foreleg.

Simultaneous Presentation

3.496: Labour does not progress beyond behavioural changes observed during first stage.

- **Key observations**: Often prolonged first stage labour. Cow appears restless.
- **Clinical examination**: Endless combinations of heads/legs.

Hypocalcaemia

3.497: Aged beef cow unable to stand during first stage labour. **3.498:** Treatment of cow (left) for hypocalcaemia with intravenous calcium borogluconate. **3.499:** Milk fever delaying calving in this recumbent dairy cow.

- **Key observations**: Weakness progressing to recumbency, mucus string visible at vulva.
- **Clinical examination**: Reduced responses, ruminal atony/bloat, increased heart rate.
- **Key history events**: Older cows at pasture during autumn.

Caesarean Operation

Paravertebral anaesthesia is preferred by many surgeons and reduces the risk of surgical site contamination with subsequent wound breakdown. While distal paravertebral anaesthesia can be achieved without any reaction from the cow, it is prudent to administer the block with the cow in cattle stocks. 3.260, 3.261, 3.262

The cow is restrained in cattle stocks or by a halter in a large well-bedded loose box. A large area of the left flank is shaved and surgically prepared while the flank analgesia takes effect. (A plastic disposable drape is fenestrated and held in position with towel clips—this may only be practical in recumbent cattle.)

Surgery is performed through a left flank incision midway between the last rib and the wing of the ilium commencing 10 to 15 cm below the level of the transverse processes of the lumbar vertebrae. A 25 cm incision is made through the skin, external abdominal oblique muscle and thick internal abdominal oblique muscle using a scalpel blade. The transversus muscle and closely adherent peritoneum near the top of the incision are grasped with forceps and raised before a small nick is made with scissors and subsequently extended. Care is necessary at this stage to avoid puncturing an underlying viscus if the cow is in lateral recumbency. A sloping incision starting 10–15 cm ventral to the tuber coxae and extending ventrally and cranially at an angle of 35° is also described. This latter method has the reported advantage that blunt dissection can be used to separate the muscle fibres of the thick internal abdominal oblique muscle that run parallel to the incision site.

The uterine incision is made starting on the greater curvature 15 to 20 cm from the tip of the uterine horn (at the level of the calf's hind fetlock joints in those calves in anterior presentation) and extended towards the cervix as necessary using blunt ended scissors (Roberts embryotomy knife or plastic disposable letter-opening device).

The uterine incision is normally extended 30 to 40 cm to just over the calf's tailhead. While the incision may be started with the uterine tip exteriorised, it is usually necessary to enter the abdomen whilst extending the incision taking care to protect the cutting edge. Once the calf's hind feet have been freed from the uterine horn, it helps to have an assistant hold the calf's hind feet as high as possible which draws the uterine incision up to the lower margin of the abdominal incision where it can be safely extended. Always make the uterine incision long enough—if not, an L-shaped tear is difficult to suture.

The uterine incision is exteriorised with the uterine horn resting on the lower margin of the abdominal incision. Some surgeons prefer farmers to hold the uterus proximal to the incision site with uterine clamps rather than a hand to avoid possible contamination. A single Cushing (inversion) suture of absorbable material closes the uterus. Approximately 2 cm of wound edges are inverted to form a tight seal with the needle passing through serosa and muscular layer and not into the lumen. The assistant is often asked to hold the suture material taut as the surgeon places the next suture. Two suture layers are recommended by some surgeons, but this approach is not necessary nor is there always sufficient time due to the contracting uterus. It is essential that the incision is water-tight and that none of the fetal membranes protrude through the suture line. The bulk of the protruding fetal membranes can be debrided before commencing uterine closure to prevent them interfering with the suturing process. If a tear has occurred, suture this first then the longer length of the uterine incision.

Flush the uterine closure site with one litre of sterile saline to remove all remains of blood contamination before returning the uterus into the abdominal cavity. Before commencing closure of the abdominal incision, any blood clots are scooped from the abdominal cavity. Intra-abdominal antibiotics are commonly used but with little evidence of their efficacy. Ensure the calf's umbilicus is treated before you leave the farm and give instructions regarding colostrum feeding if the calf is not to its feet within 30 minutes. 🎥 3.263, 3.264

3.9

Common Causes of Tenesmus and Prolapse

Growing Lambs and Adult Sheep

- Farmers only take their sheep with severe prolapse problems to the veterinary surgery. Transport could aggravate the problem with tears/rupture of the vaginal/uterine wall.
- Always administer a low extradural block of lignocaine and xylazine before proceeding.
- Intravenous NSAID injection is also indicated.
- A webbing-style harness works well if applied immediately after the vaginal prolapse occurs.
- Ultrasound examination of the bladder (diagnosis) and kidney (prognosis) is essential in suspect urolithiasis cases.
- Salvaging lambs from ewes with herniated intestines is rarely worthwhile. If the farmer insists, shoot the ewe, you then have 5 minutes to retrieve the lambs.

TABLE 3.67

Common Causes of Tenesmus and Prolapse in Growing Lambs and Adult Sheep

	Likely duration at presentation	Tenesmus	Size of prolapse	Haemorrhage	Shock/ toxaemia
Dystocia	Hours	+/+++	NAD	NAD/+	NAD
Vaginal prolapse	Hours/days	++/+++	+/+++	NAD/+++	NAD/++
Uterine prolapse	Hours	NAD/+	+++	+/++	NAD/++
Eviscerated intestines	Hours	NAD/+	+++	+/++	+++
Rectal prolapse	Days/weeks	+/++	+/++	NAD	NAD
Urolithiasis	Days	+/+++	NAD	NAD	NAD/+

TABLE 3.68

Action to Correct Common Causes of Tenesmus and Prolapse in Growing Lambs and Adult Sheep

Dystocia	Low extradural anaesthesia. Obstetrical examination
Vaginal prolapse	Low extradural block. Buhner suture/harness
Uterine prolapse	Low extradural block. Buhner suture/harness
Herniated intestines	Immediate euthanasia
Rectal prolapse	Low extradural block. Buhner suture. Amputation of prolapsed tissues should be approached with caution. Sheep are not pigs
Urolithiasis	Ultrasonography of bladder and right kidney

DOI: 10.1201/9781003106456-27

First Stage Labour/Dystocia

3.500: Powerful abdominal contractions are normal during first stage labour. **3.501:** Powerful abdominal contractions continue after excessive force to deliver lambs. Note the painful expression.

- **Key observations**: Powerful abdominal contractions are normal during first stage labour and may persist after unskilled assisted delivery.
- **Clinical examination**: Oedema and bruising of the posterior reproductive tract.
- **Key history events**: Unskilled delivery.

Vaginal Prolapse

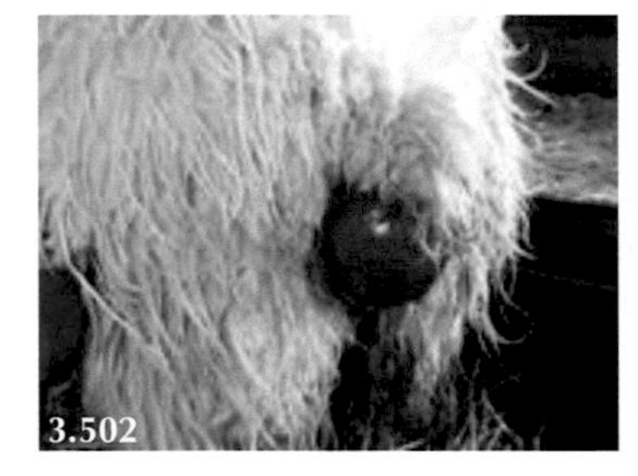

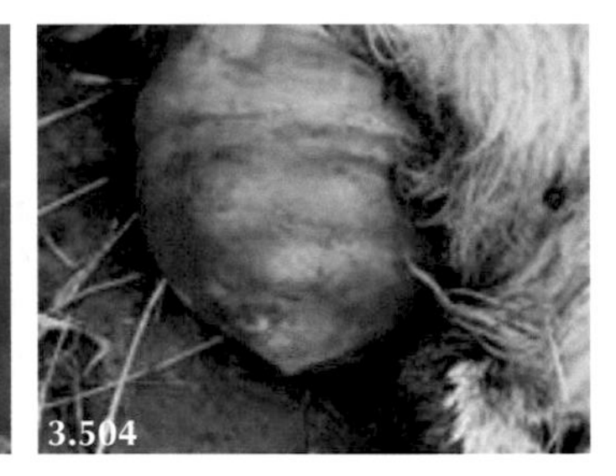

3.502–3.504: A range of vaginal prolapses are encountered in sheep practice. These prolapses are relatively clean.

- **Key observations**: Large viscus protruding from the vulva.
- **Clinical examination**: Oedematous vaginal wall, prolapse usually contains retroflexed urinary bladder.
- **Key history events**: Last three weeks of pregnancy, multiple pregnancy, fibrous diet. Tenesmus following assisted lambing. 🎥 3.265, 3.266, 3.267, 3.268

Complications Associated with Vaginal Prolapses

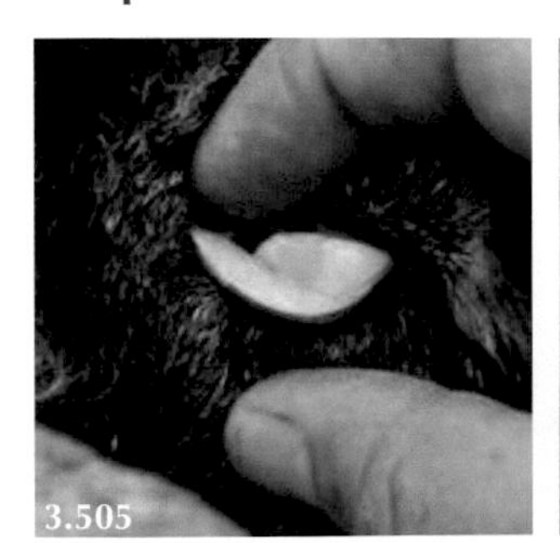

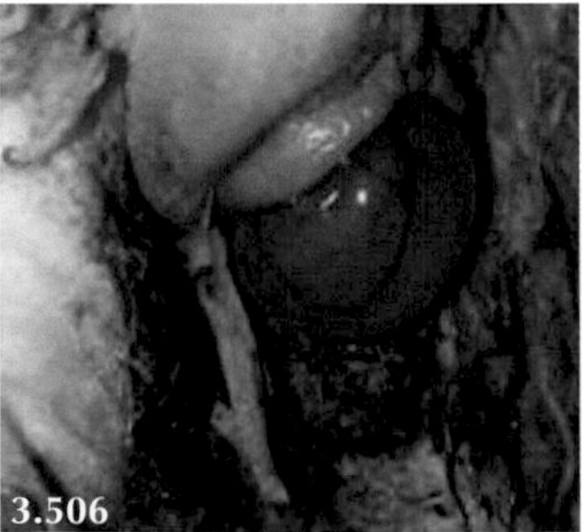

3.505: Very pale mucous membranes caused by severe internal haemorrhage. **3.506:** Allantois protruding through the vulva made narrow by the Buhner retention suture. **3.507:** Infected fetal membranes protruding through vaginal prolapse.

- **Key observations**: Complications to vaginal prolapse include superficial infection, haemorrhage, physical prevention of normal parturition (retention suture) and abortion.
- **Clinical examination**: Check for anaemia, toxaemia and abortion/lambing.
- **Key history events**: Difficulty catching ewe with large prolapse causing tears/haemorrhage, abortions on farm causing fetal death. 🎥 3.269, 3.270 3.271, 3.272

Concurrent Oedematous Vaginal and Rectal Prolapses

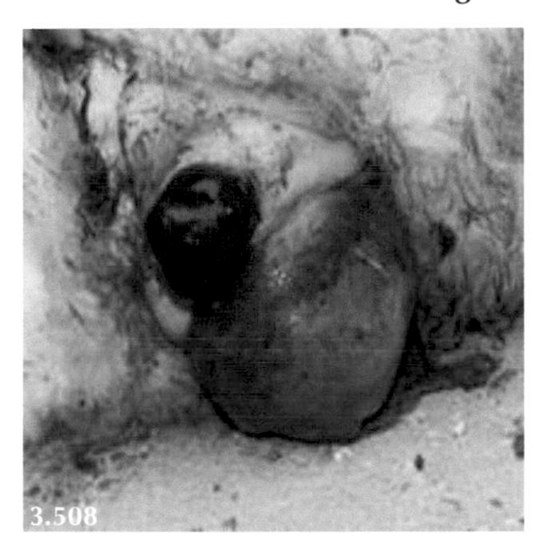

3.508 and **3.509:** Concurrent oedematous vaginal and rectal prolapses.

- **Key observations**: Two prolapses visible.
- **Clinical examination**: Large oedematous prolapses.
- **Key history events**: Vaginal prolapse causing prolonged tenesmus results in rectal prolapse. 🎥 3.273

Incorrect Management

3.510: Chronic vaginal prolapse where a "plastic intravaginal retainer" has not worked. **3.511:** Chronic vaginal prolapse in ewe showing pelvic limb weakness and skin excoriation (prolonged recumbency) likely caused by bilateral sciatic and obturator nerve paralysis.

- **Key observations**: Re-prolapse of vagina often because restraint methods have failed. Failure to control pain/tenesmus by extradural injection.
- **Clinical examination**: Prolapsed tissues often oedematous and grossly contaminated.
- **Key history events**: Poor treatment/management of vaginal prolapse. 🎥 3.274, 3.275

Uterine Prolapse

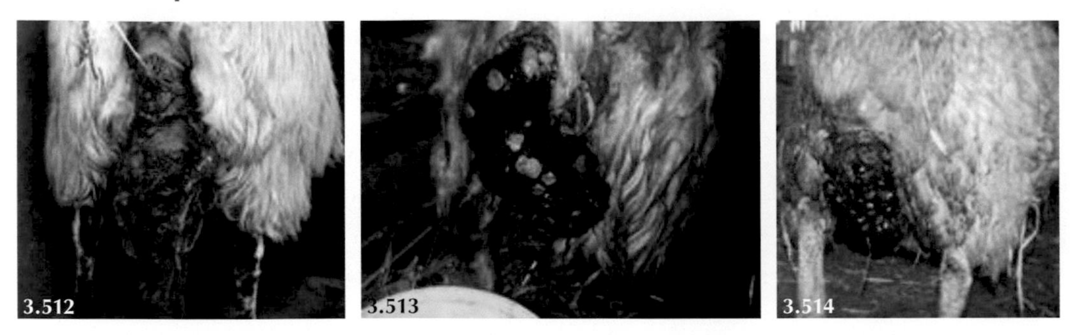

3.512: Uterine prolapse 24 hours after assisted parturition. **3.513:** Uterine prolapse 24 hours after assisted parturition. assisted parturition. **3.514:** Considerable haemorrhage from caruncles following uterine prolapse.

- **Key observations**: Very large prolapse with caruncles/fetal membranes visible.
- **Key history events**: Often immediately follows dystocia/traction or after period of tenesmus stimulated by trauma to posterior reproductive tract caused during assisted lambing.

Herniated Intestines

3.515–3.517: Evisceration through tear in vaginal wall shown at necropsy (right).

- **Key observations**: Purple discoloured intestines.
- **Clinical examination**: Shock, rapid heart rate, immediate euthanasia.
- **Key history events**: Late gestation, multiple pregnancy, within 30 minutes of concentrate feeding. 🎥 3.276

Rectal Prolapse

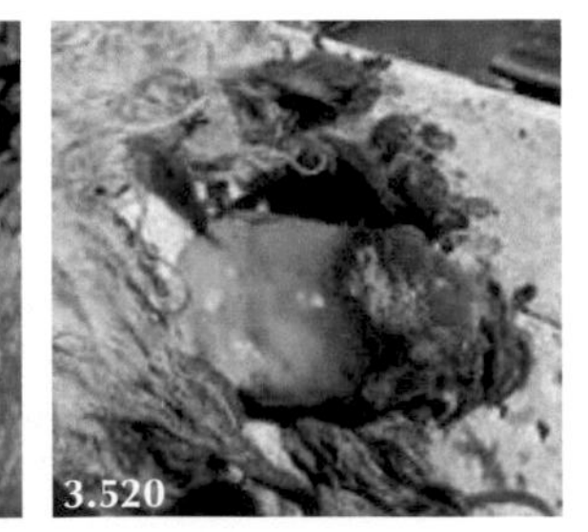

3.518: Rectal prolapse secondary to prolonged tenesmus (vaginal prolapse at lambing time). **3.519** and **3.520:** Rectal prolapse in a growing lamb.

- **Key observations**: Frequent tenesmus, raised tail, poor body condition/weight loss.
- **Clinical examination**: Tubular prolapse from rectum (intussusceptions are rare).
- **Key history events**: Usually secondary to prolonged tenesmus (vaginal prolapse). Occurs sporadically in growing lambs associated with persistent diarrhoea/coccidiosis. 🎥 3.277, 3.278

Urolithiasis

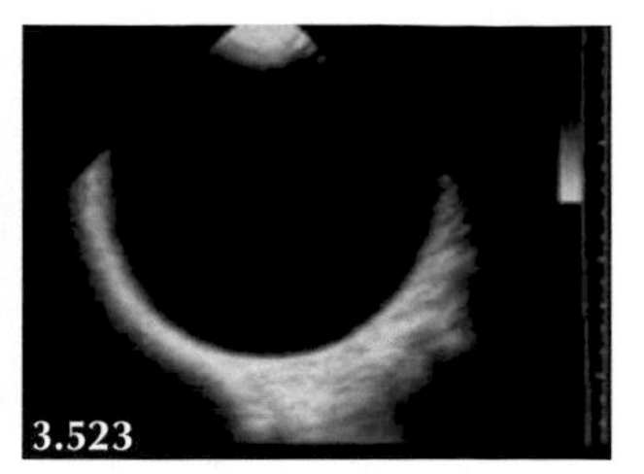

3.521: Scottish Blackface ram not grazing. **3.522:** Painful expression with curled upper lip. **3.523:** Transabdominal ultrasound reveals considerable bladder distension (18 cm diameter).

- **Key observations**: Not grazing, colic signs, straining to urinate.
- **Clinical examination**: Transabdominal ultrasound reveals considerable bladder distension. Check for hydronephrosis (right kidney).
- **Key history events**: Often fed inappropriate minerals/concentrate feeds. 3.279

Cattle

- Uterine prolapse usually occurs after prolonged second stage labour with the calf in anterior presentation, and associated with hypocalcaemia.
- Vaginal prolapse is typically seen in older, leaner beef cows during late gestation, and associated with oestrus.
- Rectal prolapse is very occasionally seen in four–nine-month-old beef calves after attempting to bull a cow in the group.

TABLE 3.69

Common Causes of Tenesmus/Prolapses in Cattle

	Likely duration at presentation	Tenesmus	Size	Haemorrhage
Uterine prolapse	Hours	NAD	+++	+/++
Vaginal prolapse	Days	NAD	++/+++	NAD
Rectal prolapse	Days/weeks	+/++	+	NAD

TABLE 3.70

Action to Correct Common Causes of Tenesmus/Prolapses in Cattle

Condition	Action
Uterine prolapse	Low extradural block. Buhner suture
Vaginal prolapse	Low extradural block. Buhner suture
Rectal prolapse	Low extradural block. Purse string suture. Amputation of prolapsed tissues rarely necessary. Castrate if immature bull calf

Vaginal Prolapse

3.524–3.526: Vaginal prolapses in older, leaner beef cattle during late gestation. Note the considerable fecal contamination suggesting the tissues have been prolapsed for several days.

- **Key observations**: Large 30–50 cm diameter viscus protruding from the vulva. No caruncles.
- **Clinical examination**: Oedematous vaginal wall with fecal contamination.
- **Key history events**: Beef cows; last few weeks of pregnancy or bulling activity several months after calving. 🎥 3.280, 3.281

Uterine Prolapse

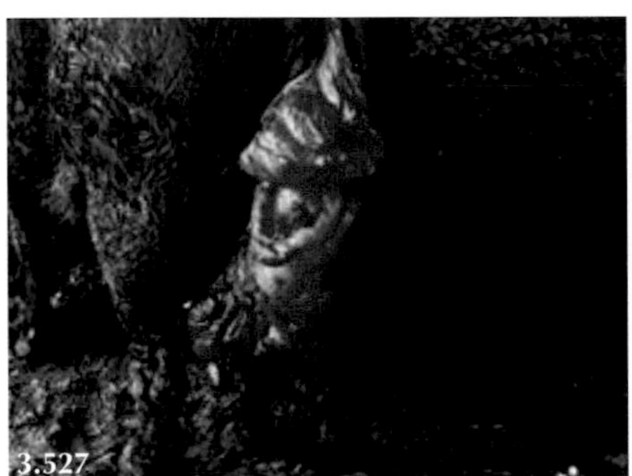

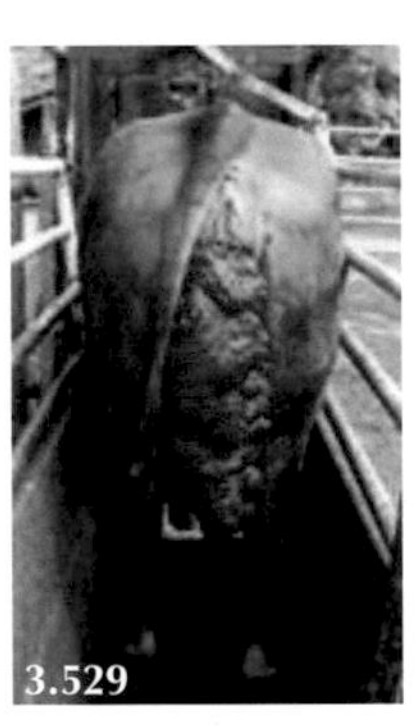

3.527–3.529: Uterine prolapse in three beef cows. Note the haemorrhage on the hindlegs in the centre cow.

- **Key observations**: Extensive pendulous prolapse to cow's hocks, caruncles/fetal membranes visible, may be trauma to caruncles causing considerable haemorrhage.
- **Clinical examination**: Sometimes very oedematous uterine wall.
- **Key history events**: Often immediately follows dystocia/excessive traction in beef cattle or associated with hypocalcaemia in dairy cows. 🎥 3.282, 3.283

Rectal Prolapse

- **Key observations**: Tubular prolapse from rectum.
- **Clinical examination**: (Intussusceptions are rare.)
- **Key history events**: Bulling activity in four- to nine-month-old male beef calves.

3.10

Lameness

Sheep—Foot Lesions

- Chronic severe foot lameness is the most common cause of poor body condition in adult sheep. This opinion is based upon handling all adult sheep in 120 flocks visited bi-annually/annually over six years totalling 300,000 sheep.
- Few farmers had received veterinary instruction on how to deal with toe fibroma and no farmer within this group had ever seen digit amputation for septic pedal arthritis.
- Foot paring remains standard practice on most farms for any lame sheep. Foot-bathing is rarely planned but used whenever sheep are gathered for other reasons.
- Duration of lameness can be estimated from limb muscle wastage, overgrown horn and bone lysis/osteophyte formation.

TABLE 3.71

Lameness in Sheep: Foot Lesions

	Duration of lameness	Lameness/ Pain	Swelling at coronary band	Hair at coronary band	Limb muscle
Footrot	Several days to months	+/+++	NAD	NAD	–/–––
Interdigital dermatitis	Several days to weeks (lambs)	+/+++	NAD	NAD	NAD
Contagious ovine digital dermatitis	Several days to weeks	+++	NAD	–/–––	––/–––
White line abscess	Several days	++/+++	NAD/+	NAD	NAD/–
Toe fibroma	Several weeks/months	+/++	NAD	NAD	–/––
Septic pedal arthritis (acute)	Several days	+++	++/+++	NAD	–/––
Septic pedal arthritis (chronic)	Several weeks/months	+++	++/+++	––/–––	–––
Septic pedal arthritis (ankylosis)	3–6 months	+	+	–/––	/––

TABLE 3.72

Ancillary Tests to Diagnose the Cause of Foot Lameness in Sheep

	Ancillary tests
Footrot	Bacteriology/PCR
Interdigital dermatitis	Response to topical treatment
Contagious ovine digital dermatitis	Bacteriology/PCR
White line abscess	Careful foot paring following black mark/separation at white line to release pus
Toe fibroma	Foot examination
Septic pedal arthritis (acute)	Radiography (abaxial soft tissue swelling, slightly increased P2/P3 joint space). Arthrocentesis unsuccessful due to pannus formation
Septic pedal arthritis (chronic)	Destruction of articular cartilage P2/P3, osteophyte formation. Dislocated P2/P3 joint
Septic pedal arthritis (ankylosis)	Fusion of considerable osteophytes

DOI: 10.1201/9781003106456-28

Footrot

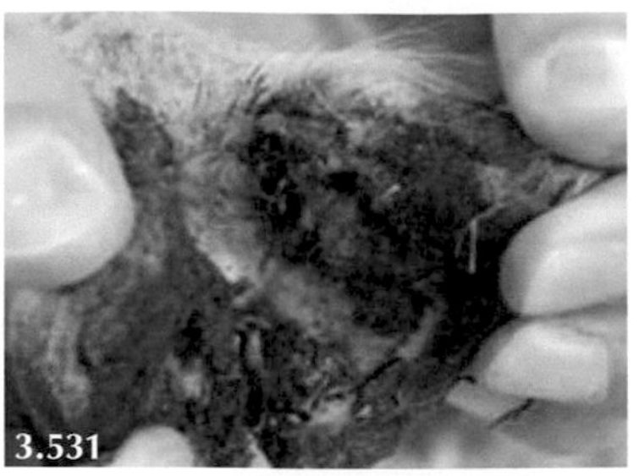
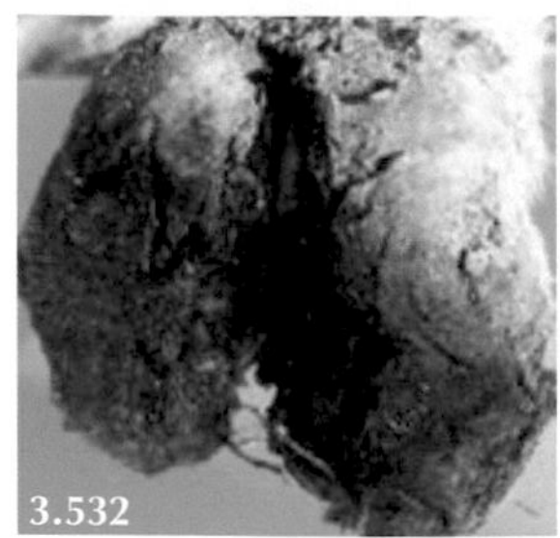

3.530: Non-weight-bearing lameness of right hind foot. **3.531:** Early separation of sole horn from the corium starting at the interdigital space. **3.532:** Extensive underrunning of the hoof horn exposing haemorrhagic corium.

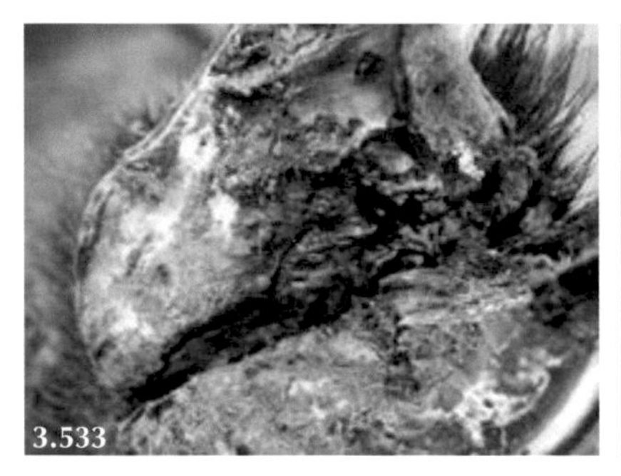
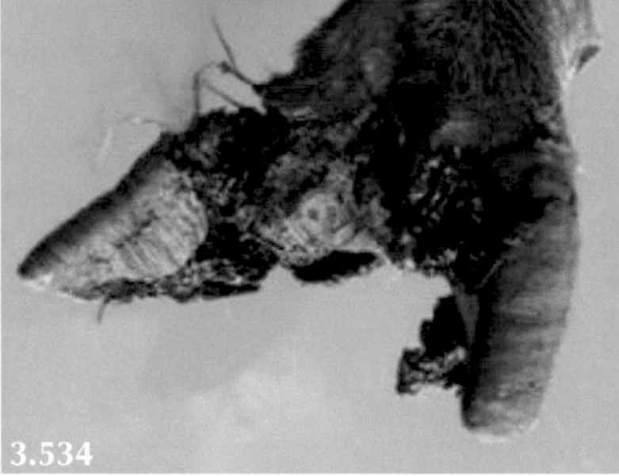
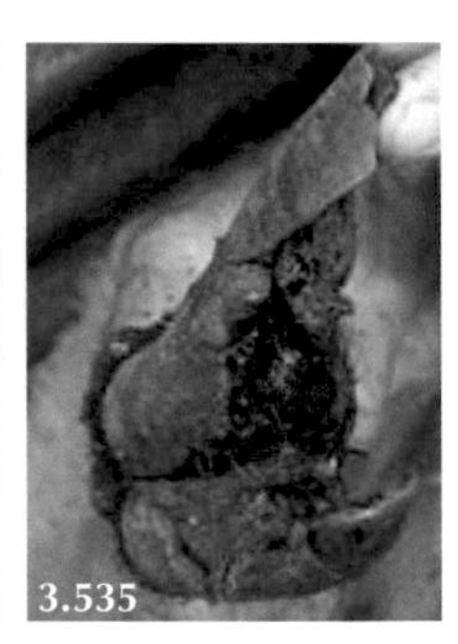

3.533: Hoof horn of the sole separated from the bleeding corium (there was no advantage to paring this foot). **3.534:** Extensive under-running of both hoof capsules exposing the bleeding corium. **3.535:** Chronic irritation of the corium to form an early toe fibroma; hoof horn overgrowth indicates duration of at least four to six weeks.

- **Key observations**: Large percentage of flock can show severe lameness.
- **Clinical examination**: Separation of hoof horn from corium starting at the interdigital space and spreading abaxially.
- **Key history events**: Recently purchased sheep. Warm wet weather. Delayed treatment exacerbates problem. 🎥 3.284, 3.285, 3.286, 3.287, 3.288, 3.289, 3.290

Interdigital Dermatitis (Lambs)

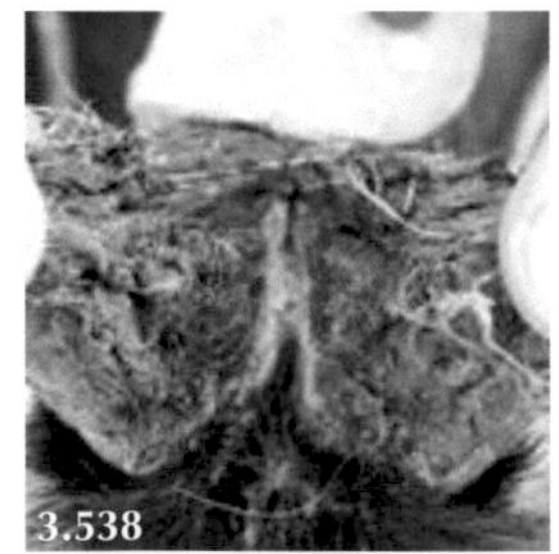

3.536: Four-month-old twin lambs grazing on their knees. **3.537:** Severe bilateral forefoot lameness. **3.538:** White diphtheritic surface to the interdigital space.

- **Key observations**: Large numbers of lambs suddenly very lame often grazing on their knees.
- **Clinical examination**: Interdigital skin hyperaemic with superficial infection. No hoof horn separation.
- **Key history events**: May follow overnight frosts, long grass. 🎥 3.291, 3.292, 3.293

Contagious Ovine Digital Dermatitis

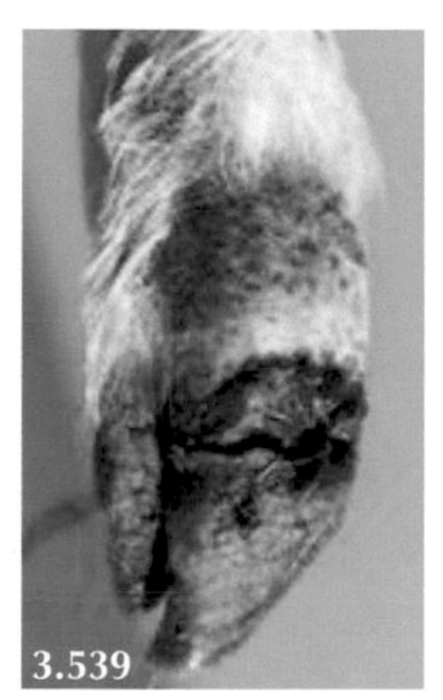
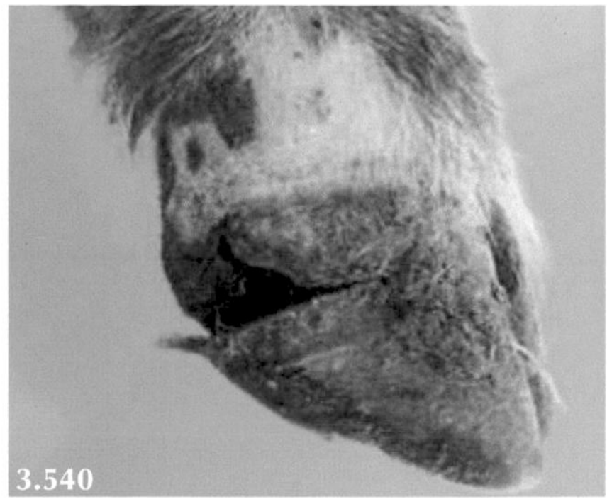
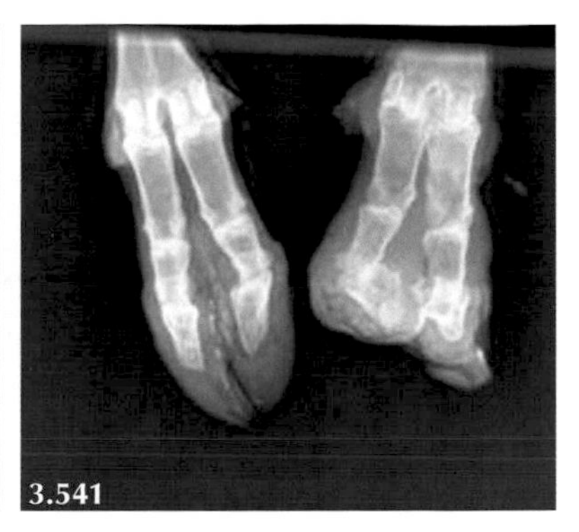

3.539 and **3.540:** Hair loss extending 2–3 cm proximal to the coronary band, separation/sloughing of hoof capsule at coronary band. **3.541:** DP radiographs of normal foot (left as viewed) compared to CODD-affected foot showing loss of hoof capsules and bony re modelling (resorption) of both P3 bones.

- **Key observations**: Large percentage of flock shows severe lameness.
- **Clinical examination**: Hair loss/bleeding extending 2–3 cm above coronary band, sloughing of hoof capsule in severe cases. Absence of granulation tissue.
- **Key history events**: Recently purchased sheep, sheep returning from rented (dairy farm) grazing. 🎥 3.294, 3.295, 3.296

White Line Abscess

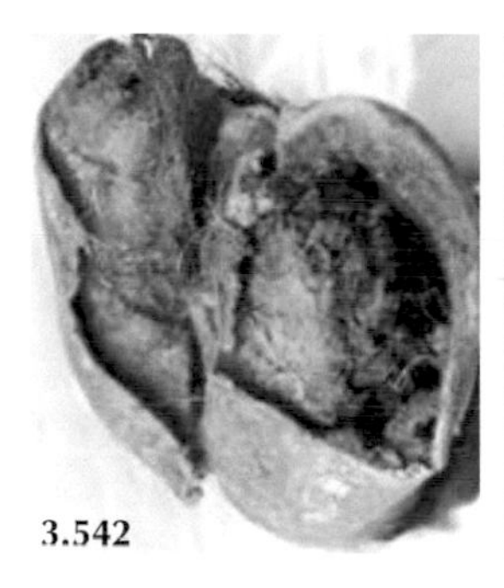
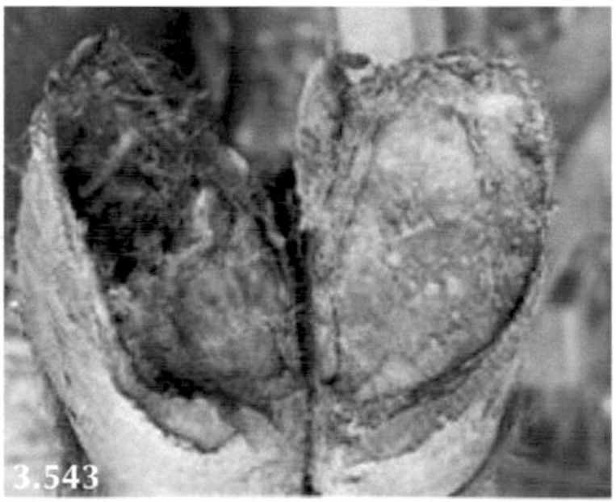
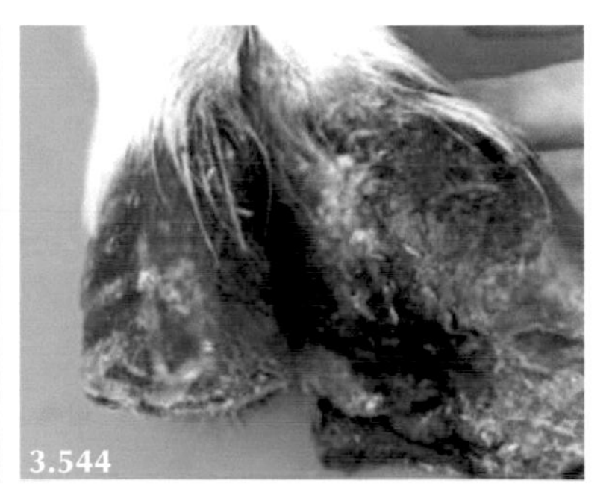

3.542 and **3.543:** Separation of hoof horn at the white line with impaction of foreign material leading to an abscess. **3.544:** Infection has tracked up the medial wall to rupture at the coronary band in the interdigital space.

- **Key observations**: Individual sheep, sudden marked lameness. White line separation with impacted dirt leading to abscess formation. If not treated, infection eventually tracks to discharge at coronary band with much reduced lameness.
- **Clinical examination**: Foot paring, taking care not to damage the corium, releases pus.
- **Key history events**: Severe lameness for 10–14 days if untreated; lameness much reduced once abscess ruptures at coronary band.

Toe Fibroma

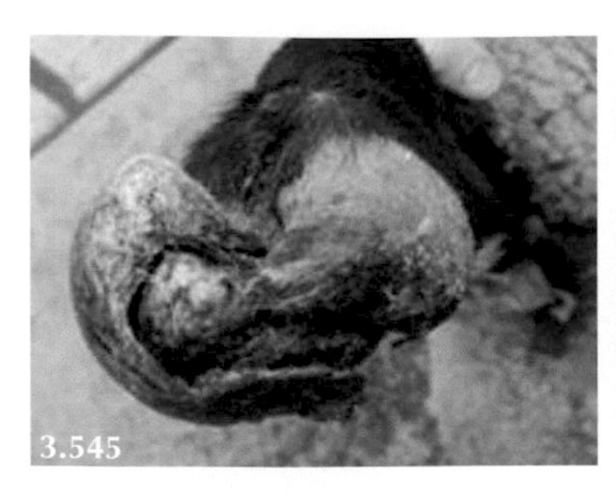
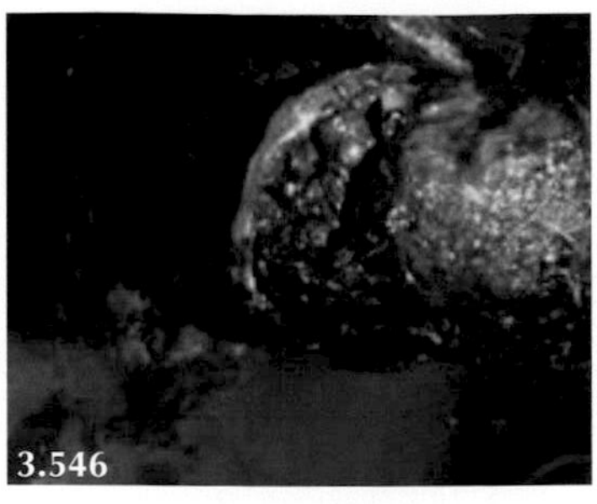
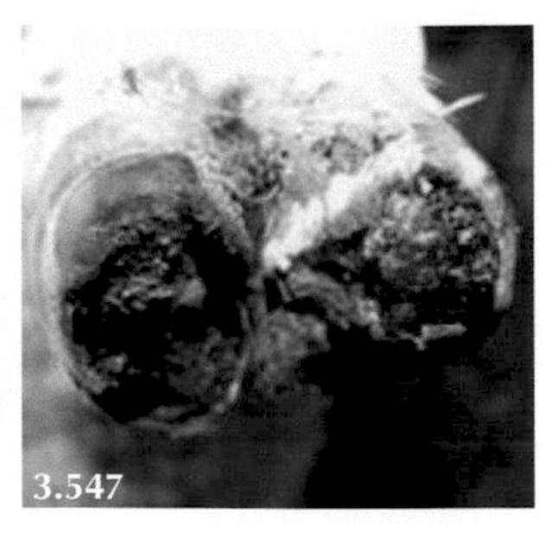

3.545: Toe granuloma. Lameness and lack of wear has resulted in overgrown hoof horn. **3.546** and **3.547:** Removal of overgrown horn to reveal the toe granuloma prior to excision.

- **Key observations**: Individual sheep, variable lameness. Approximately 10–20 mm diameter granulation tissue arising from corium often obscured by overgrown hoof horn but exposed by careful paring.
- **Clinical examination**: Granulation tissue attached to corium by narrow stalk.
- **Key history events**: Over-paring feet with damage to corium aggravated by formalin footbaths. 3.297

Septic Pedal Arthritis (Acute)

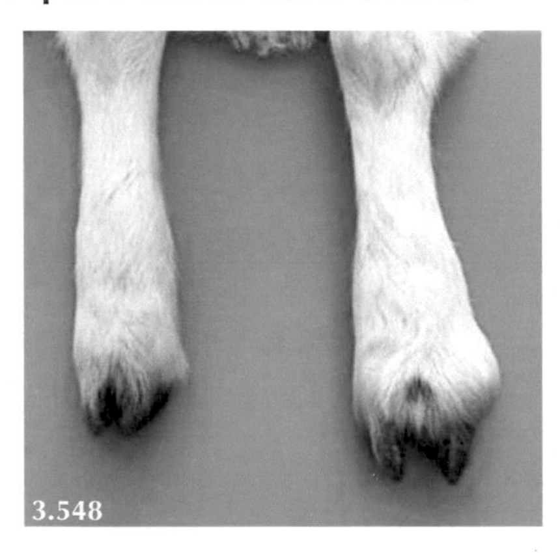
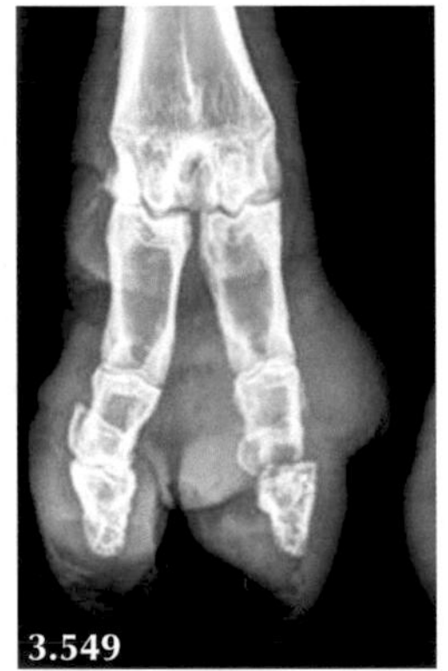

3.548: Swelling of interdigital space and proximal to the lateral claw of the left forefoot (right as viewed). **3.549:** DP radiograph of the left forefoot reveals a possible track from the interdigital space to the distal interphalangeal joint of the lateral claw, slight widening of the joint space and considerable soft tissue swelling around the coronary band. There are no conclusive bony lesions.

- **Key observations**: Individual sheep, sudden severe lameness.
- **Clinical examination**: Soft tissue swelling of interdigital area and abaxial coronary band. Thinned skin over abaxial coronary band.
- **Key history events**: Severe lameness, often seven to 10 days' duration before veterinary examination. 3.298

Septic Pedal Arthritis (Chronic)

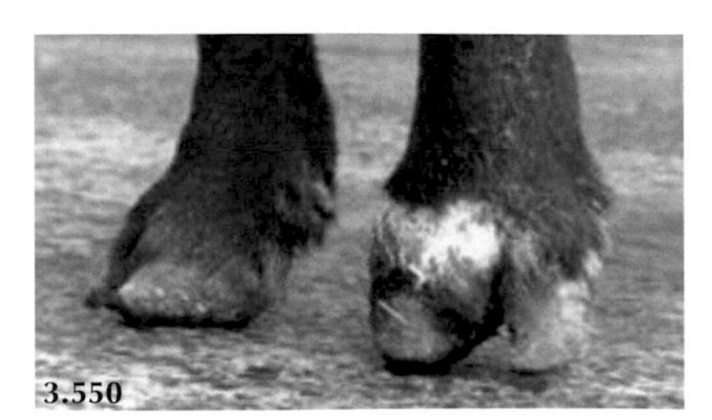
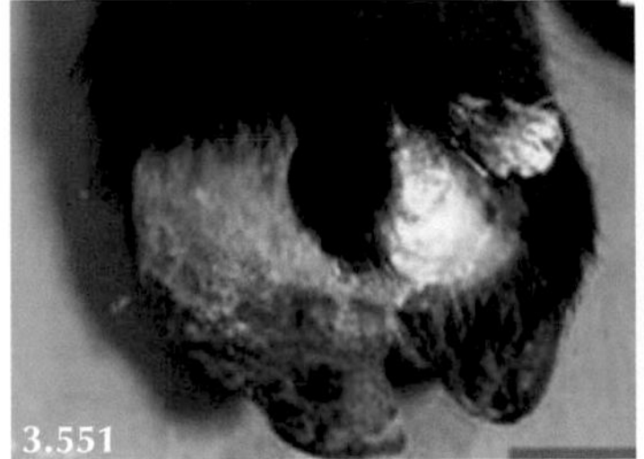
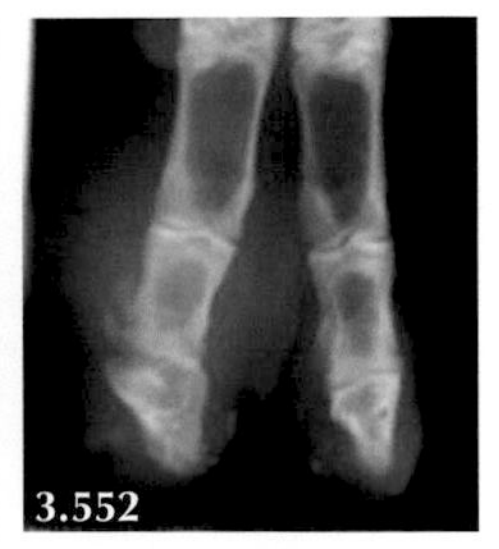

3.550: Medial claw of left forefoot shows considerable soft tissue swelling with loss of hair extending approximately 2 cm above the coronary band. **3.551:** Increased interdigital space. Soft tissue swelling and hair loss extending proximally

(CONTINUED)

from the coronary band of the lateral claw only. **3.552:** DP radiograph shows destruction of articular surface of P3 with osteophyte formation on distal P2 (left side as viewed).

- **Key observations**: Individual sheep, severe chronic lameness.
- **Clinical examination**: Soft tissue swelling of interdigital area and abaxial coronary band with overlying thinned skin showing hair loss. Severe localised pain. Markedly increased drainage lymph node.
- **Key history events**: Severe lameness of at least two to three months' duration. 3.299, 3.300, 3.301

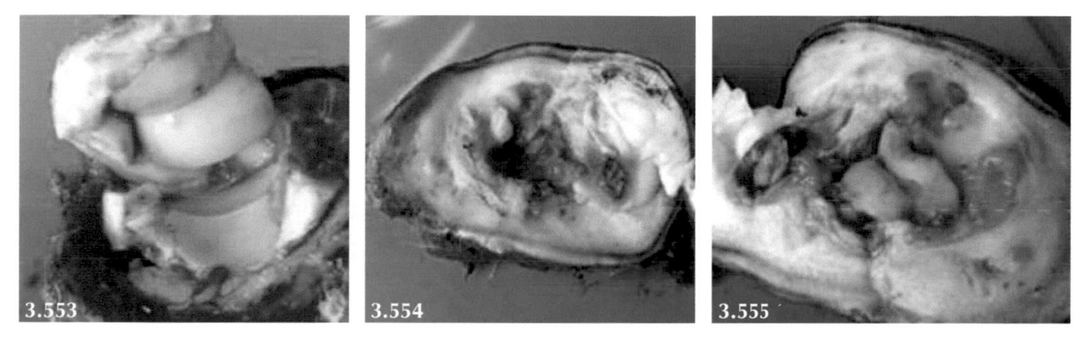

3.553: Normal distal interphalangeal joint opened at necropsy. **3.554:** Septic distal interphalangeal joint opened at necropsy—articular surface of P3 obscured by synovial hypertrophy and hyperaemia. **3.555:** Erosion of articular cartilage of distal P2 and synovial hypertrophy.

Septic Pedal Arthritis (Chronic) with Dislocation of P2/P3 Joint

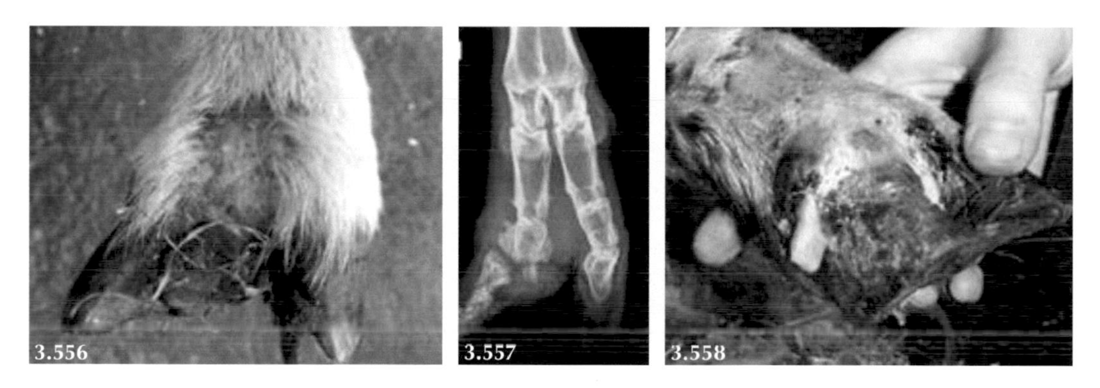

3.556: Soft tissue swelling of interdigital area with abnormal angulation of the lateral claw (left as viewed) suggests dislocation of this P2/P3 joint. **3.557:** Radiography confirms soft tissue swelling of interdigital area and dislocation of P2/P3 joint. **3.558:** Infection originating in the interdigital space has tracked across the distal interphalangeal joint to rupture above the coronary band.

Septic Pedal Arthritis (Ankylosis)

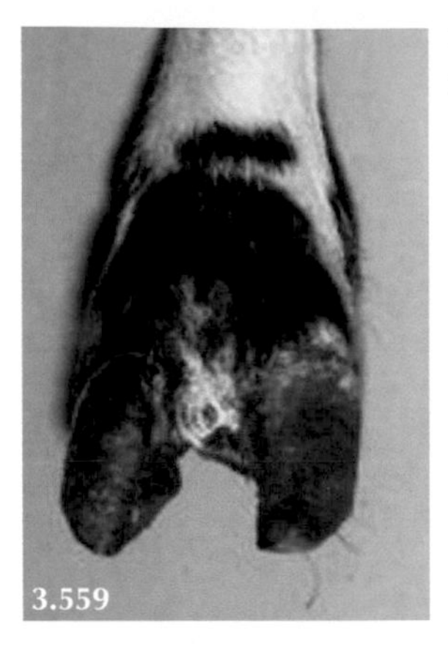

3.559: Swelling of interdigital area with abaxial skin thinned and showing hair loss of the right digit as viewed.

- **Key observations**: Chronic severe lameness has now largely resolved.
- **Clinical examination**: Individual sheep, mild/moderate lameness, swelling of interdigital area, abaxial skin thinned and shows hair loss above coronary band.
- **Key history events**: Severe lameness of several months' duration has now largely resolved.

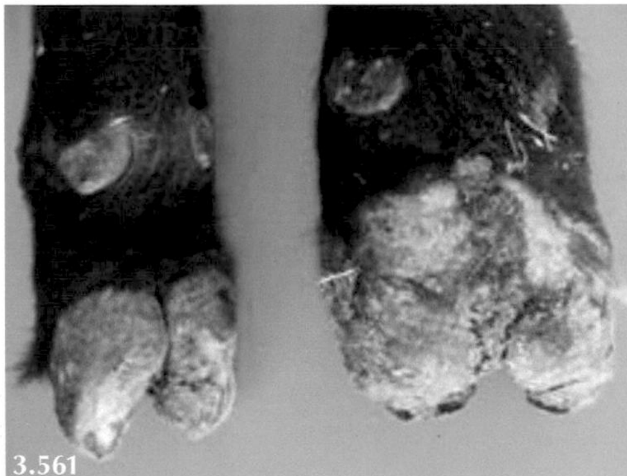

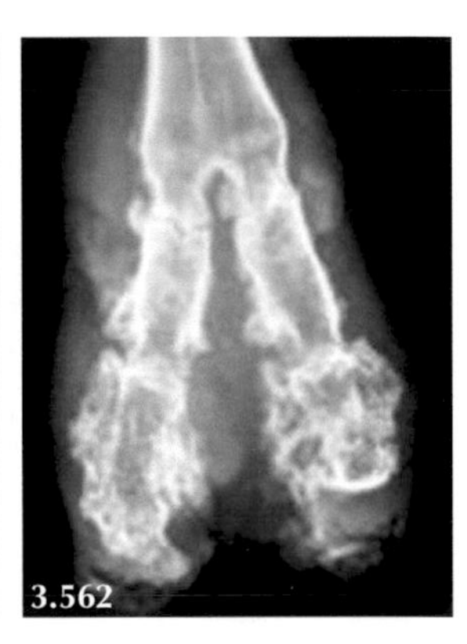

3.560: Massively swollen right hind foot with hair loss above the coronary band of both digits. **3.561:** Comparison of this Suffolk ram's hind feet. **3.562:** DP radiograph of the right hind foot reveals destruction of both distal interphalangeal joints with ankylosis and extensive osteophyte formation involving the proximal interphalangeal joints and extending to involve the fetlock joint.

- **Key observations**: Individual sheep with moderate lameness, dramatic swelling of foot.
- **Clinical examination**: Hard painful swelling extending proximal to the fetlock joint.
- **Key history events**: Severe lameness of up to six months' duration, may have improved recently. 🎥 3.302

Bilateral Distal Interphalangeal Joint Infection

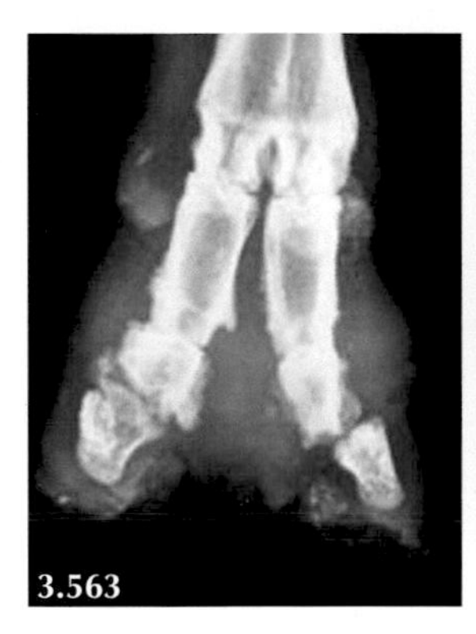

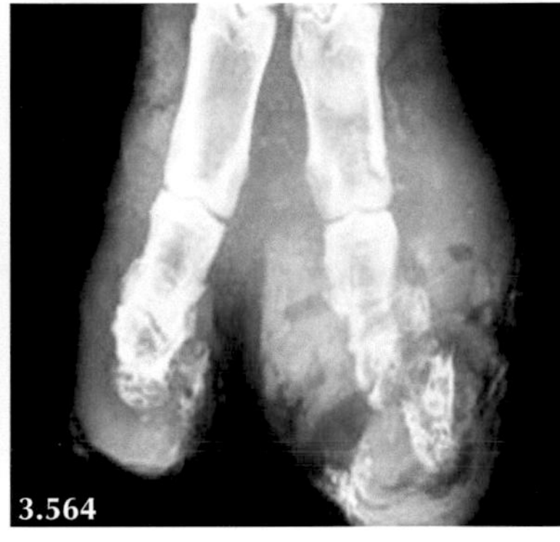

3.563: DP radiograph shows infection of both distal interphalangeal joints with considerable new osteophyte formation of the lateral joint (left as viewed) and dislocation of the medial joint; digit amputation of the medial claw is not possible.
3.564: DP radiograph one month after curettage of the medial joint and repeated flushing shows considerable soft tissue swelling and loss of bone density but early osteophyte formation. Effective pain management is essential during healing but is not easily achieved.

- **Key observations**: Individual sheep with severe lameness, swollen foot.
- **Clinical examination**: Interdigital swelling extending above the coronary band of both digits.
- **Key history events**: Severe lameness often several months' duration.

Distal Interphalangeal Joint Infection Extending to Involve the Fetlock Joint

In this unusual case infection has extended from the distal interphalangeal joint to involve the fetlock joint.

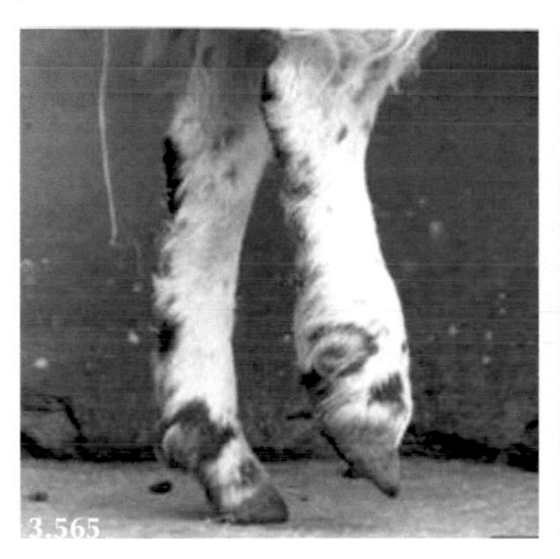

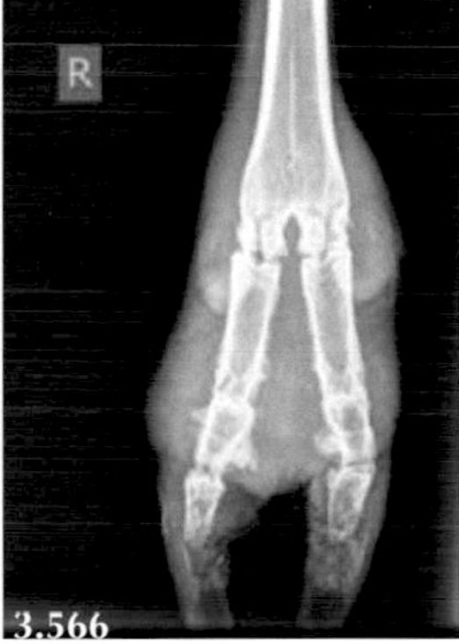

3.565: Soft tissue swelling extends from around the lateral digit and interdigital space proximally above the fetlock joint.
3.566: Extensive soft tissue swelling extends above the fetlock joint, lysis of articular surface of the distal interphalangeal joint with osteophyte formation on distal P2. 3.303

Foot Paring

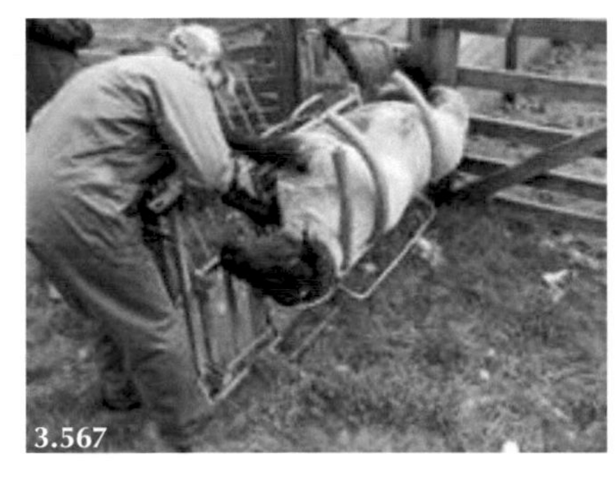

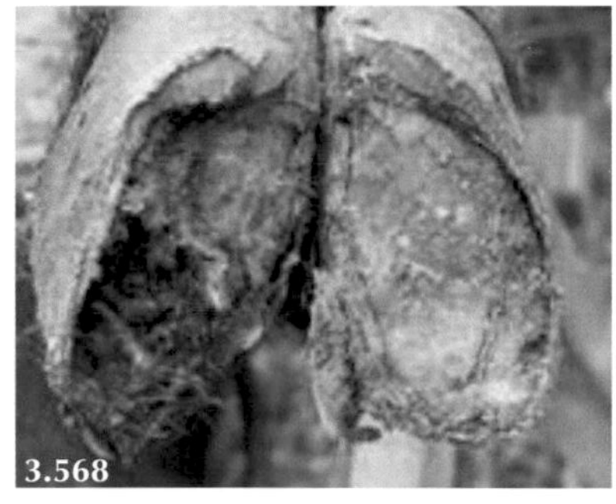

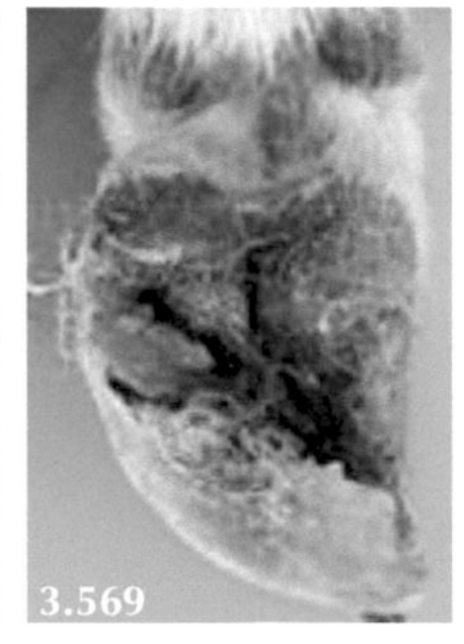

Foot paring (**3.567**) is only indicated where the hoof wall has become impacted with dirt and the degree of lameness suggests an abscess (**3.568**). Overgrown horn is caused by lack of wear due to lameness; overgrown horn (**3.569**) does not cause lameness.

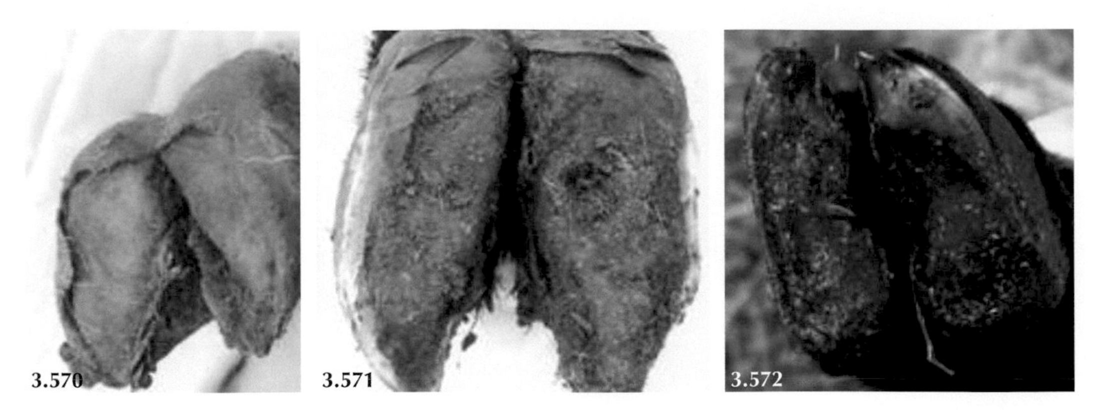

Paring the normal foot (**3.570**) is counter-productive because it transfers weight onto the sole (**3.571**). Over-paring (**3.572**) with exposure of the corium results in more severe lameness and may cause a toe fibroma. 🎥 3.304

Cattle—Foot Lesions

- Foot lesions are the major animal welfare concern for dairy cows where they also cause poor milk production, reduced fertility and considerable weight loss.
- Despite many farms using routine biannual foot paring services, a recent survey reports 36.8% mean prevalence of moderate to severe lameness across 205 UK dairy farms.
- Foul of the foot and joint trauma are major causes of lameness in beef cows.
- The general public considers year-round housing of dairy cows a major welfare concern.

TABLE 3.73

Common Causes of Foot Lameness in Cattle

	Likely duration at presentation	Lameness	Swelling	Muscle
Digital dermatitis	Days	++/+++	NAD	NAD
Sole ulcer	Days/weeks	++/+++	+	−−/−−−
Foul in the foot	Days	++/+++	++	NAD
White line abscess	Days	++/+++	NAD/+	NAD
Infected P2/P3	Weeks	+++	+	−−−
Thimbling	Months	−	NAD	NAD
Toe necrosis	Weeks	++/+++	NAD	−/−−
Overgrown claws	Months	NAD	NAD	NAD

TABLE 3.74

Ancillary Tests to Diagnose the Common Causes of Foot Lameness in Cattle

Digital dermatitis	Bacteriology, PCR
Sole ulcer	Clinical examination
Foul in the foot	Clinical examination
White line abscess	Clinical examination with careful foot paring
Infected P2/P3	Radiography
Thimbling	Clinical examination
Toe necrosis	Clinical examination with careful foot paring. Radiography
Overgrown claws	Clinical examination

Digital Dermatitis

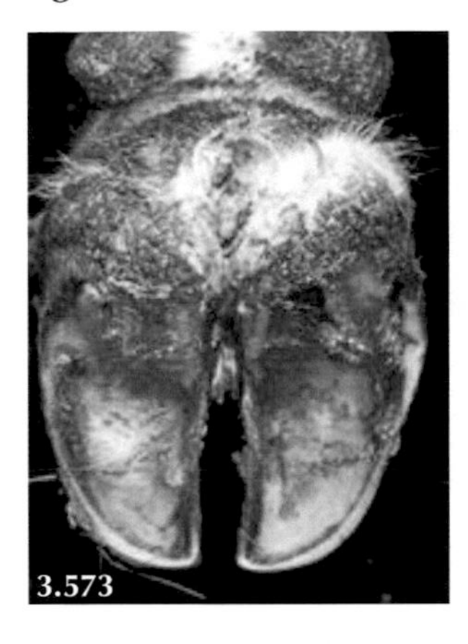

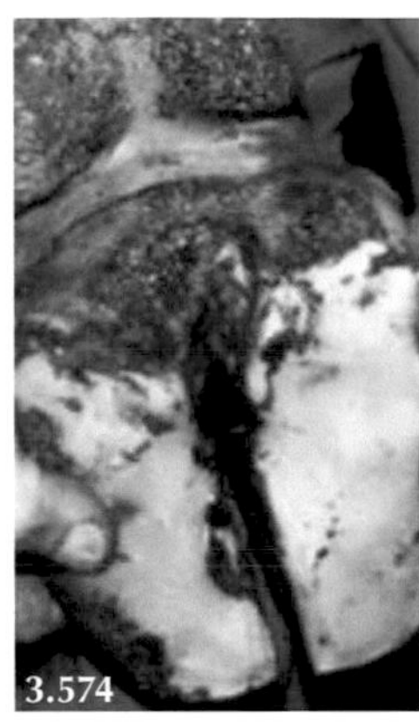

3.573 and **3.574:** Show 2–4 cm diameter diphtheritic areas between heel bulbs.

- **Key observations**: Acute moderate/severe lameness in housed cattle. Poor hygiene.
- **Clinical examination**: Painful 2–4 cm diameter diphtheritic area between heel bulbs.
- **Key history events**: Introduced cattle, high prevalence of lameness.

Sole Ulcer

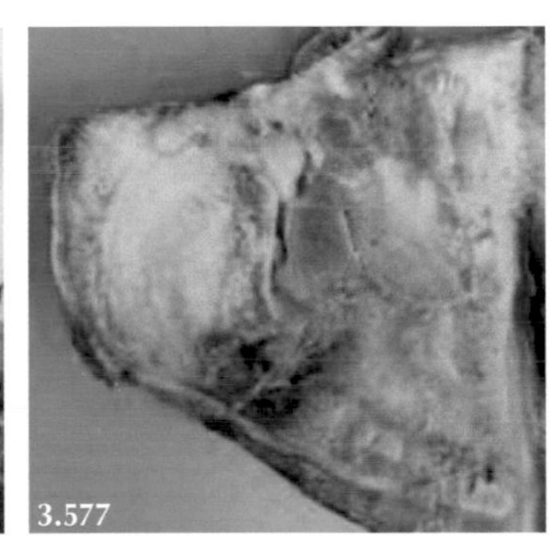

3.575: Severe muscle wastage of the left hindleg. **3.576:** Extensive under-running of sole horn, 2 cm diameter granulation tissue arising from the damaged corium. **3.577:** Sagittal section through digit after amputation reveals that infection has tracked from the sole ulcer site to the distal interphalangeal joint and deep flexor tendon sheath.

- **Key observations**: Muscle wastage, lateral claw hind foot. Housed dairy cattle.
- **Clinical examination**: Under-running of sole horn, exposed damaged corium producing 2 cm diameter exuberant granulation tissue.
- **Key history events**: Poor cubicle/housing and lack of routine foot trimming, excessive time standing. 🎥 3.305, 3.306

Foul of the Foot

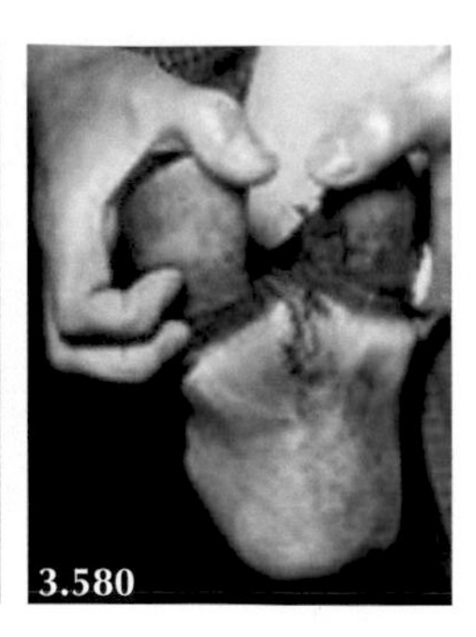

3.578: Severe non-weight-bearing lameness of the right foreleg in a beef heifer at pasture. **3.579:** Severe non-weight-bearing lameness of the right foreleg of a housed "bull beef" animal. **3.580:** Swollen interdigital space with skin break and superficial infection.

- **Key observations**: Sudden onset, severe lameness, swollen foot. Several cattle affected especially if housed.
- **Clinical examination**: Widening of the interdigital space with skin break.
- **Key history events**: Contaminated/flooded area around water trough. 3.307, 3.308, 3.309

Superfoul

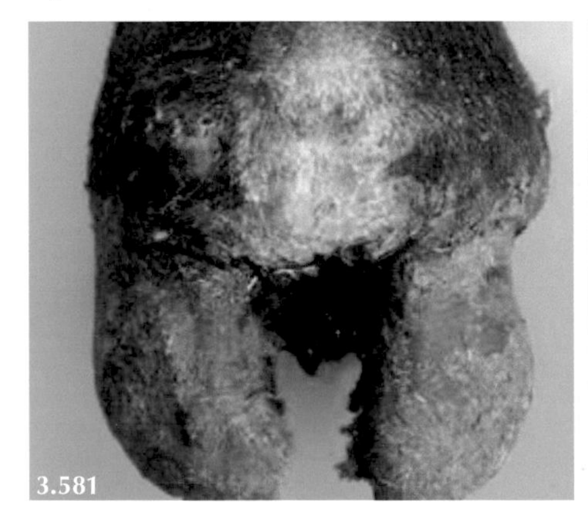
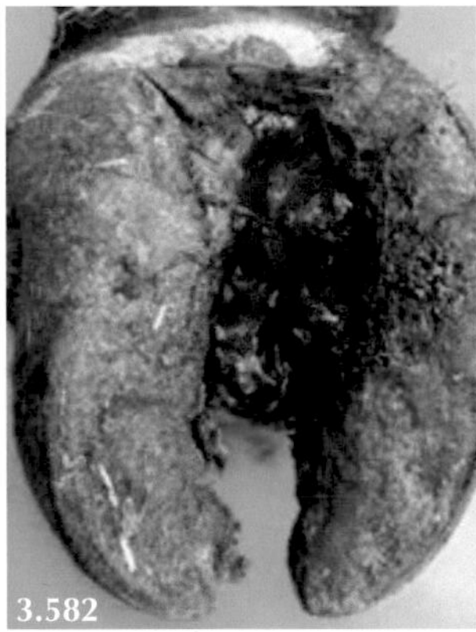
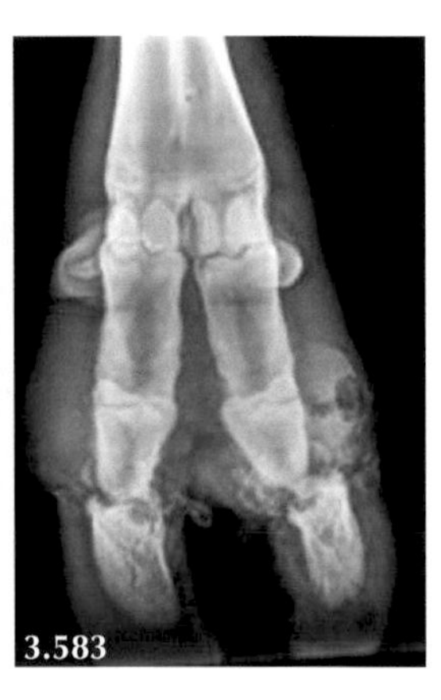

3.581: Severe and extensive infection tracking into both pedal joints and above the abaxial coronary band of both digits. **3.582:** Plantar view showing depth and extent of interdigital infection. **3.583:** DP radiograph confirming infection of both pedal joints with considerable bone lysis.

- **Key observations**: Sudden onset, severe, non-weight bearing lameness, swollen foot.
- **Clinical examination**: Widening of the interdigital space, deep tissue infection.
- **Key history events**: Infection likely to have been introduced into herd with purchased cattle. 3.310

White Line Abscess

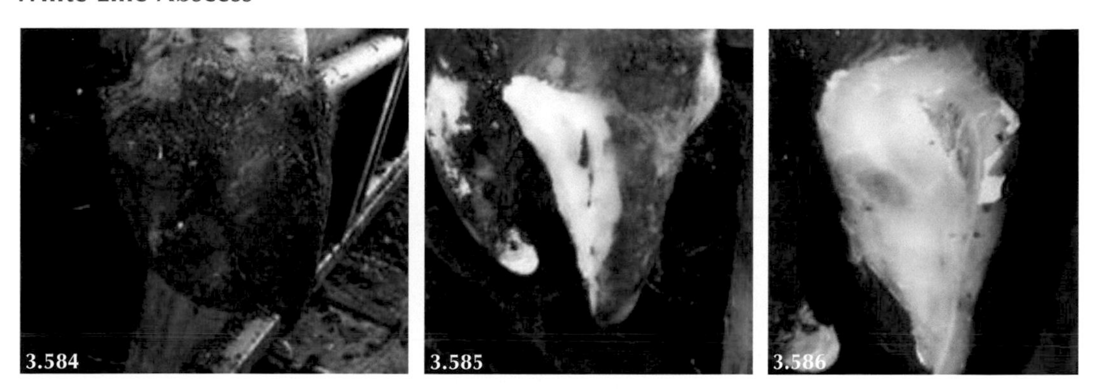

3.584: Foot raised in cattle stocks reveals overgrown hoof horn. **3.585:** Paring horn off the sole reveals a black mark in the abaxial white line of the lateral claw two-thirds along the white line towards the heel. **3.586:** White line abscess pared out. Note also the sole bruising.

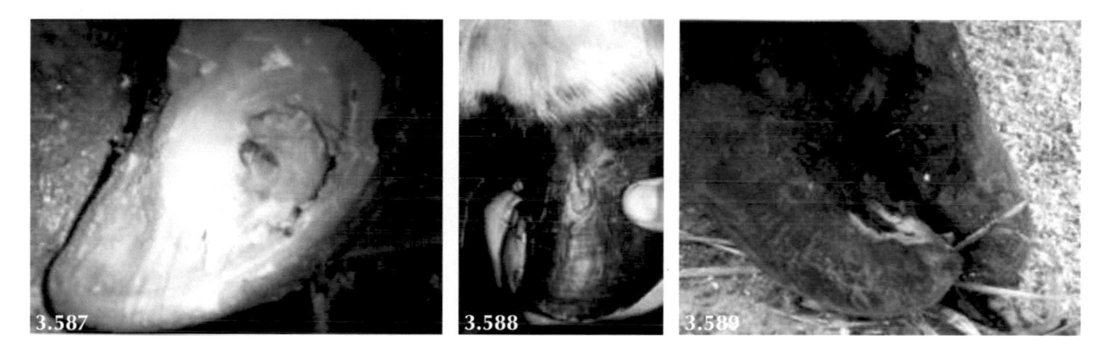

3.587: Puncture wound of sole with pus under-running sole now pared out. **3.588** and **3.589:** Infected sand cracks pared out to release pus.

- **Key observations**: Severe lameness over several days.
- **Clinical examination**: Impacted dirt/stone in the abaxial white line of the lateral claw two-thirds along the white line towards the heel leads to abscess under pressure.
- **Key history events**: Often associated with overgrown feet, poorly maintained farm tracks, worn concrete surfaces. 3.311, 3.312, 3.313, 3.314

Septic Pedal Arthritis

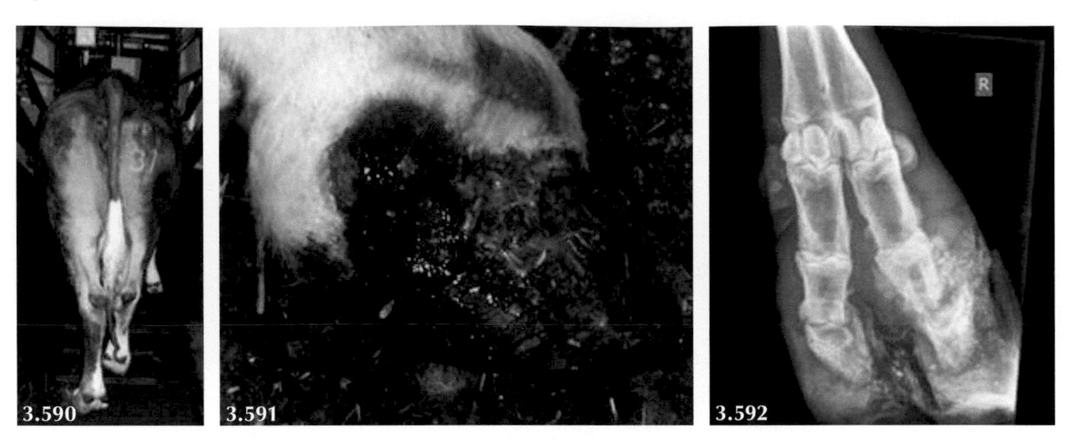

3.590: Pronounced wastage of right hindleg; very swollen lateral claw especially above coronary band on lateral aspect. **3.591:** Discharging sinus at coronary band consistent with septic pedal arthritis. **3.592:** DP radiograph of affected foot (left).

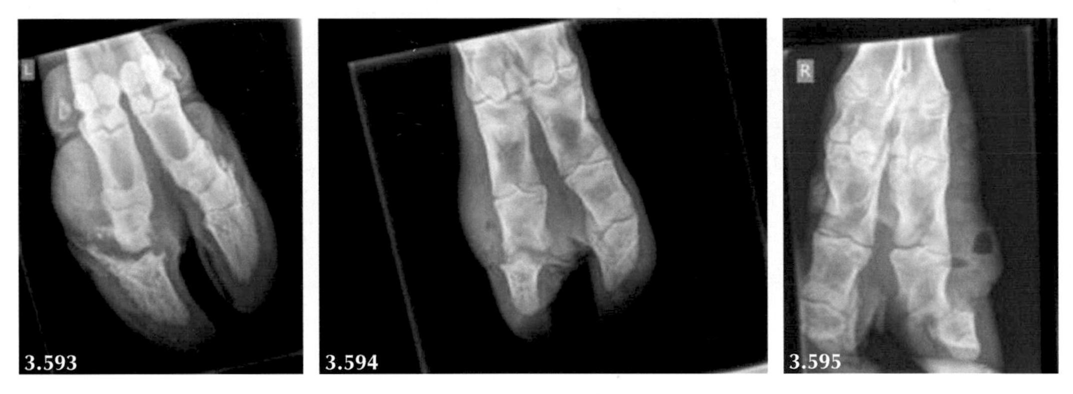

3.593–3.595: DP radiographs of infected pedal joints. Important features to note: soft tissue swelling causing widening of the interdigital space and above the coronary band, abscesses within the soft tissue above the coronary band, widening of the distal interphalangeal joint, loss of articular surfaces, osteophyte formation and disarticulation of the joint (especially **3.595**).

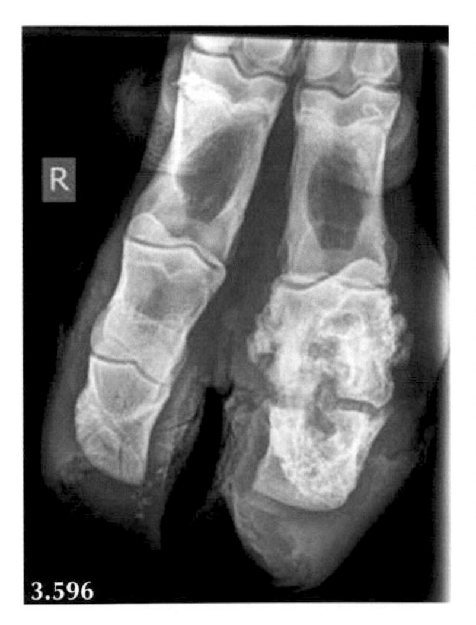

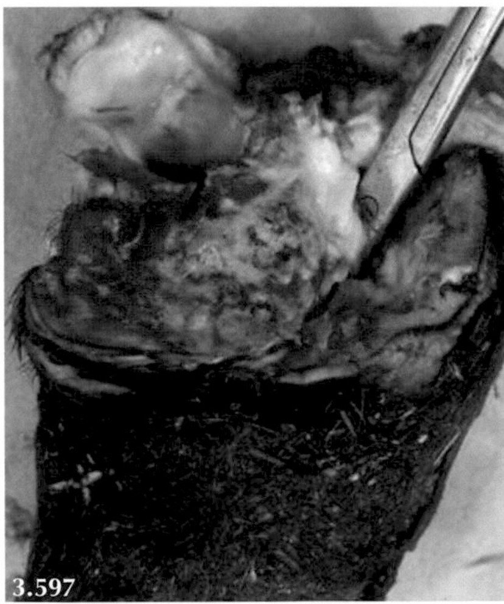

3.596: Soft tissue swelling causing widening of the interdigital space and above the coronary band, marked widening of the distal interphalangeal joint, loss of articular surfaces and extensive osteophyte formation extending proximally as far as distal P1. **3.597:** Scissor points extend into the septic distal interphalangeal joint with the distal articular surface of P1 visible (top left); the digit has been amputated through mid P1.

- **Key observations**: Severe lameness of several weeks' duration, marked muscle wastage, may extend from sole ulcer in bulls and dairy cows. Often foreign body penetration in growing animals and adult beef cattle.
- **Clinical examination**: Swollen at coronary band with possible discharging sinus above coronary band.
- **Key history events**: Treated earlier for sole ulcer but temporary/no improvement. 3.315, 3.316, 3.317, 3.318, 3.319, 3.320, 3.321

Septic P1/P2

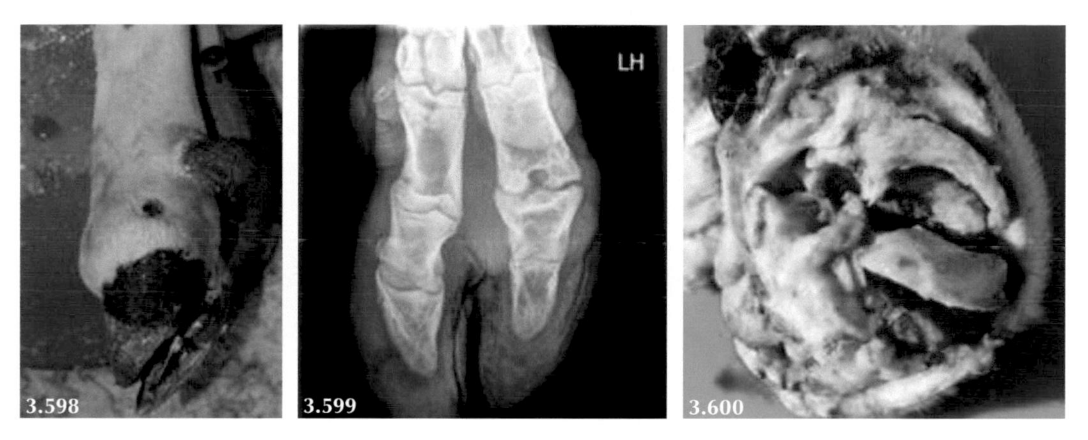

3.598: Discharging sinus 6–7 cm above coronary band. **3.599:** Radiograph reveals increased joint space of lateral P1/P2 joint. **3.600:** The digit has been amputated through mid P1.

Thimbling

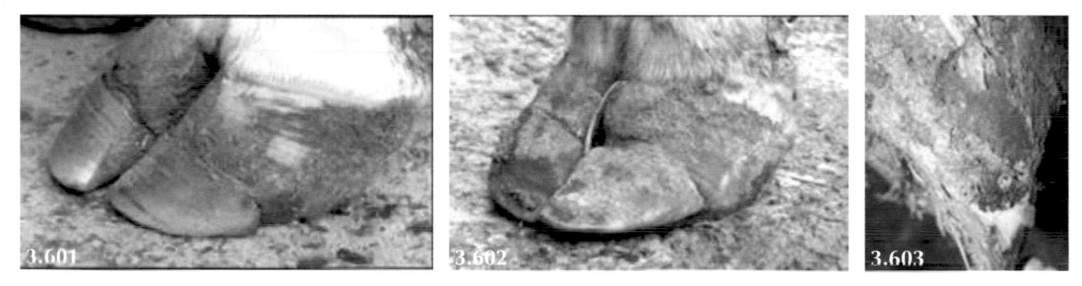

3.601 and **3.602:** Old hoof capsules form "caps" at the toes of all eight digits. **3.603:** Right: Impaction with dirt often forms abscesses putting pressure on the laminae causing marked lameness.

- **Key observations**: Replacement/regrowth of all eight hoof capsules from coronary band.
- **Clinical examination**: Old hoof capsule forms a cap at the toe often becoming impacted with dirt forming abscesses.
- **Key history events**: Often acute disease episode around calving whether coliform mastitis/septic metritis three to four months previously. 3.322, 3.323

Toe Necrosis

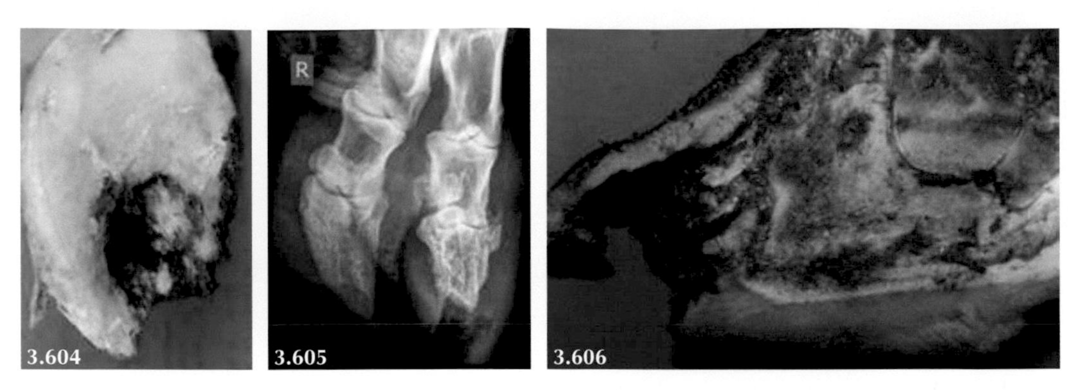

3.604: Hoof horn erosion extending onto distal P3. **3.605:** Bone lysis/resorption at toe (right digit as viewed). **3.606:** Digit amputation reveals loss of hoof horn of the sole and lysis of P3 (sagittal section).

- **Key observations**: Chronic severe lameness.
- **Clinical examination**: Paring away under-run hoof horn leads to exposed bone of P3.
- **Key history events**: Prolonged standing on concrete floor/slats, and during transport. 🎥 3.324, 3.325

Corkscrew Claws

3.607: Corkscrewing of the lateral claws of both hind feet in a bull. **3.608:** Severe corkscrewing of the lateral claw of the right hind foot. **3.609:** Radiography reveals the extent of horn overgrowth.

- **Key observations**: Obvious corkscrewing of the lateral claws.
- **Clinical examination**: Weight bearing on curved lateral wall not sole.
- **Key history events**: Possible hereditary component. 🎥 3.326, 3.327

Overgrown Claws

3.610 and **3.611:** Overgrown claw with weight borne on heels.

- **Key observations**: Toe not in contact with the ground.
- **Clinical examination**: Grossly overgrown claw.
- **Key history events**: Consequence of lameness rather than cause. Commonly encountered in "bull beef" intensive rearing systems. 🎥 3.328

Interdigital Fibromas

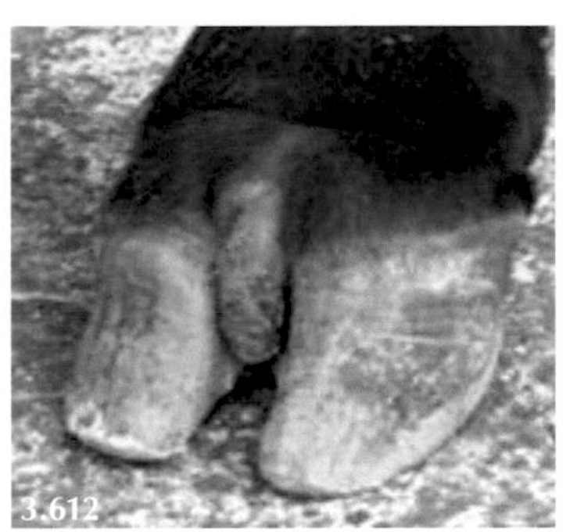

3.612: Interdigital skin hyperplasia (toe fibroma).

- **Key observations**: Interdigital skin hyperplasia often affecting all four feet in beef bulls.
- **Clinical examination**: Condition does not cause lameness unless so large that excoriation leads to superficial infection.
- **Key history events**: Often observed in certain beef bloodlines. 3.329

Sheep—Joint Lesions

- Polyarthritis in neonatal lambs is a major welfare problem. Penicillin is the antibiotic treatment of choice for *Streptococcus dysgalactiae* infections.
- A single injection of dexamethasone on the first day of treatment improves outcome—see AO joint video recordings.
- Radiography has limited value when assessing all but chronic joint lesions.
- Joint lavage does not work because of well-established pannus.
- Synovial membrane yields better bacterial culture results than joint fluid.

TABLE 3.75

Lameness in Sheep: Joint Lesions

	Likely duration at presentation	Lameness/pain	Limb muscle
Septic polyarthritis (lambs)	Several days	++/+++	-/---
Erysipelas (multiple joints)	Days to months	+++	---
Septic joint (acute)	Days	+++	-
Septic joint (chronic)	3–6 Months	+++	---
Endocarditis	Weeks to months	+/+++	-/---

TABLE 3.76

Ancillary Test to Diagnose the Cause of Ovine Joint Lesions

	Ancillary tests
Septic polyarthritis (lambs)	Arthrocentesis/necropsy. >90 per cent caused by *Streptococcus dysgalactiae*
Erysipelas	Culture from synovial membrane samples
Septic joint (acute)	Arthrocentesis, culture
Septic joint (chronic)	Radiography if >3 months' duration
Endocarditis	Ultrasonography of heart valves, necropsy

Septic Polyarthritis (*Streptococcus dysgalactiae* in Lambs)

3.613: Non-weight-bearing lameness of right hindleg in two-week-old lamb. **3.614:** Joint pain causing recumbency.

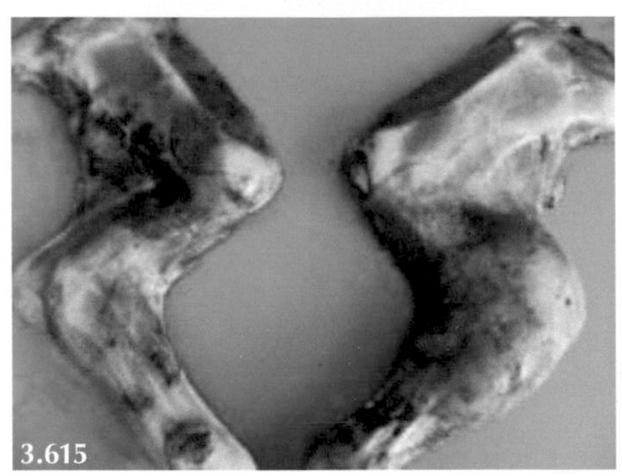

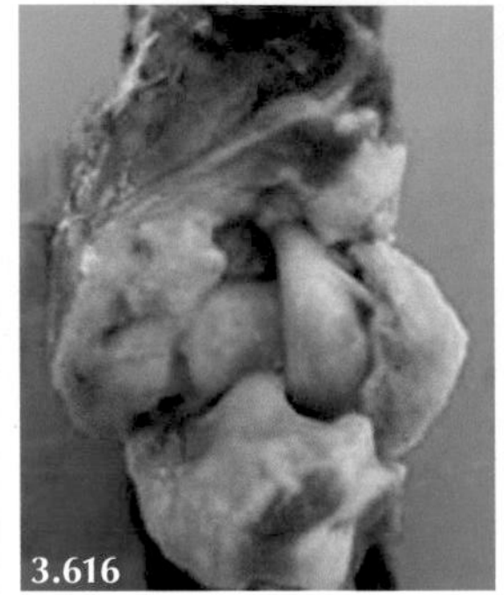

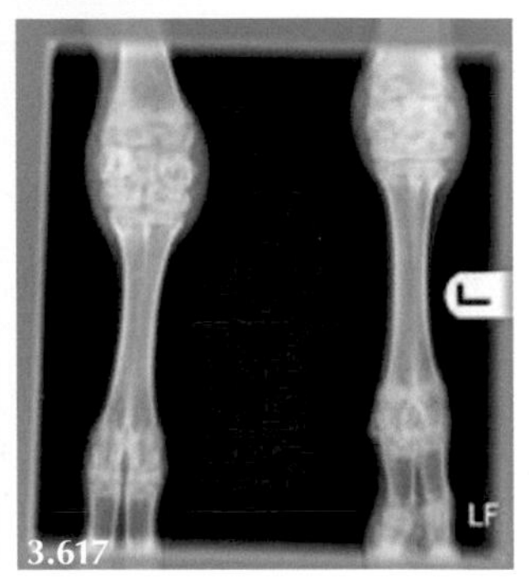

3.615: Necropsy showing enlargement of the left elbow joint (right as viewed) caused largely by accumulation of pannus (centre). **3.616:** Pannus within the elbow joint. **3.617:** Radiograph of chronically infected (six weeks) carpal joints showing soft tissue swelling but no bony lesions.

- **Key observations**: Large number of five–ten-day-old lambs show severe lameness.
- **Clinical examination**: Painful swollen joints (pannus) but usually little fluid distension.
- **Key history events**: Lambs born indoors but often good hygiene and colostrum management.
  3.330, 3.331, 3.332, 3.333, 3.334, 3.335, 3.336, 3.337, 3.338

Erysipelas

Acute joint infections cause sudden severe lameness but no joint effusion. Chronic joint infections are shown next.

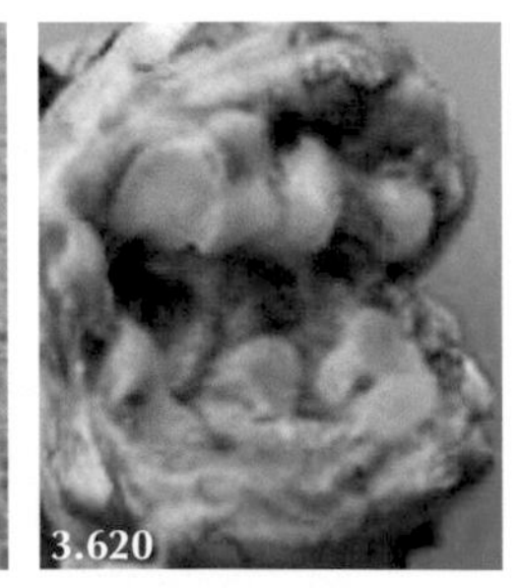

3.618: Joint pain results in recumbency for long periods. **3.619:** Pain evident as arched back, swollen carpal joints with weight borne by hindlegs. **3.620:** Note pink/red colour of joint caused by hyperaemia and proliferation of synovial membrane obscuring articular cartilage.

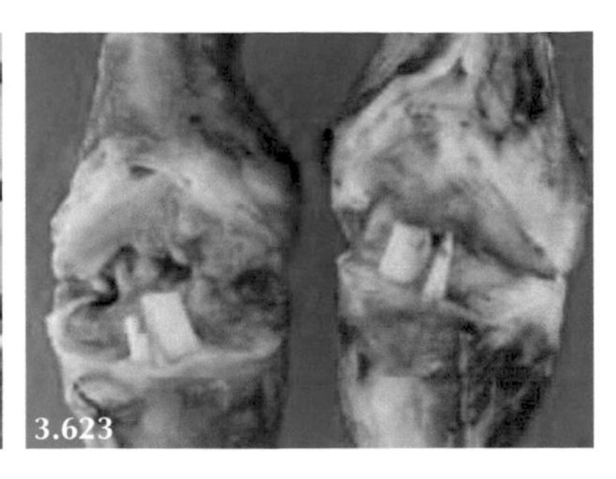

3.621: Markedly swollen carpal joints. **3.622:** Painful expression with head held lowered, hindleg lameness with weight borne by forelegs. **3.623:** Note pink colour of hock joints caused by hyperaemia and proliferation of synovial membrane; the white articular surfaces are barely visible.

- **Key observations**: Acute disease, large numbers of two–four-month-old lambs suddenly very lame.
- **Clinical examination**: Markedly thickened joint capsule with little joint effusion.
- **Key history events**: Commercial erysipelas vaccine no longer available. 🎥 3.339, 3.340, 3.341, 3.342

Septic Arthritis (Acute)

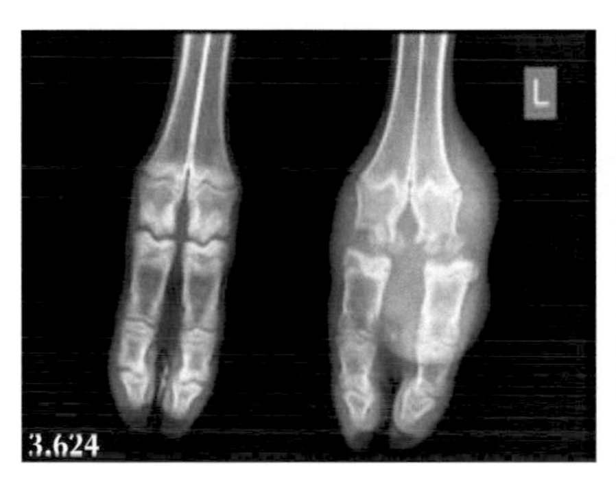

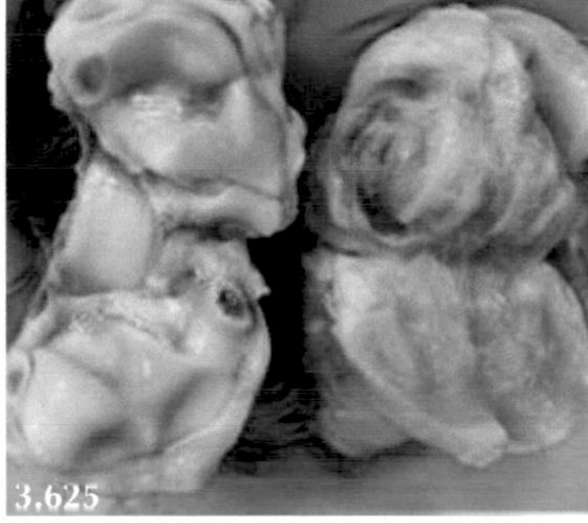

3.624: Soft tissue swelling and distension of the left fetlock joint. Joint lavage proved unsuccessful because of the severity of pathology when presented. **3.625:** Necropsy of the affected joint reveals extensive pannus and synovial membrane hypertrophy.

- **Key observations**: Individual sheep, severe lameness.
- **Clinical examination**: Very painful swollen joint, effusion during early stage quickly progressing to pannus.
- **Key history events**: Possible puncture wound. 🎥 3.343

Joint Lavage

Joint lavage is rarely successful in sheep for many reasons including duration of infection, presence of pannus and the numerous compartments of the fetlock, hock, carpal and stifle joints.

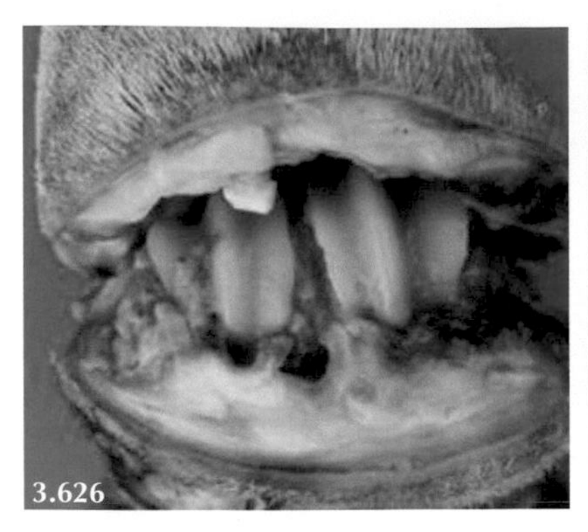

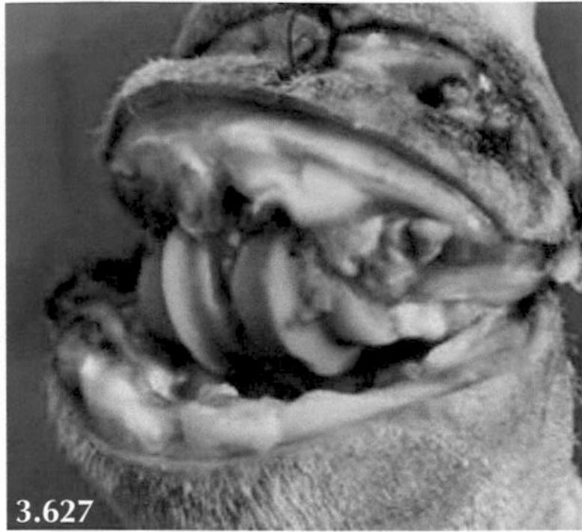

3.626 and **3.627:** Necropsy findings of an infected fetlock joint where joint lavage was attempted under general anaesthesia. Note the presence of pannus and the complex joint structure with numerous "compartments". 🎥 3.344

Septic Arthritis (Chronic)

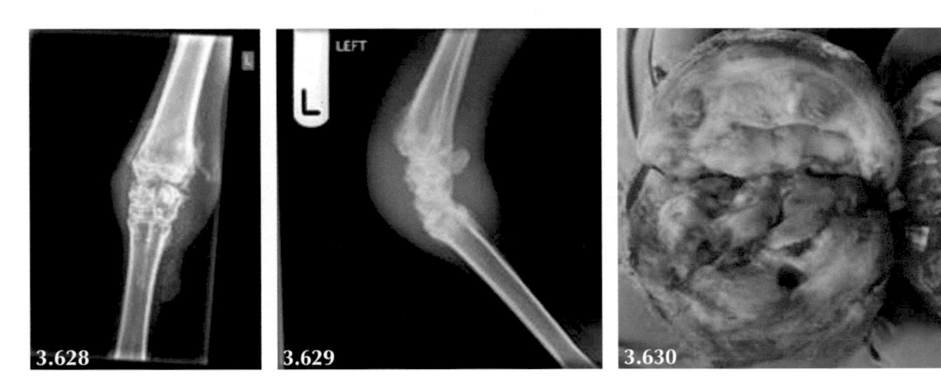

3.628 and **3.629:** DP and lateral radiographs of the left carpus showing soft tissue swelling, widening of the radiocarpal joint in particular and osteophyte formation on the distal radius. **3.630:** Affected carpus (left as viewed) showed marked thickening of the fibrous joint capsule, synovial hypertrophy and erosion of articular cartilage.

- **Key observations**: Individual sheep, severe lameness, muscle wastage over limb.
- **Clinical examination**: Thickened joint capsule, reduced joint excursion, marked muscle wastage, grossly enlarged drainage lymph node (if palpable). If multiple joints involved investigate possibility of endocarditis.
- **Key history events**: Neglect of severe lameness for several/many months. 🎥 3.345

Endocarditis

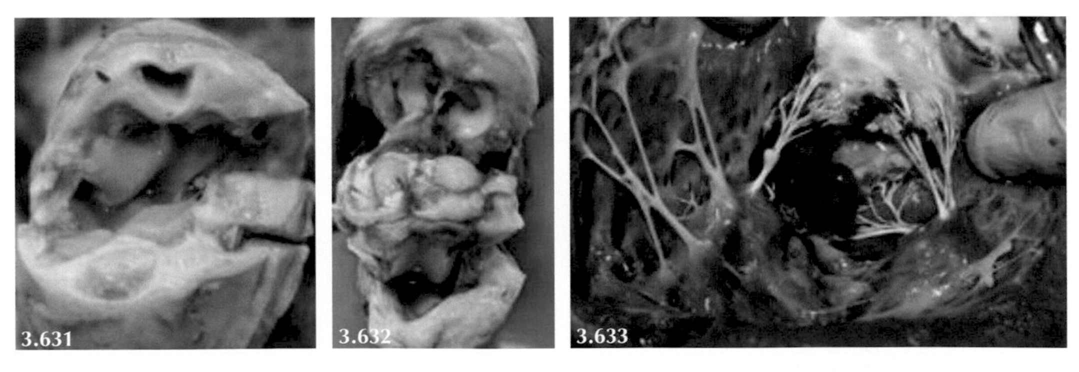

3.631: Septic carpus distended with pannus/pus and marked thickening of joint capsule. **3.632:** Stifle joint showing pus and marked synovial inflammation (note pink/red colour). **3.633:** Vegetative lesion on the tricuspid valve.

- **Key observations**: Insidious, progressive lameness affecting more than one leg.
- **Clinical examination**: Multiple joints—pain, distension and marked thickening of joint capsule. Enlargement of drainage LNs. Heart murmur often not present; need ultrasound diagnosis.
- **Key history events**: Originating septic focus rarely found. 3.346

Cattle—Joint Lesions (Excluding Joints Distal to the Fetlock [See Foot Lameness])

- The prognosis for neonatal septic polyarthritis is guarded.
- Joint lavage generally yields poor results.
- Definitive diagnosis of OCD is difficult even after multiple radiographs.
- Trauma of the stifle and hip joints are common in older beef cows.
- High extradural block using lignocaine provides excellent analgesia for joint lavage/other painful procedures of the pelvic limbs.

TABLE 3.77

Joint Lesions But Excluding Joints Distal to the Fetlock

	Likely duration of lameness at presentation	Lameness/Pain	Limb muscle
Septic polyarthritis (neonates)	1–2 Days	++/+++	-
Septic joint (acute)	Days	+++	-
Septic joint (chronic)	Weeks	++/+++	--/---
Osteochondrosis dissecans	Weeks to months	+	-
Osteoarthritis	Weeks to months	+/++	-/--
Dislocated hip	Days	++/+++	--/---

TABLE 3.78

Ancillary Test to Diagnose the Cause of Bovine Joint Lesions

	Ancillary tests
Septic polyarthritis (calves)	Arthrocentesis but often pannus. Necropsy
Septic joint (acute)	Arthrocentesis, culture
Septic joint (chronic)	Radiography if >2 months' duration
Osteochondrosis dissecans	Joint effusion. Radiography often fails to yield critical findings
Osteoarthritis	Radiography but difficult to image changes
Dislocated hip	Radiography in calf

Septic Polyarthritis (Septicaemia)

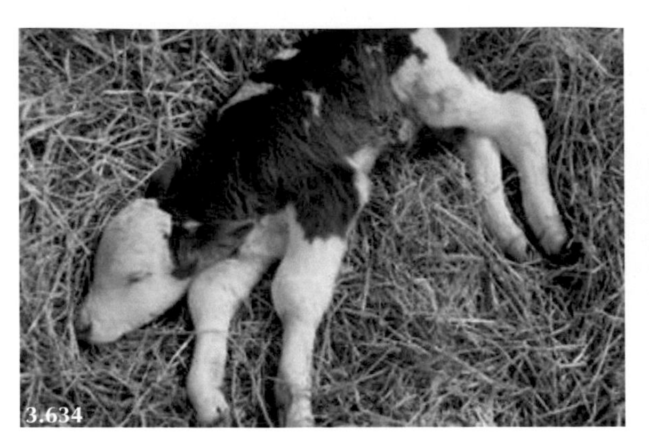

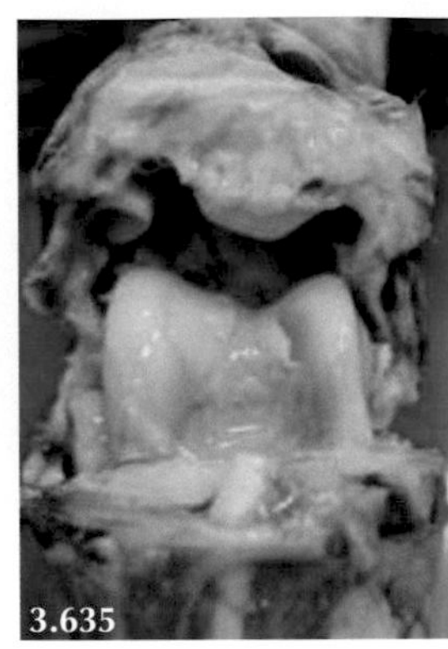

3.634: Three-day-old calf in lateral recumbency. Swollen left hock and both carpal joints. **3.635:** Pannus in the left hock joint revealed at necropsy.

- **Key observations**: Two- to three-day-old calf. Lateral recumbency.
- **Clinical examination**: Several distended painful joints.
- **Key history events**: Failure of passive antibody transfer. Unhygienic environment. 3.347, 3.348

Septic Arthritis (Acute)

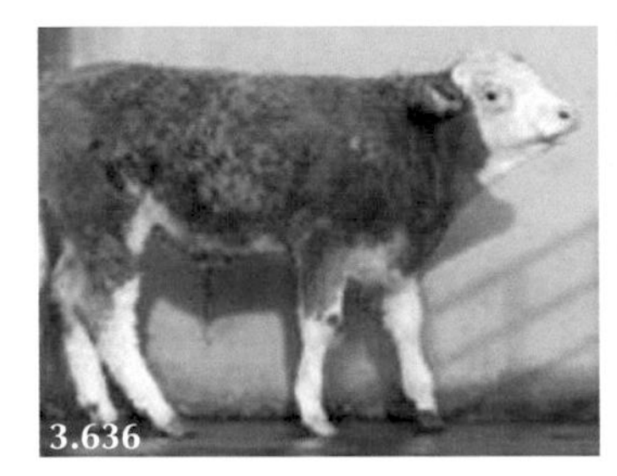

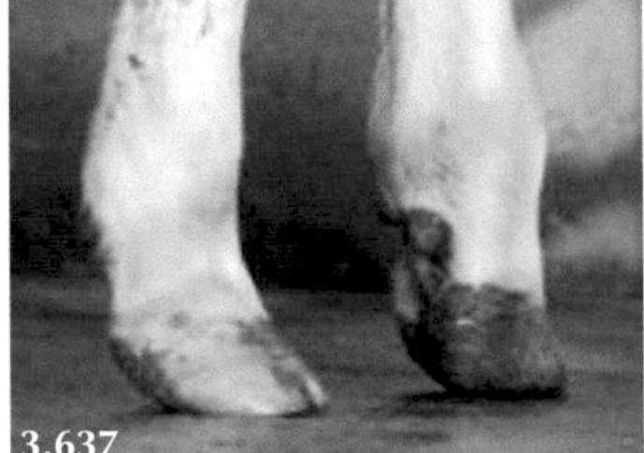

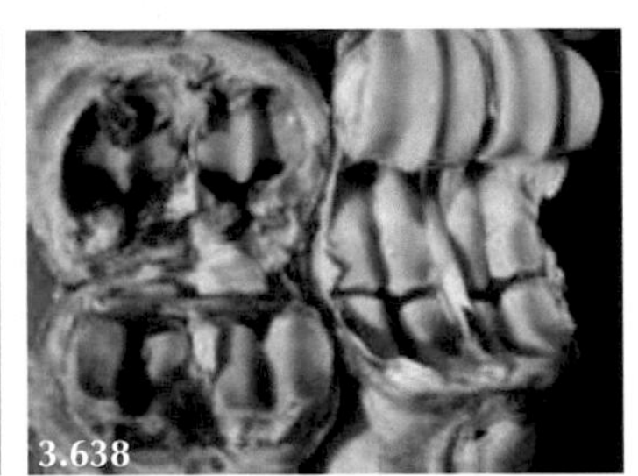

3.636 and **3.637:** Severe onset left foreleg lameness with distended fetlock joint. **3.638:** Comparison of foreleg fetlock joints at necropsy revealing pannus and hyperaemic synovia of infected joint.

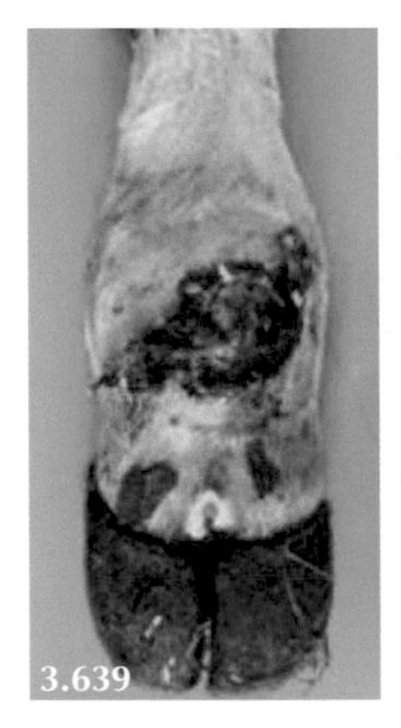

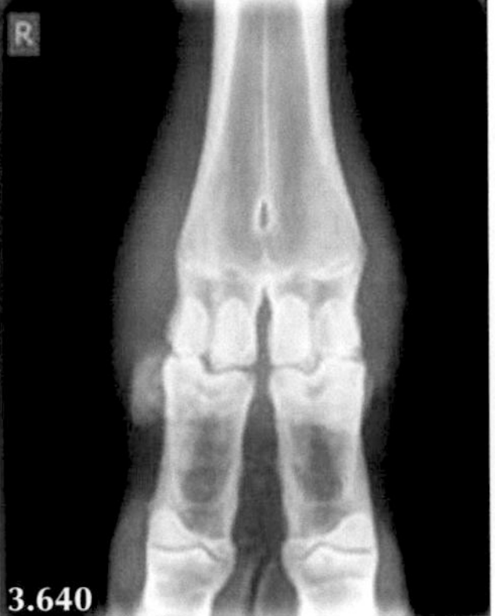

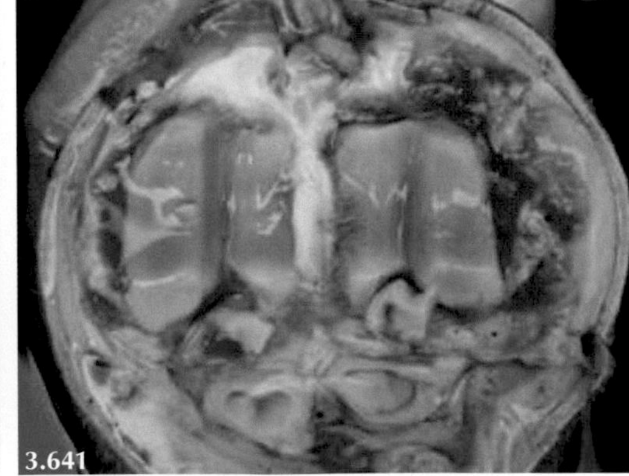

3.639: Abraded skin over swollen fetlock joint. **3.640:** DP radiograph shows only soft tissue swelling. **3.641:** Necropsy findings show early pannus formation with the joint appearing red due to the hyperaemic proliferative synovial membrane.

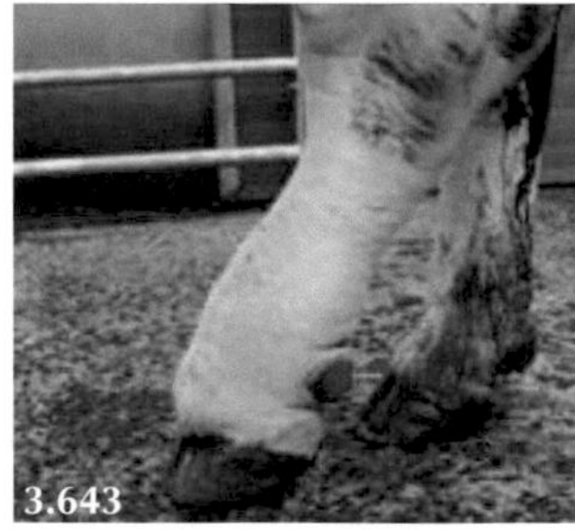
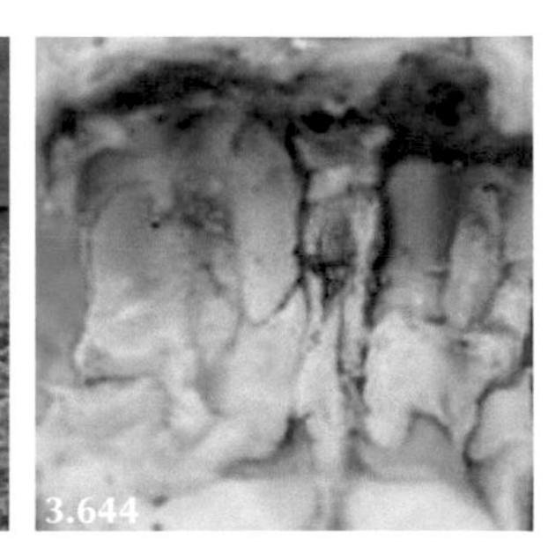

3.642 and **3.643:** Non-weight-bearing lameness of the left hindleg with extensive oedema of the distal limb most pronounced surrounding the left fetlock joint. **3.644:** Necropsy of the left fetlock joint reveals extensive pannus.

- **Key observations**: Severe lameness affecting single joint.
- **Clinical examination**: Very painful swollen joint difficult to palpate because of limb oedema.
- **Key history events**: None. Consider endocarditis if several joints affected. 3.349, 3.350, 3.351

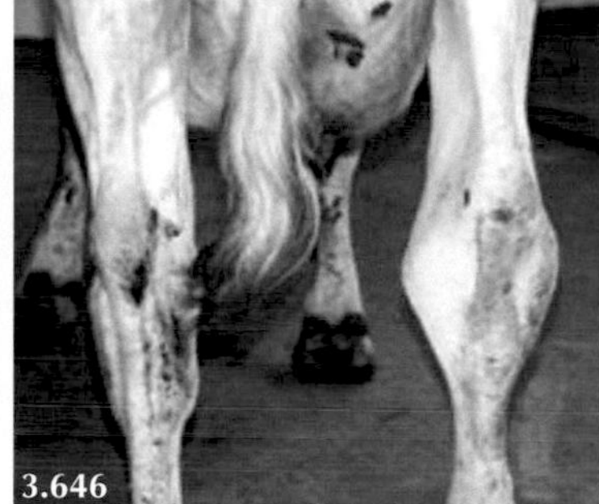
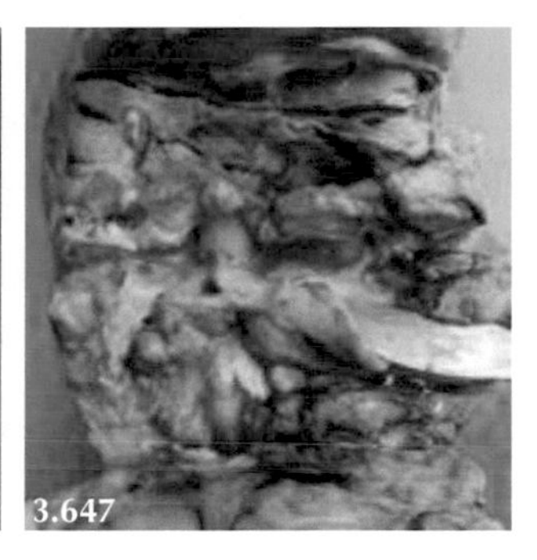

3.645 and **3.646:** Grossly swollen left hock joint. **3.647:** Necropsy reveals extensive pannus, synovial hypertrophy and fibrous tissue reaction of the joint capsule. 3.352

Septic Arthritis (Chronic)

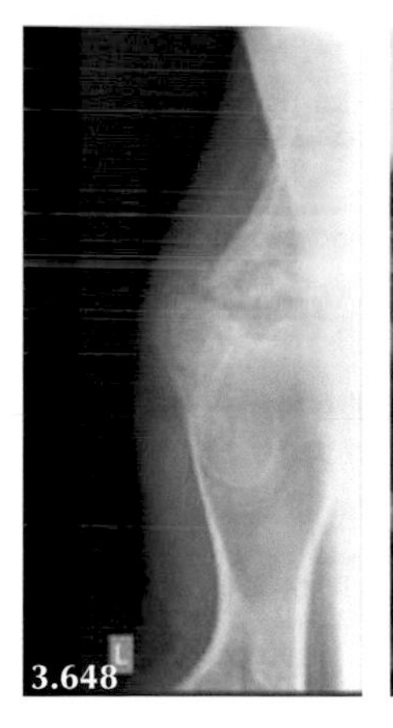
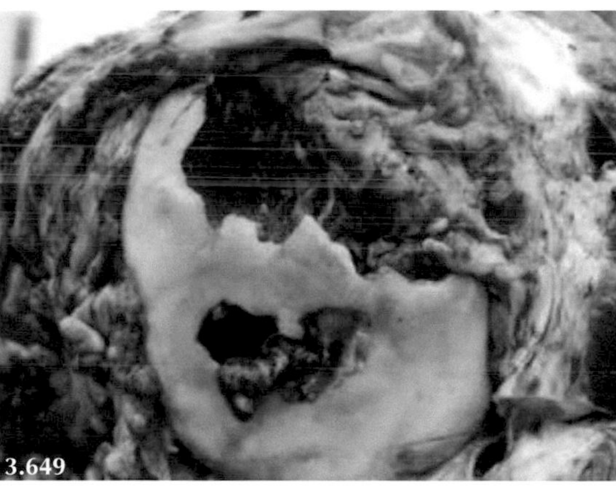

3.648: Cranio-caudal view of the left shoulder reveals extensive erosion of articular cartilage and osteophyte formation. **3.649:** Necropsy reveals extent of joint pathology.

- **Key observations**: Severe lameness, muscle wastage over limb.
- **Clinical examination**: Oedematous, painful joint with variable fluid distension, grossly enlarged drainage lymph node (if palpable). If multiple joints involved investigate possibility of endocarditis.
- **Key history events**: Severe lameness of several months' duration. 3.353

Osteochondrosis Dissecans (OCD)

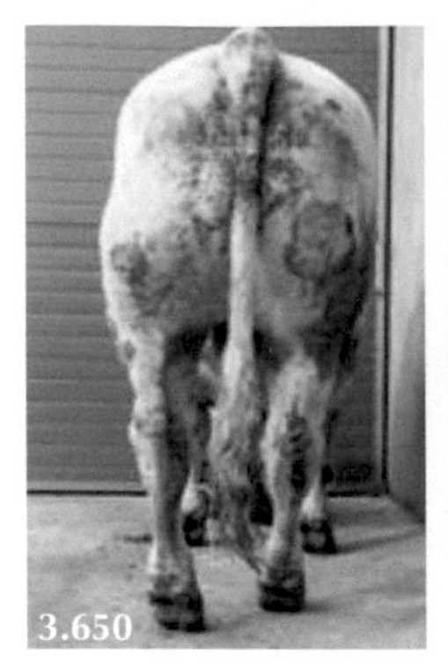
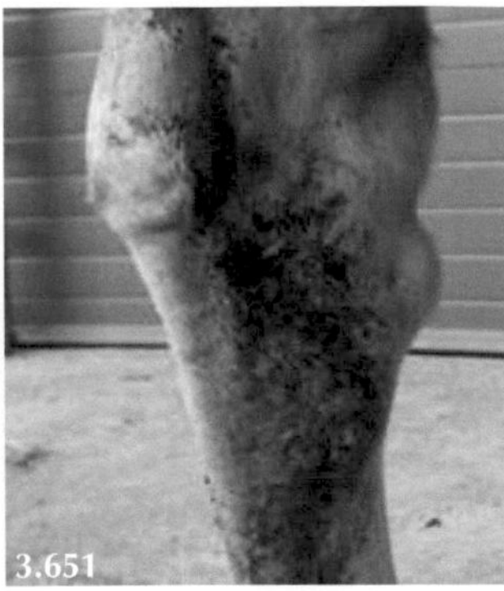
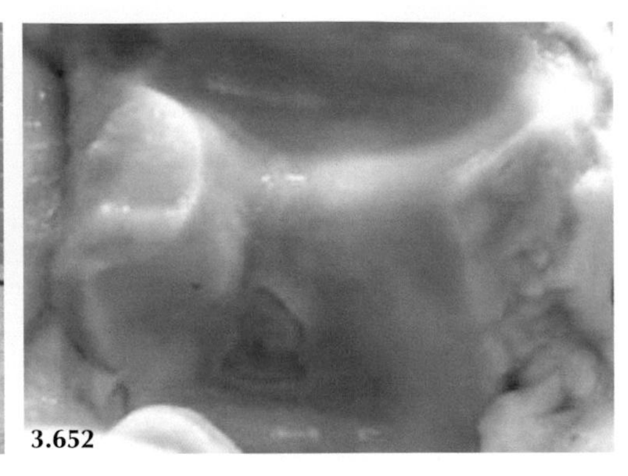

3.650: Distension of both hock joints in a rapidly growing yearling Charolais bull. **3.651:** Obvious distension of the left tarso-metatarsal joint. **3.652:** Loss of articular cartilage.

- **Key observations**: Insidious onset mild lameness with considerable joint effusion.
- **Clinical examination**: Radiography may confirm the presence of a calcified flap free within the joint ("joint mouse").
- **Key history events**: Seen most commonly in rapidly growing bulls between 1 and 2 years old typically affecting the shoulder, elbow, stifle or hock joints. 🎥 3.354

Osteoarthritis

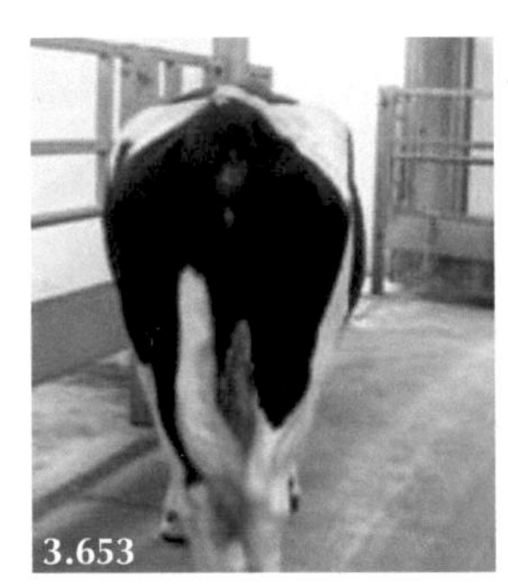
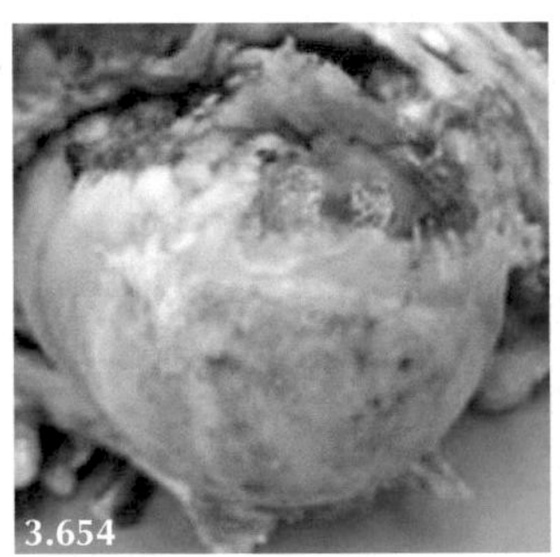
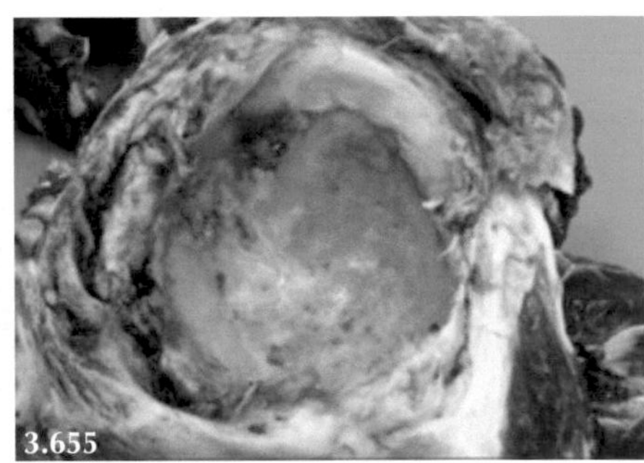

3.653: Marked muscle wastage over the right hip. **3.654** and **3.655:** Erosion of articular cartilage and pitting of subchondral bone. Eburnation refers to friction in the joint causing the reactive conversion of the sub-chondral bone to an ivory-like surface at the site of the cartilage erosion.

- **Key observations**: Mild to moderate chronic lameness.
- **Clinical examination**: Variable joint effusion and thickening of the joint capsule.
- **Key history events**: Hip and stifle joints commonly involved. 🎥 3.355, 3.356, 3.357, 3.358, 3.359

Dislocated Hip

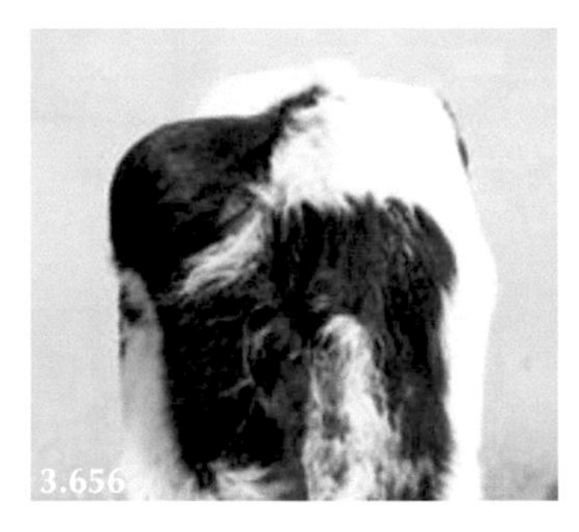
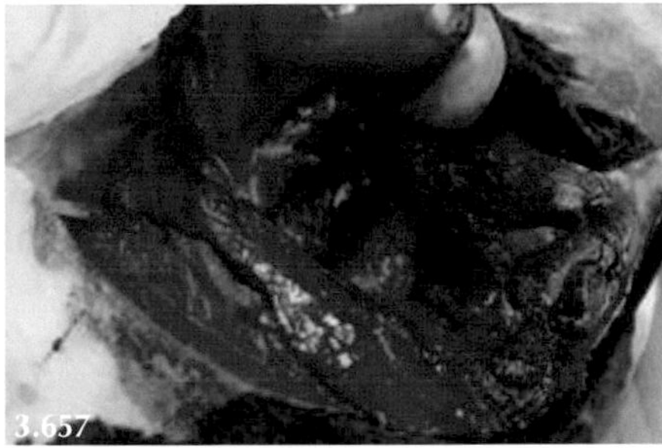
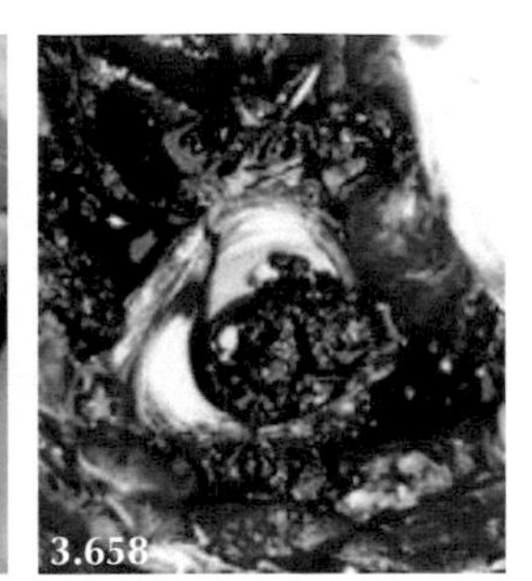

3.656: Dislocated left hip in week-old calf which is much more prominent than right hip. **3.657:** Torn muscle obscures the acetabulum. **3.658:** Clotted blood fills the acetabulum in another case of dislocated hip.

- **Key observations**: Difficulty rising from sternal recumbency. Mechanical lameness. Swelling cranial and dorsal to hip joint.
- **Clinical examination**: Cranio-dorsal femoral head displacement. Shorter leg when position of hock joints compared.
- **Key history events**: Dystocia, calf delivered with excessive traction. 🎥 3.360

Joint Lavage

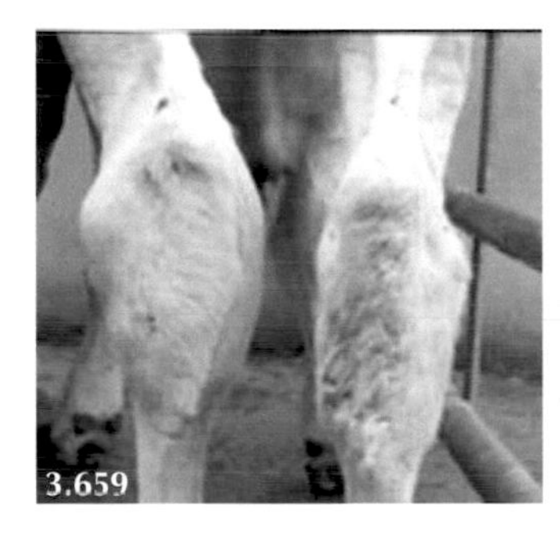
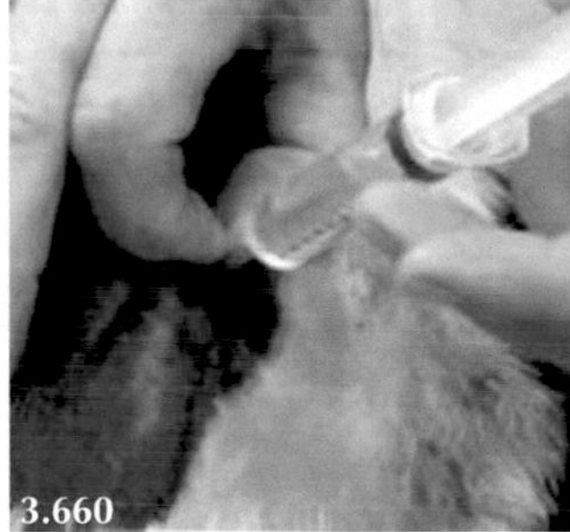
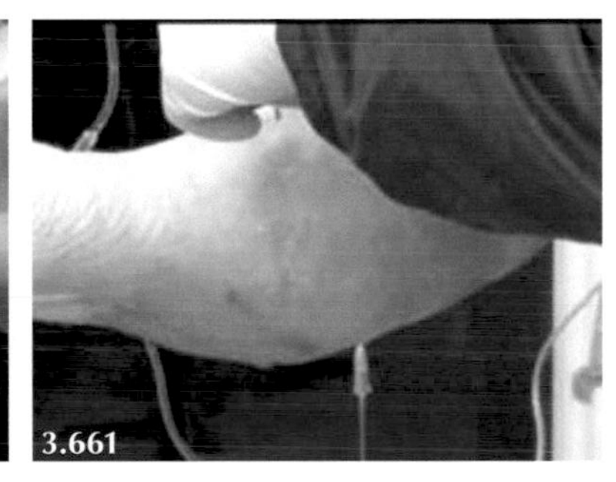

3.659: Distended left hock joint. **3.660:** Administration of high extradural block. **3.661:** Joint lavage. 🎥 3.361, 3.364

3.11

Skeletal System/Fractures

Sheep

- Splints are commonly used by shepherds to stabilise long bone fractures in neonatal lambs, but poor stabilisation leads to extensive bony callus.
- Correct reduction and alignment of foreleg fractures in growing lambs/adults often cannot be achieved by traction alone and general anaesthesia may be indicated. For pelvic limb fractures lumbosacral extradural injection causes temporary paralysis of the pelvic limbs allowing pain-free fracture alignment.
- Greenstick and growth plate fractures can only be confirmed by radiography.
- Fractures of the humerus and femur cannot be satisfactorily stabilised using a cast and require internal fixation.
- Enthesophyte formation is common in the elbow joint of adult sheep with best diagnostic images obtained from an oblique view.

TABLE 3.79

Fractures in Sheep

	Likely duration at presentation	Lameness/Pain	Swelling	Muscle
Long bone fractures (lambs)	Hours/days	+/+++	NAD/+	NAD
Rib fractures (lambs)	Hours/days	+/+++		
Greenstick fractures (lambs)	Hours/days	+	NAD	NAD
Long bone fractures (growing lambs/adults)	Hours to months	+++	+	NAD/---
Elbow arthritis (enthesophytes)	Months	+	+/++	-/--

TABLE 3.80

Ancillary Tests to Diagnose Cause(s) of Ovine Fractures

	Ancillary tests
Long bone fractures (lambs)	Radiography where necessary
Rib fractures (lambs)	Palpation
Greenstick fractures (lambs)	Radiography is essential to confirm diagnosis
Long bone fractures (adults)	Radiography is essential to confirm diagnosis
Elbow arthritis	Radiography is essential to confirm diagnosis

DOI: 10.1201/9781003106456-29

Long Bone Fractures (Lambs)

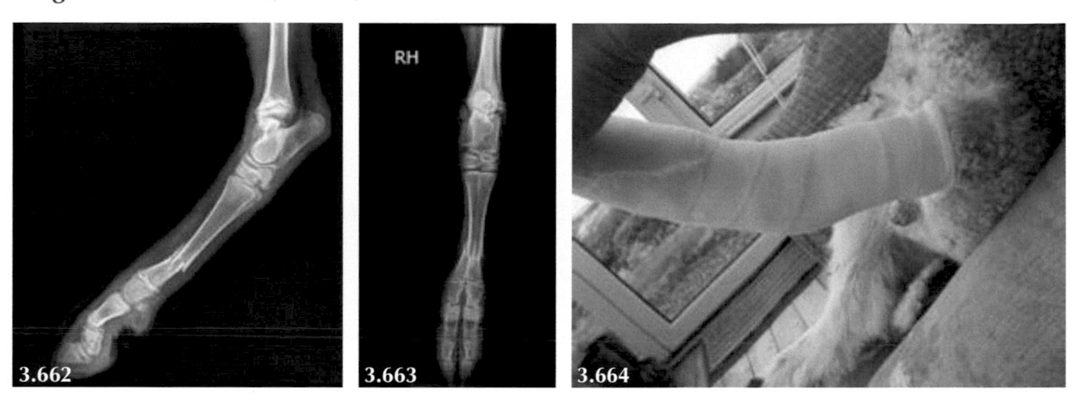

3.662 and **3.663:** Fracture of the right third metatarsal bone in a two-day-old lamb. **3.664:** Reduction/alignment/casting of similar fracture under high extradural block.

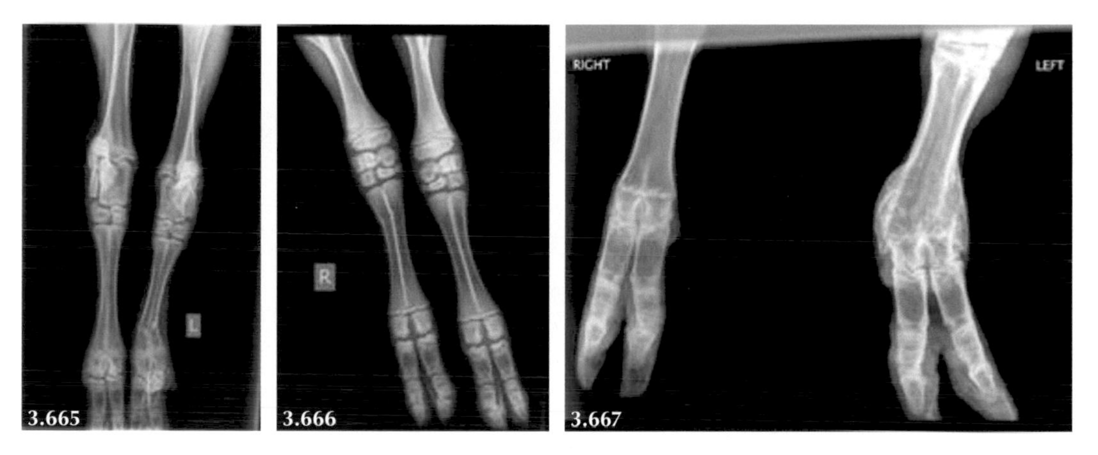

3.665: Fracture of the distal third metatarsal bone of the left hindleg in a two-day-old lamb. **3.666:** Greenstick fracture of the proximal third metacarpal bone of the right foreleg. **3.667:** Extensive osteophyte/new bone deposition (bony callus) formed at an unstable fracture site of the left third metacarpal bone.

- **Key observations**: Sudden onset lameness affecting one leg.
- **Clinical examination**: Gentle palpation/movement at fracture site (not greenstick fracture).
- **Key history events**: Often caused by dam standing on lamb during confinement/transport. 3.362

Rib Fractures

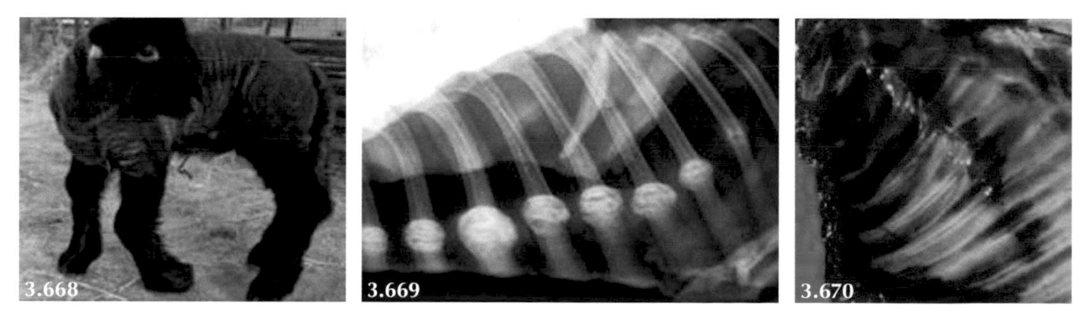

3.668: Trauma to the brachial plexus. **3.669:** Radiograph shows healing at the costochondral junctions. **3.670:** Necropsy of lamb that died during delivery. Note the sharp fracture edges which would have caused severe pain had the lamb survived.

- **Key observations**: Tachypnoea, "flattened" chest wall on one side. Brachial plexus damage may also result.
- **Clinical examination**: Gentle palpation reveals fracture through multiple costo-chondral junctions.
- **Key history events**: Dystocia with lamb in posterior presentation, trodden by dam. 3.363

Long Bone Fractures in Growing Lambs/Adult Sheep

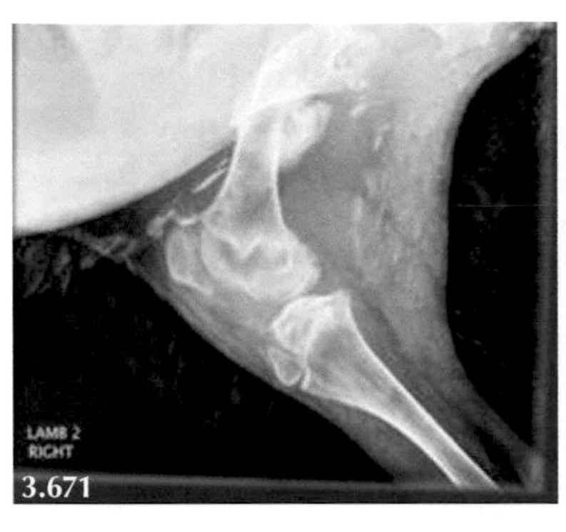

3.671: Mid-shaft femoral fracture which was not reduced/fixed correctly resulting in extensive new bone surrounding the site. **3.672:** Sagittal section through the fractured right femur (right hand side as viewed) compared to the normal left femur.

- **Key observations**: Severe lameness.
- **Clinical examination**: Difficult to palpate femoral fracture—either anesthetise or high extradural block.
- **Key history events**: Often no known incident. 3.364, 3.365

Growth Plate Infections

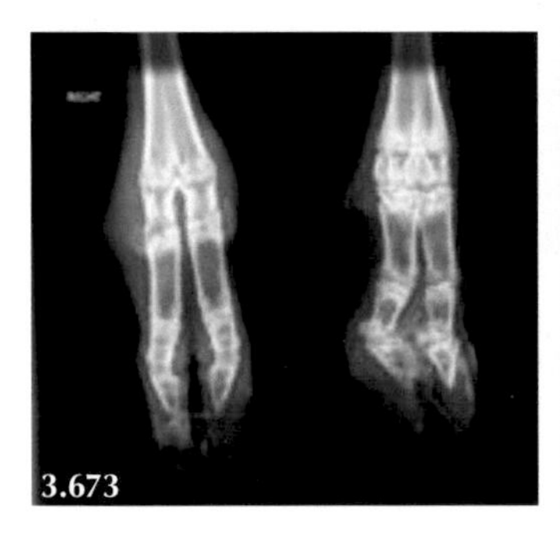

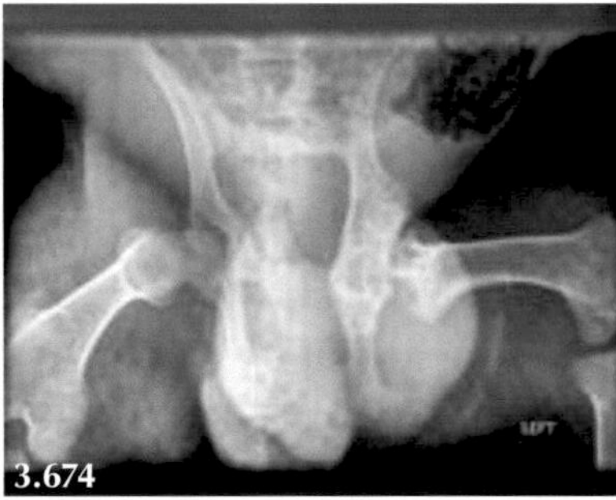

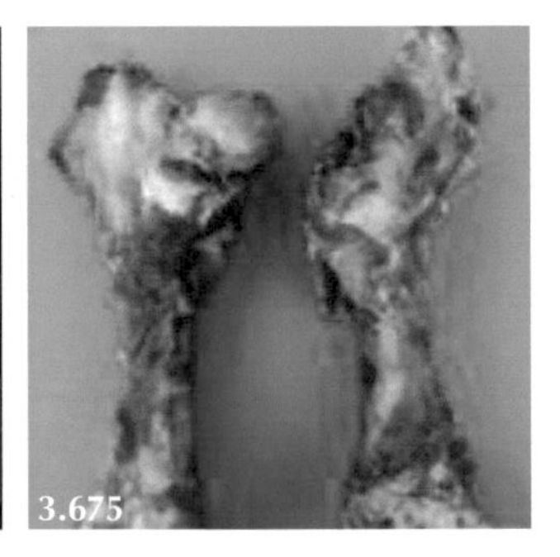

3.673: Infection of the proximal growth plate of the lateral first phalanx of the right foreleg. **3.674:** Infection of the left proximal femoral growth plate showing extensive bone destruction/remodelling surrounding the left hip joint. **3.675:** Necropsy findings of the left femur (right hand side as viewed); normal right femur on left.

- **Key observations**: Severe lameness with visible localised swelling if growth plate distal to the stifle/elbow joints.
- **Clinical examination**: In chronic cases it is difficult to distinguish from septic joint.
- **Key history events**: None.

Elbow Arthritis

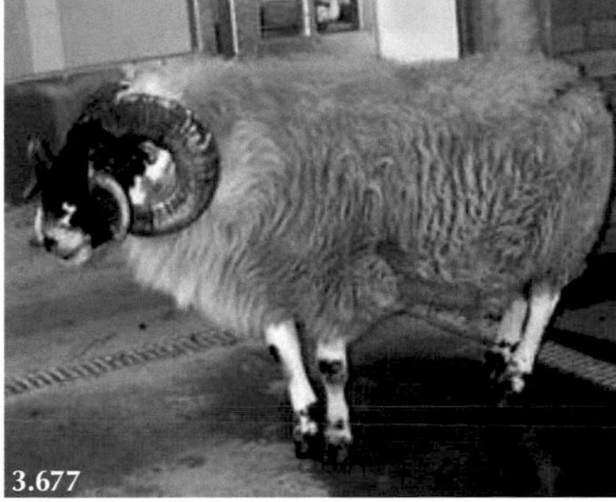
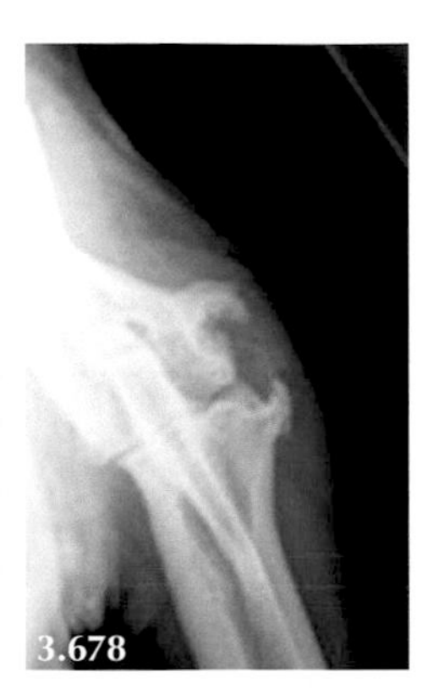

3.676: Early mild lameness of the right foreleg. **3.677:** Bilateral foreleg lameness with its weight taken more on the hindlegs. **3.678:** Oblique radiograph of the right elbow joint showing extensive enthesophyte formation.

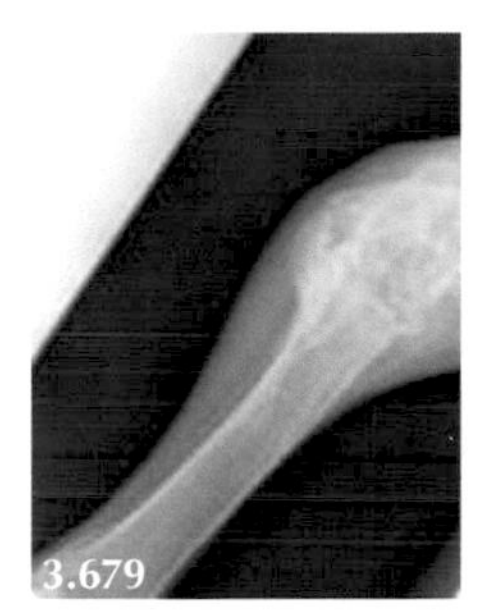
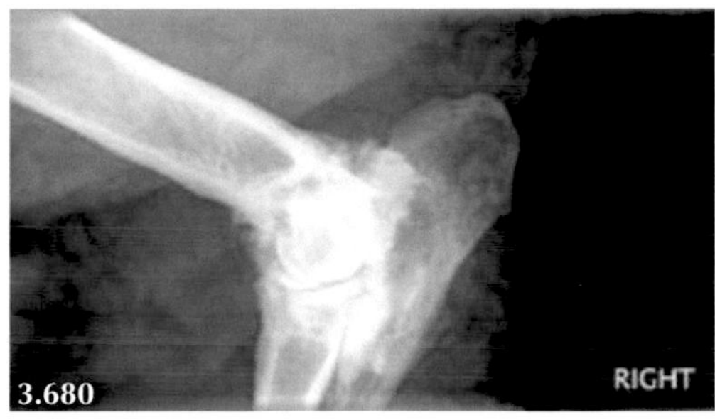

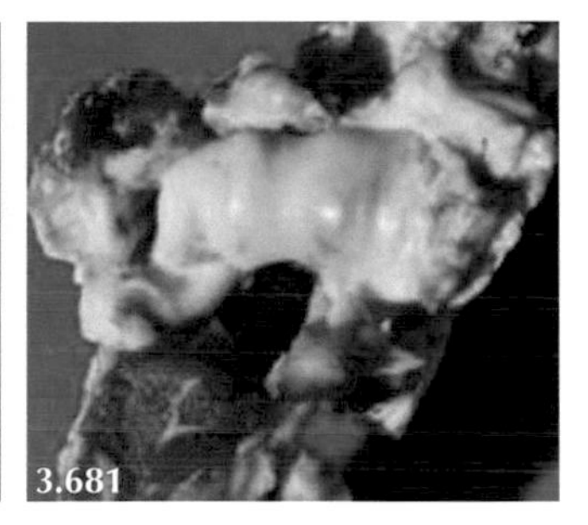

3.679: Fusion of the advancing enthesophytes. **3.680:** Lateral radiograph fails to reveal the localised enthesophyte reaction. **3.681:** Opened elbow joint at necropsy reveals extensive enthesophytes (to the left of joint as viewed).

- **Key observations**: Chronic uni- or bi lateral foreleg lameness.
- **Clinical examination**. Palpable bony swelling on lateral aspect of distal humerus and proximal radius
- **Key history events**: Trauma to the support mechanism preventing overextension of the elbow joint.

The elbow is the ovine joint most commonly affected by osteoarthritis in the UK. Unlike other joints, arthropathy of the elbow joint in adult sheep is characterised by osteophytic reaction and extensive enthesophyte formation involving the lateral ligament (*Lig. collaterale ulnae*). 3.366, 3.367

Calves

- Long bone fractures in neonates are not uncommon following excessive traction in dystocia cases; fractures of the humerus and femur cannot be fixed.
- Fractures of the radius/ulna and tibia require internal fixation.
- Growth plate fractures, commonly involving the femoral head require radiography for definitive diagnosis.
- Be aware of the possibility of rickets predisposing to multiple long bone fractures and joint lesions in intensively managed growing cattle fed vitamin D deficient diets.

TABLE 3.81

Skeletal System/Fractures: Calves

	Duration	Lameness/Pain	Swelling	Muscle
Long bone fractures (calves)	Hours	+++	NAD/+	NAD
Long bone fractures (growing cattle)	Days	++/+++	+/++	NAD
Growth plate fractures	Days/Weeks	+/++	NAD/+	––
Growth plate infections	Weeks	+/++	NAD/+	––
Long bone fractures (Rickets)	Days	++/+++	- NAD/+	NAD
Sequestrum	Months	NAD/+	++	NAD

TABLE 3.82

Ancillary Tests to Diagnose Cause(s) of Bovine Fractures

	Ancillary tests
Long bone fractures (neonates)	Radiography essential before considering fixing long bone fractures
Growth plate fractures	Radiography
Growth plate infections	Radiography
Long bone fractures (rickets)	Radiography is essential to confirm diagnosis
Sequestrum	Radiography

Long Bone Fractures (Neonates)

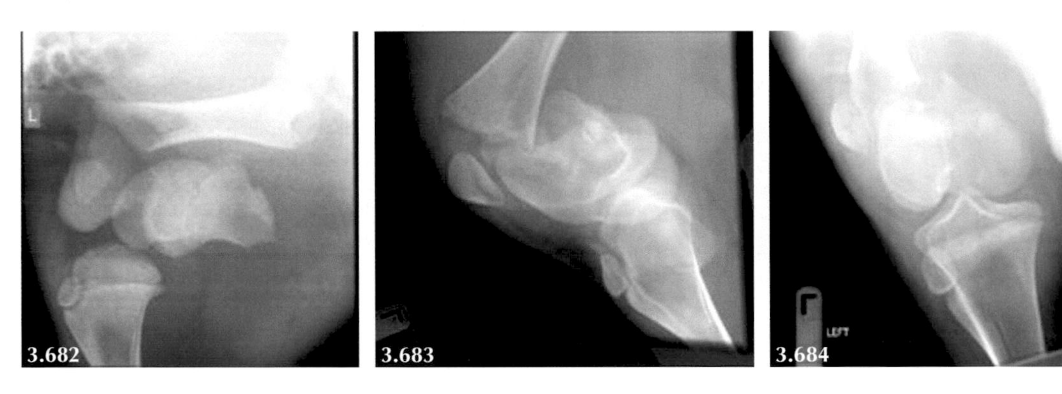

3.682: Lateral view of femoral fracture in a calf following assisted delivery. **3.683** and **3.684:** Lateral and DP views of femoral fracture following assisted delivery.

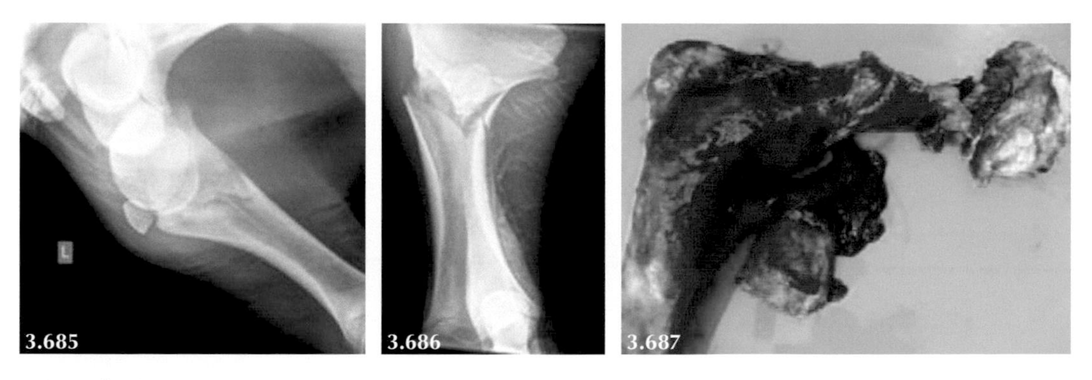

3.685 and **3.686:** Lateral and DP views of tibial fracture following assisted delivery. **3.687:** Dissection of fracture following humane destruction supports this decision.

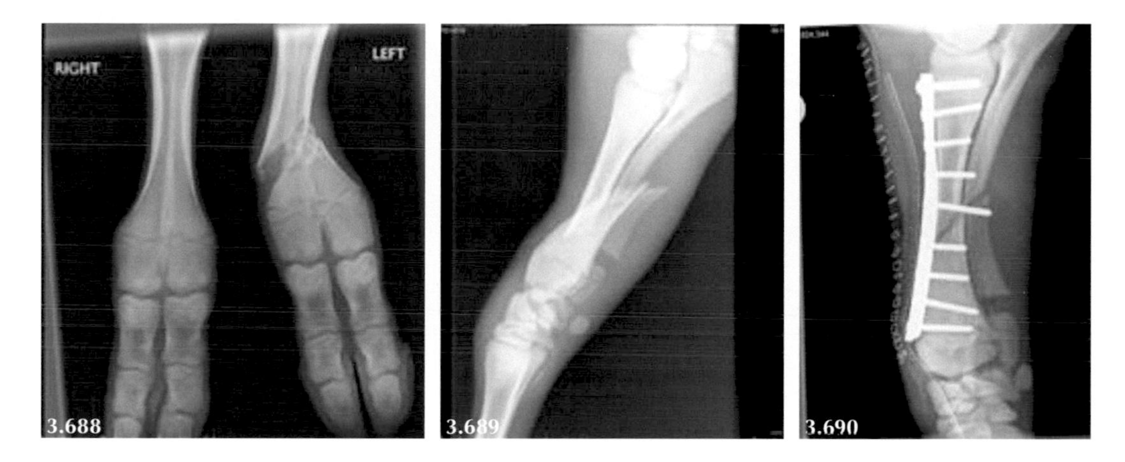

3.688: Fracture through the left distal third metacarpal bone caused by excessive traction at delivery—such fractures heal well after casting. **3.689** and **3.690:** Internal fixation of the fractured radius and ulna caused by excessive traction.

- **Key observations**: Abnormal angulation of affected limb, spontaneous crepitus when leg carefully handled.
- **Clinical examination**: Gentle palpation reveals excessive movement at fracture site.
- **Key history events**: Excessive traction to deliver calf. Cow was standing but attempts to lie down when traction is applied. 3.368, 3.369

Spontaneous Long Bone Fractures in Growing Cattle

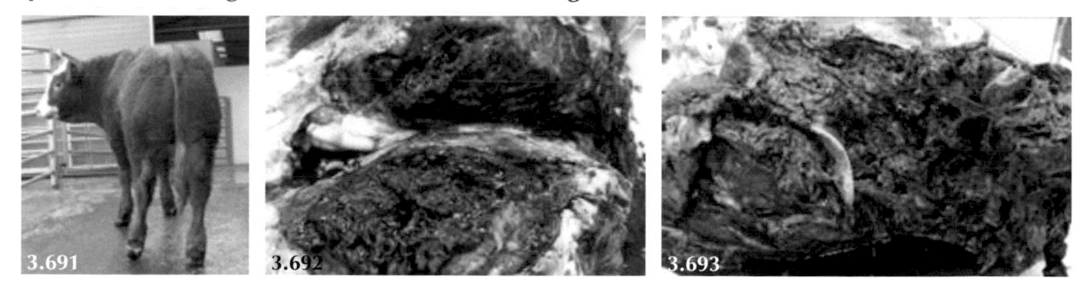

3.691: Yearling beef bull showing severe lameness of left hindleg. There is considerable swelling of the mid-femoral region. **3.692** and **3.693:** Extensive haemorrhage at the fracture site with bone fracture visible in the right image.

- **Key observations**: Sudden, severe lameness. Localised swelling due to haemorrhage.
- **Clinical examination**: Crepitus may be difficult to detect with femoral fractures due to large amounts of supporting muscle. Gentle palpation may reveal movement at fracture site.
- **Key history events**: Slipping on wet concrete surface. Bulling injury. Rickets. 3.370, 3.371

Growth Plate Fractures

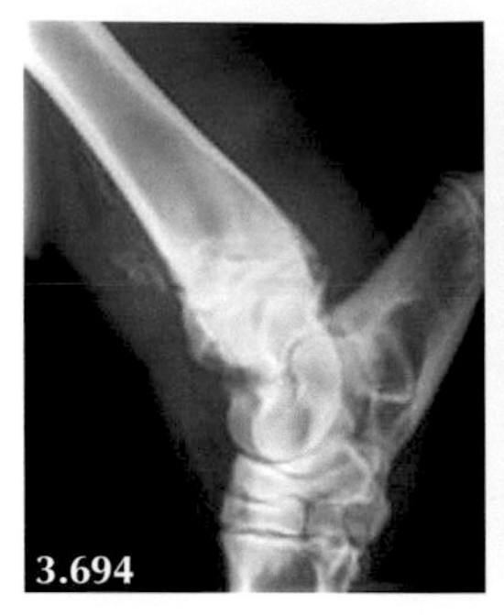

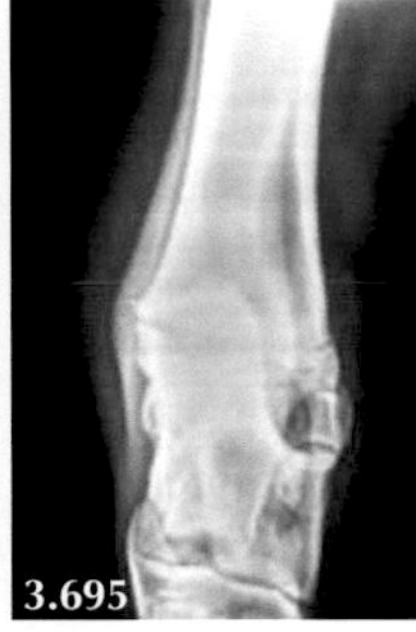

3.694 and **3.695:** Lateral and cranio-caudal radiographs reveal fracture of the distal tibial growth plate. **3.696:** Fracture revealed at necropsy; note the associated haemorrhage. 3.372

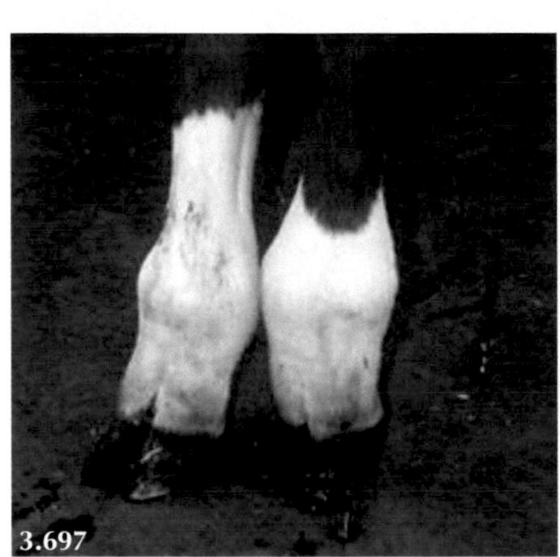

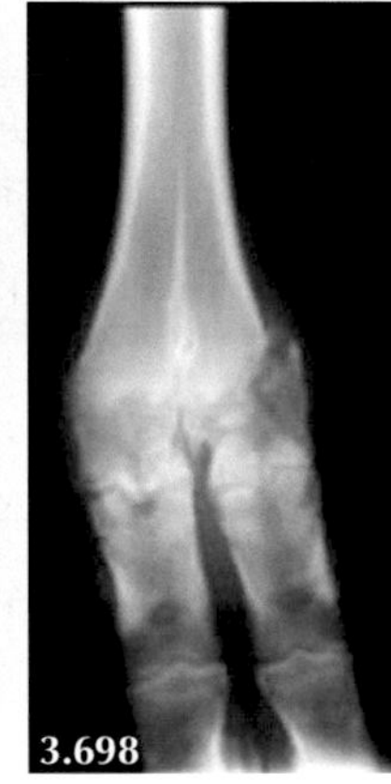

3.697: Swelling proximal to the left hind fetlock joint (right as viewed). **3.698:** DP radiograph reveals Salter Harris type III fracture through distal growth plate of the third metatarsal bone.

- **Key observations**: Chronic moderate lameness (often weeks) of affected limb. Marked muscle wastage.
- **Clinical examination**: Lesion localised to bone extremity not joint. Need (portable) radiography to confirm diagnosis.
- **Key history events**: Often no known history of trauma. 3.373

Growth Plate Infections

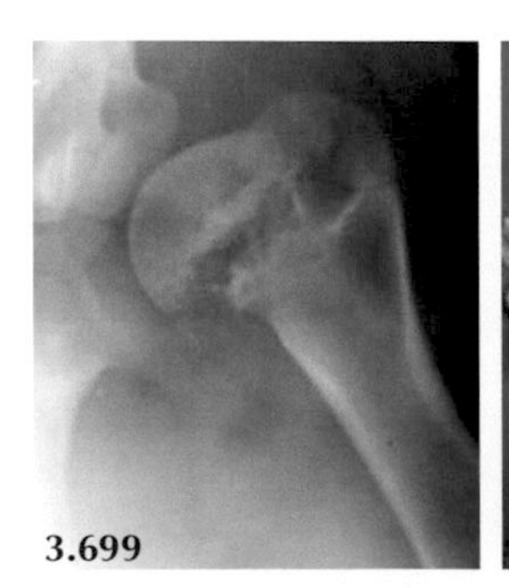

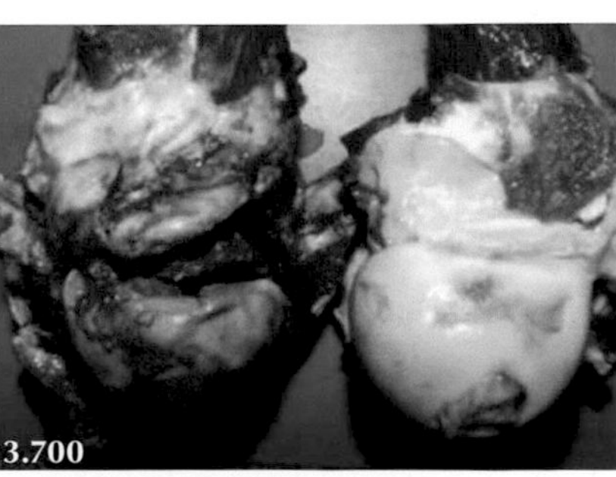

3.699: Radiograph reveals infection of the proximal femoral growth plate.
3.700: Necropsy findings with the infected growth plate shown on the left hand of the image.

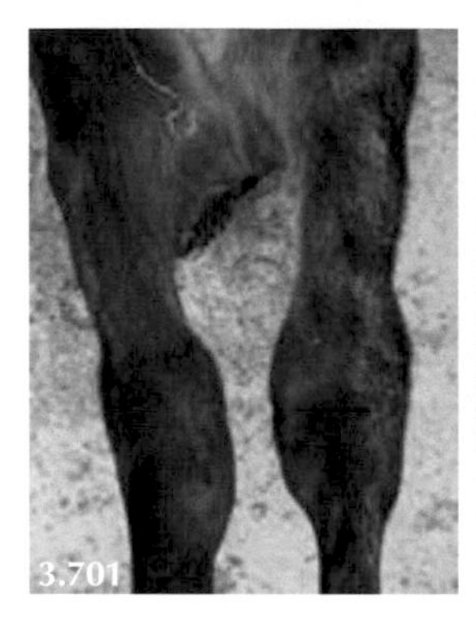
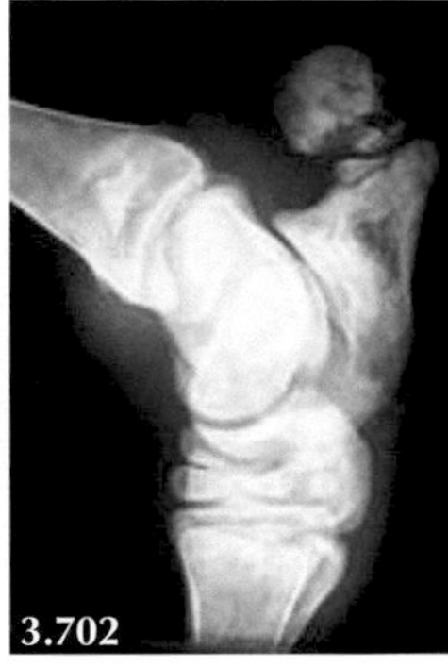
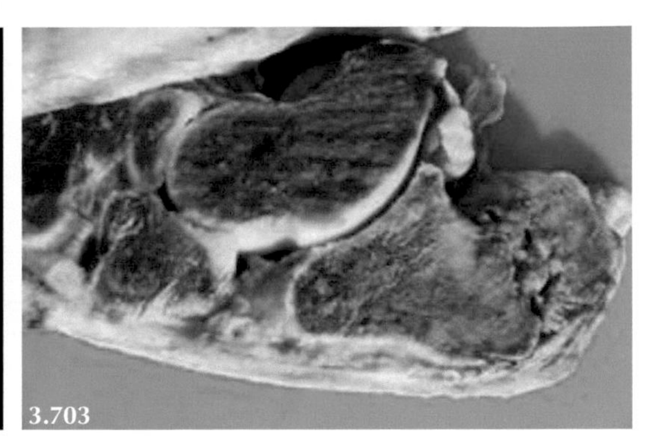

3.701: Swelling of the right tibio-tarsal area in two-month-old beef calf. **3.702:** Lateral radiograph shows infection of the calcaneal growth plate. **3.703:** Necropsy confirms radiographic observations.

- **Key observations**: Chronic lameness (weeks before presentation). Marked muscle wastage.
- **Clinical examination**: Often not possible to differentiate joint infection/growth plate fracture/ growth plate infection on clinical examination—need radiography.
- **Key history events**: Often no known history of trauma/infections. 3.374

Rickets: Long Bone Fractures in Growing Cattle Associated with Vitamin D3 Deficiency

3.704: Sudden onset recumbency and inability to rise resulting from bilateral humeral fractures associated with dietary vitamin D3 deficiency. **3.705:** Healing fracture of left femur in bull beef animal fed ration without appropriate minerals and vitamins.

- **Key observations**: Sudden onset severe lameness but can be several weeks before animal presented. Possible healing of several long bone fractures.
- **Clinical examination**: Often multiple joint effusions in chronic severe cases.
- **Key history events**: Home-mixed ration without appropriate minerals. 3.375

Sequestrum

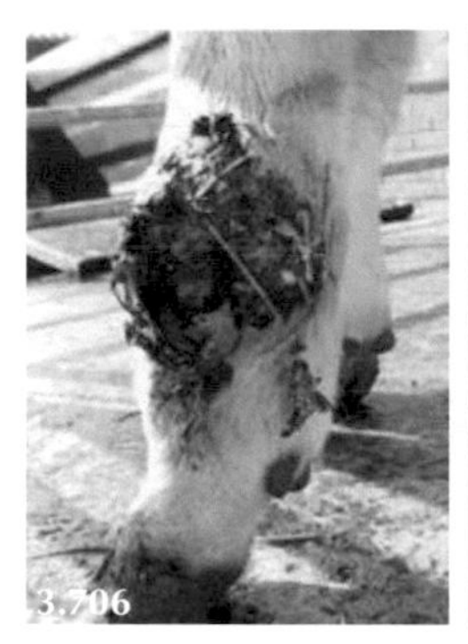
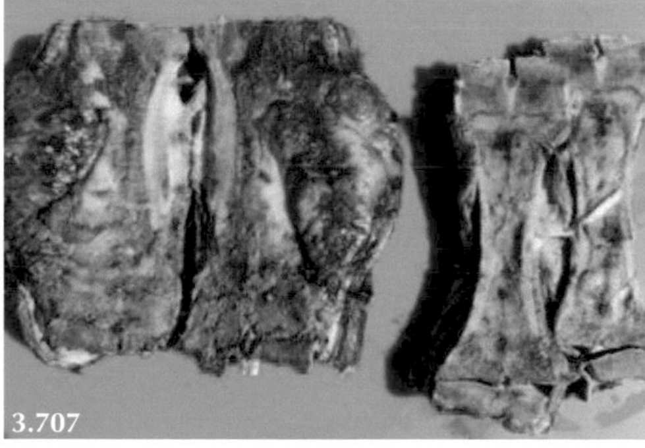

3.706: Large sequestrum of the lateral aspect of the third metacarpal bone of a six-month-old calf. **3.707:** Necropsy through the lesion demonstrates the extent of cortical bone destruction (left) with the normal third metacarpal bone on the right hand side of the image.

- **Key observations**: Large, slowly growing fibrous swelling of the distal limb. Animal may not be lame.
- **Clinical examination**: Non-painful. Enlarged drainage lymph node.
- **Key history events**: No history of trauma or penetration wound.

3.12

Common Causes of Muscle Lesions

Sheep

- Muscle atrophy reflects severe lameness.
- Subcutaneous abscess/cellulitis is surprisingly uncommon in sheep considering the unhygienic state of injection equipment on many farms.

TABLE 3.83

Common Causes of Muscle Lesions in Sheep

	Size	Lameness	Toxaemia	Rectal temperature
Muscle atrophy	–/–––	+/+++	NAD	NAD
Blackleg	++/+++	++/+++	+/++	+/++
Cellulitis/abscess	+/+++	NAD/+	NAD	NAD

TABLE 3.84

Ancillary Tests to Diagnose Common Causes of Muscle Lesions in Sheep

	Ancillary tests
Muscle atrophy	Muscle atrophy secondary to trauma, joint and/or nerve lesion
Blackleg	Response to antibiotics (penicillin). Vaccination status. Necropsy
Cellulitis/abscess	Ultrasonography, needle puncture

Muscle Atrophy

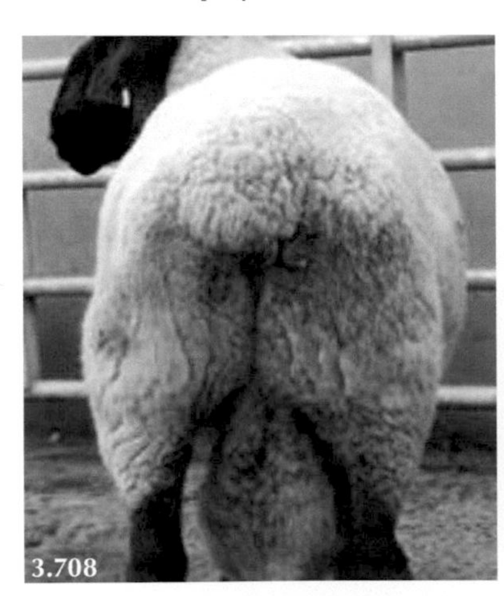
3.708

3.709

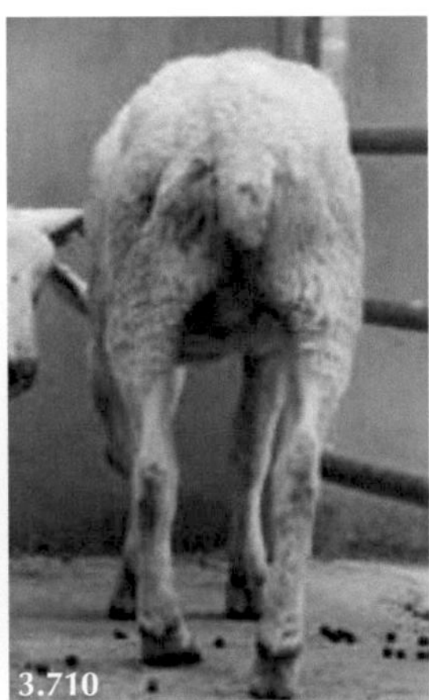
3.710

3.708 and **3.709:** Muscle atrophy of the left hindleg in a Suffolk ram; and a ewe **(3.710)**.

DOI: 10.1201/9781003106456-30

- **Key observations**: Marked limb muscle atrophy.
- **Clinical examination**: Bony prominences readily palpable, identify primary lesion leading to secondary muscle atrophy.
- **Key history events**: Often joint trauma several weeks/months earlier. 3.376

Blackleg

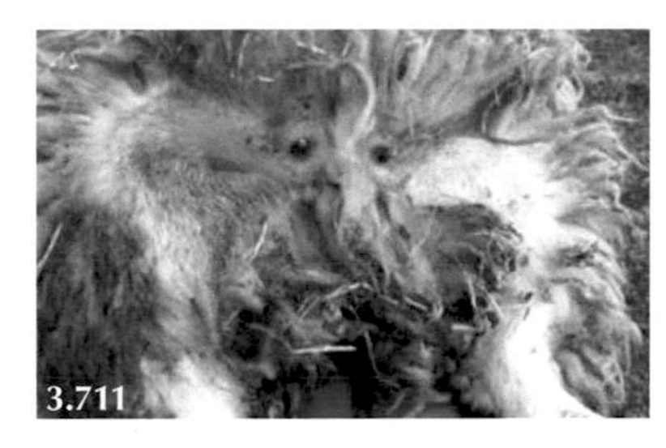

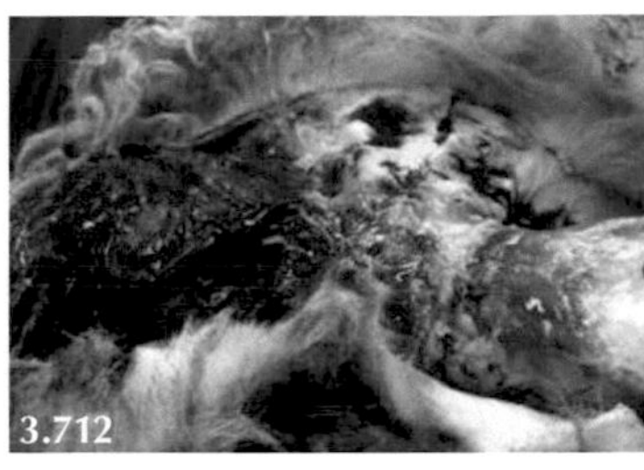

3.711: Swollen and oedematous right inguinal area showing early gangrenous change.
3.712: Necropsy findings showing subcutaneous oedema and discoloured muscle mass.

- **Key observations**: Sudden severe lameness.
- **Clinical examination**: Subcutaneous oedema.
- **Key history events**: Incomplete vaccination history. Dog bite/contaminated intramuscular injection site.

Cellulitis/Abscess

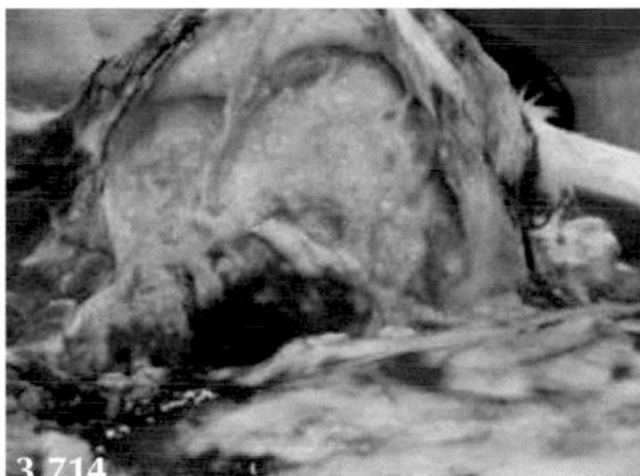

3.713: Very large swelling of the left hindleg. **3.714:** Extensive abscessation/cellulitis. The infection is deep-seated in fascial planes which is more likely explained by a contaminated deep intramuscular injection.

- **Key observations**: Lame. Large subcutaneous fluid swelling.
- **Key history events**: Dog bite/contaminated intramuscular injection. Check antibiotic storage, use of disposable syringes and needles, injection technique. 3.377, 3.378

Cattle

- Subcutaneous abscesses and cellulitis lesions are common in cattle.
- Recent earthworks often source of outbreak of blackleg in growing cattle.

TABLE 3.85

Common Causes of Muscle Lesions in Cattle

	Duration	Size	Lameness	Toxaemia	Rectal temperature
Cellulitis/abscess	Days/months	+/+++	NAD/+	NAD	+
Blackleg	Hours	++/+++	++/+++	+/+++	+/++
Spastic paresis	Months	NAD	+/+++	NAD	NAD
Barn cramps	Months	NAD	+	NAD	NAD

TABLE 3.86

Ancillary Tests to Diagnose Common Causes of Muscle Lesions in Cattle

Disease	Ancillary tests
Cellulitis/abscess	Ultrasonography (snowstorm appearance), needle puncture
Blackleg	Vaccination status. Necropsy
Spastic paresis	Response to surgical transection of tibial nerve
Barn cramps	Clinical signs

Cellulitis

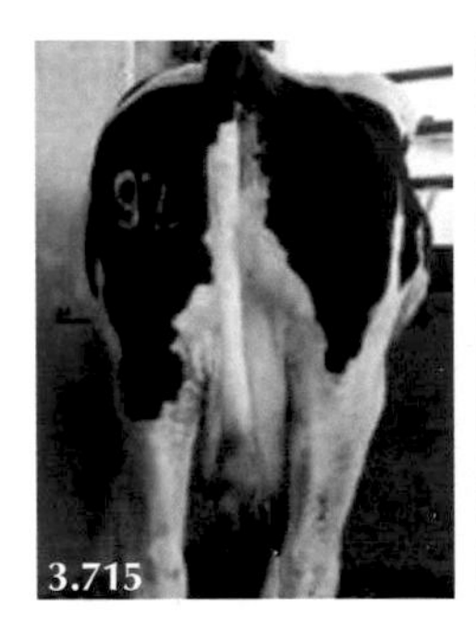

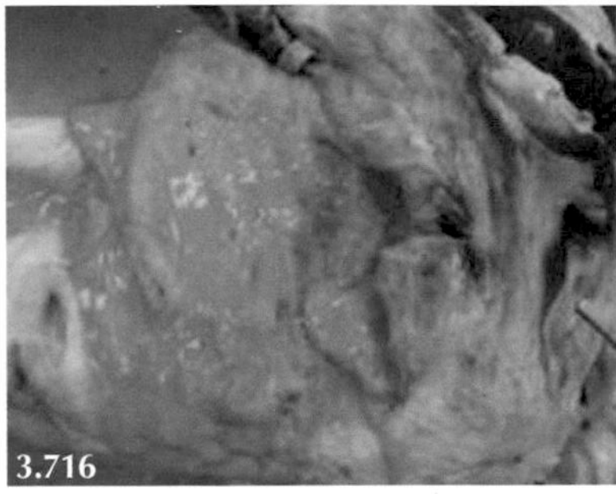

3.715: Swollen left leg and lameness after deep (contaminated) intramuscular injection. **3.716:** Deep-seated infection (cellulitis) revealed at necropsy. **3.717:** Non-weight-bearing lameness of the right foreleg caused by cellulitis from a capped knee.

- **Key observations**: Sudden moderate/severe lameness develops over several days.
- **Clinical examination**: Extensive painful swelling, subcutaneous oedema. Enlarged drainage lymph node.
- **Key history events**: Infected injection technique, contaminated solution/suspension injected. Skin penetration from abrasions/puncture over bony prominences. 3.379

Subcutaneous abscess

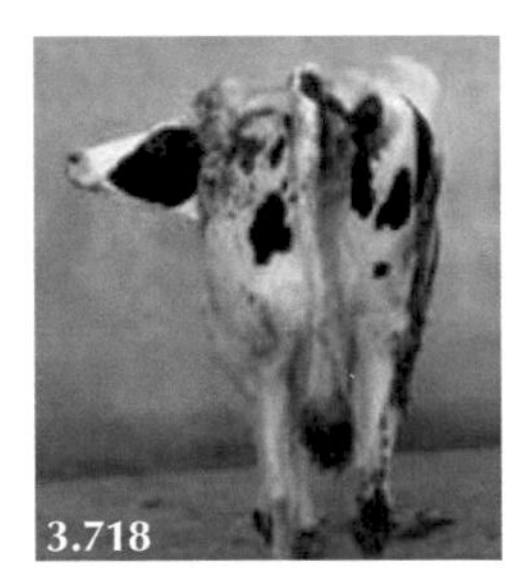

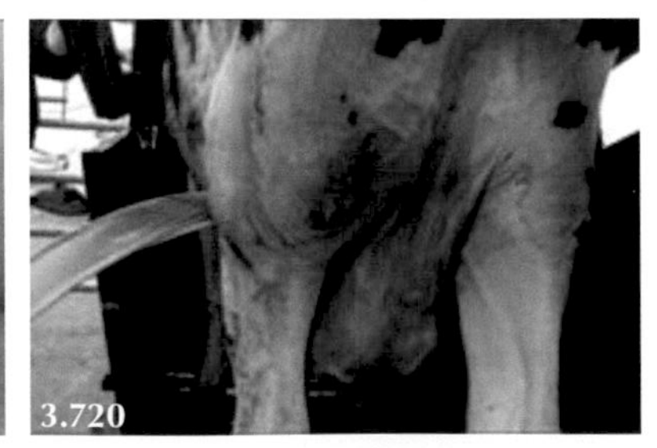

3.718 and **3.719:** Considerable swelling immediately caudal and lateral to the left stifle joint. **3.720:** Lancing the subcutaneous abscess released about 40 litres of pus. 3.380

Blackleg

3.721 and **3.722:** Severe lameness, very swollen left hindlegs in two of 15 affected Holstein heifers from a group of 85 cattle. **3.723:** The heifers had access to recent earthworks which were fenced off after the outbreak.

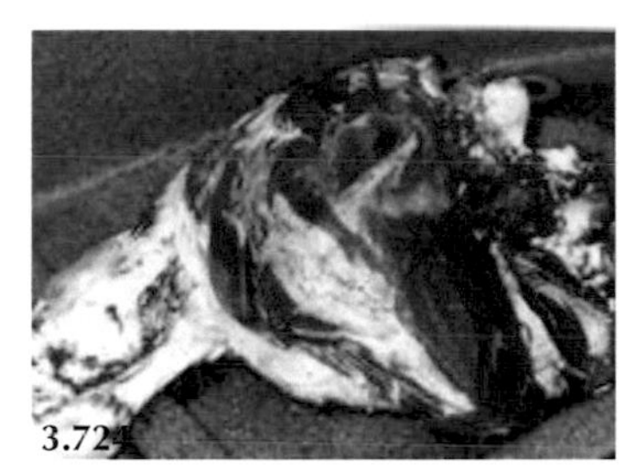
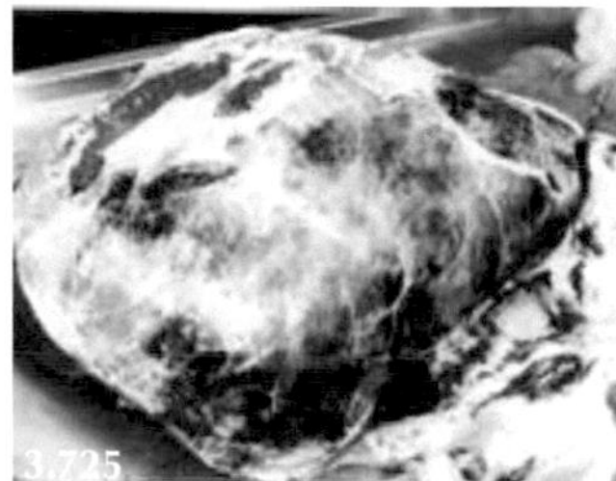
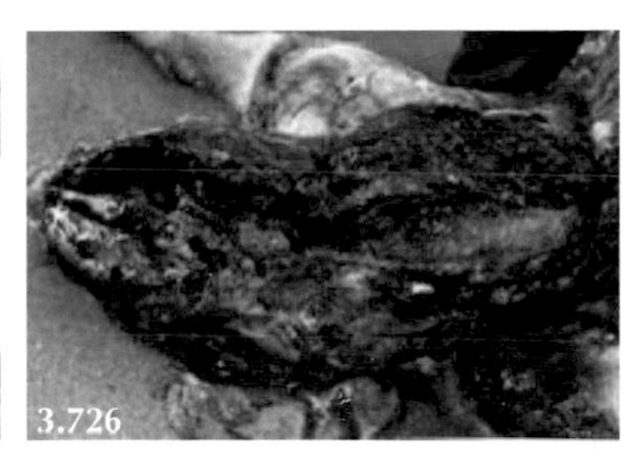

3.724: Normal hindleg at necropsy compared to contra-lateral hindleg affected by blackleg (**3.725**). **3.726:** Necropsy revealing blackleg affecting the ventral neck musculature causing rapid autolysis.

- **Key observations**: Sudden severe lameness. Usually found dead without premonitory signs.
- **Clinical examination**: Extensive swelling, subcutaneous oedema and crepitus.
- **Key history events**: No clostridial vaccination history. Recent earthworks/exposed riverbanks. Bulling injury in heifers. 3.381

Spastic Paresis

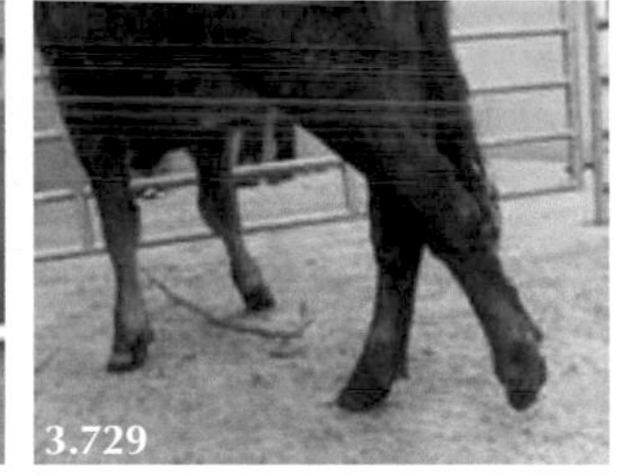

3.727: Too straight hindlegs caused by over-extended stifle and hock joints. **3.728** and **3.729:** Over-extended hindleg with hind foot held caudally and clear of the ground.

- **Key observations**: Insidious onset, over-extension of stifle and hock joints, foot may not contact ground when walking in advanced cases.
- **Clinical examination**: Altered stance and gait otherwise NAD.
- **Key history events**: May be several cattle affected sired by same bull. 3.382

Barn Cramps

3.730: "Saw horse" appearance caused by overextension of the hindleg joints.

- **Key observations**: Insidious onset, excessive extension of both hindleg(s).
- **Clinical examination**: Altered stance otherwise NAD.
- **Key history events**: May be hereditary component. 3.383

Hygromas

- Carpal hygromas are very common in meat breed rams reflecting their susceptibility to foot lameness.
- Carpal hygromas simply reflect poor management. It is depressing to see carpal hygromas in so many purchased rams.

	Duration	Size	Lameness
Carpal hygromas	Weeks/months	+/++	NAD

Carpal Hygromas

3.731–3.733: Rams with carpal hygromas; note the sale numbers in centre and right images. Why would a farmer buy a ram with a proven history of chronic lameness?

- **Key observations**: Do not cause lameness. Large swellings over the carpal joints. Often white hairs replace black ones resulting from damage to skin.
- **Clinical examination**: Some fluid "under skin" but not within joint capsule. Not painful.
- **Key history events**: Previous chronic foreleg lameness causing sheep to graze on its knees. 3.384, 3.385

Bursal Lesions

- Capped knees and hocks indicate poor dairy cow cubicle design and maintenance.

TABLE 3.87

Common Bursal Lesions in Cattle

	Duration	Lameness/pain	Limb muscle
Capped hock	Weeks/months	NAD	NAD
Capped knee	Weeks/months	NAD	NAD

TABLE 3.88

Ancillary Tests for Common Bursal Lesions in Cattle

	Ancillary tests
Capped hock	Not lame. No additional tests necessary
Capped knee	Not lame. No additional tests necessary

Capped Hock (Tarsal Bursitis)

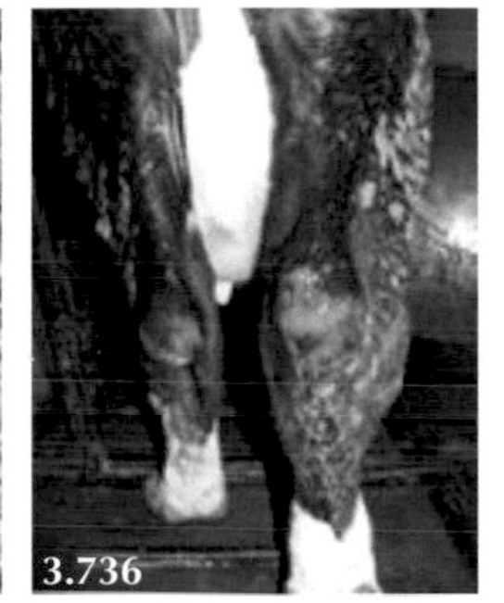

3.734: Large capped right hock. **3.735:** Superficial infection of the left tarsal bursa. **3.736:** Large capped right hock.

- **Key observations**: Large swelling of lateral hock. Often abraded but not lame.
- **Clinical examination**: Firm fibrous swelling on lateral aspect; no joint involvement.
- **Key history events**: Poorly designed/maintained cubicles/slats. 3.386, 3.387

Capped Knee (Carpal Bursitis)

3.737: Left: Capped and abraded knees.

- **Key observations**: Large swollen knee(s). Often abraded but not lame.
- **Clinical examination**: Firm swelling on dorsal aspect; no joint involvement.
- **Key history events**: Poorly designed/maintained cubicles. 3.388

3.13

Common Causes of Peripheral Oedema, Jugular Distension and Ascites

Sheep

Causes of circulatory failure with extensive peripheral oedema, jugular distension and ascites are rare in sheep. Submandibular oedema occurs secondary to anaemia in chronic fascioliasis and haemonchosis, and hypoalbuminaemia in some cases of paratuberculosis.

TABLE 3.89

Common Causes of Peripheral Oedema in Sheep

Disease	Duration	Oedema	Appetite
Fasciolosis	Weeks	+/++	–/––
Haemonchosis	Weeks	+	NAD/–

Cattle

- Septic pericarditis and DCM are common causes of oedema.
- Right-sided heart failure commonly arises from thymic lymphosarcoma and DFA and other lung pathologies.
- Cardiac tumours can only be diagnosed in the live animal using ultrasonography.

TABLE 3.90

Common Causes of Peripheral Oedema, Jugular Distension and Ascites in Cattle

Disease	Duration	Ascites	Oedema	Appetite	Pyrexia
Periparturient oedema	Days	NAD	+	NAD	NAD
Septic pericarditis	Days/weeks	+/++	+/+++	–/––	+
Dilated cardiomyopathy	Weeks/months	+/++	+++	–/––	NAD
Thymic lymphosarcoma	Months	+/++	+/+++	–/––	NAD
Diffuse fibrosing alveolitis	Days/weeks	NAD/+	+/++	NAD	NAD
Cardiac tumours	Weeks/months	+/++	+/+++	–/––	NAD/+

DOI: 10.1201/9781003106456-31

TABLE 3.91

Ancillary Tests for Common Causes of Peripheral Oedema, Jugular Distension and Ascites in Cattle

Disease	Ancillary tests
Periparturient oedema	Pitting oedema, resolves within 2–3 days
Septic pericarditis	Auscultation, ultrasonography. Necropsy
Dilated cardiomyopathy	Examination of heart at necropsy
Thymic lymphosarcoma	Necropsy
Diffuse fibrosing alveolitis	Response to corticosteroid injection. Necropsy
Cardiac tumours	Ultrasonography. Necropsy

Periparturient Oedema

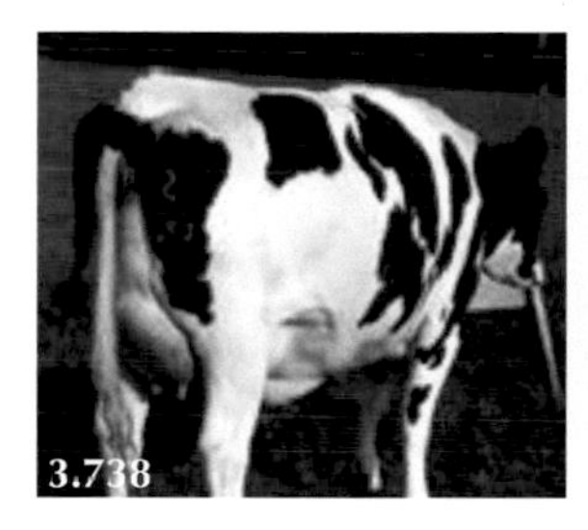

3.738–3.740: Udder and ventral oedema in periparturient heifers.

- **Key observations**: Pitting oedema of the ventral udder extending along the ventral body wall.
- **Clinical examination**: No clinical significance.
- **Key history events**: Typically seen in heifers from two to three days before calving.

Septic Pericarditis

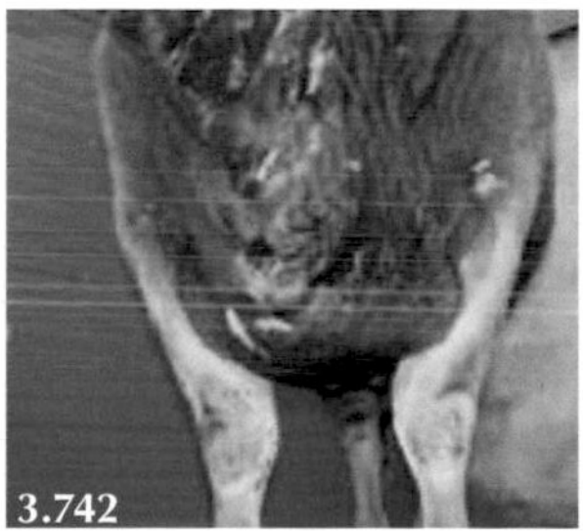

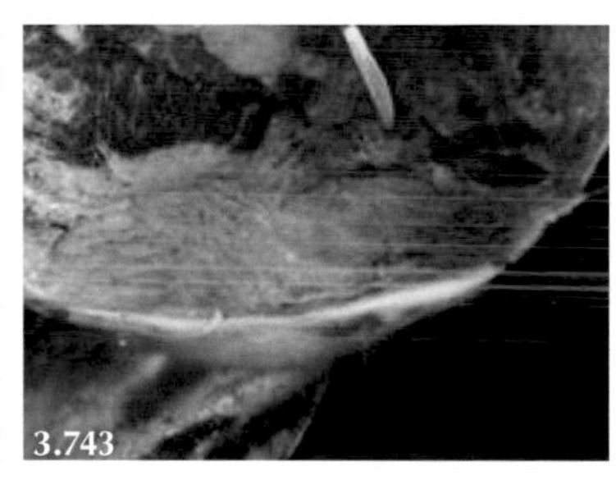

3.741: Early (mild) brisket oedema. **3.742:** Late (severe) brisket oedema. **3.743:** Necropsy reveals severe brisket oedema extending for 15 cm (see centre image).

- **Key observations**: Mild to severe pitting oedema, submandibular and ventral oedema also possible.
- **Clinical examination**: "Splashing/tinkling" sounds extending halfway up chest wall.
- **Key history events**: Earlier traumatic reticulitis not noted by farmer especially in beef cattle. 3.389, 3.390, 3.391

Dilated Cardiomyopathy

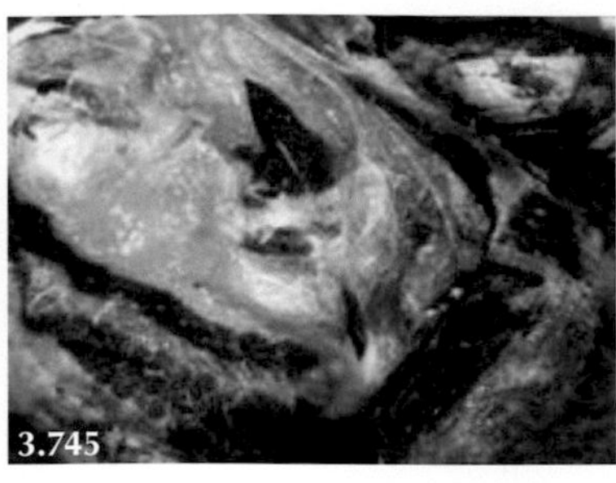

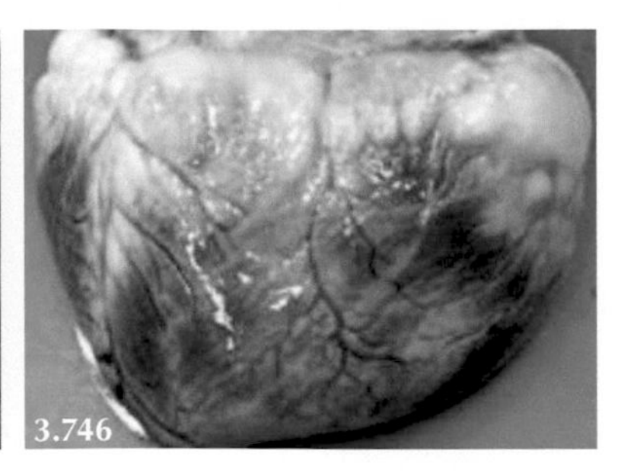

3.744: Extensive brisket and ventral oedema in a twenty-two-month-old Holstein heifer. **3.745:** Subcutaneous oedema revealed at necropsy. **3.746:** Globose-shaped heart.

- **Key observations**: Extensive brisket and ventral oedema develop over several months.
- **Clinical examination**: Widespread pitting oedema, often pleural effusion.
- **Key history events**: Genetic effect—several animals in cohort sired by same bull.

Thymic Lymphosarcoma

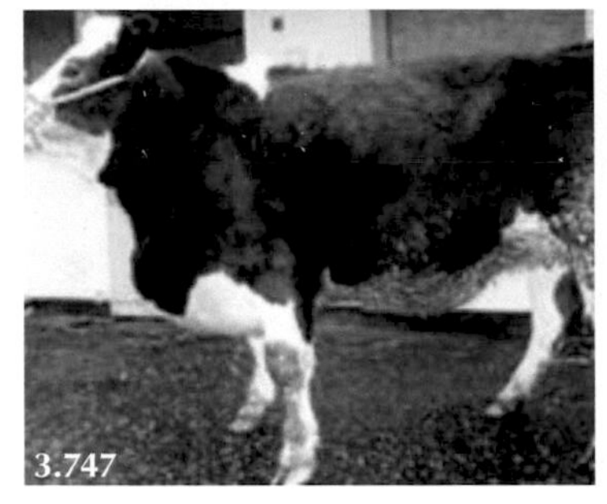

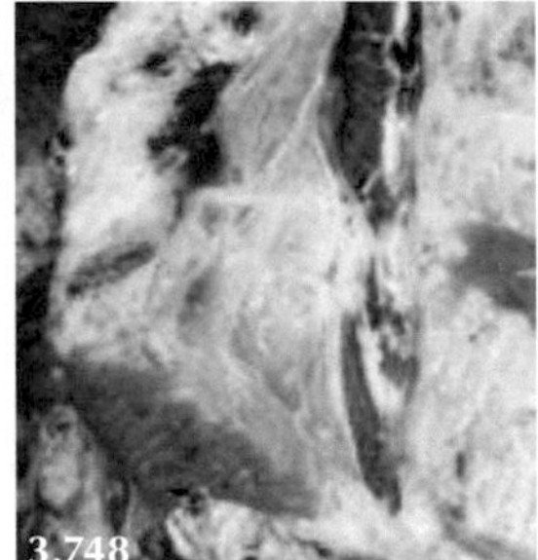

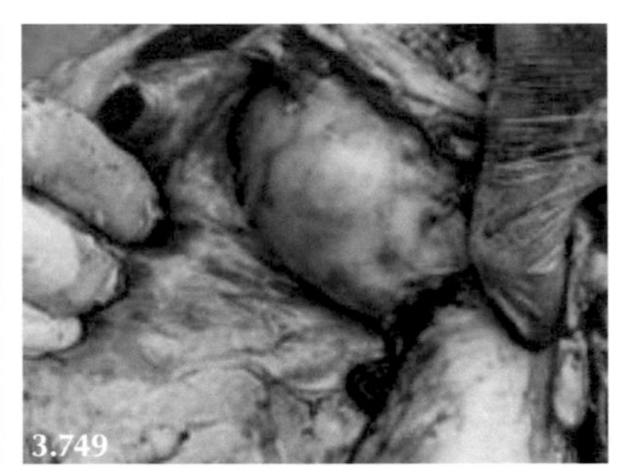

3.747: Brisket oedema in twenty-one-month-old beef heifer. **3.748:** 8–10 cm oedema in the brisket at necropsy. **3.749:** Caudal margin of mass in the mediastinum.

- **Key observations**: Often recurrent bloat (compression of oesophagus), development of brisket oedema.
- **Clinical examination**: Muffled heart and lung sounds.
- **Key history events**: Chronic weight loss over several months. Sporadic occurrence in absence of EBL virus. 🎥 3.392

Diffuse Fibrosing Alveolitis

3.750: Aged beef cow with extensive brisket oedema. **3.751:** Same cow two days later with much reduced oedema after corticosteroid injection.

- **Key observations**: Increased respiratory rate, frequent coughing.
- **Clinical examination**: Marked brisket oedema.
- **Key history events**: Older beef cows, exposure to mouldy hay. 3.393

Pericardial Mass/Tumour

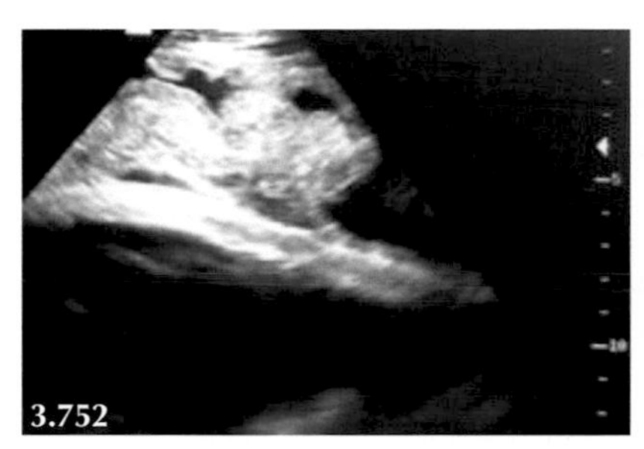

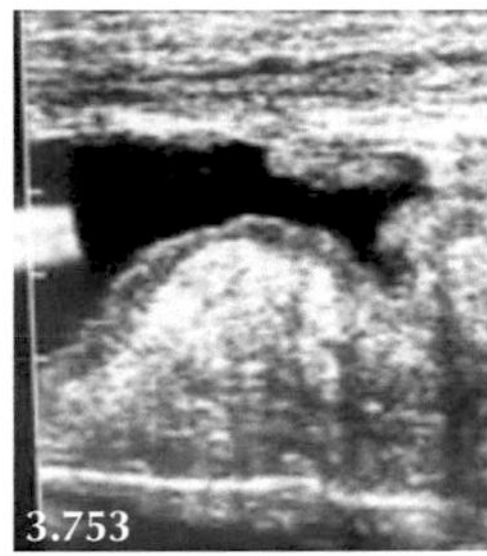

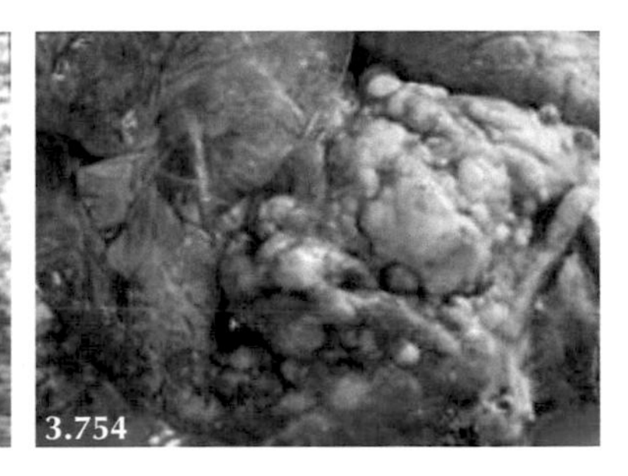

3.752: 6.5 MHz microarray probe reveals 5 cm deep multi-lobulated mass on the surface of the pericardium. There are numerous 1–2 cm diameter masses attached to the parietal pleura throughout the chest cavity. **3.753:** Similar findings are revealed using a 5 MHz linear probe. **3.754:** Mass revealed at necropsy.

- **Key observations**: Cow presents in poor body condition and an obvious jugular pulse.
- **Clinical examination**: Much reduced heart sounds caused by the pleural effusion. Masses noted during ultrasound examination.
- **Key history events**: None noted. 3.394

Tumours (Heart Base Tumour)

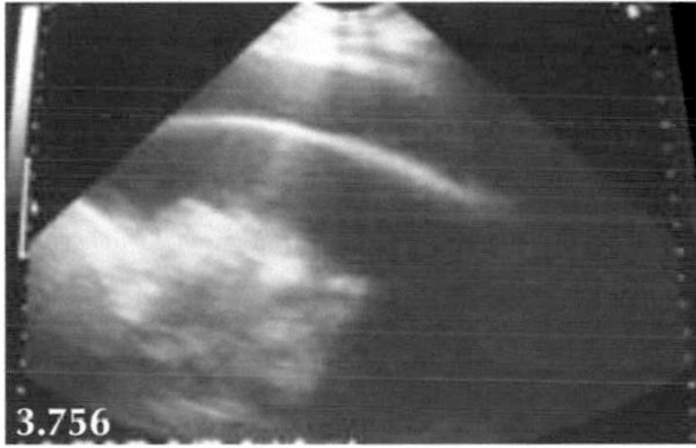

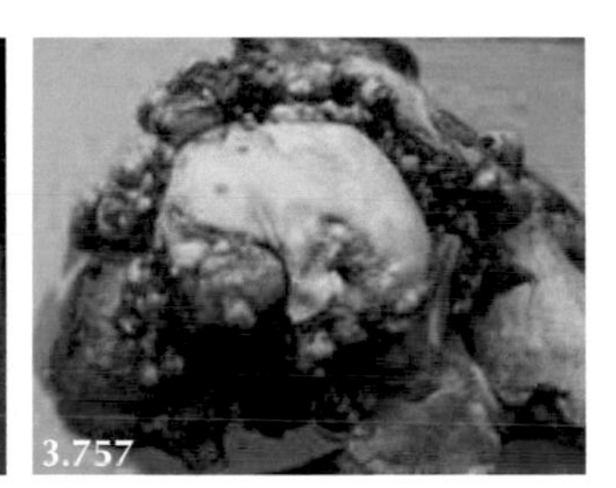

3.755: Extensive brisket and ventral oedema in an aged beef cow. **3.756:** Ultrasound examination reveals fluid within the pleural space and pericardial sac, and a >10 cm diameter multi lobulated mass attached to heart. **3.757:** Necropsy reveals approximately 20 cm diameter heart base tumour.

- **Key observations**: Subcutaneous oedema slowly increasing over several months.
- **Clinical examination**: Inaudible heart sounds.
- **Key history events**: None. 3.395

3.14

Veterinary Investigation of Perinatal Mortality

Lambs

Rather than necropsy dead lambs to investigate the major causes of perinatal lamb losses on a farm, aim to prevent mortality by providing detailed advice on correct ewe nutrition during late pregnancy and provide standard operating procedures for critical tasks in the veterinary flock health plan.

- Be aware that skilled labour will likely be a critical limiting factor when 75–90 per cent of ewes lamb within two weeks.
- Low lamb birthweight and failure of passive antibody transfer are regarded as the most important factors causing lamb losses. Environmental hygiene can be improved on most sheep farms.
- 75 per cent of fetal growth occurs during the last six weeks of pregnancy. Measuring lamb birth-weights from a representative sample is essential information.
- Energy underfeeding during late pregnancy results in low lamb birthweight and poor colostrum supply.
- The mean 3-OH butyrate concentration of multi-gravid ewes determined one month before lambing allows immediate correction of energy underfeeding.
- 50 ml of colostrum per kg birthweight is essential for all lambs in the first 4–6 hours, preferably within 30 minutes. Measured after 24 hours old, the total plasma protein concentration for lambs which have not sucked sufficient colostrum is <45 g/l compared to >60 g/l for lambs which have.
- Abortion is a separate category of fetal loss with recognised pathogens, such as EAE and toxoplasmosis, many of which can be effectively controlled by vaccination.

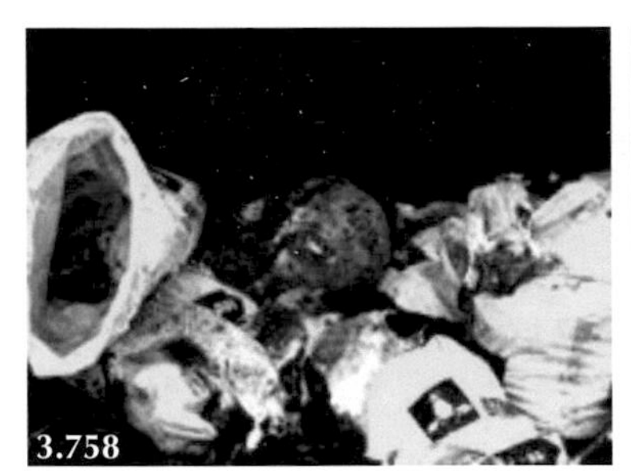

3.758–3.760: There have been few detailed studies of all lamb losses on sheep farms. Random post-mortem examinations are unlikely to provide the complete answer.

DOI: 10.1201/9781003106456-32

The Effects of Nutrition during Pregnancy on Lamb Viability

Second trimester	Placental development. One possible consequence of poor placentation is chronic intra-uterine growth retardation (IUGR)
Third trimester	Fetal growth and development. Poor energy supply results in low lamb birthweight

Placental Development

Weather and/or grazing conditions need to be severe for more than seven to 10 days during the second trimester to seriously impair placental development. This condition is more usually seen in ewes with a chronic disease such as paratuberculosis.

3.761: Weather and feed restriction need to be severe for more than seven to 10 days to seriously impair placental development. **3.762** and **3.763:** Chronic IUGR; twin lamb birthweight less than 3.0 kg when a Greyface ewe should produce lambs around 5.5 kg. 3.396, 3.397, 3.398

Reduced lamb birthweights can also occur when placental development has been limited by competition for caruncles resulting in a reduced number of placentomes per fetus. This situation is thought to occur in multiple litters where the birth of twins with disproportionate weights e.g. 6 kg and 3.5 kg probably indicates that three embryos implanted and underwent early fetal development but one fetus failed to develop further and was resorbed. The limited number of caruncles of the remaining fetus in the ipsilateral horn results in poor placental nutrient supply and a much reduced birthweight (e.g. 3.5 kg) compared to the co-twin which develops without competition in the contralateral horn and had the normal number of caruncles (e.g. 6 kg).

3.764 and **3.765:** Twins with disproportionate weights e.g. 6.0 kg versus 3.5 kg probably indicates that three embryos implanted but one fetus failed to develop further and was resorbed.

Blood Sampling Lowground Flocks during Late Gestation to Determine Nutritional Status

Adequate ewe nutrition during the last six weeks of pregnancy, when 75 per cent of fetal growth occurs, is essential to ensure appropriate lamb birthweights. Dietary energy supply relative to metabolic demands can be accurately determined during late gestation by measuring ewes' serum (or plasma) 3-OH butyrate concentration. Increased 3-OH butyrate values reflect inefficient fatty acid utilisation caused by high glucose demand from the developing fetuses not matched by dietary propionate or glucogenic amino acid supply. The reader is directed to the excellent article written almost 40 years ago by Dr Angus Russel which appeared in *In Practice* (vol 7, pp 23–28; 1985) which describes the interpretation of the serum 3-OH butyrate concentration in relation to dam nutritional requirements. This article is essential reading for all farm animal veterinary surgeons and undergraduate students.

Farm Visit

Ewes due to lamb during the first week should be body condition scored and blood sampled four weeks before the start of lambing time thereby allowing sufficient time to implement dietary changes. An equal number of twin and triplet-bearing ewes should be sampled; there is little benefit in collecting samples from ewes with singletons other than to establish reference values. Details of the diet, forage analyses and future alterations should be noted, and feed allocations checked on weigh scales.

3.766: Feed storage could be improved by using sealed bins (**3.767**). **3.768:** Snacker feeding is now used to feed large numbers of sheep on clean ground every day. Calibration of the dispenser is critical to accurate rationing.

3.769: Be aware than feeding whole grain using a snacker does not work very well—here barley has been trodden into soft ground by the feeding ewes. **3.770** and **3.771:** Grain may attract seagulls, geese and crows which pose a possible abortion risk from fecal contamination of the feeding areas. 🎥 3.399

Results and Interpretation

The target mean 3-OH butyrate concentration is below 1.0 mmol/l; values above 1.6 mmol/l in individual ewes represents severe energy underfeeding with the likelihood of pregnancy toxaemia developing as pregnancy advances and fetal energy requirements increase unless dietary changes are implemented. 3-OH butyrate concentrations greater than 3.0 mmol/l are consistent with a diagnosis of ovine pregnancy toxaemia. Once the mean 3-OH butyrate concentration has been determined, any alteration in the ration can be made with reference to Dr Russel's *In Practice* article.

3.772: Feeding small groups of sheep in troughs has long been abandoned in favour of snacker feeding to save time. **3.773** and **3.774:** Feeding in troughs can present its own hygiene risk.

Evaluation of Protein Status

Blood samples can also be analysed for blood urea nitrogen (BUN) which indicates short-term protein intake, and albumin which reflects longer term protein status. Recent feeding can greatly affect BUN concentration; blood samples should be collected either before concentrate feeding or at least 4 hours later to avoid post-prandial increases. Low BUN concentrations usually indicate a shortage of rumen degradable protein. Serum albumin concentrations fall during the last month of gestation as immunoglobulins are manufactured and accumulate in the udder thus serum albumin concentrations in the region of 26 to 30 g/l are "normal" during the last month of gestation.

Evaluation of Housing Arrangements for Lambing Time

Straw used for bedding sheep must be stored indoors from harvest to lambing time.

3.775 and **3.776:** Exemplary standards of hygiene in the lambing shed during late pregnancy/lambing. Lambing pens should be custom-built (**3.777**); the use of wooden pallets and old gates tied with string simply wastes time. 🎥 3.400, 3.401, 3.402

3.778: Lambing time requirements including stored bovine colostrum. Note that routine hoof paring would not be recommended; the hoof shears are likely to be redundant. A medicine record book should be available. **3.779:** Wearing disposable arm-length gloves should be mandatory when correcting a dystocia. Arm length gloves cost 8 pence each. 🎥 3.403

3.780 and **3.781:** Adequate staffing is essential during the main lambing period (up to 90 per cent lambing within two weeks) because mis-mothering can happen very quickly when sheep are housed in tight groups.

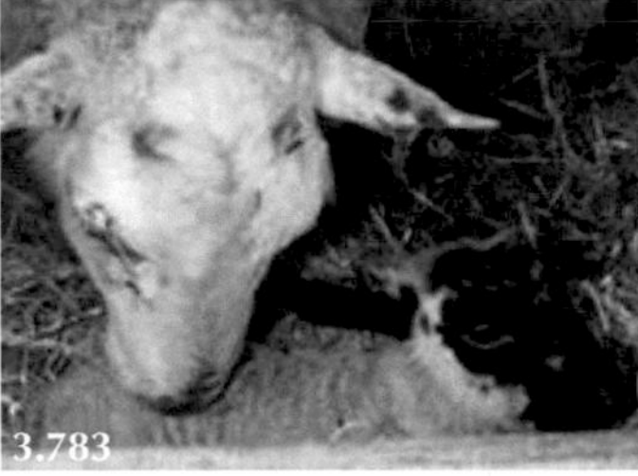

3.782: Pen hygiene is essential to prevent neonatal infections. **3.783:** Confinement of the ewe with her lamb in individual pens for 24 hours ensures the development of a strong maternal bond. **3.784:** Provision of fresh water and roughage, with concentrates as appropriate, are essential.

3.785 and **3.786:** Poor pen hygiene—the pens are wet and dirty with no fresh straw. There is no food or fresh water for the ewes. Shepherds will tell you that well-fed sheep make better mothers. 3.404

Colostrum Ingestion

The total plasma protein concentration for lambs which have not sucked sufficient colostrum is <45 g/l compared to >60 g/l for lambs which have.

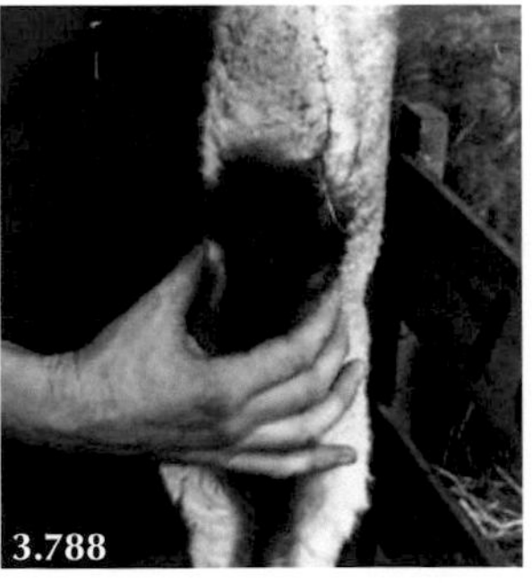

3.787: Inspection would suggest that this lamb has not sucked colostrum. **3.788:** After sucking, the lamb's full abomasum drops below the costal arch and is easily seen or palpated when the lamb is held by both forelegs. 3.405, 3.406

3.789–3.791: When lambs have not sucked, supplementary colostrum feeding—whether stripped from the dam, another ewe or bovine colostrum pooled from several cows—is essential within the first 4–6 hours after birth.

Navel Dressing

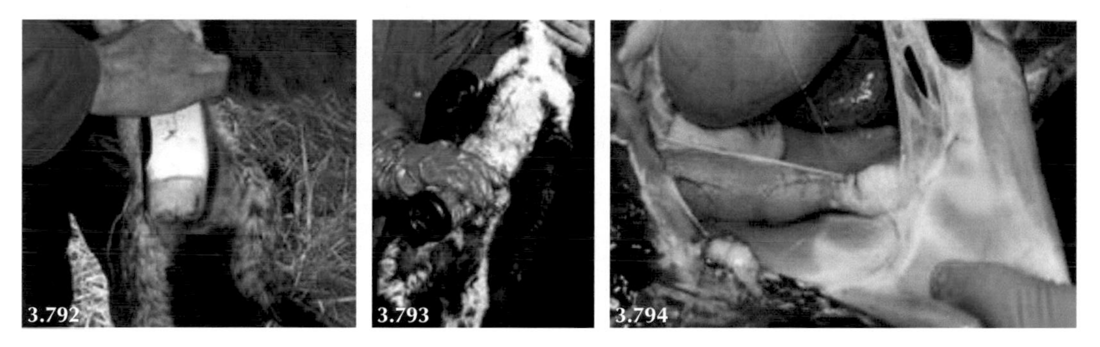

3.792 and **3.793:** Navel dressing with strong veterinary iodine BP is an essential routine procedure to prevent localised umbilical infections. **3.794:** Umbilical abscess extending via the urachus to the bladder. However, bacteraemia largely originates via the gut or tonsil.

The use of prophylactic oral antibiotics within 15 minutes of birth to prevent/control watery mouth disease is inconsistent with responsible use of antibiotics and is not recommended unless a problem arises and animal welfare is compromised.

Watery Mouth Disease

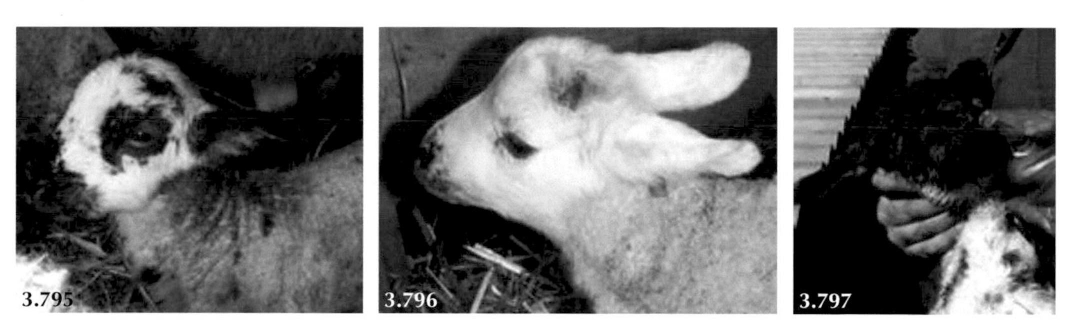

3.795 and **3.796:** Excess salivation during the early stages of watery mouth. **3.797:** Oral antibiotic prophylaxis.

3.798: Exemplary hygiene standards for housed ewes and their new-born lambs. **3.799** and **3.800:** Turn-out to pasture when lambs are 36–48 hours old. 🎥 3.407, 3.408

Lambing Outdoors

3.801: Lambing outdoors has many benefits but requires good weather and excellent shepherding.
3.802: Note that mis-mothering can also occur in outdoor lambing systems. 🎥 3.409

Triplets

3.803: It is possible for ewes to nurse triplets, but they require good grazing and generous concentrate allowance including a creep feed for the lambs. Despite all efforts, mastitis is much more likely in ewes rearing triplets.

Foster Crates

There are various methods used to foster triplet lambs onto ewes that have a singleton or have lost one or more lambs. Foster crates are popular but are high maintenance. Removal of the dead lamb's skin and transferring it onto the foster lamb is another commonly used method for fostering.

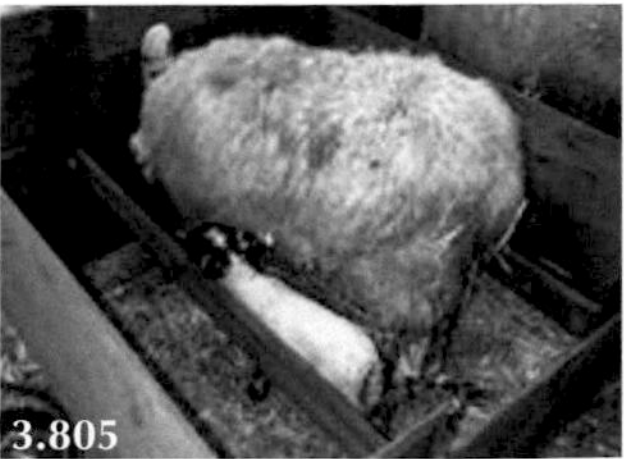

3.804: No food and fresh water for these ewes in foster crates mean that they will not lactate well and are less inclined to accept fostered lambs. **3.805:** The lamb's view of the foster crate. **3.806:** This image likely reflects the management on this farm whereby the foster lamb has also died (the foster lamb's skin is evident at the base of the other lamb's neck). 3.410

Orphan Lambs

All sheep farms have orphan lambs and the rearing system must be organised well before lambing commences. An automatic feeder provides excellent results, albeit financially costly, because milk replacer is very expensive.

3.807: A fully automated milk bar for orphan lambs. **3.808:** Alternatively, buckets with multiple teats can be used but bloat is a constant risk (site the teats at the correct height—at lamb shoulder height). 3.411, 3.412

Calves

There are many definitions of perinatal calf mortality; by choosing birth to 24 hours old this period largely excludes infections acquired in the immediate post-partum period and factors determining adequate passive antibody transfer.

- Perinatal dairy calf mortality (birth to 24 hours old) is estimated to be 6–7 per cent.
- Approximately 75 per cent of perinatal dairy calf mortality occurs within 1 hour of calving.
- 90 per cent of calves which die in the perinatal period were alive at the start of calving.
- The majority of perinatal mortality has been attributed directly to difficult calving particularly in heifers.
- Increased duration of second stage calving beyond 2 hours, poor abdominal contractions, use of mechanical calf pullers and changes in the calving supervision all increase significantly the risk of perinatal mortality.
- Micro-nutrients (iodine, selenium, copper and zinc) deficiencies have been associated with high stillbirth rates but results from randomised clinical trials have not always supported a causal relationship.
- The causes of calf loss, as opposed to the risk factors associated with such losses, are best diagnosed from necropsy examination.

Major causes of bovine perinatal mortality in necropsy studies are dystocia (evidenced by traumatic lesions such as fractures, 35%), anoxia (necropsy findings of pulmonary atelectasis and sub-serosal haemorrhages, 30%), infections (5%) and congenital defects (approximately 5%); 25% of deaths have no diagnosed cause.

Dystocia

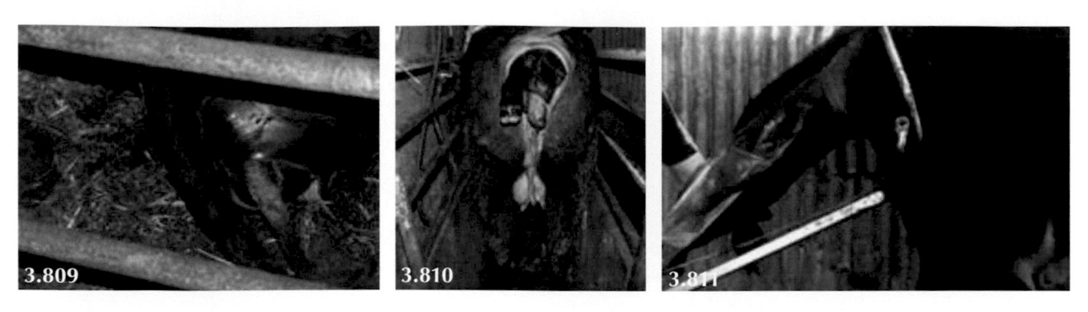

3.809: Breech presentation. **3.810:** Unilateral shoulder flexion (leg back). **3.811:** Absolute fetal oversize.

3.812: Excessive traction is difficult to define afterwards. **3.813** and **3.814:** Excessive traction and prolonged dystocia are management causes of calf death.

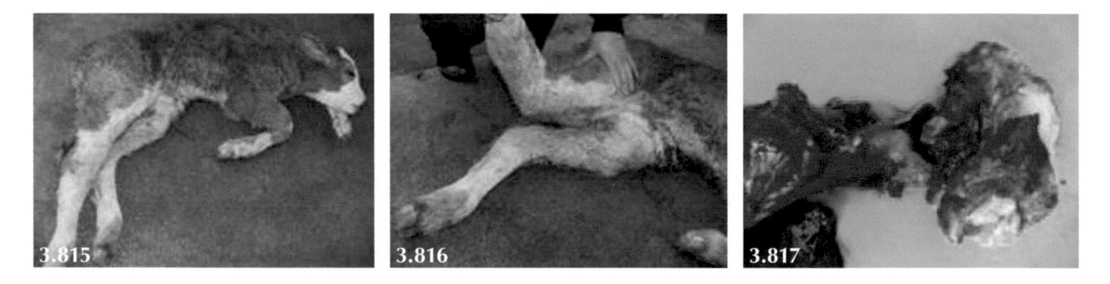

3.815: Calf died during assisted delivery by farmer. **3.816:** Proximal tibial fracture. **3.817:** Tibial fracture confirmed at necropsy.

- **Key observations**: Malposture/malpresentation, absolute fetal oversize.
- **Clinical examination**: Fractures of limbs, ribs and/or vertebral column.
- **Key history events**: Prolonged forced extraction.

Anoxia

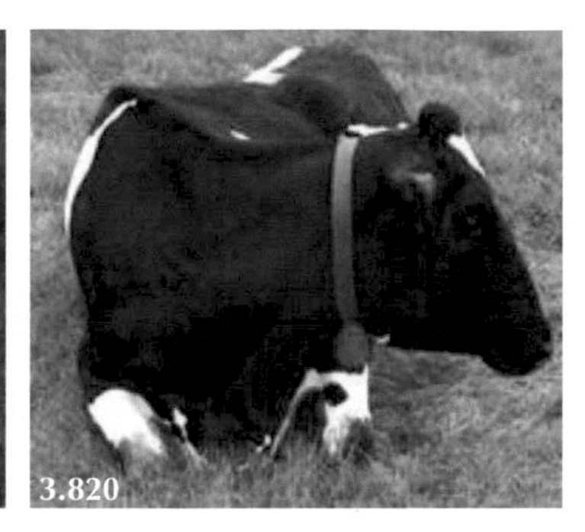

3.818: Early first stage labour—cow requires close supervision over next 2–4 hours. **3.819** and **3.820:** Hypocalcaemia, parturition does not progress.

- **Key observations**: Cow often presented with twins and/or hypocalcaemia.
- **Clinical examination**: Necropsy findings of pulmonary atelectasis, sub-serosal haemorrhages and meconium-staining but rarely undertaken.
- **Key history events**: Prolonged second stage labour.

Septicaemia

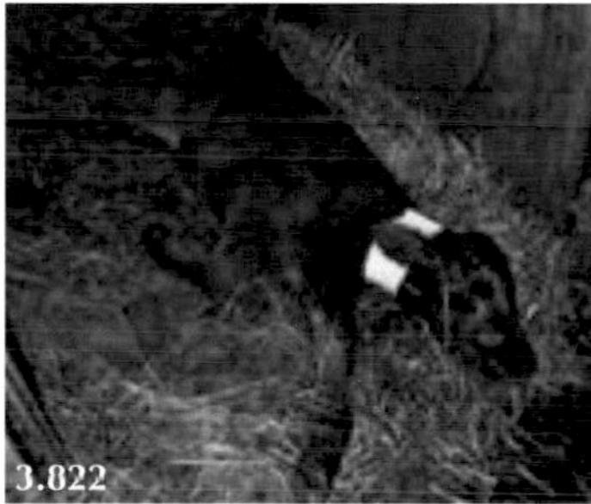

3.821 and **3.822:** Low birthweight calves showing early seizure activity/opisthotonos.

- **Key observations**: Low birthweight, poorly fleshed, cow likely to have retained placenta.
- **Clinical examination**: Injected scleral vessels, not always febrile, rapidly progress to seizures/meningitis.
- **Key history events**: Slow to stand, failure of passive antibody transfer.

Congenital Defects

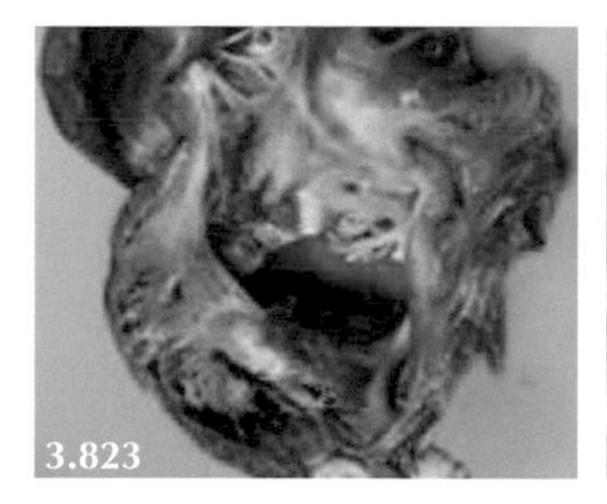
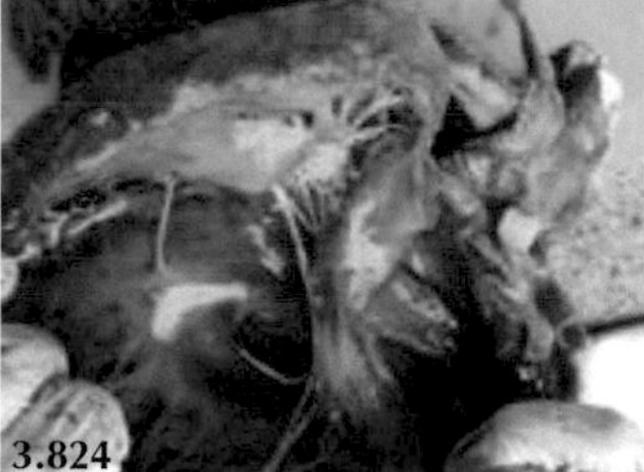

3.823: Ventricular septal defect. **3.824:** Tetralogy of Fallot. **3.825:** Distended intestines as a consequence of atresia coli.

- **Key observations**: Calf with atresia coli may appear normal for the first 36 hours.
- **Clinical examination**: Audible murmur/palpable thrill depend upon size of cardiac defect.
- **Key history events**: May appear as small cluster if genetic effect.

3.15

Poor Growth Rate

Lambs

- Be aware of the farmer's aims regarding profitable lamb production on his/her farm.
- Lambs can be marketed at 38–40 kg liveweight from 8 weeks to 12 months old.
- Financial impact = percentage of lamb crop affected × intensity × duration.
- Preventive measures should be detailed in the veterinary flock health plan.

Some farmers target early lamb markets which necessitate rapid (>400g/day) growth rate often achieved using creep feeding while other farmers sell store lambs produced at minimal cost soon after weaning; it is essential to discuss your client's ambitions for his/her flock's performance. It could therefore be reasoned that lambs reaching slaughter weight beyond 4 months old have either been underfed or suffered from production-limiting disease(s).

The diseases causing poor lamb growth rates are ranked next but the relative importance will vary greatly between flocks depending upon breed, management/targets, soil type (trace element deficiencies), location (grass growing conditions) etc. Conditions affecting the whole lamb crop even to a small degree will have a much greater financial impact than several sick/dead lambs.

The intensity × duration (severity) are critical factors affecting growth rate. Timely correction will reduce the overall disease impact.

- Lambs reaching slaughter weight beyond 4 months old have either been grossly underfed or suffered parasitic infestation/chronic disease.
- PGE is the most common cause of sub-optimal growth; SCOPS policy/advice is readily available on line but not universally adopted.
- It will be many years before farmers fully engage targeted selected anthelmintic treatment and automatically calibrated dosing guns.
- There are a wide range of trace element supplements (copper, selenium and cobalt) including drenches, boluses, long-acting injections etc.

DOI: 10.1201/9781003106456-33

TABLE 3.92

Common Causes of Poor Growth Rate in Lambs

	Age/Season	Growth rate	Appetite	Feces
PGE (Mixed helminth species)	>2 months	–/––	NAD	Diarrhoea
Cobalt deficiency	>4 months	–/––	–	NAD
Liver fluke	September onwards	–/––	–/––	NAD
Inadequate grazing post-weaning	>5 months	–	NAD	NAD
Poor dam milk supply/inadequate grazing	1–5 months	–	NAD	NAD
PGE (Trichostrongylosis)	>7–8 months, late autumn	–/––	NAD	Diarrhoea
PGE (Nematodirosis)	1–3 months	–/–––	–	Diarrhoea
Sheep scab	Winter	–/–––	–/––	NAD
Selenium deficiency	>4 months	–/––	–/––	NAD
Coccidiosis	>4 weeks	–/––	NAD	Mucoid diarrhoea
Scald	Any age	–		NAD
Myiasis/head fly	June–September	–/––	–/––	NAD
Severe joint lesion	Any age	––/–––	––/–––	NAD

TABLE 3.93

Ancillary Tests to Determine Cause(s) of Poor Growth Rate in Lambs

Disease	Ancillary test
PGE (Mixed species)	FWEC >400–1,000 epg although larval stages can cause diarrhoea before patent infestation develops
Cobalt deficiency	Serum and liver vitamin B12 concentrations
Liver fluke	Serum GLDH, GGT concentrations. Coproantigen ELISA, fecal egg count (patent)
Inadequate grazing post-weaning	Check pasture sward/catch crops such as kale etc
Poor dam milk supply/inadequate grazing	Check for mastitis, udder fibrosis/induration due to MVV
PGE (Trichostrongylosis)	Usually very high FWECs
PGE (Nematodirosis)	FWEC although larval stages can cause diarrhoea before patent infestation develops
Sheep scab	Skin scraping. Serology
Selenium deficiency	Plasma glutathione peroxidase below 20 IU/L packed red blood cells at 30°C
Coccidiosis	Fecal oocyst count and speciation
Scald	Clinical examination
Myiasis/head fly	Clinical examination
Severe joint lesion	Clinical examination. Radiography

PGE (Mixed Species)

3.826: Chronic diarrhoea with fecal contamination of the tail and hindquarters in ten-week-old lambs, and weaned lambs (**3.827** and **3.828**).

- **Key observations**: Chronic diarrhoea with fecal contamination of the tail and hindquarters.
- **Clinical examination**: Check for myiasis. FWEC >400–1,000 generally considered significant and worthy of anthelmintic treatment.
- **Key history events**: Larval challenge continues from contaminated "dirty" pasture despite anthelmintic treatment. 🎥 3.413, 3.414, 3.415

3.829 and **3.830:** "Cell grazing" with 1–2 hectare plots grazed intensively for one to two days, then rotated, can work very well.

Cobalt Deficiency

3.831–3.833: Weaned lambs dull with reduced appetite/abdominal fill, low condition scores and poor fleece quality.

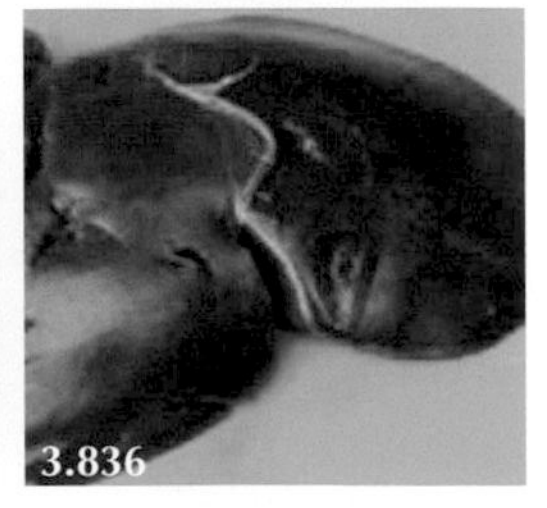

3.834 and **3.835:** Very lethargic weaned lambs with poor abdominal fill, very low condition scores and poor fleece quality. **3.836:** Pale, fatty and friable liver (white liver disease) in severe cobalt deficiency.

- **Key observations**: Very poor growth rates affecting weaned lambs despite adequate grazing.
- **Clinical examination**: Lethargic, weak, often have serous ocular discharges. Pale, fatty and friable liver (white liver disease) in severely affected lambs.
- **Key history events**: Specific geological areas. Deficiency problems occur every year unless supplemented. 🎥 4.416, 3.417

Liver Fluke

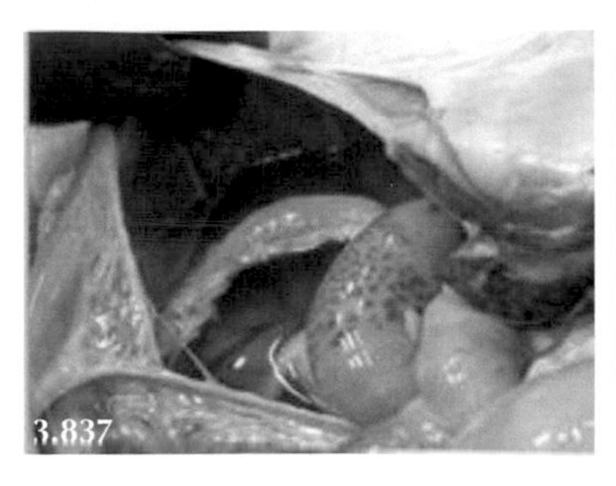

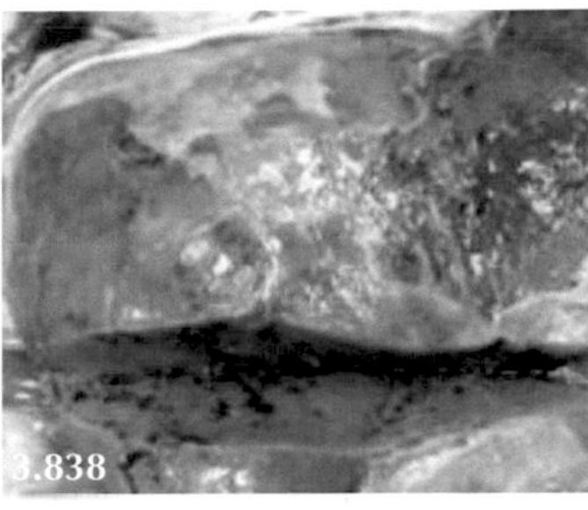

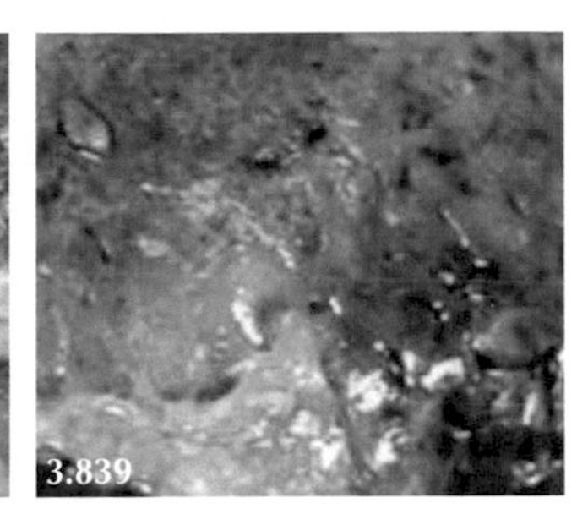

3.837: Large volumes of peritoneal exudate at necropsy. **3.838:** Migrating stages cause haemorrhagic tracts throughout liver parenchyma. **3.839:** Flukes can be measured to determine age and thereby when (and where) metacercaria were ingested.

- **Key observations**: Rapid weight loss, difficult to gather/exercise intolerance.
- **Clinical examination**: Reduced grazing, low condition scores, anaemia, abdominal pain.
- **Key history events**: Failure to administer correct preventive flukicides, purchased infested stock with no quarantine drenching. Triclabendazole resistance. 🎥 3.418

Inadequate Grazing Post-Weaning

3.840 and **3.841:** Poor grazing compounding PGE. **3.842:** High stocking density but with abundant grazing on 6–8 cm long silage aftermath sward.

- **Key observations**: Overstocked/very short grass. Heavy fecal contamination. Grazed all year round by sheep.
- **Clinical examination**: Poor grazing compounding PGE.
- **Key history events**: Poor grass growth/delayed marketing of lambs.

Poor Dam Milk Supply

3.843 and **3.844:** Four-month-old lambs which are poorly grown compared to (**3.845**) excellent lambs with clean tight fleece and good body condition. Note that both dams are fat so unlikely to be ewe's diet.

- **Key observations**: Low bodyweight, poor quality open fleece, often chronic diarrhoea.
- **Clinical examination**: NAD except much poorer bodyweight and condition than peers.
- **Key history events**: Same management only ewe has poor milk yield. 3.419

PGE (Trichostrongylosis)

3.846–3.848: Diarrhoea, often foetid, during late autumn affecting high percentage of group.

- **Key observations**: Foetid diarrhoea.
- **Clinical examination**: Often very high FWECs.
- **Key history events**: Grazing permanent pasture during late autumn. No recent anthelmintic treatment. 3.420, 3.421

PGE (Nematodirosis)

3.849–3.851: Profuse diarrhoea affecting six-week-old lambs; the ewes are not affected.

- **Key observations**: Profuse diarrhoea affecting large percentage of lambs, sudden deaths may occur.
- **Clinical examination**: Disease may be caused by larval stages; FWEC may be zero.
- **Key history events**: On pasture grazed by lambs the previous year. 3.422, 3.423, 3.424

Sheep Scab

3.852–3.854: Neglect of fattening lambs with severe chronic (>8–10 weeks' duration) sheep scab lesions.

- **Key observations**: Intense pruritus, fleece loss, rapid weight loss.
- **Clinical examination**: Excoriated and corrugated skin, serum exudation. Take numerous skin scrapings at periphery of fleece loss.
- **Key history events**: Purchased/feral sheep, common grazing areas introduce infection. Failure to implement routine preventive measures. 🎥 3.425, 3.426

Selenium Deficiency

3.855–3.857: Poorly grown lambs. Nutrition, parasites and trace elements may all play a role, and interpreting the most significant cause is often very difficult.

- **Key observations**: Poor growth, low body condition scores.
- **Clinical examination**: Compounded by other trace element deficiency states and PGE.
- **Key history events**: Selenium-responsive poor growth in lambs has been reported in certain geographic areas.

Coccidiosis

3.858: Poor condition, empty abdomen and tenesmus. **3.859:** Diarrhoea with fresh blood clots. **3.860:** Rapid condition loss.

- **Key observations**: Condition loss, tenesmus, diarrhoea with fresh blood.
- **Clinical examination**: Fecal oocyst counts very variable and not diagnostic without speciation.
- **Key history events**: Intensive (housed) rearing. Contaminated feeding/creep areas. No coccidiostat in feed/preventive treatment.

Scald

3.861 and **3.862:** Marked lameness with lambs grazing on their knees caused by interdigital dermatitis. **3.863:** Interdigital dermatitis showing diphtheritic membrane covering interdigital skin.

- **Key observations**: Sudden onset, marked lameness.
- **Clinical examination**: Diphtheritic membrane covering interdigital skin.
- **Key history events**: Rapid spread in lambs causing high prevalence of lameness. 3.427

Myiasis/Head Fly

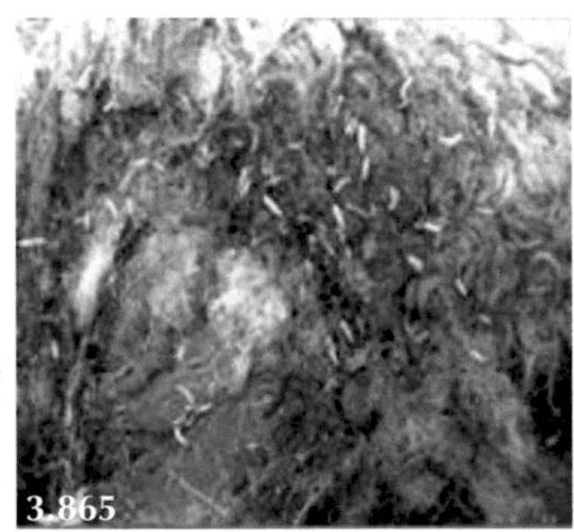

3.864: Lamb isolated from group, not grazing, blackened dry skin around tailhead, fecal contamination of tail. **3.865:** Blackened dry skin after moist fleece peeled back, maggots. **3.866:** Damaged skin grows back black wool (lamb shown three months after myiasis).

- **Key observations**: Isolated from group, not grazing, fecal contamination of fleece, attempting to nibble at lesions usually around the tailhead.
- **Clinical examination**: Foul-smelling, moist fleece, blackened dry skin, maggots.
- **Key history events**: Fecal contamination of fleece and dermatophilosis lesions attract blowflies. Lack of preventive measures. 3.428

Severe Joint Lesion

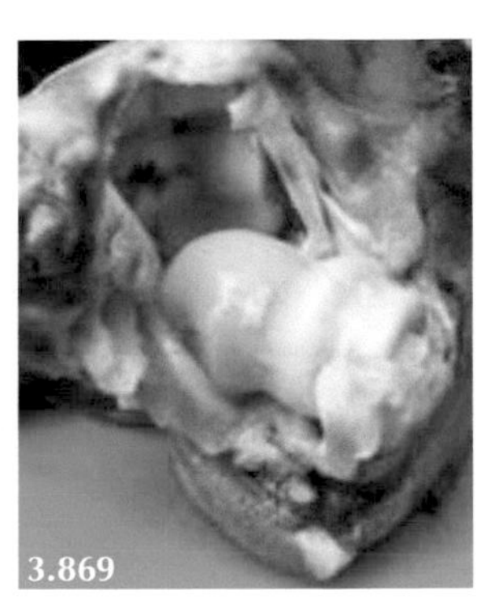

3.867: Traumatic right stifle injury causing reduced grazing/poor rumen fill. **3.868:** Lameness caused by septic elbow. **3.869:** Necropsy of lamb (left) reveals pannus and erosion of articular cartilage.

- **Key observations**: Poor fleece, low condition score, much poorer weight gain than peers.
- **Clinical examination**: Enlarged drainage lymph node, thickened fibrous joint capsule.
- **Key history events**: Chronic joint infection persisting from neonatal polyarthritis, erysipelas or a traumatic event. 3.429

Calves

- Trace element deficiencies are not commonly reported in growing cattle in the UK. Supplementary feeding is much more common in cattle than lambs.
- Outbreaks of lungworm, type I ostertagiosis and fasciolosis have severe adverse effects on growth.

Management of growing cattle for beef production has changed from "store" rations where growth was stalled during the more expensive winter feeding period until the following spring grazing when "compensatory" growth resulted. Now, many male beef calves are kept entire and reared as bulls, and an intensive cereal ration is the most common method to rear "bull beef" from the dairy industry. Dairy heifers have target growth rates for all critical stages to first calving with regular liveweight measurements. Poor growth in dairy and beef calves is most likely to occur either due to failures of routine parasite control allowing PGE, lungworm and possibly liver fluke to develop or a lack of grazing during hot summer weather in certain areas of the country when supplementary feeding will be necessary. Unlike lambs, trace element deficiencies are not commonly reported in growing cattle in the UK. Relative to the group, individual disease problems, such as chronic pneumonia, do not impact upon farm profits.

3.870: Intensively reared entire bull calves from a beef herd. **3.871:** Intensive "barley beef" production of Holstein bulls. **3.872:** Maintaining growth rates at pasture by feeding supplementary concentrates.

TABLE 3.94

Common Causes of Poor Growth Rate in Calves

	Age/Season	Growth rate	Feces
Lungworm	Late summer/autumn	–/–––	NAD
PGE	Late summer/autumn	–/––	Diarrhoea
Inadequate grazing/feeding	Summer months	–	NAD
Chronic pneumonia/BVD PI	>4 weeks	–/––	NAD
Coccidiosis	>4 weeks	–/––	Mucoid diarrhoea
Liver fluke	September onwards	–/––	NAD
Bloat		–	NAD
Liver abscesses	Intensive ration	–/––	NAD

Lungworm *(Dictyocaulus viviparus)*

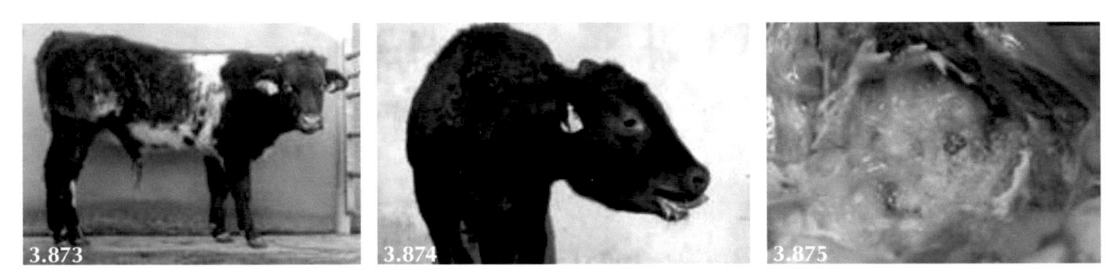

3.873 and **3.874:** Frequent coughing and increased respiratory rate in beef calves. **3.875:** Large numbers of mature lungworm in the large airways.

- **Key observations**: Sudden onset frequent coughing in large percentage of group. Mouth breathing in severely affected calves.
- **Clinical examination**: Adventitious lung sounds over whole lung field. Possible subcutaneous emphysema.
- **Key history events**: Non-vaccinated calves grazing pasture grazed by growing infested cattle last year or earlier this season. No preventive anthelmintic programme.

PGE (Ostertagiosis Type I)

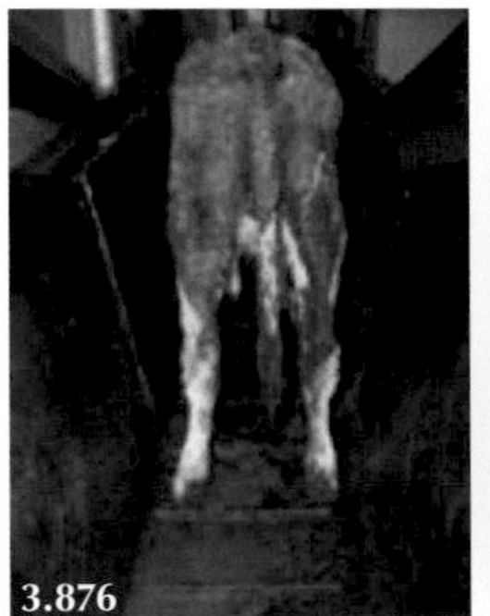

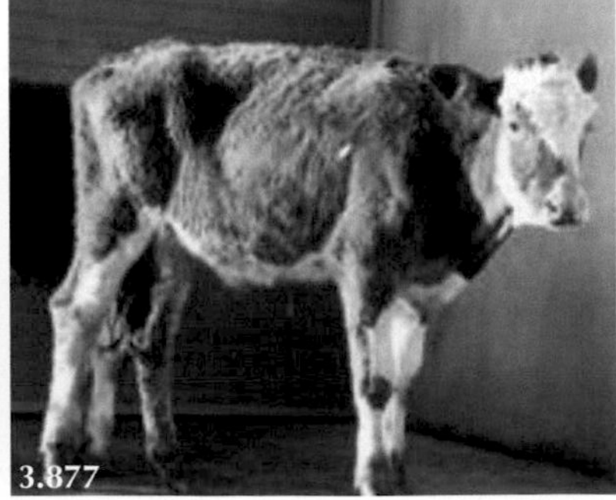

3.876 and **3.877:** Profuse diarrhoea and severe weight loss over several weeks caused by ostertagiosis type I. **3.878:** Profuse diarrhoea at pasture in September.

- **Key observations**: Profuse diarrhoea with rapid weight loss affecting a large percentage of calves in group.

- **Clinical examination**: NAD other than profuse diarrhoea.
- **Key history events**: Grazing contaminated pasture without preventive parasite control measures.

Inadequate Grazing/Feeding

3.879 and **3.880:** Poor grass growth during hot dry weather can impact on liveweight gain.

- **Key observations**: Poor liveweight gain at pasture.
- **Clinical examination**: Often NAD.
- **Key history events**: Insufficient grass growth due to very dry/hot weather necessitating supplementary feeding.

Chronic Pneumonia

3.881–3.883: Single BVD PI calf on left side of all images as viewed.

- **Key observations**: Much reduced bodyweight compared to peer group.
- **Clinical examination**: Often signs of chronic respiratory disease identified ultrasonographically.
- **Key history events**: Often have experienced several episodes of respiratory disease. Often persistently infected BVD animals. 🎥 3.430, 3.431

Coccidiosis

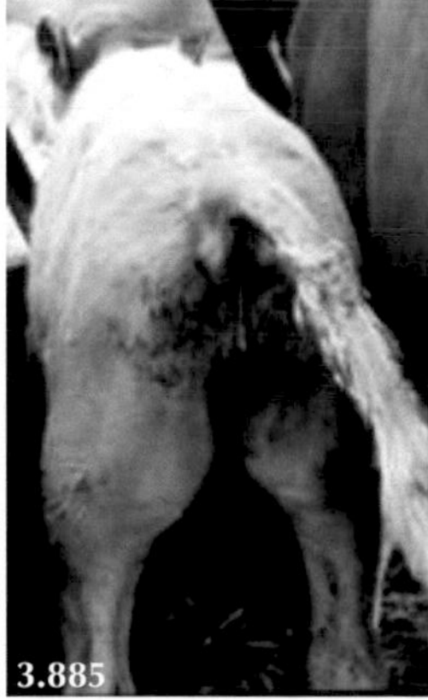

3.884: Contaminated surface water. **3.885:** Tenesmus with cocked tailhead, mucoid diarrhoea containing small blood clots. **3.886:** Gaunt appearance, fecal staining of hindquarters, tenesmus.

- **Key observations**: Rapid weight loss, gaunt appearance.
- **Clinical examination**: Rapid dehydration/weight loss. Frequent tenesmus, mucoid diarrhoea.
- **Key history events**: Often young beef calves at grass with contaminated surface water as only supply.

Liver Fluke

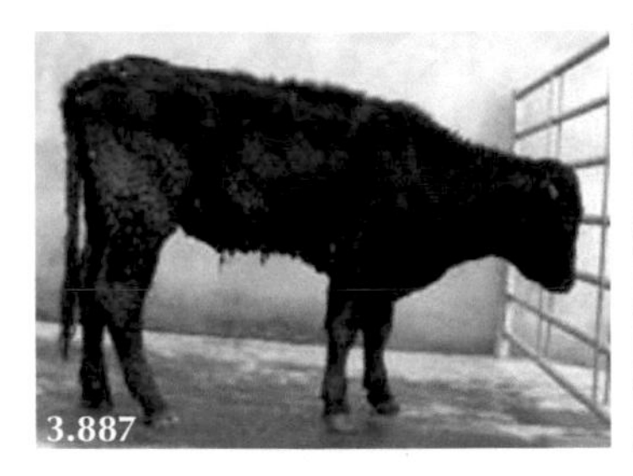

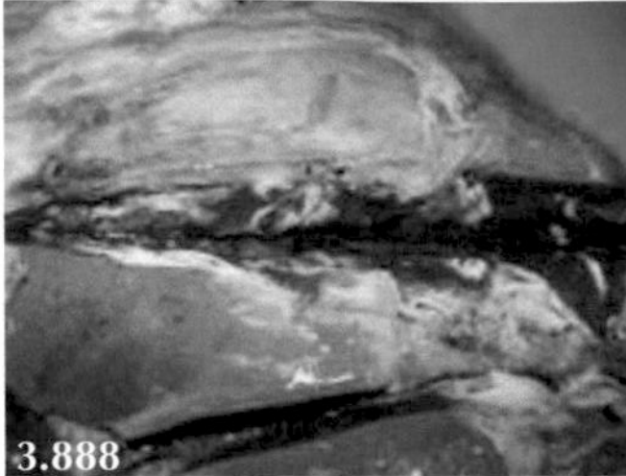

3.887: Poor growth in a group of fifteen-month-old steers. **3.888** and **3.889:** Fibrosis of bile ducts containing mature flukes.

- **Key observations**: Poor liveweight gain/loss of body condition over several weeks.
- **Clinical examination**: Poor condition.
- **Key history events**: Grazing infested (wet) pastures from September onwards. 🎥 3.432

Chronic Bloat

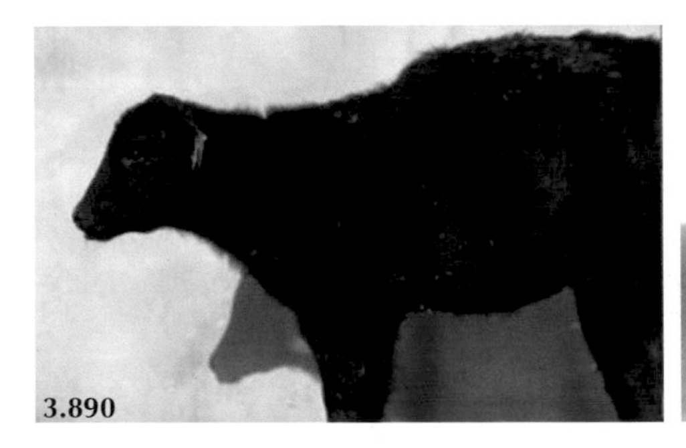

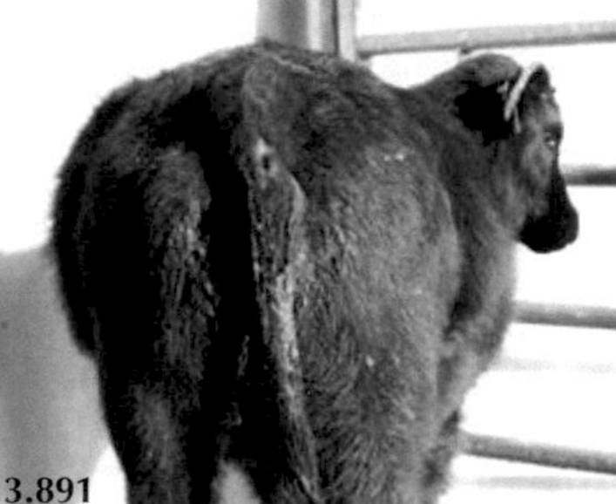

3.890 and **3.891:** Free gas bloat and very poor body condition.

- **Key observations**: Recurrent bloat often over several weeks, poor body condition.
- **Clinical examination**: Often no specific cause.
- **Key history events**: None. 🎥 3.433

3.16

Poor Conception/Pregnancy Rate in Sheep

- Investigating the cause of poor pregnancy rates within a sheep flock several months after the breeding season presents many problems not least that records are poor and rams are moved between groups during the breeding season.
- Breeding soundness examinations using electro-ejaculation have their limitations.
- Liver fluke is a major cause of poor pregnancy rate. Ask para-professionals about fetal death and resorption at ultrasound pregnancy scanning. Assess farm's fluke management programme especially during "high risk" year.

TABLE 3.95

Common Causes of Poor Conception/Pregnancy Rates in Sheep

Cause	Findings
Severe lameness affecting ram	Ram lying down for long periods. Rapid condition loss. Reduced number/no more ewes keeled after onset of lameness
Previous illness affecting ram	History of illness in previous two months. High percentage of ewes return to service at 17 days
Adverse weather	Ewes not mated during adverse weather, greater problem in extensive hill systems
Testicles	High percentage of ewes return to service at seventeen-day intervals. Small scrotal circumference, soft testicles. Immature ram lamb
Epididymitis	Enlarged very firm unilateral scrotal swelling, unable to distinguish testicle and tail of epididymis
Orchitis	Testicles not moveable in the scrotum caused by adhesions
Fasciolosis	Failure of quarantine treatment of ram, incorrect treatment of ewes in high-risk year. Reports of fetal death and resorption at ultrasound pregnancy scanning
Bluetongue (not UK at present)	Infertility, abortion, mummification, stillbirths and congenital anomalies depending when contracting infection
Brucella ovis (not UK)	Multiple returns to service. Very high barren rate

Ancillary Tests

TABLE 3.96

Ancillary Tests to Determine Common Causes of Poor Conception/Pregnancy Rates in Sheep

Disease	Ancillary tests
Severe lameness	Clinical examination
Previous illness/fever	History of treatments/veterinary examinations. Semen examination after A/V collection
Adverse weather	Meteorological data although weather events may be localised
Testicles	Determine maximum scrotal circumference (<32 cm). Semen examination after A/V collection
Epididymitis	Ultrasonography. Culture of ejaculate
Orchitis	Ultrasonography
Fasciolosis	Coproantigen ELISA as often caused by subacute fasciolosis (immature flukes)
Bluetongue	Serology, virus isolation
Brucella ovis (not UK)	Serology. White cells in ejaculate and sperm defects (both non-specific)

DOI: 10.1201/9781003106456-34

The relative importance of the conditions listed will vary between farms. In Scotland, in high-risk liver fluke years such as 2020, subacute fasciolosis was a common cause of fetal resorption reported by paraprofessional sheep pregnancy scanners. Many farmers have their veterinary surgeon undertake a prebreeding check of rams when all scrotal, testicular and epididymal physical abnormalities should be detected. Bluetongue is included low down in the table because the virus has not significantly affected UK flocks since 2007 but this could change as the virus is widespread in France.

Further veterinary investigation of the rams may include ultrasound examination of palpable abnormalities and semen collection by electroejaculation. Semen collection and examination of all rams was advocated for many years by Sheep Vet Soc, but in a ram which has no palpable abnormality this risks condemning a ram purely on poor collection technique, sample processing and incomplete microscopic examination.

Lameness

3.892: Extended periods in sternal recumbency is unusual behaviour when there are ewes in oestrus. **3.893** and **3.894:** Severe hindleg lameness may well prevent service.

- **Key observations**: Extended periods in sternal recumbency. Severe lameness.
- **Clinical examination**: White line abscesses are common. Fighting injury from another ram in group may also cause lameness, especially to stifle injury.
- **Key history events**: Foot lesion may be a chronic problem such as toe fibroma. 3.434, 3.435, 3.436, 3.437, 3.438, 3.439, 3.440, 3.441, 3.442, 3.443, 3.444

Previous Illness

3.895: Laryngeal chondritis affecting a Suffolk ram. **3.896** and **3.897:** Rams may have been ill and treated several weeks before sale.

- **Key observations**: Laryngeal chondritis is common in certain breeds necessitating treatment.
- **Clinical examination**: Stertor caused by laryngeal oedema.
- **Key history events**: Treatment may include a corticosteroid which could affect fertility (much debated). Other infections occurring before sale could cause a fever affecting spermatogenesis. 3.445

Adverse Weather

3.898 and **3.899:** The effects of adverse weather on conception rates are not known. **3.900:** Ewes cannot be mated if they are separated from the ram by bad weather.

- **Clinical examination**: Sheep may be in lower body condition score as consequence of poor weather.
- **Key history events**: Prolonged periods of deep snow but also gales with heavy rain are thought to reduce pregnancy rates (largely speculative). 🎥 3.446

Testicles

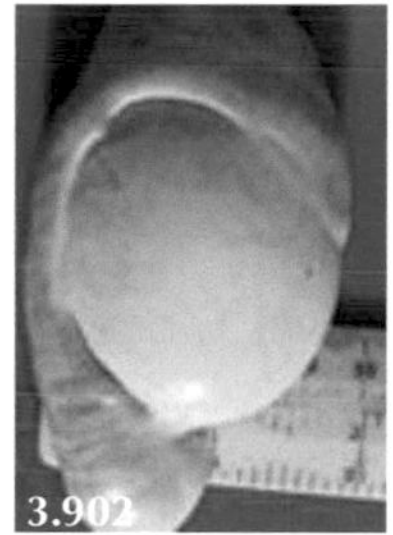
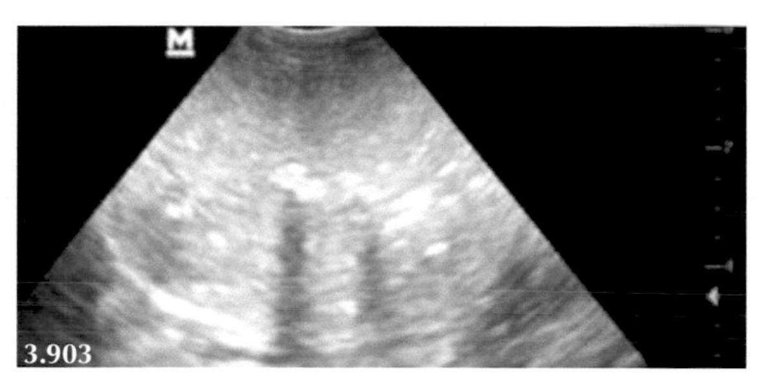

3.901: Small testicles are more common in immature ram lambs. **3.902:** Testicle less than 5 cm diameter. **3.903:** Ultrasound examination reveals abnormal heterogenous testicle with shadowing consistent with atrophy.

- **Key observations**: There are few ram sales with pre-sale veterinary examination.
- **Clinical examination**: Ultrasound examination. Semen collection using A/V and examination.
- **Key history events**: Vendors will check rams before sales, but some rams that palpate normally prove sub-fertile. Some conditions may develop/deteriorate after sale.

Orchitis/Epididymitis

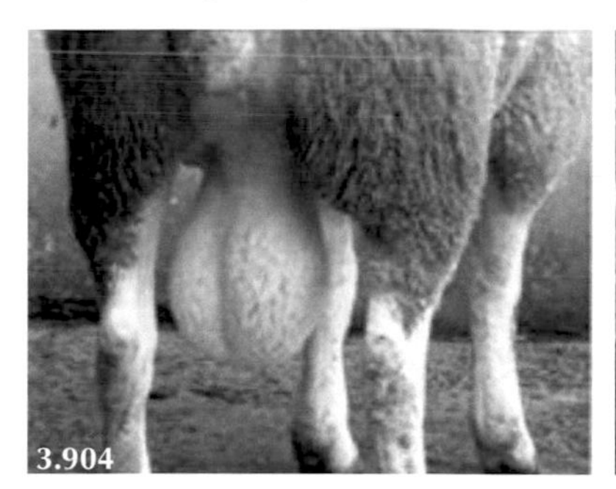
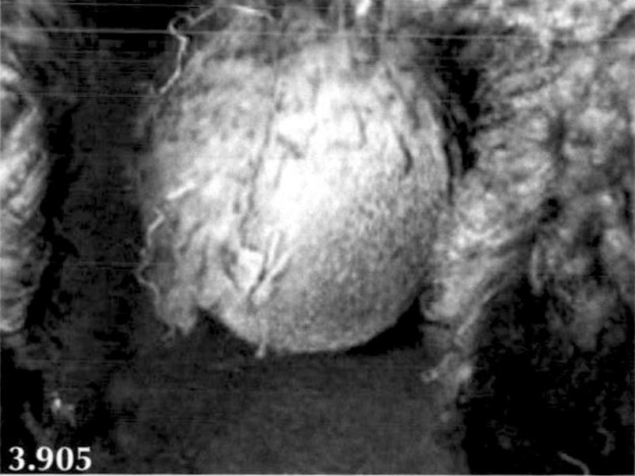
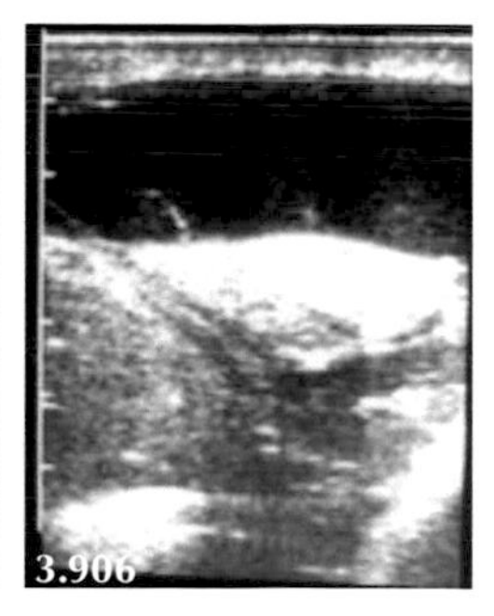

3.904: Swollen scrotum with testicles adherent to scrotum. **3.905:** Unilateral scrotal swelling (right testicle) with wool loss. **3.906:** 2 cm deep fibrinous exudate within the vaginal tunics in this ram with orchitis.

- **Key observations**: Acute testicular/epididymal lesions are painful and often cause hindleg lameness.
- **Clinical examination**: Ultrasound examination readily identifies inflammatory exudate/epididymal lesions.
- **Key history events**: Palpable changes in scrotum/testicles can occur after purchase.

Fasciolosis

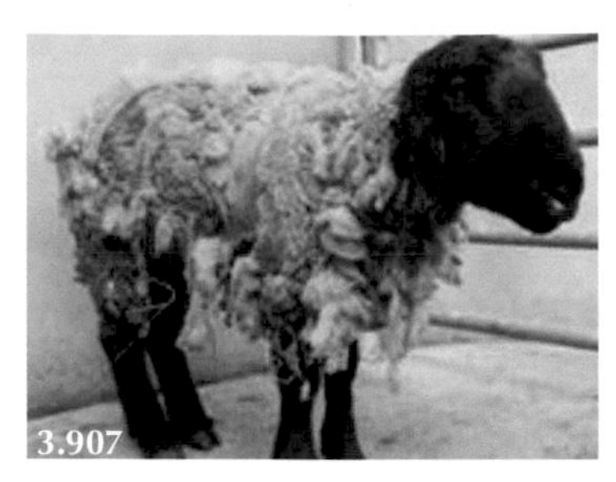

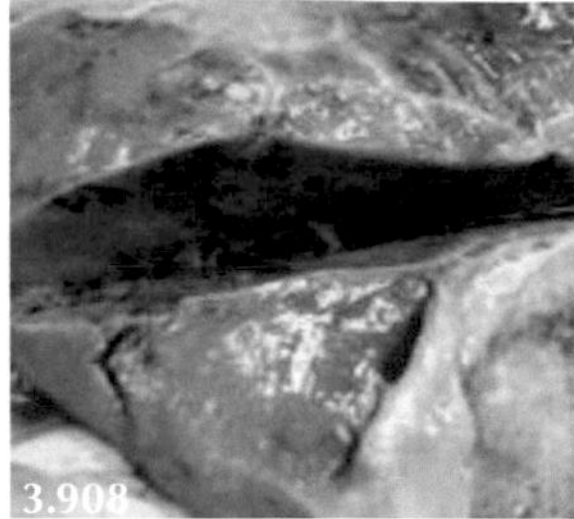

3.907 and **3.908:** Failure of quarantine treatment of ram. Liver affected by severe subacute fasciolosis. **3.909:** Fetal death (lower) and eventual resorption which would have caused lower litter size.

- **Key observations**: High barren rate, lower litter size, rapid loss of ram/ewe body condition. Possible ewe deaths.
- **Clinical examination**: Low condition scores, likely anaemia.
- **Key history events**: Failure of quarantine treatment for purchased sheep. Incorrect flukicide and/or timing. Failure to take account of high-risk year. 3.447

3.17

Common Causes of Tachypnoea and/or Coughing

Adult Sheep

- Chest auscultation fails to identify lung pathology except for advanced disease.
- Ultrasonographic examination of the lungs/pleurae is essential to accurately define pathology and takes only 20–60 seconds. 5 MHz linear scanners work perfectly well but 6.5 MHz microarray probes allow you to work much faster.
- All pathologies should be video recorded (e.g. using Elgato software) for future comparison and assessment of treatments.
- Fibrinous pleurisy and large pleural abscesses are common in sheep.
- Studies have demonstrated the gross under-reporting of OPA in the UK.
- "Wheelbarrow testing" of sheep for suspect OPA is not sensitive.

DOI: 10.1201/9781003106456-35

TABLE 3.97

Common Causes of Tachypnoea and/or Coughing in Adult Sheep

	Likely duration at presentation	BCS	Tachypnoea	Coughing	Rectal temperature	Toxaemia
Septicaemia	Hours	NAD	+/+++	NAD/+	++/+++	+++
OPA	Months	-/--	+/++	NAD/+	NAD	NAD
Laryngeal chondritis	Hours/days	-	++/+++	NAD	NAD/+	NAD
Pleurisy	Days/weeks	-/--	+/++	+/++	+	NAD/+
Large pleural abscesses >6 cm	Weeks	-	+	+	NAD	NAD
Bronchopneumonia	Weeks/months	-/--	+	+	NAD/+	+
Lungworm	Weeks	NAD	NAD	+/++	NAD	NAD
Inhalation pneumonia	Days	-	+/++	+	++	+++
ENT tumours	Months	-/--	+/+++	NAD	NAD	NAD

TABLE 3.98

Ancillary Tests to Investigate Common Causes of Tachypnoea and/or Coughing in Adult Sheep

Disease	Diagnostic tests
Septicaemia	Response to antibiotics (oxytetracycline or macrolide). Necropsy, histology and possible bacteriology of lung lesions
OPA	Ultrasonography then euthanasia, necropsy and histology. Only 60 per cent of moderate/ severe OPA cases are wheelbarrow-positive
Laryngeal chondritis	Endoscopy, response to single corticosteroid injection on day one and antibiotics (typically clavulanic acid/amoxycillin combination)
Fibrinous pleurisy	Ultrasonography, response to antibiotics (penicillin)
Pleural abscesses	Ultrasonography (well encapsulated—no treatment necessary)
Bronchopneumonia	Altered serum protein concentrations: elevated globulin, lowered albumin (chronic bacterial infection). Ultrasonography, slow response to antibiotics (penicillin)
Lungworm	Baermann technique positive in patent infestation. Response to anthelmintic (levamisole)
Inhalation pneumonia	Guarded response to treatment. Necropsy
Tumours	Nasal tumour—radiography. Note widespread presence of enzootic nasal tumour in much of southern Europe

Septicaemia

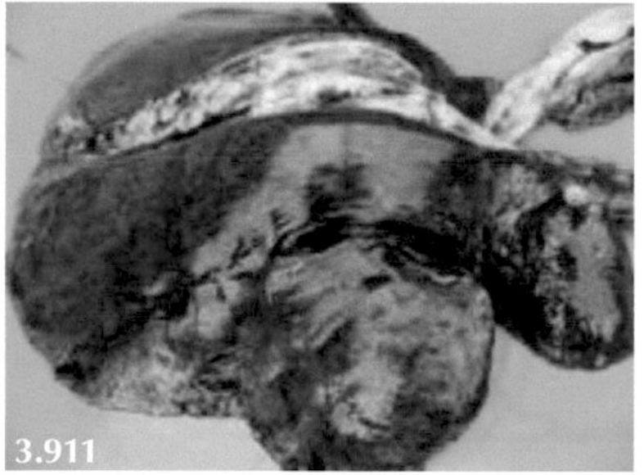

3.910: Severe illness (septicaemia) caused by *Mannheimia haemolytica*. **3.911:** Necropsy of ewe (left) reveals lung congestion, consolidation of the right cardiac lung lobe (predisposing OPA) and widespread petechiae and ecchymoses on the visceral pleura. **3.912:** Autolytic change may confuse gross lung interpretation—diseased lung always sinks.

- **Key observations**: Sudden onset, severe depression, weakness.
- **Clinical examination**: Fever >41°C, toxic mucous membranes. Tachypneoa.
- **Key history events**: Acute respiratory disease caused by *Mannheimia haemolytica* often with predisposing problem such as OPA. 3.448, 3.449

OPA

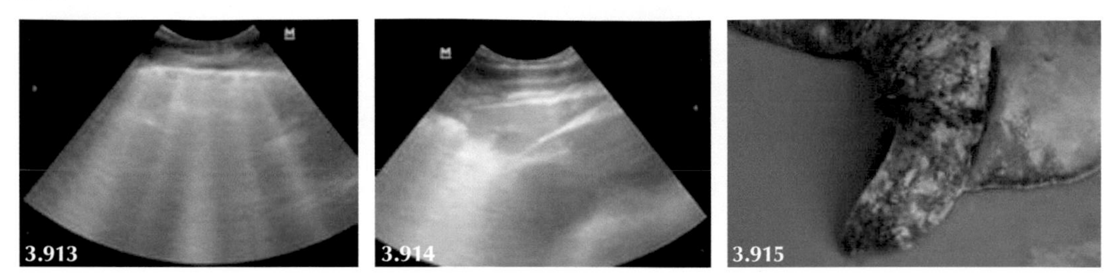

3.913: Thickened visceral pleural with many B lines. **3.914:** 1 cm deep, sharply demarcated hypoechoic area represents a very early OPA lesion at the ventral margin of the lung lobe displacing the heart from its normal position against the chest wall (see necropsy sample right). **3.915:** Early OPA lesions (centre image) affecting many lobules before becoming confluent.

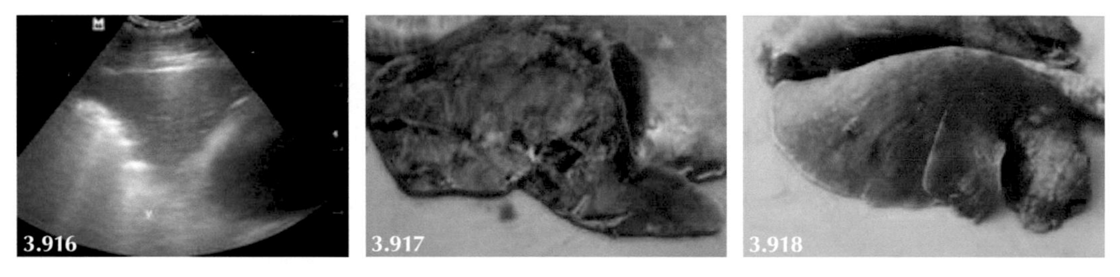

3.916: 3 cm sharply demarcated lung consolidation representing large OPA lesion displacing heart from chest wall. **3.917:** OPA affecting the apical and cardiac lobes of the left lung. There is some overlying fibrinous pleurisy. **3.918:** There are no visible OPA lesions in the right lung. Such unilateral occurrence is not unusual.

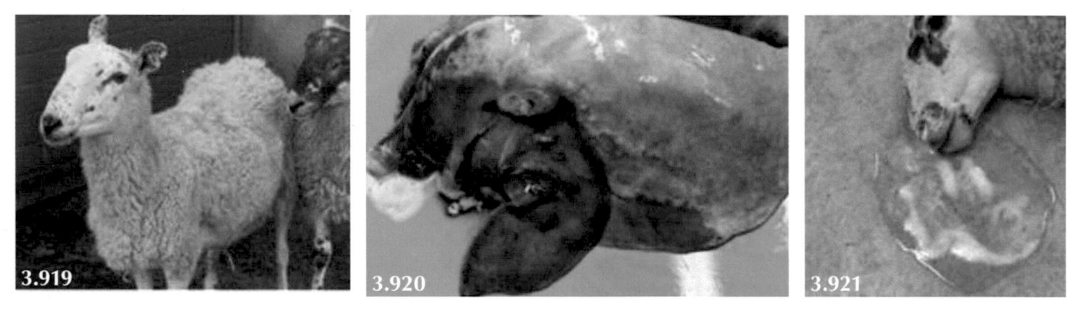

3.919: Chronic weight loss in advanced OPA. **3.920:** Necropsy finding of advanced OPA affecting the left lung. **3.921:** A positive "wheelbarrow" test (note this ewe had been euthanased following a "positive" ultrasound scan before this test was undertaken).

- **Key observations**: Bright and alert. Weight loss with lower body condition and exercise intolerance only in advanced cases.
- **Clinical examination**: Tachypnoea, widespread crackles ventrally but only in advanced cases. Wheelbarrow test detects only 60 per cent of sheep with moderate/severe OPA.
- **Key history events**: Disease present on farm and/or affects sheep (often rams) purchased from infected premises. 3.450, 3.451, 3.452, 3.453, 3.454, 3.455, 3.456, 3.457, 3.458, 3.459

Laryngeal Chondritis

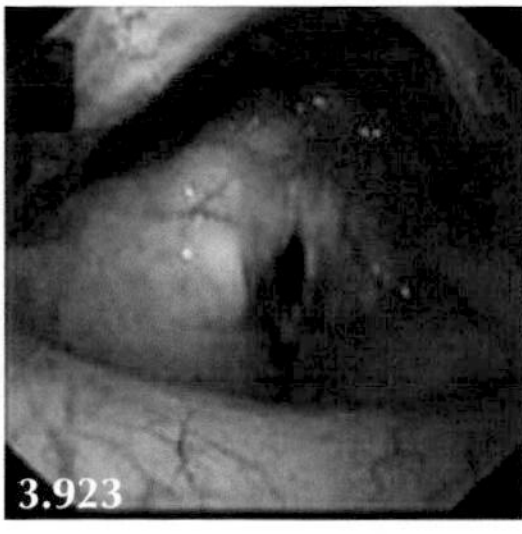

3.922: Sudden onset open-mouth breathing in a Texel ram. **3.923:** Endoscopy reveals swollen oedematous arytenoid cartilages greatly reducing airway. **3.924:** Necropsy of another case reveals abscessation of the left arytenoid cartilage with enlarged left retropharyngeal lymph node (dorsal view of longitudinal opening of larynx [top] and trachea).

- **Key observations**: Sudden onset, often severe dyspnoea, open mouth breathing ("honking").
- **Clinical examination**: Marked stertor/stridor transmitted over entire lung field.
- **Key history events**: Texel and Beltex shearling rams most commonly affected. 3.460, 3.461, 3.462

Fibrinous Pleurisy

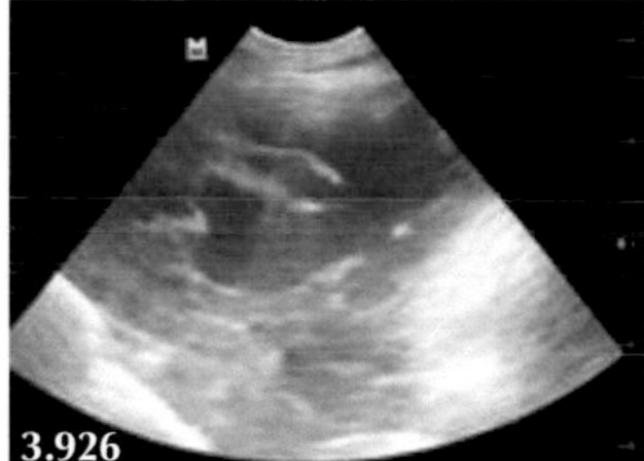
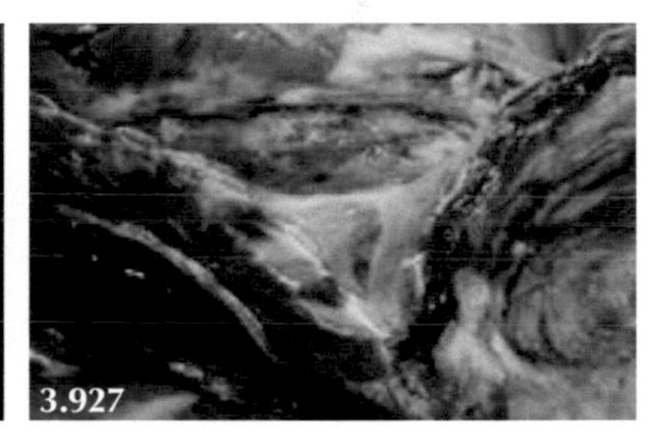

3.925: Sick ewe showing much reduced rumen fill and body condition loss. **3.926:** Ultrasound examination of sheep (left) shows unilateral 6–7 cm organised fibrinous exudate within the pleural space. This sheep responded well to treatment. **3.927:** Organised exudate within pleural space (different case).

- **Key observations**: Poor appetite/reduced abdominal fill, dull, head lowered, ears back, coughing.
- **Clinical examination**: Raised rectal temperature, unilateral pleurisy reduces lung/heart sounds however there are no friction rubs. Exudate readily detected on ultrasound scan.
- **Key history events**: May have been "off colour" for several days but not detected because of irregular/cursory flock inspection. 3.463, 3.464, 3.465

Pleural Abscess

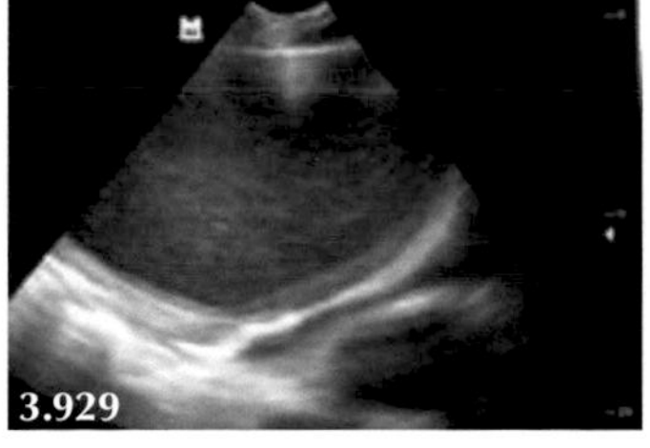
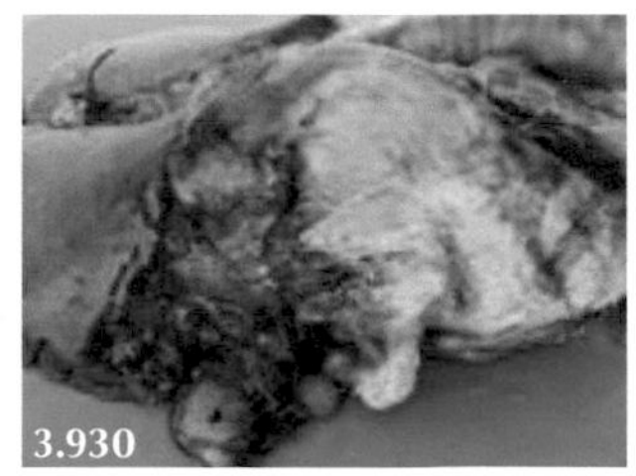

3.928: Ram in poorer body condition than peers. **3.929:** Ultrasound examination shows 6–7 cm diameter well-encapsulated abscess within the right pleural space. **3.930:** Necropsy confirms the ultrasound diagnosis of a pleural abscess with some lung consolidation and fibrinous pleurisy caudally (ram culled for an unrelated reason).

- **Key observations**: Sheep with large abscess (12–20 cm diameter) often in poorer body condition than peers and may show exercise intolerance.
- **Clinical examination**: Large (unilateral) abscess reduces lung sounds and displaces the heart, greatly reducing heart sounds.
- **Key history events**: Previous fibrinous pleurisy often goes unnoticed. 3.466, 3.467, 3.468

Bronchopneumonia

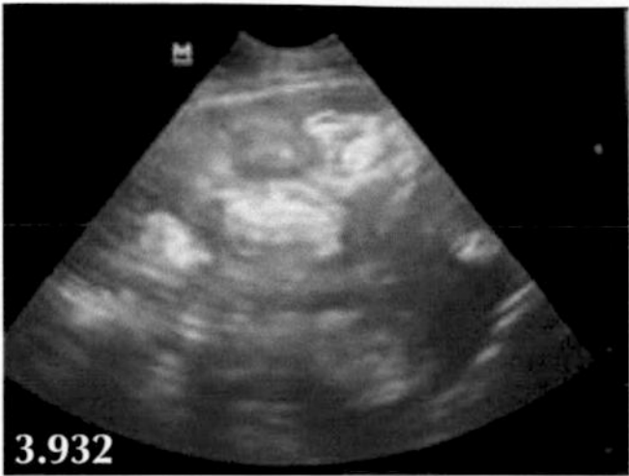
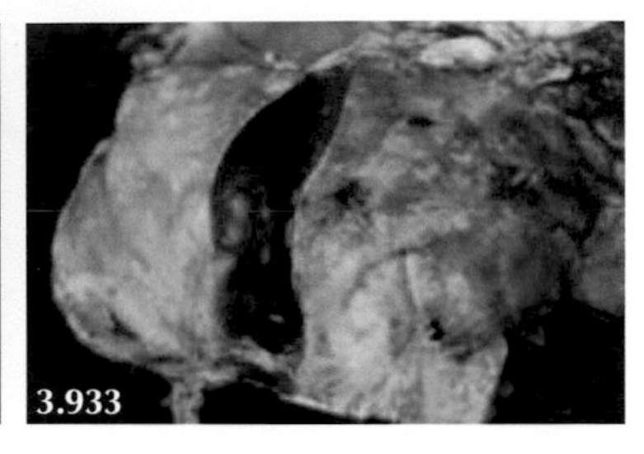

3.931: Dull ram, reduced abdominal fill and poorer body condition than peers. **3.932:** Ultrasound reveals multiple 1–3 cm diameter well-encapsulated abscesses within lung consolidation/fibrosis extending to 6–7 cm deep from chest wall. **3.933:** Necropsy reveals fibrinous pleurisy, lung consolidation and multiple abscesses throughout the lung parenchyma.

- **Key observations**: Poor appetite/abdominal fill, condition score loss.
- **Clinical examination**: Reduced/absence of normal lung sounds. Ultrasound examination needed to establish accurate diagnosis.
- **Key history events**: Rams much more often affected than ewes (related to housing management and increased risk of viral/bacterial pneumonia). 3.469, 3.470, 3.471

Lungworm

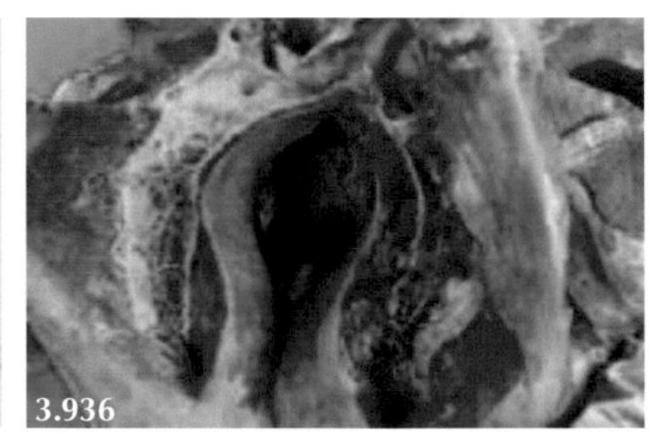

3.934: Ewe in poor condition caused by paratuberculosis. **3.935:** Fecal examination of ewe (left) reveals coincidental lungworm larvae. **3.936:** Severe lungworm infestation in this ewe suffering from paratuberculosis.

- **Key observations**: Anecdotal reports of increased coughing in large numbers of sheep in autumn which stops after anthelmintic treatment. Severe infestations in individual sheep with paratuberculosis due to immunosuppression.
- **Clinical examination**: Often emaciated (primary paratuberculosis), occasional coughing.
- **Key history events**: No recent anthelmintic treatment. 3.472, 3.473

Inhalation Pneumonia

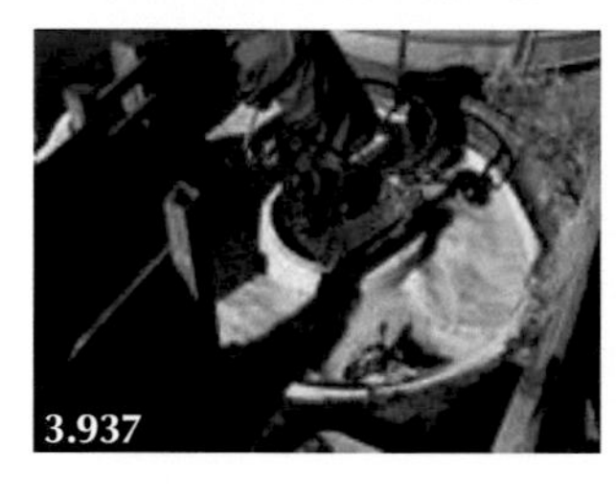
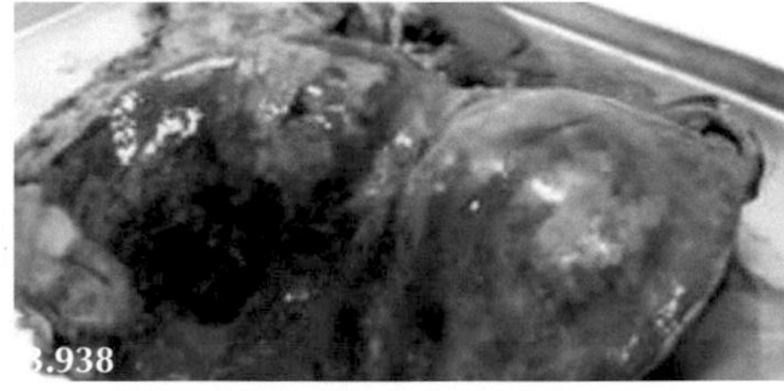
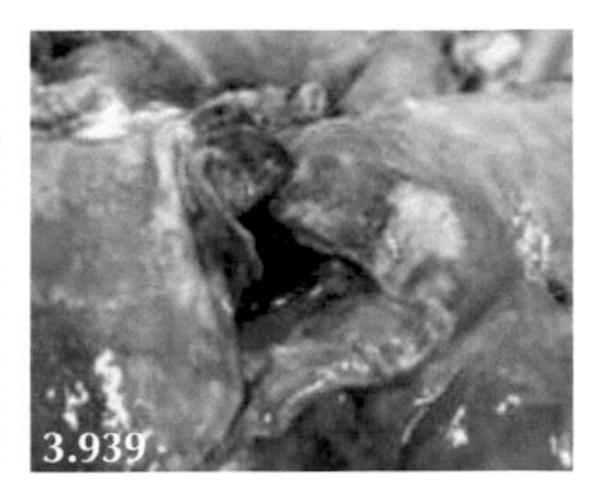

3.937: Incorrect plunge dipping may cause inhalation pneumonia. **3.938** and **3.939:** Large gas-filled necrotic lesions in the dorsal lung lobes.

- **Key observations**: Onset of illness within 24 hours of dorsal recumbency/regurgitation of rumen contents after becoming cast. Possible inhalation of plunge dip solution.
- **Clinical examination**: Sudden onset, laboured respirations, marked toxaemia.
- **Key history events**: Found in dorsal recumbency with head pointed uphill. Possible drenching error. 3.474

Nasal Tumour

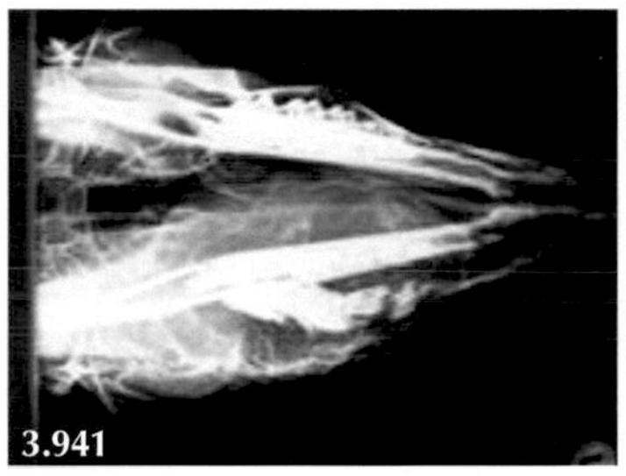
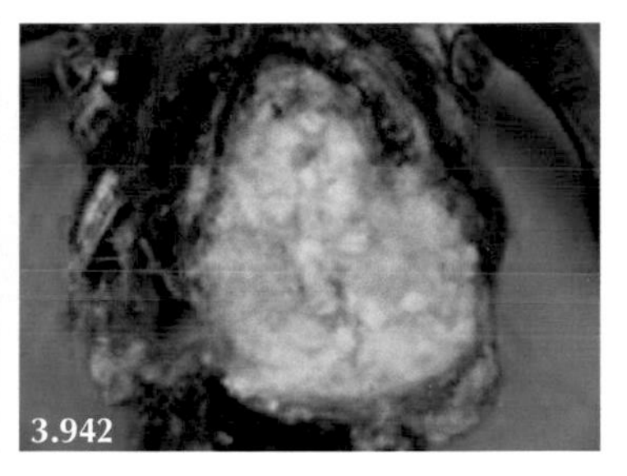

3.940: Profuse left-sided mucopurulent nasal discharge. **3.941:** Radiograph reveals soft tissue mass and much bone remodelling of left frontal and maxillary sinuses. **3.942:** Sagittal section through the frontal sinuses reveals a squamous cell carcinoma.

- **Key observations**: Facial distortion, profuse unilateral nasal discharge.
- **Clinical examination**: Marked abdominal respiratory effort. Unilateral nasal blockage. Swollen frontal sinus.
- **Key history events**: Chronic nasal discharge, weight loss over several months. Note enzootic nasal tumour not present in UK but endemic in much of southern Europe. 3.475

Lambs

- Pasteurellosis in growing lambs is identified as a per-acute disease with fever and toxaemia.
- Outbreaks of pasteurellosis in recently weaned lambs can kill 2–5 per cent of the group.
- Always submit lung samples to confirm your diagnosis of respiratory disease.
- While slaughterhouse reports of fat lambs identify "pneumonia" and "pleurisy" during winter such observations overestimate their clinical significance.

TABLE 3.99

Common Causes of Tachypnoea and/or Coughing in Lambs

	Likely duration at presentation	BCS	Tachypnoea	Coughing	Rectal temperature	Toxaemia
Systemic pasteurellosis (*Pasteurella trehalosi*)	Hours	NAD	+/+++	NAD	+++	+++
Septicaemic pasteurellosis (*Mannheimia haemolytica)*	Hours	NAD	+/+++	NAD	+++	+++
Bronchopneumonia	Days/weeks	-/--	+/++	+/++	NAD/+	+
Enzootic pneumonia	Weeks	NAD	NAD/+	+/++	NAD	NAD
Lungworm	Days/weeks	NAD	NAD	+/++	NAD	NAD

TABLE 3.100

Ancillary Tests to Investigate Common Causes of Tachypnoea and/or Coughing in Lambs

	Ancillary tests
Systemic pasteurellosis (*Pasteurella trehalosi*)	Histology and possible bacteriology of sample collected from lung pathology
Septicaemic pasteurellosis (*Mannheimia haemolytica)*	Diffuse fibrinous pleurisy often observed at necropsy. Histology and possible bacteriology (lung, liver, kidney, spleen, thoracic fluid, and heart blood)
Bronchopneumonia	Altered serum protein concentrations; elevated globulin, lowered albumin—chronic bacterial infection. Ultrasonography
Enzootic pneumonia	Commonly noted in slaughterhouse reports
Lungworm	Baermann technique for patent infections. Response to anthelmintic

Systemic Pasteurellosis *(Pasteurella trehalosi*)

The disease is most common in recently weaned lambs occurring in the UK from August to December.

3.943: Lambs with systemic pasteurellosis may simply be found dead. **3.944** and **3.945:** Recently weaned five-month-old lambs, recumbent, dull and unresponsive.

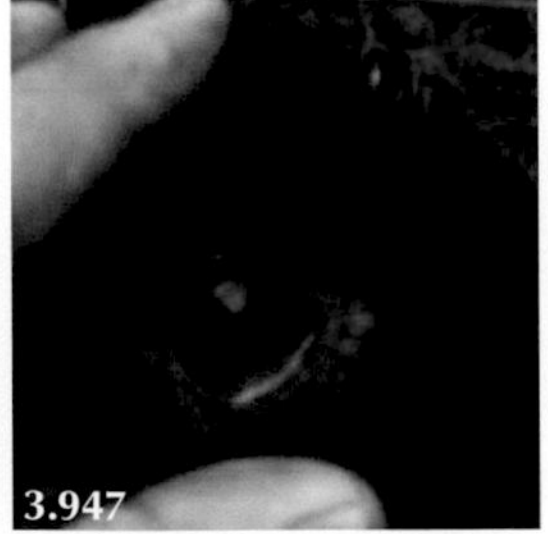
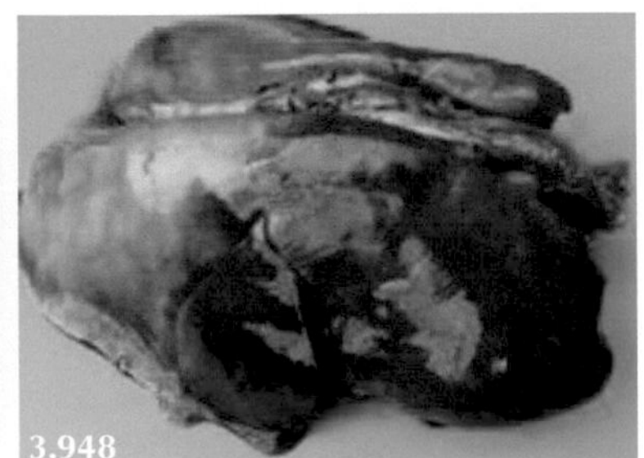

3.946: Recently weaned five-month-old lamb, recumbent, dull and unresponsive, head and ears down. **3.947:** Toxic ocular mucous membranes. **3.948:** Oedematous and congested lungs at necropsy.

- **Key observations**: Isolated from group, severely depressed, may simply be found dead.
- **Clinical examination**: Febrile (41–42°C), marked toxaemia. Auscultation fails to reveal convincing adventitious sounds.
- **Key history events**: Weaning/moved to aftermath two to three weeks previously. Often best lambs in group affected. 3.476, 3.477, 3.478

Septicaemic Pasteurellosis (*Mannheimia haemolytica*)

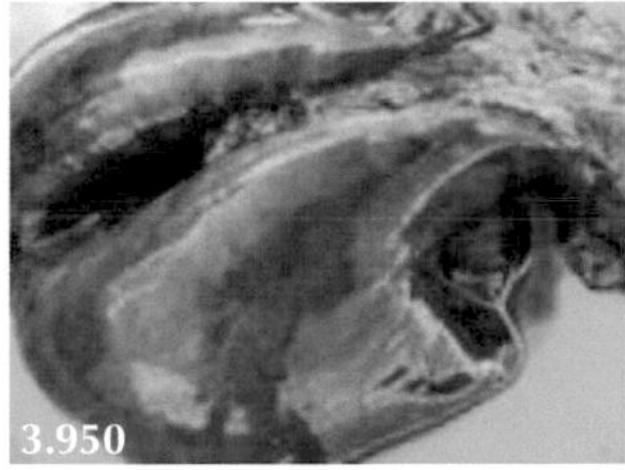
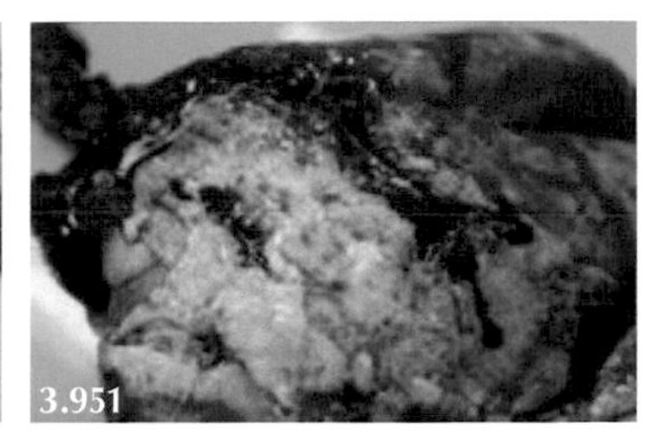

3.949: Severely depressed/unresponsive lamb. **3.950:** Very oedematous and congested lungs. **3.951:** Extensive unilateral fibrinous pleurisy.

- **Key observations**: Isolated from dam, severely depressed, may simply be found dead.
- **Clinical examination**: Febrile (41–42°C), marked toxaemia.
- **Key history events**: Often best lambs in group affected. Widespread petechiae, congested and oedematous lungs; less acute cases are characterised by fibrinous pleurisy. 3.479, 3.480, 3.481

Bronchopneumonia

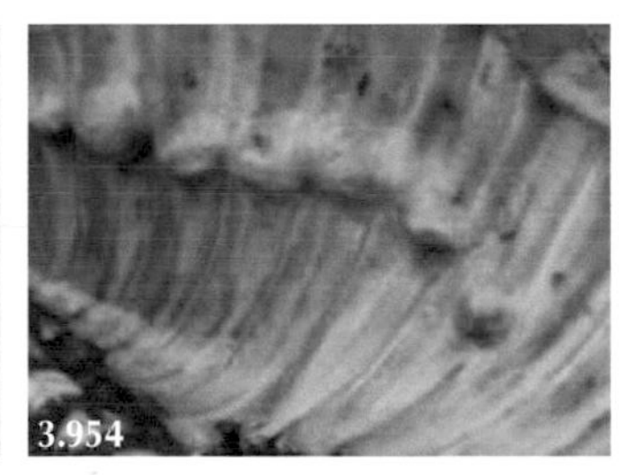

3.952: Bronchopneumonia is often confined to orphan/housed lambs. **3.953:** Dull orphan lamb with head lowered, salivation and ears back. **3.954:** There are anecdotal reports that rib fractures may predispose to respiratory disease.

- **Key observations**. Often orphan/housed lambs, dull, poor growth and body condition, poor appetite.
- **Clinical examination**: Tachypnoea, frequent coughing, muco-purulent nasal discharge, salivation.
- **Key history events**: Poor lambs from early age. 3.482, 3.483, 3.484, 3.485

Enzootic Pneumonia

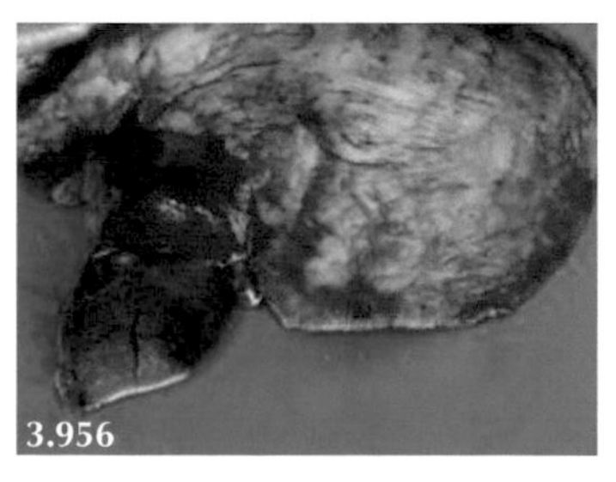

3.955: Housed in poorly ventilated buildings over past four to eight weeks. Mixing groups of lambs, housed with/near adult sheep. **3.956:** Antero-ventral lung consolidation.

- **Key observations**: Reduced appetite for several days, frequent coughing in group.
- **Clinical examination**: Often NAD.
- **Key history events**: Housed in poorly ventilated buildings over past four to eight weeks. Mixing groups of lambs, housed with/near adult sheep. Frequently noted in slaughterhouse reports often with "pleurisy". 🎥 3.486

Lungworm

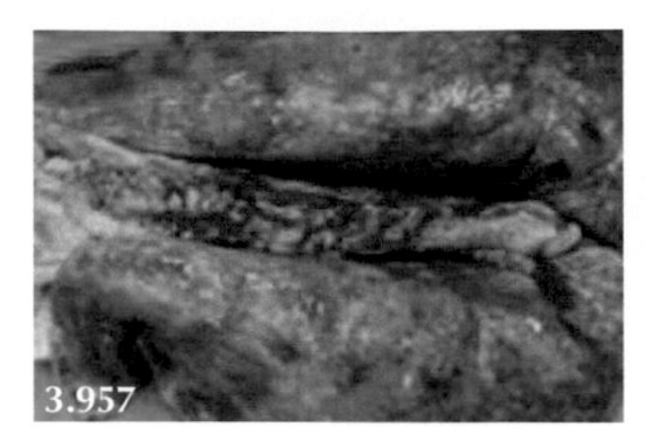
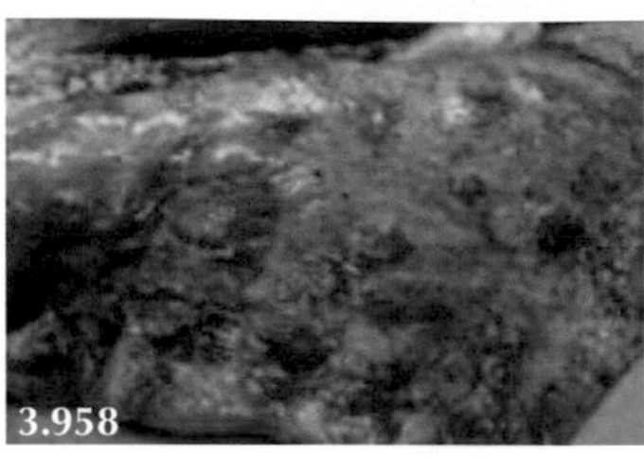
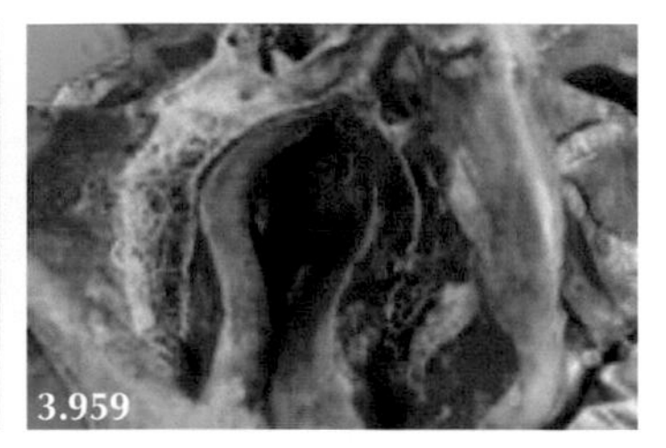

3.957 and **3.958:** Widespread superficial lung lesions caused by *Muellerius capillaris* infestation. **3.959:** Adult *Dictyocaulus filaria*. Significant infestations in adult sheep are rarely seen except for sheep with paratuberculosis.

- **Key observations**: Frequent coughing is reported from mid-summer onwards in young sheep at pasture with apparent response to anthelmintic treatment.
- **Clinical examination**: Coughing could be caused by pre-patent infestation therefore Baermann test may be negative.
- **Key history events**: No recent anthelmintic treatment. Significant infestations in individual adult sheep affected by paratuberculosis presumably due to poor host immune function.

Adult Cattle

- Adventitious lung sounds do not indicate the extent or nature of lung pathology.
- Ultrasonographic examination is essential to accurately define lung/pleural pathology and takes only 1–2 minutes. 5 MHz linear scanners work perfectly well.
- All pathologies should be recorded for future comparison and assessment of treatments. Allows interpretation of recorded findings by telemedicine services.
- Always check cases for the possibility of lungworm in your practice lab.

TABLE 3.101

Common Causes of Tachypnoea and/or Coughing in Adult Cattle

	Likely duration at presentation	Tachypnoea	Coughing	Rectal temperature	Nasal discharge	Toxaemia
Broncho-pneumonia/ bronchiectasis	Days/weeks	+	+/++	NAD/+	+/++	+
IBR	Days	+	+/++	+++	+++	+/++
Lungworm	Days	+/++	++/+++	NAD	NAD	NAD
Hepato-caval thrombosis	Days	+	+/++	+	+/+++	+
Endocarditis	Weeks	+	NAD/+	+	NAD	NAD
Inhalation pneumonia	Days	+/++	+/++	+/++	+/++	+++
Pleural abscess/ pyothorax	Weeks	+	NAD	NAD/+	NAD	NAD/+
Fog fever	Days	++/+++	NAD	NAD	NAD	NAD
Diffuse fibrosing alveolitis	Weeks/months	+/++	+/+++	NAD	+	NAD
Tumours	Weeks/months	+/++	NAD/+	NAD	NAD/+	NAD

TABLE 3.102

Ancillary Tests to Investigate Common Causes of Tachypnoea and/or Coughing in Adult Cattle

	Ancillary tests
Bronchopneumonia/bronchiectasis	Ultrasonography, good response to penicillin treatment. Necropsy
IBR	FAT on conjunctival swab. Necropsy
Lungworm	Baermann technique positive in patent infestations. Response to anthelmintic (levamisole)
Hepato caval thrombosis	Ultrasonography—difficult to be certain of dilated vena cava, hepatic vessels. Necropsy
Endocarditis	Ultrasonography—vegetative lesion on heart valve. Necropsy
Inhalation pneumonia	Ultrasonography. Necropsy
Pleural abscess/pyothorax	Ultrasonography, thoracocentesis
Fog fever	No diagnostic test. Necropsy
Diffuse fibrosing alveolitis	No diagnostic test. Response to corticosteroid injection (if not >3-months pregnant) Necropsy
Tumours	Lung tumours—ultrasonography. Nasal tumour—endoscopy, radiography

Bronchopneumonia/Bronchiectasis

3.960: Dairy cow with bronchiectasis at presentation.
3.961: Six weeks later after change of antibiotic treatment from marbofloxacin to procaine penicillin.

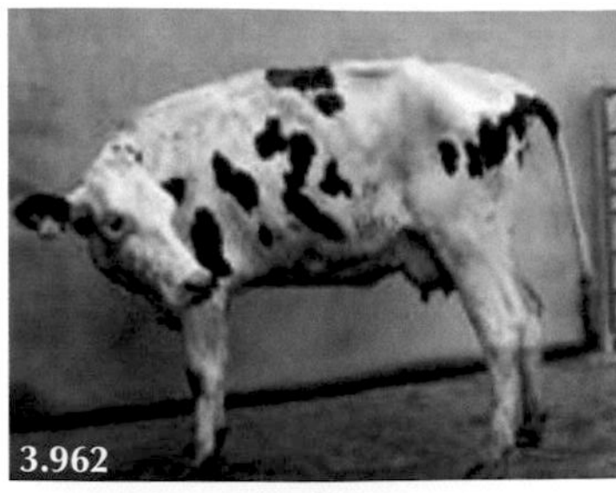

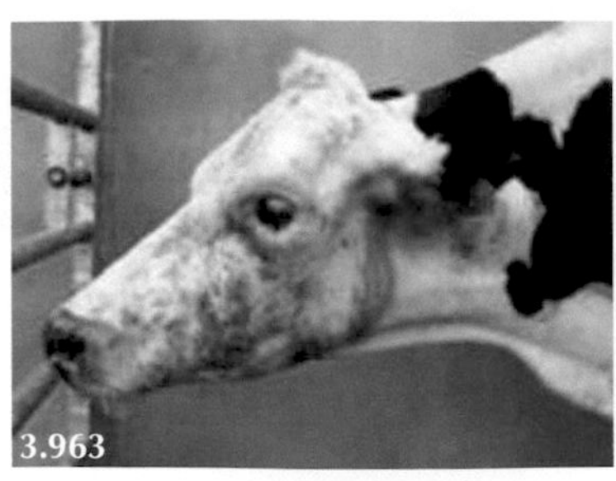

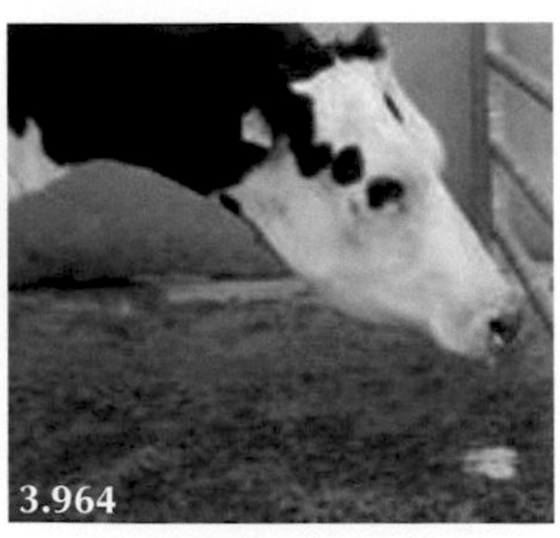

3.962 and **3.963:** Severe case at first presentation to veterinarian. Note arched back, gaunt, painful expression, neck extended with head lowered and ears back. Low body condition and reduced abdominal fill. **3.964:** Purulent nasal discharge more evident when head lowered. 3.488, 3.489, 3.490, 3.491

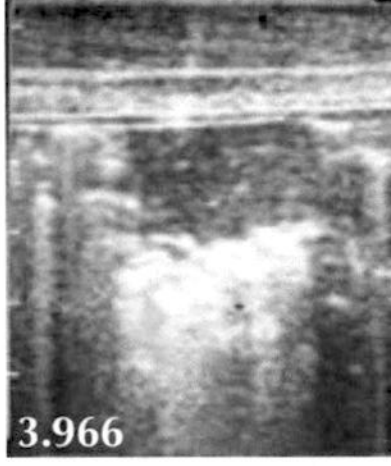

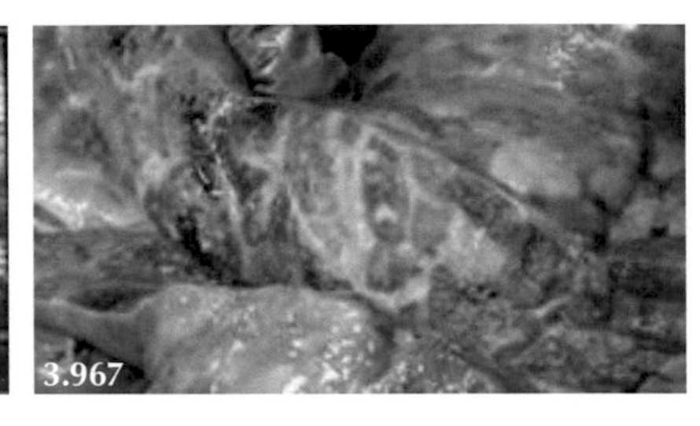

3.965: Painful expression, neck extended with head lowered, ears back, poor body condition and reduced abdominal fill. **3.966:** 2 cm hypoechoic area represents lobular lung pathology in the dorsal lung. **3.967:** Necropsy confirms lobular distribution in the dorsal lung lobes before extensive pathology ventrally (see the following).

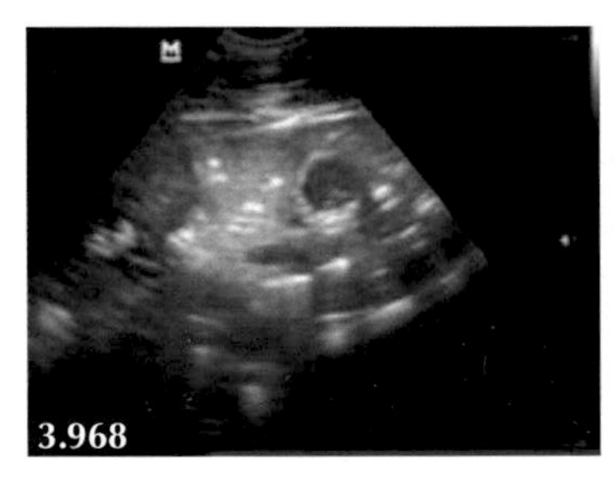

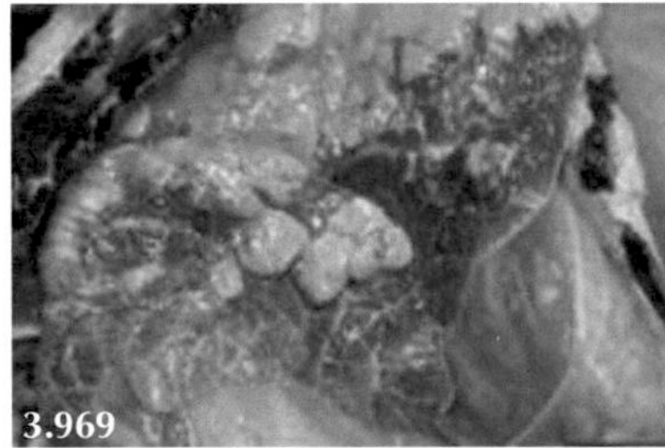

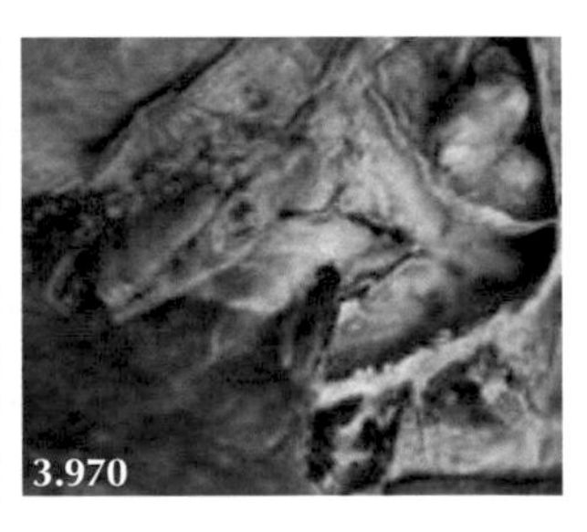

3.968: Extensive hypoechoic area extending 8–10 cm from chest wall with several 2 cm diameter anechoic circles containing hyperechoic dots which may represent abscesses. **3.969:** Necropsy reveals extensive ventral lung consolidation. **3.970:** Pus within the incised bronchi (bronchiectasis) at necropsy. 3.492

- **Key observations**: At presentation, the cow has an arched back, painful expression, neck extended with head lowered, ears back, intermittent purulent nasal discharge (when head lowered), cough and reduced abdominal fill (poor appetite/rumen fill) with a low body condition score and dry coat.
- **Clinical examination**: 39.0°C, halitosis, often reduced adventitious sounds caused by lung consolidation. Ultrasound examination reveals widespread hepatoid change (bronchiectasis).
- **Key history events**: Calved two to six weeks previously, no response to certain antibiotics. History of respiratory disease as calf. 3.487

Response of Cows with Bronchiectasis to Treatment

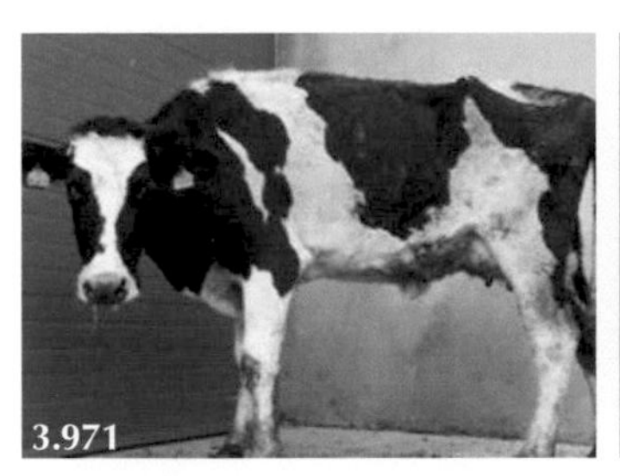

Moderate case. At admission (**3.971**) and six days later after change of antibiotic treatment from marbofloxacin to procaine penicillin (**3.972**). Note the fuller abdomen (rumen), no nasal discharges, much brighter demeanour. The cleaner coat is the result of self-grooming and straw bedding replacing dirty cubicles. Response of cows with bronchiectasis to treatment—four cases shown in Videos 🎥 3.493, 3.494, 3.495, 3.496.

IBR

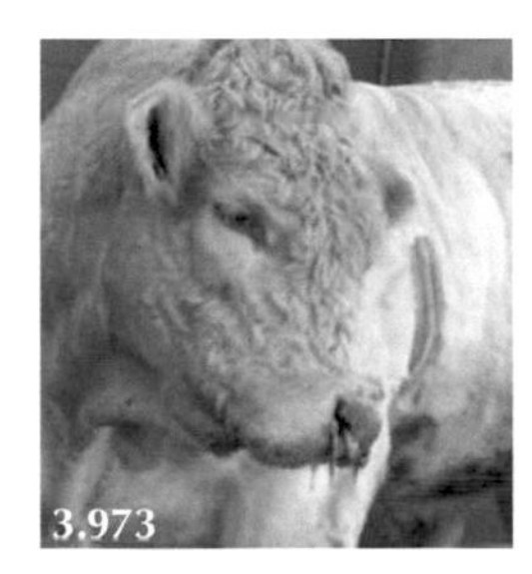

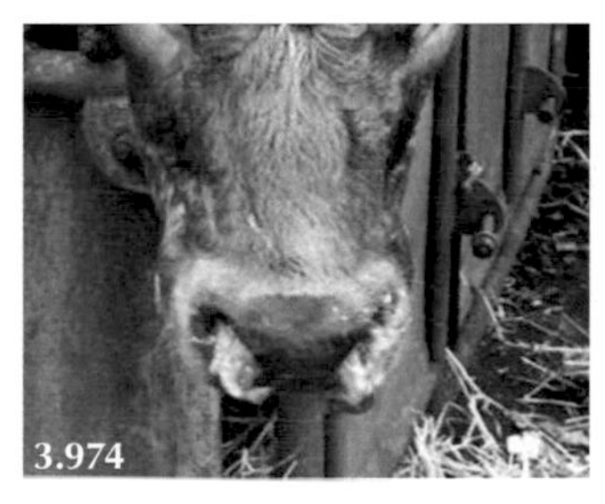

3.973 and **3.974:** Pronounced bilateral purulent nasal discharge.

- **Key observations**: Bilateral purulent ocular and nasal discharge. Frequent coughing. Large numbers of cattle affected.
- **Clinical examination**: Often >41°C, dull, not eating.
- **Key history events**: Recently purchased unvaccinated cattle introduced into infected herd or carrier animals introduced into naïve herd. 🎥 3.497

Lungworm

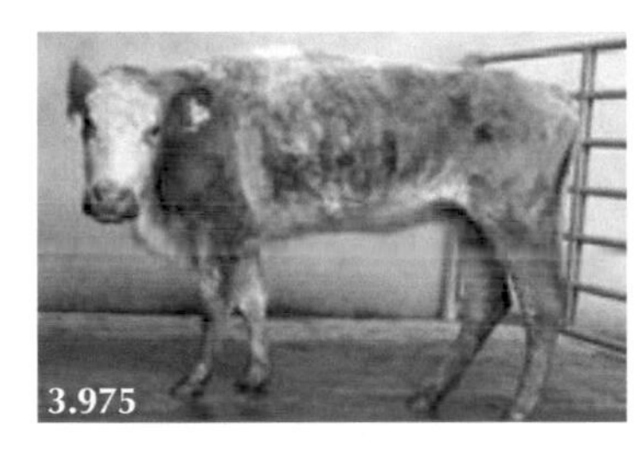

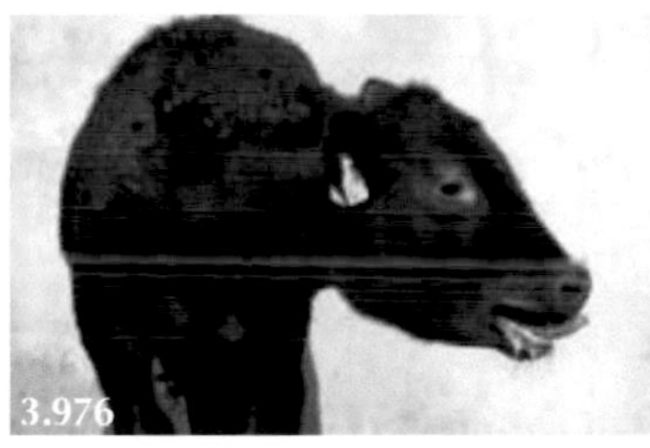

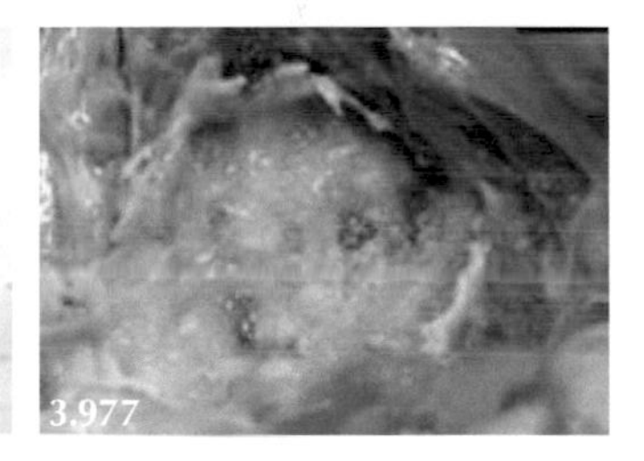

3.975: Chronic severe weight loss with frequent coughing. **3.976:** Acute onset, frequent coughing, neck extended. **3.977:** Patent lungworm infestation with adult lungworms in the major airways.

- **Key observations**: At pasture, several/many animals affected with frequent coughing, neck extended, often dyspneoic.
- **Clinical examination**: Afebrile unless secondary bacterial infection. Baermann examination but clinical signs can be caused by prepatent lungworm infestations.
- **Key history events**: Unvaccinated, pasture previously grazed by infected young stock, rapid deterioration. 🎥 3.498, 3.499, 3.500

Hepatocaval Thrombosis

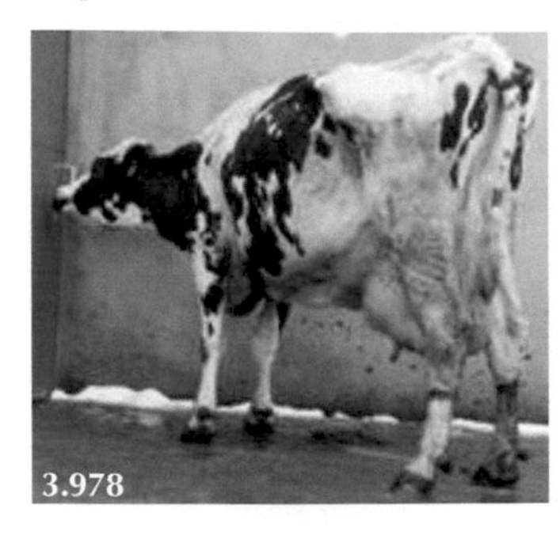

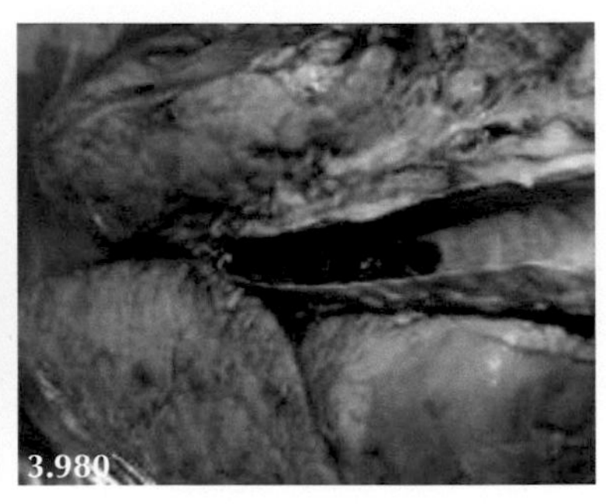

3.978: Chronic weight loss and poor milk production. **3.979:** Mild epistaxis (typically followed by fatal bleed several days/ weeks later). **3.980:** Blood in the trachea at necropsy.

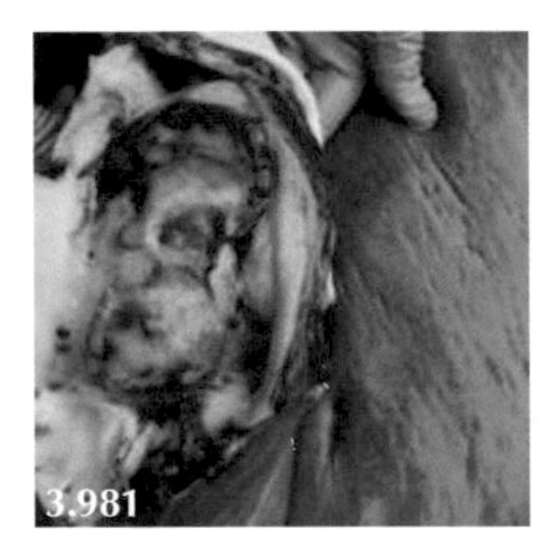
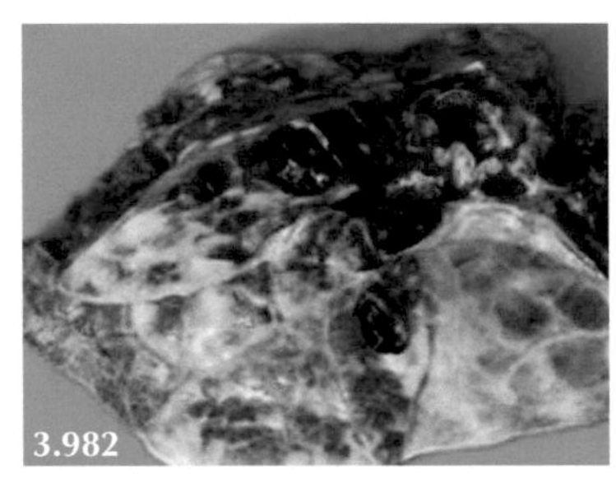

3.981: Very large abscess (15 cm) juxtaposed the caudal vena cava at necropsy. **3.982:** Haemorrhage throughout the lung.

- **Key observations**: Similar presentation as chronic suppurative pneumonia until epistaxis.
- **Clinical examination**: Rapid deterioration/sudden death some days after first mild epistaxis episode.
- **Key history events**: Liver abscesses are very common in dairy cows fed intensive rations.

Endocarditis

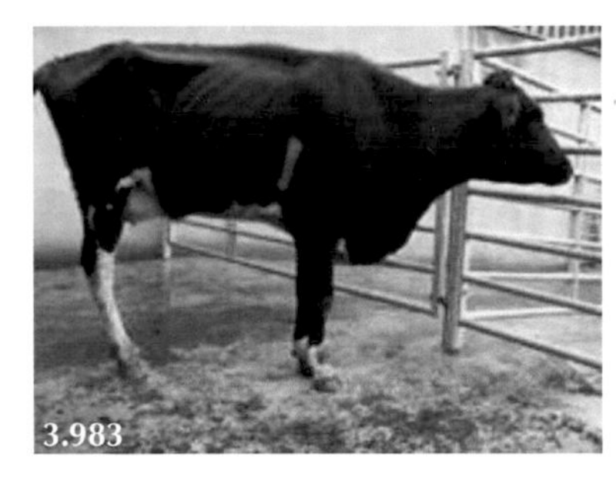
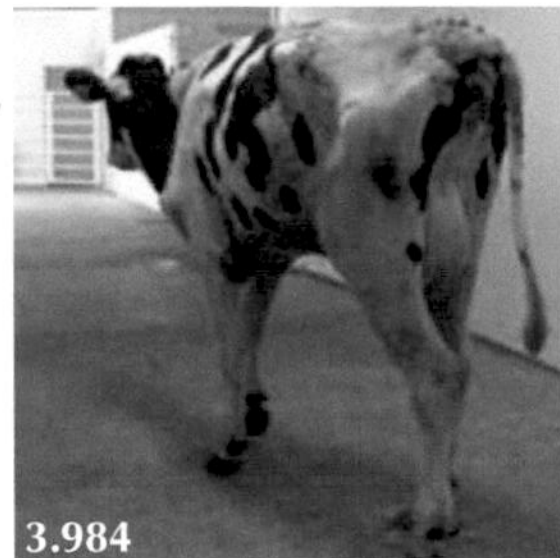
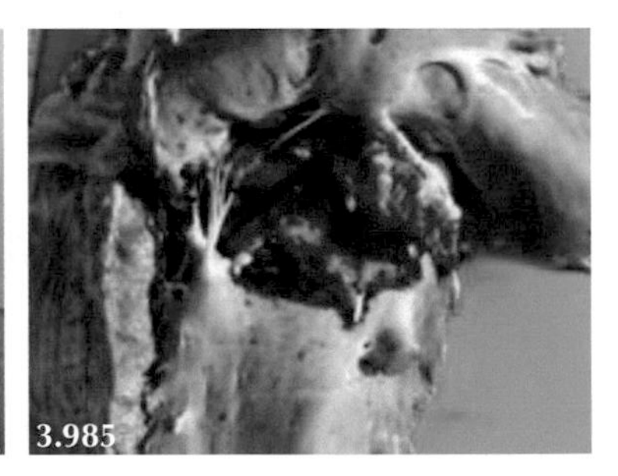

3.983: Arched back, head held lowered poor production/weight loss. **3.984:** Loss of body condition, lameness arising from joint effusion/infection. Pain causing increased respiratory rate. **3.985:** Large vegetative lesion affecting the mitral valve.

- **Key observations**: Chronic weight loss, poor abdominal fill, arched back, often lame with fetlock and hock joint effusions with pain causing increased respiratory rate.
- **Clinical examination**: Auscultation fails to detect significant heart murmur in most cases.
- **Key history events**: None.

Inhalation Pneumonia

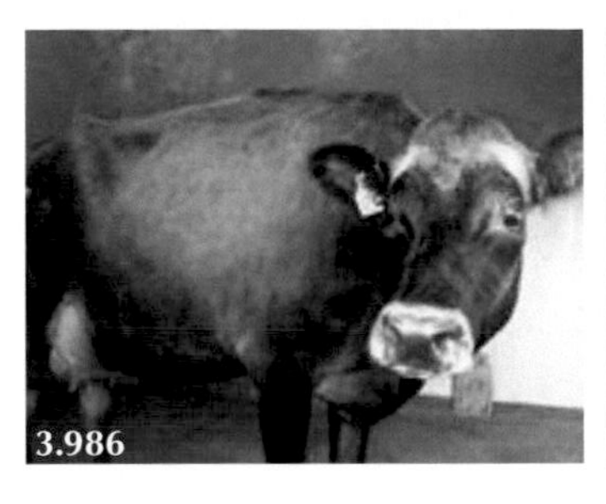
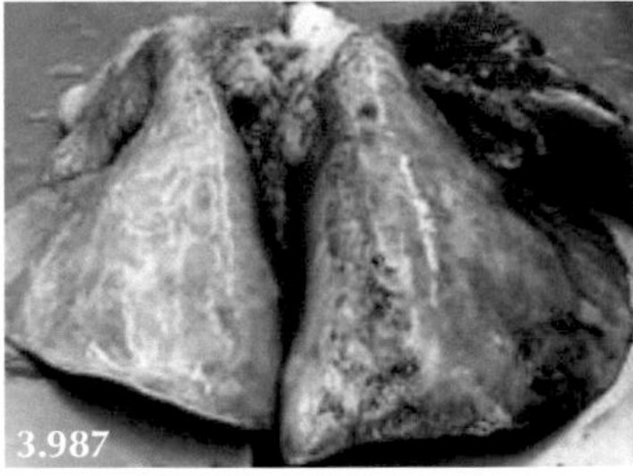
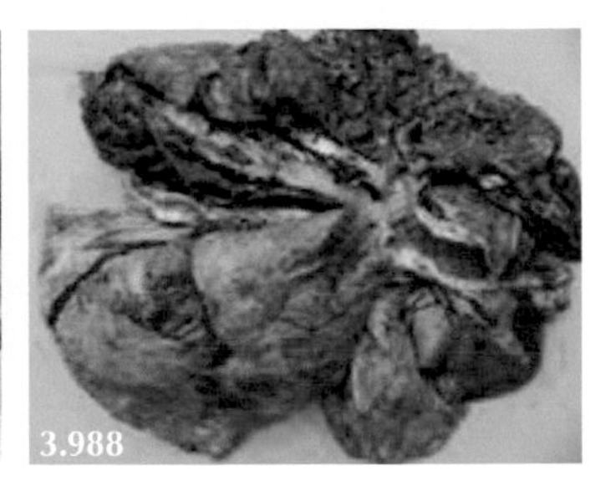

3.986: Arched back, painful expression, no appetite/reduced abdominal fill, purulent nasal discharge. **3.987:** Right lung reveals consolidation and extensive fibrinous pleurisy. **3.988:** Extensive left lung destruction following inhalation of rumen fluid when cow cast.

- **Key observations**: Painful expression/arched back, reluctant to move, profound toxaemia, no appetite/ milk production.
- **Clinical examination**: Necropsy varies from (unilateral) per-acute pleurisy to extensive pleural abscess.
- **Key history events**: Usually hypocalcaemia and becoming cast in dorsal recumbency. 3.501

Fog Fever

3.989: Severe dyspnoea audible over large distance, open mouth breathing/flared nostrils.

- **Clinical examination**: Auscultation often prevented by widespread subcutaneous emphysema over thorax.
- **Key history events**: Summer months, lush pasture/silage aftermath for past seven to 10 days, beef cattle most often affected. 3.502

Pleural Abscess/Pyothorax

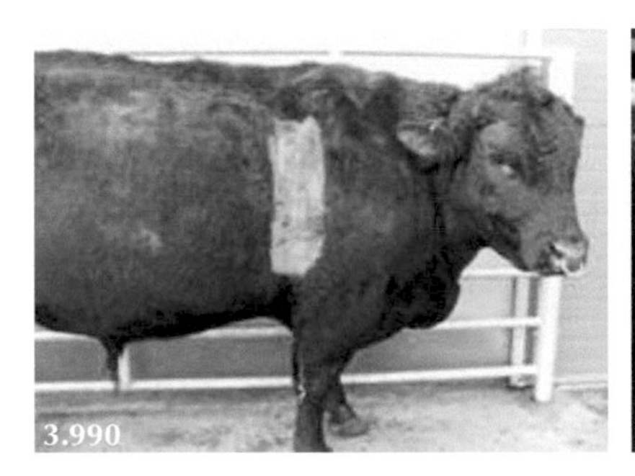
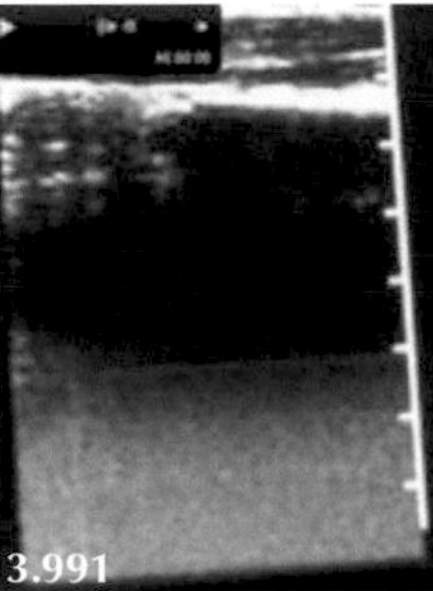
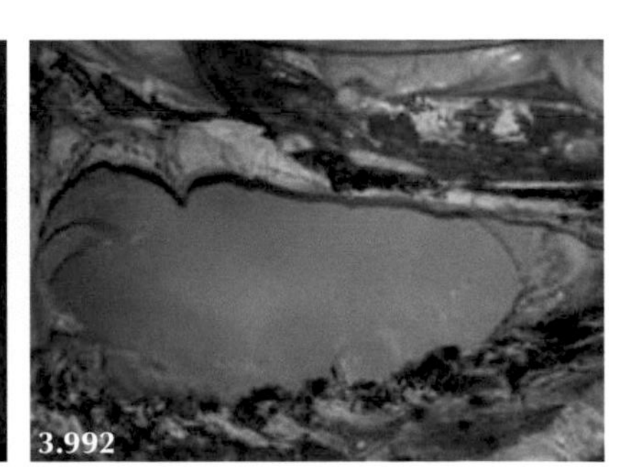

3.990: Pyothorax secondary to traumatic reticulitis in a bull which had caused right sided heart failure seen as brisket oedema. **3.991:** Ultrasonography shows anechoic area (pus) extending beyond 10 cm field depth (right) of 5 MHz linear probe. **3.992:** True extent of pyothorax revealed at necropsy.

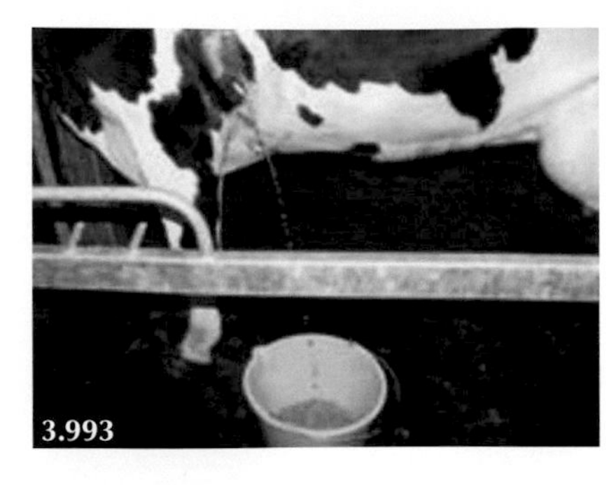

3.993: Attempted drainage is not often successful.

- **Key observations**: Painful expression, respiratory signs depend upon size of space-occupying abscess/ pyothorax. Often signs of right-sided heart failure.
- **Clinical examination**: Absence of normal lung/heart sounds on affected side.
- **Key history events**: Secondary to traumatic reticulitis causing pyothorax.

Diffuse Fibrosing Alveolitis (Extrinsic Allergic Alveolitis)

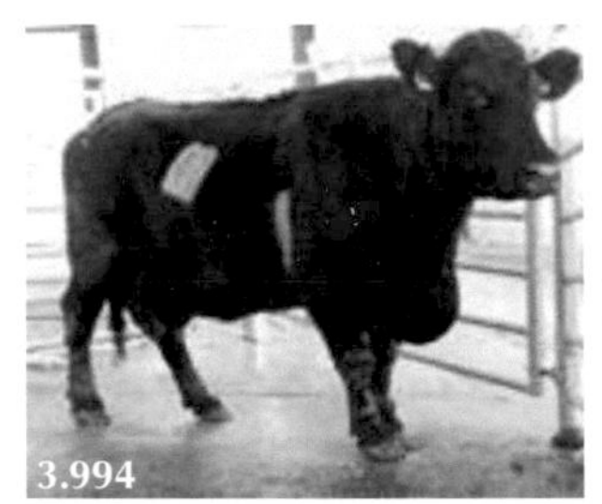

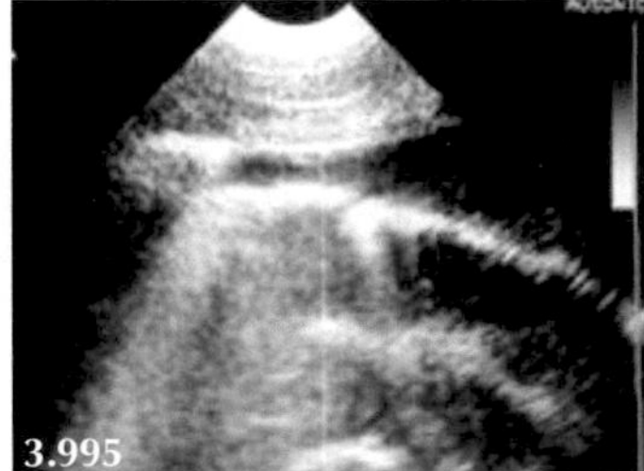

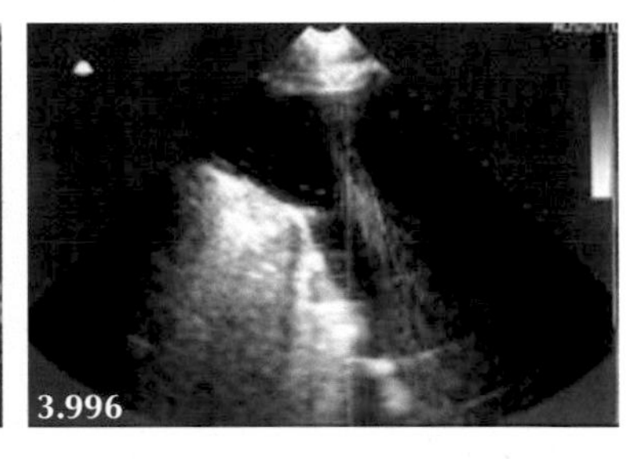

3.994: Bright, alert, older beef cow with brisket oedema. **3.995** and **3.996:** Pleural effusion (right-sided heart failure) increases in depth to 20 cm as ultrasound probe moves down the chest wall (dorsal to the left, probe head at top of images).

- **Key observations**: Bright and alert older beef cow. Increased respiratory rate, frequent coughing over weeks/months although acute episodes occur when suddenly exposed to allergens (often at housing).
- **Clinical examination**: Development of pleural effusion/right-sided heart failure in chronic severe cases quantified ultrasonographically.
- **Key history events**: Prevalence depends upon risk factors including poor ventilation, allergens in poorly conserved hay but now considered very uncommon/rare condition. 🎥 3.503

Primary Lung Tumours/Metastatic Spread

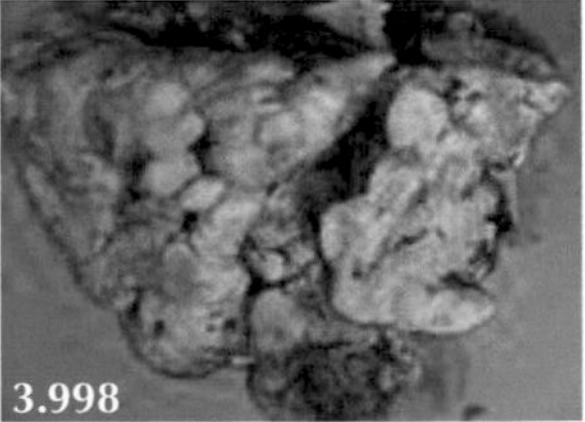

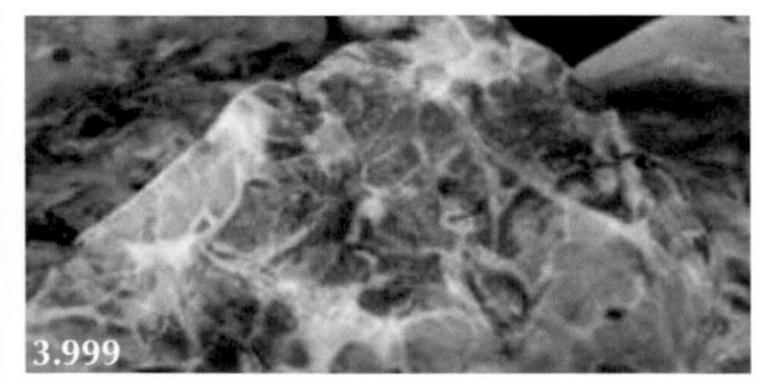

3.997: Respiratory difficulty shown by extended neck and lowered head carriage. **3.998:** Tumour occupies much of the lung. **3.999:** Distinct lobular tumour distribution.

- **Key observations**: Chronic weight loss, increasing respiratory signs.
- **Clinical examination**: Respiratory signs depend upon loss of functional lung capacity. Lobular distribution identified on ultrasound examination.
- **Key history events**: Primary lung tumours/metastatic spread to the lungs are rare in cattle. 🎥 3.504

Nasal Tumours

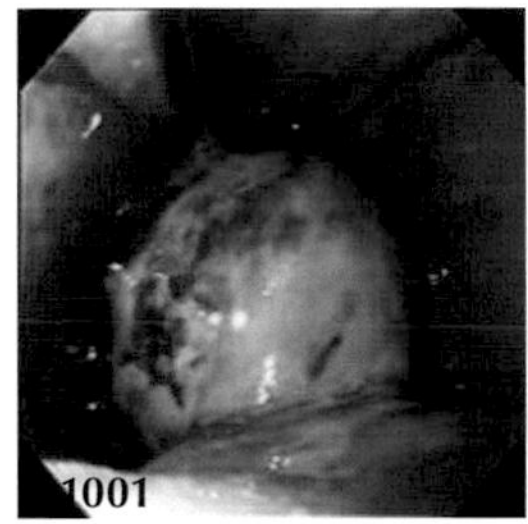

3.1000: Note nasal occlusion (breath expelled only from the cow's right nostril); exophthalmos and ocular discharge on left side.
3.1001: Mass (tumour) identified on endoscopic examination of left nasal passage.

- **Key observations**: Rare, occlusion of nasal passage, possible bony changes to skull.
- **Clinical examination**: Localised signs depend upon size of tumour.
- **Key history events**: None. 🎥 3.505

Squamous Cell Carcinoma

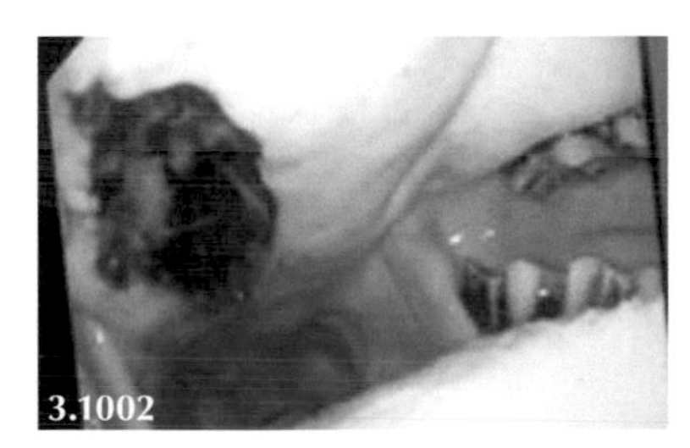

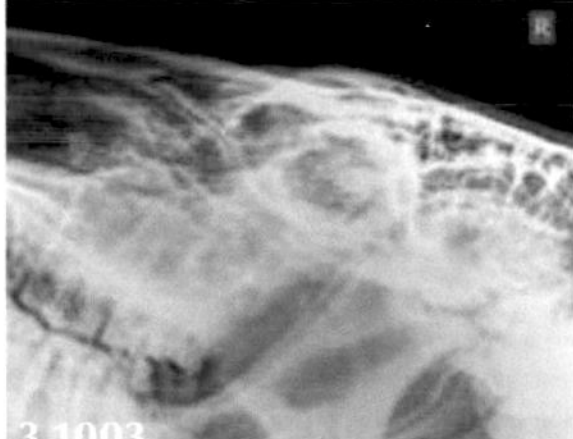

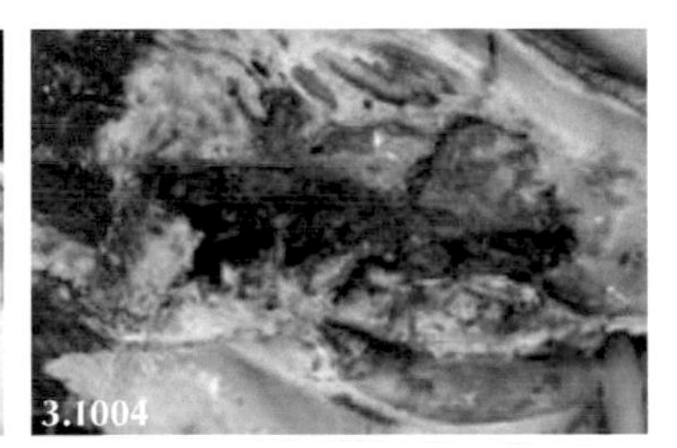

3.1002: Endoscopic examination reveals an oval-shaped 3 cm × 6 cm erosion through the hard palate into the nasal passages. **3.1003:** Destruction of the bony turbinates of the nasal passages. **3.1004:** Impacted food material in the nasal passages. 🎥 3.506

Calves and Growing Cattle

- There are several clinical scoring charts for detecting acute respiratory disease in calves but none work as well as rectal temperature. Recordings >39.6°C are frequently used as the sole indicator for antibiotic treatment during outbreaks allowing calves to be examined at a rate of >100 animals per hour; a fall to below 39.2°C after 48 hours is considered a positive treatment response.
- Recurrence of respiratory disease indicated by pyrexia greater than 39.6°C after four or more days' interval should be treated with the same antibiotic—this indicates re-infection of compromised respiratory tract defence mechanisms rather than rapid development of antibiotic resistance.
- Chest ultrasound examination reporting lung consolidation measuring only 1–2 cm^2 in young calves is often claimed to be clinically significant but this is not consistent with scans from diseased calves in first opinion practice.
- BRSV and IBR vaccines administered before the challenge period, particularly when given intranasally, work very well in disease prevention.

TABLE 3.103

Common Causes of Tachypnoea and/or Coughing in Growing Cattle

	Duration	Tachypnoea	Coughing	Rectal temperature	Nasal discharge	Toxaemia
Pasteurellosis	Days	+	+	+++	+	+/++
BRSV	Hours/days	+++	+/++	++/+++	NAD/+	+/++
IBR	Days	+	++/+++	+++	+++	+/++
Mycoplasma bovis	Days	+	+	++	+	+
Chronic pneumonia	Weeks	+/++	+	+	+	NAD/+
Lungworm	Days	+/+++	++/+++	NAD/+	NAD	NAD

TABLE 3.104

Ancillary Tests to Investigate Common Causes of Tachypnoea and/or Coughing in Growing Cattle

	Ancillary tests
Pasteurellosis	Commensal of upper respiratory tract therefore bacteriology of limited use. Response to antibiotic. Necropsy
BRSV	FAT/virus isolation on BAL fluid. Paired serology
IBR	FAT on conjunctival swab. Paired serology. Necropsy
Mycoplasma bovis	Often present with concurrent otitis media and/or joint infections
Chronic pneumonia	Ultrasonography. Necropsy
Lungworm	Baermann technique positive. Response to anthelmintic. Necropsy

Pasteurellosis

3.1005–3.1007: Dull, head lowered, ears down, muco-purulent nasal discharges, pyrexic 40–41°C, following housing spring-born beef calves in autumn.

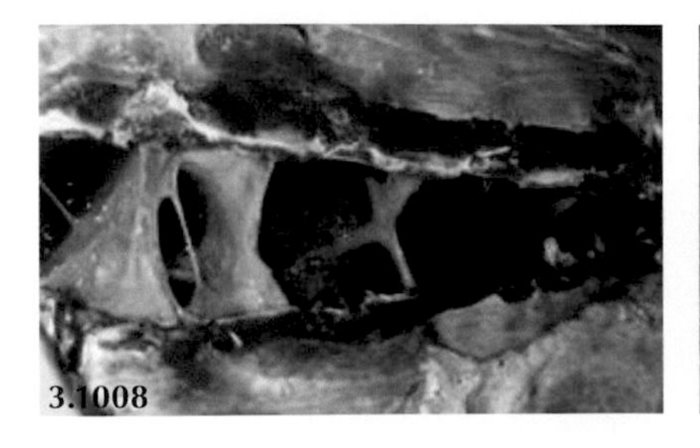
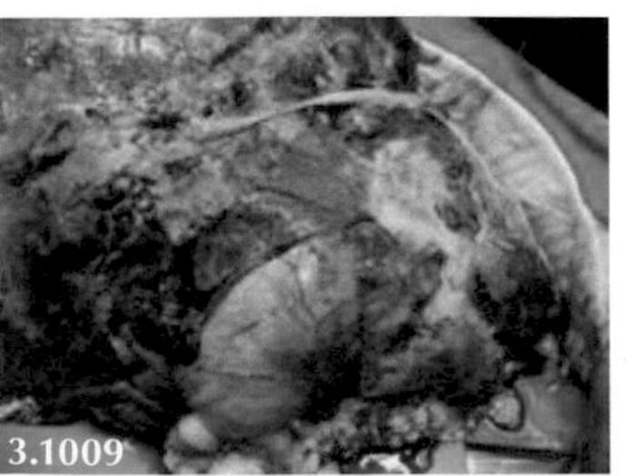

3.1008: Acute pleurisy is an uncommon finding associated with pasteurellosis in the UK.
3.1009: Extensive pathology at necropsy is difficult to attribute to specific bacterial pathogens especially after antibiotic therapy.

3.1010 and **3.1011:** Detection of affected calves based on inspection is not easy especially when outdoors in poor weather. **3.1012:** Stocking density is an important predisposing factor for respiratory disease in young calves.

- **Key observations**: Dull, neck extended with head held lowered, ears down, variable mucopurulent nasal discharge.
- **Clinical examination**: Febrile 40–41°C. Few adventitious lung sounds.
- **Key history events**: Stressful event(s) 10–14 days previously including weaning, sale, transport, co-mingling, housing, over-stocking buildings, castration/dehorning, prolonged adverse weather. 🎥 3.507, 3.508, 3.509, 3.510, 3.511, 3.512, 3.513

BRSV

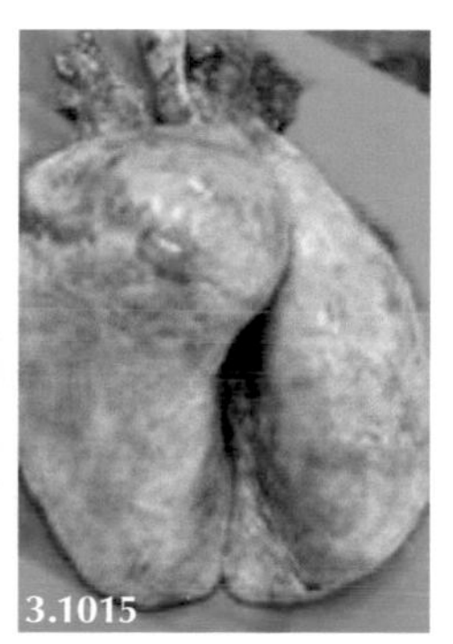

3.1013 and **3.1014:** Anxious expression, mouth breathing, flared nostrils, salivation, no observed ocular/nasal discharges. Eyes appear sunken suggesting dehydration. **3.1015:** Widespread interlobular emphysema/bullae at necropsy.

- **Key observations**: Sudden onset respiratory disease most severe in first calves affected. Mouth breathing, flared nostrils, salivation, neck extended with head held lowered, ears down, usually no ocular/nasal discharges.
- **Clinical examination**: Tachypnoea, febrile ca. 41°C. May be subcutaneous emphysema over dorsal chest.
- **Key history events**: Stressful event(s) 10–14 days previously including weaning, sale, transport, co-mingling, housing, castration/dehorning. Necropsy reveals widespread bullae formation explaining poor treatment response. 🎥 3.514

IBR

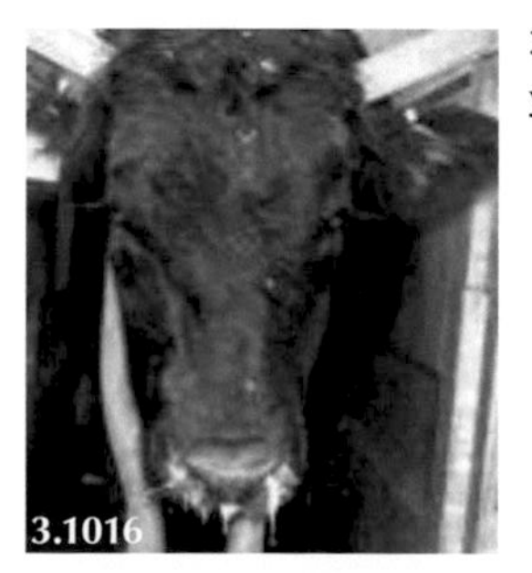

3.1016: Dull demeanour, purulent ocular and nasal discharges in recently purchased yearling steer.

- **Key observations**: Purulent nasal discharges. Frequent coughing.
- **Clinical examination**: Febrile-41°C. Inflamed conjunctivae. Halitosis.
- **Key history events**: Stressful event(s) 10–14 days previously, unvaccinated cattle, weaning, sale, transport, co-mingling, housing, castration/dehorning and calving.

Mycoplasma bovis

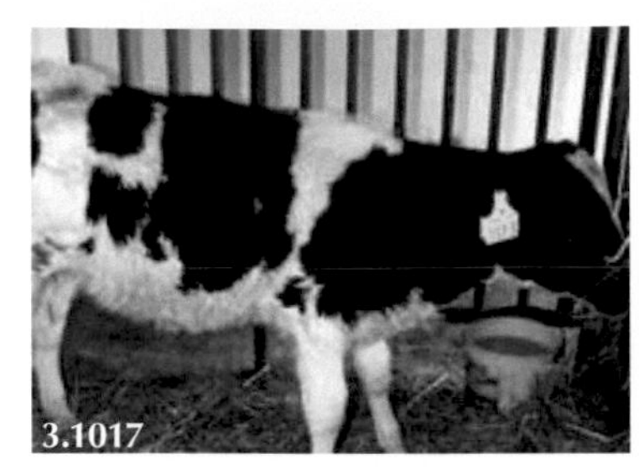

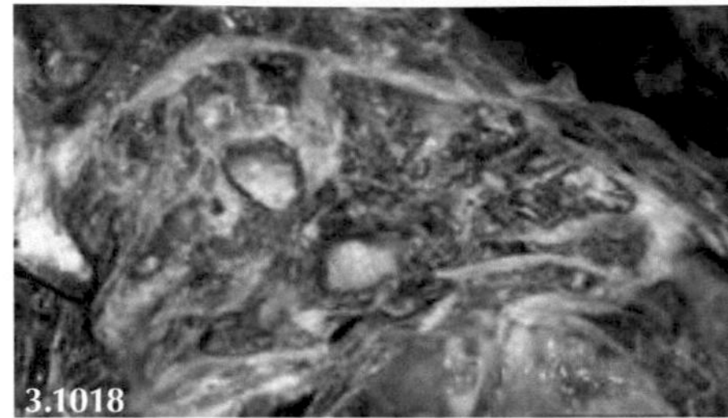

3.1017: Poor growth in dairy calf due to chronic pneumonia. **3.1018:** Extensive lung pathology, including widespread abscessation, at necropsy.

- **Key observations**: Insidious onset poor growth/respiratory signs.
- **Clinical examination**: Extensive antero-ventral consolidation on ultrasonographic examination.
- **Key history events**: Ongoing (dairy) herd problem with mastitis and joint infections, otitis media in calves.

Chronic Pneumonia/Bronchiectasis

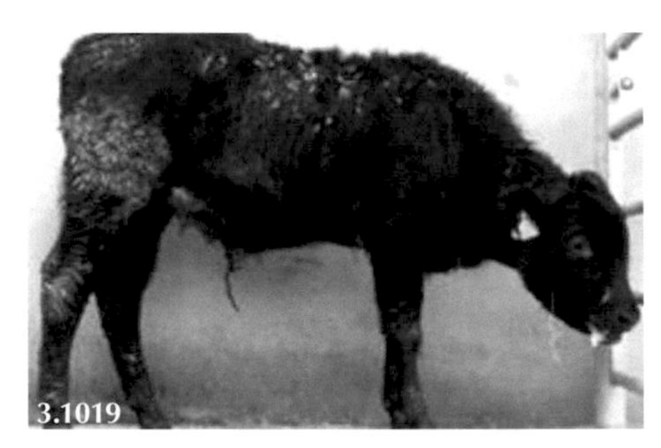

3.1019: Poor condition, frequent coughing. **3.1020** and **3.1021:** Poor growth and body condition caused by persistent infection with BVD virus and secondary chronic pneumonia (PI calf on right in **3.1021**).

- **Key observations**: Chronic weight loss/poor condition. Poor quality coat. Occasional purulent nasal discharge and frequent coughing.
- **Clinical examination**: Rectal temperature ca. 39°C. Reduced heart/lung sounds ventrally indicate lung consolidation, crackles dorsally arise from pus in larger airways.
- **Key history events**: Three or more antibiotic treatments with only temporary improvement in appetite. Often persistently infected with BVD virus. 🎥 3.515, 3.516, 3.517, 3.518, 3.519, 3.520, 3.521

Lungworm

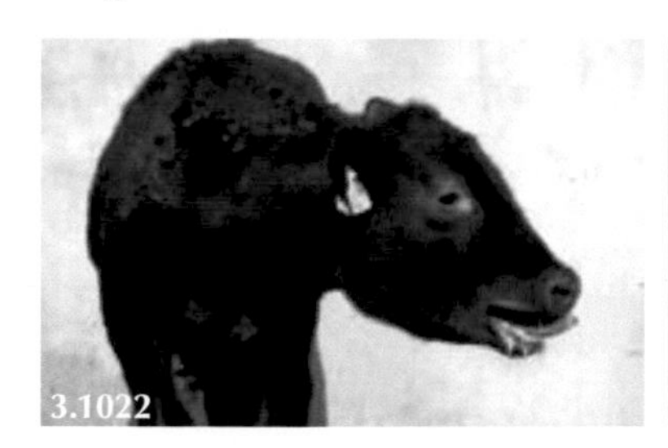
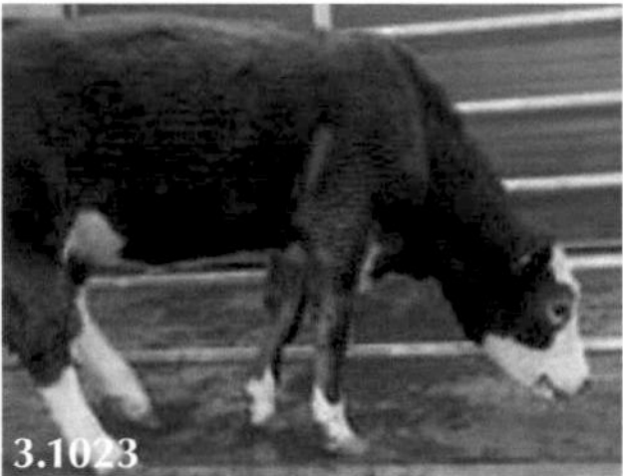
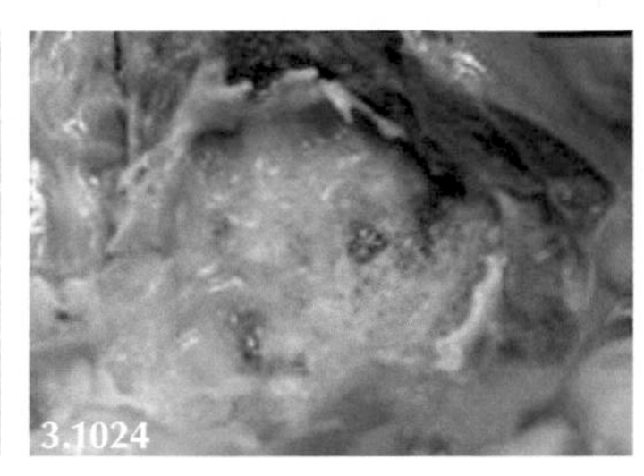

3.1022 and **3.1023:** Frequent coughing in calves infested with lungworm. **3.1024:** Necropsy reveals severe patent lungworm infestation.

- **Key observations**: Several/many cattle in group affected. Acute onset frequent coughing at pasture, variable dyspnoea/mouth-breathing.
- **Clinical examination**: Rectal temperature around 39°C. Widespread wheezes.
- **Key history events**: First season at pasture. No vaccination/preventive anthelmintic strategy.

Laryngeal Diphtheria

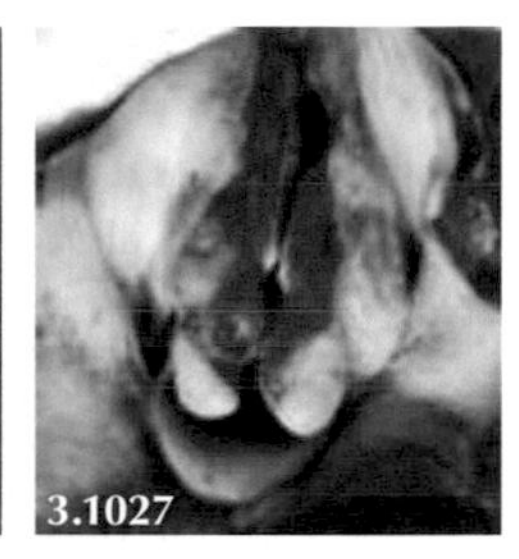

3.1025 and **3.1026:** Acute respiratory signs (stridor) with necks extended and flared nostrils. **3.1027:** Endoscopy reveals oedema of the arytenoid cartilages.

- **Key observations**: Two three month old beef calf. Acute onset stridor (a harsh vibrating noise when breathing, caused by obstruction of the larynx), variable inspiratory effort/dyspnoea.
- **Clinical examination**: Rectal temperature ca. 39°C. Sounds from upper airway transmitted over whole lung fields.
- **Key history events**: Often best calf in group. 3.522, 3.523, 3.524

3.18

Common Causes of Changes in Scrotal Size

Rams

- Ram breeders operate a strict culling policy regarding scrotal/testicular swellings with few problems encountered in sale rams as a consequence.
- Testicular hypoplasia is common in immature ram lambs.

TABLE 3.105

Common Causes of Changes in Scrotal Size in Rams

	Likely duration at presentation	Size	Consistency	Rectal temperature
Epididymitis (chronic)	Weeks/months	++/+++	+++	NAD
Epididymitis (acute)	Days	+/++	+	+/++
Orchitis	Days	+/++	+	+/++
Sperm granuloma	Months/years	–	+++	NAD
Testicular hypoplasia	Weeks/years	–/––	–	NAD
Scrotal hernia	Weeks/months	+++	–––	NAD

TABLE 3.106

Ancillary Tests to Investigate Cause of Changes in Scrotal Size in Rams

	Ancillary tests
Epididymitis (chronic)	Ultrasonography
Epididymitis (acute)	Ultrasonography
Orchitis	Ultrasonography
Sperm granuloma	Palpation, ultrasonography
Testicular hypoplasia	Maximum scrotal circumference, ultrasonography
Scrotal hernia	Palpation, ultrasonography

Epididymitis (Chronic)

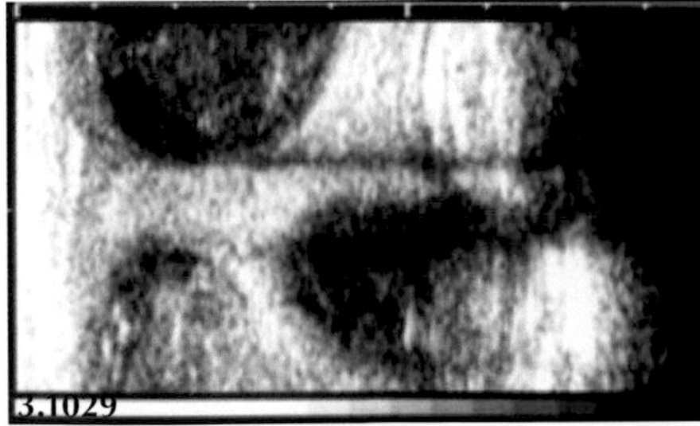

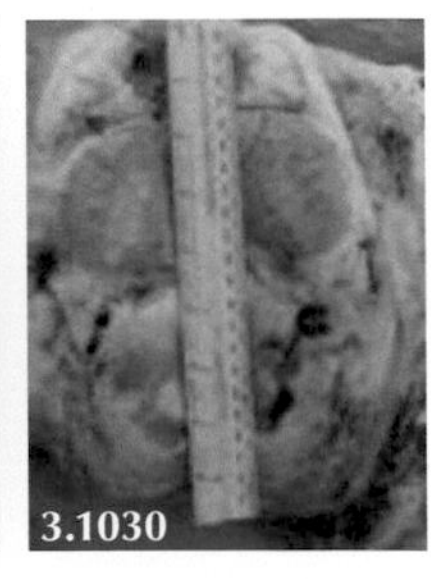

3.1028: Very large right side of scrotum. **3.1029:** Thick-walled abscesses in the tail of the epididymis. **3.1030:** Size (ruler 300 mm) and location of abscesses in the right epididymis.

DOI: 10.1201/9781003106456-36

- **Key observations**: Large/very large, usually unilateral, fibrous swelling of the scrotal content.
- **Clinical examination**: Firm fibrous lesion, adherent to scrotal skin, unable to identify normal testicular and epididymal morphology.
- **Key history events**: Lesion develops over several/many months. 🎥 3.525, 3.526

Epididymitis (Acute)

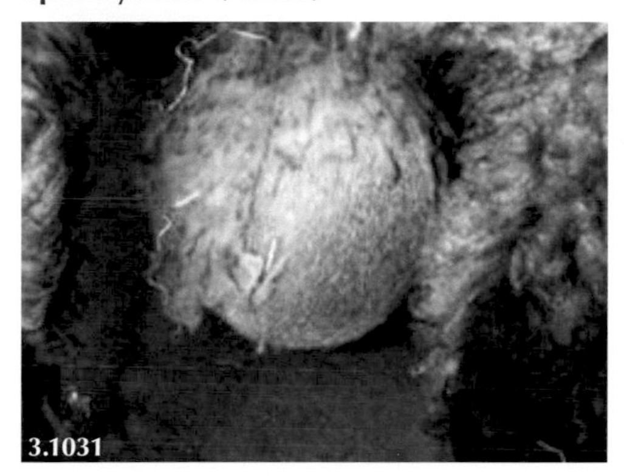

3.1031: Enlarged right side of scrotum with hair loss. Enlarged epididymis identified on ultrasound examination.

- **Key observations**: Ipsilateral hindleg lameness.
- **Clinical examination**: Oedematous painful scrotum. Possible concurrent bacteraemia.
- **Key history events**: Sudden onset swollen scrotum, often affecting ram lambs. 🎥 3.527

Orchitis

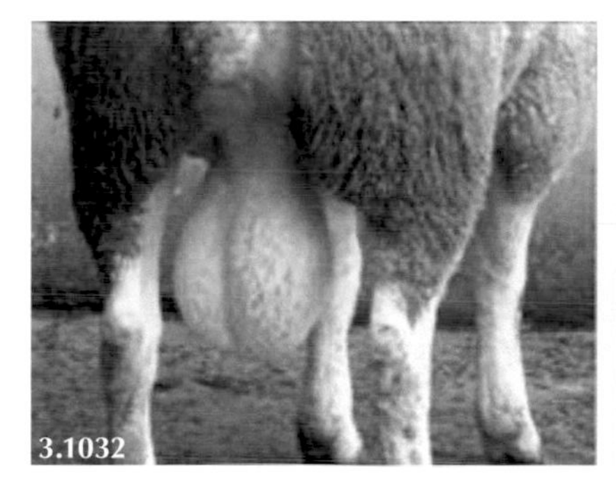

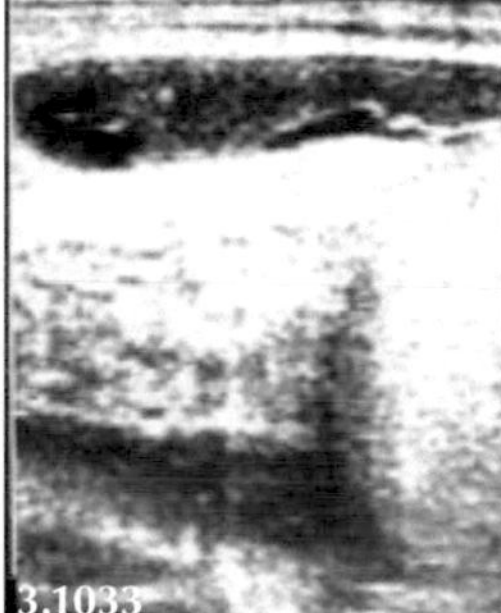

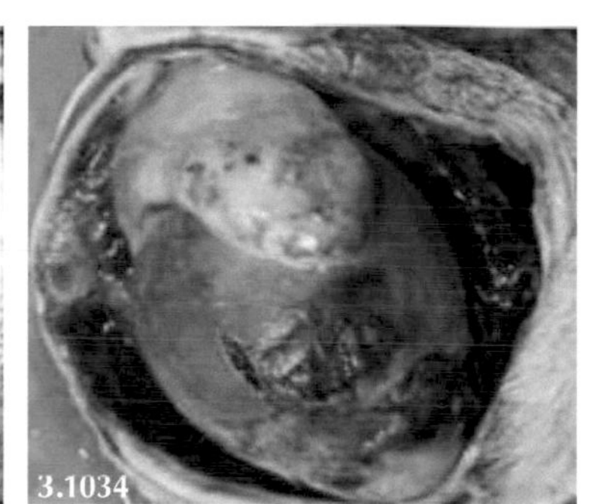

3.1032: Swollen right side of scrotum especially proximally around the pampiniform plexus. **3.1033:** Ultrasonography reveals inflammatory exudate containing fibrin between the vaginal tunics. **3.1034:** Necropsy confirms the sonographic findings with the fibrinous exudate more marked surrounding the pampiniform plexus and swollen head of the epididymis. 🎥 3.528

Sperm Granuloma

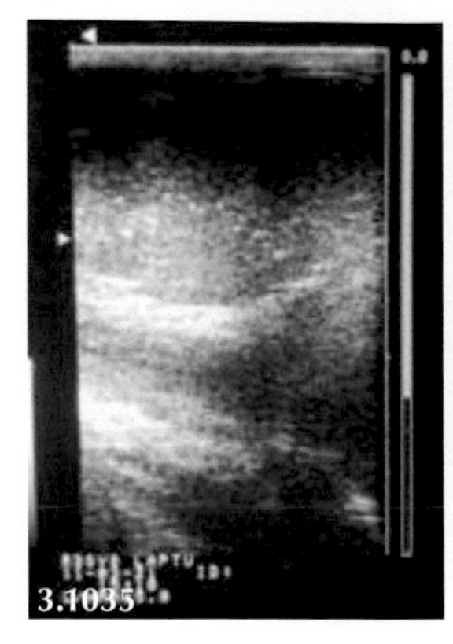

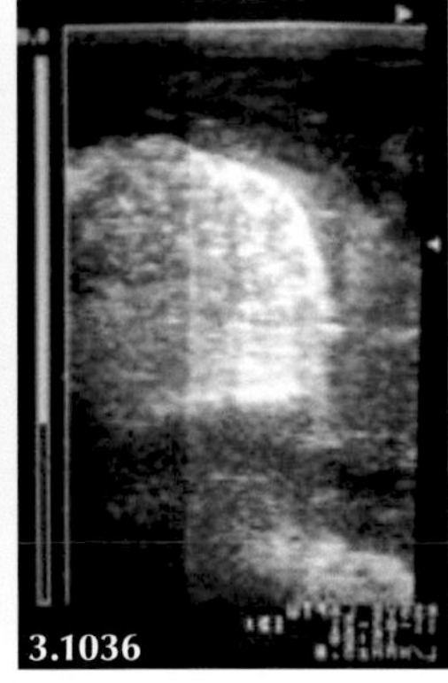

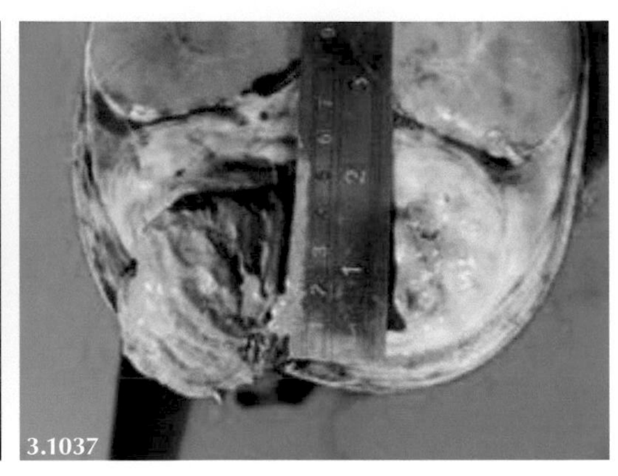

3.1035: Sonogram of testicle reveals hypoechoic appearance but containing multiple hyperechogenic dots consistent with testicular atrophy (proximal to the left in both sonograms). **3.1036:** Sonogram shows a 4 cm diameter thick-walled abscess ("snowstorm appearance"). **3.1037:** 4 cm diameter abscess within the tail of the epididymis at necropsy (proximal at top of image).

- **Key observations**: None until detected at routine pre-breeding check.
- **Clinical examination**: Bi-lobed appearance with enlarged tail of epididymis same diameter as (atrophied) testicle.
- **Key history events**: Most commonly observed several years after vasectomy.

Testicular Atrophy

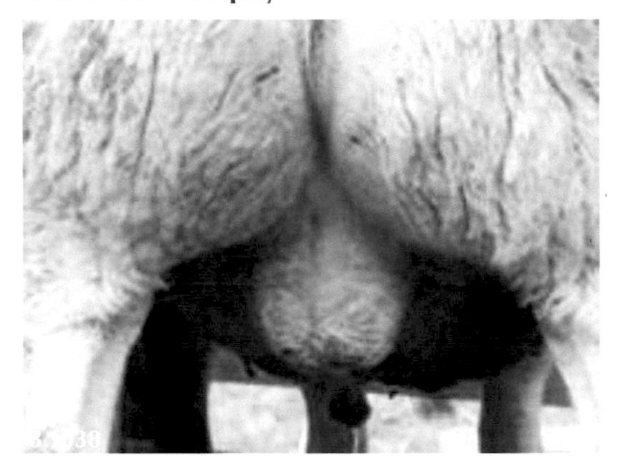

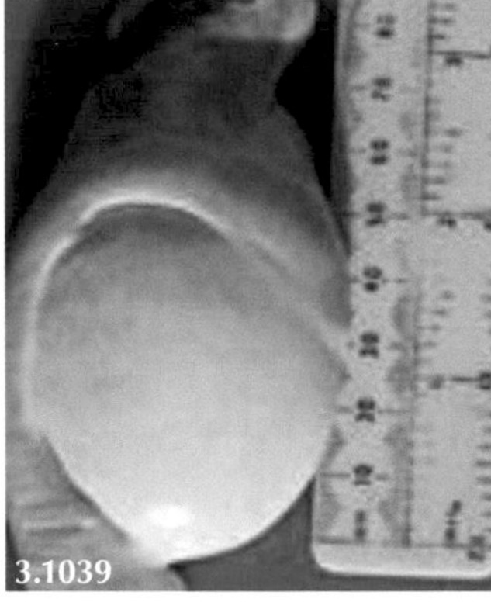

3.1038: Small scrotum. **3.1039:** The atrophied testicle measures <5 cm diameter.

- **Key observations**: Small scrotum/testicles detected at pre-breeding check.
- **Clinical examination**: Maximum scrotal circumference <34 cm. Testicles 4–5 cm diameter; normal >7 cm.
- **Key history events**: Poor reproductive performance in previous years, recent significant weight loss/debility. 3.529

Scrotal Hernia

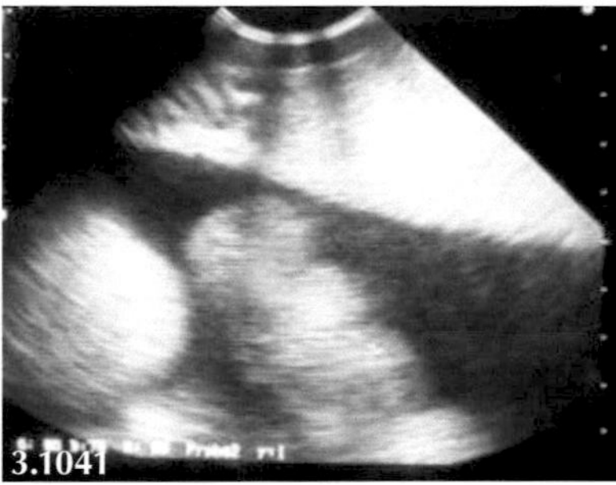

3.1040: Swollen right side of scrotum. **3.1041:** Sonogram revealing pampiniform plexus and testicle across top of image with fluid and loops of small intestine beneath.

- **Key observations**: Fluid distension of scrotum.
- **Clinical examination**: Able to reduce lesion when ram cast on back.
- **Key history events**: Increasing scrotal distension over weeks/months. 3.530, 3.531

Bulls

TABLE 3.107

Common Causes of Changes in Scrotal Size in Bulls

	Duration	Size	Consistency	Rectal temperature
Epididymitis	Days	+/++	+	+/++
Seminal vesiculitis	Days/weeks	NAD	NAD	NAD
Varicocoele	Weeks/months	++/+++	---	NAD

TABLE 3.108

Ancillary Tests to Investigate Cause of Changes in Scrotal Size in Bulls

Disease	Ancillary tests
Epididymitis	Ultrasonography
Seminal vesiculitis	Trans-rectal ultrasonography
Varicocoele	Ultrasonography

Epididymitis

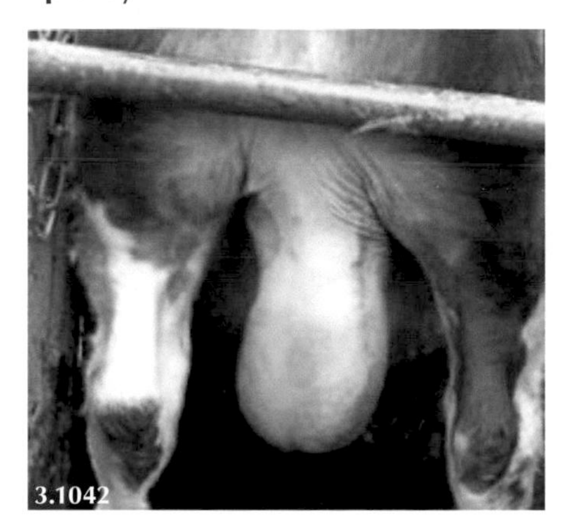

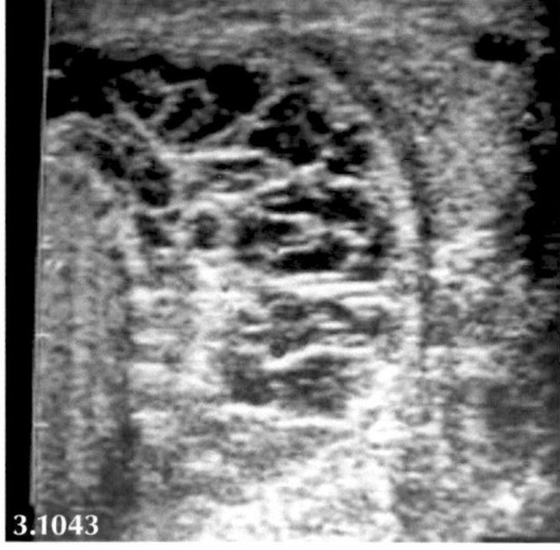

3.1042: Enlarged oedematous scrotum. **3.1043:** Enlarged fluid-filled tail of the epididymis identified on ultrasound examination; 5 MHz linear scanner, proximal to the left. 3.532

Seminal Vesiculitis

- **Key observations**: Discomfort during urination. Possible increased returns to service.
- **Clinical examination**: Rectal palpation reveals swollen, painful, firm vesicles with loss of lobulation. Ultrasound examination may reveal abscessation and dilation of the glands.
- **Key history events**: Poor bull fertility.

Varicocoele

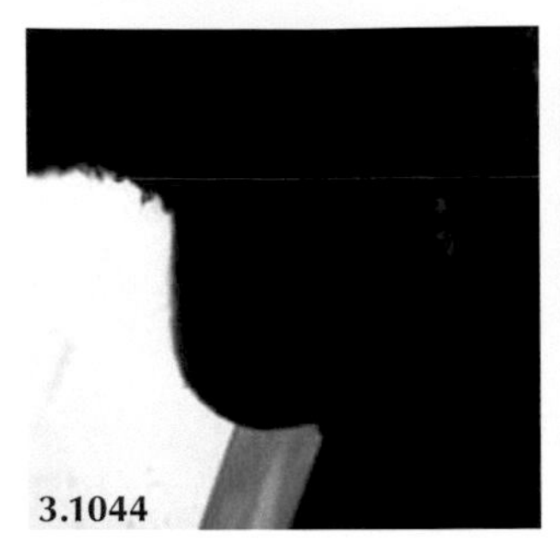
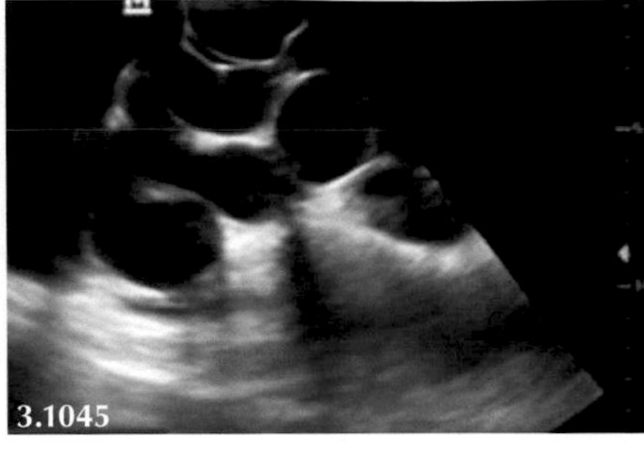
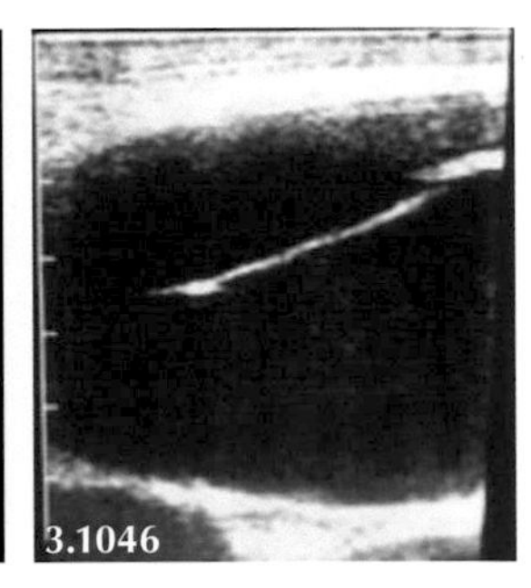

3.1044: Distended, fluid-filled scrotum. **3.1045:** Cross section of scrotum using a 6.5 MHz microarray probe reveals thin-walled fluid-filled circles. **3.1046:** Longitudinal section using a 5 MHz linear probe reveals thin-walled fluid-filled tubes.

- **Key observations**: Distended, fluid-filled scrotum.
- **Clinical examination**: Ultrasonography confirms a varicocoele.
- **Key history events**: Possible consequence of late Burdizzo castration but this cannot be confirmed. 🎥 3.533

3.19

Common Causes of Skin Lesions

Where Pruritus Is the Major Presenting Sign—Sheep

- Ectoparasite infestations in sheep are becoming an increasingly serious and widespread welfare concern in the UK.
- A return to compulsory plunge dipping is long overdue which could eliminate sheep scab and lice in one year.
- The widespread use of macrocyclic lactones to control sheep scab increases the selection pressure for resistant gut parasites.
- Sheep with headfly and other open skin lesions must be housed to prevent further nuisance and resultant self-trauma.

DOI: 10.1201/9781003106456-37

TABLE 3.109

Common Causes of Skin Lesions in Sheep Where Pruritus Is the Major Presenting Sign

	Duration of lesions at presentation	Pruritus	Fleece	BCS
Pediculosis	Weeks/months	+/++	–/––	NAD/–
Sheep scab	Weeks/months	+/+++	–/–––	–/–––
Cutaneous myiasis	Days	+/+++	–	–––
Headfly	Days/weeks	+/++	NAD	–/––

TABLE 3.110

Ancillary Test to Investigate Causes of Pruritus in Ovine Skin Lesions

	Ancillary tests
Pediculosis	Magnifying glass, >5 lice per skin parting significant
Sheep scab	Skin scraping, ×100 magnification. Pso o 2 ELISA
Cutaneous myiasis	Observation of larvae
Headfly	Observing flies on skin lesions

Pediculosis

3.1047: Pruritus with sheep nibbling at skin. **3.1048** and **3.1049:** Ewes in low body condition with large numbers of lice; fleece plucks/loss over flanks and shoulders.

3.1050: Pruritus caused by lice. **3.1051:** Poor quality wool with fleece breaks. **3.1052:** A pathetic emaciated hill ewe with a huge louse burden it could well do without.

- **Key observations**: Pruritus with wool loss. Nibble and kick at skin.
- **Clinical examination**: No serum exudation/skin hyperaemia. Louse burden more severe in emaciated sheep rather than pediculosis causing body condition loss (consequence rather than effect). More than five lice visible to naked eye in fleece parting over tailhead considered significant.
- **Key history events**: Purchased sheep, poor biosecurity, no effective quarantine treatment, no plunge dipping, misdiagnosis scab versus lice. 🎥 3.534, 3.535, 3.536, 3.537

Sheep Scab

3.1053: Intense pruritus, fleece plucked/rubbed out over flanks. **3.1054:** Serum exudation/sticky fleece. **3.1055:** Almost complete loss of fleece in neglected sheep with some wool re-growth over flank; thickened skin.

3.1056–3.1058: Unacceptable neglect of sheep scab infested sheep. The estimated duration of infestation would be around 16 weeks.

- **Key observations**: Intense pruritus, extensive fleece loss if infestation more than two to three months.
- **Clinical examination**: Hyperaesthetic to touch, positive "nibble response", serum exudation, skin hyperaemia, excoriation of skin/thickening. Handling may precipitate seizures in some hypersensitive cases.
- **Key history events**: Purchased sheep, poor biosecurity, no effective quarantine treatment, misdiagnosis scab versus lice. 🎥 3.538, 3.539, 3.540

Cutaneous Myiasis (Flystrike)

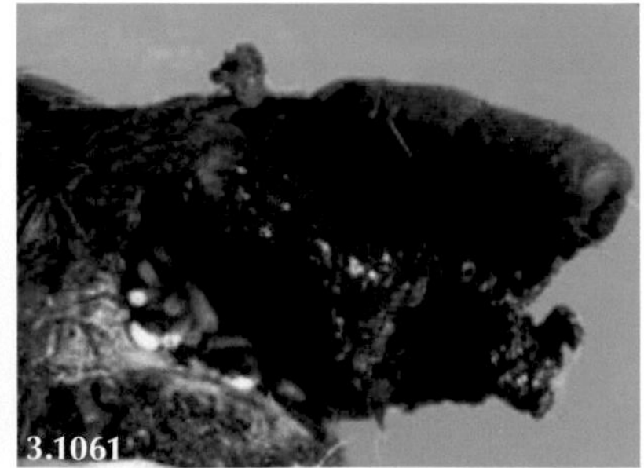

3.1059: Fecal contamination of tail and perineum, lamb nibbling at lesion around tailhead. **3.1060:** Extensive flystrike lesion over sheep's back. **3.1061:** Maggots infecting a proliferative footrot lesion.

- **Key observations**: Disturbed grazing, frequent tail swishing, attempts to nibble at lesion.
- **Clinical examination**: Foul putrid smell, matted wet overlying fleece, large numbers of maggots feeding on blackened skin surface.
- **Key history events**: Warm wet weather, full fleece, fecal contamination, proliferative foot and dermatophilosis lesions, no preventative measures (insecticides) adopted, poor shepherding. 🎥 3.541, 3.542, 3.543, 3.544

Head Fly

3.1062: Sheep tormented by headflies attempts to avoid contact by sheltering in bank in the field. **3.1063:** Skin lesions aggravated by sheep repeatedly kicking head with hind foot. **3.1064:** Photosensitisation lesion likely initiating cause in this case.

- **Key observations**: Sheep not grazing, "hiding" in banking, head on ground, frequent kicking at head. These sheep must be housed to prevent further fly contact.
- **Clinical examination**: Self-trauma, presence of feeding headflies.
- **Key history events**: Often horned sheep breeds, nearby woodland, predisposing lesions such as ear tag wounds, photosensitisation. No preventative strategy adopted. 3.545, 3.546, 3.547

Other Common Causes of Skin Lesions in Sheep (Where Pruritus Is Not a Major Presenting Sign)

- Dermatophilosis, as bottle-brush lesions on the face and ears, is seen in debilitated sheep.
- Photosensitisation occurs sporadically with the cause rarely identified.
- Caseous lymphadenitis of the parotid lymph node follows fighting injury to the poll. CLA is not a major cause of chronic weight loss/emaciation in the UK.

TABLE 3.111

Common Causes of Skin Lesions in Sheep (Where Pruritus Is Not a Major Presenting Sign)

	Likely duration	Hair/Fleece loss	Localised swelling
Dermatophilosis	Weeks/months	NAD/–	NAD
Orf	Weeks/months	NAD	+/++
Photosensitisation	Days	NAD	+/++
Woolslip	Weeks	–––	NAD
Scrotal mange	Weeks/months	–	NAD/+
Ringworm	Weeks/months	–	NAD
Actinobacillosis	Weeks/months	–	+/++
Caseous lymphadenitis	Months	–	+
In-growing horns	Weeks	NAD	NAD/+
Staphylococcal folliculitis	Days/weeks	–	+
Strawberry footrot	Weeks/months	–	+/++
Keloid	Months/years	NAD	+

TABLE 3.112

Ancillary Tests to Investigate Common Causes of Skin Lesions in Sheep (Where Pruritus Is Not a Major Presenting Sign)

Disease	Ancillary tests
Dermatophilosis	Impression smear of pus. Bacteriology
Orf	Direct EM observation of pox virus
Photosensitisation	Clinical signs. Causal plant rarely identified
Woolslip	Clinical examination. Failure to demonstrate ectoparasites
Scrotal mange	Skin scraping
Ringworm	Bacteriology
Actinobacillosis	Clinical examination. Bacteriology rarely justified
Caseous lymphadenitis	Clinical examination. Serology
In-growing horns	Inspection
Staphylococcal folliculitis	Bacteriology
Strawberry footrot	Direct EM observation of pox virus in impression smear
Keloid	Clinical examination

Dermatophilosis, Rain Scald

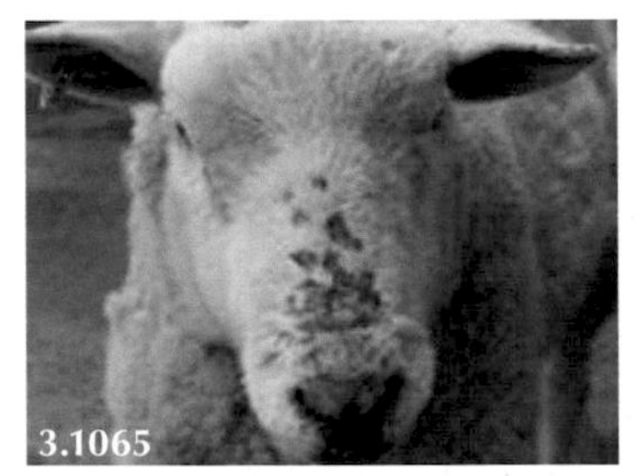

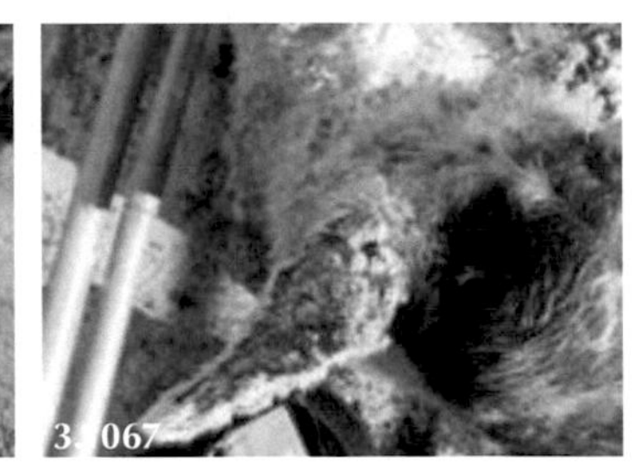

3.1065: "Bottle brush" lesions on face. **3.1066:** "Bottle brush" lesions which have bled when dislodged. **3.1067:** Larger lesions along the pinnae.

- **Key observations**: "Bottle brush" lesions on face, ears and along dorsum.
- **Clinical examination**: 3–5 mm scabs with purulent skin surface underneath. Skin may become keratinsed.
- **Key history events**: Very common in debilitated sheep. Following shearing, lesions often develop along the dorsum during prolonged wet weather. 🎥 3.548

Orf (Contagious Pustular Dermatitis)

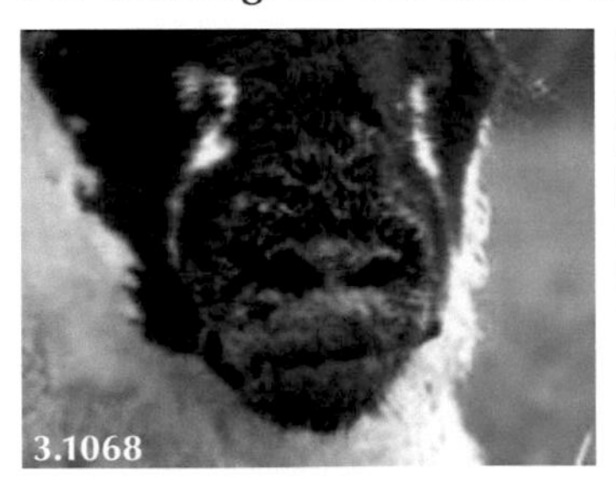

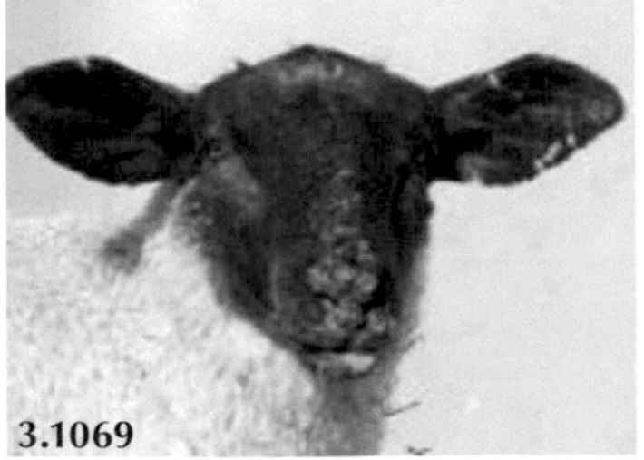

3.1068 and **3.1069:** Proliferative skin lesions on gum margins which bleed readily. **3.1070:** Concurrent infection with *Dermatophilus congolensis* causing severe superficial skin infection extending over muzzle.

- **Key observations**: Granulomatous lesions surrounding mouth and on muzzle.
- **Clinical examination**: Proliferative skin lesions on gum margins which bleed readily.
- **Key history events**: Affects orphan lambs most severely. Outbreaks may result from grazing rough pastures causing superficial skin trauma. 🎥 3.549

Photosensitisation

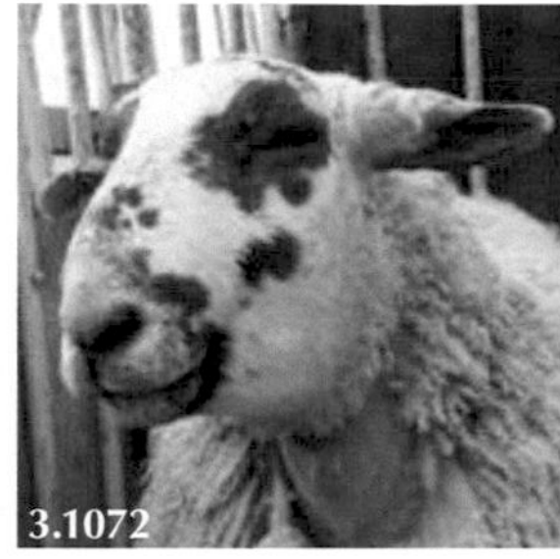

3.1071 and **3.1072:** Dullness with oedematous head and ears in early stages of photosensitisation. **3.1073:** Chronic stage—ears are often sloughed. Skin lesions attract head flies leading to self-trauma.

- **Key observations**: Sudden onset dullness, not eating, swollen head. May induce vigorous head-rubbing, self-trauma.
- **Clinical examination**: Marked pitting oedema of the head and ears.
- **Key history events**: Often only one sheep affected, predominantly white-faced breeds. 🎥 3.550, 3.551, 3.552

Woolslip

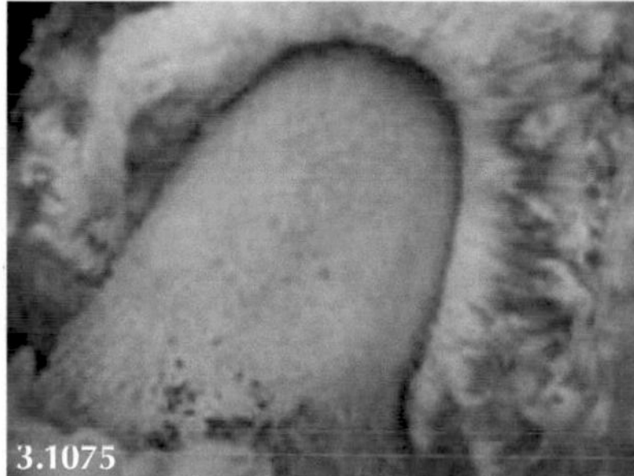

3.1074: Woolslip occurring about four to six weeks after an acute disease event. **3.1075:** Note the lack of skin reaction. **3.1076:** Woolslip showing wool re-growth about three months after disease event.

- **Key observations**: Premature fleece loss. No parasitc involvement.
- **Clinical examination**: No pruritus/skin inflammation. Appears similar to normal fleece rise before shearing.
- **Key history events**: Process starts four to six weeks after severe illness such as pregnancy toxaemia or abortion.

Scrotal Mange

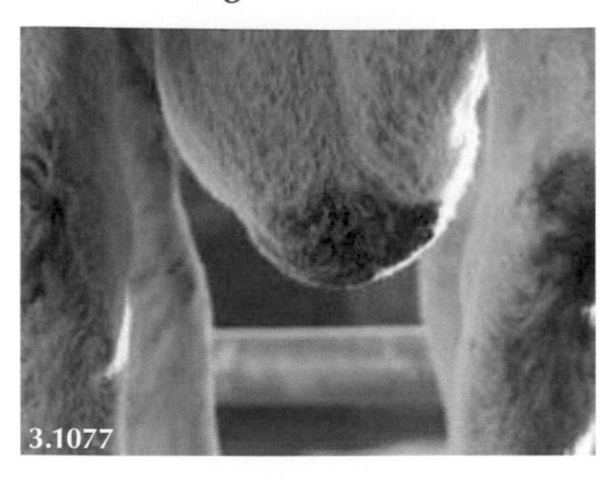

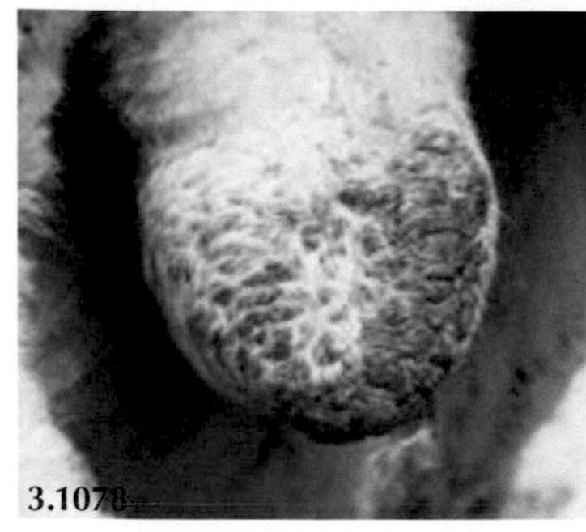

3.1077 and **3.1078:** Thickened scrotal skin forming keratotic nodules.

- **Key observations**: Thickened scrotal skin.
- **Clinical examination**: Keratinised ventral skin area of scrotum.
- **Key history events**: Common finding in many rams without apparent adverse effects on fertility.

Ringworm

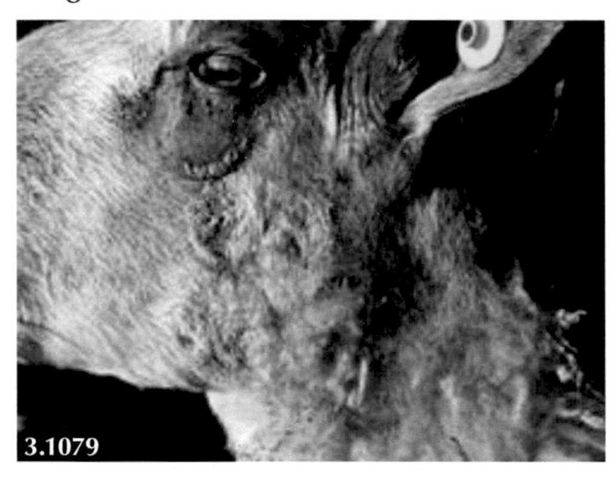

3.1079: 2–4 cm diameter lesions on the haired skin areas, usually the face/ears.

- **Key observations**: Circular areas with dry flaking skin.
- **Clinical examination**: Crusting lesions leading to hair loss. Non-pruritic.
- **Key history events**: Sheep housed in building with cattle, only haired areas affected.

Actinobacillosis

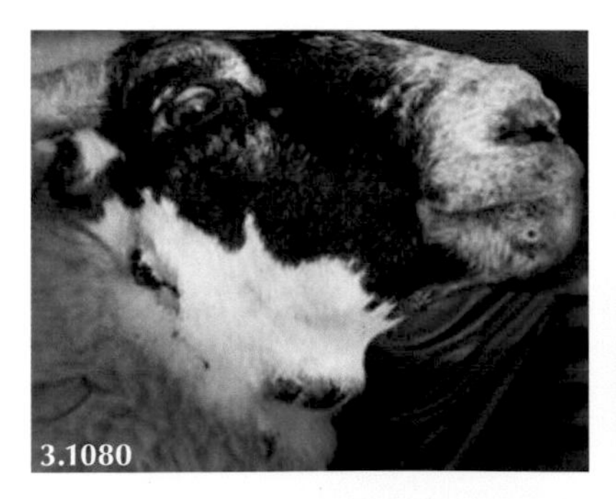

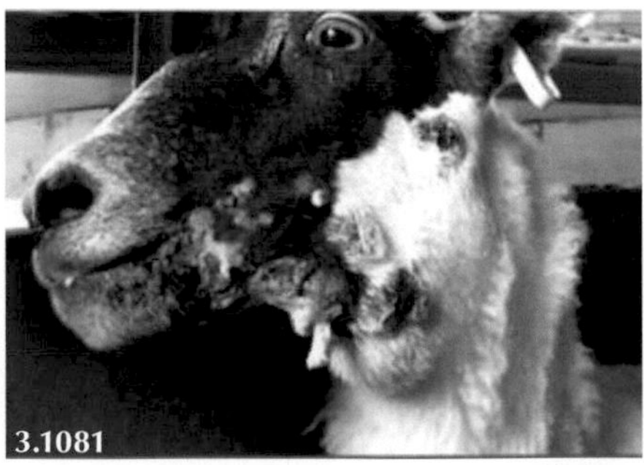

3.1080 and **3.1081:** Multiple subcutaneous abscesses, some ruptured discharging green toothpaste-like pus.

- **Key observations**: Abscesses on face and neck, some discharging viscous green pus.
- **Clinical examination**: Subcutaneous abscesses, enlarged drainage lymph nodes.
- **Key history events**: Present for weeks/months. Sheep grazing pastures with thorn bushes but only one or two sheep in group affected. 🎥 3.553

Caseous Lymphadenitis

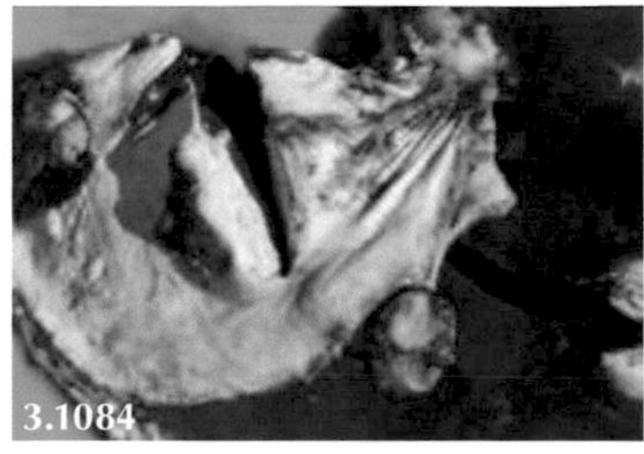

3.1082: CLA affecting the right parotid lymph node at the base of the ear. **3.1083:** Lamellar CLA abscess affecting the mediastinal lymph node. **3.1084:** Lamellar CLA abscess in the liver.

- **Key observations**: Affected carcase lymph nodes swollen and may discharge pus. Loss of hair/fleece overlying lymph node. May have concurrent head/poll wound from head butting (rams).
- **Clinical examination**: In the UK, parotid lymph node most commonly affected. Visceral CLA is rare.
- **Key history events**: Purchase of infected sheep. Poor quarantine programme. Poor biocontainment once identified in purchased stock.

In-Growing Horns

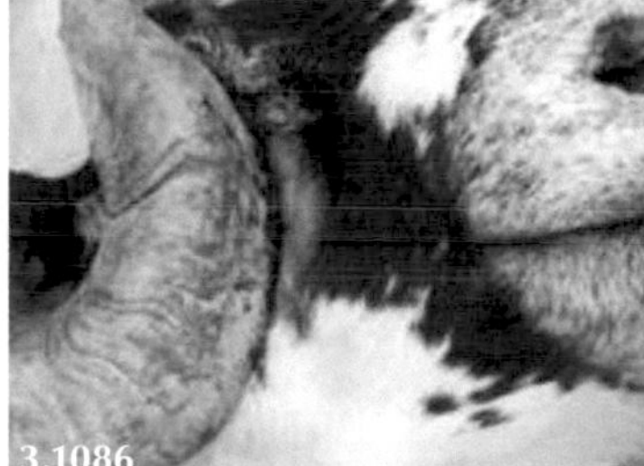

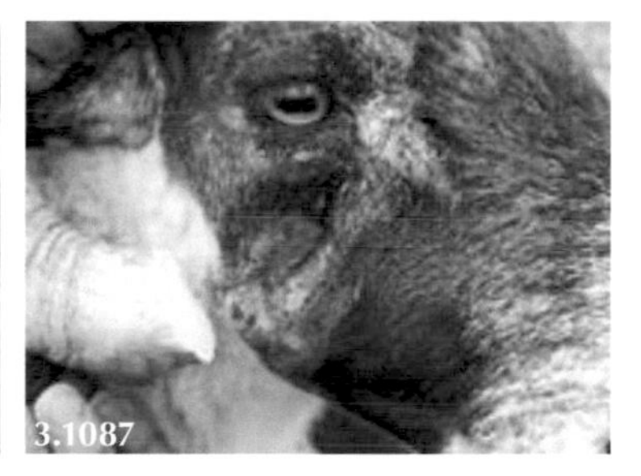

3.1085 and **3.1086:** Horn pressing on skin. **3.1087:** Removal of horn tip using saw or embryotomy wire revealing early skin abrasion.

- **Key observations**: Horn turned into face.
- **Clinical examination**: Horn contacting/penetrating skin.
- **Key history events**: Broken horn may grow into face, most commonly observed in rams with tight horns and their progeny. 3.554, 3.555

Keloid

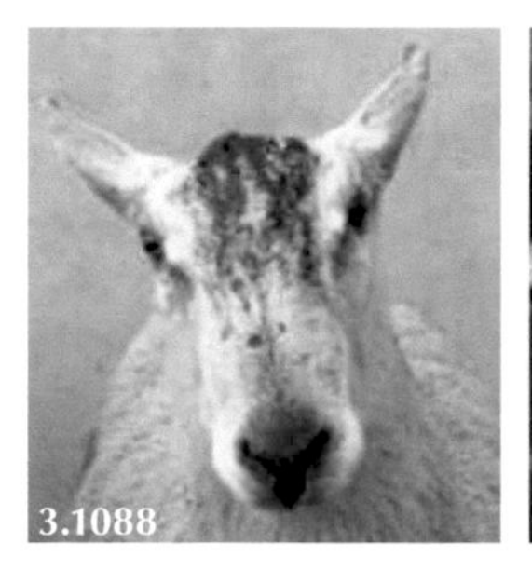

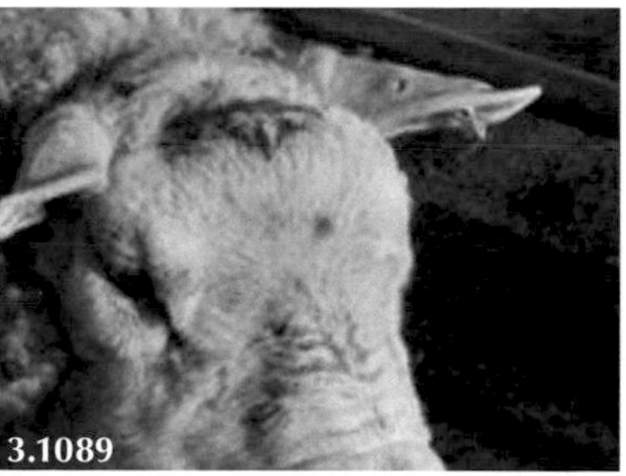

3.1088: Persistent skin trauma from fighting likely leads to keloid formation. **3.1089:** Horn growth from the poll in this Texel ram. **3.1090:** Keloid likely originating from persistent ear tag injury/infection.

- **Key observations**: Horn growing from poll/skin injury site.
- **Clinical examination**: Varies from keratinised skin proliferation to horn-like structure.
- **Key history events**: Repeated skin trauma to poll by fighting. May also arise from skin trauma sites caused by chronic dermatophilosis and ear tags. 3.556, 3.557

Staphylococcal Folliculitis

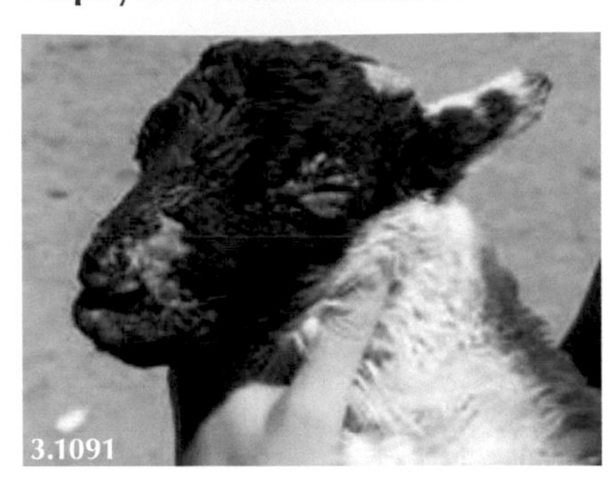

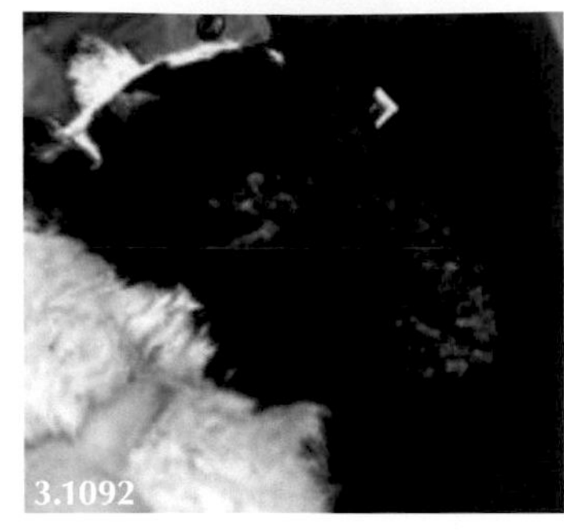

3.1091 and **3.1092:** Proliferative skin lesions with areas of hair loss in this orphan Scottish Blackface ram.

- **Key observations**: Nodular facial swellings.
- **Clinical examination**: Proliferative/granulomatous skin lesions with over-lying hair loss.
- **Key history events**: Usually artificially reared orphan lambs.

Strawberry Footrot

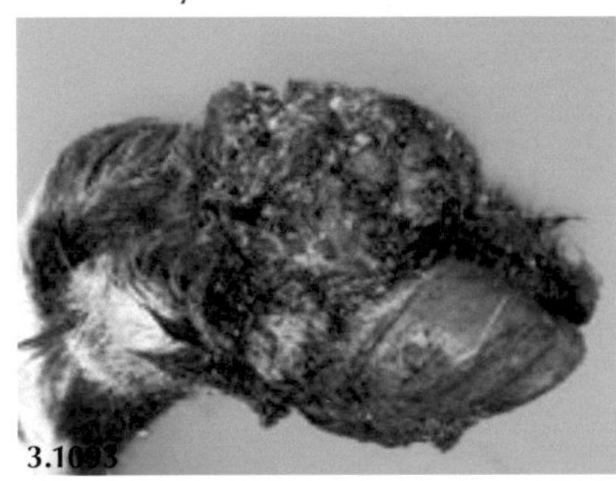

3.1093: Proliferative painful skin granuloma likely caused by orf virus and secondary bacterial infection.

- **Key observations**: Large spherical granulomatous lesion on distal limb in growing lambs.
- **Clinical examination**: Marked drainage lymph node reaction.
- **Key history events**: History of orf on farm. Grazing stubbles or rough grazing.

Skin Conditions of Cattle

- Some beef farmers appear not to recognise lice with almost total body hair loss especially in bulls.
- Papillomata occur sporadically but can become severe and widespread in individual animals for no obvious reason.

TABLE 3.113

Common Skin Conditions of Cattle

	Likely duration	Pruritis	Hair	Body condition
Lice	Weeks/months	++/+++	–/–––	NAD/–
Chorioptic mange	Weeks/months	+	––/–––	NAD
Ringworm	Weeks	NAD	–	NAD
Papillomata	Weeks/months	NAD	NAD	NAD
Photosensitisation	Weeks/months	NAD	–/––	–
Rain scald	Weeks/months	NAD	–	NAD

TABLE 3.114

Ancillary Tests to Investigate the Common Causes of Skin Lesions in Cattle

	Ancillary tests
Lice	X10 Microscopic identification
Chorioptic mange	Location. Microscopic identification in skin scrapings
Ringworm	Microscopic identification in hair pluck. Culture
Papillomas	Direct electron microscopic examination
Photosensitisation	Clinical examination
Rain scald	Impression smear. Culture of scab material

Lice

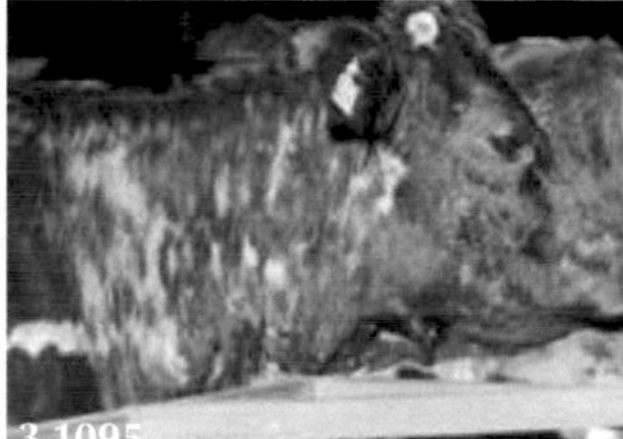

3.1094–3.1096: Extensive hair loss caused by heavy louse infestation.

- **Key observations**: Extensive hair loss, frequent rubbing/licking.
- **Clinical examination**: Biting and chewing lice visible to naked eye/magnifying glass.
- **Key history events**: No lice control, much more prevalent during winter housing period, and in bulls. 🎥 3.558

Chorioptic Mange

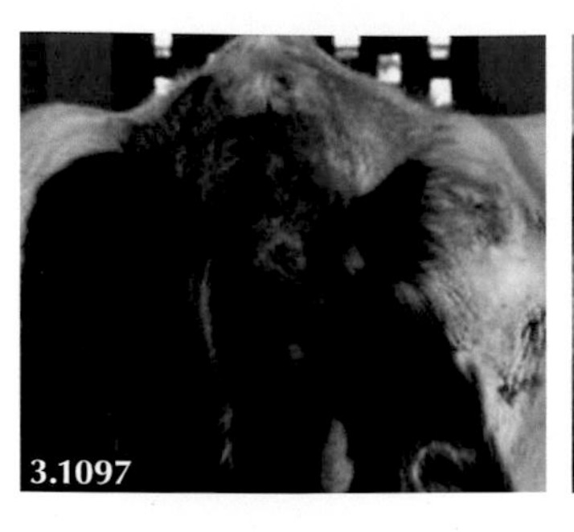

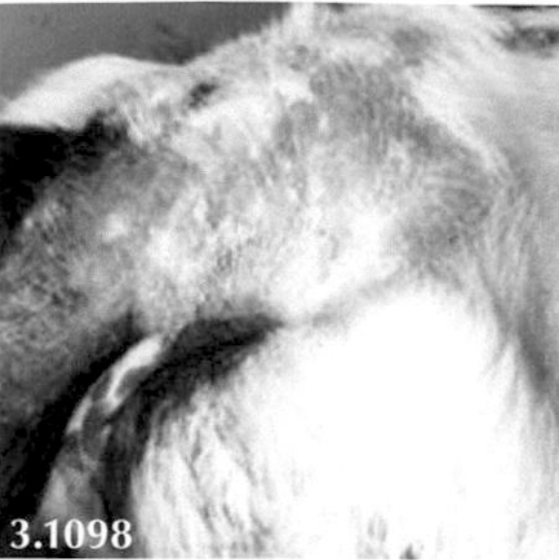

3.1097 and **3.1098:** Hair loss and keratinised skin on tailhead of dairy cow caused by chorioptic mange.

- **Key observations**: Focal areas of highly keratinised skin around tailhead with hair loss.
- **Clinical examination**: Dry, fissured and flaky skin around tailhead.
- **Key history events**: Observed more often in dairy cows during winter months.

Ringworm

3.1099–3.1101: Areas of hair loss on the head and neck with dry flaking skin.

- **Key observations**: Many circular skin lesions on head and neck becoming confluent, not pruritic.
- **Clinical examination**: Dry scaling lesions, slight skin thickening.
- **Key history events**: Growing calves, usually during housing periods. More severe in cattle in poor body condition although does not cause debility. 🎥 3.559

Papillomata

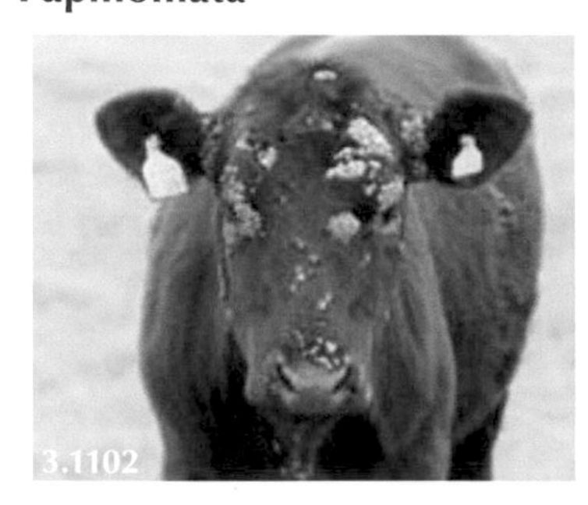

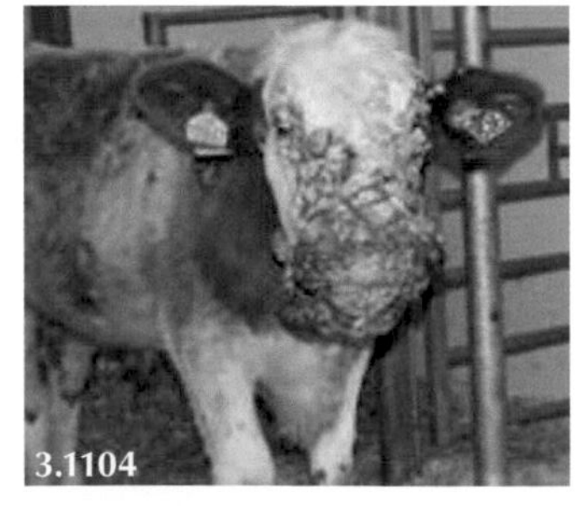

3.1102 and **3.1103:** Multiple small papillomata on the head and neck. **3.1104:** Extensive lesions are rare.

- **Key observations**: Multiple small skin growths affecting most young animals in group.
- **Clinical examination**: Small papillomatous growth usually confined to head and neck. Unusually these can become extensive.
- **Key history events**: Slow growth over several months before spontaneous resolution. 🎥 3.560, 3.561

Photosensitisation

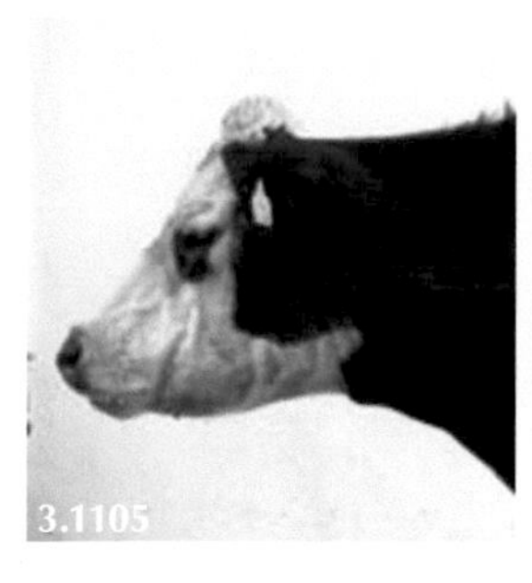
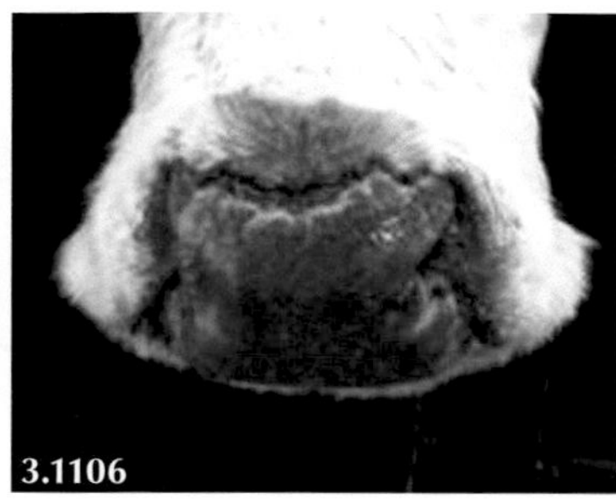
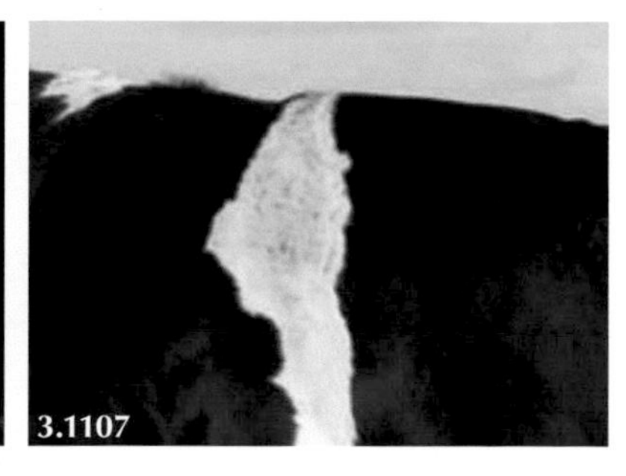

3.1105: Oedematous skin folds affecting white skin of head. **3.1106:** Sharply demarcated skin lesion; muzzle is normal. **3.1107:** Serum exudation and hair loss over white skin area only.

- **Key observations**: Swollen face and eyelids, only white skin areas affected.
- **Clinical examination**: Initial oedema with eventual sloughing of serum exudation and hair.
- **Key history events**: Individual animal, summer months, likely ingestion of photodynamic agent from toxic plant. 🎥 3.562, 3.563

Rain Scald

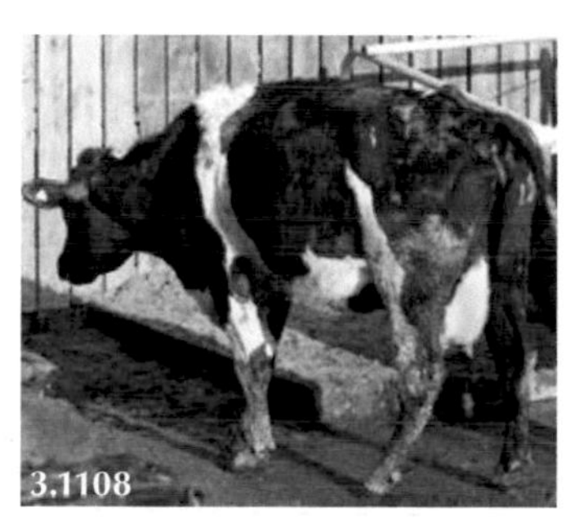

3.1108: Widespread superficial skin infection appears along the dorsum causing extensive hair loss in out-wintered cattle during wet winter weather.

- **Key observations**: Loss of hair over dorsum.
- **Clinical examination**: Superficial skin infection.
- **Key history events**: Out-wintered cattle in persistent wet weather.

3.20

Common Causes of Facial/ Mandibular Swellings in Cattle

- Poor milk feeding hygiene can result in an outbreak of calf diphtheria in dairy calves.
- Actinomycosis lesions in beef cows are often too advanced for successful treatment.
- Farm machinery striking an animal with its head through a feed barrier can cause a mandibular fracture.

TABLE 3.115

Common Causes of Facial/Mandibular Swellings in Cattle

	Duration	Size	Salivation	Rectal temperature
Calf diphtheria	Days	+/++	+	NAD/+
Actinobacillosis	Days	++	++/+++	NAD/+
Actinomycosis	Weeks/months	++/+++	+/++	NAD/+
Fractured mandible	Hours	+	+/++	NAD
Tooth root abscess	Weeks/months	+	+/++	NAD/+
Photosensitisation	Days	+/++	NAD	NAD
Pharyngeal abscess	Days	+/++	NAD	NAD

TABLE 3.116

Ancillary Test to Investigate the Common Causes of Facial/Mandibular Swellings in Cattle

	Ancillary tests
Calf diphtheria	Mouth gag and torch
Actinobacillosis	Palpation using mouth gag
Actinomycosis	Radiography
Fractured mandible	Radiography
Tooth root abscess	Radiography
Photosensitisation	None
Pharyngeal abscess	Ultrasonography. Radiography?

DOI: 10.1201/9781003106456-38

Calf Diphtheria

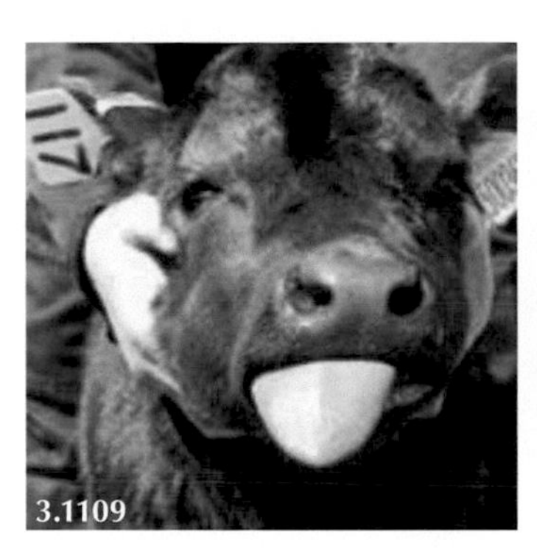

3.1109 and **3.1110:** Swelling of right cheek in young beef calves caused by calf diphtheria.

- **Key observations**: Discrete swelling of cheek.
- **Clinical examination**: Using mouth gag and light source reveals well-circumscribed diphtheritic membrane with erosion of papillae. Enlarged submandibular LN.
- **Key history events**: Related to poor hygiene/dirty feeding equipment in dairy calves. 🎥 3.564

Actinobacillosis

- **Key observations**: Profuse salivation, tongue protrudes 10–15 cm.
- **Clinical examination**: Very firm tongue.
- **Key history events**: Occurs sporadically often when cattle are fed roughages containing very fibrous plant stems.

Actinomycosis

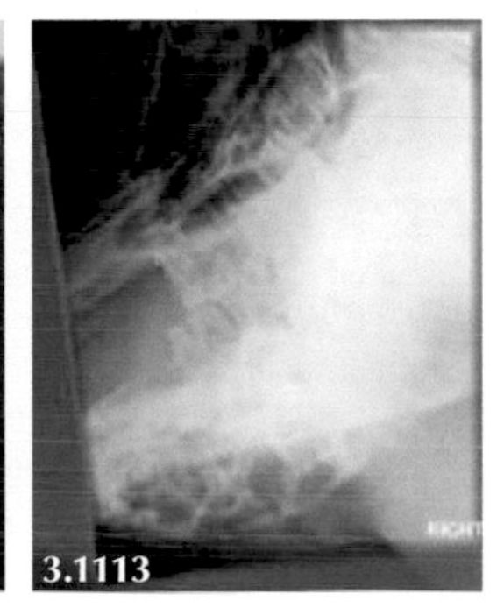

3.1111: Large bony swelling of left mandible. **3.1112:** Large bony swelling of right mandible. **3.1113:** Lateral radiograph of the left mandible.

- **Key observations**: Salivation, loss of body condition, extensive bony swelling of horizontal ramus of one mandible.
- **Clinical examination**: Gag—soft tissue swelling of mandible/gum, possible loose and maligned premolar and molar teeth.
- **Key history events**: Insidious onset over weeks/months most commonly adult beef cattle. 🎥 3.565

Fractured Mandible

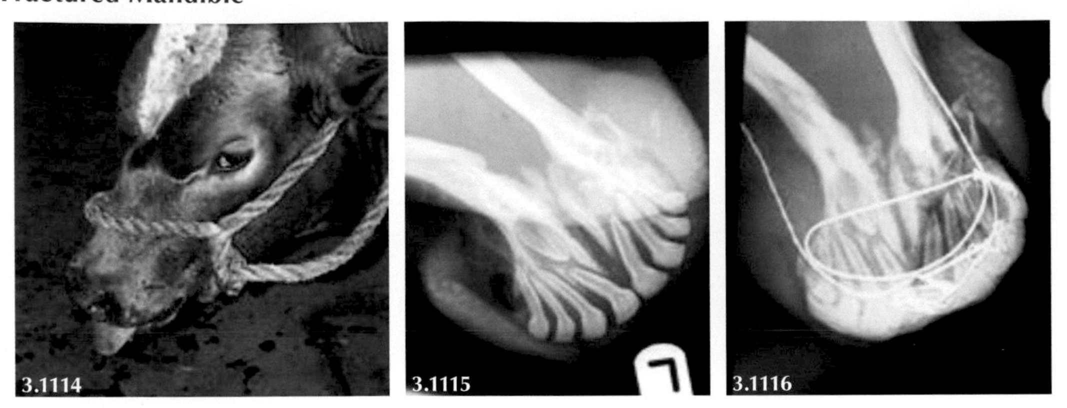

3.1114: Protruding tongue and profuse salivation. **3.1115:** Radiograph reveals fracture of the mandibular symphysis. **3.1116:** Stabilisation of fracture.

- **Key observations**: Salivation, quidding.
- **Clinical examination**: Gag—mal-aligned premolar and molar teeth if affects horizontal ramus.
- **Key history events**: Animal's head through feed barrier struck by tractor wheel.

Tooth Root Abscess

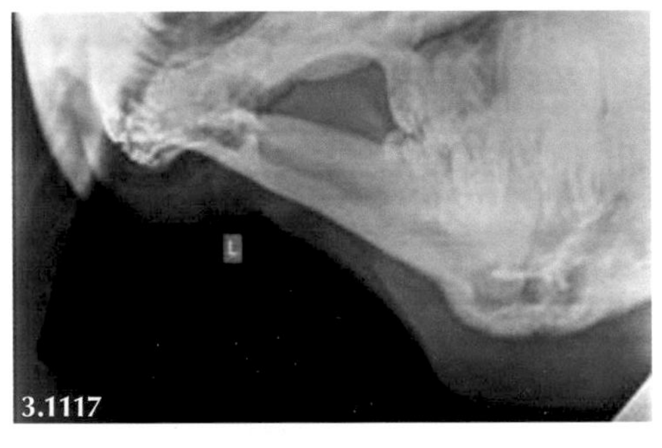

3.1117: Lateral radiograph of the left horizontal ramus of the mandible showing considerable bone lysis involving the tooth roots, and new bone formation.

- **Key observations**: Poor appetite, possible quidding, ventral swelling of horizontal ramus.
- **Clinical examination**: Bony and soft tissue swelling of one mandible, fistula is uncommon, swelling of the submandibular lymph node.
- **Key history events**: Insidious onset over weeks/months.

Photosensitisation

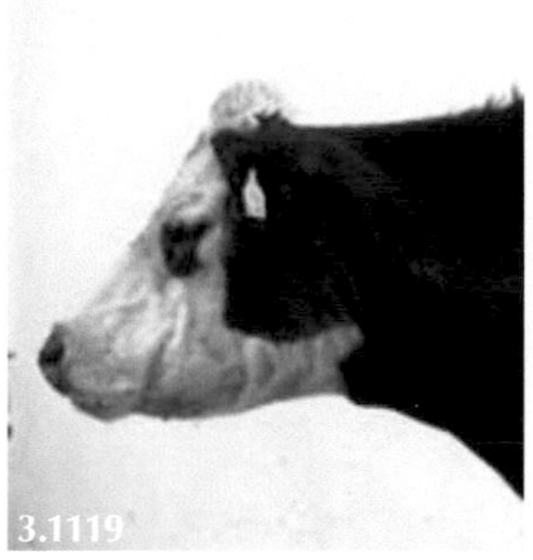
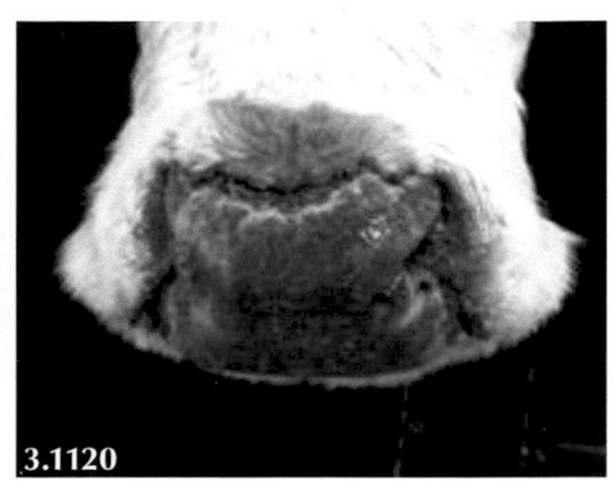

3.1118: Pendulous ears caused by oedema, slightly swollen face. **3.1119:** Oedematous folds of non-pigmented skin. **3.1120:** Same cow as centre, showing sharply demarcated area of non-pigmented skin affected.

- **Key observations**: Swollen oedematous face and ears, painful with frequent rubbing/kicking at head.
- **Clinical examination**: Only non-pigmented skin affected. Pitting oedema in early stages.
- **Key history events**: Sunny days, often caused by plant toxin (primary photosensitisation). 🎥 3.566

Pharyngeal Abscess

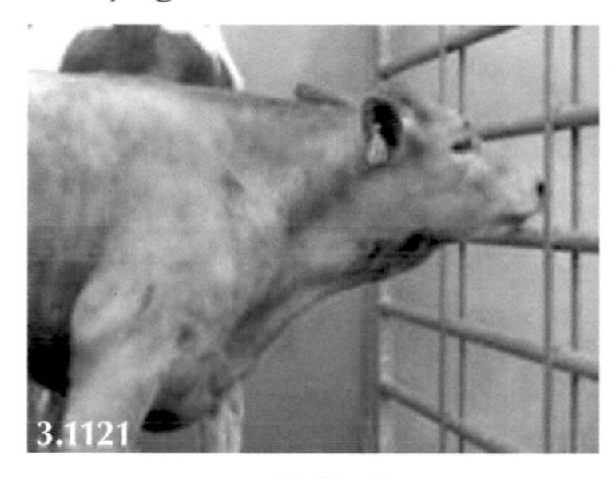
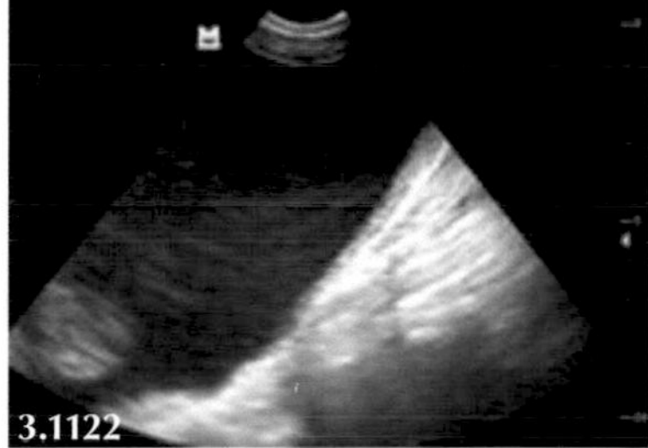
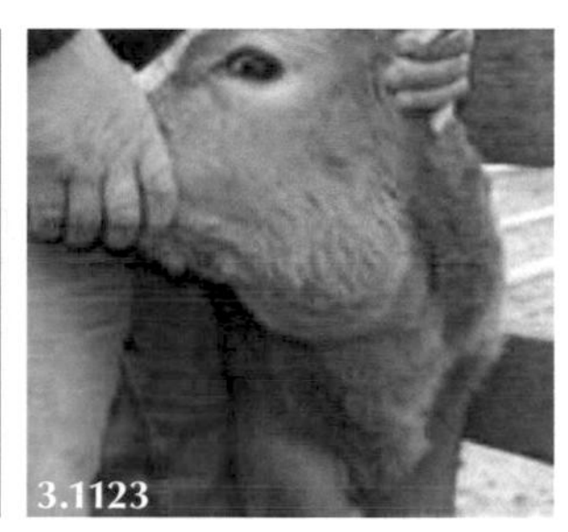

3.1121: Stirk presents with an extended neck and a firm subcutaneous swelling situated immediately caudal to the angle of the right mandible. **3.1122:** Ultrasound examination reveals the swelling extends for 12 cm with a distinct capsule; the contents appear anechoic, containing multiple hyperechoic dots ("snowstorm appearance") consistent with an abscess. **3.1123:** Large abscess in the pharyngeal area in a month-old calf.

- **Key observations**: Localised firm swelling.
- **Clinical examination**: Ultrasound examination with a "snowstorm appearance", followed by needle aspiration to yield pus.
- **Key history events**: Often none noted. 🎥 3.567

3.21

Common Causes of Subcutaneous Swellings in Cattle

- Large abscesses and haematomata are common in cattle and are readily differentiated on ultrasound examination.
- Injection site reaction/infections are common because many farmers do not use new needles and syringes for each procedure. Flutter valves are rarely correctly sterilised after use.

TABLE 3.117

Common Causes of Subcutaneous Swellings in Cattle

	Size	Lameness	Toxaemia
Abscess	+/+++	NAD	NAD
Haematoma	+/+++	NAD/+	NAD
Injection sites	+/++	NAD	NAD
Oedema	+/++	NAD	NAD
Urine	+/++	NAD	NAD/+

TABLE 3.118

Ancillary Tests to Investigate the Common Causes of Subcutaneous Swellings in Cattle

	Ancillary tests
Abscess	Ultrasonography then drain
Haematoma	Ultrasonography. Do not needle puncture or attempt drainage
Injection sites	Ultrasonography will identify abscess
Oedema	Palpation, identification of primary cause
Urine	Location. Palpate distended bladder

Subcutaneous Abscess

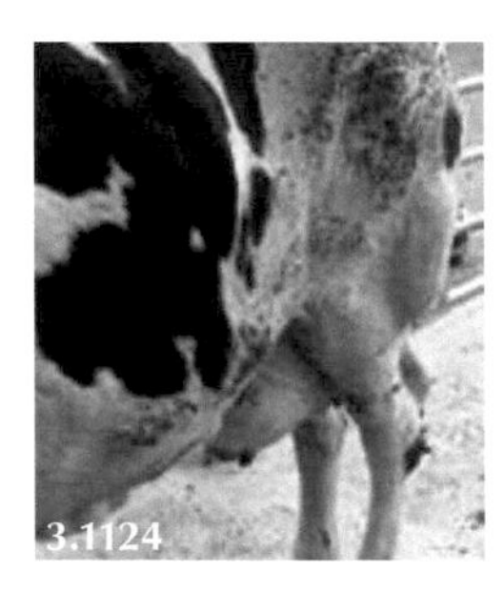

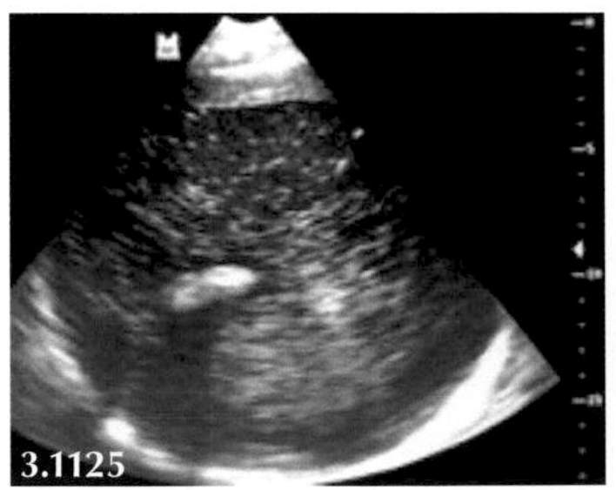

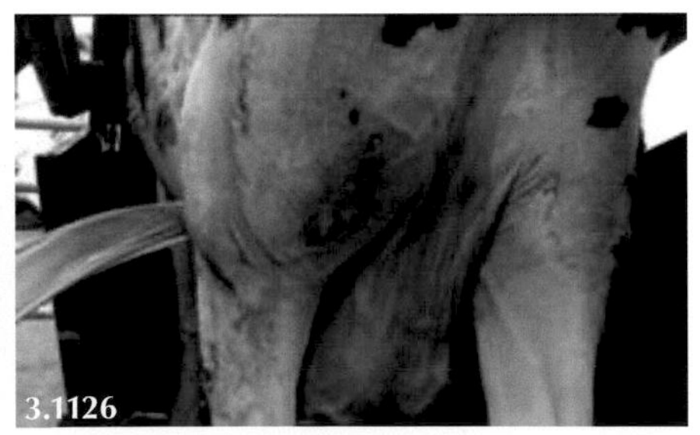

3.1124: Large subcutaneous fluid swelling. **3.1125:** Ultrasound reveals 16 cm deep subcutaneous abscess. **3.1126:** Lancing the abscess at the lowest point.

- **Key observations**: Large subcutaneous fluid swelling develops over several weeks/months.
- **Clinical examination**: Large subcutaneous fluid accumulation. Not painful.

DOI: 10.1201/9781003106456-39

- **Key history events**: Contaminated intramuscular injection. Check antibiotic storage, use of disposable syringes and needles, injection technique. 3.568, 3.569

Haematoma

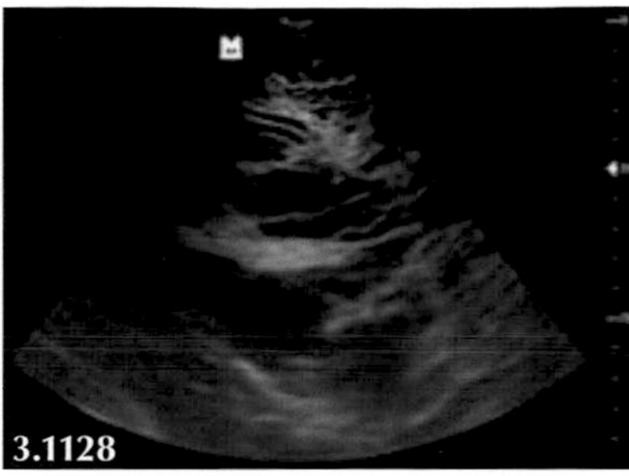

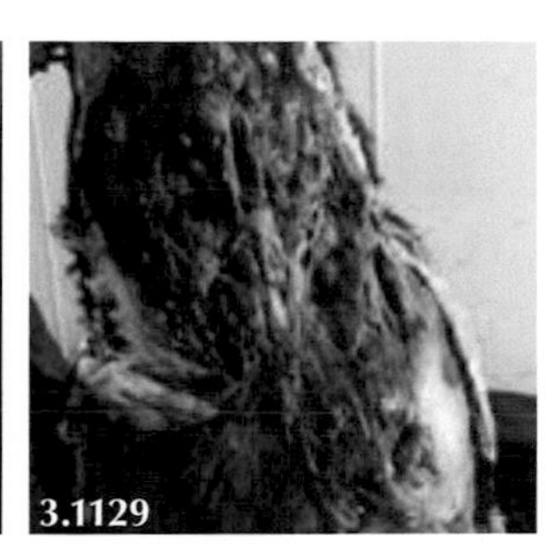

3.1127: Large haematoma of the neck. **3.1128:** Ultrasound scan of a haematoma extending to 30 cm. The hyperechoic fibrin matrix is clearly visible within the anechoic fluid. **3.1129:** Necropsy from another case shows an extensive haematoma over most of the chest and abdominal wall. 3.570

- **Key observations**: Large/very large subcutaneous fluid swellings.
- **Clinical examination**: Oedematous, non-painful.
- **Key history events**: Blunt trauma, develop very rapidly then gravitate over several weeks/months.

Injection Sites

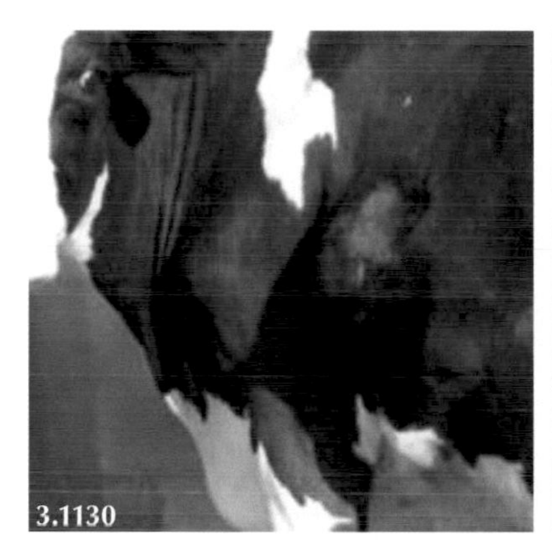

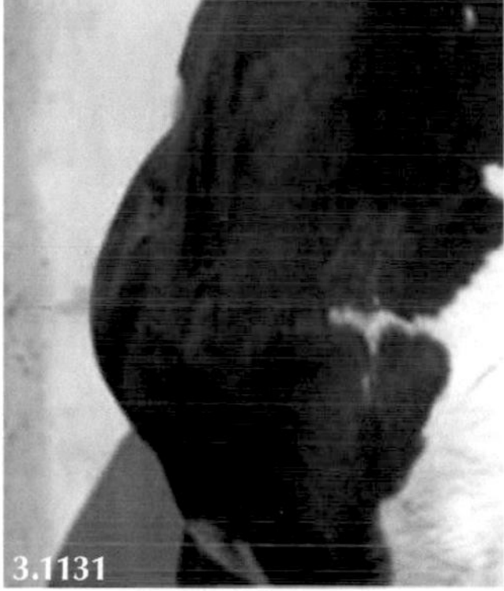

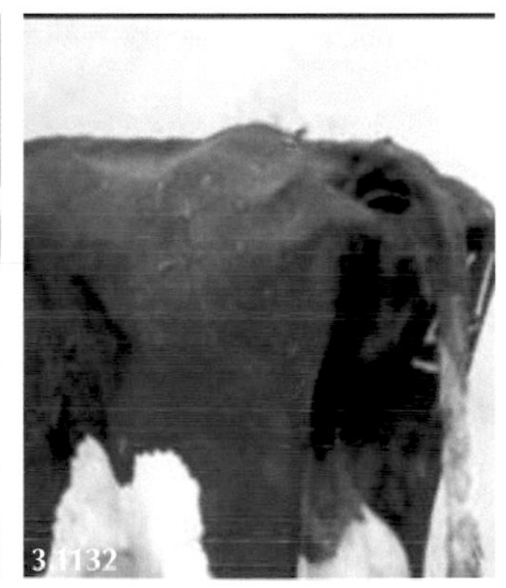

3.1130: 30 cm diameter swelling at site of subcutaneous calcium borogluconate injection (lateral view). **3.1131:** Caudal view of subcutaneous calcium borogluconate injection site. **3.1132:** Large (>15 cm diameter) painful swelling over the left hip area after a long-acting oxytetracycline injection.

- **Key observations**: Large, firm and painful swellings following subcutaneous/intramuscular injection.
- **Clinical examination**: Often surrounding oedema, painful.
- **Key history events**: Within days of subcutaneous/intramuscular injection using contaminated equipment/incorrectly stored product. 3.571, 3.572

Oedema

3.1133 and **3.1134:** Both cows show extensive brisket oedema and variable submandibular oedema caused by traumatic pericarditis. **3.1135**: Dilated cardiomyopathy. 3.573

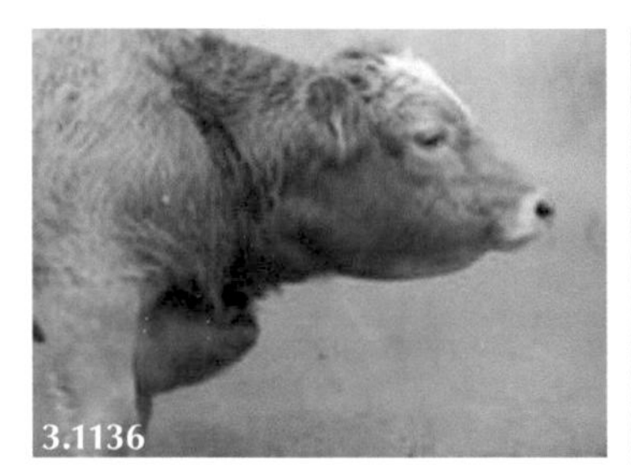

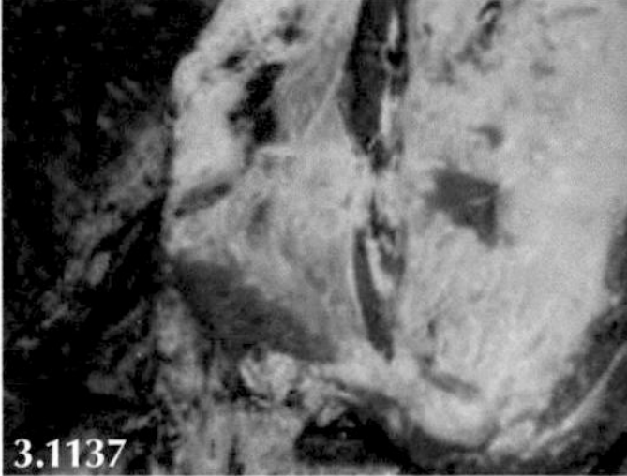

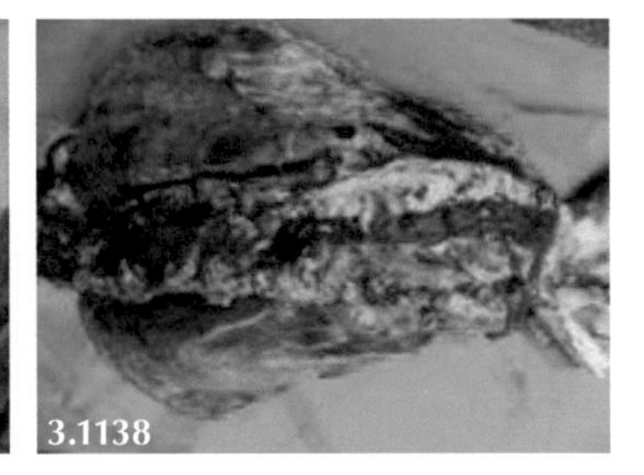

3.1136: Extensive brisket oedema caused by thymic lymphosarcoma. **3.1137:** Oedema shown at necropsy. **3.1138:** Thymic lymphosarcoma.

- **Key observations**: Large/very large subcutaneous fluid swellings which pit under finger pressure.
- **Clinical examination**: Oedematous, non-painful.
- **Key history events**: Secondary to right-sided heart failure.

Urine

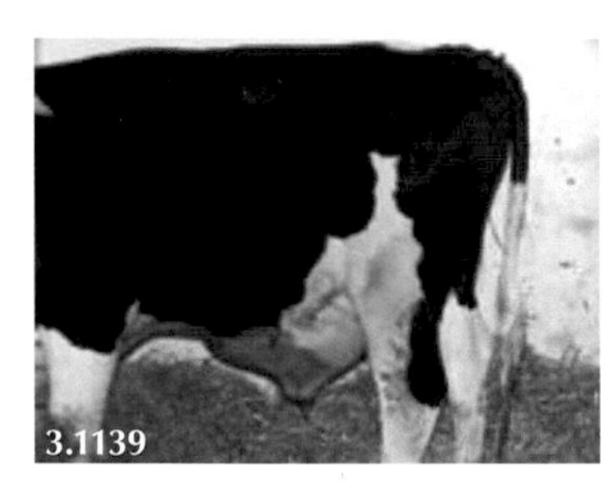

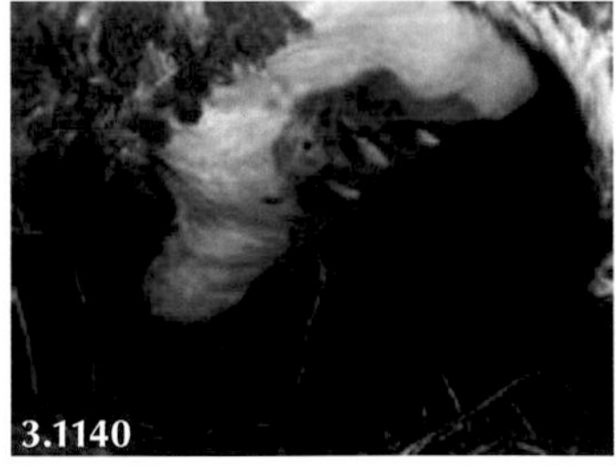

3.1139: Large swelling surrounding the prepuce and extending caudally along the penis.
3.1140: Subcutaneous urine accumulation causing skin necrosis.

- **Key observations**: Large/very large subcutaneous fluid swellings which may pit under finger pressure.
- **Clinical examination**: Overlying skin necrosis in neglected cases. Distended urinary bladder on rectal palpation. Note stab incisions do not work to drain urine.
- **Key history events**: Secondary to urolithiasis.

3.22

Common Causes of Sudden Death

Adult Sheep

- Not all sheep reported as "sudden death" died within 24 hours of onset of illness.
- Single on-farm necropsies are of limited use in determining the important infectious diseases on a farm as they fail to quantify annual incidence.
- Differentiating pulmonary congestion from acute respiratory disease necessitates histopathology.
- It is not possible to diagnose OPA on gross examination alone.

While it is a legal requirement that sheep under intensive management are inspected at least once daily, not all sheep are managed to this standard. Not all sick individuals are easily detected once they have gathered into a large mob of several hundred sheep. Without veterinary examination, not all sick sheep are treated correctly. Sudden death has therefore been defined as "has died within 48–72 hours of last close inspection" however "sudden death" may follow illness of up to one week. In some situations, euthanasia is necessary for welfare reasons and some of these conditions are featured in the video recordings with affected sheep killed immediately afterwards. A veterinary service that examines all farm mortalities at a knackery collection centre is providing very practical advice which is highly regarded by livestock farmers.

TABLE 3.119

Necropsy Findings and Laboratory Tests Used to Determine the Common Causes of Sudden Death in Adult Sheep

	Necropsy findings	Laboratory tests
Cast on back	As found	None necessary
Gangrenous mastitis	Purple/black udder, thrombus of mammary vein	Culture rarely necessary
Respiratory disease	Septicaemia/fibrinous pleurisy often secondary to OPA	Histology essential to identify potential OPA involvement
Acute fasciolosis	Swollen liver, haemorrhagic tracks. Fibrinous peritonitis	Immature flukes visible
Abortion/some dystocias	Purple black uterus. Emphysematous fetuses	Culture fetal stomach contents
Prolapses	Anaemic carcass, large blood clot(s)	None necessary
Eviscerated intestines through vaginal tear	Small intestines prolapsed through vaginal tear	None necessary
Assisted dystocia	Ruptured uterus, haemorrhage	None necessary
Hypocalcaemia	Bloat, otherwise NAD	
Torsion of small intestine	Torsion around root of mesentery	None necessary
Haemonchosis	Anaemia	FWEC. Adult parasites in abomasum
Acidosis	Foetid diarrhoea containing whole grain. Large amounts grain in rumen	Rumen liquor pH<4.5
Copper poisoning	Jaundiced carcase	Kidney copper concentrations often in excess of 3,000 μmol/kg DM (normal <314 μmol/kg DM)
Clostridial disease (blackleg)	Rapid autolyisis. Often related to dystocia	FAT for *C chauvoei*
Cervical fracture	Fighting injury in rams, fracture C4/C5 at necropsy	None necessary

DOI: 10.1201/9781003106456-40

Cast on Back

3.1141: Cast for less than 2 hours with moderate bloat. **3.1142:** Ewe alive but severely bloated, head up hill. Note the amount of feces behind ewe. **3.1143:** Ewe dead with head pointed downhill.

- **Necropsy examination**: Severe bloat.
- **Key history events**: Usually occurs during summer months before shearing and immediately after heavy rain showers. 🎥 3.574

Gangrenous Mastitis

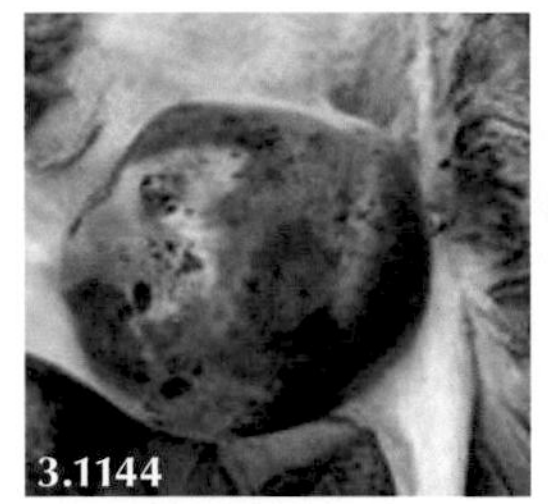

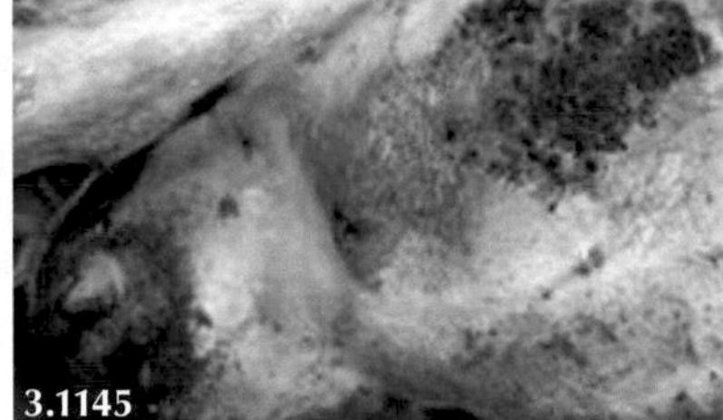

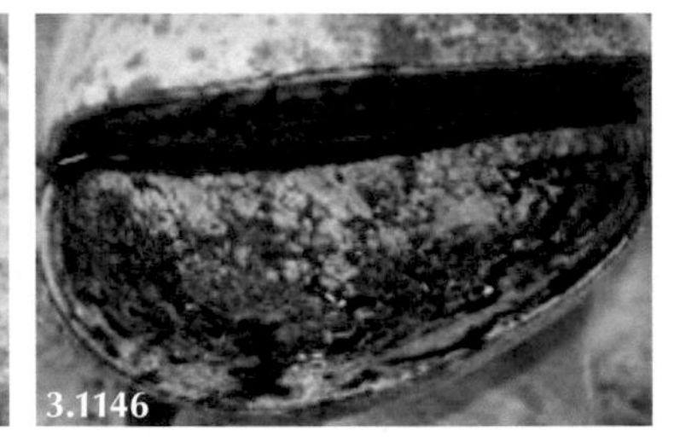

3.1144: Gangrenous quarter. **3.1145:** Thrombosis extending along right mammary vein. **3.1146:** Early gangrenous changes in the mammary tissue.

- **Key history events**: Lambs often 4–6 weeks old. May be traumatic lesions on medial aspect of teats. 🎥 3.575

Chronic ("Recovered") Case of Gangrenous Mastitis

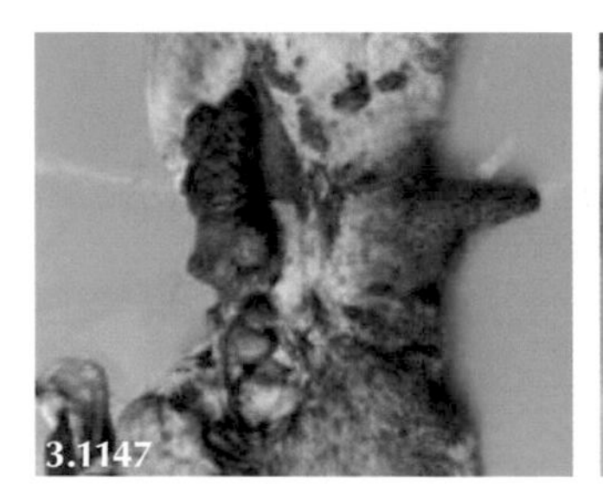

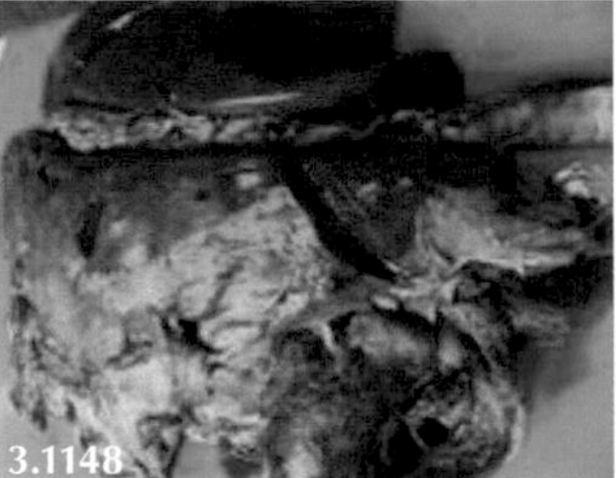

3.1147: Udder three months after acute episode of gangrenous mastitis; the right mammary gland has been sloughed off. Necropsy of this ewe revealed extensive chronic fibrinous pleurisy (**3.1148**) and bacterial endocarditis involving the tricuspid valve (**3.1149**).

Respiratory Disease

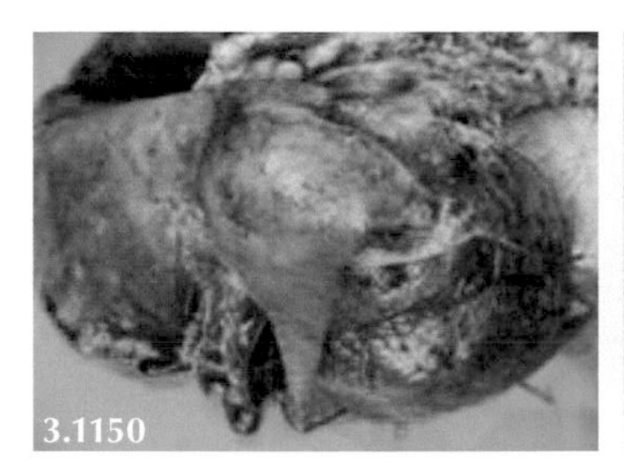
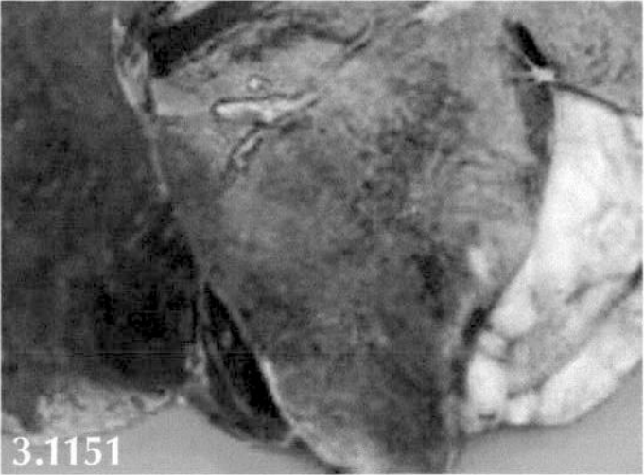
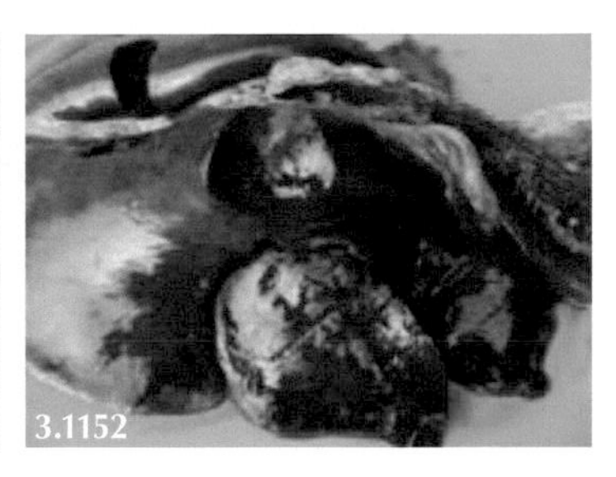

All images OPA and secondary septicaemia. **3.1150:** Purple/black swollen apical, cardiac and accessory lung lobes with adherent fibrin. **3.1151:** Adherent fibrin on serosal surface of lung. **3.1152:** Oedematous and congested lungs with OPA lesion of cardiac lung lobe.

- **Necropsy examination**: Swollen, oedematous purple lungs with variable inflammatory exudate in chest.
- **Key history events**: Rapid deterioration although likely to have had OPA lesions for several months. Histology (and immunostaining) essential for OPA diagnosis. 3.576, 3.577

Acute/Subacute Fasciolosis

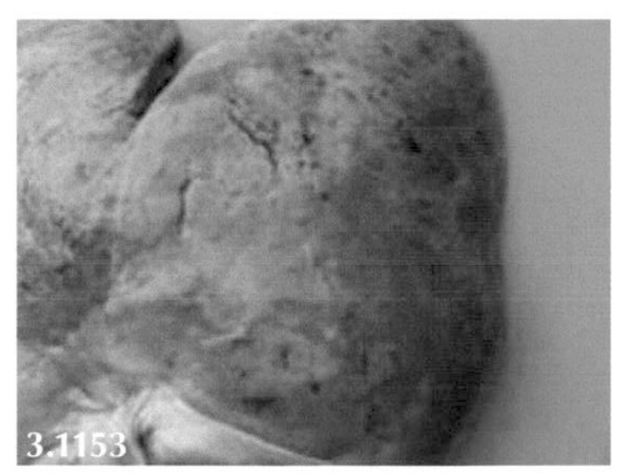
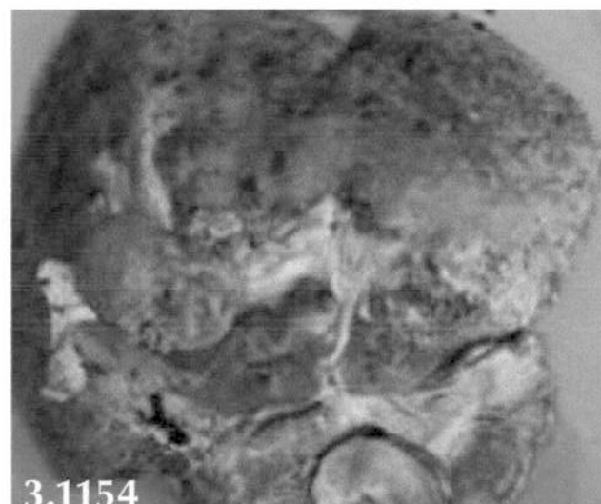
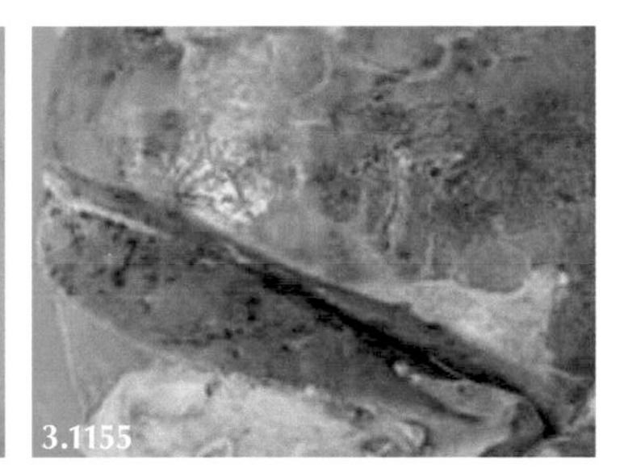

3.1153: Diaphragmatic surface of the liver showing many haemorrhagic tracts. **3.1154:** Visceral surface showing enlarged hepatic lymph nodes. **3.1155:** Many haemorrhagic tracts visible on cut section of liver with fibrin on diaphragmatic surface.

- **Necropsy examination**: Pale liver with multiple haemorrhagic tracts.
- **Key history events**: Failure of liver fluke control measures, high-risk year based upon meteorological data. 3.578

Abortion

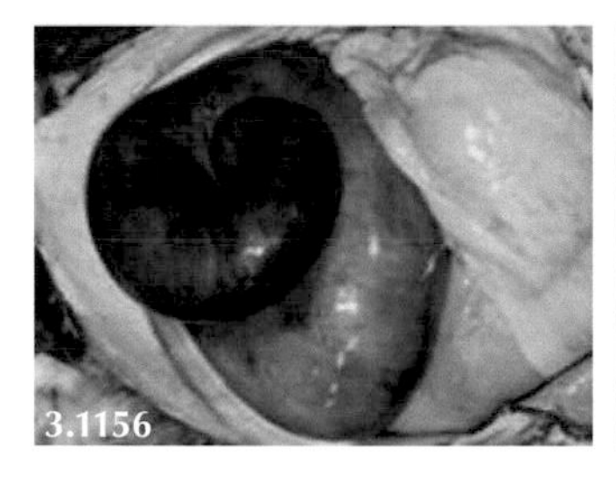
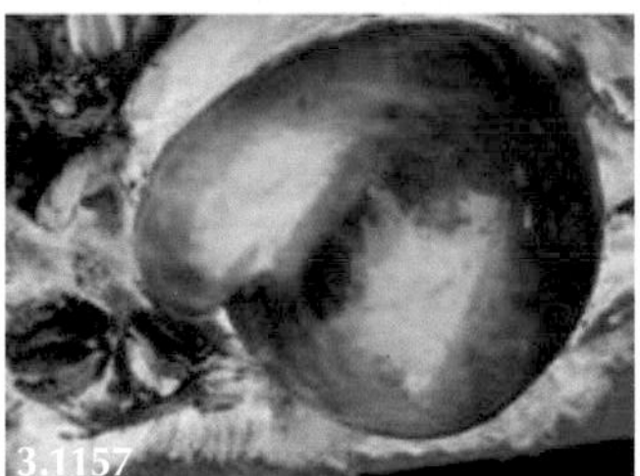

3.1156 and **3.1157:** Fetal infection/death producing severe toxaemia and pressure necrosis of uterine wall. **3.1158:** Delivery of rotten lamb; metritis likely to develop.

- **Necropsy examination**: Purple/black pressure necrosis of uterine wall.
- **Key history events**: Often associated with certain Salmonella serotypes.

Vaginal Prolapse

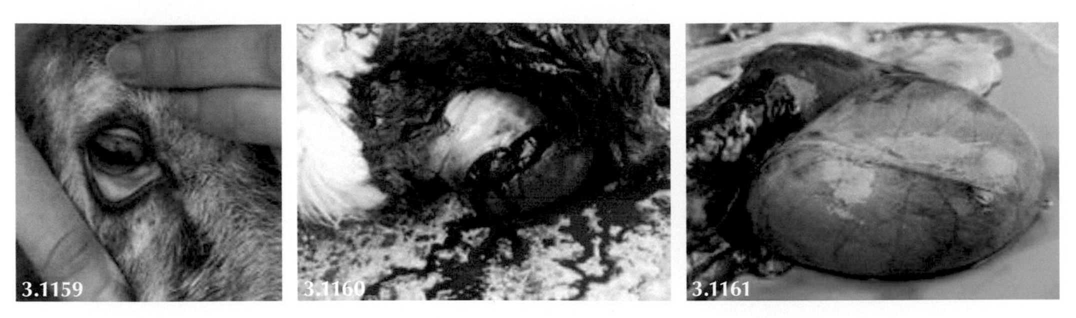

3.1159: Anaemia resulting from vaginal tear. **3.1160:** Arterial bleeding from ruptured vaginal prolapse. **3.1161:** Vaginal prolapse contained retroflexed bladder. 3.579

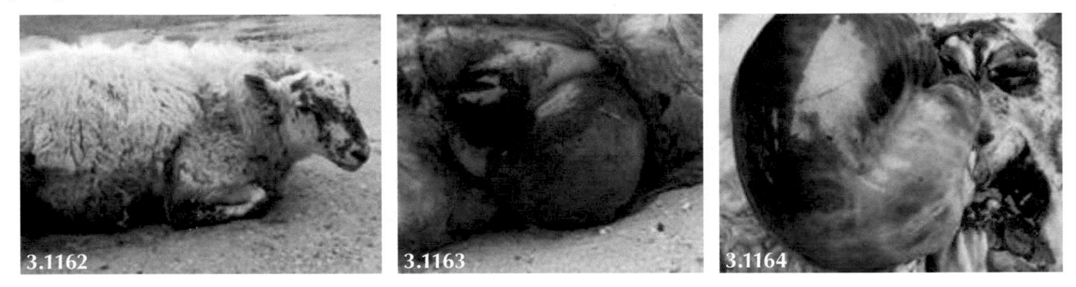

3.1162: Profound toxaemia in ewe with rectal and vaginal prolapses (**3.1163**). **3.1164:** After euthanasia for welfare reasons, necropsy reveals fetal death and necrosis of uterine wall.

- **Necropsy examination**: Haemorrhage/toxaemia.
- **Key history events**: Tenesmus causing rupture of uterine artery. Fetal death causing profound toxaemia. 3.580

Eviscerated Intestines through Vaginal Tear

3.1165 and **3.1166:** Evisceration through tear in vaginal wall. **3.1167:** Tear in vaginal wall (top right).

- **Necropsy examination**: Evisceration through tear in vaginal wall.
- **Key history events**: Occurs within 30 minutes of concentrate feeding without premonitory signs. Most often housed, heavily pregnant crossbred ewes in excellent body condition. 3.581

Dystocia

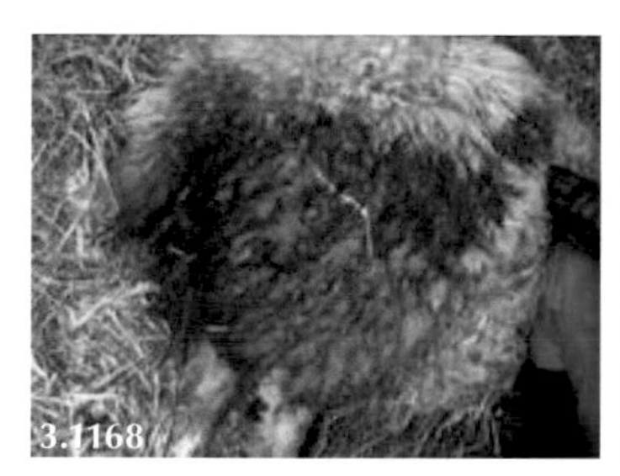

3.1168: Haemorrhage caused by unskilled attempted correction of dystocia. **3.1169:** Dead ewe with lamb's forefeet presented. **3.1170:** Second stage labour could not progress due to vaginal prolapse suture present.

- **Necropsy examination**: Lamb engaged in maternal pelvis.
- **Key history events**: Parturition cannot progress beyond second stage labour. More common in extensively managed sheep with infrequent supervision.

Hypocalcaemia

3.1171–3.1173: Heavily pregnant ewes, dull, weak and unable to stand. Bloat develops, followed by coma and death without appropriate treatment (all three ewes shown responded within 5 minutes to intravenous calcium borogluconate injection).

- **Necropsy examination**: No pathognomonic signs.
- **Key history events**: Movement/housing/dietary change in last three weeks of pregnancy. Several animals affected. 3.582

Torsion of Small Intestine

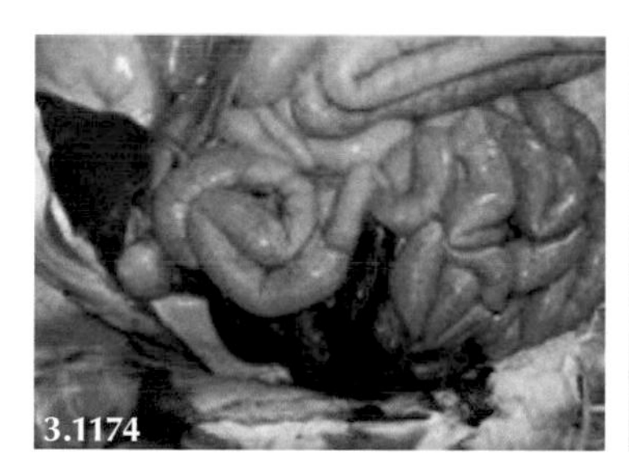
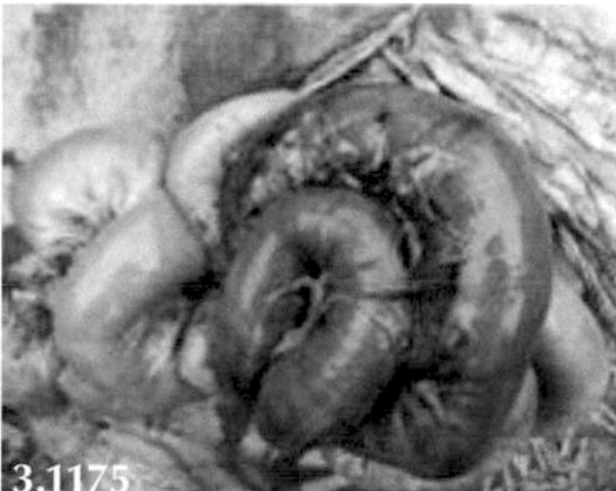
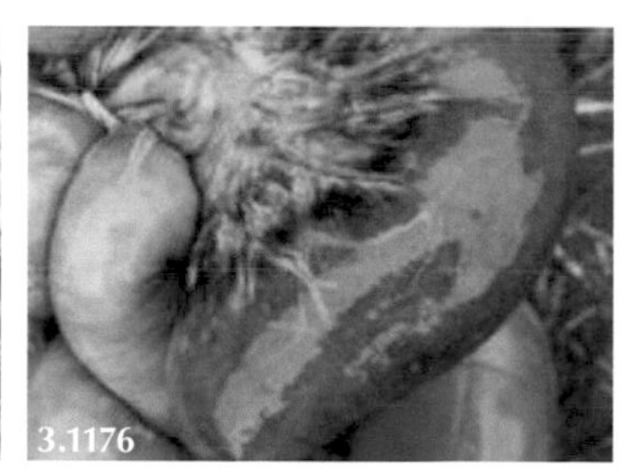

3.1174: Normal gut contrasts with purple/black gut involved in the torsion. **3.1175:** Normal gut contrasts with red/purple congested gut with adherent fibrin in ram with small intestine torsion around root of mesentery. **3.1176:** Torsion clearly visible at root of mesentery (top left of image; see Video 3.583).

- **Necropsy examination**: Sharply demarcated necrotic gut affected by torsion.
- **Key history events**: Sudden change in ration.

Haemonchosis

- **Necropsy examination**: Anaemic carcase. Presence of large numbers of larval stages and mature *H. contortus* in abomasum.
- **Key history events**: No recent effective anthelmintic treatment, noting likely multiple resistance in this worm species.

Acidosis

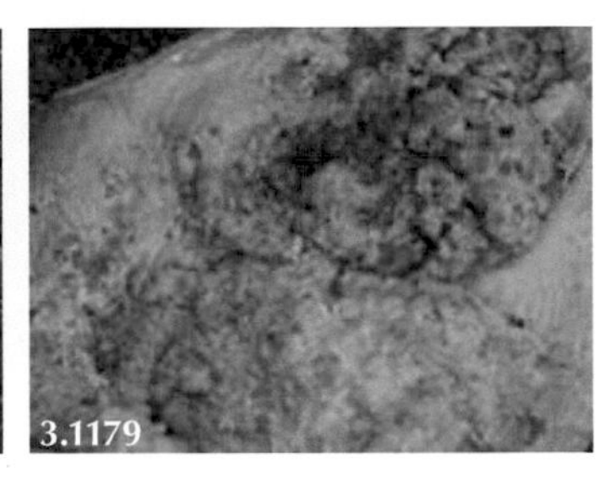

3.1177: Accidental access to grain store. **3.1178:** Large amounts of grain in rumen. **3.1179:** Rumen mucosa stripping off (this must be differentiated from post-mortem change).

- **Necropsy examination**: Large amounts of grain in foul-smelling rumen contents. Rumen liquor pH<4.5. Rumen mucosa strips easily.
- **Key history events**: Accidental access to grain store.

Chronic Copper Toxicity

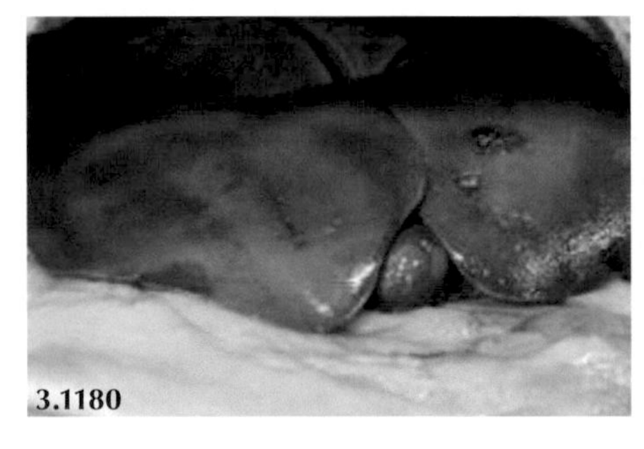

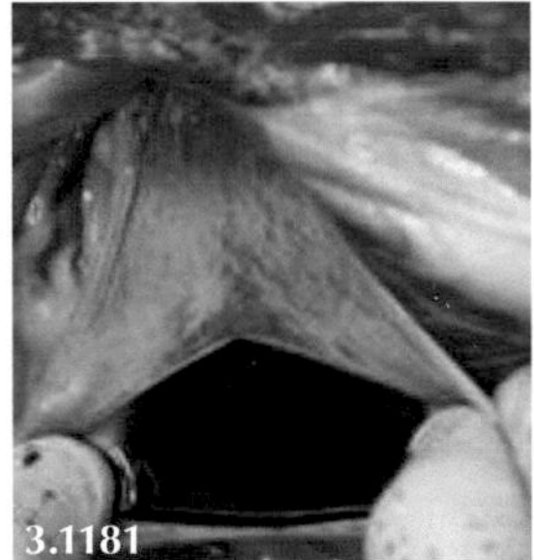

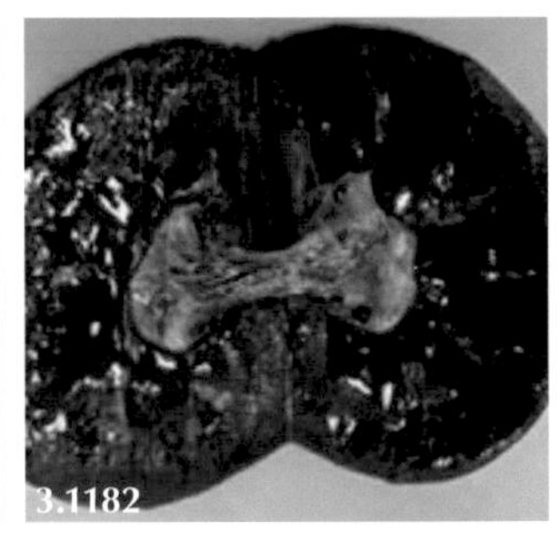

3.1180: Yellow liver. **3.1181:** Opened urinary bladder reveals haemoglobinuria. **3.1182:** Black kidneys.

- **Necropsy examination**: Jaundiced carcase, yellow liver, black kidneys and haemoglobinuria.
- **Key history events**: Often susceptible breed such as Texel, inappropriate concentrate feeding/supplementation. 3.584

Blackleg

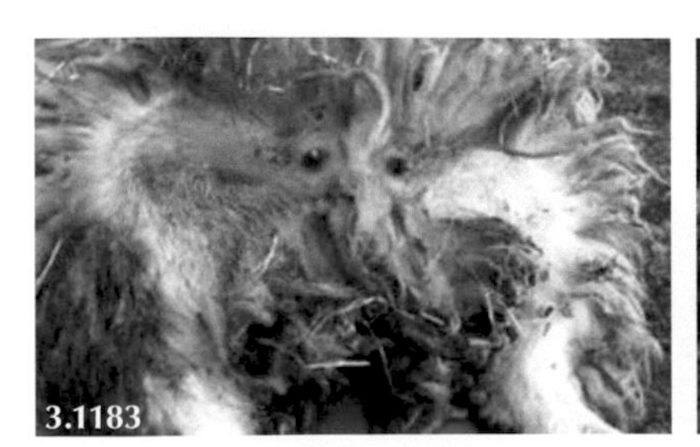

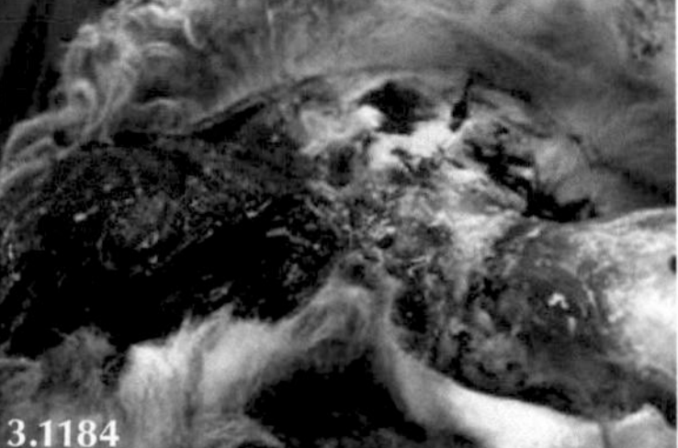

3.1183: Swollen right hindleg with purple/black discolouration of the skin at necropsy. **3.1184:** Swollen muscles of the right hindleg are dark red/black and crepitant. There is subcutaneous oedema extending to the right hand side.

- **Key history events**: No clostridial vaccination programme.

Lambs Less Than One Year Old

TABLE 3.120

Necropsy Findings and Laboratory Tests Used to Determine the Common Causes of Sudden Death in Lambs Less Than One Year Old

	Necropsy findings	Laboratory tests
Respiratory disease	Purple/red, oedematous, swollen lungs. Fibrinous pleural exudate. Petechiation, septicaemia	Histology of lung. Culture *Mannheimia haemolytica or Bibersteinia trehalosi* in septicaemic distribution
Clostridial disease (lamb dysentery)	Focal necrosis of small intestine. Rapid carcase autolysis. Effusion in pericardial sac	Clostridial toxins B and E in small intestinal contents
Clostridial disease (pulpy kidney)	Rapid carcase autolysis. Effusion in pericardial sac	Clostridial toxin E. Glucosuria
Acute fasciolosis	Anaemic carcase, peritoneal exudate. Haemorrhagic tracts	Immature flukes in liver
Nematodirosis	Large numbers of larvae in small intestine	FWEC may be nil as disease caused by larval stages
Braxy	Rapid carcase autolysis. Effusion in pericardial sac	
Redgut (SI torsion)	Torsion around root of mesentery	None required
PEM	No gross findings	Autofluorescence under Wood's lamp
Acidosis	Large amounts of grain in rumen/abomasum. Stripping of rumen mucosa	Rumen pH<4.5
Brassica poisoning	Blood and tissues appear brown due to methaemoglobin	
Chronic copper toxicity	Jaundice, haemoglobinuria. Yellow liver, black kidneys	Kidney copper concentrations often in excess of 3,000 µmol/kg DM (normal <314 µmol/kg DM)
Bloat/cast on back	Caudal oesophagus pale compared to the congested cervical oesophagus ("bloat line")	
White muscle disease (cardiac)	Focal muscle pallor may be difficult to detect	Histology. Liver selenium and vitamin E concentrations
Plant poisons including yew and rhododendron	Identification of leaves in rumen contents	

Respiratory Disease (*Mannheimia haemolytica*)

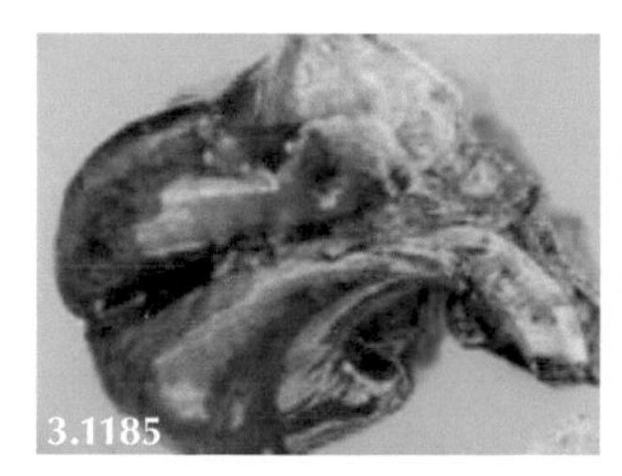
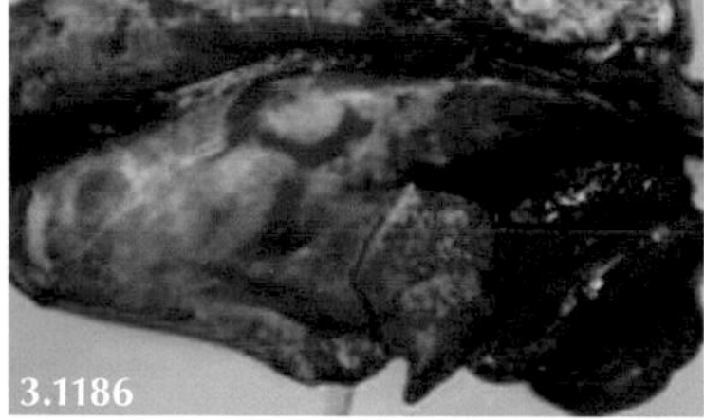
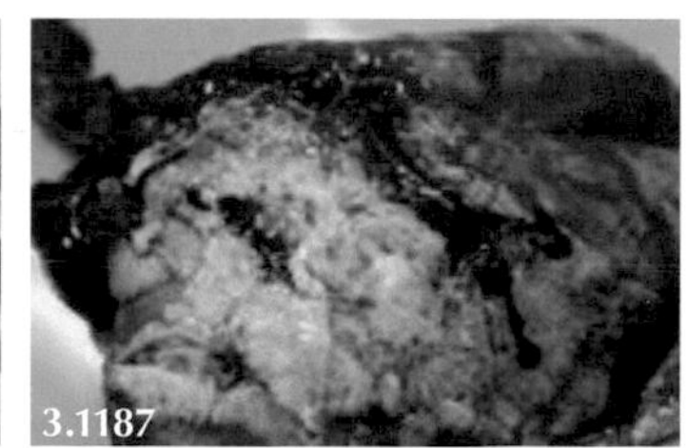

3.1185: Purple/red, oedematous, swollen lungs. **3.1186:** Antero-ventral lung consolidation. **3.1187:** Right: Unilateral fibrinous pleural exudate.

- **Necropsy examination**: Purple/red, oedematous lungs—requires histology to confirm disease. Often extensive fibrinous pleural exudate.
- **Key history events**: Often no recognised risk factors. Often vaccinated lambs. 🎥 3.585

Respiratory Disease (*Bibersteinia trehalosi*)

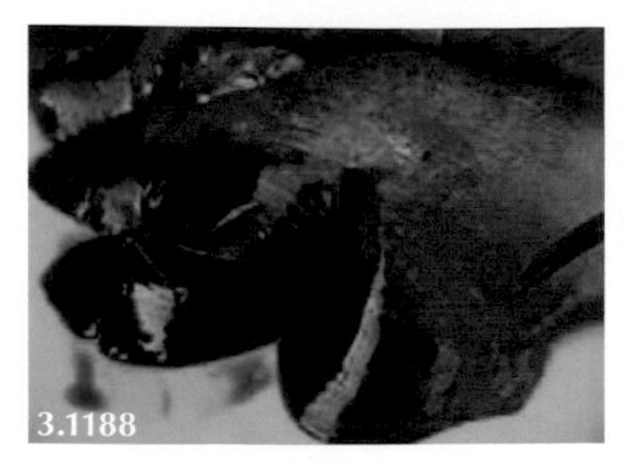
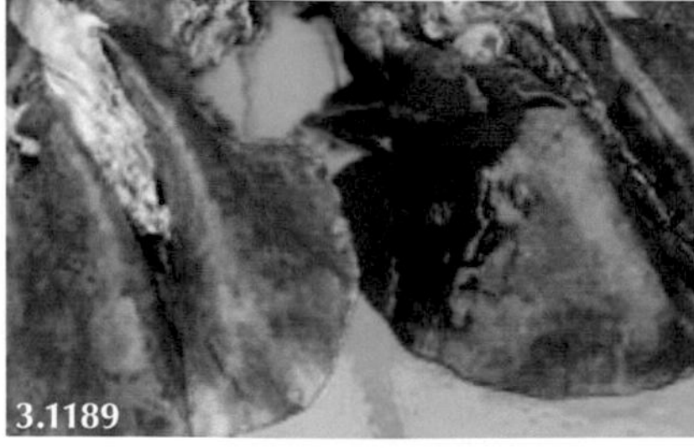
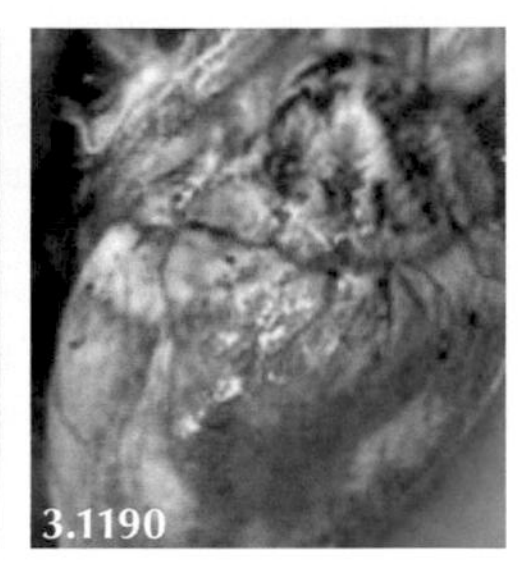

3.1188: Antero-ventral consolidation. **3.1189:** Comparison of affected lungs (right) with normal lungs (left). **3.1190:** Petechiae on epicardium.

- **Necropsy examination**: Antero-ventral lung consolidation, petechiation of heart and lungs.
- **Key history events**: Two to three weeks after weaning/sale/transport. 3.586

Lamb Dysentery

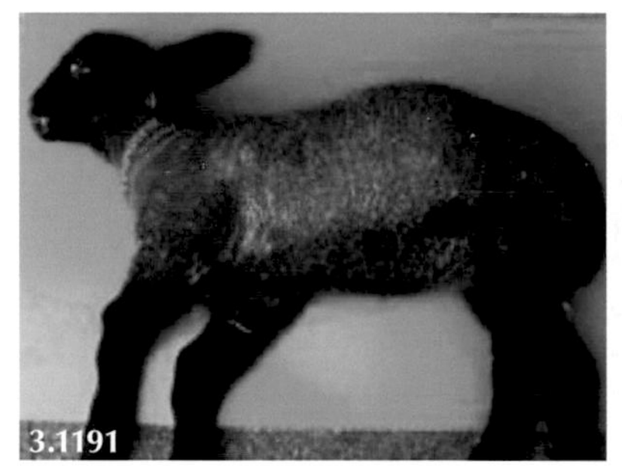
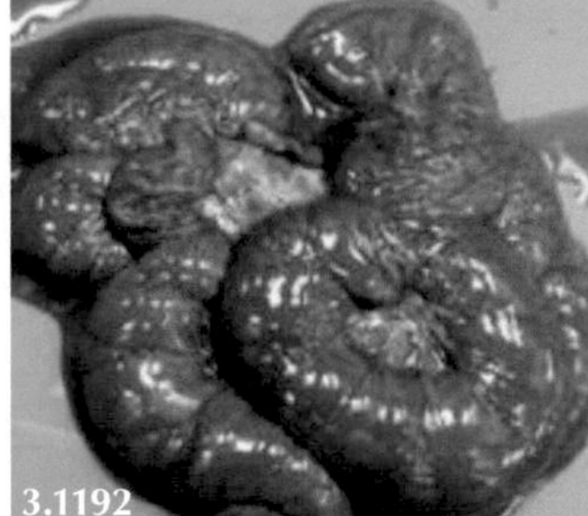
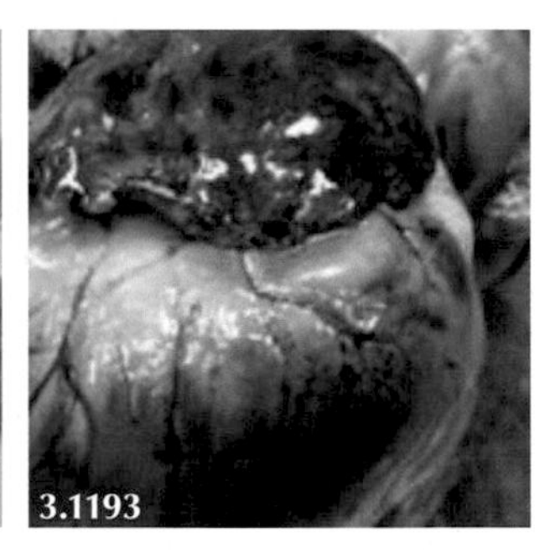

3.1191: Collapsed week-old lamb with distended abdomen—euthanised for welfare reasons. **3.1192:** Necropsy reveals congested and fluid-distended small intestine. **3.1193:** Widespread petechiae on the epicardium.

- **Necropsy examination**: Focal necrosis of small intestine. Rapid carcase autolysis. Effusion in pericardial sac.
- **Key history events**: Unvaccinated dam, failure of passive antibody transfer. 3.587

Pulpy Kidney

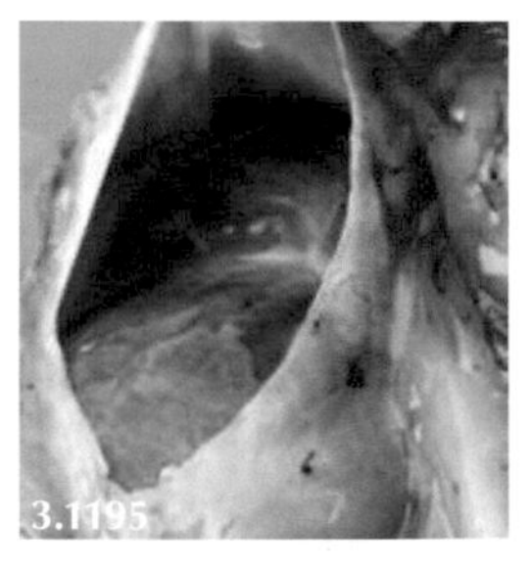

3.1194: "Pulpy" kidneys from confirmed case (left) compared to normal kidney at necropsy. **3.1195:** Large amount of pericardial effusion.

- **Necropsy examination**: Often difficult to appreciate rapid kidney autolysis. Effusion in pericardiac sac. Glucosuria.
- **Key history events**: Failure to vaccinate weaned lambs.

Acute/Subacute Fasciolosis

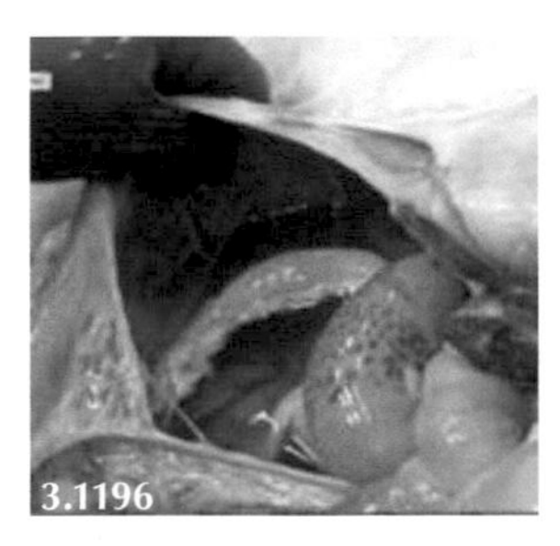
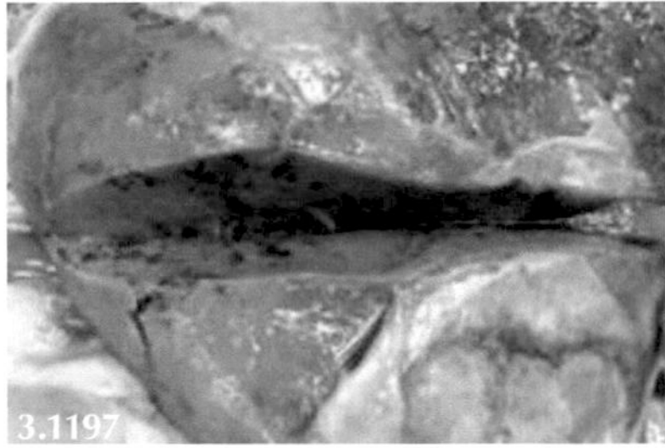
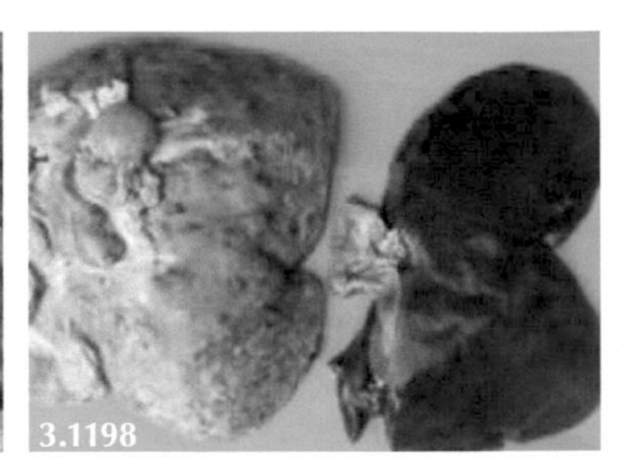

3.1196: Large volume of peritoneal exudate with fibrin tags. **3.1197:** Haemorrhagic tracts in liver. **3.1198:** Comparison of sub-acute liver fluke (left) with normal liver (right).

- **Necropsy examination**: Anaemic carcase, peritoneal exudate. Haemorrhagic tracts throughout swollen liver.
- **Key history events**: High-risk year. No prophylactic drench or treatment failure (e.g. triclabendazole resistance). 🎥 3.588

Nematodirosis

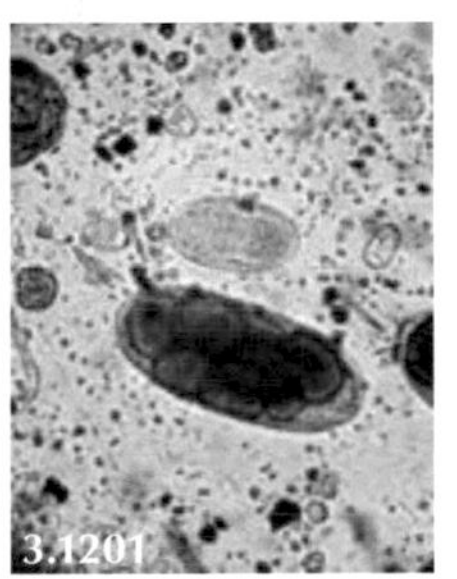

3.1199: Collapsed lamb with profuse diarrhoea. **3.1200:** Dead lamb from same field. **3.1201:** *Nematodirus battus* egg (note per-acute disease may be caused by larval stages).

- **Necropsy examination**: Rapid digestion of larvae/adult worms after lamb death so may not be detected. Death may occur before adult worms become patent with a negative FWEC.
- **Key history events**: Grazing pasture grazed by young lambs the previous spring. Specific weather conditions delaying egg hatching resulting in high risk.

Braxy

3.1202: Sudden death following frosty weather.

- **Necropsy examination**: Rapid carcase autolysis may be affected by weather conditions. Effusion in pericardiac sac.
- **Key history events**: Unvaccinated against clostridial diseases. 🎥 3.589

Redgut (SI Torsion)

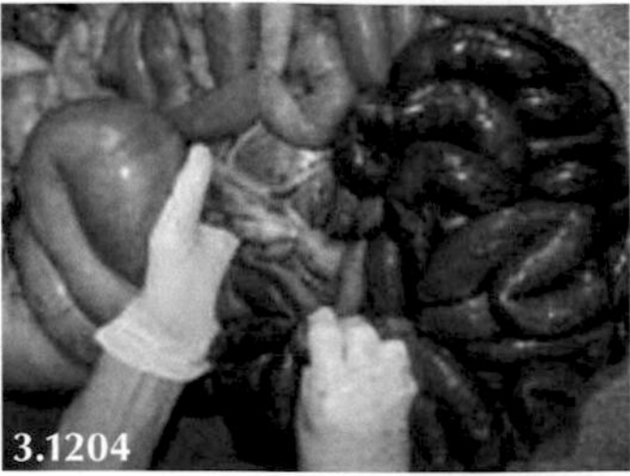

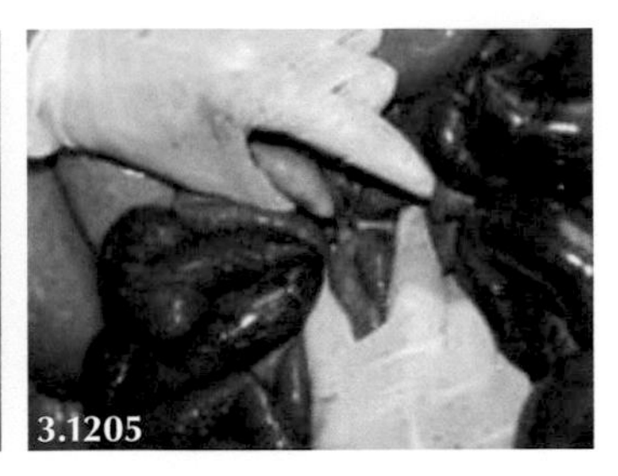

3.1203 and **3.1204:** Sharply demarcated necrotic section of small intestine. **3.1205:** Torsion around the root of the mesentery clearly demonstrated.

- **Necropsy examination**: Dramatic differentiation between normal and necrotic small intestine.
- **Key history events**: Often occurs following move to lush pasture/brassica crop.

PEM

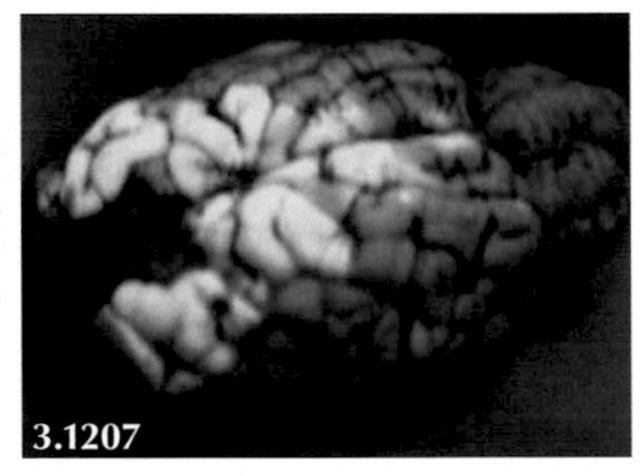

3.1206: Seizure activity in advanced PEM. **3.1207:** Autofluorescence under Wood's lamp at necropsy.

- **Necropsy examination**: No gross brain abnormality. Autofluorescence under Wood's lamp variable so requires brain histology.
- **Key history events**: Later stages of disease cause opisthotonos and seizures over one to two days before death.

Acidosis

3.1208: Grazing grain stubbles with spilt grain. **3.1209:** Sudden access to *ab lib* concentrates in creep feeder. **3.1210:** Large amounts of grain in rumen, foetid rumen contents.

- **Necropsy examination**: Rumen liquor pH<4.5.
- **Key history events**: Accidental access to feed stores.

Brassica Poisoning

3.1211–3.1213: Prolonged access to kale and other brassicas.

- **Necropsy examination**: Blood and tissues appear brown due to methaemoglobin.
- **Key history events**: Prolonged sole access to kale and other brassicas.

Chronic Copper Toxicity

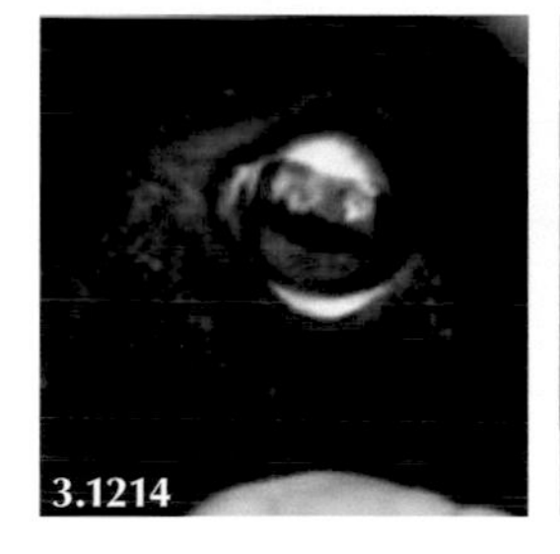
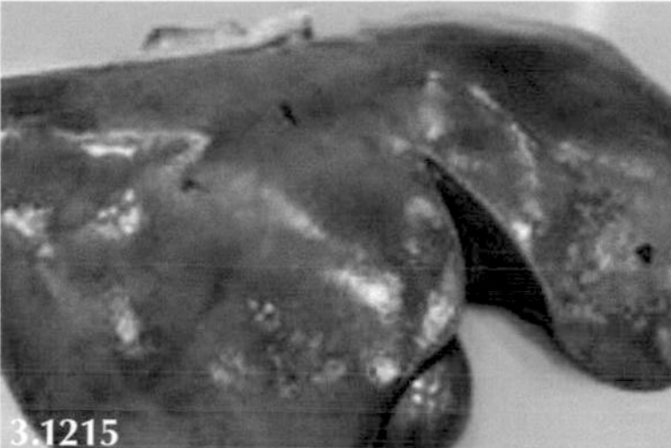
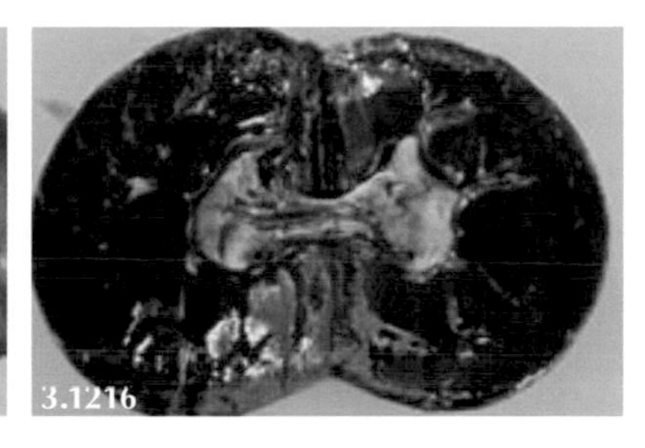

3.1214: Jaundice. **3.1215:** Yellow liver. **3.1216:** Black kidneys.

- **Necropsy examination**: Jaundice, haemoglobinuria. Yellow liver, black kidneys.
- **Key history events**: Incorrect minerals too high in copper. Susceptible breed such as Texel. 🎥 3.590

Bloat

- **Necropsy examination**: Caudal oesophagus pale compared to the congested cervical oesophagus ("bloat line").
- **Key history events**: Bloat is rare in lambs except for abomasal conditions in artificially reared lambs. Weaned lambs do occasionally become cast on their back. 🎥 3.591

White Muscle Disease

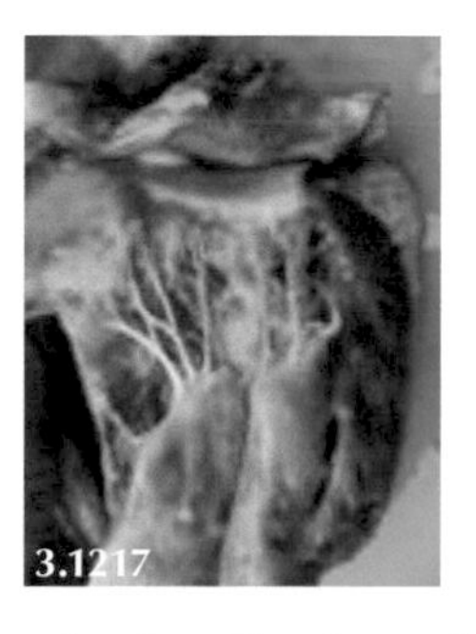

3.1217: Focal muscle pallor of the ventricular wall.

- **Necropsy examination**: Focal muscle pallor difficult to detect and interpret without histology.
- **Key history events**: Often rapidly growing, four–eight-week-old meat breed, ram lambs.

Plant Poisons Including Yew, *Rhododendron* and *Pieris* spp.

- **Necropsy examination**: Identification of leaves in rumen contents.
- **Key history events:** Often associated with heavy snowfall/access to gardens and churchyards.

Cattle More Than One Year Old

With the common exception of hypomagnesaemia, there are few true causes whereby adult cattle die within 24 hours without premonitory signs. The term "acute death" has been adopted here to take into account those cattle which had been ill for up to 24–48 hours before death but not noted by the farmer and/or not treated correctly. The causes of acute death may differ for beef and dairy cattle but have been included together in the table that follows. This table does not include those animals euthanased for accidents/acute trauma including peripheral nerve and spinal injuries, long bone fractures etc.

TABLE 3.121

Necropsy Findings and Laboratory Tests Used to Determine the Common Causes of Sudden Death in Adult Cattle

	Necropsy findings	Laboratory findings
Hypomagnesaemia	Found in lateral recumbency. Signs of seizure activity	Aqueous humor magnesium <0.25 mmol/l
Dystocia/ruptured uterus	Haemorrhage, large uterine tear	
Toxic mastitis	Gangrenous mastitis—cold, black gland. Coliform mastitis—serum-like secretion	Culture milk sample
Prolapsed uterus	Prolapsed uterus	
Clostridial disease—blackleg, tetanus	Dry, crepitant, black muscle masses; leg, neck, diaphragm, heart	
Bronchiectasis	Extensive lung consolidation. Chronic pathology but acute recurrence of disease	Culture, histology
Grain overload	Foetid diarrhoea, large volume grain in rumen	Rumen liquor pH <4.5
Metritis/abortion	Retained placenta(e). Foetid uterine contents	
Abomasal ulceration/perforation	Leakage of contents causing acute septic peritonitis	
Hepatocaval thrombosis	Anaemia, arterial blood in lungs, trachea and rumen	
Inhalation pneumonia	Extensive unilateral lung pathology including necrosis	Histology
Botulism	None	Botulinum toxin in gut contents
Bloat/choke	Object in oesophagus. Bloat line	
Dilated cardiomyopathy	Often extensive oedema, pleural and peritoneal transudate	
Lightning strike	Local storms. Scorch marks on skin. Widespread petechiae	
Anthrax	Historical occurrence of anthrax on farm/area. Widespread petechiae	Blood smear, McFadyean stain

Hypomagnesaemia

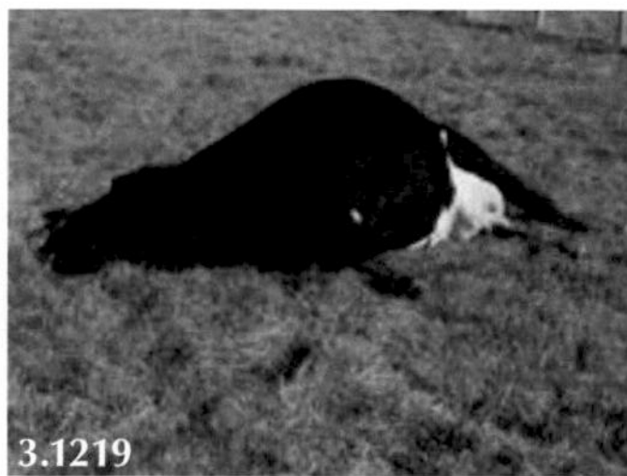

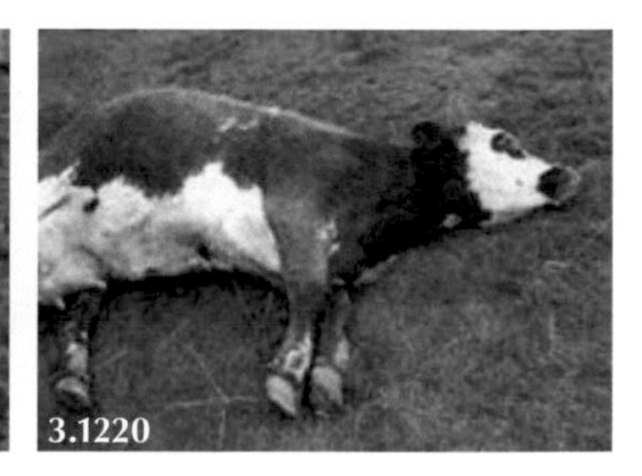

3.1218–3.1220: There are no specific gross findings in adult cattle that have died from hypomagnesaemic tetany.

- **Necropsy examination**: No specific findings.
- **Key history events**: Often older beef cows, early lactation, lush pasture with limited magnesium-supplemented feeding. Recent storms. 3.592

Dystocia/Ruptured Uterus

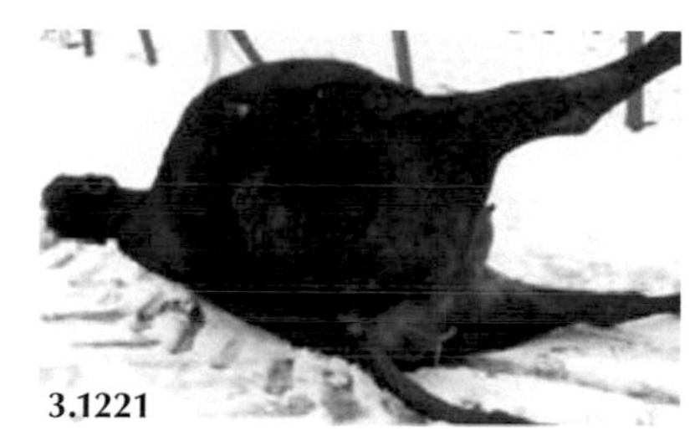

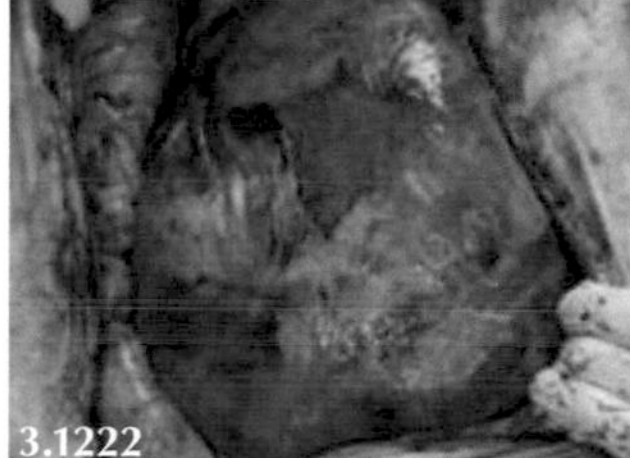

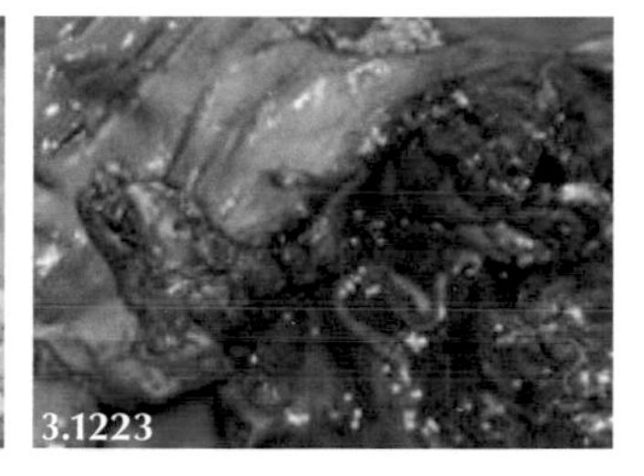

3.1221: Death caused by mal-presentation of twin pregnancy. **3.1222:** Devitalised uterine wall with fibrin deposited on serosal surface. **3.1223:** Uterine tear revealed at necropsy. Only the serosal surface should be viewed but large tear reveals placenta.

- **Necropsy examination**: Dead/autolytic fetus *in utero* with compromised uterus, uterine tear.
- **Key history events**: Cow with twin pregnancy not noted calving under extensive management. Uterine tear following unskilled intervention most often attempted correction of breech calving.

Toxic Mastitis

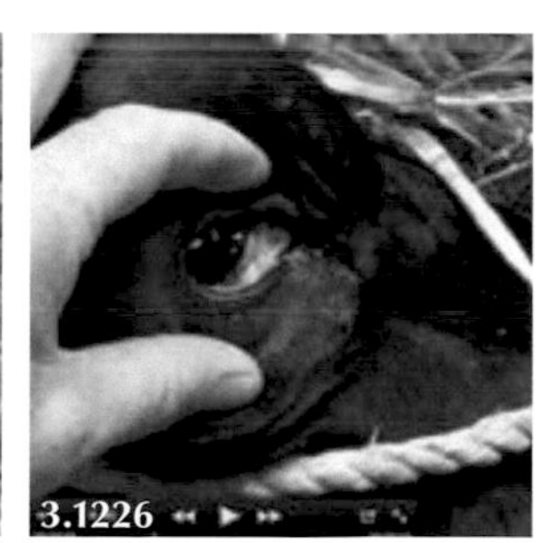

3.1224 and **3.1225:** Weak, recently calved dairy cows. **3.1226:** Marked dehydration and toxaemia.

- **Necropsy examination**: Dehydration, toxaemia, udder flaccid with serum-like secretion in one quarter.
- **Key history events**: Often within first 36 hours after calving. 3.593

Uterine Prolapse

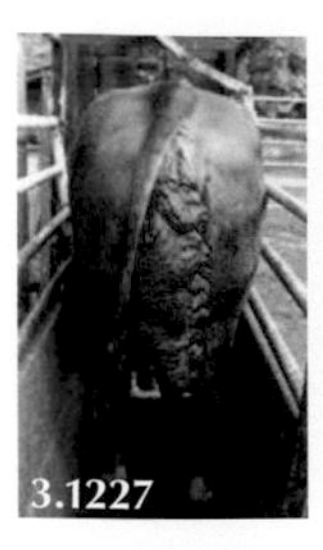

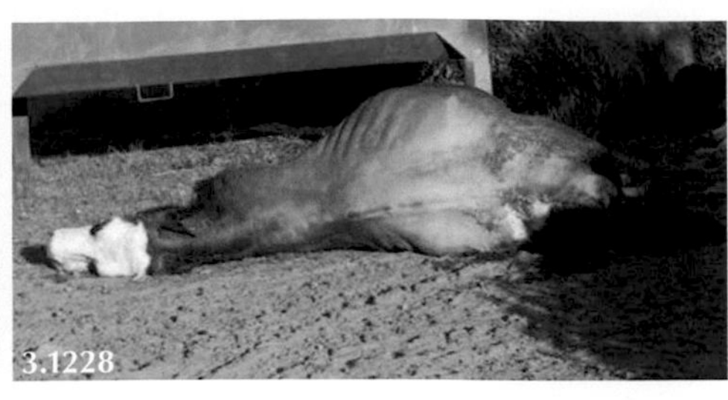

3.1227: Uterine prolapse post-calving. **3.1228:** Death due to internal haemorrhage after rupture of middle uterine artery.

- **Necropsy examination**: External haemorrhage from damaged caruncles, internal haemorrhage from middle uterine artery.
- **Key history events**: Following dystocia caused by absolute fetal oversize; uterus follows immediately after delivery of calf. Also associated with hypocalcaemia. 3.594

Blackleg

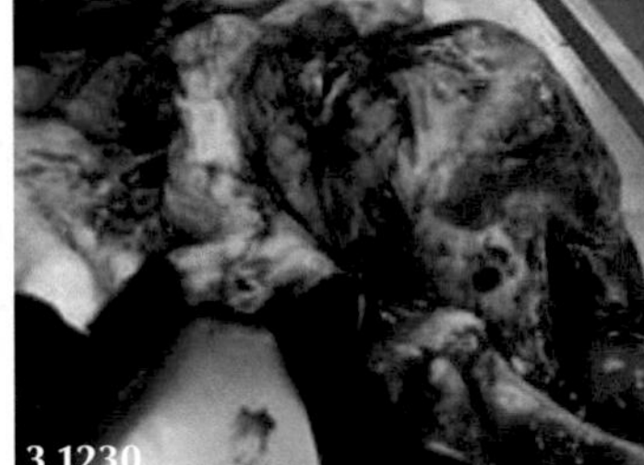

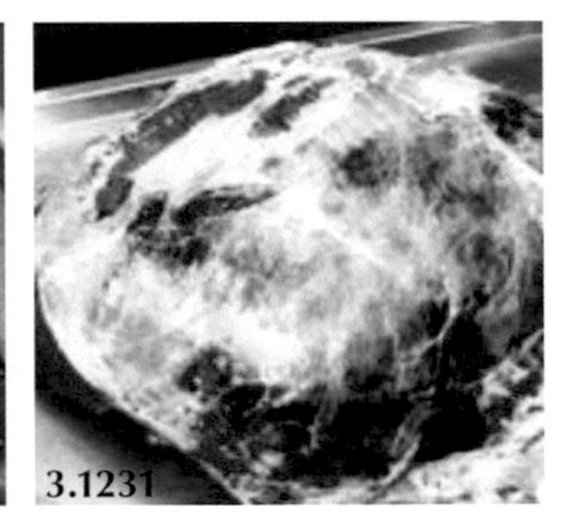

3.1229: Haemorrhage, oedema, emphysema of ventral neck revealed at necropsy (animal in dorsal recumbency). **3.1230:** Animal on its back and viewed from tail. **3.1231:** Very swollen, oedematous and crepitant left hindleg.

- **Necropsy examination**: Rapid carcase autolysis, local changes.
- **Key history events**: Skeleton muscle—bulling activity, neck—barrier feeding. Myocardium—no known association.

Bronchiectasis

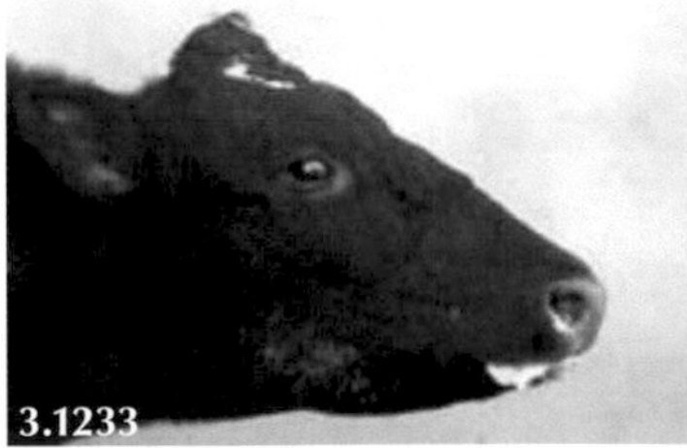

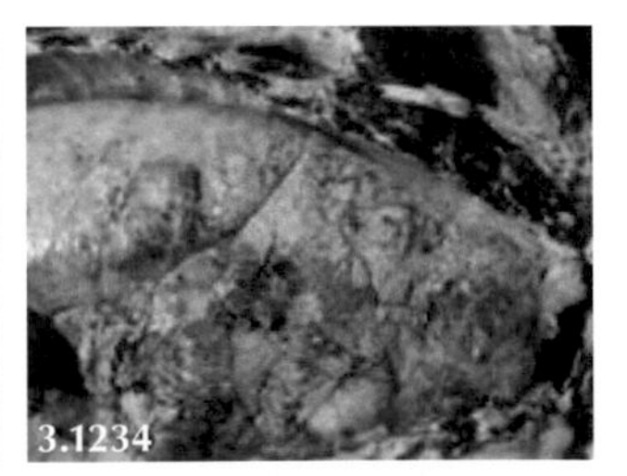

3.1232 and **3.1233:** Sudden flare-up of severe respiratory distress with open mouth breathing and flared nostrils. **3.1234:** Extensive lung pathology of bronchiectasis.

- **Necropsy examination**: Extensive lung consolidation with pus within airways (bronchiectasis).
- **Key history events**: Acute exacerbation of chronic disease process following stressful event such as dystocia.

Grain Overload

3.1235: Recumbent heifer (note abnormal hindleg position) with abdominal distension and diarrhoea. **3.1236:** Very depressed bloated heifer in same group. **3.1237:** Rumen mucosa strips easily at necropsy.

- **Necropsy examination**: Large volumes of grain/concentrate in rumen, foetid "soup-like" consistency, pH<4.5. Rumen mucosa strips off easily.
- **Key history events**: Sudden access to excessive quantities of grain/concentrates. 3.595

Metritis

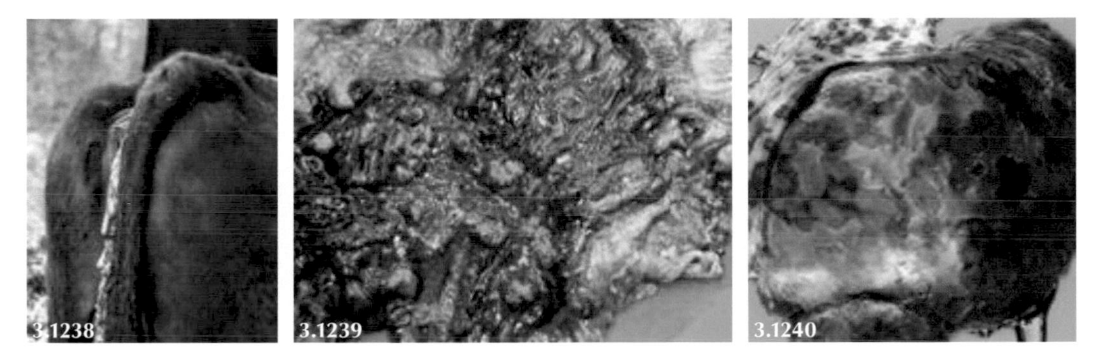

3.1238: Placenta(e) retained more than three to four days. **3.1239:** Lining of uterus with accumulation of necrotic debris. **3.1240:** *Fusobacterium necrophorum* lesions in liver following haematogenous spread from uterus.

- **Necropsy examination**: Retained placenta, foetid uterine discharge.
- **Key history events**: Twins, assisted dystocia, hypocalcaemia, fatty liver.

Abomasal Ulceration/Perforation

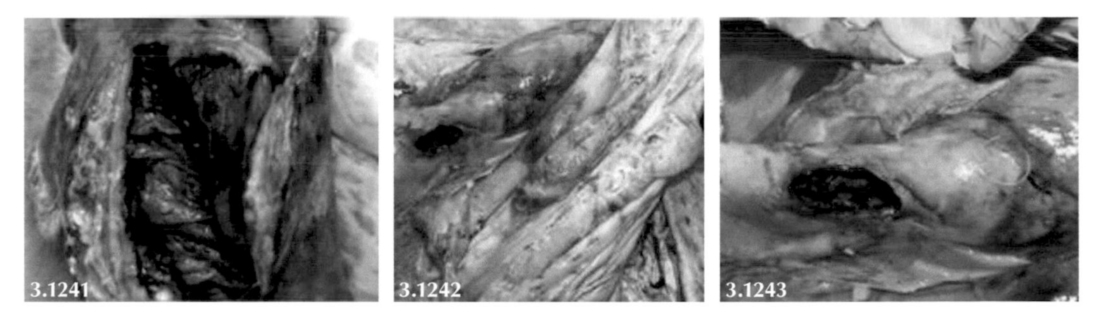

3.1241: 4–6 cm fibrin deposited on serosal surface of abomasum, melaena. **3.1242:** Abomasal perforation. **3.1243:** Close-up showing fibrin deposition on serosal surface of abomasum.

- **Necropsy examination**: Perforated abomasal ulcer, extensive fibrinous exudate/peritonitis.
- **Key history events**: Melaena for several days before rapid deterioration. 3.596

Hepatocaval Thrombosis

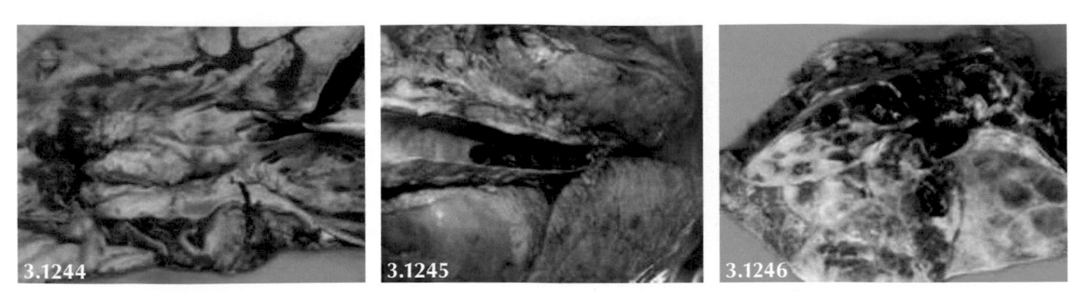

3.1244: Large liver abscess excised through the wall of the caudal vena cava. **3.1245:** Large blood clot within trachea. **3.1246:** Blood within lobules throughout the lung.

- **Necropsy examination**: Arterial blood at nostrils, within larger airways and in rumen. Often very large septic thrombus within anterior vena cava.
- **Key history events**: Fatal haemorrhage often preceded by minor bleeding events over previous seven to 10 days. 🎥 3.597

Inhalation Pneumonia

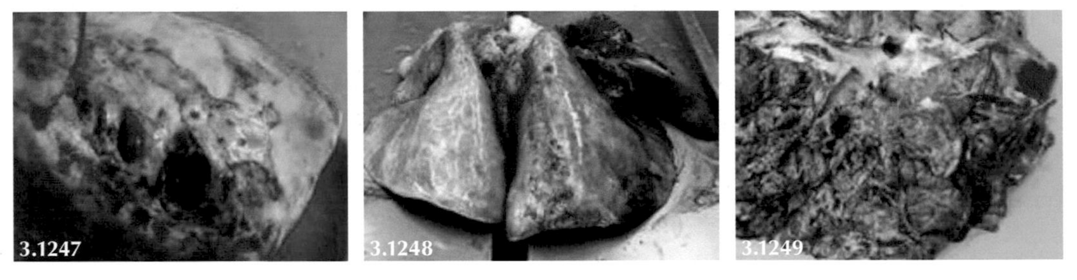

3.1247: Aspiration (anaerobic) pneumonia lesion. **3.1248:** Swollen oedematous right lung with adherent fibrin following cow becoming cast and aspiration pneumonia. **3.1249:** Same as **3.1248** but necrosis of left lung parenchyma.

- **Necropsy examination**: Necrosis of dependent lung after inhalation of rumen contents.
- **Key history events**: Recumbent/cast due to hypocalcaemia. Drenching accident.

Botulism

Recumbency (**3.1250**) with flaccid paralysis of tongue (**3.1251**). **3.1252:** Poultry manure spread on pasture as a fertiliser.

- **Necropsy examination**: NAD.
- **Key history events**: Access to poultry carcases/manure.

Choke

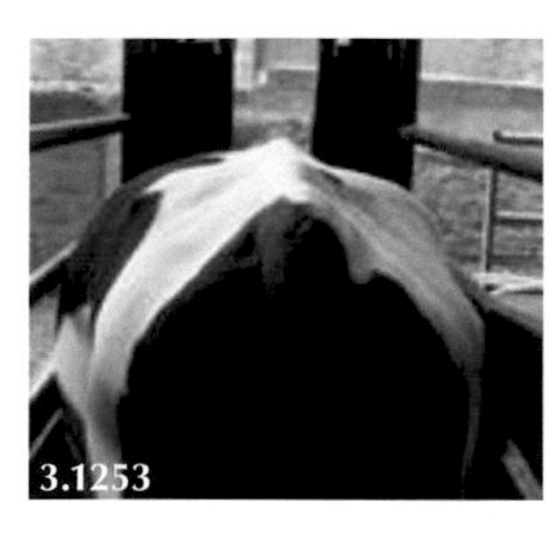

3.1253: Free gas bloat can quickly lead to death most commonly caused by an oesophageal foreign body. **3.1254:** A lodged potato causing bloat has resulted in death. **3.1255:** Surplus potatoes are a cheap and commonly used cattle feed.

- **Necropsy examination**: Oesophageal foreign body. Oesophagus may be perforated if aggressive attempts made to dislodge using probang.
- **Key history events**: Feeding potatoes.

Dilated Cardiomyopathy

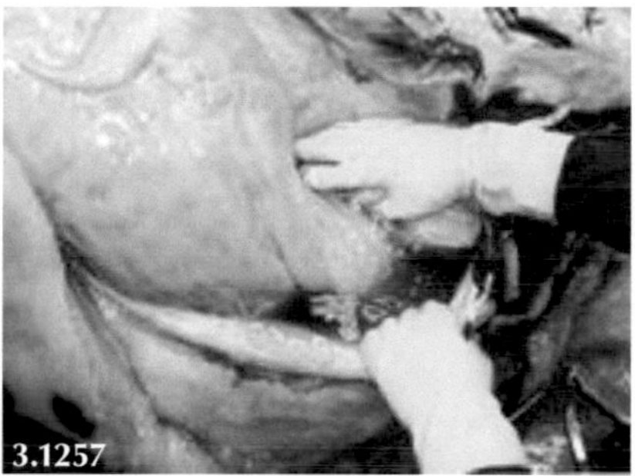

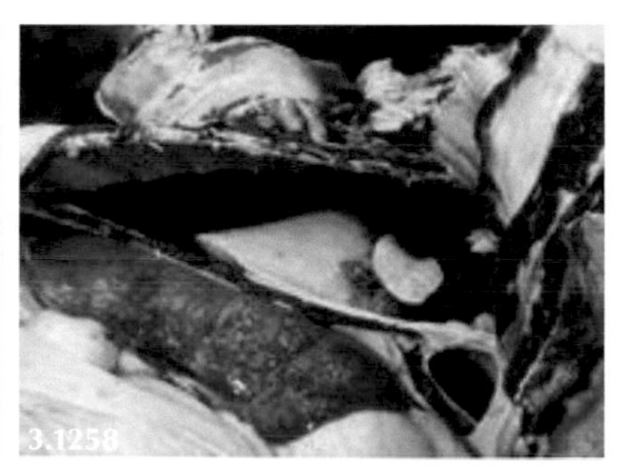

3.1256: Extensive brisket oedema. **3.1257:** Ascites. **3.1258:** Pleural transudate.

- **Necropsy examination**: Signs of right-sided heart failure.
- **Key history events**: Cluster of related cases.

Lightning Strike

- **Necropsy examination**: Singe marks on hide, otherwise NAD.
- **Key history events**: Several dead cattle found under tree or similar circumstantial findings.

Anthrax

- **Necropsy examination**: The carcase must not be opened. Septicaemic changes.
- **Key history events**: History of anthrax on farm/area.

Young Calves and Growing Cattle

- Acute respiratory disease is considered the most important cause of sudden death in this age category.
- An outbreak of rota-/coronavirus in calves from unvaccinated dams can cause serious losses.

TABLE 3.122

Age Profile, Necropsy Findings and Laboratory Tests Used to Determine the Common Causes of Sudden Death in Young Calves and Growing Cattle

Disease/Disorder	Usual age prevalence	Necropsy findings	Laboratory tests
BRSV	2–6 months	Extensive lung emphysema	FAT on lung section, histology
Septicaemia	2–7 days	Widespread petechiae	Blood/tissue cultures such as heart, liver, kidney. Failure passive transfer
Neonatal diarrhoea	8–21 days	Evidence of chronic diarrhoea, fluid-distended, atonic abomasum and intestines	Isolation of entero-pathogen does not necessarily confirm causation
Enterotoxaemia	4–8 weeks	Dysentery, dehydrated carcase	*Clostridium perfringens* Type A from gut contents
Blackleg	>6 months	Dry, crepitant muscle masses; leg, neck, diaphragm, heart	
ETEC	1–3 days	Intestines distended with fluid.	Culture of gut contents.
Grain overload	>6 months	Foetid diarrhoea, large volume of grain in rumen	Rumen liquor pH <4.5
Abomasal perforation	4–8 weeks	1 cm diameter abomasal perforation, peritonitis	None required
Peritonitis	3–5 days	Diffuse septic peritonitis	None required
Congenital abnormalities	At birth	Ventricular septal defect	None required
Bloat/choke	None	Object in oesophagus. Bloat line	None required
Tetanus	None	N/A	N/A

Acute Respiratory Disease Including BRSV

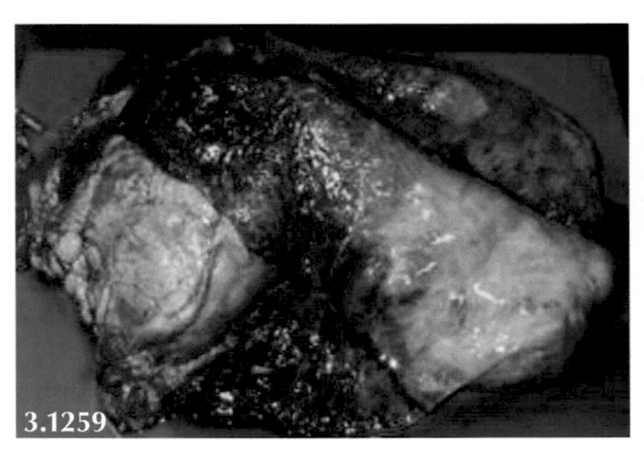

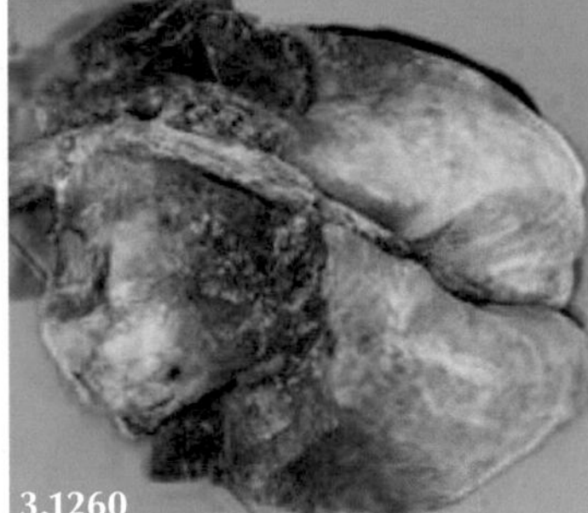

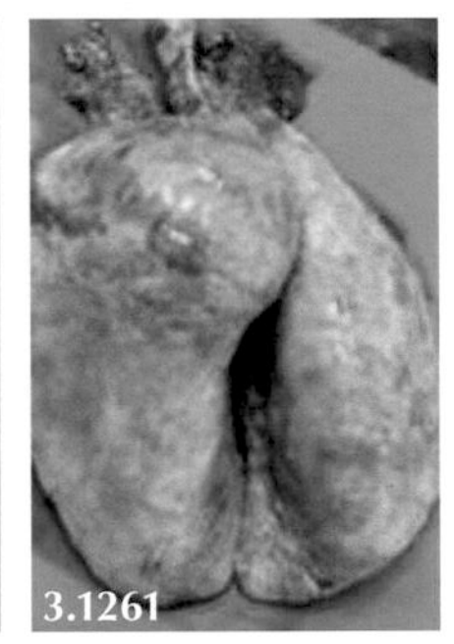

3.1259–3.1261: Extensive antero-ventral consolidation with caudo-dorsal emphysema most marked in the left lung of the right image.

- **Necropsy examination**: Antero-ventral consolidation, caudo-dorsal emphysema with bullae formation (some emphysema can occur agonally).
- **Key history events**: Sudden severe dyspnoea although antero-ventral consolidation present for several days.

Septicaemia

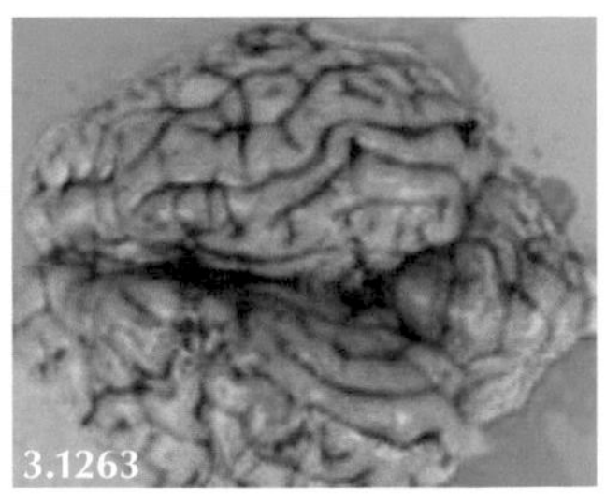
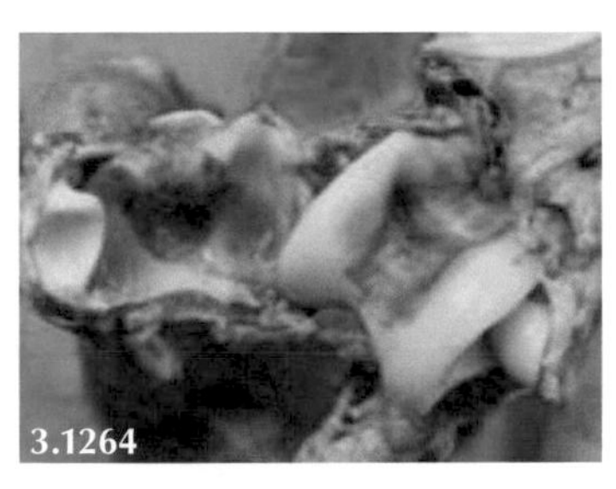

3.1262: Small dairy calf with failure of passive antibody transfer. **3.1263:** Congested meningeal vessels. **3.1264:** Pannus and synovial reaction within the hock joint.

- **Necropsy examination**: Petechial and ecchymotic haemorrhages on pleural and serosal surfaces, pulmonary congestion, meningeal and joint lesions.
- **Key history events**: Contaminated calving environment, failure of passive antibody transfer.

Neonatal Diarrhoea

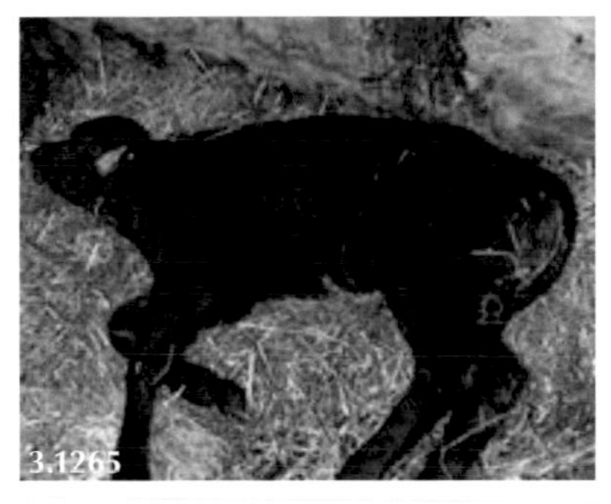

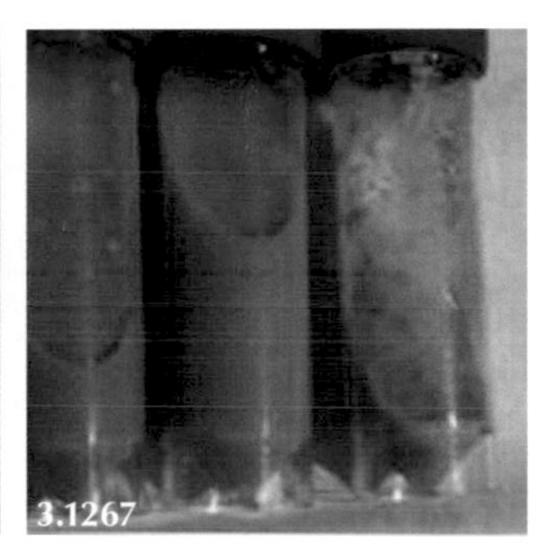

3.1265 and **3.1266:** Collapsed acidotic calves. **3.1267:** Fecal samples from other calves in group for virus testing.

- **Necropsy examination**: Atonic, fluid-distended loops of small intestine.
- **Key history events**: No dam vaccination.

Clostridial Enterotoxaemia

3.1268: Profuse dysentery in neonatal calf with clostridial enterotoxaemia

- **Necropsy examination**: Dysentery.
- **Key history events**: Acute dysentery, abdominal pain, seizures, and opisthotonos.

Blackleg

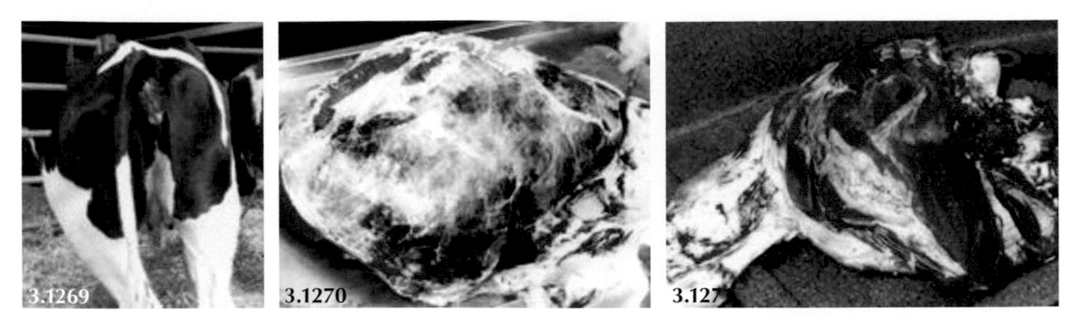

3.1269: Very painful and swollen left hindleg. Unusual to be found alive. **3.1270:** Purple/black leg muscle compared to normal necropsy specimen (**3.1271**).

- **Necropsy examination**: Purple/black muscle compared to normal necropsy specimen.
- **Key history events**: Most often found dead. Often bulling activity in heifers causing muscle damage activating spores. Recent earthworks/excavations.

ETEC

- **Necropsy examination**: Sudden death of calves less than 3 days old. Atonic, fluid-distended intestines.
- **Key history events**: Cluster of cases. No dam vaccination, failure of passive antibody transfer.

Grain Overload

- **Necropsy examination**: Rancid rumen contents with high grain contents, pH<4.5, no protozoa.
- **Key history events**: Sudden, unaccustomed access to grain/feed.

Abomasal Perforation

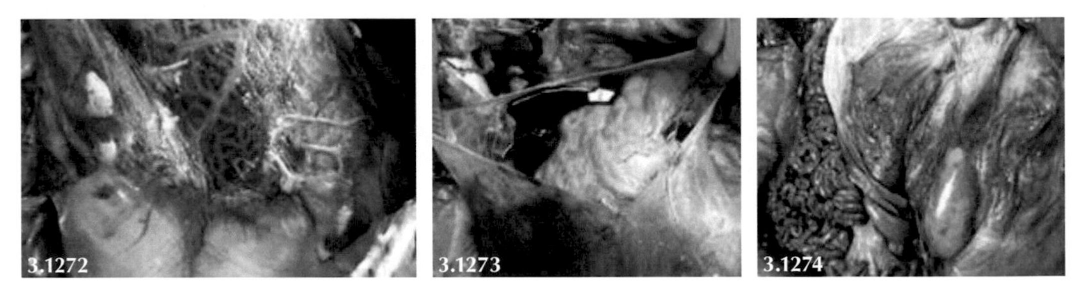

3.1272: Fibrin deposited on omentum. **3.1273:** Omental bursitis with abomasal contents. **3.1274:** Fibrin adherent to the omentum but discoloured by abomasal contents.

- **Necropsy examination**: Omental bursitis containing abomasal contents.
- **Key history events**: Often best calf in group, 6–8 weeks old, sudden onset illness/sudden death. 🎥 3.598

Peritonitis

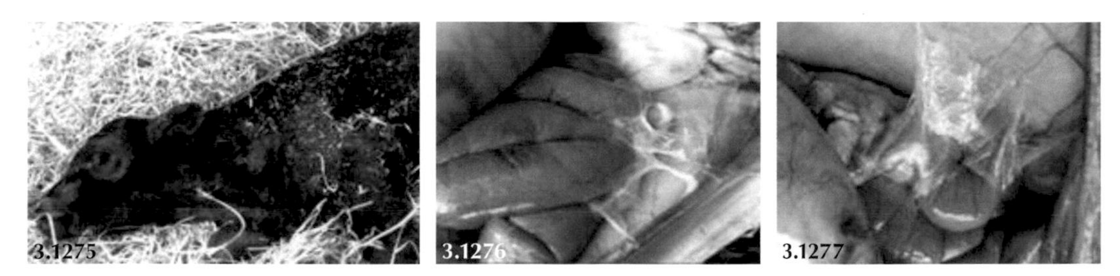

3.1275: Moribund three-day-old calf with distended abdomen. **3.1276:** Distended atonic loops of inflamed intestine with localised large fibrin tags. **3.1277:** Focal peritoneal exudate.

- **Necropsy examination**: Focal accumulation of inflammatory exudate impairing gut motility causing sequestration and dehydration.
- **Key history events**: Rapid deterioration over 12 hours, often moribund when presented. 3.599

Congenital Abnormalities

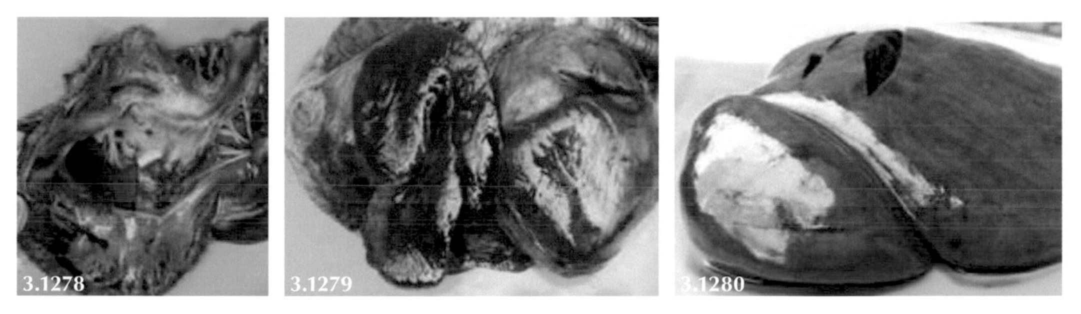

3.1278: Ventricular septal defect. **3.1279:** Globose-shaped heart and congested lungs. **3.1280:** Chronic venous congestion of liver.

- **Necropsy examination**: Globose-shaped heart, variably sized ventricular septal defect, congested lungs and chronic venous congestion of liver.
- **Key history events**: Large defect-respiratory signs present from birth. 3.600

Tetanus

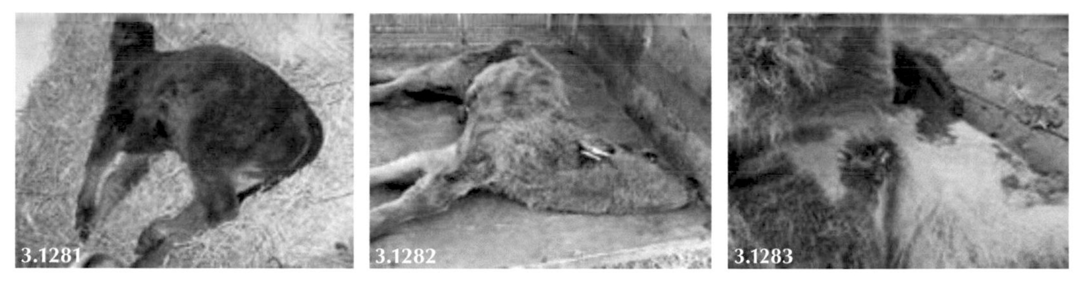

3.1281 and **3.1282:** Calves 4–8 weeks old presented in opisthotonos necessitating euthanasia for welfare reasons. **3.1283:** Necrotic scrotum removed to reveal localised swelling and likely source of infection.

- **Necropsy examination**: NAD.
- **Key history events**: Usually arises from rubber ring castration site (calf too old when applied). 3.601

3.23

Udder Lesions in Cattle

TABLE 3.123
Udder Lesions in Cattle

	Duration	Udder	Secretion	Toxaemia	Rectal temperature
Acute coliform mastitis	Hours	NAD	Serum-like	+/+++	–/––
Chronic suppurative mastitis (Summer mastitis)	Days	+/++	Pus	+/++	+/++
Gangrenous mastitis (per-acute)	Hours	Cold	Bloody	+++	++/+++
Mammary tumour	Months	Firm mass	NAD	NAD	NAD

TABLE 3.124
Ancillary Tests to Determine Cause of Udder Lesions in Cattle

Disease	Ancillary tests
Acute coliform mastitis	Bacteriology
Summer mastitis	Bacteriology
Gangrenous mastitis (per-acute)	Cold to touch, with sharply defined skin necrosis
Mammary tumour	Necropsy, histology

Acute Coliform Mastitis in Dairy Cows

3.1284 and **3.1285:** Very dull, weak cow with coliform mastitis. **3.1286:** Sunken eyes consistent with 5–7 per cent dehydration; toxic mucous membranes.

- **Key observations**: Profound dullness, weak, diarrhoea.
- **Clinical examination**: 5–7 per cent dehydration, elevated heart rate, serum-like secretion from affected flaccid mammary gland.
- **Key history events**: Several days after calving. 3.602

DOI: 10.1201/9781003106456-41

Chronic Suppurative Mastitis (Summer Mastitis) in Beef Cattle

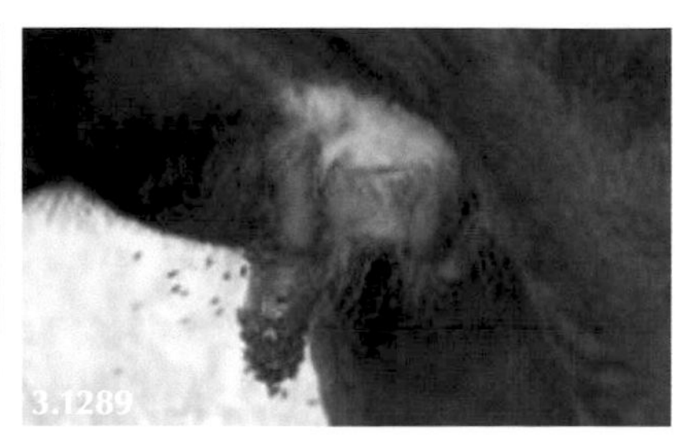

3.1287 and **3.1288:** Very dull, non-lactating beef cow with swollen right fore mammary gland and teat affected by summer mastitis. **3.1289:** Note the swollen right fore teat compared to other teats and many flies which may play a role in disease transmission.

- **Key observations**: Dry cow isolated from others in group with markedly swollen teat(s) and gland(s). Very occasionally affects non-pregnant heifer, steer.
- **Clinical examination**: Oedematous, painful mammary gland with scant foul-smelling purulent secretion.
- **Key history events**: Non-lactating cow, summer months, wooded areas nearby, no preventative teat sealant/intramammary antibiotic infusion. 🎥 3.603, 3.604

Gangrenous Mastitis (Per-Acute)

- **Key observations**: Sick cow with very swollen, cold mammary gland/udder.
- **Clinical examination**: Cold mammary gland with bloody secretion.
- **Key history events**: No recognised predisposing factor.

Mammary Tumour

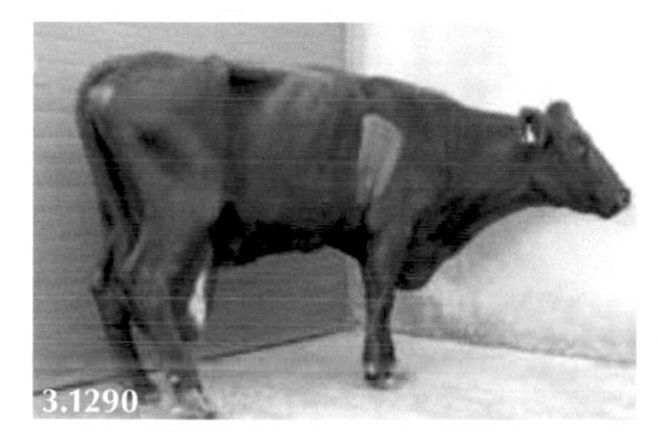
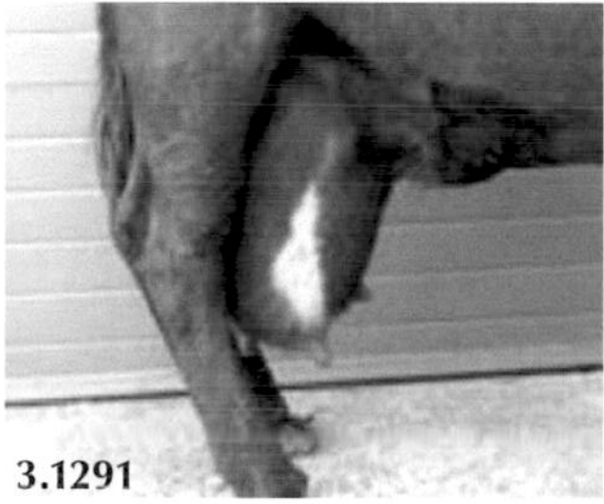
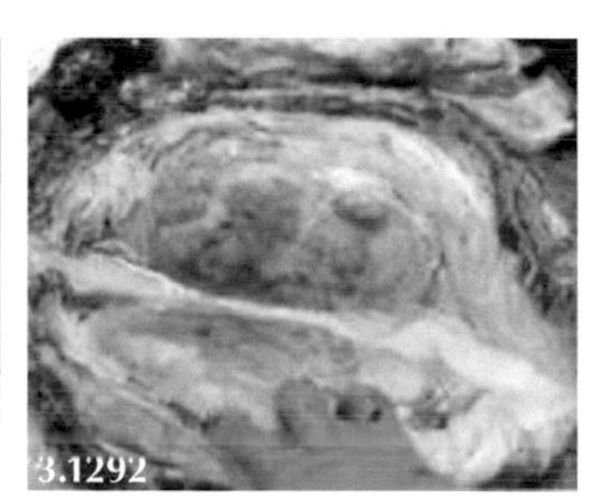

3.1290 and **3.1291:** Pendulous udder with very firm/hard, non-painful 30 cm diameter centre. **3.1292:** Necropsy reveals a mammary tumour (possibly a fibrosarcoma) but no metastases.

- **Key observations**: Very firm/hard, non-painful, uniform mass within the udder.
- **Clinical examination**: The lack of oedema would exclude mastitis.
- **Key history events**: Tumours of the mammary gland are rare. 🎥 3.605

3.24

Common Causes of Mastitis in Sheep

- Gangrenous mastitis is a major cause of death in low-ground sheep nursing four–eight-week-old twin lambs.
- Sloughing of the gangrenous mammary glands occurs four to six weeks later, leaving fibrotic nodules which bleed readily, attracting nuisance flies and causing serious welfare concerns.
- Indurative mastitis caused by MVV may be more common than current diagnostic tests suggest.

TABLE 3.125

Common Causes of Mastitis in Sheep

	Likely duration at presentation	Udder	Toxaemia
Gangrenous mastitis (per-acute)	Hours	++/+++	+++
Gangrenous mastitis (chronic)	Months	Slough	NAD
Acute mastitis	Days	+/++	+/++
Chronic suppurative mastitis/abscess	Months	+	NAD
Indurative mastitis	Months	Fibrotic	NAD
Teat lesions	Days	NAD	NAD

TABLE 3.126

Ancillary Tests to Determine Cause of Mastitis in Sheep

	Ancillary tests
Gangrenous mastitis (per-acute)	Euthanasia for animal welfare reasons
Gangrenous mastitis (chronic)	Clinical examination
Acute mastitis	Bacteriology if necessary
Chronic suppurative mastitis	Ultrasonography confirms abscess(es) within fibrous tissue
Indurative mastitis	Visna-maedi infection—serology or histology
Teat lesions	Electron microscopy for parapox virus if orf suspected

Gangrenous Mastitis (Per-Acute)

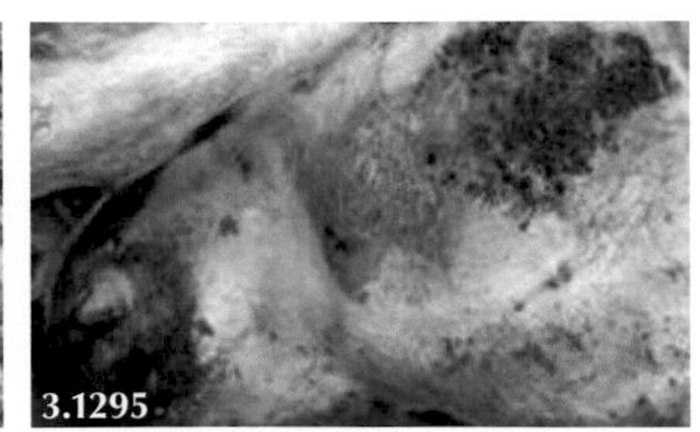

3.1293: Profound, sudden onset toxaemia. **3.1294:** Sharply demarcated gangrenous lesion affecting left mammary gland. **3.1295:** Thrombosis extending cranially along mammary vein.

DOI: 10.1201/9781003106456-42

- **Key observations**: Sudden onset illness with ipsilateral hindleg lameness.
- **Clinical examination**: Sharply demarcated gangrenous mastitis with thrombosis extending cranially along mammary vein.
- **Key history events**: Nursing four–eight-week-old twin lambs, insufficient grazing. 🎥 3.606, 3.607

Gangrenous Mastitis (Consequences)

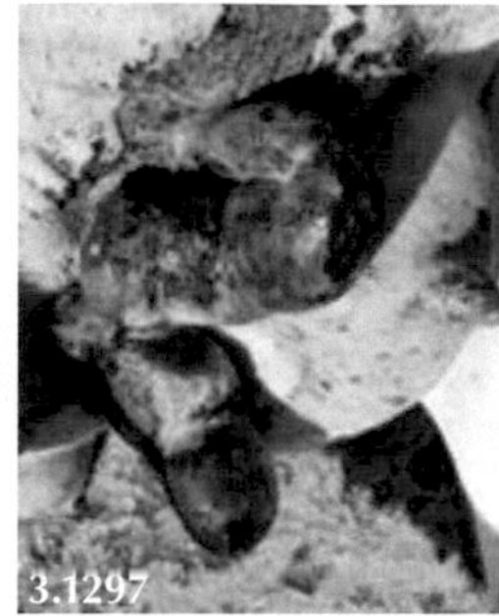
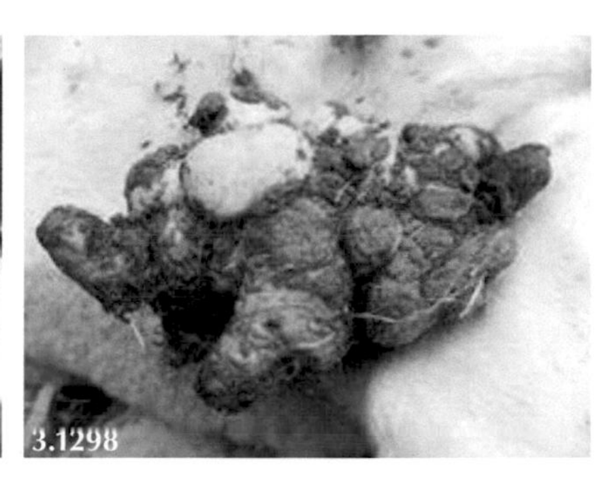

3.1296: Ewe four months after gangrenous mastitis; note blood on hindlegs from udder. **3.1297:** Fibrous tissue proliferation from sloughed udder site. **3.1298:** Keratinised fibrous stumps of remaining udder tissue.

- **Key observations**: Bright, alert ewe with proliferative fibrous stumps which bleed profusely.
- **Clinical examination**: Mammary tissue sloughed two to four months earlier leaving proliferative fibrous tissue which bleeds readily.
- **Key history events**: Gangrenous mastitis some months earlier. 🎥 3.608, 3.609, 3.610

Acute Mastitis

3.1299: Ewe isolated from flock, reduced rumen fill, swollen teat. **3.1300:** Swollen right mammary gland. **3.1301:** Swollen left mammary gland.

- **Key observations**: Ewe not grazing, lambs not allowed to suck.
- **Clinical examination**: Oedematous mammary gland often with lesion on medial aspect of teat caused by lamb's incisor teeth/orf.
- **Key history events**: Possible over-sucking/orf with lamb's teeth causing lesion on medial aspect of teat. Inadequate grazing. 🎥 3.611, 3.612, 3.613

Chronic Suppurative Mastitis

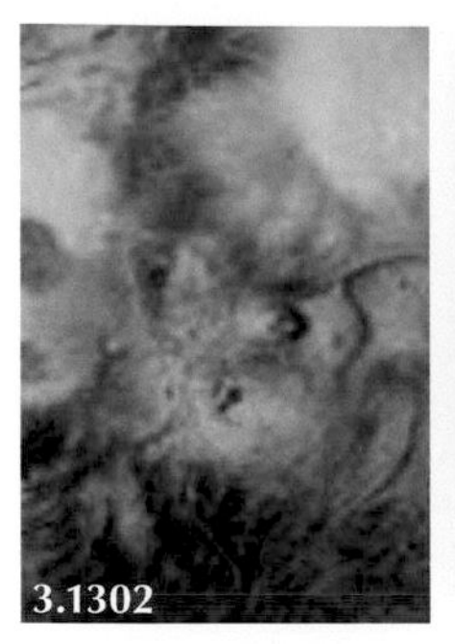

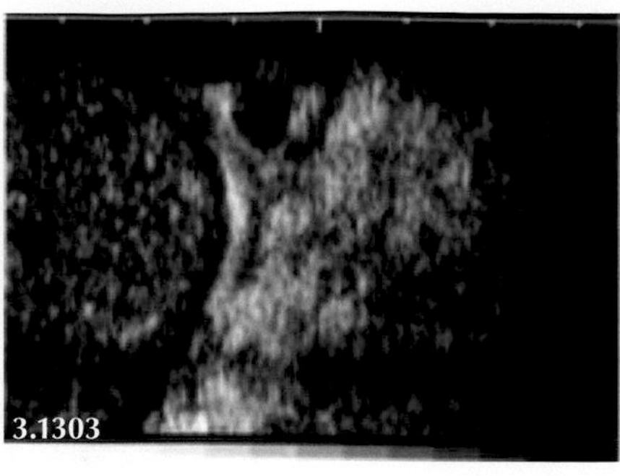

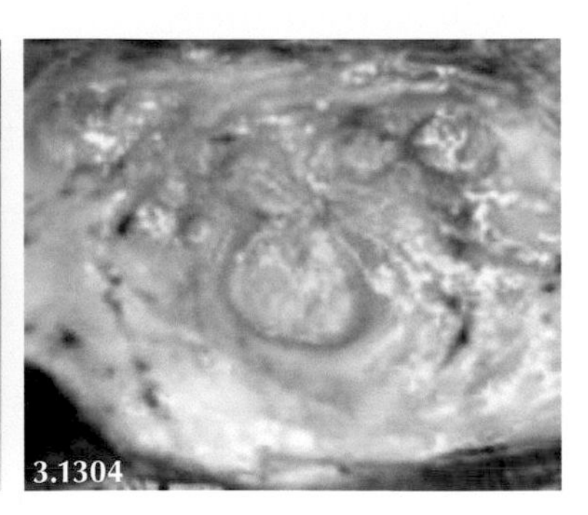

3.1302: Swollen left teat, several 2–3 cm diameter spherical swellings within mammary gland. **3.1303:** Ultrasound examination reveals 3 cm diameter abscess within the mammary gland (probe on skin on right hand side, scale top with cm gradations). **3.1304:** Necropsy reveals abscesses encapsulated by considerable fibrous tissue.

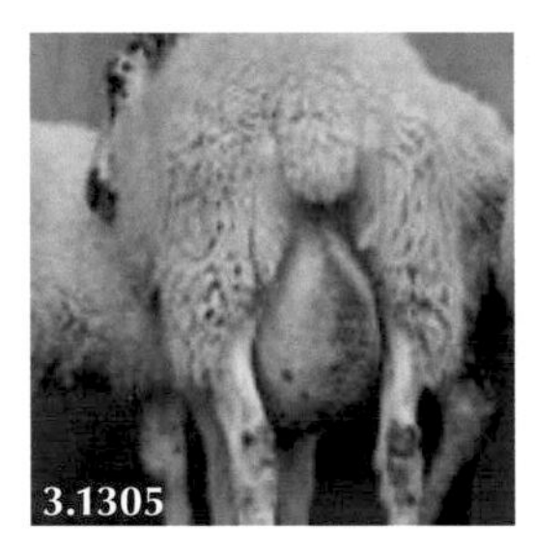

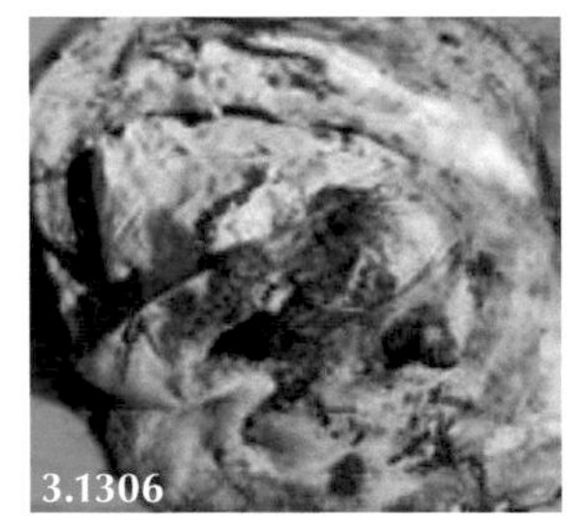

3.1305: Swollen left mammary gland. **3.1306:** Chronic mastitis with thick-walled abscess.

- **Key observations**: Swollen firm teat/mammary gland.
- **Clinical examination**: Firm swollen mammary gland containing numerous 2–3 cm diameter abscesses often with thick fibrous capsules.
- **Key history events**: Mastitis during lactation/post-weaning. Lesions usually found at pre-breeding check in autumn.

Visna-Maedi Infection

3.1307: Chronic indurative mastitis with poor milk production leading to very hungry lambs despite ewe in good BCS and adequate grazing/energy supply.

- **Key observations**: Hungry, poorly grown lambs.
- **Clinical examination**: Large firm/fibrous udder.
- **Key history events**: No routine maedi/visna monitoring. 3.614

Teat Lesions

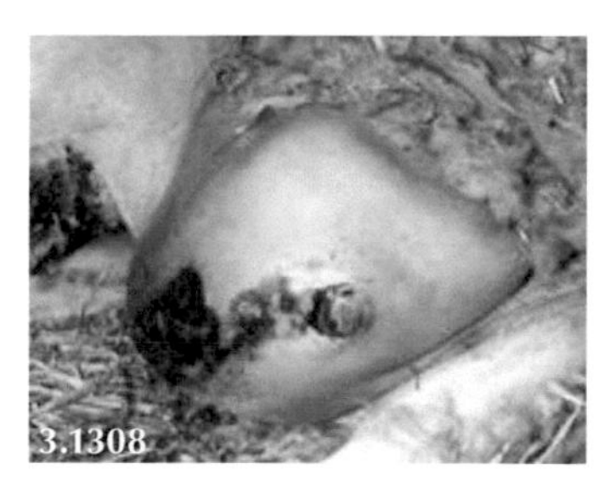

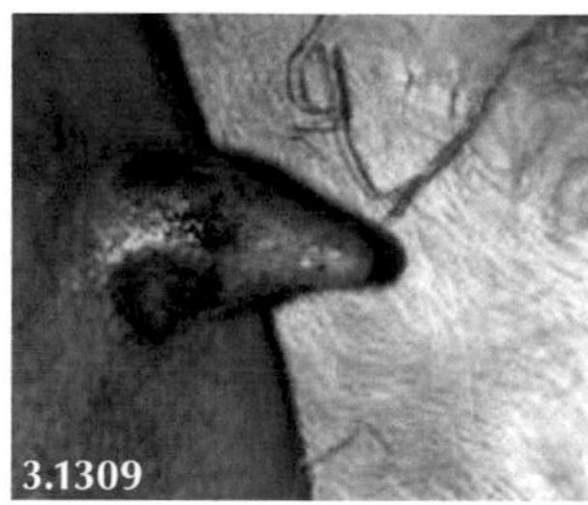

3.1308: Oedematous teat with lesions on the medial aspect causing by over-sucking by lambs.
3.1309: Gangrenous mastitis of the left mammary gland (ewe positioned on hindquarters).

- **Key observations**: Ewe not allowing lamb to suck causing hungry lamb.
- **Clinical examination**: Oedematous teat with lesion on medial aspect. Possible orf virus involvement.
- **Key history events**: Poor milk supply leading to over-sucking and damage by lambs' incisor teeth (to medial aspect of teat). Typically, four–six-week-old lambs at foot.

3.25

Common Causes of Abnormal Urination in Sheep

- Urolithiasis is a concern for farmers breeding pedigree rams and when fattening lambs on intensive cereal rations.
- Immediate demonstration of bladder distension is possible using ultrasonography.
- Ultrasonographic assessment of the right kidney for hydronephrosis aids prognosis.

TABLE 3.127

Common Causes of Abnormal Urination in Sheep

Disease	Likely duration at presentation	Urine voided	Tenesmus/colic	Appetite
Urolithiasis	Days	--/---	+/+++	---
Urolithiasis—ruptured urethra	Days	---	+	--/---
Nephrosis	Weeks	NAD/+	NAD	-/--
Tumour	Weeks/months	+	NAD	NAD

DOI: 10.1201/9781003106456-43

TABLE 3.128

Ancillary Tests to Determine the Common Causes of Abnormal Urination in Sheep

Disease	Ancillary tests
Urolithiasis	Ultrasonography—massively-distended bladder
Urolithiasis—ruptured urethra	Clinical examination
Nephrosis	Elevated BUN and creatinine. Pale swollen kidneys at necropsy. Histology
Tumour	Ultrasonography

Urolithiasis

3.1310 and **3.1311:** Male sheep showing colic signs—lying down, curled upper lip and ears directed caudally. **3.1312:** Ultrasound examination of sheep (centre image) reveals >12 cm diameter bladder in 40 kg sheep.

- **Key observations**: Colic signs—not eating, lying down, bruxism, curled upper lip and ears back.
- **Clinical examination**: Ultrasound demonstration of distended urinary bladder.
- **Key history events**: Often none but always check correct feed and mineral content (rams often fed ewe ration in error). 🎥 3.615

Ruptured Urethra

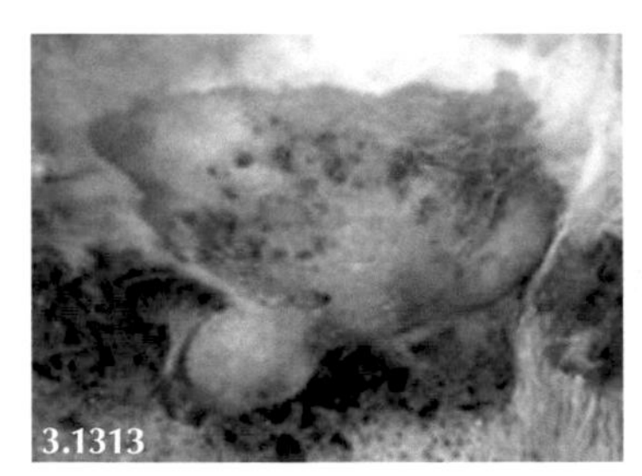

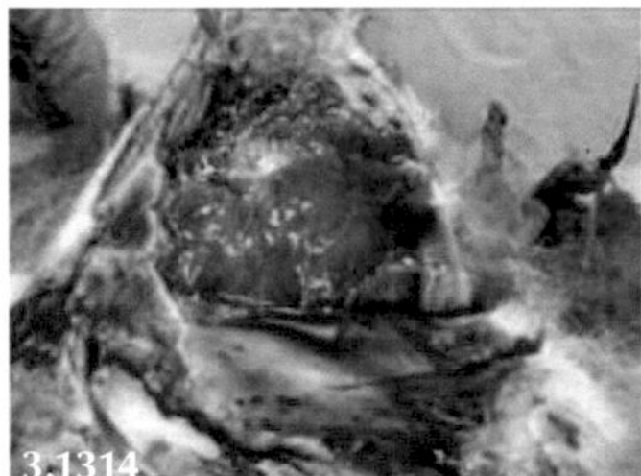

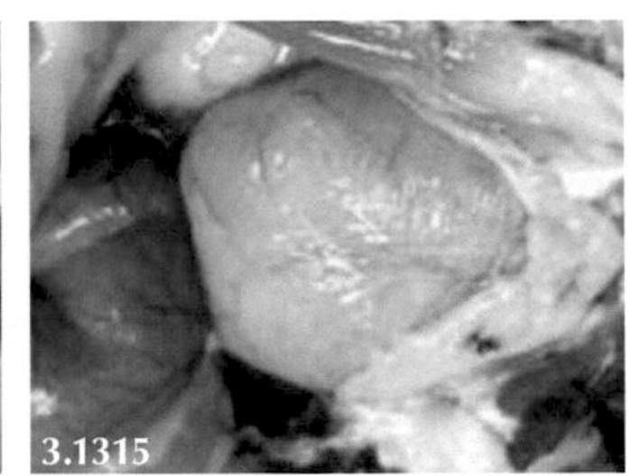

3.1313: Sharply demarcated skin necrosis and oedematous prepuce. **3.1314:** Subcutaneous urine accumulation. **3.1315:** Distended intact bladder.

- **Key observations**: Sharply demarcated skin necrosis and oedematous prepuce.
- **Clinical examination**: Distended bladder on ultrasound examination.
- **Key history events**: Castrated lambs fed incorrectly mineralised ration.

Nephrosis

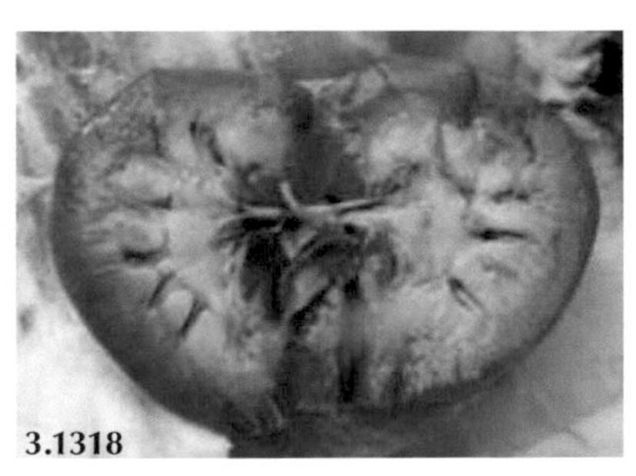

3.1316: Lamb frequently seen drinking at field drain. **3.1317:** Low condition score, poor wool quality and chronic diarrhoea. **3.1318:** Pale swollen kidney at necropsy.

- **Key observations**: Poor lambs, chronic diarrhoea, often seen standing over water troughs.
- **Clinical examination**: Low condition scores and poor wool quality consistent with chronic problem. Pale swollen kidneys at necropsy.
- **Key history events**: Previous nematodirosis and/or coccidiosis. 3.616, 3.617, 3.618

Bacteraemic Spread to Kidneys

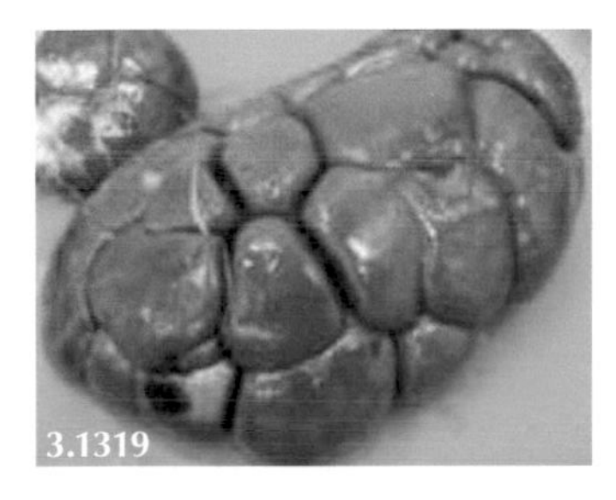

3.1319: Infarcts in the kidney secondary to vegetative endocarditis.

- **Clinical examination**: Lesions found at necropsy. There are no specific signs attributable to bacteraemic spread to kidneys.
- **Key history events**: Primary septic focus rarely identified.

Tumour (Renal Carcinoma)

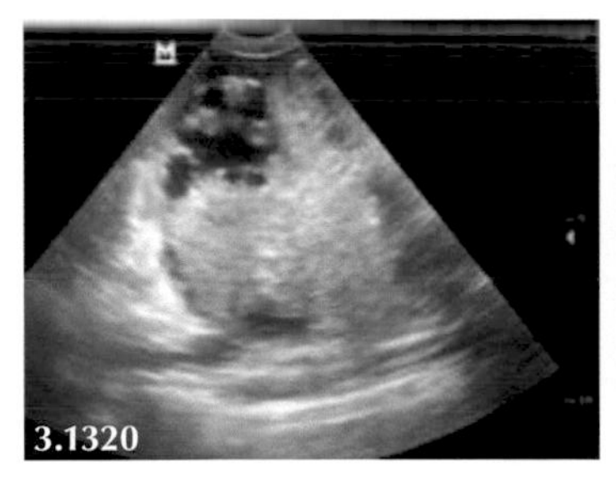
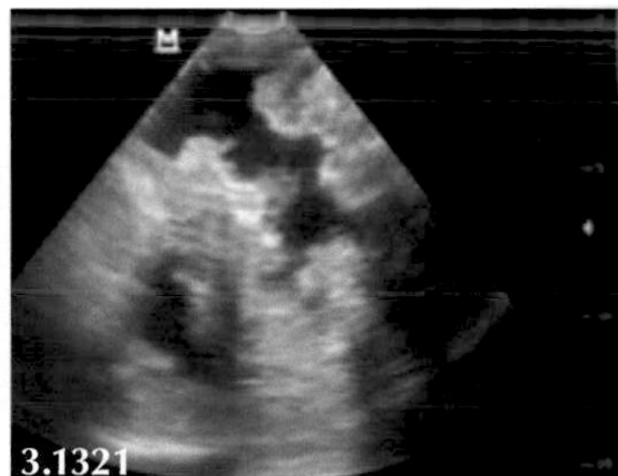
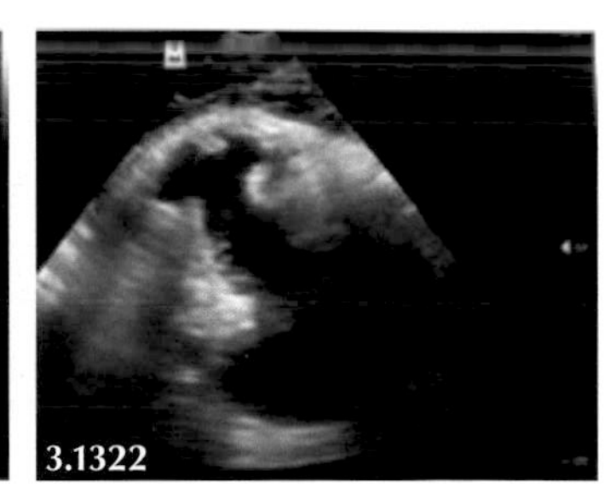

3.1320: Renal carcinoma when first examined, two months later (**3.1321**) and three months after first examination (**3.1322**). Note the change in size by referring to field depth. The ewe was clinically normal during this period, eating well and increased her body condition score.

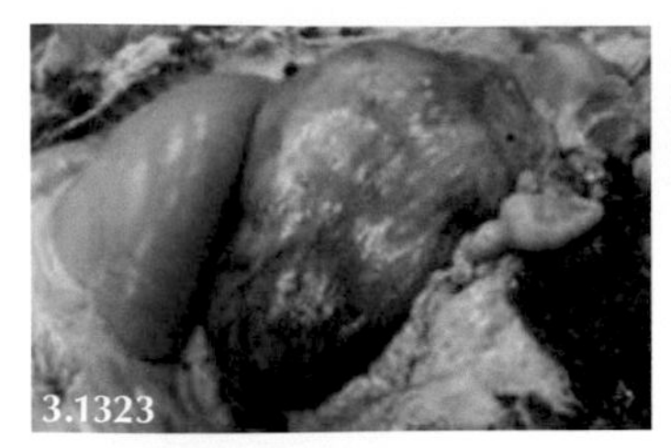
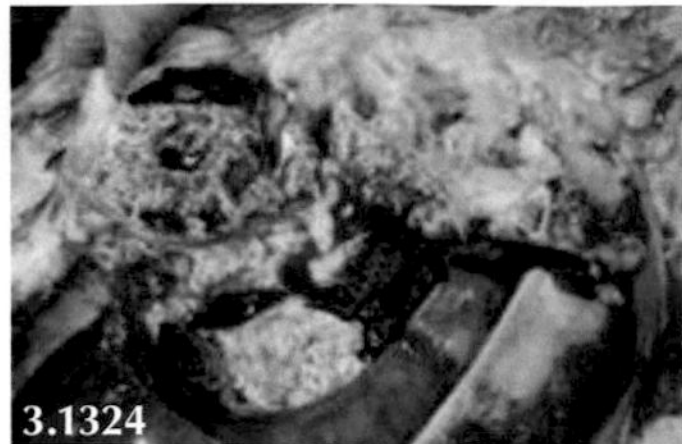
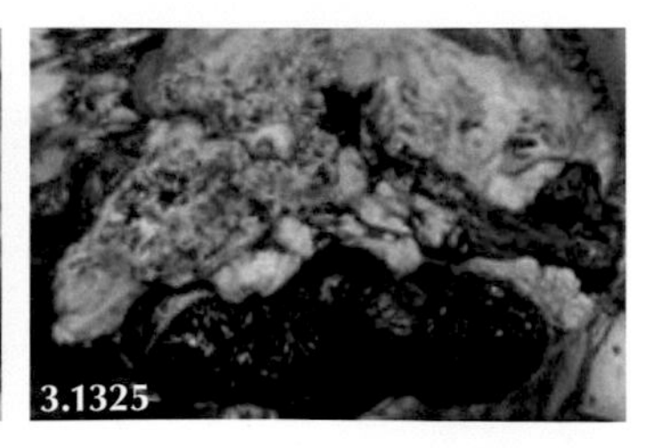

3.1323: Confinement of the tumour by the greater omentum (omental bursitis). Note the large amount of omental fat. **3.1324:** Necropsy confirmation of the renal carcinoma with adjacent haemorrhage. **3.1325:** Euthanasia after identification of significant haemorrhage.

- **Key observations**: Blood in urine.
- **Clinical examination**: Ultrasound reveals abnormal kidney.
- **Key history events**: Renal carcinomas are very rare in sheep. While uterine leiomyomas are common in goats, they have not been reported in sheep. 3.619

Uterine Leiomyoma (in Goats)

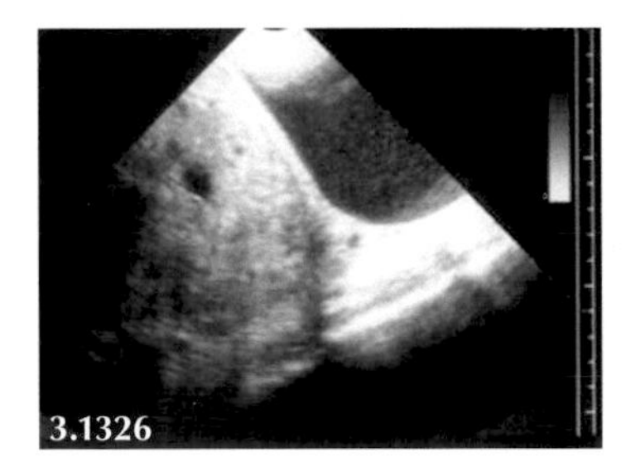
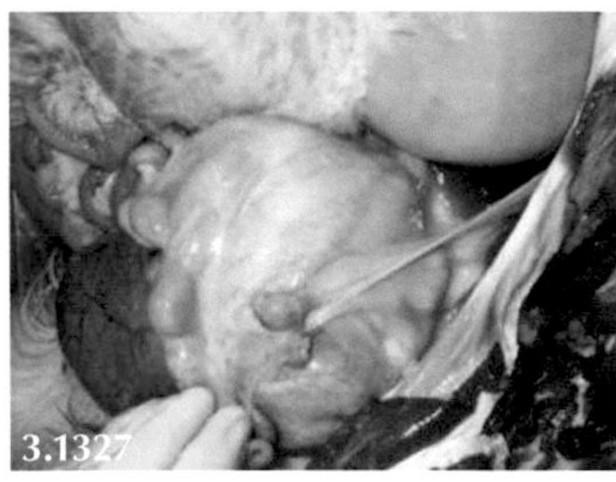

3.1326: Approximately 16 cm diameter soft tissue mass dorsal (to the left) of distended urinary bladder. **3.1327:** Confirmation of uterine leiomyoma at necropsy.

3.26

Common Conditions of the Penis in Rams

- Balanoposthitis is a very common condition of rams associated with very high protein levels in grass.

TABLE 3.129

Common Condition of the Penis in Rams

	Likely duration at presentation	Urination	Localised swelling	Pain
Balanoposthitis	Days	+/++	+	+
Paraphimosis	Days/weeks	NAD	+	NAD/+
Ruptured penis	Days	---	+	+/++
Ruptured glans	Days/weeks	NAD/+	NAD/+	+

DOI: 10.1201/9781003106456-44

TABLE 3.130

Ancillary Tests to Investigate the Common Penile Lesions in Rams

	Ancillary tests
Balanoposthitis	Culture *Corynebacterium renale*
Paraphimosis	Clinical inspection
Ruptured penis	Necropsy. Ultrasound examination demonstrates distended bladder and subcutaneous fluid accumulation. Surgery is not indicated because of large skin slough
Ruptured glans	Clinical examination

Balanoposthitis

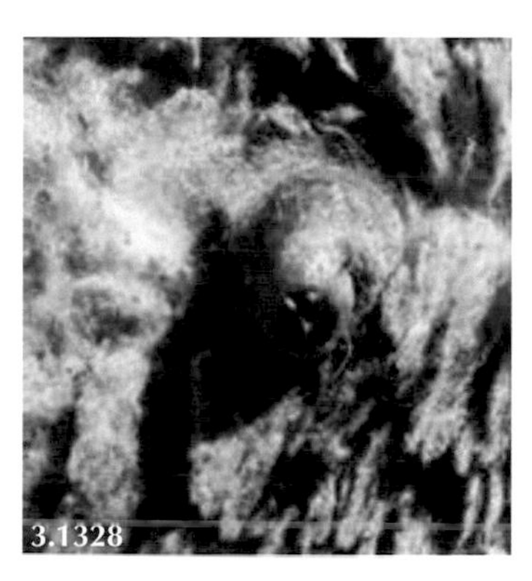

3.1328: Ulcerated and oedematous preputial skin.

- **Key observations**: Frequent dribbling of urine and tail swishing suggestive of localised irritation/pain.
- **Clinical examination**: Oedematous and frequently ulcerated skin surrounding the prepuce which bleeds readily.
- **Key history events**: High protein intake with proliferation of *C. renale*. Sternal recumbency secondary to lameness traumatises preputial skin.

Paraphimosis

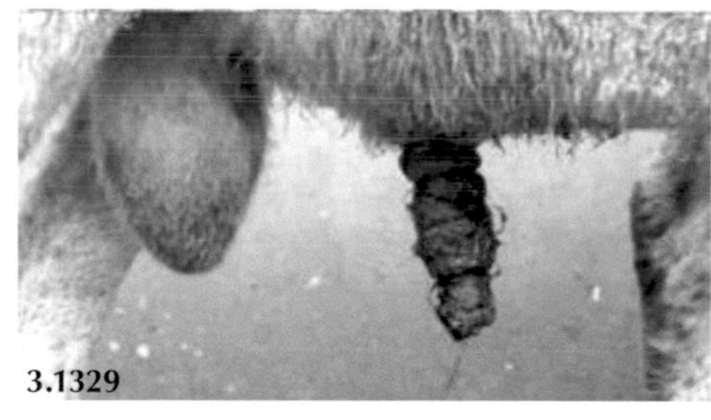

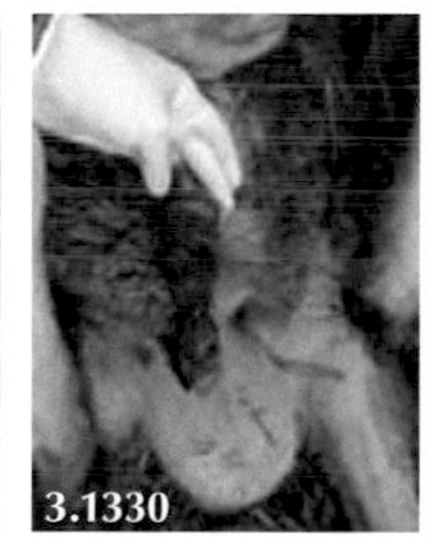

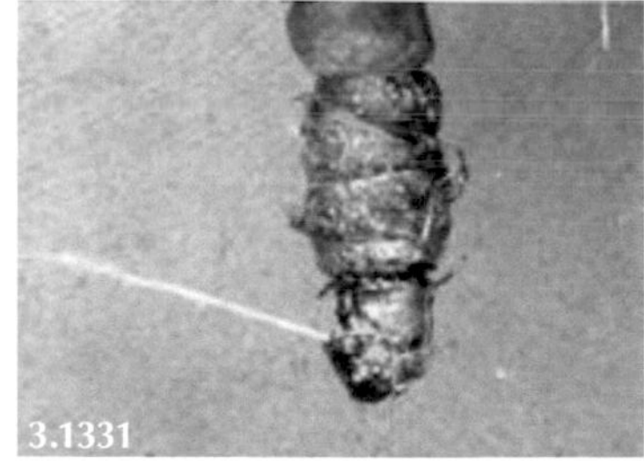

3.1329 and **3.1330:** Considerable oedema and superficial contamination of the prepuce and glans penis causing paraphimosis. **3.1331:** The vermiform appendage is difficult to identify when covered by scab material.

- **Key observations**: Superficial contamination and infection of the prepuce and glans penis.
- **Clinical examination**: Observing urination and/or ultrasound bladder examination essential.
- **Key history events**: Mating period with wool wrapped around penis. 🎥 3.620

Ruptured Penis

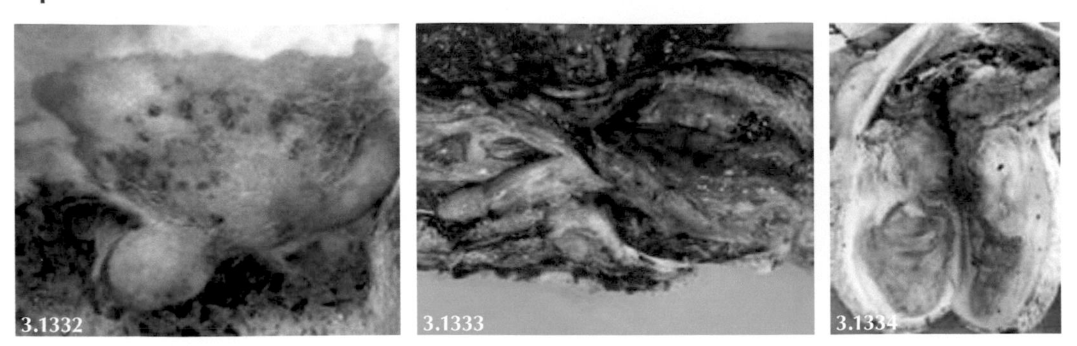

3.1332: Subcutaneous urine accumulation causing skin necrosis. **3.1333:** Subcutaneous urine accumulation surrounding the penis. **3.1334:** Urine accumulation within the scrotum.

- **Key observations**: Swollen prepuce. Purple/black skin on ventral abdomen/scrotum.
- **Clinical examination**: Bladder distension. Uraemic.
- **Key history events**: Intensively fed male sheep; often fattening castrated male lambs. 🎥 3.621, 3.622

Ruptured Glans

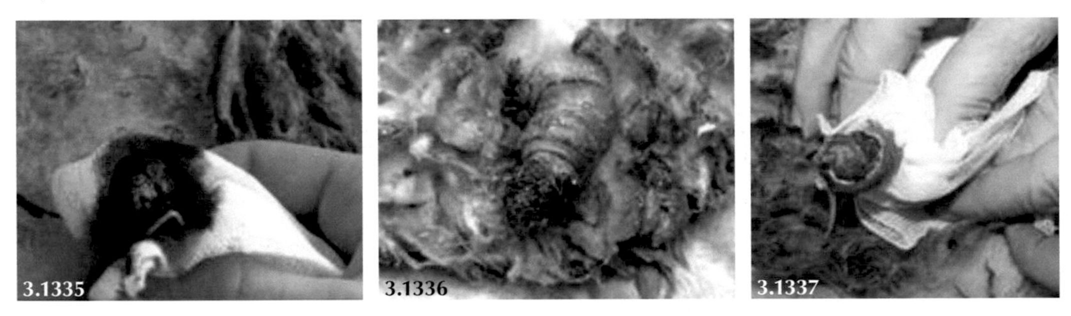

3.1335: Haemorrhage from the glans penis. **3.1336:** Dried scabs firmly adherent to the glans penis; the vermiform appendage cannot be identified. **3.1337:** The torn surface of the glans penis is clearly visible.

- **Key observations**: Blood on the fleece surrounding the prepuce and on the tail of served ewes.
- **Clinical examination**: Fresh blood/dried scabs on the glans penis.
- **Key history events**: Trauma during service. Possible spontaneous event. 🎥 3.623, 3.624

3.27

Common Causes of Weakness/Recumbency

Adult Sheep

The ranking that follows will vary considerably between farm management systems and possibly breeds of sheep. Diseases/disorders that cause weakness/recumbency, such as hypocalcaemia, are strongly associated with the production cycle. Culling policy will influence the number of ewes that become weak/recumbent due to paratuberculosis or dental problems.

- Hypocalcaemia affects ewes during late gestation often following a stressful event such as movement, change of forage/grazing, concentrate mineralisation, handling/vaccination etc.
- Weakness and depression are more prominent than cranial nerve V and VII deficits in some cases of listeriosis.
- Weakness/recumbency, as a result of emaciation, should not occur; these sheep should be culled much sooner.

TABLE 3.131

Common Causes of Weakness/Recumbency in Adult Sheep

	Strength	Mental state	Toxaemia	Pain	Abdomen	Rectal temperature
Hypocalcaemia	---	-	NAD	NAD	+/+++	NAD
Listeriosis	/	--/---	NAD	NAD	NAD	NAD/+
Ovine pregnancy toxaemia	-/--	--/---	NAD	NAD	+/++	NAD
Polioencephalo-malacia	NAD	-/+++	NAD	NAD	NAD	NAD
Endocarditis	-	-	+	+++	-	+
Joint lesions	NAD	NAD	NAD	+++	-/--	NAD/+
Mastitis	-	--	+++	+	-	+++
Fasciolosis			--- (Anaemia)	++	---	NAD
Paratuberculosis (end stage)	-/---	-	NAD	NAD	-/--	NAD
Emaciation (dentition)	-/--	-	NAD	NAD	-/--	NAD
Sarcocystosis	--/---	NAD	NAD	NAD	NAD	NAD
Basilar empyema	--/---	--/---	NAD	NAD	NAD	NAD/+

DOI: 10.1201/9781003106456-45

TABLE 3.132

Ancillary Tests to Investigate the Common Causes of Weakness/Recumbency in Adult Sheep

	Ancillary tests
Hypocalcaemia	Serum calcium <1.4 mmol/l. Response to intravenous calcium borogluconate within 5–10 minutes
Listeriosis	Increased lumbar CSF protein and white cell count (monocytes)
Ovine pregnancy toxaemia	Serum 3-OH butyrate concentration >3–4 mmol/l
Polioencephalomalacia	Response to intravenous vitamin B_1. Wood's lamp autofluorescence at necropsy can be variable. Histology
Endocarditis	Often no audible murmur, ultrasonography of heart valves
Mastitis	California mastitis test, bacterial culture
Fasciolosis	Raised serum enzymes esp. GLDH, copro-antigen ELISA
Paratuberculosis (end stage)	Profound hypoalbuminaemia, normal globulin concentration. ELISA. Necropsy
Emaciation (dentition)	Necropsy
Sarcocystosis	Increased lumbar CSF eosinophil count and percentage
Basilar empyema	Multiple cranial nerve deficits. Increased CSF protein and white cell count

Hypocalcaemia

3.1338: Housed 4-crop Greyface ewe due to lamb within two weeks. The ewe is weak and unable to stand. **3.1339:** Five minutes after intravenous administration of 20 ml of 40 per cent calcium borogluconate.

3.1340: 4-crop Greyface ewe due to lamb within two weeks at pasture. The ewe is weak and unable to stand. **3.1341:** Five minutes after intravenous administration of 20 ml of 40 per cent calcium borogluconate; the ewe is able to stand but is still bloated.

- **Key observations**: Sternal recumbency, weak/unable to stand, abdominal distension due to gut stasis/bloat, dull, no cranial nerve abnormalities.
- **Clinical examination**: Afebrile, normal mucous membranes, no ruminal activity, atonic rectum.
- **Key history events**: Late gestation, handled/housed (stressed) one or two days previously. 🎥 3.625, 3.626, 3.627, 3.628, 3.629

Listeriosis

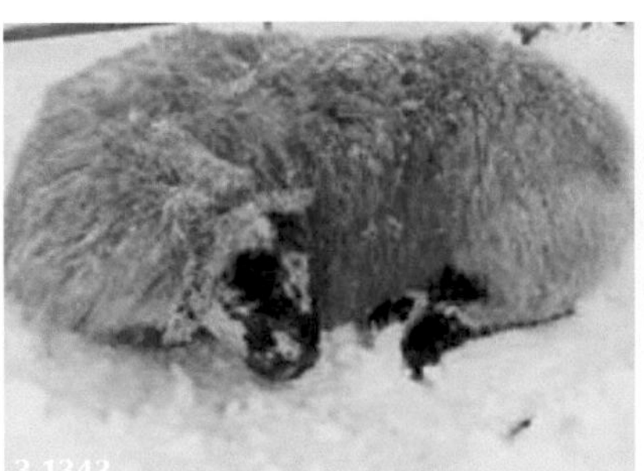

3.1342 and **3.1343:** Obtunded ewes unable to stand. **3.1344:** Left-sided hemiparesis with ipsilateral V and VII deficits.

- **Key observations**: Frequently obtunded and unable to stand. Ipsilateral hemiparesis in some cases.
- **Key clinical findings**: Unilateral cranial nerve V and VII deficits.
- **Key history events**: Access to poorly conserved silage 10–14 days ago. 3.630, 3.631, 3.632, 3.633

Ovine Pregnancy Toxaemia (Advanced Stages)

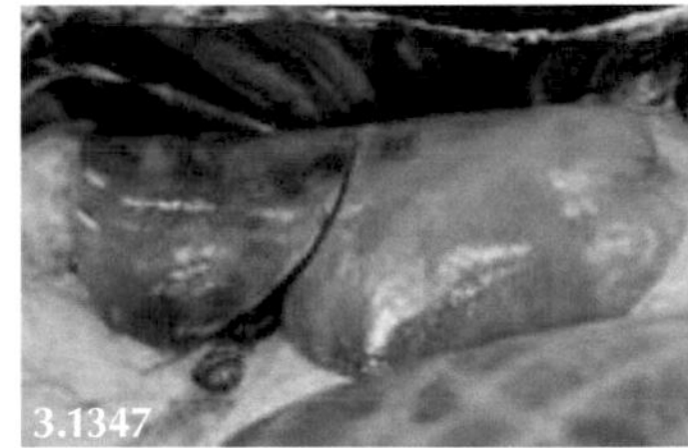

3.1345: Debilitated ewe unable to stand. **3.1346:** Debility caused by energy deficiency during prolonged winter storm. **3.1347:** Development of severe fatty liver.

- **Key observations**: Weak, unable to stand, large abdomen due to advanced multiple pregnancy.
- **Key clinical findings**: Blind, muscle fasciculations of face/head, very poor body condition.
- **Key history events**: Multiple pregnancy, chronic energy underfeeding. Severe winter storms during critical last few weeks of pregnancy. 3.634, 3.635

Polioencephalomalacia

3.1348 and **3.1349:** These two ewes are unable to stand, are blind and show dorsiflexion of the neck. **3.1350:** Recumbent ewe shows early seizure activity.

- **Key observations**: Dorsiflexion of neck, blind.
- **Key clinical findings**: Progress to hyperaesthesia, handling often induces seizure activity, blind but normal pupillary light reflexes.
- **Key history events**: Change in ration/grazing around 14 days ago. 3.636

Endocarditis

3.1351–3.1353: Weakness associated with pain arising from endocarditis and associated (poly)arthritis.

- **Key observations**: Painful expression; loss of body condition; difficulty rising; and chronic, severe lameness.
- **Clinical examination**: Painful joint(s), thickened joint capsule. Often no audible murmur.
- **Key history events**: Chronic lameness unresponsive to treatment. Original pyaemia not noticed or caused no clinical illness. 3.637

Joint Lesions (Erysipelas)

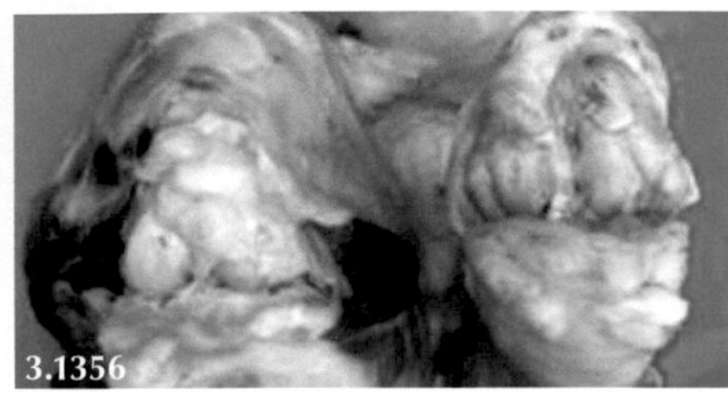

3.1354: Poor body condition. **3.1355:** Painful expression with long periods in sternal recumbency. **3.1356:** Comparison of stifle joints (right joint as viewed) shows a thickened joint capsule, synovial membrane hypertrophy and erosion of articular cartilage.

- **Key observations**: Painful expression; loss of body condition; chronic, severe lameness with long periods in sternal recumbency.
- **Clinical examination**: Painful joint(s), thickened joint capsule.
- **Key history events**: Chronic lameness unresponsive to treatment. 3.638

Endotoxaemia

3.1357–3.1359: Weakness and toxaemia resulting from severe mastitis.

- **Key observations**: Weak, obtunded, profuse salivation.
- **Clinical examination**: Febrile (41.5°C), toxic mucous membranes, often arising from gangrenous mastitis.
- **Key history events**: Nursing four–eight-week-old twins/triplets. 3.639

Subacute Fasciolosis

3.1360 and **3.1361:** Rapid weight loss leading to profound weakness and emaciation.

- **Key observations**: Very weak, dull, painful expression, poor appetite and little abdominal fill.
- **Clinical examination**: Afebrile, pale mucous membranes.
- **Key history events**: Autumn, no appropriate flukicide treatment. 3.640

Paratuberculosis (Agonal)

3.1362–3.1364: Weakness during agonal stages of neglected Johne's disease cases.

- **Key observations**: Sternal recumbency, dull, very weak, emaciated, empty abdomen.
- **Clinical examination**: Emaciated, afebrile, normal mucous membranes, no ruminal activity, atonic rectum.
- **Key history events**: Should have been culled months previously when poor condition first noted. 3.641, 3.642

Basilar Empyema

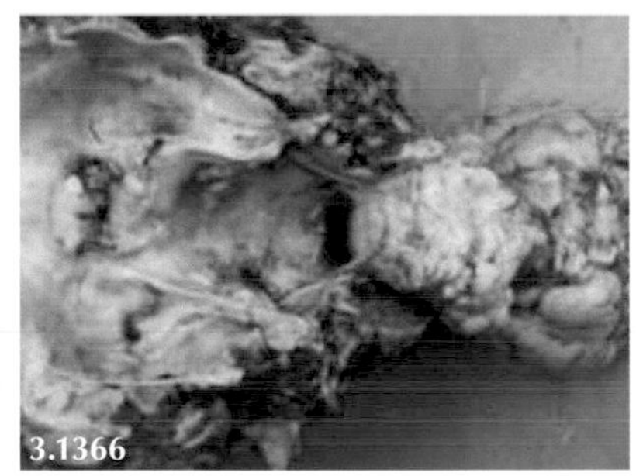

3.1365: Basilar empyema.
3.1366: Extent of lesion in relation to cranial nerve roots.

- **Key observations**: Obtunded, weak.
- **Clinical examination**: Variable but multiple bilateral cranial nerve deficits of II, III, V and VII.
- **Key history events**: NAD. 3.643

Sarcocystosis

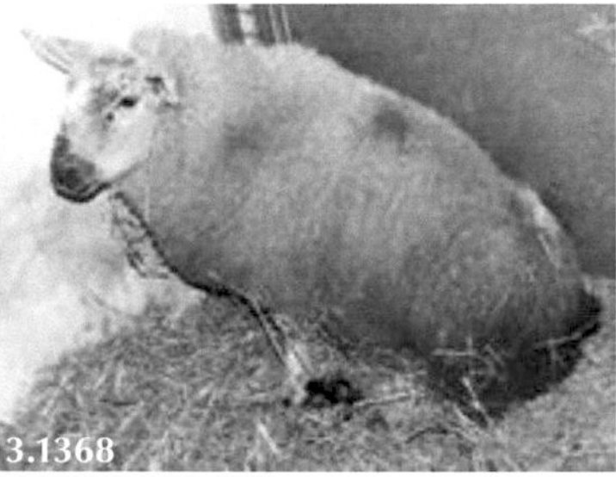

3.1367 and **3.1368:** Affected sheep appear bright and alert but weak most commonly affecting the hindlegs. **3.1369:** Weakness progressing to sternal recumbency.

- **Key observations**: Bright and alert but weak most commonly affecting the hindlegs.
- **Clinical examination**: Myelitis can cause weakness of all four legs or, more often, both hindlegs.
- **Key history events**: Access to heavily contaminated (dog feces) pastures.

Lambs (See Also Spinal Lesions)

- Atlanto-occipital joint infection causing tetraparesis is common but often misdiagnosed.
- Lumbar CSF analysis is the best means to differentiate vertebral empyema T2-L3 from delayed swayback.

TABLE 3.133

Common Causes of Weakness/Recumbency in Lambs

	Duration	Muscle tone	Mentation	Toxaemia
Atlanto-occipital infection	Hours/days	-/---	NAD	NAD
Delayed swayback	Days/weeks	-/--	NAD	NAD
Watery mouth disease	Hours	-/--	--	++/+++
Polyarthritis	Days	NAD	NAD	NAD/+
Omphalophlebitis/peritonitis	Days	-	-/--	+/++
Meningitis	Hours	Seizures	---/+++	+/++
Muscular dystrophy	Days	-/--	NAD	NAD
Severe cobalt deficiency	Weeks	-/--	-/--	NAD

TABLE 3.134

Ancillary Tests to Investigate the Common Causes of Weakness/Recumbency in Lambs

Disease	Ancillary tests
Atlanto-occipital infection	Elevated lumbar CSF protein concentration. *Streptococcus dysgalactiae* on culture
Delayed swayback	Histology of spinal cord. Blood/tissue copper levels of little diagnostic value
Watery mouth disease	Bacteriology of limited value
Polyarthritis	Joint tap likely to be unsuccessful—pannus. Necropsy
Omphalophlebitis/peritonitis	Ultrasonography. Necropsy
Meningitis	Lumbar cerebrospinal fluid analysis—very high neutrophil count and percentage, and increased protein concentration
Muscular dystrophy	Increased serum CK (>10,000 iu/L). Response to vitamin E/selenium
Severe cobalt deficiency	Plasma vitamin B_{12} concentration is <250pg/ml

Atlanto-Occipital Infection

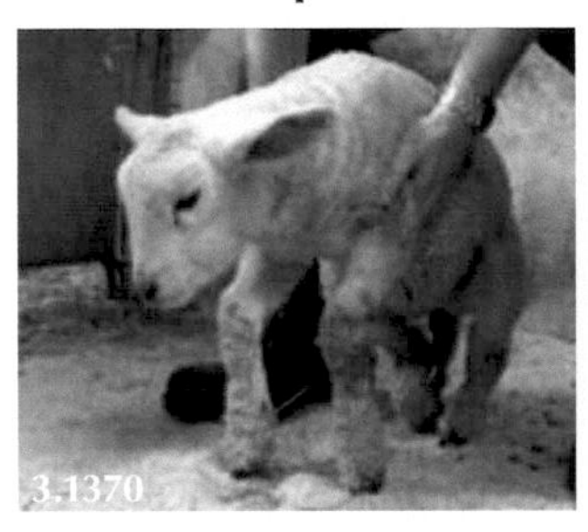

3.1370–3.1372: Profound weakness of all four legs caused by atlanto-occipital joint infection causing spinal cord compression. Possible cervical pain in right hand image.

- **Key observations**: Weakness progressing to recumbency.
- **Clinical examination**: Upper motor neuron signs in all four legs.
- **Key history events**: Entry via ears tags/elastrator rings coupled with failure of passive transfer suspected but no convincing evidence yet. 3.644, 3.645

Delayed Swayback

3.1373 and **3.1374:** Weakness of the hind legs eventually progressing to recumbency.

- **Key observations**: Weakness of the hindlegs eventually progressing to recumbency.
- **Clinical examination**: Upper motor neuron signs in hindlegs.
- **Key history events**: Several two–four-month-old lambs affected. Often farm with improved pastures. 3.646

Watery Mouth Disease

3.1375: Early watery mouth disease progresses to weakness and recumbency within 6 hours (**3.1376**).

- **Key observations**: Salivation, abdominal distension, weakness progressing to recumbency.
- **Clinical examination**: Ileus, dehydration.
- **Key history events**: Failure of passive transfer suspected but no convincing evidence yet. Contaminated environment. 3.647

Polyarthritis

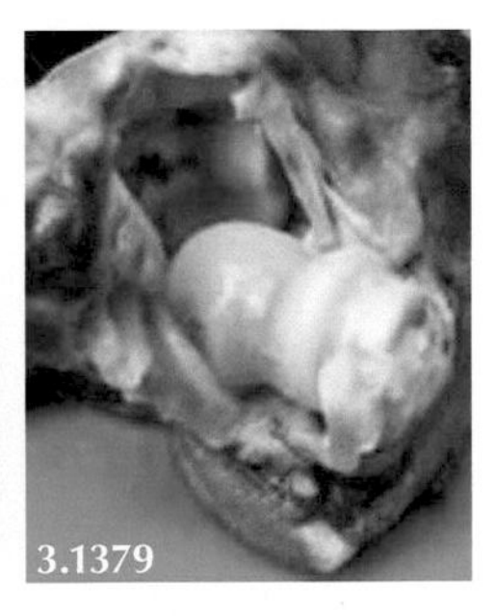

3.1377 and **3.1378:** Severe joint pain of the right elbow causing lameness, progressing to recumbency. **3.1379:** Pannus in the elbow point.

- **Key observations**: Severe joint pain causing lameness with long periods in sternal recumbency.
- **Clinical examination**: Swollen joint(s) caused by pannus and inflamed joint capsule.
- **Key history events**: Contaminated environment, gut versus navel as most important portal, failure of passive transfer suspected but no convincing data. 3.648, 3.649

Omphalophlebitis/Peritonitis

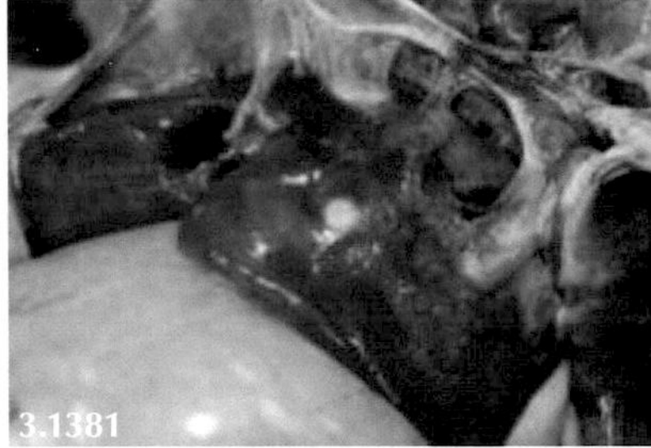

3.1380: Weight and condition loss in recumbent lamb caused by peritonitis. **3.1381:** Peritonitis associated with liver abscesses.

- **Key observations**: Weak, not sucking, loss of condition.
- **Clinical examination**: Anterior abdominal pain.
- **Key history events**: Born in contaminated environment, no umbilical treatments.

Meningitis

3.1382 and **3.1383:** Obtunded, weak three-week-old lambs with early meningitis. **3.1384:** Profound dullness and weakness in a three-day-old lamb with septicaemia/meningitis.

- **Key observations**: Profound dullness and blindness progressing rapidly to seizures.
- **Clinical examination**: Dullness/weakness rapidly progresses to hyperaesthesia, dorso-medial strabismus, injected episcleral blood vessels.
- **Key history events**: Failure of passive antibody transfer, contaminated lambing environment.

Muscular Dystrophy

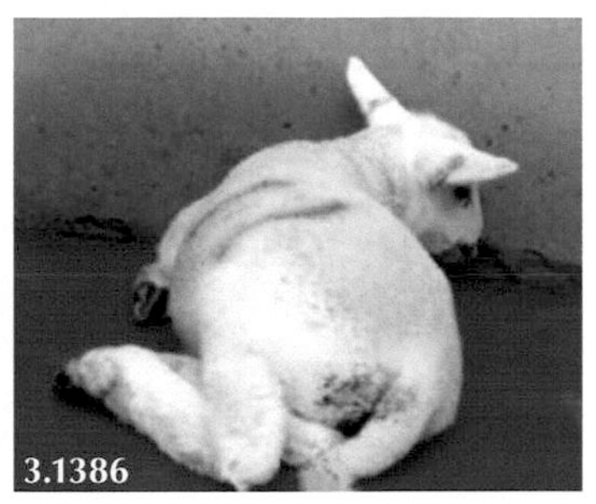

3.1385 and **3.1386:** Weakness affecting four-week-old rapidly growing Suffolk and Texel ram lambs.

- **Key observations**: Profound weakness in rapidly growing four–eight-week-old meat breed lambs.
- **Clinical examination**: Bright and alert but weak.
- **Key history events**: Dam and lamb rations very low in selenium/Vit E.

Severe Cobalt Deficiency/White Liver Disease

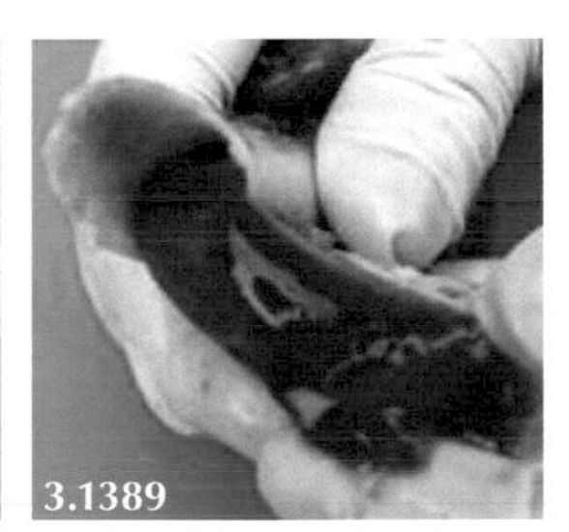

3.1387: Weakness in a five-month-old weaned lamb caused by severe cobalt deficiency. **3.1388:** Recumbency associated with white liver disease **3.1389:** Bone marrow from lamb with white liver disease.

- **Key observations**: Poor growth/body condition progressing to weakness.
- **Clinical examination**: Serous ocular discharge, anaemia.
- **Key history events**: Poor growth in many lambs after weaning. 3.650

Adult Cattle

- Most, if not all, recumbent recently calved dairy cows will have received multiple bottles of Ca, Mag and Phos administered intravenously and subcutaneously before veterinary attendance is requested.
- Testing spinal reflexes and peripheral nerves in adult cattle is difficult.
- Regular turning, milking and feeding recumbent dairy cows takes time.

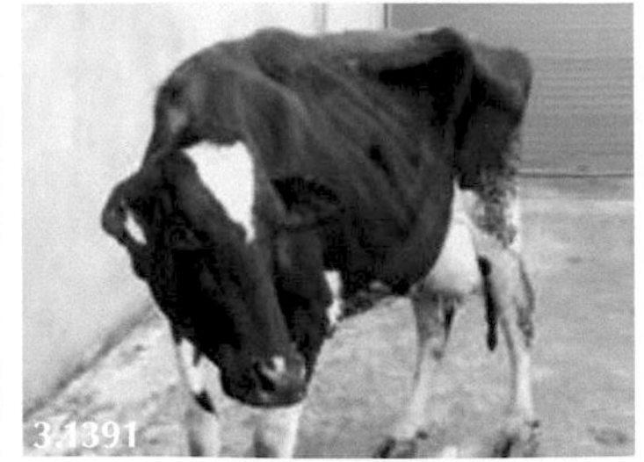

3.1390–3.1392: Emaciation may complicate other diseases that cause recumbency in dairy cattle.

TABLE 3.135

Common Causes of Recumbency in Adult Cattle

	Likely duration at presentation	Muscle tone	Mentation	Toxaemia
Hypocalcaemia	Hours	--/---	+/--	NAD
Hypomagnesaemia	Hours	+/+++	+/+++	NAD
Obturator nerve	Days	-/--	NAD	NAD
Coliform mastitis	Hours	-/--	--/---	+++
Listeriosis	Days	NAD	-/--	NAD
Endocarditis	Weeks	NAD	NAD	NAD/+
Botulism	Days	---	-	NAD
Hypophosphataemia	Days	-/--	NAD	NAD
Thoraco-lumbar spinal lesion	Days	-/---	NAD	NAD
Cervical lesion	Days	-/---	NAD	NAD
Tetanus	Days	+++	NAD	NAD
Pregnancy toxaemia	Days	-/--	--	++/+++

TABLE 3.136

Ancillary Tests for Recumbent Adult Cattle

Disease	Ancillary tests
Hypocalcaemia	Serum calcium concentration less than 1.4 mmol/l
Hypomagnesaemia	Serum magnesium concentration less than 0.4 mmol/l
Obturator nerve	Clinical presentation.
Coliform mastitis	Culture of milk sample
Listeriosis	Increased lumbar CSF protein concentration and absolute monocytosis
Endocarditis	Ultrasound demonstration of vegetative lesion
Botulism	No test available
Hypophosphataemia	Low blood phosphorus concentration not necessarily diagnostic
Thoraco-lumbar spinal lesion	Increased lumbar CSF protein concentration
Cervical lesion	UMN in all legs
Tetanus	No test available
Pregnancy toxaemia	Serum 3–0H butyrate concentration greater than 4.0 mmol/l

Hypocalcaemia

3.1393: Recently calved dairy cow with head averted against flank. **3.1394:** Dairy cow showing "S" bend in neck. **3.1395:** Recently calved beef cow unable to rise.

- **Key observations**: Weak/unable to stand, constipation, increasing bloat.
- **Clinical examination**: Flaccid muscles, tachycardia, no ruminal activity.
- **Key history events**: Older dairy cows (beef crosses from the dairy herd) within 24 hours after calving but can occur beforehand, and associated with oestrus for several months after calving. Much more common in cattle at pasture. 3.651, 3.652, 3.653

Hypomagnesaemia

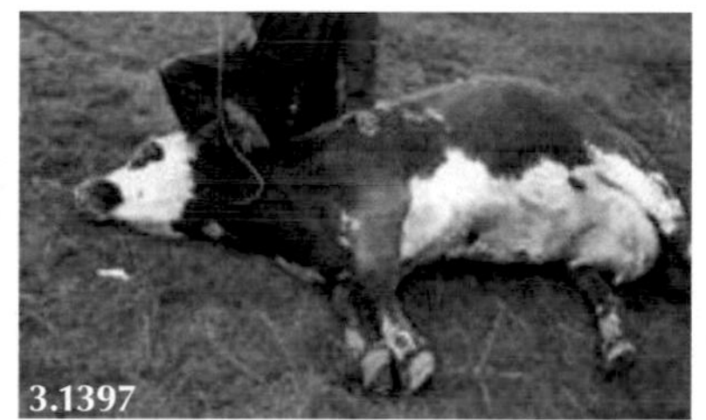

3.1396: Beef cow unsteady on her legs has been walked into a shed for treatment during which time she became very aggressive. Salivation is the result of constant bruxism. **3.1397:** Treatment for hypomagnesaemia; the cow has been heavily sedated and no longer has seizures. **3.1398:** Cow sedated for hypomagnesaemia now appears normal.

- **Key observations**: Beef cattle unsteady on legs (ataxic) progressing quickly to seizures.
- **Clinical examination**: Restricted by aggressive nature of cow; do not want to induce seizures.
- **Key history events**: Lush pasture, calved one to two months, sudden change to stormy weather, no preventive measures in place. 3.654, 3.655

Obturator and Sciatic Nerve Damage

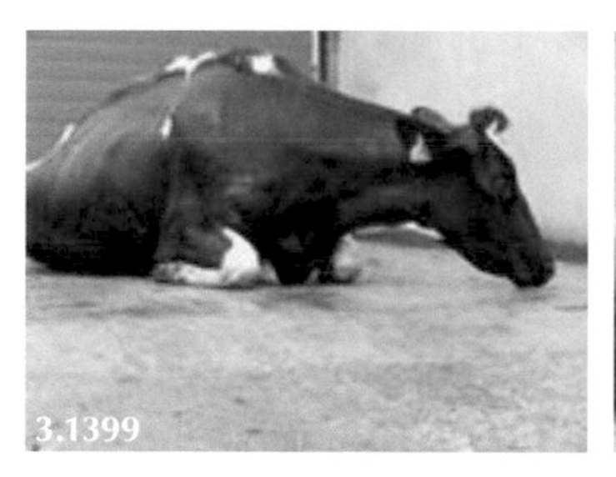

3.1399 and **3.1400:** Dairy cow unable to rise; treated for hypocalcaemia then "did splits" on smooth wet concrete. **3.1401:** Note the abnormal stifle joint extension and abduction of the upper hindleg of this cow in sternal recumbency.

- **Key observations**: Unable to adduct hindlegs and rise to feet otherwise BAR.
- **Clinical examination**: Difficult to assess.

- **Key history events**: Slipping on wet smooth concrete soon after recovering from bout of hypocalcaemia. Absolute fetal oversize and excessive traction causing "hip lock". 3.656

Coliform Mastitis

3.1402–3.1404: Dairy cow weak and unable to stand several days after calving.

- **Key observations**: Acute onset, mammary secretion often "serum-like".
- **Clinical examination**: Toxic mucous membranes, dehydration, rapid pulse.
- **Key history events**: Often within several days of calving. 3.657

Listeriosis

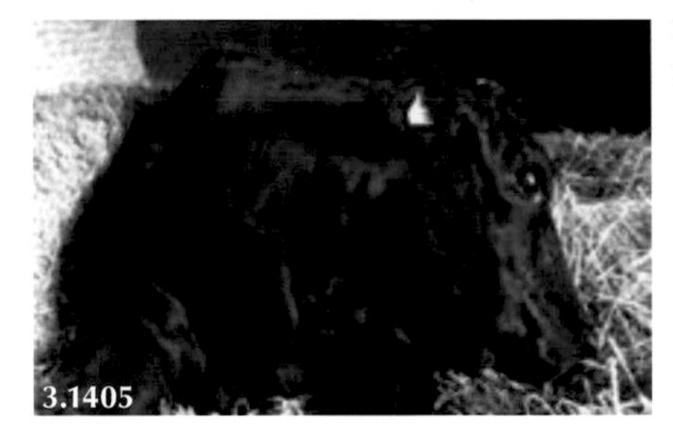

3.1405: The animal is dull and weak with unilateral trigeminal and facial nerve paralysis.

- **Key observations**: Dull and weak (hemiparesis). May propel themselves across cubicles, into corners, under gates etc.
- **Clinical examination**: Unilateral V and VII (if other deficits—consider basilar empyema).
- **Key history events**: Access to spoiled silage but cattle much less susceptible than sheep. 3.658

Endocarditis (Painful Joints)

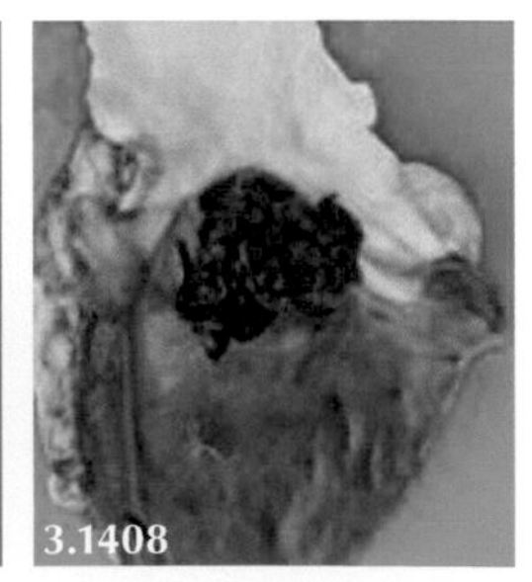

3.1406: Steer with multiple joint swellings. **3.1407:** Pannus within the hock joint. **3.1408:** Vegetative lesion on pulmonary valve.

- **Key observations**: Distended painful joints, often reluctant to stand.
- **Clinical examination**: No audible heart murmur in most cases.
- **Key history events**: Insidious onset, chronic weight loss.

Botulism

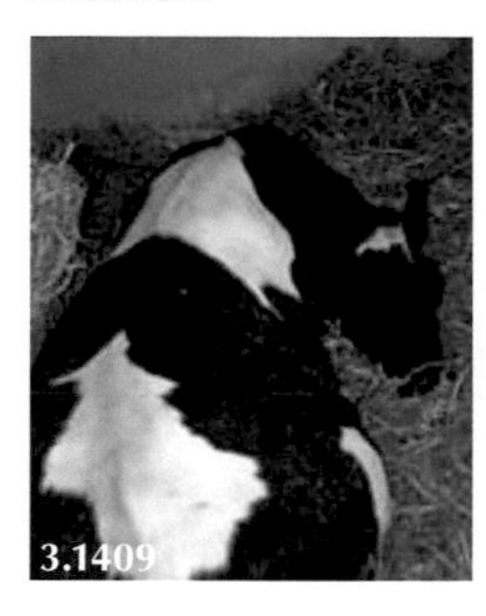

3.1409 and **3.1410:** Cattle very weak and unable to raise their head. **3.1411:** Tongue paralysis.

- **Key observations**: Weak, sternal recumbency, unable to eat/drink/swallow, protruding tongue.
- **Clinical examination**: Progressive motor paralysis over two to six days.
- **Key history events**: Several to many cattle affected over short period. Rapidly fatal motor paralysis caused by ingestion of the toxin produced by *Clostridium botulinum* often associated with spreading poultry manure (containing carcases) on pasture.

Hypophosphataemia

3.1412 and **3.1413:** Dairy cows unable to rise after several treatments for hypocalcaemia.

- **Key observations**: Periparturient hypophosphatemia of dairy cattle has been associated with muscle weakness but this is controversial. The term "crawlers" is often used to describe weak periparturient dairy cattle after treatment for hypocalcaemia.
- **Clinical examination**: No specific clinical signs other than weakness.
- **Key history events**: Often dairy cows treated for hypocalcaemia which do not rise.

Thoracolumbar Spinal Lesion

3.1414 and **3.1415:** Beef cows adopt a dog-sitting posture which is abnormal. **3.1416:** Size disparity between Charolais bull and beef cow highlights the risk of trauma during mating.

- **Key observations**: Adopt dog-sitting posture as unable to rise.
- **Clinical examination**: Normal forelegs; upper motor signs in hindlegs but difficult to assess reflex arcs in adult cattle.
- **Key history events**: Possible bulling injury in beef cattle. EBLv not present in UK.

Cervical Spinal Lesion

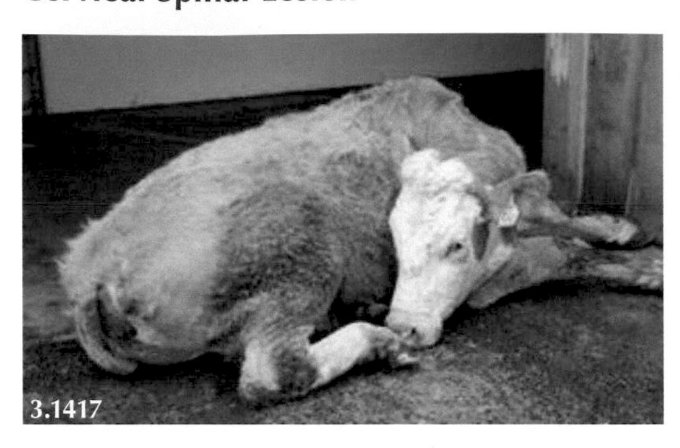

3.1417: Weakness on all four legs causing falls and abnormal leg position.

- **Key observations**: Weak on all four legs with difficulty getting onto feet.
- **Clinical examination**: Difficult to appreciate increased reflexes in all four legs in adult cattle.
- **Key history events**: Often none noted. 3.659

Tetanus

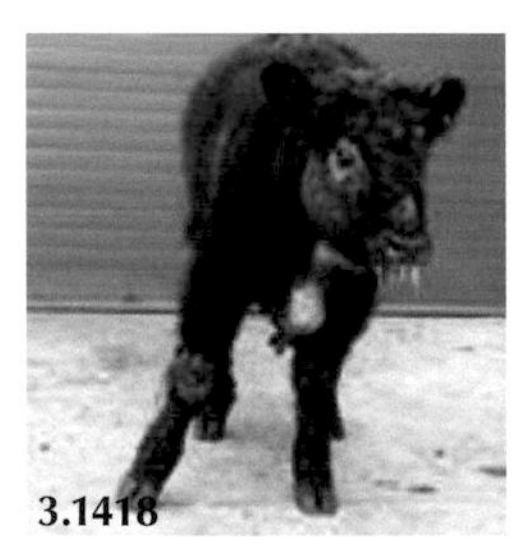

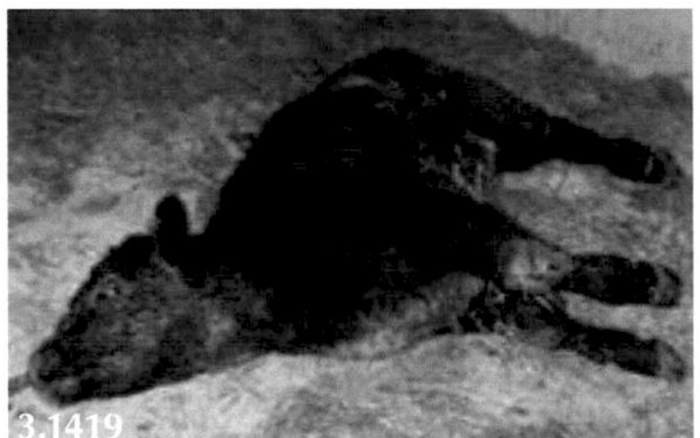

3.1418: Spastic gait and continuous salivation caused by tetanus. **3.1419:** Disease has progressed to opisthotonos. **3.1420:** When startled cattle often fall to the ground and show opisthotonos.

- **Key observations**: Lockjaw, startled appearance, prominent third eyelid.
- **Clinical examination**: Unable to open mouth, examination very limited if animal shows opisthotonos.
- **Key history events**: Infection of open castration and other wounds. Ingestion of pre-formed toxin causes bloat. 3.660

Pregnancy Toxaemia/Fatty Liver (Beef Cows)

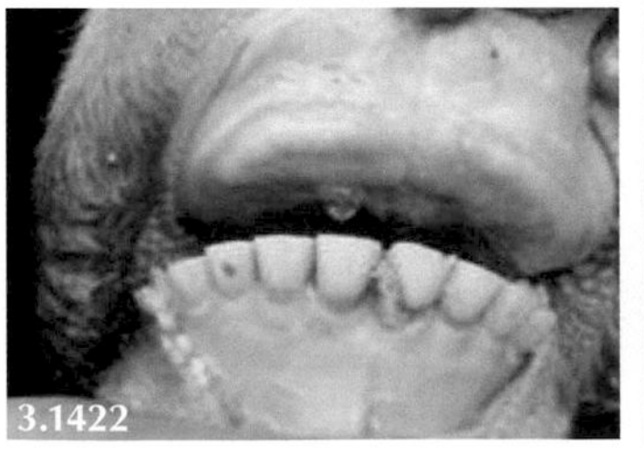

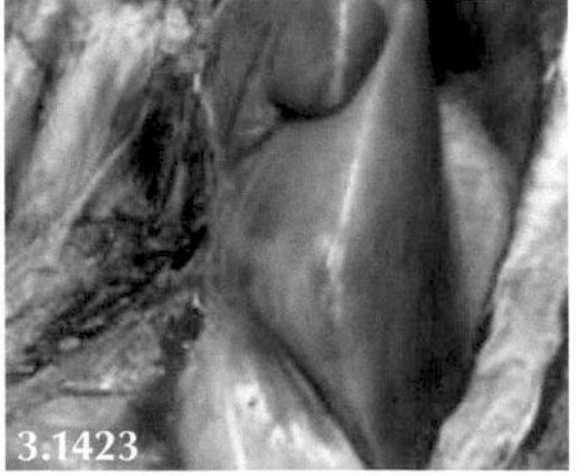

3.1421: Beef cow—weak, inappetant and unresponsive. Early brisket oedema. **3.1422:** Jaundiced oral mucous membranes. **3.1423:** Jaundice noted in colour of omental fat with enlarged very fatty liver with rounded margins at necropsy.

- **Key observations**: Dull, inappetant and unable to rise.
- **Clinical examination**: Jaundice and subcutaneous oedema.
- **Key history events**: In this case dystocia caused by absolute fetal oversize lead to severe and sudden dietary energy reduction to all pregnant cows one week ago. This cow was carrying twins. 3.661

Young Calves

- Pain arising from polyarthritis is a common cause of recumbency in neonatal calves.
- Acidosis following virus-induced diarrhoea is a common cause of weakness/recumbency in eight–fourteen-day-old beef calves.

TABLE 3.137

Common Causes of Weakness/Recumbency in Young Calves

	Likely duration at presentation	Muscle tone	Mentation	Toxaemia
Septicaemia	Hours	-	---	++/+++
Viral diarrhoea with acidosis	Hours/days	--/---	--/---	NAD
Omphalophlebitis/peritonitis	Days	-	-/--	+/++
Meningitis	Hours	-/+++	---/+++	+/++
Polyarthritis	Days	NAD	NAD	NAD/+
Dystocia-related e.g. fracture	At birth	NAD	NAD	NAD
Diarrhoea (ETEC)	Hours	-/--	NAD	NAD
Cervical spinal lesion	Days	--/---	NAD	NAD
Congenital defects	At birth	-	NAD	NAD
Bovine neonatal panleucopaenia	Days	-	NAD	NAD
Aortic thrombus	Hours	Hindlegs ---	NAD	NAD

TABLE 3.138

Ancillary Tests for Weak/Recumbent Young Calves

Disease	Ancillary tests
Septicaemia	Blood culture rarely undertaken. Necropsy
Viral diarrhoea with acidosis	Calf-side test kits for causal virus. Blood gas analysis. Response to intravenous bicarbonate
Omphalophlebitis/peritonitis	Ultrasonography. Necropsy
Meningitis	Lumbar cerebrospinal fluid analysis—very high neutrophil count and percentage, and increased protein concentration
Polyarthritis	Joint tap may be unsuccessful—pannus. Necropsy
Dystocia-related e.g. fracture	Radiography
Diarrhoea (ETEC)	Calf-side test kits. Culture
Cervical lesion	Radiography. Increased lumbar CSF protein concentration
Bovine neonatal panleucopaenia	Bone marrow analysis
Aortic thrombus	Necropsy

Septicaemia

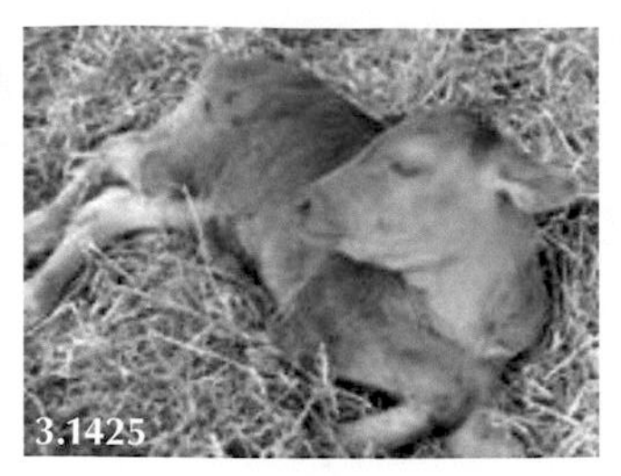

3.1424: Dull, ears down, weakness leading to recumbency. **3.1425:** Note profound weakness and dullness, abnormal head position. **3.1426:** Agonal seizure activity.

- **Key observations**: Dullness in one–three-day-old calves rapidly progressing to recumbency and possible seizure activity (meningitis).
- **Clinical examination**: Normal hydration status but injected scleral vessels. Umbilicus often normal—not the major portal of bacterial invasion.
- **Key history events**: Failure of passive antibody transfer, possibly related to dystocia. Poor calving box/area hygiene. 3.662, 3.663, 3.664, 3.665, 3.666

Acidosis Secondary to Viral-Induced Diarrhoea

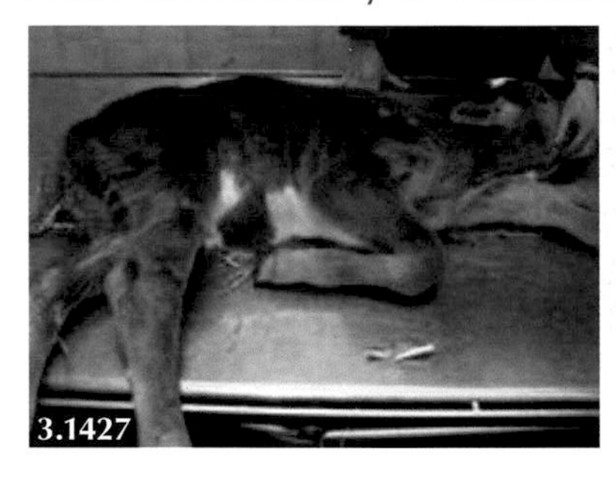
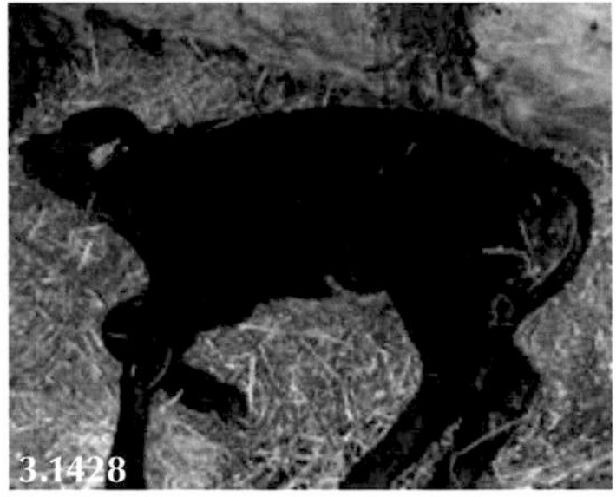
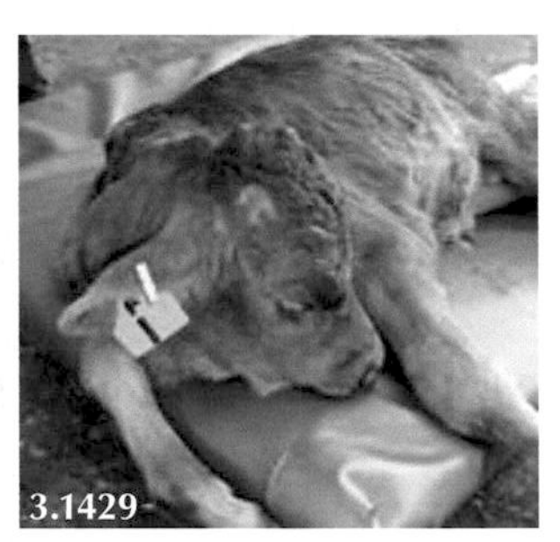

3.1427–3.1429: All three calves are very weak and unable to maintain sternal recumbency. Profuse diarrhoea is evident in the centre image.

- **Key observations**: Typically seven–ten-day-old beef calves. Diarrhoea for previous one to three days.
- **Clinical examination**: Profound weakness, dullness progressing to coma, often have normal hydration status due to oro-gastric administration of rehydration solution by farmer.
- **Key history events**: High prevalence of diarrhoea with some losses occurring towards the end of calving period with rapid spread amongst housed beef calves. 3.667

Omphalophlebitis/Peritonitis

3.1430: Dull, head and ears down, weak three-day-old dairy calf which has not been sucking. **3.1431:** Necropsy reveals extensive fibrinous peritonitis; the distended fluid-filled intestines explain the normal "abdominal fill" in the left hand image. 3.668

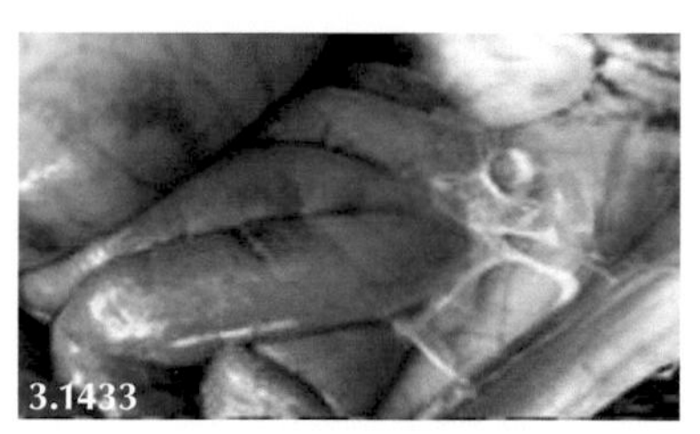

3.1432: Collapsed four-day-old beef calf with a distended abdomen with widespread fluid sounds on succussion. Treatment, including intravenous fluid therapy, was unsuccessful and the calf was euthanased for welfare reasons. **3.1433:** Fibrinous peritonitis causing ileus and distended fluid-filled intestines.

- **Key observations**: Dullness, not sucking over previous 24–36 hours. Painful, swollen umbilicus.
- **Clinical examination**: Abdominal distension. Fluid-filled intestines on succussion.
- **Key history events**: Unhygienic conditions, inadequate navel treatment. 3.669, 3.670

Meningitis

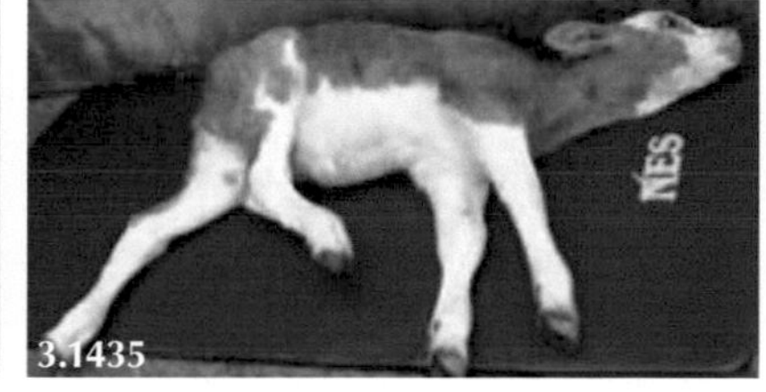
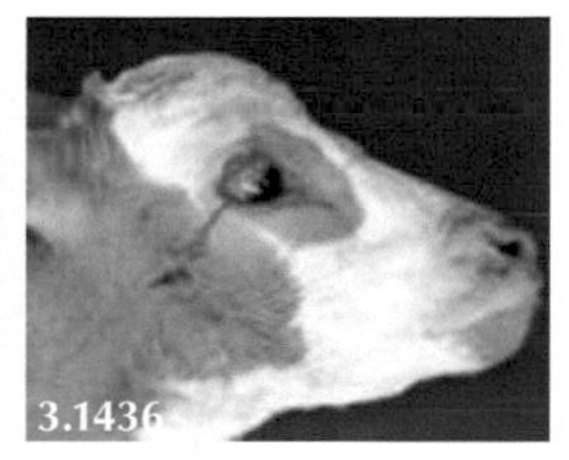

3.1434: Profound dullness, weakness in a three-day-old calf. **3.1435:** Seizure activity, dorsi-flexion of neck, extension of fore legs and flexion of hindlegs. **3.1436:** Dorso-medial strabismus, injected episcleral blood vessels.

- **Key observations**: Profound dullness and blindness progressing rapidly to seizures.
- **Clinical examination**: Hyperaesthesia, dorso-medial strabismus, injected episcleral vessels.
- **Key history events**: Unhygienic calving environment, failure of passive antibody transfer. 3.671

ETEC

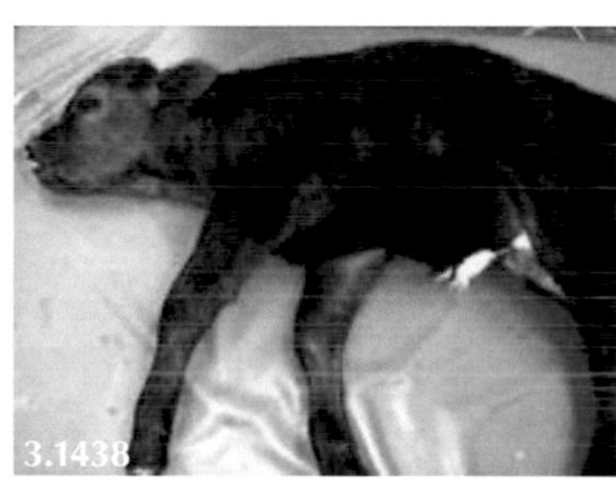

3.1437: Weakness affecting a 24-hour-old calf. **3.1438:** Collapse in a 36-hour-old beef calf.

- **Key observations**: Very rapid progression to recumbency and death within first one to three days of life. May be no evidence of profuse diarrhoea.
- **Clinical examination**: Usually found dead by farmer.
- **Key history events**: Non-vaccinated dam, often several cases over short time period.

Polyarthritis

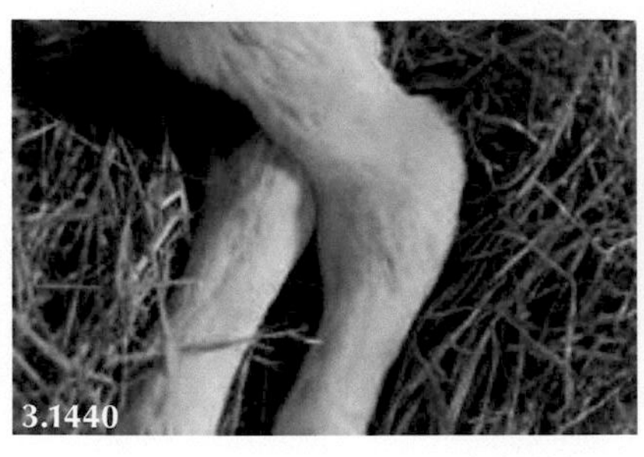

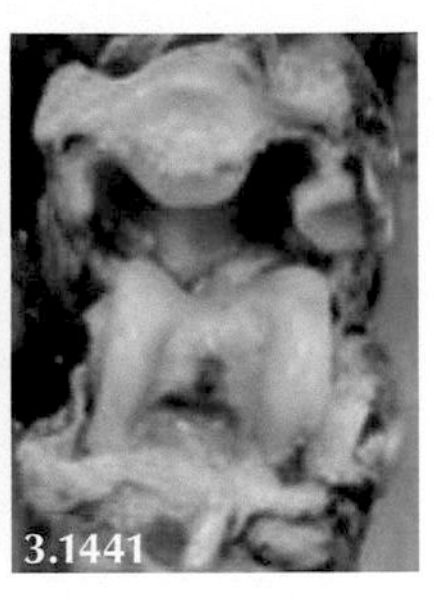

3.1439: Four-day-old calf unwilling to stand with swollen carpal and hock joints. **3.1440:** Distended left hock joint. **3.1441:** Marked pannus in the left hock joint despite the short illness.

- **Key observations**: Recumbent animal unwilling to stand.
- **Clinical examination**: Several distended joints.
- **Key history events**: Failure of passive antibody transfer. 3.672, 3.673 3.674

Limb Fracture Caused by Dystocia

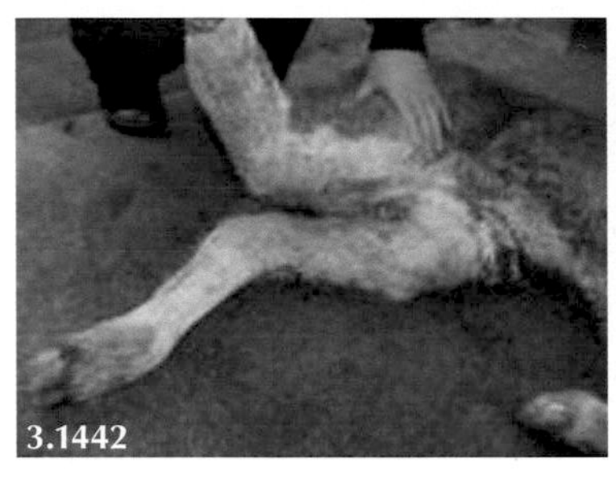

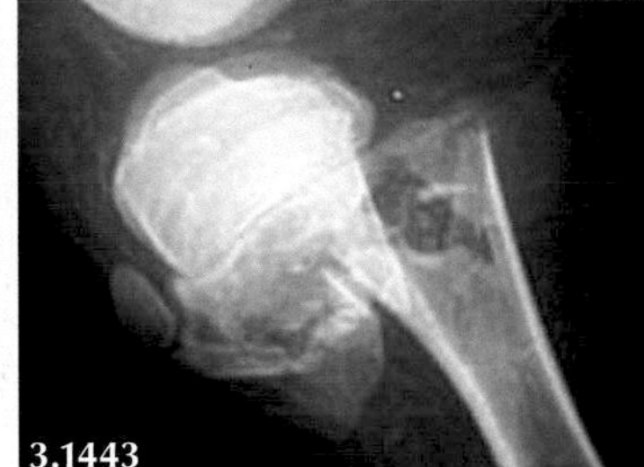

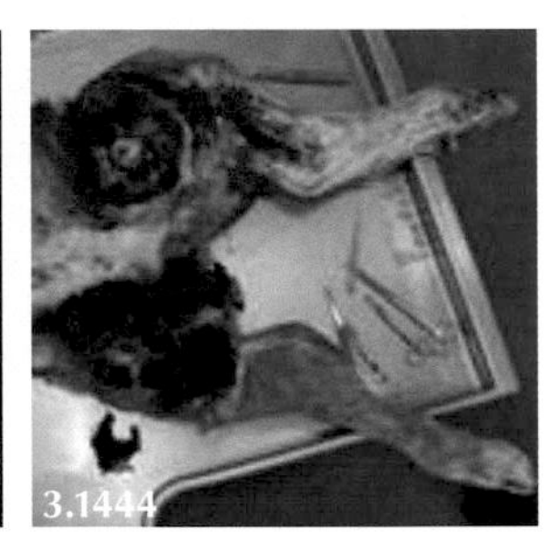

3.1442: Calf at necropsy after confirmation of tibial fracture through proximal growth plate. **3.1443:** Radiograph confirming tibial fracture. **3.1444:** Note that localised haemorrhage and soft tissue swelling may limit movement at fracture site. 3.675

Limb Fracture Unrelated to Dystocia

3.1445: Recumbent calf with metatarsal fracture. **3.1446:** Abnormal angulation of left (upper leg as viewed) third metatarsal bone at fracture site. **3.1447:** Pain-free fracture alignment after high extradural lignocaine block. 3.676

- **Key observations**: Recumbent calf unable to move affected leg.
- **Clinical examination**: High extradural lignocaine block used to achieve pain-free clinical examination. Femoral head fractures can be difficult to appreciate.
- **Key history events**: Problem during assisted delivery of calf.

Bovine Neonatal Panleucopaenia

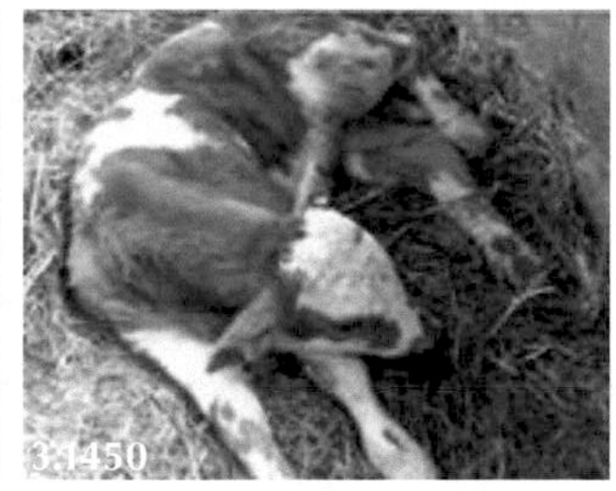

3.1448: Bleeding from the nares and jugular vein injection site. **3.1449** and **3.1450:** Recumbency secondary to colic from mesenteric bleeding.

- **Key observations**: Normal for first three to five days, then stop sucking, weak, recumbent, may show colic.
- **Clinical examination**: Widespread petechiae of mucous membranes, bleeding from needle puncture sites/fly bites. Haemorrhage into omentum causes colic signs.
- **Key history events**: Associated with ingestion of colostrum from cow vaccinated with a particular BVD vaccine (no longer available). 🎥 3.677, 3.678

Congenital Heart Defects

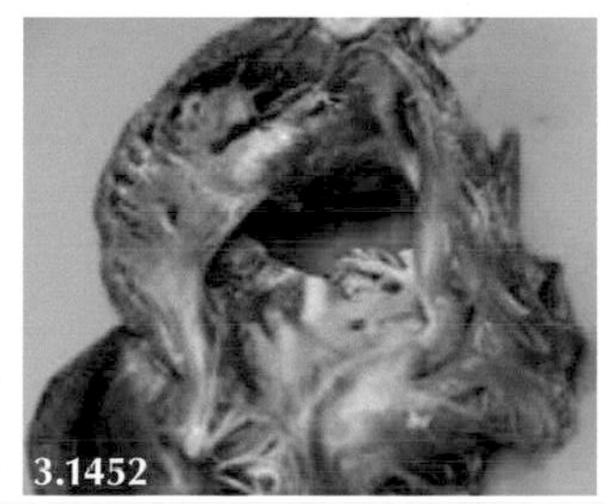

3.1451: Weak, lethargic calf, reluctant to walk. **3.1452:** Large ventricular septal defect. **3.1453:** Grossly enlarged liver with rounded margins.

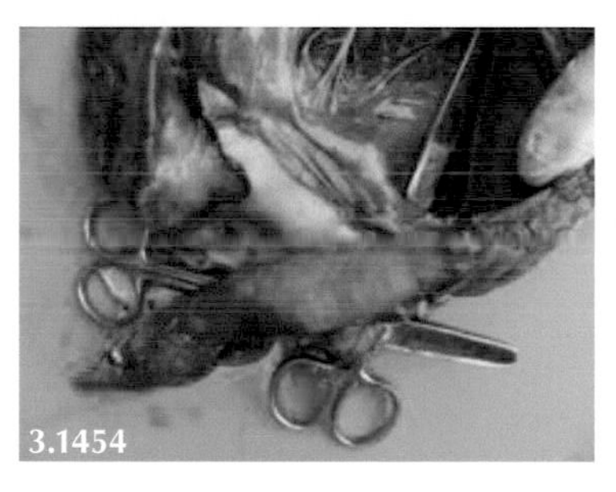
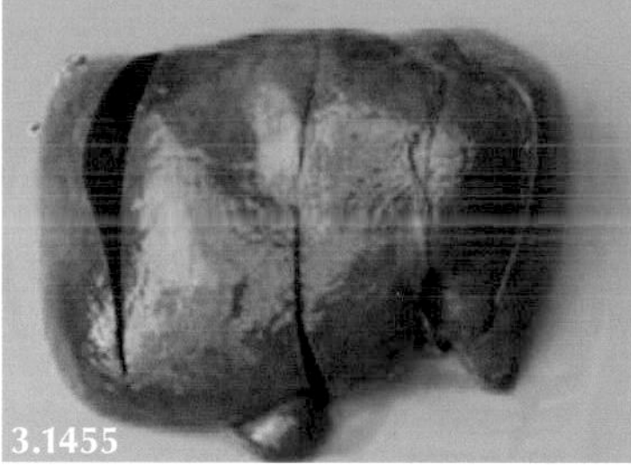

3.1454: Tetralogy of Fallot.
3.1455: Grossly fatty liver.

- **Key observations**: Larger defect—weak since birth.
- **Clinical examination**: Murmur may not be audible. Calves can survive with smaller defects which cause pansystolic murmur.
- **Key history events**: Clinical signs present since birth. 🎥 3.679

Aortic Thrombus

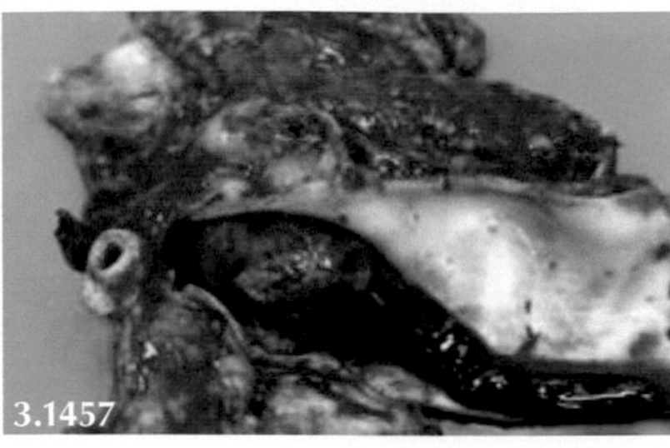

3.1456: Ten-day-old calf displaying hindleg weakness.
3.1457: Necropsy (longitudinal section of aorta) reveals thrombus lodged in the iliac bifurcation.

- **Key observations**: Sudden onset weakness affecting hindlegs leading to dog-sitting and sternal recumbency.
- **Clinical examination**: Cold hindlegs with no palpable femoral pulse.
- **Key history events**: Other septic focus may not be evident. 🎥 3.680

3.28

Common Causes of Chronic Weight Loss/Emaciation

Adult Sheep

The diseases/management practices that can result in chronic weight loss/emaciation are listed later in this chapter in order of their likely impact on the flock. Inadequate nutrition/energy supply, especially during late gestation, can affect the whole flock and is therefore ranked as the most important cause.

Ranking = percentage of flock affected × intensity × duration

Cull ewe screens, whereby four to eight low condition score ewes are culled and necropsied, are helpful but can sometimes yield misleading results e.g. this author has scanned flocks where one ewe had been diagnosed with OPA at a necropsy screen but there were no other cases in the remaining four to 600 adult sheep. While a useful first step, a small sample does not necessarily predict prevalence/annual incidence of disease and likely benefit:cost if control measures are adopted. Flock inspection by the farmer's veterinary surgeon can quickly identify lean sheep which can then be handled to quantify body condition scores allowing further investigations. 🎥 3.681

Whole flock OPA screening has allowed this author to handle every sheep on over 100 large commercial farms (150 to 2,400 ewes) and has emphasised the importance of lameness as one of the major causes of poor body condition.

Be aware that some of the conditions listed next respond to veterinary treatment (most lameness causes, parasitism) and/or nutritional changes (undernutrition and dental problems). Other causes are invariably fatal and necessitate immediate culling to remove animal welfare concerns and salvage any slaughter value (Johne's disease and OPA). Parasitic diseases such as fasciolosis, sheep scab and lice simply should not occur in a well-managed flock. An effective veterinary flock health plan will markedly reduce the annual incidence of many of the conditions listed. See the end of this section for demonstration of the clinical application of serum protein analysis in the investigation of chronic weight loss/poor condition.

- If the farmer is serious about investigating causes of chronic weight loss/emaciation, the veterinary surgeon should condition score every sheep and separate for further investigation.

DOI: 10.1201/9781003106456-46

- Cull ewe screens are helpful but cannot measure annual incidence.
- Necropsy studies of all deaths is very useful but does not include involuntary culls.
- Farmers do not understand the liver fluke cycle nor which active molecules to use when—this advice must be communicated via practice newsletters, emails, WhatsApp groups etc.
- Demonstration of hypoalbuminaemia <15 g/l (serum sample) and normal globulin concentration are much more useful than serological testing to highlight the possibility of paratuberculosis.
- Defining lung pathology necessitates ultrasound examination.
- Molar dentition problems are easily overlooked.

TABLE 3.139

Common Causes of Chronic Weight Loss/Emaciation in Adult Sheep

	BCS	Appetite	Demeanour	Toxaemia	Pain	Feces	Rectal temperature
Inadequate nutrition	-/--	NAD	NAD	NAD	NAD	NAD	NAD
Severe foot lameness	-/---	-/--	NAD	NAD	++/+++	NAD	NAD
Chronic liver fluke	-/---	-/--	-/--	NAD	NAD	NAD	NAD
Paratuberculosis	--/---	NAD/-	NAD	NAD	NAD	Agonal diarrhoea	NAD
PGE	-/--	-	NAD	NAD	NAD	Diarrhoea	NAD
OPA	-/---	-	-	NAD	NAD	NAD	NAD
Sheep scab	-/---	NAD	-/--	NAD	+	NAD	NAD
Dentition	-/---	NAD	NAD	NAD	Quidding	Fibrous	NAD
Severe joint lesion	--/---	-/--	-	NAD	++/+++	NAD	NAD
Myiasis/head fly	/	/--	-/--	NAD	NAD	NAD	NAD
Chronic pneumonia	-	-	-	NAD/+	NAD	NAD	NAD/+
Endocarditis	-/---	-/--	--/---	+	++/+++	-	NAD/+
Maedi-visna	-/--	NAD	NAD	NAD	NAD	NAD	NAD
Lice	-	NAD	NAD	NAD	NAD	NAD	NAD

TABLE 3.140

Ancillary Tests to Investigate the Common Causes of Chronic Weight Loss/Emaciation in Adult Sheep

Disease	Ancillary test
Inadequate nutrition	Serum 3-OH butyrate concentration during late gestation, much less useful during lactation
Severe foot lameness	Clinical examination. Radiography to identify chronic septic joint
Chronic liver fluke	Fecal egg count. Coproantigen ELISA. Serum GLDH, GGT concentrations
Paratuberculosis	Hypoalbuminaemia <15 g/l and normal globulin concentration (35–50 g/l). Serum ELISA. Fecal culture. Necropsy and gut/LN histopathology. Be aware FWEC may be very high due to anergy
PGE	Fecal worm egg count >1,000 epg
OPA	Ultrasonography. Necropsy and lung histopathology
Sheep scab	Skin scraping. Serology
Dentition	Mouth gag and torch. Radiography
Severe joint lesion	Clinical examination. Radiography
Myiasis/head fly	Clinical examination
Chronic pneumonia	Ultrasonography
Endocarditis	Ultrasonography
Maedi-visna	Serology. Necropsy and histology, virus isolation
Lice	Microscopic identification on fleece plucks

Inadequate Nutrition

3.1458: Pregnant ewes stranded without food in deep snow. **3.1459** and **3.1460:** Too many sheep at feed ring, which is barely visible, causing inadequate access to roughage.

3.1461–3.1463: Emaciated rams due to inadequate nutrition after breeding season.

- **Key observations**: Large percentage of group/flock affected.
- **Clinical examination**: Poor body condition scores in twin- and triplet-bearing ewes.
- **Key history events**: Usually results from prolonged periods (at least several weeks) of severe weather (usually extensively managed hill sheep). 3.682

Severe Foot Lameness

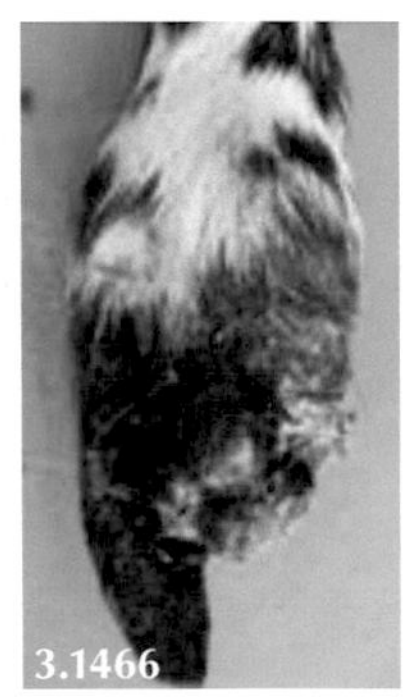

3.1464: 10/10 lame on right hind foot with considerable muscle wastage. There is no visible overgrown hoof because the hoof capsule has been sloughed off due to CODD. **3.1465** and **3.1466:** The duration of lameness is indicated by lack of hoof horn wear (likely >6–8 weeks).

- **Key observations**: Much reduced grazing leads to rapid body condition loss.
- **Clinical examination**: A wide range of foot pathologies cause severe lameness.
- **Key history events**: While the cause of lameness may be simple to diagnose, the resolution, even with veterinary attention, can prove difficult and take several months. 3.683, 3.684, 3.685, 3.686

Chronic Liver Fluke

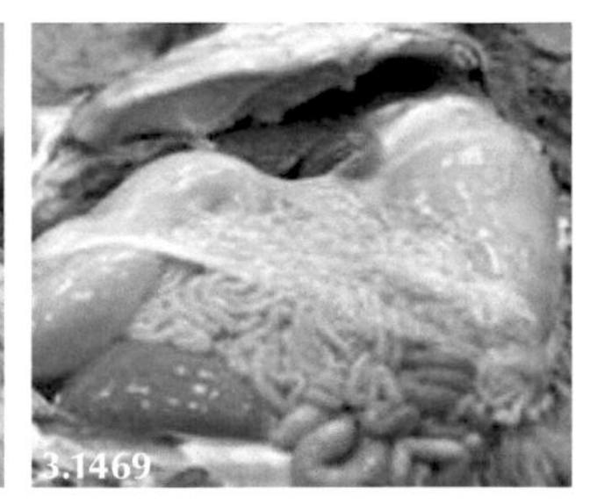

3.1467: Emaciated sheep caused by chronic liver fluke infestation. **3.1468:** Submandibular oedema ("bottle jaw"). **3.1469:** Emaciated carcase—note absence of omental fat.

- **Key observations**: Condition score loss during winter months. Occasional deaths.
- **Clinical examination**: Anaemia and submandibular oedema in severe cases.
- **Key history events**: Large numbers in group affected, failure to correctly implement veterinary flock health control programme. Possible triclabendazole resistance. 3.687, 3.688

Paratuberculosis

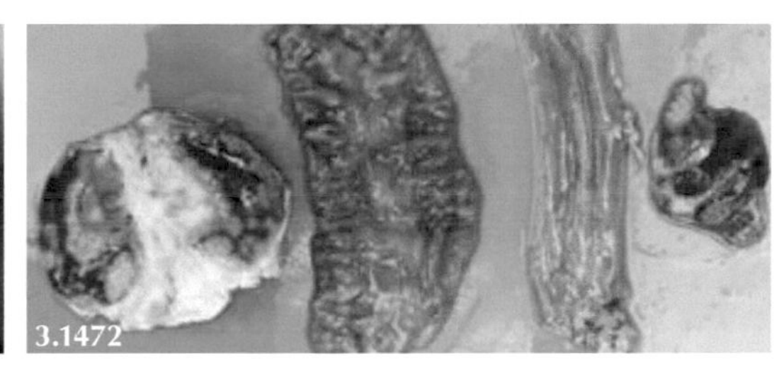

3.1470 and **3.1471:** Cheviot ewe which is bright and alert but in very poor condition. **3.1472:** Necropsy shows reactive (nodular cortex) ileo-caecal lymph node and thickened corrugated ileum of paratuberculosis (left samples) compared to normal tissues (right).

- **Key observations**: Sheep bright and alert but in very poor condition.
- **Clinical examination**: May be agonal diarrhoea due to parasitism.
- **Key history events**: Insidious weight loss over four to six months. 3.689, 3.690, 3.691, 3.692

PGE (Haemonchosis)

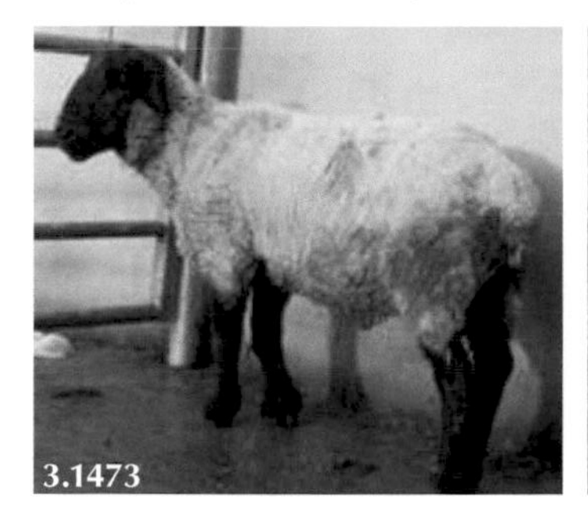

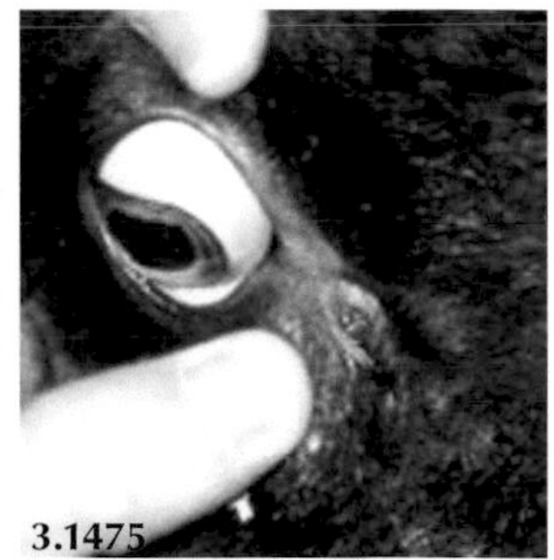

3.1473 and **3.1474:** Failure to use an effective anthelmintic has led to rapid loss of condition in these purchased rams. **3.1475:** Anaemic ocular mucous membranes of ram (centre).

- **Key observations**: Condition score loss when adequate grazing. Occasional deaths.
- **Clinical examination**: Anaemia and possibly submandibular oedema in severe cases. May not be signs of diarrhoea.
- **Key history events**: Often large numbers in group affected, failure to correctly implement veterinary flock health control programme. Possible anthelmintic resistance.

Ovine Pulmonary Adenocarcinoma

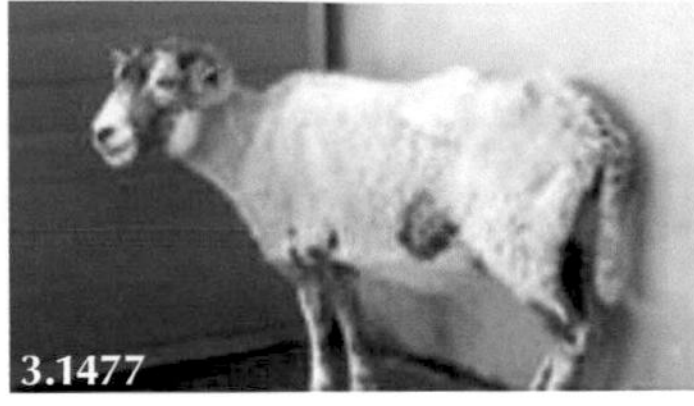

3.1476–3.1478: Emaciated ewes caused by advanced OPA.

- **Key observations**: Chronic weight loss over three to four months and poor fleece quality. Exercise intolerance in advanced OPA not always detected by farmer.
- **Clinical examination**: Increased respiratory rate and effort especially after gathering/handling stress. Widespread crackles only in advanced disease.
- **Key history events**: Endemic in many flocks at low levels (1–3 per cent). Often diagnosed in purchased sheep especially rams. 3.693, 3.694, 3.695, 3.696, 3.697, 3.698

Sheep Scab

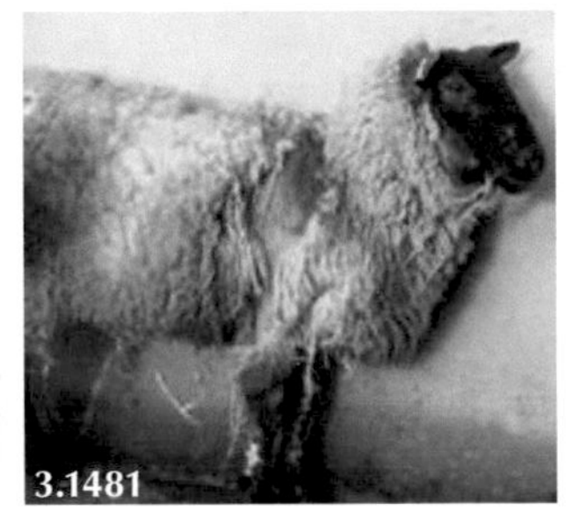

3.1479: Sheep constantly rub against fences/pen partitions. **3.1480:** Broken fleece caused by scratching/kicking with hind feet. **3.1481:** Fleece loss, serum exudation, adjacent fleece wet and sticky, skin folds/thickening/excoriation.

- **Key observations**: Frequent nibbling at fleece, rubbing against fences etc. Winter months.
- **Clinical examination**: Fleece abrasion/loss over flanks and shoulders, serum exudation. Skin scrape at periphery—take several scrapes from four or five affected sheep.
- **Key history events**: No preventive control. Common grazing, feral sheep/stray sheep introduce sheep scab. 3.699

Dentition

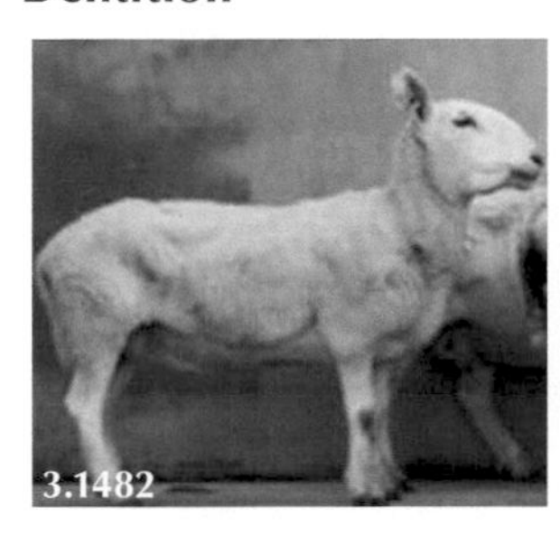

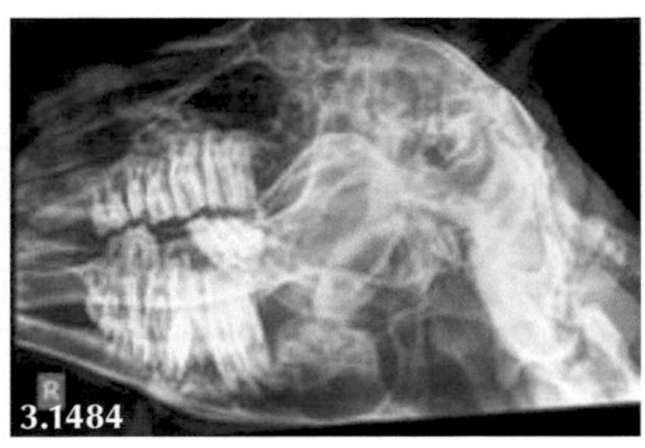

3.1482: Emaciated ewe caused by poor molar dentition. **3.1483:** Necropsy reveals missing/grossly overgrown mandibular cheek teeth. **3.1484:** Oblique radiographs can be used to investigate dental problems in valuable sheep.

- **Key observations**: Condition loss, quidding, often drool saliva, stained lower jaw.
- **Clinical examination**: Irregular dental arcade. Examine using mouth gag and torch.
- **Key history events**: Older sheep, may also present with "broken mouth". 3.700

Severe Traumatic Joint Lesion

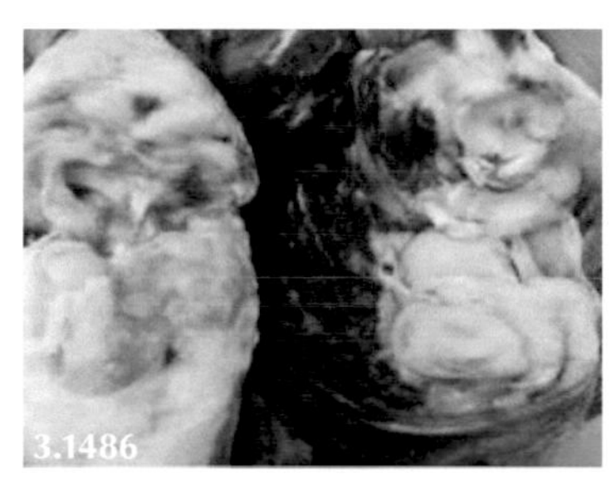

3.1485: Emaciated ewe with chronic stifle injury. **3.1486:** Left stifle shows marked fibrous tissue reaction of the joint capsule.

- **Key observations**: Chronic lameness not involving the foot.
- **Clinical examination**: Thickened joint capsule, little/no joint effusion.
- **Key history events**: Likely traumatic event but rarely known history of specific event. 3.701

Head Fly

3.1487: Head fly causing disrupted grazing. **3.1488:** Aggravation of head wound by kicking. **3.1489:** Ear notching to identify year/age group at the wrong time of year can cause headfly problems.

- **Key observations**: Disrupted grazing. Lie with head on ground often seeking shelter in banking. Frequent kicking at skin lesions.
- **Clinical examination**: Open wounds.
- **Key history events**: Shearing cuts, horned sheep, septic ear tag wounds. 3.702

Myiasis

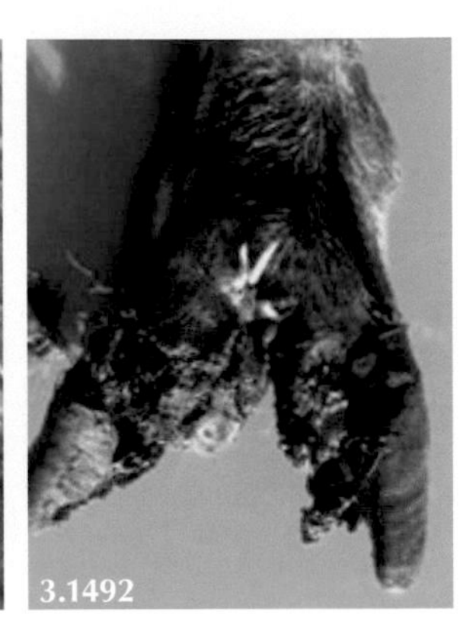

3.1490: Neglected myiasis case. **3.1491** and **3.1492:** Emaciated Scottish Blackface ewe as the consequence of severe footrot and secondary myiasis on grazing behaviour.

- **Key observations**: Dull, isolated from group, fecal staining of fleece, not grazing, frequent nibbling at affected area.
- **Clinical examination**: Putrid smell, maggots feeding on blackened skin, wool easily detached.
- **Key history events**: Poor endoparasite control, no dagging, no preventive ectoparasite application. 🎥 3.703, 3.704, 3.705

Chronic Pneumonia

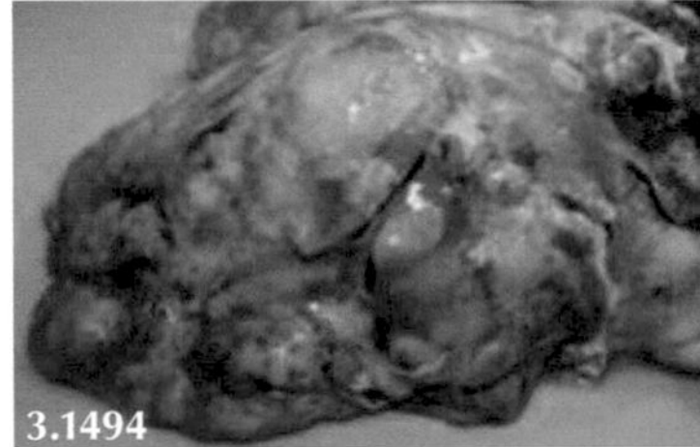

3.1493: Weight loss caused by chronic bacterial pneumonia. **3.1494** and **3.1495:** Necropsy findings of lung abscessation and pleurisy.

- **Key observations**: Poor appetite and low body condition score.
- **Clinical examination**: Increased respiratory rate with abdominal component. Much reduced lung sounds over affected lung; no adventitious sounds and no audible pleuritic rubs.
- **Key history events**: Often encountered in shearling rams probably associated with prolonged housing and respiratory disease during first winter. 🎥 3.706

Endocarditis

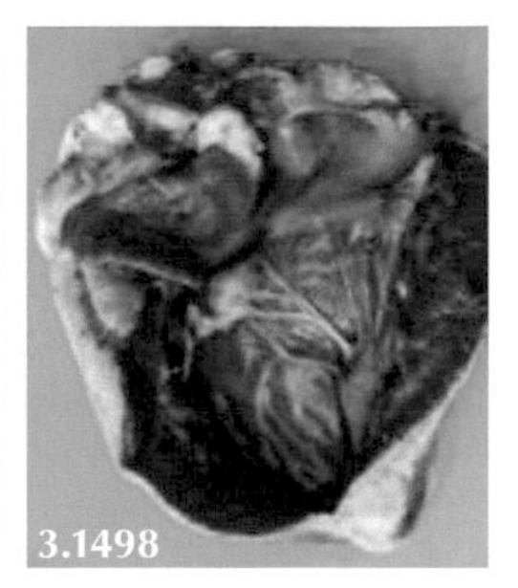

3.1496: Chronic left foreleg lameness in a Suffolk ram. **3.1497:** Multiple painful joints causing ewe to assume sternal recumbency. **3.1498:** Vegetative lesion on heart valve.

- **Key observations**: Shifting leg lameness, several swollen painful joints.
- **Clinical examination**: Thickened joint capsule but little palpable joint effusion. Often no audible heart murmur.
- **Key history events**: Usually no specific source of pyaemia identified. 3.707

Maedi-Visna

3.1499: Visna causing unilateral (right) hindleg weakness.

- **Key observations**: Clinical cases of maedi and visna are rarely described in the UK while recent seroprevalence studies show the virus is widespread.
- **Clinical examination**: Chronic weight loss, chronic respiratory disease (maedi), myelitis causing hindleg weakness (visna).
- **Key history events**: Imported sheep. Purchased sheep from infected flock.

Pediculosis

3.1500: Frequent nibbling at fleece. **3.1501:** Fleece looks "plucked". **3.1502:** Lice affects sheep in poor body condition during the winter months.

- **Key observations**: Frequent nibbling at fleece with wool "plucked".
- **Clinical examination**: >5 lice per skin parting. Always check for sheep scab irrespective of observing lice.
- **Key history events**: No preventive (plunge dipping) control. Pour-on preparations not 100 per cent effective when treated with full fleece. Winter months. Affects sheep in poor body condition. 3.708

General Findings of Emaciation at Necropsy

At necropsy, emaciation is evident as gross loss of muscle mass and absence of subcutaneous fat; and serous atrophy of omental, perirenal, epicardial, and bone marrow fat.

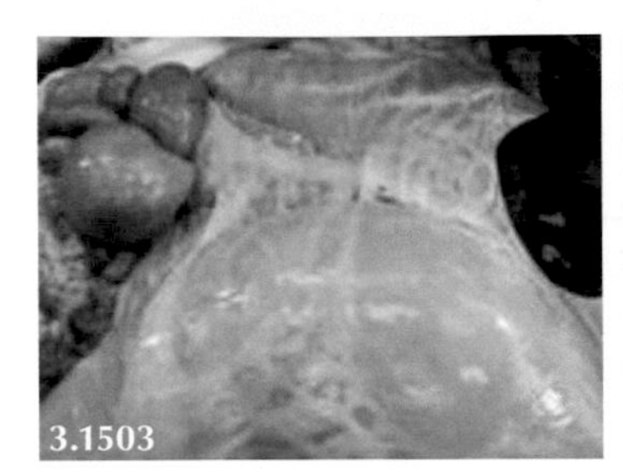

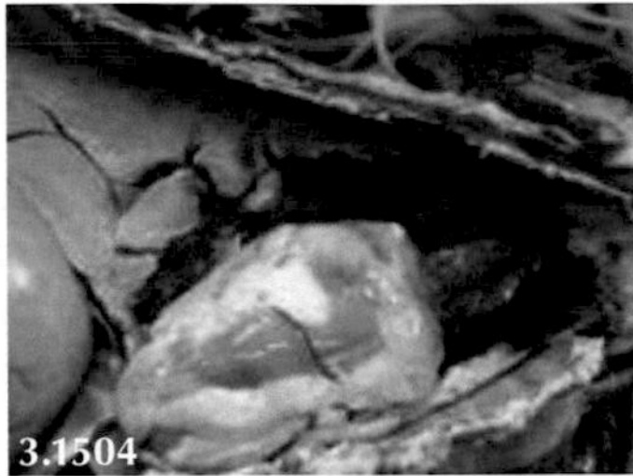

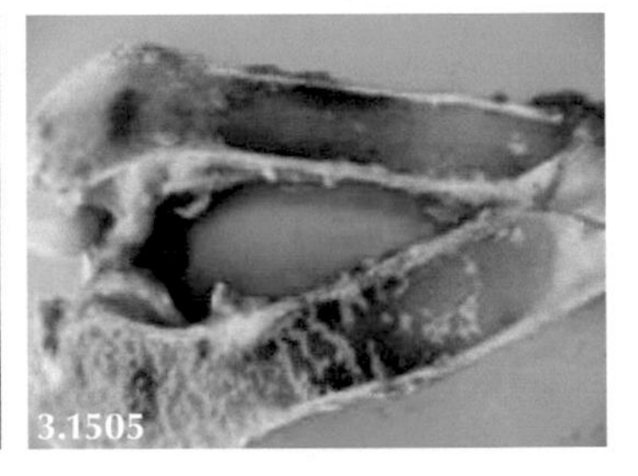

Serous atrophy of fat evident in the omentum (**3.1503**), pericardial sac (**3.1504**), and bone marrow (**3.1505**).

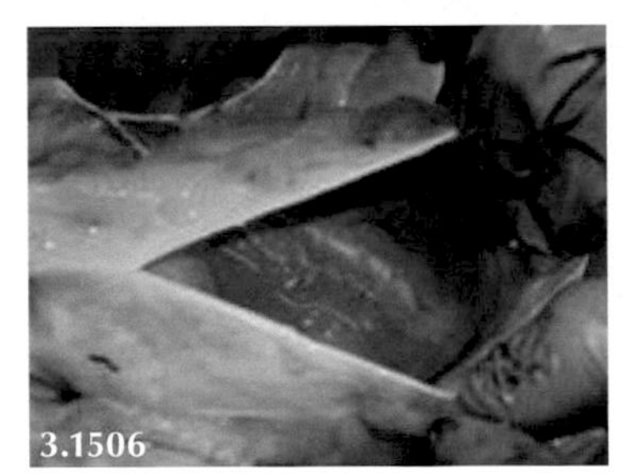

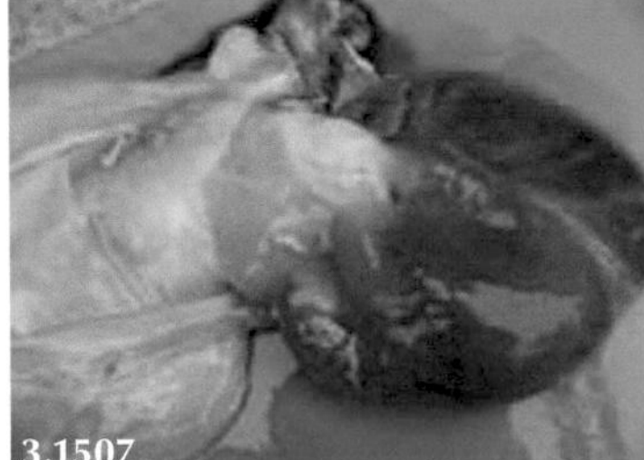

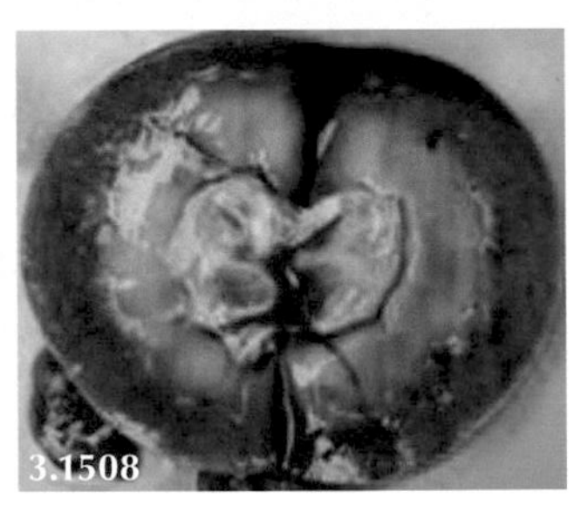

Serous atrophy of fat evident in the pericardial sac (**3.1506**), epicardial groove (**3.1507**), and kidney (**3.1508**).

Serum Protein Analysis in the Investigation of Weight Loss

Albumin reflects the balance between hepatic synthesis from dietary nitrogenous intake and endogenous demands/losses. Serum globulin is a long-term indicator of the body's response to bacterial infections and certain parasite infestations, most notably liver fluke. Other proteins, such as haptoglobin or fibrinogen, are more useful as indicators of acute disease with significant increases within one to three days of bacterial infection and greater than four to seven days, respectively. 3.709

Sheep with paratuberculosis have profound hypoalbuminaemia (serum concentration <15 g/l but as low as 6 to 10 g/l in advanced disease; normal range greater than 30 g/l) and a normal globulin concentration. These changes result from loss of albumin across the damaged intestinal mucosa (protein-losing enteropathy). These serum protein concentrations may also be encountered in cases of severe chronic intestinal parasitism such as haemonchosis, but such infestations are generally group problems. Fasciolosis may result in hypoalbuminaemia, but these sheep usually show additional clinical signs such as anaemia and submandibular oedema. In addition, many sheep with sub-acute and chronic fasciolosis have greatly increased serum globulin concentrations often exceeding 60 g/l (normal value <45 g/l).

In the absence of recent anthelmintic treatment, it must be appreciated that adult sheep with paratuberculosis often have very high fecal egg counts (>5–10,000 epg) due to immunosuppression of the host. Detailed necropsy examination typically fails to reveal significant populations of adult nematodes; the high egg output is the result of increased fecundity of a small population of adult parasites. It is not uncommon to find

Nematodirus battus eggs and lungworm large (*Dictyocaulus filaria*) on the McMaster slide in these adult sheep; these parasitological findings should alert the clinician to the likelihood of paratuberculosis. Caution is needed when interpreting pooled fecal samples from adult sheep because a single paratuberculosis case can cause a significant increase in the mean worm egg count leading to a misdiagnosis of parasitism.

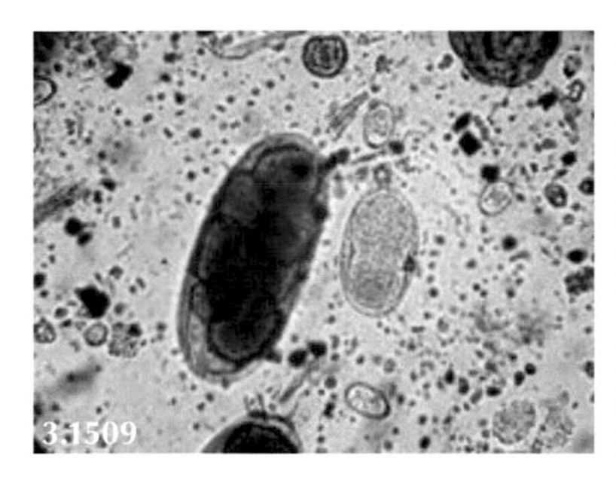

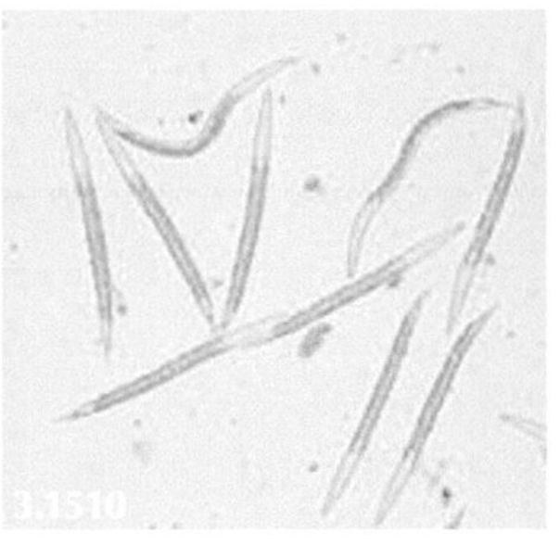

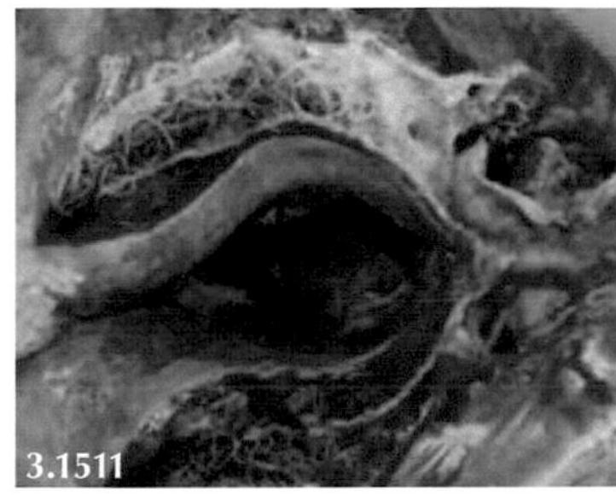

It is not uncommon to find *Nematodirus battus* eggs (**3.1509**) and *Dictyocaulus filaria* larvae (**3.1510**) in fecal samples from sheep with paratuberculosis. **3.1511:** Severe lungworm infestation affecting only one ewe in group—this ewe had paratuberculosis.

Chronic bacterial infections causing weight loss result in significant increases in serum globulin concentration (often greater than 55 g/l) and low serum albumin concentration (often around 18 to 25 g/l but rarely below 15 g/l). Low albumin/high globulin concentrations indicate a probable chronic suppurative disease process/focus such as chronic suppurative pneumonia, endocarditis, liver abscessation, mastitis, infectious polyarthritis, cellulitis etc. Once the possibility of a chronic suppurative focus has been highlighted, the sheep should be re-examined. Further specific tests can then be selected based upon the organ system suspected of being involved, e.g. serum gamma glutamyl transferase (GGT) concentrations and fecal fluke egg count for chronic fasciolosis, chest ultrasonography if chronic suppurative pneumonia is suspected, ultrasonography for peritonitis etc.

Low serum albumin/normal globulin concentrations (<25 g/l and around 45 g/l, respectively) suggest that chronic bacterial infection would be unlikely. A dietary effect such as low protein intake is one possible explanation. A lowered serum albumin concentration is also often encountered in ewes during late pregnancy fed poor quality rations at a time when protein metabolism is geared toward immunoglobulin production and transfer into the colostrum in the udder.

Example

Results of serum and fecal samples collected as part of a preliminary investigation of poor condition score in a group of six ewes (2–7 sheep; sheep 1 normal control ewe) are presented in the following table.

TABLE 3.141

Example of Serum Protein Analysis in the Investigation of Chronic Weight Loss

	1	2	3	4	5	6	7
Albumin (g/l)	34.1	28.3	12.5	14.1	21.1	24.3	22.5
Globulin (g/l)	44.1	39.1	37.1	40.1	62.7	42.1	64.1
Fecal worm egg count (epg)	100	400	50	6,100	100	150	150

Sheep 1 has normal serum albumin and globulin concentrations >30 g/l and 35–45 g/l, respectively. Sheep 2 has a serum protein profile typically observed in emaciated ewes without obvious bacterial infection or parasite infestation. Causes include poor molar dentition, lameness etc. Sheep with a protein-losing enteropathy, notably paratuberculosis (very common; sheep 3 and 4), and nephropathy (rare) have profound hypoalbuminaemia (serum concentration <15 g/l) and a normal globulin concentration.

Due to immunosuppression, sheep with paratuberculosis often have high fecal egg counts >1,000 epg (sheep 4) resulting from increased fecundity of those mature nematodes present; diarrhoea is uncommon. It is exceptional for serum albumin concentrations to fall so low in cases of chronic intestinal parasitism such as haemonchosis. Chronic severe bacterial infections (sheep 5) causing weight loss/illthrift result in significant increases in serum globulin concentration (often greater than 55 g/l) and low serum albumin concentration (18 to 25 g/l).

If fed marginal protein levels during late gestation, serum albumin concentrations can fall as immunoglobulins accumulate in colostrum (sheep 6).

Fasciolosis (sheep 7) results in hypoalbuminaemia, but these sheep usually show additional clinical signs such as anaemia and submandibular oedema. In addition, many sheep with sub-acute and chronic fasciolosis have increased serum globulin concentrations often exceeding 65 g/l. Further specific tests then can be selected based upon the organ system e.g. serum GLDH and GGT concentrations and fecal fluke egg count for fasciolosis.

Adult Cows

The ranking of diseases causing weight loss is broadly the same for individual beef and dairy cows, but there are major differences in the management of beef and dairy herds not least with their feeding. High yielding dairy cows have *ad libitum* easy access to a high energy total mixed ration whereas spring-calving beef cows often have to compete for relatively poor quality silage at a feed ring throughout winter. Such restricted feeding during late gestation of beef cows results in generally low body condition scores at calving to reduce the prevalence of dystocia. Such management affects the whole herd, and therefore restricted nutrition has been ranked first as a cause of poor body condition; this does not apply in general to dairy cows and autumn-calving beef cows where over-conditioned animals is the more common problem.

- Poor winter nutrition of spring-calving beef cows affects all cattle in the group.
- Severe foot lesions cause rapid condition loss in dairy cows.
- Fasciolosis can be difficult to control in dairy herds calving year-round.
- Paratuberculosis remains a major problem in beef and dairy herds.
- Confirmation of significant lung pathology, peritonitis, endocarditis, liver abscesses, pericarditis, pyelonephritis and hepatocaval thrombosis can usually be achieved using ultrasonography. 3.710

Restricted Nutrition in Pregnant Beef Cows over Winter

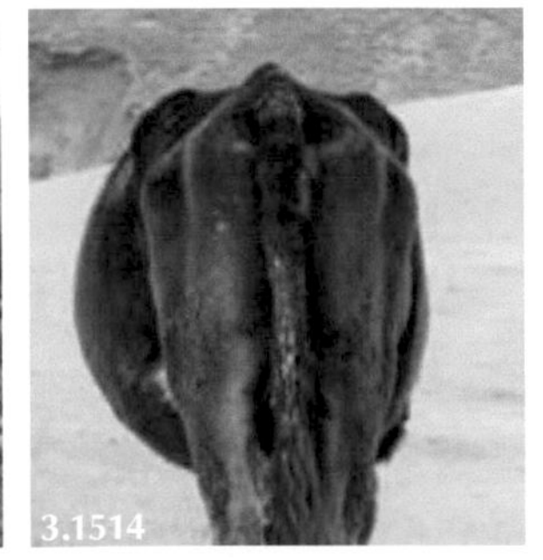

3.1512: Spring-calving cows in good body condition at weaning in early winter. **3.1513** and **3.1514:** Considerable loss of body condition by spring and calving time; abdominal distension due largely to near-term fetus.

TABLE 3.142

Common Causes of Weight Loss/Emaciation in Adult Cows

	BCS	Appetite	Demeanour	Toxaemia	Pain	Feces	Rectal temperature
Restricted nutrition (beef)	-/--	NAD	NAD	NAD	NAD	NAD	NAD
Severe foot/joint lameness	--/---	-	-/--	NAD	+++	NAD	NAD
Fasciolosis	/	NAD	NAD	NAD	NAD	NAD	NAD
Paratuberculosis	---	NAD/-	NAD	NAD	NAD	Profuse diarrhoea	NAD
Chronic pneumonia	-/--	-	-	NAD/+	NAD	NAD	+
LDA	-	-/--	-	NAD	NAD	Constipated	NAD
Peritonitis	-/---	---	-/---	++/+++	+/+++	-/---	+
Endocarditis	-/---	-/--	--/---	-+	++/+++	-	+
Liver abscesses	-/--	-	NAD/-	NAD/+	NAD/+	NAD	+
Pericarditis	-/--	-/---	-/--	+	++/+++	-	+
Hepatocaval thrombosis	--	---	--/---	+	++/+++	-	+
Pyelonephritis	-/---	--	-/--	+/++	NAD	-	+

TABLE 3.143

Ancillary Tests to Determine Cause of Weight Loss/Emaciation in Adult Cows

Disease	Ancillary tests
Restricted nutrition (beef)	Ration evaluation
Severe lameness	Radiography
Paratuberculosis	Serology. Fecal PCR. Culture. Necropsy
Chronic pneumonia	Ultrasonography
LDA	Surgical correction
Peritonitis	Ultrasonography. Abdominocentesis
Endocarditis	Ultrasonography
Liver abscesses	Ultrasonography
Hepatocaval thrombosis	Ultrasonography
Pyelonephritis	Ultrasonography. Urine analysis and culture

Restricted Nutrition

3.1515 and **3.1516:** Spring-calving cows in low condition after calving but will quickly improve on good grazing.

- **Key observations**: Low condition scores of many spring-calving beef cows at calving.
- **Clinical examination**: Restricted nutrition during winter months often to reduce dystocia caused by absolute fetal oversize (evolution for many species).
- **Key history events**: Common farm management practice. 3.711

Chronic Severe Lameness

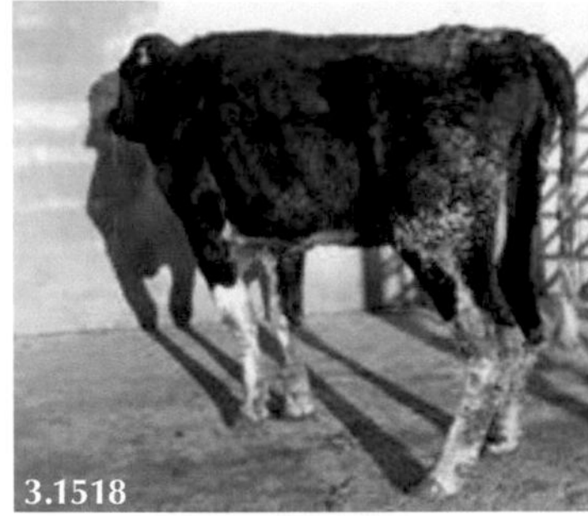

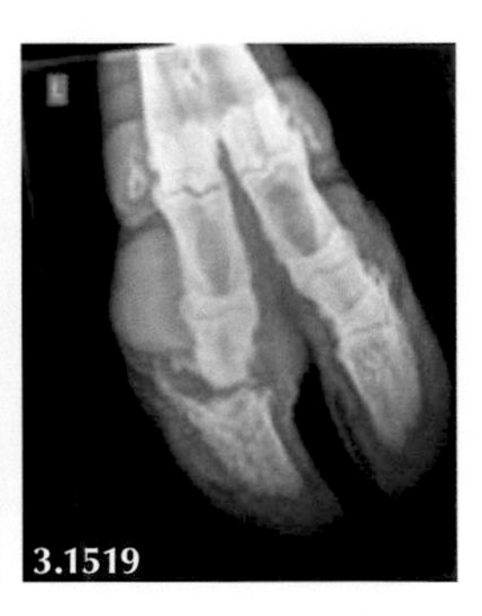

3.1517 and **3.1518:** Dirty coats and very low body condition indeed caused by chronic and severe lameness. **3.1519:** DP radiograph of the distal interphalangeal joint showing chronic infection with considerable bone lysis and osteophyte formation.

3.1520 and **3.1521:** Emaciated, chronically lame dairy cow. **3.1522:** Arched back, lowered head carriage. Dreadful state of walking track to the field.

- **Key observations**: Rapid weight loss caused by severe lameness.
- **Clinical examination**: Often sole ulceration, check for swelling around coronary band.
- **Key history events**: Often sole ulcer possibly extending to septic pedal joint. 3.712, 3.713, 3.714, 3.715, 3.716, 3.717

Fasciolosis

3.1523–3.1525: Late November when these beef cows should be in condition score >3.5 but are <2.0 caused by fasciolosis. Note the cow on the right has submandibular oedema.

- **Key observations**: Low conditions scores in many cattle but otherwise bright and alert and eating well.
- **Clinical examination**: Poor condition scores and very occasionally submandibular oedema.
- **Key history events**: Problem on many upland farms operating mixed grazing with sheep. 3.718

Paratuberculosis

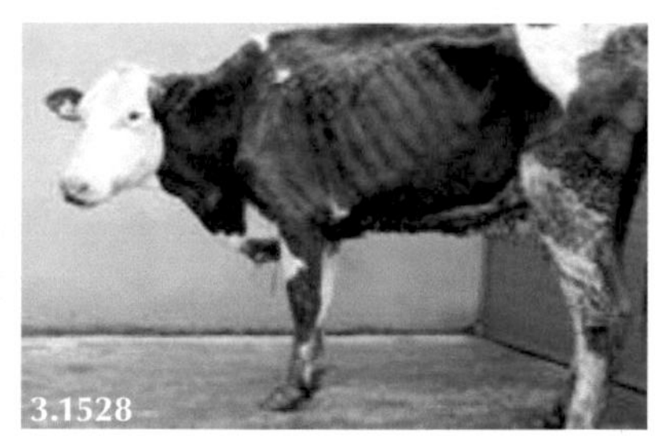

3.1526–3.1528: Individual emaciated adult cattle with "hose-pipe" diarrhoea.

- **Key observations**: Chronic weight loss, "hose-pipe" diarrhoea but otherwise bright and alert and eating well.
- **Clinical examination**: Only clinical findings are emaciation and severe diarrhoea.
- **Key history events**: Persistent problem on many farms. 3.719, 3.720, 3.721 3.722

Chronic Pneumonia

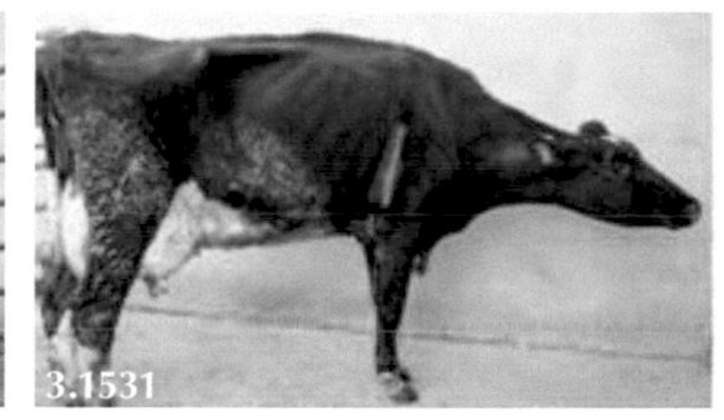

3.1529–3.1531: Chronic weight loss, neck extended, head lowered.

- **Key observations**: Poor appetite, frequent coughing. Painful expression and arched back. Occasional purulent nasal discharge.
- **Clinical examination**: Ultrasound examination essential to quantify pathology.
- **Key history events**: Flare-up of chronic disease following stressful event such as calving . 3.723, 3.724, 3.725

LDA

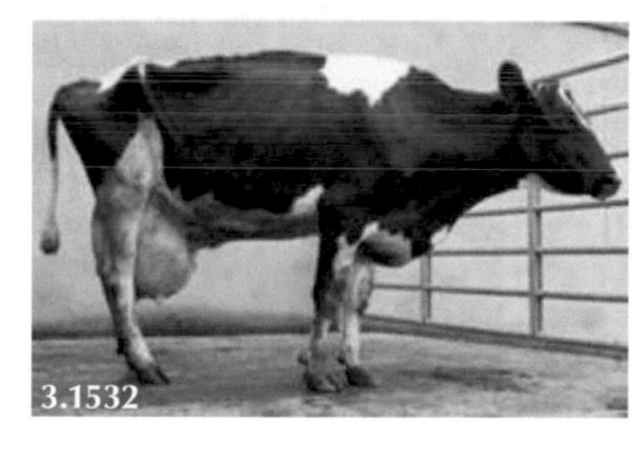

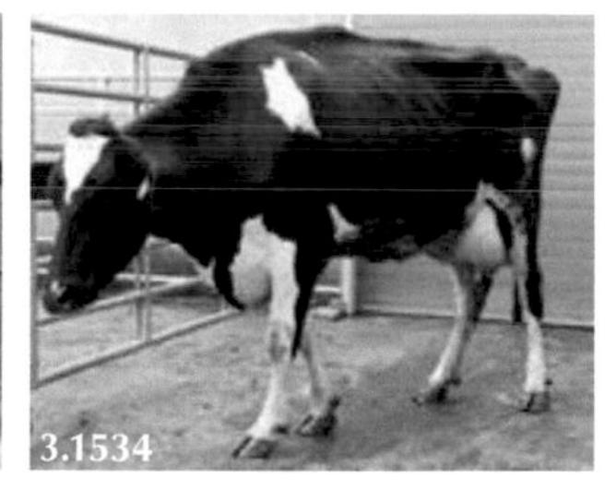

3.1532–3.1534: Chronic weight loss and poor abdominal fill associated with left displaced abomasum.

- **Key observations**: Poor appetite and reduced milk production. Constipated.
- **Clinical examination**: High-pitched resonant "pings" in upper left sublumbar fossa. May develop secondary ketosis if not corrected immediately.
- **Key history events**: Two to three weeks post-partum. Twin pregnancy, hypocalcaemia and metritis are recognised risk factors for LDA. 3.726, 3.727, 3.728

Peritonitis

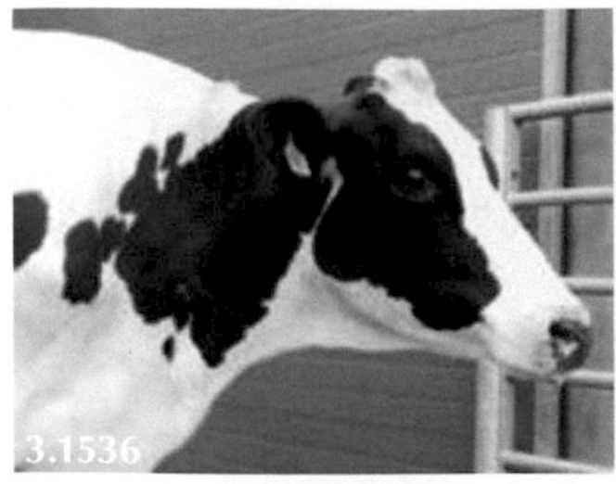
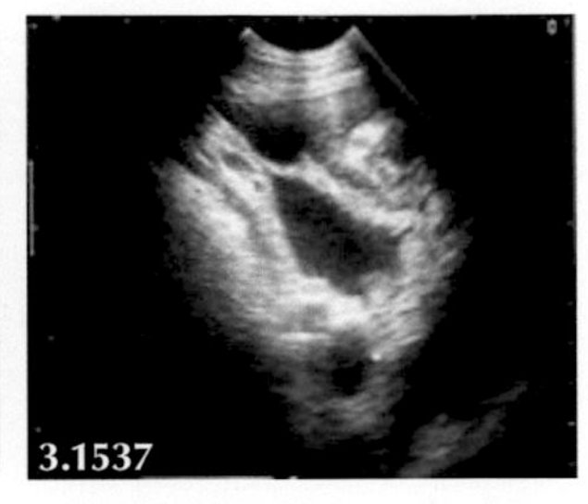

3.1535: Reduced abdominal fill. **3.1536:** Painful, fixed expression with the ears directed caudally. **3.1537:** Ultrasound examination reveals 6–8 cm inflammatory exudate with large fibrin tags within the peritoneal cavity.

- **Key observations**: Poor appetite and milk production. Painful expression.
- **Clinical examination**: Taut abdomen, reduced rumen motility. T39.0°C.
- **Key history events**: Traumatic reticulitis cases may appear as cluster if source such as access to bonfire sites. 3.729, 3.730

Endocarditis

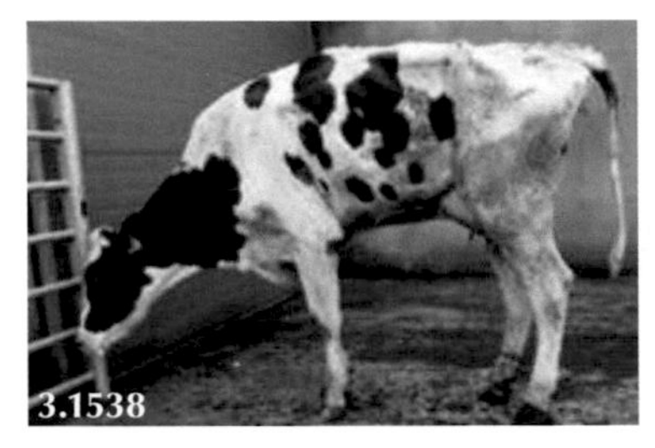

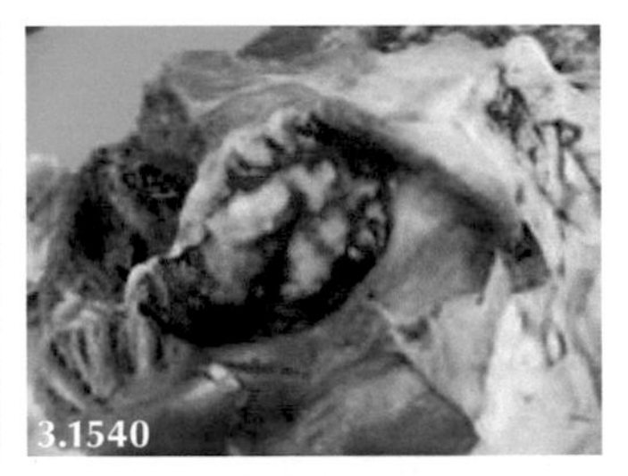

3.1538 and **3.1539:** Chronic severe weight loss, variable lameness. **3.1540:** Vegetative lesion on the tricuspid valve.

- **Key observations**: Variable lameness often affecting all four legs, arched back, painful expression.
- **Clinical examination**: Effusion of multiple joints, often no significant heart murmur. Microarray probe to identify vegetative lesion ultrasonographically as usually no murmur.
- **Key history events**: Source of pyaemia rarely identified. 3.731, 3.732

Liver Abscesses

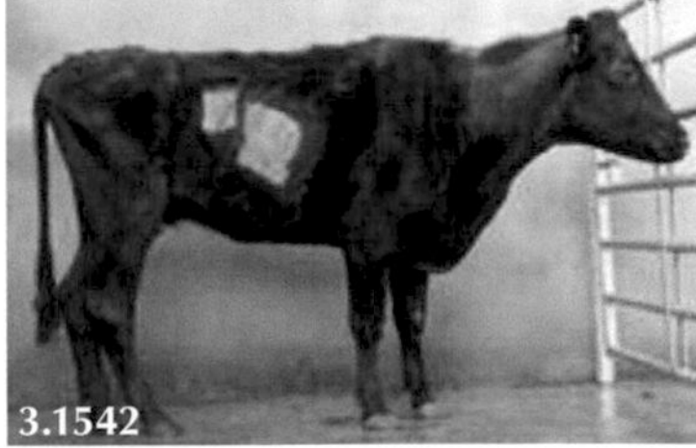
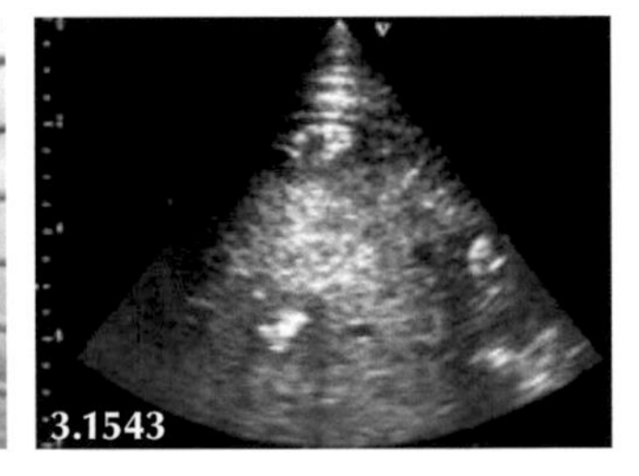

3.1541 and **3.1542:** Poor BCS and much reduced appetite/abdominal fill. **3.1543:** Multiple 1–3 cm diameter abscesses in the liver identified ultrasonographically.

- **Key observations**: Poor appetite, weight loss.
- **Clinical examination**: Often no specific clinical signs—value of ultrasound examination.
- **Key history events**: Often bull beef cattle or other intensive feeding system. Common necropsy finding in dairy cows. 3.733, 3.734

Pericarditis

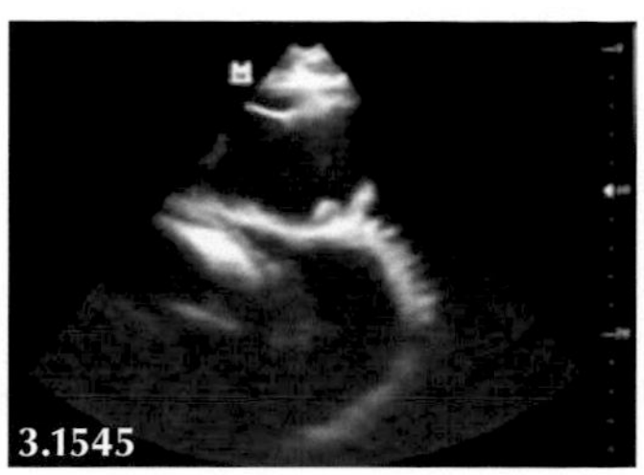
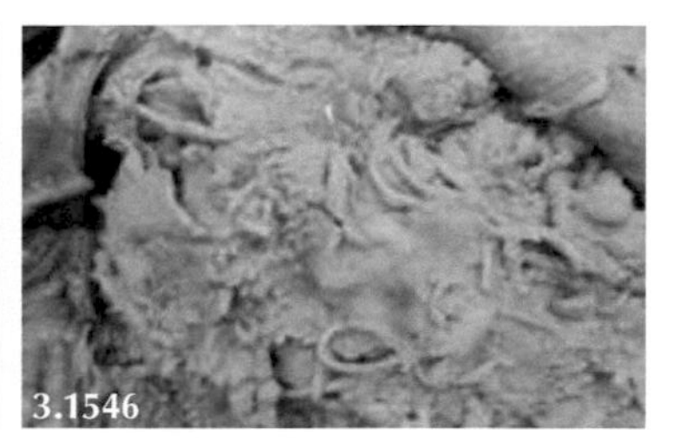

3.1544: Chronic weight loss with moderate brisket oedema. **3.1545:** Ultrasound examination reveals up to 20 cm fibrinous exudate within the pericardial sac. **3.1546:** Necropsy confirms the sonographic findings (fluid has been drained).

- **Key observations**: Weight loss, development of brisket oedema, jugular distension.
- **Clinical examination**: Normal lung sounds only heard dorsally, replaced by splashing sounds audible over large area of ventral chest.
- **Key history events**: Often a consequence to a migrating wire following traumatic reticulitis. 3.735, 3.736

Hepatocaval Thrombosis

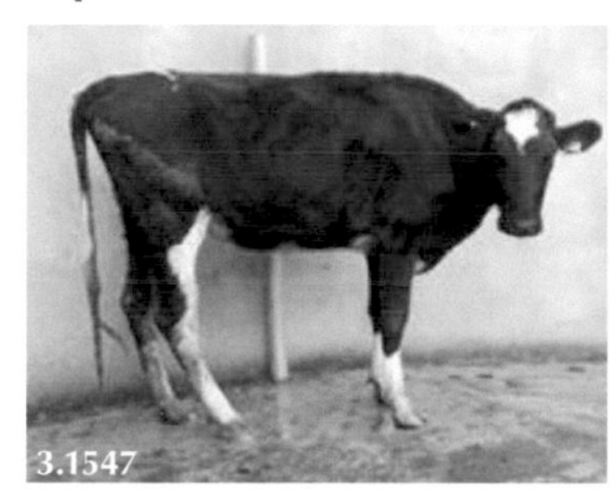

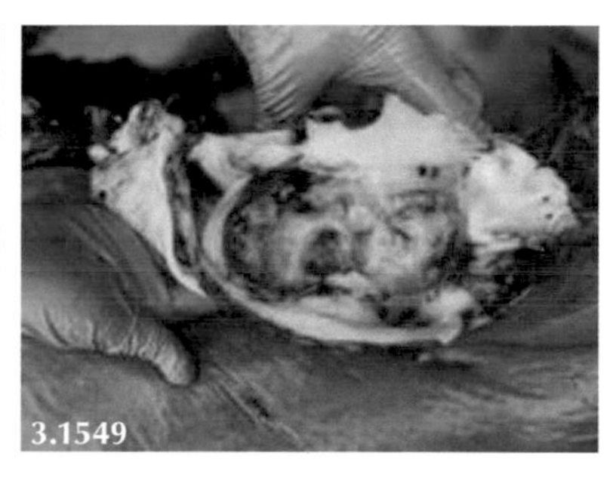

3.1547: Chronic weight loss. **3.1548:** Fatal epistaxis. **3.1549:** Large liver abscess within the wall of the opened vena cava.

- **Key observations**: Often sudden fatal bleed after period of poor appetite and weight loss.
- **Clinical examination**: Specific clinical diagnosis only really possible when epistaxis observed.
- **Key history events**: May have minor epistaxis events days before fatal bleed. 3.737

Pyelonephritis

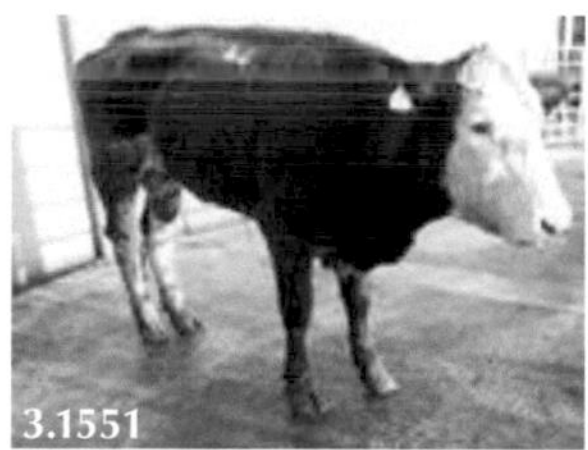
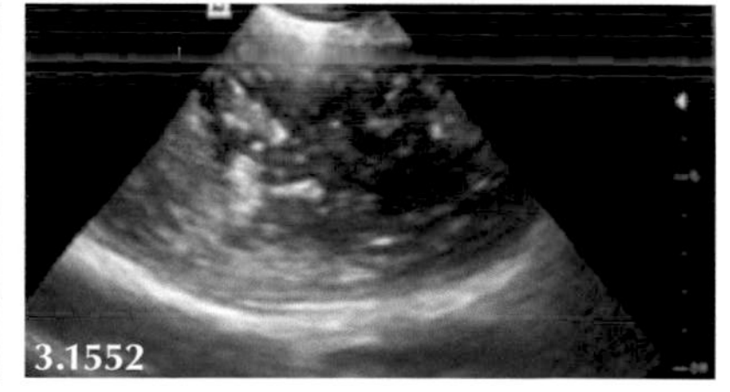

3.1550 and **3.1551:** Chronic weight loss in two beef cows. **3.1552:** Ultrasound examination reveals loss of normal renal architecture.

- **Key observations**: Chronic weight loss.
- **Clinical examination**: Frequent urination, flecks of blood/pus in urine and on tail. Rectal examination identifies thickened bladder/ureters.
- **Key history events**: Usually affects beef cows some months after normal calving. 3.738

Index

Page numbers in *italics* indicate a figure and page numbers in **bold** indicate a table on the corresponding page.

C

L

M

N

O

R

T

X

Y